STRUCTURE
of the
NUCLEUS

by

M. A. PRESTON

and

R. K. BHADURI
McMaster University

T0318485

CRC Press
Taylor & Francis Group
Boca Raton London New York

CRC Press is an imprint of the
Taylor & Francis Group, an **informa** business

First published 1975 by Westview Press

Published 2018 by CRC Press
Taylor & Francis Group
6000 Broken Sound Parkway NW, Suite 300
Boca Raton, FL 33487-2742

© 1975 by Taylor & Francis Group, LLC
CRC Press is an imprint of Taylor & Francis Group, an Informa business

No claim to original U.S. Government works

ISBN 13: 978-0-201-62729-9 (pbk)

Visit the Taylor & Francis Web site at
http://www.taylorandfrancis.com

and the CRC Press Web site at
http://www.crcpress.com

Library of Congress Cataloging in Publication Data

Preston, Melvin Alexander.
  Structure of the nucleus.

  Includes bibliographical references.
  1. Nuclear physics. I. Bhaduri, R. K., joint
author. II. Title.
QC776.P74   539.7   74-13242

Structure of the Nucleus

Contents

Preface

This book is the result of activity generated by the publisher's request for a revision of M. A. Preston's 1962 book, *Physics of the Nucleus*. As we worked at that task, we gradually realized that the intervening decade had seen such a substantial increase of knowledge of the atomic nucleus that a book at this level would perforce have to be much more voluminous or, alternatively, more concentrated in its emphasis. The result is this essentially new book, emphasizing nuclear structure. Some parts of it are unchanged or are revisions of material in the earlier book, but most of it is new.

In the preface to the earlier work, the view was expressed that nuclear physics had advanced almost far enough that one could attempt to write a book on a deductive basis, stating first the fundamental laws determining the interaction of nucleons and then deducing the properties of nuclei. If that hope were realistic in 1960, it should be even more capable of fulfilment at the present time. But physicists now have a more precise appreciation of the extent of the difficulties encountered in completing some of the links in this chain of deductive reasoning. If a fairly satisfying structure exists for calculating some properties of many-body systems, we also know that there are gaps in our current ability to perform some steps in the calculation. If our knowledge of the two-nucleon force is much better than it was fifteen years ago, we also realize more clearly the importance of the gaps in our current knowledge about the nature and role of many-body forces and the structure of the nucleons themselves. It is still true that many fascinating aspects of the behavior of nuclei are best discussed in terms of models whose limits of validity are not fully understood in terms of a more fundamental theory. Thus, although in this book we tend to emphasize the fundamental approach which should unify the concepts used in describing nuclei, this work, like its predecessor, remains semideductive.

Although we intend the book to be useful as a reference volume for research physicists, while writing most parts of it we pictured the reader as a graduate student engaged in his first detailed examination of nuclear physics. We make essential use of the material normally covered in a first course in quantum mechanics. The pace of the text also assumes that a reader has had a prior introduction to nuclear phenomenology, but in fact the discussion of each topic proceeds *ab initio* and makes no essential use of unstated facts of nuclear physics.

The book is intended to be neither "experimental" nor "theoretical"; it deals with the structure and physics of the nucleus, and not primarily with either laboratory or mathematical techniques. Although there is more discussion of theories than of experiments, we have tried to restrict the presentation of theoretical technique to that necessary for physical understanding.

The book falls into three parts. The first part, entitled "Fundamental Properties of Nuclei," contains a detailed discussion of the two-nucleon interaction and of basic model-independent nuclear properties. The description of nuclear force at a simple level begins in Chapter 2. A detailed discussion of its noncentral character, however, can be appreciated only after acquiring a background knowledge of electromagnetic moments and form factors. It is therefore natural to continue the discussion of nuclear force at a more sophisticated level in Chapter 5 after the topics of moments and sizes have been covered in Chapters 3 and 4.

The first part of the book prepares the way for the second, in which we discuss nuclear models and their agreement with experimental results and then explain their bases in fundamental theory. Chapter 7 describes the simple single-particle model, while Chapter 8 attempts to justify the independent-particle picture by developing the theory of nuclear matter and finite nuclei in a unified manner, and making a connection between the free two-nucleon force obtained from scattering experiments and the effective two-nucleon force used in nuclear structure calculations. Ordinarily, one would expect the description of microscopic calculations in finite nuclei, which is the content of Chapter 10, to immediately follow Chapter 8. We believe, however, that topics like deformed Hartree-Fock orbitals can be much better appreciated after a thorough exposition of the phenomenological collective model, which is so successful in correlating experimental data. That is why Chapter 9 intercepts the logical continuation of the materials in Chapters 8 and 10.

The third part of the book deals in some detail with two processes that cast considerable light on nuclear structure, alpha decay and fission. Our understanding of the phenomenon of fission has improved tremendously in the last six years with the advent of the double hump in fission barriers, and we have attempted to incorporate these exciting developments in Chapter 12 without cluttering up the material with too much detail.

Throughout this book, the underlying emphasis is on how a nucleus is constituted through the interaction between the nucleons. This emphasis and limitations on the size of the book have led us to exclude explicit accounts of the theory of reactions and of weak interactions and beta decay, although we do refer to results from the whole range of nuclear physics experiments which elucidate features of structure. When we use a reaction or decay experiment, we give sufficient theoretical background to make the reference understandable. We hope this choice of topics makes possible a one-volume work that not only fits the requirements of many current graduate courses but also is of some use to research physicists. In giving a course, certain sections of the book may be omitted without loss of coherence. In our opinion these are Sections 5-9, 5-11, 10-4, more detailed parts of 8-3 and 8-4, and possibly Chapter 11.

Problems for the student should be an integral part of a course in nuclear physics. Most instructors like to design their own problems, but, as in the earlier book, we include some fairly instructive problems in the text. Although most are original, over the years we have collected problems from friends and now we are not always sure of their sources.

An important function of a graduate course is to introduce the student to the periodical literature. For this reason numerous references are given to original and review articles. No attempt has been made at historical completeness; rather we have referred to an article whenever it seemed natural to do so in our presentation of a subject. The result should be a representative sampling of the literature of nuclear structure.

The preface to *Physics of the Nucleus* acknowledged indebtedness to Dr. L. G. Elliott and to Professor Sir Rudolf Peierls for their indirect but very significant influence on the book. We reiterate that sentiment now. One of the authors (R. K. B.) is grateful to Dr. Kailash Kumar for initiating him into nuclear physics. We also wish to express appreciation to our colleagues Yuki Nogami and Don Sprung for useful discussions on certain sections of the book, to Subal Das Gupta and Carl Ross for reading parts of the manuscript and giving us the benefit of their suggestions, and to Hazel Coxall and Erie Long for their patience in typing the manuscript. Finally, our thanks go to our wives and families for cheerfully tolerating our excessive allocation of time to this task.

M. A. Preston
R. K. Bhaduri

Part I
Fundamental Properties of Nuclei

Chapter 1

The Constituents of Nuclei

1-1 NUCLEONS AND NUCLEI

The basic aim of any fundamental science is the understanding of observed phenomena in terms of general underlying concepts and laws. Scientific progress is made in two ways. One is the gradual reduction of the number of fundamental hypotheses needed to explain all the results of observation and experiment. In physics this is achieved by the formulation of theories and the performing of experiments intended to confirm or reject them. The other method of progress is the discovery, often not motivated by a theory, of unexpected phenomena. Existing theories must then be revised to include these newly found facts in the structure of the science. The interplay of these two facets of development is quite evident in the history of nuclear and atomic physics.

We can begin our account with what was one of the most important unifying steps in the advance of our knowledge of matter, the recognition that all material substances are composed of "atoms" of the chemical "elements," and that the number of elements is limited. We now have knowledge of about 100 elements. At the time of these early developments, individual atoms were thought of as indivisible and the atoms of any one element were indistinguishable; in effect, there were about 100 "elementary particles." The observation that the atomic weights of the different elements are roughly integral multiples of the atomic weight of hydrogen led Prout, in about 1815, to suggest that all heavier atoms were in fact combinations of hydrogen atoms. If this hypothesis had been correct—and we now know that it was a step in the right direction—it would have brought about a great simplification in basic concepts. There would have been only one type of primary particle, from which all matter was constructed. However, Prout's hypothesis was discarded, largely because accurate determinations of atomic weights showed that the ratios differ from integers by amounts which, although usually small, are definitely not zero. Nevertheless, the pronounced regularities in the properties of the elements, as exemplified in the periodic table, led to continued speculation that atoms were indeed structures composed of some more fundamental particles.

The next unifying discovery, largely due to the work of J. J. Thomson, was that all atoms can be made to emit identical electrons. Since electrons are very much lighter than atoms, and since atoms are electrically neutral whereas electrons are negatively charged, the suggestion arises that the bulk of the mass of the atom is made up of particles with a net positive charge. One might imagine a sphere of positively charged material, the size of the atom, in which the electrons are embedded. Alternatively, the positively charged mass might be concentrated in a small region at the center of the atom, with the electrons moving about it, the extreme excursions of their paths defining the atomic radius ($\sim 10^{-8}$ cm). The decision between these two atomic models was made by a famous experiment by Rutherford in 1911 [1],[†] which showed that the positive charge is confined to a region with radius of about 10^{-11} cm or less. This constituted the first experimental demonstration of the existence of the nucleus.

An atom, then, consists of a nucleus of small volume containing almost all the mass of the atom and bearing a positive charge, Ze, surrounded by a cloud of Z electrons moving in the much larger atomic volume. The number Z is called the charge number or atomic number. We have since learned that all atomic and chemical properties of the elements are determined essentially by the motions of the extranuclear electrons, so that all atoms with the same atomic number have the same chemical properties. However, there exist nuclei with the same Z but different masses. (One can estimate the mass of the nucleus by subtracting the mass of Z electrons from the mass of the atom.) The existence of atoms of the same element differing in mass could be inferred from the different atomic weights of the lead (and other elements) produced in the natural radioactive chains. More directly, by the deflection of ionized atoms, J. J. Thomson [2] demonstrated in 1913 the existence of chemically identical atoms with different masses. These atoms are called isotopes, from Greek words meaning "same place," a reference

---

[†] Numbers enclosed in brackets refer to the references listed at the end of each chapter.

to the fact that all the isotopes of an element appear in the same place in the periodic table. It was soon found that the masses of all nuclei are very nearly, but not precisely, integral multiples of a mass very close to that of a proton, the hydrogen nucleus. For a given nucleus, this integer is called its mass number and is denoted by A. It is easy to understand why Prout's law failed so long as it was interpreted in terms of chemically determined atomic weights; in a quantity of chlorine, for example, 75% of the atoms have $A = 35, 25\%$ have $A = 37$, and standard chemical procedures suggest an effective A of 35.5.

Because of the nearly integral masses, it would seem tempting to assume that nuclei are made up of protons. Strong direct evidence that protons do exist in nuclei was forthcoming in 1919 when Rutherford [3] found that nitrogen, bombarded by sufficiently energetic α-particles, emitted hydrogen nuclei. It is clear, however, that a nucleus cannot be composed solely of protons, for this would imply $Z = A$. In fact, Z is less than A; for most stable nuclei Z is not far from $\frac{1}{2}A$. For example, the common isotope of oxygen has $Z = 8, A = 16$; the two stable isotopes of chlorine have $Z = 17$, $A = 35$ and 37; and the stable isotopes of lead are $Z = 82, A = 204, 206, 207,$ 208. A natural assumption, particularly in view of the emission of electrons in β-radioactivity, is that there are also electrons in a nucleus, in just sufficient number, namely, $A - Z$, to produce the observed charge. For example, the ³⁷Cl nucleus would consist of 37 protons and 20 electrons. This hypothesis involves a number of difficulties, and in 1931 it was shown to be in disagreement with the observed spectrum of the nitrogen molecule, which reveals information about the statistics of the ¹⁴N nucleus [4, 5, 6].[†] An alternative hypothesis which had been advanced from time to time suggests that the nucleus consists of Z protons and $A - Z$ particles which approximate the proton mass but are electrically neutral. This was merely speculation until such a particle, the neutron, was found in 1932 by Chadwick [7] in England and by Curie and Joliot [8] in France. Specifically, they found that beryllium, bombarded by α-particles from radium, emitted an electrically neutral radiation which in turn imparted considerable energy to nuclei with which it collided. It was shown that these recoils could not be explained on the assumption that the radiation was electromagnetic, but were consistent with the hypothesis that it consisted of neutral particles of proton mass.

[†] An odd number of particles of spin $\frac{1}{2}$, such as 14 protons plus 7 electrons, has a total angular momentum of half an odd integer and has Fermi-Dirac statistics, whereas an even number, 7 protons plus 7 neutrons, has integral angular momentum and Bose-Einstein statistics.

A nucleus, then, is specified by two numbers: A, the total number of nucleons, and Z, the number of protons, or $N = A - Z$, the number of neutrons.

The same reasoning which shows that there could not be electrons in the nucleus shows that a neutron must not be thought of as a composite particle, say a proton and an electron tightly bound, but must be treated as a fundamental particle on the same footing as a proton. Indeed, it is useful to consider the neutron and the proton as simply two different states of a single particle, the nucleon. If a nucleon changes from one of its states to another, it of course changes its electrical charge. Charge can be conserved if the nucleon's change of state is always accompanied by the creation of an electron, positively or negatively charged, as may be required. This provides an explanation of β-radioactivity without necessitating the presence of electrons in the nucleus.

The properties of individual nucleons have been carefully measured. Determination of the masses of the proton and neutron is based on energy measurements of certain nuclear reactions and requires a knowledge of other fundamental numbers such as the electron mass, Avogadro's number, and the fine-structure constant. The preferred values [9, 10] are as follows:

$$M_p = 938.2592 \pm 0.0052 \text{ MeV}/c^2$$

$$= 1836.109 \pm 0.011m$$

$$= (1.672614 \pm 0.000011) \times 10^{-24}\text{g}$$

$$= 1.00727661 \pm 0.00000008u \quad \text{(atomic mass units)}$$

$$M_n = 939.5527 \pm 0.0052 \text{ MeV}/c^2$$

$$= 1837.64 \pm 0.01m$$

$$= (1.674920 \pm 0.000011) \times 10^{-24}\text{g}$$

$$= 1.00866520 \pm 0.00000010u$$

where m is the mass of the electron. An accurate measurement of the mass difference yields

$$M_n - M_p = 1.29344 \pm 0.00007 \text{ MeV}/c^2 = 2.531m$$

so that sufficient energy is available for a free neutron to change to a proton, creating an electron and still providing 0.7825 MeV of kinetic energy. This makes feasible the explanation of β-decay suggested in the preceding paragraph. The half-life of the free neutron is 10.8 min [11], but to change a neutron bound in a nucleus (A, Z) into a proton bound in $(A, Z + 1)$ may require more energy than is available because of the differing binding energies of different nuclear species.

Both neutron and proton are fermions, that is, they have intrinsic spin angular momentum $\frac{1}{2}\hbar$ and obey Fermi-Dirac statistics and the exclusion principle. This means that a wave function must change sign if we interchange all the coordinates of two protons or of two neutrons. For this purpose we must think of four coordinates: the three coordinates of ordinary space and the spin component along some fixed direction. The latter can take only two values, $\pm \frac{1}{2}$.

1-2 ISOSPIN

We have suggested that the neutron and the proton are two different quantum states of a single entity, the nucleon. To give mathematical expression to this idea we must assign an additional quantum number. This quantum number can take only two values since there are only two states, and it follows that the associated variable has the same algebra as spin-$\frac{1}{2}$ operators. For this reason, it is customary to call the observable associated with this degree of freedom by the rather inept name "isotopic spin," now usually shortened to "isospin" or i-spin. In particular, we say that the nucleon has i-spin $\frac{1}{2}$ and that, just as the values $+\frac{1}{2}$ and $-\frac{1}{2}$ for s_z tell us the actual spin directions of the nucleon, there is a variable t_3 with two possible values, $+\frac{1}{2}$ and $-\frac{1}{2}$, which tell us the charge state of the nucleon. We take $t_3 = +\frac{1}{2}$ for the proton state and $t_3 = -\frac{1}{2}$ for the neutron state.[†] In complete analogy with angular momentum, we introduce $\tau_3 = 2t_3$ (corresponding to $\sigma_z = 2s_z$) and write

$$\tau_3|p\rangle = |p\rangle, \qquad \tau_3|n\rangle = -|n\rangle. \tag{1-1}$$

It is sometimes convenient to use the abbreviations $\gamma \equiv |p\rangle$ and $\delta \equiv |n\rangle$:

$$\tau_3\gamma = \gamma, \qquad \tau_3\delta = -\delta. \tag{1-2}$$

As with spin, it is possible to introduce a two-dimensional matrix representation, as described in Appendix A.

It is tempting to complete this formalism by introducing the other two components τ_1 and τ_2, with properties analogous to those of σ_x and σ_y. We

[†] The opposite sign convention (making T_3 positive for nuclei with neutron excess) is used by some nuclear physicists, but the convention given here is always used in elementary particle work and relates i-spin to charge and hypercharge in the same way for nucleons as for other particles. We recommend this usage for all purposes; the continued existence of two conventions causes confusion.

would then have a vector τ in a three-dimensional "i-spin space." This mathematical exercise would seem pointless unless some physical significance could be attached to the additional two dimensions. In fact, τ_1 and τ_2 can be connected with operators which transform a nucleon from its neutron state into its proton state, and vice versa. Thus they may be useful to describe β-decay. To see this, we recall (Appendix A) that $\sigma_x \pm i\sigma_y$ is the operator that increases (decreases) the z-component of spin by 1 unit. Similarly, if we define $\tau_\pm = \tau_1 \pm i\tau_2$, we have

$$\tau_+|p\rangle = \tau_+\gamma = 0, \qquad\qquad \tau_+|n\rangle = \tau_+\delta = 2\gamma = 2|p\rangle, \qquad (1\text{--}3)$$

$$\tau_-|p\rangle = \tau_-\gamma = 2\delta = 2|n\rangle, \qquad \tau_-|n\rangle = \tau_-\delta = 0. \qquad\qquad (1\text{--}4)$$

Thus τ_+ annihilates a proton state and converts a neutron into a proton, while τ_- has the complementary effect.

The nucleon is a particle obeying Fermi-Dirac statistics and requiring five "coordinates" to specify its position and state—the space coordinates, the component of σ in a specified direction, and the value of τ_3. Any nuclear state must have a wave function which is antisymmetric for the exchange of all coordinates of any two nucleons. Consequently, if the space and spin part of the wave function is symmetric, the i-spin part must be antisymmetric, and vice versa.

When dealing with a two-nucleon system, it is of course convenient to introduce the singlet and triplet spin states, which are, respectively, $S = 0$ and antisymmetric and $S = 0$ and symmetric. The possible i-spin states are as follows:

Symmetric i-spin triplet:
$\quad {}^3(\tau)_1 = \gamma(1)\gamma(2),\quad$ the proton-proton state,
$\quad {}^3(\tau)_{-1} = \delta(1)\delta(2),\quad$ the neutron-neutron state,
$\quad {}^3(\tau)_0 = 2^{-1/2}[\gamma(1)\delta(2) + \gamma(2)\delta(1)],\quad$ a neutron-proton state;
Antisymmetric i-spin singlet:
$\quad {}^1(\tau)_0 = 2^{-1/2}[\gamma(1)\delta(2) - \gamma(2)\delta(1)],\quad$ a neutron-proton state.

This shows explicitly that the need to have specified symmetry against two-particle interchange has led us to a formalism in which the i-spins of two nucleons (and hence of any number) are added by the same rules as are angular momenta. In the two-nucleon case, if we define $T = t(1) + t(2)$, we denote that ${}^3(\tau)_m$ is the state with $T = 1$, $T_3 = m$, and ${}^1(\tau)_0$ has $T = T_3 = 0$.

In so far as one works only with nucleons, the use of i-spin is simply a formalism which does not introduce any physical consequences different from the results of the treatment of neutrons and protons as distinct particles. Sometimes one method is more convenient, sometimes the other.

If the i-spin concept were simply a mathematical convenience, it would not be so important as it is. But in fact i-spin appears to be a physical observable of wide applicability. We introduced it in connection with the nucleon, which can be called a charge doublet; it has two charge states, and we described these by analogy with the two-component state vectors of spin $\frac{1}{2}$ and the operators $\boldsymbol{\sigma}$. Now, there are other particles that have more than one charge state. For example, the π-meson or pion can have charges $+ e, - e$, and zero, and its other properties, like those of the nucleon, depend only slightly on its charge. Can it be assigned an i-spin analogous to angular momentum of 1 unit, writing $t_\pi = 1, t_{3,\pi} = + 1, - 1, 0$ for the three states? Clearly we can do this as a mathematical device, but it turns out to be much more than that. In studying the scattering of pions from nucleons, one finds that i-spin is conserved if one adds the i-spins of the pion and nucleon vectorially, analogous to addition of two angular momenta of magnitudes 1 and $\frac{1}{2}$. The π^+-proton system has the resultant $T_3 = \frac{3}{2}$, and therefore $T = \frac{3}{2}$, whereas π^--proton and π^+-neutron systems are mixtures of $T = \frac{1}{2}$ and $\frac{3}{2}$, with $T_3 = \pm \frac{1}{2}$, and the π^--neutron system is $T_3 = - \frac{3}{2}, T = \frac{3}{2}$. The interaction at low energies is greater in the $T = \frac{3}{2}$ state than in the $T = \frac{1}{2}$ state and does not depend (except for electromagnetic forces) on the value of T_3. Hence the scattering cross sections of $(\pi^+, p), (\pi^-, p), (\pi^+, n)$ and (π^-, n) are all connected by certain relationships, permitting some to be predicted if others are known. These expectations are confirmed by experiment.[†] There are many other examples of particles that constitute charge multiplets, and there is ample confirmation that i-spin is one of the important physical observables associated with a conservation law.

Actually i-spin is not a precise constant of motion in any situation where electromagnetic forces act, since the magnitude of this force depends on the third component of the i-spin. For example, the charge Q_N of a nucleon is given by

$$Q_N = \tfrac{1}{2} + t_3, \tag{1-5}$$

and the Coulomb force between two nucleons is

$$\frac{e^2}{r_{12}} [\tfrac{1}{2} + t_3(1)][\tfrac{1}{2} + t_3(2)].$$

---

† Further details are given in a number of books. In, for example, G. Källén, *Elementary Particle Physics*, Addison-Wesley, Reading, Mass., 1964, pp. 71–79.

The algebra of i-spin being that of angular momentum, t_3 does not commute with the magnitude T of the total i-spin of a system of more than two nucleons; therefore, when the Hamiltonian contains the Coulomb force, T cannot be a constant. However, if the electromagnetic force is weak compared to other forces present, the total i-spin may well have a nearly fixed value with only small components of other values in the wave function. Of course T_3 is always conserved, because it is related to total charge.

For pions the charge

$$Q_\pi = t_3. \tag{1-6}$$

This and the expression for Q_N are special cases of a general relationship,

$$Q = t_3 + \tfrac{1}{2}Y, \tag{1-7}$$

where Y is a quantity, called the hypercharge, which must be assigned to each elementary particle.

1-3 MESONS AND EXCITED NUCLEONS

The most important force acting between nucleons is the strong nuclear interaction. Not only is this interaction associated with nucleons, but also it is the principal force applied to mesons. The meson most significant for low-energy phenomena is the lightest strongly interacting particle, the pion.

Pions are produced in nucleon-nucleon collisions:

$$N + N \rightarrow N + N + \pi. \tag{1-8}$$

There must be sufficient bombarding energy (about 290 MeV) to supply the rest-mass of the pion. Charge must be conserved, so that, although the collision of two protons will permit creation of π^+ and π^0, one of the initial nucleons must be a neutron in order to produce π^-.

Pions have zero spin and negative intrinsic parity. They have masses $m_\pm = 139.6$ MeV, $m_0 = 135.0$ MeV. In the free state they decay, the charged pions predominantly to muons and neutrinos with a mean life of 2.6×10^{-8} sec, the neutral pions electromagnetically and mainly to two γ-rays with a mean life of 0.84×10^{-16} sec. Both these lifetimes are very long compared with typical times for periods of motions of nucleons bound in nuclei (10^{-22} sec), and elementary-particle physicists refer to pions as stable particles.

Pions can also be absorbed by nucleons:

$$\pi + N \rightarrow N. \tag{1-9}$$

It will be seen that pions are created and destroyed by the strong interaction of nucleons, much as photons are created and destroyed by the electromagnetic interaction of charged particles. In the same sense that the electromagnetic force is attributed to the exchange of photons, the nuclear force is attributed to the exchange of mesons. This is discussed in considerable detail in Chapter 5, but some preliminary remarks here will be useful.

A reaction in which one particle absorbs another and does not change its mass is in fact impossible, since the conservation of momentum demands that the initial state have greater energy than the final state. But there can be short-lived fluctuations in the energy of a system, a fluctuation of amount ΔE lasting for a time

$$\Delta t \sim \frac{\hbar}{\Delta E}. \tag{1-10}$$

Hence a nucleon can emit or absorb a pion (or an electron, a photon) provided there is a mechanism for rapidly readjusting the energy of the total system of interacting particles. Let us consider a nucleon that emits a pion with momentum q. Then the energy of the system will have increased by the rest-mass of the pion and the change in the kinetic energy. If q is small $(q \ll m_\pi c)$, then the fluctuation in energy is about $m_\pi c^2$, which corresponds to $\Delta t \sim 5 \times 10^{-24}$ sec. If a mechanism exists to reduce the energy to its original value in this time, such "virtual" mesons will exist. In fact, the emitting nucleon can itself reabsorb the pion, and since this process can occur *in vacuo* it is always present and its effects are a part of the properties of the "physical" nucleon. When we use this phrase, we are contrasting the "physical" nucleon with an idealized nucleon which is "undressed," that is, is not accompanied by the cloud of virtual pions we have just described. It is of course only the "dressed" nucleon which can be observed. The physical and undressed nucleons have different masses; the difference is the self-energy contribution of the meson cloud (and also the similar electromagnetic self-energy).

The emitted meson could also be absorbed if another nucleon were sufficiently close that the emitted meson could reach it within the time Δt. It would need to be within a distance $c \, \Delta t \sim 1.5$ fm.† Then the second nucleon would acquire the extra momentum q carried by the meson. The net effect

---

† The International Union of Pure and Applied Physics authorizes the abbreviation f for femto, signifying 10^{-15}. Thus a femtometer (fm) is 10^{-13} cm. (The full recommendations on symbols, units and nomenclature may be found in *Nucl. Phys.* 81, 677 (1966).) There is a widespread unofficial usage which refers to the length 10^{-13} cm as a Fermi.

is that two nucleons are seen to have scattered off each other, with conservation of momentum and energy. In experiments that do not examine times as short as 10^{-24} sec, the whole interaction process appears as a nucleon-nucleon force, and, if the particles are moving at nonrelativistic speeds, the force can be ascribed to a potential. We see that the nucleon-nucleon potential will be expected to have a range of the order of 1–2 fm.

There is, however, a circumstance in which one particle can absorb another. Photons can be absorbed by a single atom, but of course the atom is raised to an excited state, to conserve both momentum and energy, and only photons of suitable energy can be absorbed. Similarly, a nucleon can absorb a pion if there are excited states of nucleons:

$$\pi + N \rightarrow N^*. \tag{1-11}$$

Photons are elastically scattered from atoms by a process that can be described as absorption and re-emission:

$$\gamma + A \rightarrow A^* \rightarrow A + \gamma.$$

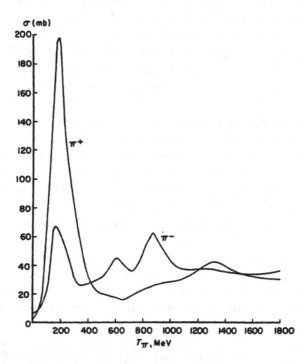

Fig. 1-1 Total cross section for π^+ and π^- scattering from protons from 0- to 1800-MeV pion kinetic energy.

In this case, photons of any energy may be scattered, since the excited state of the atom lives for only a short time and hence some energy fluctuations occur. The maximum scattering occurs for an energy-conserving photon—one with energy near $E(A^*) - E(A)$—and the lifetime of the state A^* determines for how wide a region of energy around this value photons will experience appreciable scattering. In other words, the scattering cross section has a "resonance" for a certain value of the photon energy corresponding to the excited state of the atom.

The total cross section for pions incident on protons is shown in Fig. 1–1. A number of peaks appear in the figure and are highly suggestive of resonances due to excited states of the nucleon. Thorough analysis bears out this interpretation. Some resonances are obvious on visual inspection, and the bumpy shape of the curves implies others that are wide and relatively weak. The lowest few excited states are listed in Table 1–1. The state π^+ with proton has $T_3 = \frac{3}{2}$ and hence has only $T = \frac{3}{2}$ states, whereas π^- with proton has both $T = \frac{1}{2}$ and $T = \frac{3}{2}$ components. The symbol Δ is used for $t = \frac{3}{2}$, and N for $t = \frac{1}{2}$, states.

TABLE 1-1 Resonances of the Pion-Nucleon System: Excited State of the Nucleon. These resonances are well established. The recommended mass is that shown in the name, but the possible experimental range is also indicated. Corresponding ranges are shown for the width. A width of 100 MeV corresponds to $\Delta t = 6.6 \times 10^{-24}$ sec.

Name	T_π MeV	Mass	Width Γ MeV	J^P	t
$\Delta(1236)$	195	1230–1236	110–122	$\frac{3}{2}+$	$\frac{3}{2}$
$\Delta(1650)$	830	1615–1685	130–200	$\frac{1}{2}-$	$\frac{3}{2}$
$\Delta(1670)$	870	1650–1720	175–300	$\frac{3}{2}-$	$\frac{3}{2}$
$\Delta(1890)$	1280	1840–1920	135–350	$\frac{5}{2}+$	$\frac{3}{2}$
$\Delta(1910)$	1330	1780–1935	230–420	$\frac{1}{2}+$	$\frac{3}{2}$
$\Delta(1950)$	1410	1930–1980	140–220	$\frac{7}{2}+$	$\frac{3}{2}$
$N(1470)$	530	1435–1505	165–400	$\frac{1}{2}+$	$\frac{1}{2}$
$N(1520)$	610	1510–1540	105–150	$\frac{3}{2}-$	$\frac{1}{2}$
$N(1535)$	640	1500–1600	50–160	$\frac{1}{2}-$	$\frac{1}{2}$
$N(1670)$	870	1655–1680	105–175	$\frac{5}{2}-$	$\frac{1}{2}$
$N(1688)$	900	1680–1692	105–180	$\frac{5}{2}+$	$\frac{1}{2}$
$N(1700)$	920	1665–1765	100–400	$\frac{1}{2}-$	$\frac{1}{2}$
$N(1780)$	1070	1650–1860	50–450	$\frac{1}{2}+$	$\frac{1}{2}$
$N(1860)$	1220	1770–1900	180–330	$\frac{3}{2}+$	$\frac{1}{2}$

The lowest state, the $\Delta(1236)$, is also by far the most readily excited. It is a $t = \frac{3}{2}$ state, can be found with charge $- 1, 0, 1,$ or 2, and has spin $j = \frac{3}{2}$, parity positive. Because of the j- and t-values it is often called the (3, 3) resonance. Its energy lies just 300 MeV above the proton mass. The shape of the $\pi^- - p$ scattering cross section above 350 MeV is seen to be consistent with a weak peak at 530 MeV due to $N(1470)$, which is lost in the rise to the $N(1520)$ and $N(1535)$ peaks.

A nucleon-nucleon collision at an energy of around 300 MeV can produce a $\Delta(1236)$. Over 500 MeV is needed before the next state, $N(1470)$, can be formed, and it has a much lower formation cross section because of certain angular-momentum considerations, which we shall not discuss here. As we shall suggest in a moment, this energy difference makes the $\Delta(1236)$ about the only excited nucleon state important for nuclear structure.[†]

1-4 MESONIC EFFECTS IN NUCLEI

We began this chapter with the early concept of the indivisible atom and gradually introduced successive levels of complication which brought us to the picture of atomic electrons moving outside a small nucleus composed of massive, but structureless, neutrons and protons, interacting with each other through a nuclear potential, undergoing β-decay, and subject to electro-magnetic interactions. This model has been the main conceptual framework for the remarkable progress of nuclear physics from 1930 to the present. But in Section 1-3 we mentioned the virtual meson clouds which give a nucleon structure and the existence of excited states of nucleons. Are we about to enter an era in which still a further level of complication of our model will be needed? Does the nucleus contain some $\Delta(1236)$'s some of the time? Are the nucleons of the nucleus "dressed" with mesons differently than in free space, because of the presence of the other nucleons? Should we think of a nucleus as nucleon kernels in a meson soup?

Nuclear physics has made immense strides in the experimental examination of fine details of the structure and of the dynamics of the nucleus. In Fig. 1-2 we show the spectrum produced by 70-MeV oxygen ions on ^{188}Os as determined in two experiments, representative of the best techniques available at two times about 10 years apart. This dramatically illustrates the attainment of a new level of detail. But the experimental refinements of recent years have not required any basic change in our theoretical nuclear model; on

---

[†] The baryons Λ and Σ lie at 175 and 250 MeV above the nucleons; however, because strange particles must be produced in pairs, the energy required to have them as virtual particles in the nucleus is greater than these values by the mass of the K-meson (i.e., 500 MeV).

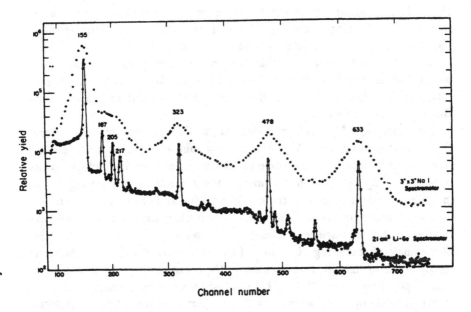

Fig. 1-2 Comparison of the γ-radiation spectra, from Coulomb excitation of ^{188}Os nuclei by a 70-MeV beam of oxygen ions, as measured with a sodium iodide spectrometer and with a lithium-germanium semiconductor spectrometer. (From *Physics in Perspective*, Vol. I, National Academy of Science, Washington, D.C., 1972.)

the contrary, confidence in it has grown as increasingly precise experiment has been understood in terms of increasingly precise calculations.

Have we now, however, reached the point at which a new model should be introduced? The answer depends on the facts to be understood. Clearly, if one contemplates some types of experiment at several hundred MeV, where very short times are important, the structure of the nucleons and the mesonic "components" of the nucleus will be important, as will other aspects of high-energy elementary particle physics. But if one wishes to examine the structure of nuclei in their ground states and other low-lying states and processes up to 100 MeV or so, what model should be used?

There is no single answer to this question. The $\Delta(1236)$ clearly is present in every nucleus for some fraction of the time—how large a fraction? If we think of the wave function of a nucleus as

$$\psi = \psi(A) + \psi(A - 1, \Delta) \qquad (1\text{--}12)$$

with $\psi(A)$ representing the wave function of an A-nucleon system and

$\psi(A - 1, \varDelta)$ a system with one $\varDelta(1236)$ present, we know, from perturbation theory,[†] that the amplitude of $\psi(A - 1, \varDelta)$ is a transition matrix element, $\langle\psi(A)|h(N \to \varDelta)|\psi(A - 1, \varDelta)\rangle$, divided by an average excitation energy. The quantity $h(N \to \varDelta)$ symbolizes that part of the Hamiltonian of the system which changes a nucleon to a $\varDelta(1236)$. It is clear that in order to estimate this quantity we need details of both meson field theory and nuclear structure. But some indications exist.

For example, if ^{208}Pb is bombarded with 24-GeV protons, one can observe the protons coming out. A proton of such high energy has a wavelength short in comparison with the spacing between nucleons in the nucleus; hence it is a good approximation to consider the reaction as proceeding by an individual encounter between the incident proton and one in the nucleus. In such a calculation one uses the measured free nucleon cross section; the wave function of the Pb nucleus is thought to be well understood. As a function of the angle of scattering, the value of the cross section alters by a factor of nearly 10^6, and through this large range the results of the calculation agree with experiment to 15–20%. This impressive agreement suggests that Pb is indeed made up of nucleons, with the properties of free physical nucleons, to at least 85% of its contents. But there is presumably some fraction of $\varDelta(1236)$, and the ground-state wave function presumably also contains a component consisting of a pion moving in an "orbit" in the nucleus, since the lowest state of the pionic Pb atom is only about 100 MeV above the ground state. Since the Pb nucleus has 1600-MeV binding energy, it is easy to realize that the continual random allocation of the kinetic energy among the nucleons could lead to the creation for short times of a state 100 MeV up. Experiments involving pions (e.g., pion production) would be sensitive to a pionic component in the wave function.

At the other end of the periodic table, calculations have been made of the amount of $\varDelta(1236)$ in the wave functions of ^{3}H and ^{3}He [12, 13]. It is found that the probability of the presence of a $\varDelta$ is a few per cent; more exactly, if we write

$$\psi = a\psi(N^3) + b\psi(N^2\varDelta) + c\psi(N\varDelta^2),$$

where we allow for the presence of either one or two $\varDelta$'s, it is found that $|b|^2$ is 0.04–0.06 and $|c|^2$ is 0.02–0.03. These small percentages may, however, have important consequences, particularly in any effect which would involve b or c linearly. (These small terms are of the same order as the important

---

[†] Or from a generalized Bethe-Goldstone equation (Chapter 8).

inclusion of the D-state in the deuteron.) Such percentages in three-body systems permit the otherwise difficult explanation of the β-decay rate of ^{3}H and the magnetic moments of ^{3}He and ^{3}H.

We shall briefly describe the latter point. The magnetic moments of the free neutron and proton are [9, 14] as follows:

$$\mu_p = (2.792782 \pm 0.000017)\mu_0, \qquad \mu_n = -(1.913148 \pm 0.000066)\mu_0,$$

where $\mu_0 = \hbar e/2M_p c$ is the nuclear magneton $= 3.152526 \times 10^{-18}$ MeV/G. A spin-$\frac{1}{2}$ particle of charge e would have magnetic moment μ_0. Hence the proton has an anomaly, $1.79\mu_0$, and the neutron's whole moment, $-1.91\mu_0$, is anomalous. The contributions are to be attributed to the meson clouds. A rough calculation would proceed as follows. The physical proton can be described as a mixture of a bare proton, a π^+ around a bare neutron, and a π^0 around a bare proton, and terms involving more than one pion and other particles. However, the first few terms indicate the idea. All three terms of the physical proton state contribute magnetic moments of varying amounts. The charged pion, with spin 1, with lower mass, and also with some orbital angular momentum, makes a much greater contribution than μ_0; hence a bare neutron with a π^+ will have almost the same moment (except for sign) as a bare proton with a π^-, and each will be much greater than the moment of the bare proton alone. If these two states are about equally admixed in a small amount into the physical proton and neutron, respectively, the magnetic-moment values will be understood.

But if the magnetic moment is so importantly determined by the pionic cloud, we can expect significant meson effects when we put nucleons together to form a nucleus. The first effect will be that some virtual pions are re-absorbed, not by the nucleon that emitted them, but by another. This gives "exchange" currents of charged mesons and therefore leads to "exchange magnetic moments." Also we note that the excited nucleon states, the $\varDelta$'s and N's, will each have a magnetic moment different from that of the proton or neutron, since they have different spins and different internal structures. Given some model of nucleon structure, we can calculate these values and find to what extent the presence of a $\varDelta$(1236) in the nucleus will alter the magnetic moment. Such a calculation [13] substantiates the wave function of Eq. 1–12.

In the rest of this book, there will be only a few references to mesons or $\varDelta$'s as constituents of the nucleus. This reflects the current state of knowledge (or, rather, of ignorance) concerning their role but is not a serious defect, since, as we have seen, these particles are not expected to be present with high probability. Nevertheless, this hitherto largely unexplored aspect of nuclear structure does make it clear that the best parameters for the "pure nucleon" model of the nucleus may sometimes differ somewhat from those derived from the properties of free nucleons. This may have particular significance in connection with "effective" forces between nucleons in nuclei.

REFERENCES

1. E. Rutherford, *Phil. Mag.* **21**, 669 (1911).
2. J. J. Thomson, *Rays of Positive Electricity*, Longmans Green, London, 1913.
3. E. Rutherford, *Phil. Mag.* **37**, 537 (1919).
4. P. Ehrenfest and J. R. Oppenheimer, *Phys. Rev.* **37**, 333 (1931).
5. H. Heitler and G. Herzberg, *Naturwissenschaften* **17**, 673 (1929).
6. F. Rasetti, *Z. Phys.* **61**, 598 (1930).
7. J. Chadwick, *Nature* **129**, 312, and *Proc. Roy. Soc. (London)* **A136**, 692 (1932).
8. M. Curie and F. Joliot, *C. R. Acad. Sci.* **194**, 273, 708, 876, 2208 (1932).
9. B. N. Taylor, W. H. Parker, and D. N. Langenberg, *Rev. Mod. Phys.* **41**, 375 (1969).
10. J. H. E. Mattauch, W. Thiele, and A. H. Wapstra, *Nucl. Phys.* **67**, 1 (1965).
11. C. J. Christenson, A. Nielsen, A. Bahnsen, W. K. Brown, and B. M. Rustad, *Phys. Lett.* **26B**, 11 (1967).
12. A. M. Green and T. H. Shucan, *Nucl. Phys.* **A188**, 289 (1972).
13. M. Ichimura, H. Hyuga, and G. E. Brown, *Nucl. Phys.* **A196**, 17 (1972).
14. V. W. Cohen, N. R. Corngold, and N. F. Ramsey, *Phys. Rev.* **104**, 283 (1956).

PROBLEMS

1-1. Consider $N + N \rightarrow N + N + \pi$. Let the initial state be $p + p$, that is, $T = 1$, $T_3 = 1$. The final state must also have $T = 1$. Therefore the final state is

$$A[(1110|11)|pp\pi^0\rangle + (1101|11)|(pn)^3\pi^+\rangle] + B|(pn)^1\pi^+\rangle$$

where the superscripts on (pn) refer to i-spin triplet and singlet. The quantity A is related to a triplet-triplet production amplitude; B, to a triplet-singlet amplitude. What angular-momentum states are possible for the two final nucleons?

1-2. Calculate the connection between the masses in column 1 and the kinetic energies in column 2 of Table 1-1.

1-3. Calculate the $t = \frac{1}{2}$, $t_3 = \frac{1}{2}$ state of the pion and bare nucleon and find its magnetic moment. Assume the pion to be in an $l = 1$ state. You will need to use an arbitrary value for pion-nucleon distance.

Chapter 2

Internucleon Forces: I

2-1 INTRODUCTION

In order to investigate the forces which act between nucleons, it would seem advisable to study first the simplest systems, namely, those in which only two nucleons are interacting. Actually, only the main features of these systems are necessary background for a discussion of heavier nuclei. Indeed, the finer details of the two-nucleon force are to some extent irrelevant in such discussions because there is no reason to suppose that the force between two nucleons is unaffected by the presence of other nucleons. In other words, the force on a nucleon need not be the simple vector sum of the elementary two-body forces due to the other nucleons in its neighborhood, but may involve terms that arise only when third, fourth, etc., particles are present and that are typical of groups of three, four, etc., nucleons. Such forces are referred to as "many-body forces." Meson theories predict these forces. In spite of many theoretical attempts, however, little is known definitively about many-body forces, except the long-range part of the three-body force. Furthermore, experimentally it is not possible to isolate the effects of these from the effects of two-body forces. Indeed much progress has been made in understanding features of nuclear behavior in terms of two-body forces alone. Nevertheless, we certainly cannot expect the force between two isolated protons to be precisely the same as that between two protons the same distance apart in a nucleus (see Chapter 8).

The two-nucleon systems are, of course, the proton-proton, neutron-neutron, and neutron-proton systems. The last of these, the neutron-proton system, can exist in a stable bound state, the deuteron. The deuteron is a quite simple nuclide in that no excited bound states are known, that is, it exists only in its ground state. Apart from the deuteron, we can study the neutron-proton system also in a state of positive energy in which the particles are free. Such experiments are normally performed by letting a beam of neutrons from a nuclear pile or an accelerator impinge on a substance containing many essentially free protons. The simplest substance is hydrogen gas, but in many experiments it is advantageous to use other substances that contain many hydrogen nuclei, such as thin nylon sheets or paraffin. Corrections must be made for scattering from the other nuclei in these substances, and hence their scattering cross sections must be known. When such a beam of neutrons impinges on protons, a few will be captured to form deuterons, with the extra energy emitted as a γ-ray, but the great majority will be scattered elastically (i.e., without any change in the total kinetic energy of neutron and proton). Of course, the neutron and proton "feel" each other's presence through the nuclear forces which change the magnitude and direction of their velocities. Hence the angle through which a neutron is scattered is determined by nuclear forces, or, conversely, experimental determination of the angular distribution of scattered neutrons and protons gives evidence concerning the nuclear force. In addition to the various properties of the stable state of the deuteron and the observations on nuclear scattering, there is one more source of information on the neutron-proton system, namely, a study of the photodisintegration of the deuteron; since the properties of the electromagnetic field are well known, these experimental results allow us to draw certain conclusions about the nuclear force.

When we turn to the system of two protons, we see an immediate difference in that there are no bound states: ^{2}He does not exist. This means that the nuclear force between two protons is not sufficiently strong to bind the protons in the face of the Coulomb repulsion, e^2/r^2. As we shall see, an analysis of the proton-proton scattering data reveals that the system would be unbound even if there were no electrostatic repulsion. In any case, the experimental evidence on the two-proton system comes entirely from scattering experiments.

There is no direct experimental evidence on the neutron-neutron system. Since neutrons decay to protons in a few minutes, and since it would, in any event, be difficult to devise a suitable box for neutrons, direct neutron-neutron scattering would seem of necessity to involve the intersection of two neutron beams. The intensity of scattered particles would be low. Such experiments have not been performed. There is considerable indirect evidence, however, on the neutron-neutron force, which will be described in this chapter, and which leads to the general conclusion that it is very similar to the force between two protons, excluding the electrostatic interaction. The question of a possible bound state, the dineutron, has been raised from time to time. Experiments have been performed in which it might be a possible product, but have met with negative results. If the *n-n* force were exactly the same as the *nuclear p-p* force, the bound state would not exist. The evidence favors the conclusion that there is no dineutron. An excellent review of the historical development of the subject is given in Brink's book [1].

2-2 POSSIBLE NUCLEON FORCES

Before examining in detail the possible form of the nucleon interaction, one might find it worth while to recall the conservation and invariance laws. The physical behavior of a system is not affected by one's arbitrary choice of coordinate systems. If the energy is independent of the choice of origin or, in other words, if the system behaves identically wherever it is located in space, then the total momentum of the system is a constant of motion. Similarly, if there are no physically preferred directions in space or if the energy is independent of the orientation of the coordinate axes, then the total angular momentum of the system is conserved. There is apparently no "center of the universe" and no preferred direction, or, as is sometimes said, space is homogeneous and isotropic. We therefore expect the two momentum-conservation laws to hold for any isolated system.

A third conservation law, that of parity, is connected with physical invariance under the change from a right-handed to a left-handed coordinate system or, equivalently, under a reflection of axis. In this case the wave function is either odd or even when all coordinates x, y, and z are replaced by $-x$, $-y$, and $-z$. However, if there should be a fundamental process in nature which distinguishes between left- and right-handedness, the wave function describing systems subject to such processes need not have definite parity.

It will be assumed that all nuclei are in states of definite parity and that in all reactions involving only nuclear or electromagnetic forces the total parity of the system is conserved. Strictly, nuclear states are not of definite parity because there is a force between nucleons associated with the β-decay interaction which does possess left-right asymmetry. However, these forces are approximately 10^{-20} of nuclear forces, and it has been estimated that they should produce a term of opposite parity with a relative amplitude of order 10^{-7} in the wave function [2]. Experimentally, the effect of such admixtures in the wave function has been detected, for example, in the α-decay of the 8.8-MeV 2⁻ state of ^{16}O to the 0⁺ ground state of ^{12}C [3] and in the observed left-right asymmetry in the angular distribution of γ-radiation from polarized ^{180m}Hf [4].

We shall begin our discussion with the neutron-proton system, in which there are no electrostatic forces. There is a force due to the two magnetic moments, but this is small enough to be treated as a slight correction to the nuclear forces. As Rutherford's experiment in 1911 on α-particle scattering showed, the nuclear forces must be of very short range, certainly less than 10^{-12} cm. We therefore consider a situation in which no appreciable force acts on the nucleons so long as they are far enough apart, but a reasonably strong force is effective when the distance between them falls below a certain value. Such a situation could be represented, for example, by postulating a potential energy between neutron and proton, given by

$$V = - V_0 \exp(- r/r_n), \qquad (2-1)$$

where r is the distance between the nucleons, r_n is a fixed (short) length, and V_0 is a fixed positive energy. This potential, of course, leads to a force

$$F = - \frac{\partial V}{\partial r} = - \frac{V_0}{r_n} \exp(- r/r_n).$$

The minus sign indicates that the force is attractive, and this is associated with the fact that the potential is negative. This form of potential satisfies our condition that both the potential and the force be essentially zero for sufficiently large r—in this case, for r greater than a few times r_n. Of course, many other potentials are also *a priori* admissible. Forms commonly used are

$$V = - V_0 \exp(- r^2/r_n^2) \quad \text{(Gaussian)}, \qquad (2-2)$$

$$V = - V_0 \frac{\exp(- r/r_n)}{r/r_n} \quad \text{(Yukawa)}, \qquad (2-3)$$

$$\begin{aligned} V &= - V_0, \quad r < r_n \\ &= 0, \quad\quad r > r_n \end{aligned} \quad \text{(square well)}. \qquad (2-4)$$

The exponential, Gaussian, and square-well potentials are employed because they offer mathematical simplicity and because they represent a range of different shapes. In particular, they have different cutoffs at the larger distances: the square-well potential represents an extremely strong force just at $r = r_n$ with zero force elsewhere, while the Gaussian potential, although falling fairly abruptly in a short distance, nevertheless has a tail outside this region. The exponential has a still longer tail (see Fig. 2–1). The Yukawa potential has some theoretical significance and was predicted by the earliest form of meson theory. It provides an example of a potential that is strongly attractive at very small distances of separation.

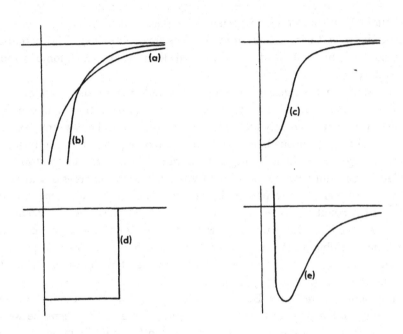

Fig. 2-1 Various proposed nuclear potentials. The shapes shown are (a) exponential, (b) Yukawa, (c) Gaussian, (d) square-well, and (e) repulsive core followed by exponential.

We shall use the expression "shape" of the potential in the following sense. If the constants, such as V_0 and r_n, in the potential are varied but the functional form remains unchanged, we say that the shape is unchanged. It will be realized that a potential shape must have at least two constants: one, V_0, determining the strength of the potential, and the other, r_n, fixing its range. It could, however, have more than two parameters. An example is the so-called repulsive core potential, which we shall have to consider later. This is shown in Fig. 2–1e and is defined by

$$V = \infty, \qquad r < r_c,$$

$$= - V_0 f\left(\frac{r}{r_n}\right), \quad r > r_c, \qquad (2\text{--}5)$$

where $f(r/r_n)$ is any short-range function, such as one of the four shapes already considered. This potential implies a very strong (strictly infinite) repulsive force at $r = r_c$, with an attractive force outside. In effect, it is as though the nucleons were impenetrable spheres of diameter r_c, since the distance between their centers cannot become smaller than r_c. This potential may, however, be thought of as a simplified description of a region of strong repulsive forces.

In describing the nuclear force by such simple potentials we are making great assumptions. In the first place, the concept of potentials is, in general, an approximation, valid only for nonrelativistic particles. Thus we may expect it to be valid for nucleons bound in nuclei or moving with reasonably small kinetic energy, but we may begin to doubt its validity for nucleons of 300 MeV. Certainly some form of relativistic correction is necessary at such energies. Secondly, these potentials are central, that is, they depend only on r and the resulting forces are directed along the join of the two particles. It is true that for point particles it is hard to imagine how dependence on an angular variable could arise, since there is no direction in space other than the join of the particles determined by the physics of the problem. (This is not true if we admit velocity-dependent forces, since the relative velocity provides another direction.) But nucleons have intrinsic spin, and hence there seems to be no reason why the force could not depend on the angles between the two spin directions, or on the angles between the spin directions and the join **r**. Indeed, we shall see that it depends on all these angles.

Let us examine the dependence on the angle between the two spin vectors. Since the energy of the system is assumed to be invariant under translation, rotation, and reflection of the coordinate system, all terms in the energy must be scalars. Consequently, the vectors σ_1 and σ_2 must appear in the potential in the form of the scalar product $\sigma_1 \cdot \sigma_2$. This, then, is the nature of the spin dependence. Now $\sigma_1 \cdot \sigma_2$ is an operator with definite eigenvalues for the singlet and triplet spin states. Indeed,

$$\sigma_1 \cdot \sigma_2 \, {}^1(\sigma)_0 = - 3 \, {}^1(\sigma)_0,$$

$$\sigma_1 \cdot \sigma_2 \, {}^3(\sigma)_m = {}^3(\sigma)_m.$$

Hence, if we write

$$V = - V_0 f\left(\frac{r}{r_n}, \sigma_1 \cdot \sigma_2\right), \qquad (2\text{--}6)$$

we have defined two potentials, one for the singlet state and the other for the triplet state:

$$V_s = - V_0 f\left(\frac{r}{r_n}, -3\right), \qquad (2\text{--}7a)$$

$$V_t = - V_0 f\left(\frac{r}{r_n}, 1\right). \qquad (2\text{--}7b)$$

As an important example, take

$$V = - V_0(A + B\sigma_1 \cdot \sigma_2) f\left(\frac{r}{r_n}\right). \qquad (2\text{--}8)$$

This gives potentials of the same shape in singlet and triplet states, but with different strengths, namely,

$$V_s = - (A - 3B) V_0 f\left(\frac{r}{r_n}\right), \qquad (2\text{--}9a)$$

$$V_t = - (A + B) V_0 f\left(\frac{r}{r_n}\right). \qquad (2\text{--}9b)$$

More generally, of course, A and B could be functions of r, so that it is possible to have quite different forces in the singlet and triplet states. However, in each state, the force is central, so that Eq. 2–6 is usually referred to as a central potential.

A case of particular interest is Eq. 2–8 with $A = B = \frac{1}{2}$, that is,

$$V = - V_0(\tfrac{1}{2} + \tfrac{1}{2}\sigma_1 \cdot \sigma_2) f\left(\frac{r}{r_n}\right), \qquad (2\text{--}10)$$

$$V_s = V_0 f\left(\frac{r}{r_n}\right), \qquad (2\text{--}11a)$$

$$V_t = - V_0 f\left(\frac{r}{r_n}\right). \qquad (2\text{--}11b)$$

The singlet and triplet potentials are equal but have opposite signs, the singlet corresponding to a repulsive force. As explained in Appendix A, $P_\sigma = \frac{1}{2}(1 + \sigma_1 \cdot \sigma_2)$ is called the spin-exchange operator, since it exchanges the spins of the two particles in the wave functions. Consequently Eq. 2–10 describes a "spin-exchange force." For historical reasons, this is often called a Bartlett force [5].

In addition to the angles discussed above, there is another variable on which the potential might depend. This is the *parity* of the state. We can best see how to introduce this dependence by considering another exchange operator. We define P_x as the operator that exchanges the position coordinates of the two particles:

$$P_x\psi(\mathbf{r}_1, \mathbf{r}_2) = \psi(\mathbf{r}_2, \mathbf{r}_1). \tag{2-12}$$

For a two-particle system, this is the same as changing the sign of the relative coordinates. If the parity of the state is even, the interchange of space coordinates leaves the wave function unaltered, whereas if the parity is odd, the sign of the wave function is changed. For a two-particle system, in a state of definite orbital angular momentum, the parity is even or odd, according as the quantum number l is even or odd. Hence

$$P_x \equiv 1 \qquad (l \text{ even}), \tag{2-13a}$$

$$P_x \equiv -1 \qquad (l \text{ odd}). \tag{2-13b}$$

A potential

$$V = -V_0 f\left(\frac{r}{r_n}\right) P_x \tag{2-14}$$

is called a "space-exchange" potential or a Majorana potential [6]. We can also introduce a Heisenberg exchange force [7] defining the operator

$$P_H = P_\sigma P_x, \tag{2-15}$$

and write

$$V = -V_0 f\left(\frac{r}{r_n}\right) P_H. \tag{2-16}$$

It will be realized that the Heisenberg operator, which exchanges both spin and space coordinates, is equivalent to an exchange of the charges of the nucleons. Indeed, $P_H = -P_\tau$, where P_τ is a charge-exchange operator analogous to P_σ but based on i-spin.

To complete the assignment of physicists' names to types of forces and to commemorate one of the early workers in this field, it is customary to refer to an ordinary force derived from $V = -V_0 f(r/r_n)$ as a Wigner force [8]. Originally this name was applied to any short-range force, but it has now come to be used exclusively for exchange-independent forces.

To summarize this discussion, we see that it is possible to write a potential which, while still essentially central, displays spin and parity dependence in the form

$$V = - V_0[W(r) + B(r)P_\sigma + M(r)P_x + H(r)P_H], \qquad (2\text{--}17)$$

where the various functions of r are arbitrary. Since Eq. 2–17 implies different potentials for (a) triplet and singlet states, and (b) states of even and odd parity, we can equally well speak of four different potentials: one for triplet even states (such as $^3S_1$, $^3D_1$); one for triplet odd states ($^3P_0$, $^3P_1$, $^3P_2$,...); one for singlet even states ($^1S_0$, $^1D_2$,...); and one for singlet odd states ($^1P_1$, $^1F_3$,...). This last method of description corresponds more closely to the experimental situation (at least at low energies), where individual states may be studied with the hope of finding the form of the potential acting. Of course, if we knew the potentials in the four types of states, we could easily calculate the functions in Eq. 2–17 (see Problem 2–5).

2-3 SCATTERING EXPERIMENTS

Consider a neutron beam incident on a target. As suggested in Fig. 2–2, many of the neutrons may not come close enough to any nucleus to interact with it and therefore pass straight through the target. Some, however, do come within nuclear force range of a nucleus and are scattered. We must be a little more precise. The de Broglie wavelength of a neutron is determined by its momentum through the formula $\lambda = \hbar/p$. The position of a neutron cannot be considered to be fixed more closely than λ. Hence, if λ is of the order of magnitude of the distance between nuclei in the target, a neutron cannot avoid a collision. There will still be some probability of its continuing undeflected, but a large number of unscattered neutrons (as in Fig. 2–2)

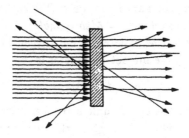

Fig. 2-2 Scattering of a neutron beam incident from the left.

implies that $\lambdabar$ is small in comparison with internuclear distances. In such a case, we can consider each neutron in interaction with at most one nucleus, the one that it approaches most closely. Then the neutron is well localized so far as the target nuclei are concerned. However, since the distance between two nuclei, even in a molecule, is never much less than 1 Å, $\lambdabar$ can be much smaller than internuclear distance, but at the same time considerably greater than the range of nuclear forces. This is the situation for neutrons of energy up to about 10 MeV.

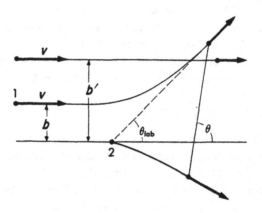

Fig. 2-3 Various parameters in a scattering experiment.

Let such a neutron approach a proton with energy $E = \frac{1}{2}M_n v^2$. Since it is fairly well localized, let us apply a mixture of classical and quantum mechanics. Let its impact parameter be b (Fig. 2–3). Then its angular momentum about the proton is $L = Mvb = \hbar(b/\lambdabar)$. Quantum-mechanically, we know that $L^2 = l(l + 1)\hbar^2$, where l is an integer. Hence, for a neutron with quantum number l, the impact parameter is

$$b \sim [l(l + 1)]^{1/2}\lambdabar. \qquad (2\text{--}18)$$

The neutron will not be scattered unless it is acted upon by the nuclear field of the proton, that is, unless $b \lesssim r_n$, the nuclear force range. Hence, for a given neutron energy and therefore $\lambdabar$, only those neutrons are scattered whose quantum number l is given by

$$l(l + 1)\lambdabar^2 < r_n{}^2. \tag{2-19}$$

Thus, so long as $\lambdabar/r_n$ is greater than about $2^{-1/2}$, only neutrons of zero angular momentum will be scattered. For sufficiently low energies ($\lesssim 10$ MeV), only the S-state potential need be considered; as the energy increases, the P-, D-, etc., states progressively become of importance. (This is, of course, a semiclassical argument, and the result is not exact.)

We set our problem up in the coordinate system in which the center of mass is at rest. Figure 2–3 shows the situation in the laboratory coordinate system in which particle 2 (the proton) is initially at rest. The center of mass G is moving, and θ_{lab} is the laboratory scattering angle. Particle 1 has initial velocity v in the laboratory system. Figure 2–4 shows the same scattering process in the system in which G is at rest. Particles 1 and 2 have initial (and final) speeds v_1 and v_2, and θ is the scattering angle. The following relationships hold:[†]

$$\tan \theta_{\text{lab}} = \frac{\sin \theta}{\cos \theta + M_1/M_2}, \tag{2-20a}$$

$$v_1 = \frac{M_2}{M_1 + M_2} v, \tag{2-20b}$$

$$v_2 = \frac{M_1}{M_1 + M_2} v, \tag{2-20c}$$

$$E = \tfrac{1}{2}M_1 v_1{}^2 + \tfrac{1}{2}M_2 v_2{}^2 = \frac{1}{2}\frac{M_1 M_2}{M_1 + M_2} v^2 = \frac{1}{2}\frac{M}{M_1} E_{\text{lab}}, \tag{2-20d}$$

where

$$M = \frac{2M_1 M_2}{M_1 + M_2}. \tag{2-20e}$$

The quantity E is the center-of-mass energy or the kinetic energy of relative motion, E_{lab} is the energy in the laboratory, and M is twice the reduced mass of the system. If we ignore the difference between proton and neutron masses, M is the "mass of a nucleon," and

$$\theta_{\text{lab}} = \tfrac{1}{2}\theta, \tag{2-21a}$$

$$v_1 = v_2 = \tfrac{1}{2}v, \tag{2-21b}$$

$$E = \tfrac{1}{2}E_{\text{lab}}. \tag{2-21c}$$

[†] For example, ref. [26], p. 86.

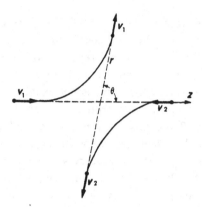

Fig. 2-4   The scattering in the center-of-mass system.

The momentum of relative motion is $\frac{1}{2}Mv$, and we shall define

$$k = \frac{1}{2}\frac{Mv}{\hbar} = \left(\frac{ME}{\hbar^2}\right)^{1/2}. \qquad (2\text{-}22)$$

It is clear, of course, that a uniform beam of neutrons incident on a proton contains neutrons of many angular momenta; if the beam could be considered of infinite cross section, it would contain all angular momenta. This means that the wave function which represents a beam of particles of momentum $k\hbar$ traveling in the z-direction, that is, e^{ikz}, can be expanded in a series of the wave functions appropriate to all angular momenta. Mathematically,

$$e^{ikz} = \sum_{l=0}^{\infty} (2l + 1)i^l j_l(kr)P_l(\cos\theta). \qquad (2\text{-}23)$$

The geometry is indicated in Fig. 2-4: P_l is the Legendre polynomial (the angular-momentum wave function), and $j_l(kr)$ is a Bessel function, whose properties are summarized in Appendix B.   Roughly speaking, we can interpret $|(2l + 1)i^l j_l(kr)|^2$ as the probability that a neutron of angular momentum l will be found at a distance r. If $\lambda > r_n$, the motion of particles with $l > 0$ is undisturbed; they are not scattered, and the terms of Eq. 2-23 for $l > 0$ are the appropriate wave functions for the complete problem. To examine the scattering, we must study the wave equation for $l = 0$.

For the S-state, we can take as function $(1/r)u(r)$, since for $l = 0$ there is no angular dependence. The Schrödinger equation in the center-of-mass system gives

$$\frac{d^2u}{dr^2} + [k^2 - V(r)]u = 0, \qquad (2\text{–}24)$$

where $V(r)$ is the neutron-proton potential in the S-state in question, which can be either singlet or triplet, depending on the spin orientations of the incident and target nucleons. For very large values of r, the nuclear potential is zero, and Eq. 2–24 becomes

$$\frac{d^2u}{dr^2} + k^2u = 0, \qquad r \gg r_n. \qquad (2\text{–}25)$$

Since the two independent solutions of this equation are e^{ikr} and e^{-ikr}, the wave function at large r has the form†

$$u \sim A \sin(kr + \delta) = \frac{Ae^{-i\delta}}{2i} (e^{2i\delta}e^{ikr} - e^{-ikr}). \qquad (2\text{–}26)$$

The first term in the parentheses represents particles traveling outward from the scattering center; the second term, incoming particles. But in a scattering experiment the only incoming particles are those in the incident beam. Therefore we must adjust the value of A so that the incoming particles of Eq. 2–26 are just those of the $(l = 0)$ part of the plane wave (of Eq. 2–23), namely, $j_0(kr)$:

$$j_0(kr) = \frac{\sin kr}{kr}$$

$$= \frac{1}{2ikr} (e^{ikr} - e^{-ikr}). \qquad (2\text{–}27)$$

Thus, if we take $A = e^{i\delta}/k$, $u(r)/r$ will contain as incoming particles only the particles in the incident beam. Then $u \sim (e^{ikr}e^{2i\delta} - e^{-ikr})/2ik$. Comparing this with Eq. 2–27, we see that the effect of scattering on the asymptotic wave function is to modify the outgoing part of the wave by a factor $e^{2i\delta(k)}$, which is denoted by the symbol

$$S(k) = e^{2i\delta(k)}. \qquad (2\text{–}28)$$

---

† The symbol $\sim$ means "equal for large r." The constant δ is real. If it were not, the amplitudes of the outgoing and incoming waves would be different; this would mean that inelastic collisions were occurring, effectively removing particles of wave number k. The only inelastic $n\text{-}p$ process is radiative capture, which has a very small cross section.

In this simple example, $S(k)$ is just a k-dependent number. In more complicated cases with noncentral forces present, the Schrödinger equation has two channels which are coupled (see Chapter 5), and then $S(k)$ is a 2×2 matrix. It is for this reason that $S(k)$ is generally called the S-matrix. Writing

$$u \sim \frac{1}{2ik} (e^{ikr} - e^{-ikr}) + \frac{1}{2ik} (e^{2i\delta} - 1)e^{ikr}$$

$$\sim u_{\text{inc}} + u_{\text{sc}},$$

where u_{inc} is the incident and u_{sc} is the scattered wave, we see that the scattered particles are represented by

$$u_{\text{sc}} \sim \frac{1}{2ik} (e^{2i\delta} - 1)e^{ikr} = \frac{e^{i\delta}}{k} \sin \delta \, e^{ikr}. \qquad (2\text{–}29)$$

Hence we can write

$$u \sim u_{\text{inc}} + f_k e^{ikr}, \qquad (2\text{–}30)$$

where

$$f_k = \frac{e^{i \, \delta(k)}}{k} \sin \delta(k). \qquad (2\text{–}31)$$

The quantity f_k is called the scattering amplitude. For S-waves, as we see, f_k is independent of θ, but for higher partial waves it is θ dependent.

Consider a sphere S of large radius R. In one second the scattered particles which were on the surface of S will be on a sphere of radius $R + v$, and during this second all particles that lie between R and $R + v$ will have passed through the surface of S. From Eq. 2–29, we see that there are $\sin^2\delta/k^2 r^2$ scattered particles per unit volume. Hence the number of scattered particles that pass through the sphere per second is $(\sin^2\delta/k^2 R^2)4\pi R^2 v = (4\pi v/k^2) \sin^2\delta$. This is the number of particles scattered per second.

By similar arguments, the wave function e^{ikz} represents a beam of v incident particles per unit area per second. Since the cross section for any event is defined as the number of desired events per second divided by the number of particles incident per unit area per second, we see that the cross section for scattering in the S-state is

$$\sigma_S(E) = \frac{4\pi}{k^2} \sin^2\delta = 4\pi \lambdabar^2 \sin^2\delta. \qquad (2\text{–}32)$$

We remark that S-wave scattering is isotropic, since u has no angular dependence. The scattering in states of higher angular momentum is not isotropic; this makes it possible, by measuring the angular distribution experimentally, to decide at what energy the effects of higher angular momenta are becoming important.

Let us consider very low velocities, indeed the limit of zero energy. Since

$$u \sim e^{i\delta} \frac{\sin(kr + \delta)}{k} ,$$

it is clear that, if u is to remain finite at large but finite values of r (as it must), δ must approach zero at least as fast as k does. Thus $\sin \delta/k$ must be finite. This can also be seen from Eq. 2-29 for the scattered amplitude u_{sc}. If we define the *scattering length* a by the equation

$$\lim_{k \to 0} \left(-\frac{\sin \delta}{k} \right) = a, \tag{2-33}$$

we have

$$\sigma_S(0) = 4\pi a^2 \tag{2-34}$$

and

$$u \sim \frac{e^{i\delta}}{k} (\sin kr \cos \delta + \cos kr \sin \delta) \to r - a. \tag{2-35}$$

Equations 2-32 and 2-34 show that the phase shifts δ (or, for very low energy, the scattering length a) are the quantities that may be found by measuring the scattering cross section. The quantities δ and a are in turn related to the potential causing the scattering by asymptotic expressions for u. This is seen as follows.

The Schrödinger equation (Eq. 2-24) must be satisfied by a quantity $u(r)$ with the boundary condition

$$u(0) = 0. \tag{2-36}$$

[Otherwise the wave function $u(r)/r$ would be infinite.] This initial condition fixes $u(r)$ except for an arbitrary multiplication constant, A, which was determined by consideration of the incoming wave in u. Hence, if we solve the Schrödinger equation subject to $u(0) = 0$, we shall be able to establish the asymptotic form of u and therefore find δ. Similarly, for $E = 0$, the equations

$$\frac{d^2 u_0}{dr^2} - V(r)u_0 = 0 \tag{2-37}$$

and

$$u_0(0) = 0 \qquad (2\text{-}38)$$

will yield a solution which asymptotically is proportional to $r - a$. Hence a can be found if V is known.

It is clear, then, that if a potential $V(r)$ is suggested as the nuclear potential, its agreement with experiment (at least so far as low-energy scattering is concerned) can be checked by solving Eq. 2–24 subject to 2–36 and thus determining δ at various energies. The converse procedure, that is, deducing $V(r)$ from a knowledge of δ at various energies, is not as straight-forward; indeed, it can be shown that the values of δ define a unique potential function only under certain restrictive assumptions. In particular, low-energy scattering does not reveal very much about the shape of the potential, as we shall now proceed to show.

Let us introduce the symbol $\psi(r)$ for the asymptotic form of u. The quantity ψ is determined by solving Eq. 2–24. Thus, for a fixed nuclear potential V, ψ depends only on r and k^2, apart from a normalizing constant; this means that $(1/\psi)(d\psi/dr)$ depends only on r and k^2. Therefore, since $\psi = A \sin(kr + \delta)$, we have, setting $r = 0$,

$$k \cot \delta = \left(\frac{1}{\psi} \frac{d\psi}{dr} \right)_{r=0}. \qquad (2\text{-}39)$$

However, if we are concerned only with low energies, we may expand this function in powers of k^2. Hence we see that, in general,

$$k \cot \delta = \alpha + \beta k^2 + \cdots. \qquad (2\text{-}40)$$

We can fix α by considering zero energy, for by Eq. 2–33 $\lim_{k \to 0} (k \cot \delta) = -1/a$. Also, it is clear that β must have the dimension of length. We set $\beta = \frac{1}{2}r_0$ and write

$$k \cot \delta = \frac{-1}{a} + \tfrac{1}{2}r_0 k^2, \qquad (2\text{-}41)$$

which is valid, provided k is sufficiently small so that higher powers may be neglected. To a good approximation, this condition allows energies up to about 10 MeV. It is shown in Appendix B that the length r_0 is given by

$$r_0 = 2 \int_0^\infty \left[\left(\frac{\psi_0}{a} \right)^2 - \left(\frac{u_0}{a} \right)^2 \right] dr, \qquad (2\text{-}42)$$

where u_0 is the zero-energy wave function and ψ_0 is its asymptotic form.

From Eq. 2–35 we have

$$r_0 = 2 \int_0^\infty \left[\left(1 - \frac{r}{a} \right)^2 - \left(\frac{u_0}{a} \right)^2 \right] dr. \tag{2–43}$$

Since $\psi_0(r) = u_0(r)$ when r is appreciably outside the range of nuclear forces, we see that r_0 is a length of the same order as the range of the nuclear potential. It is called the *effective range*. Since the result of a scattering experiment is simply the determination of the phase shifts, it follows that, as long as Eq. 2–41 is valid, the only data available from scattering are the values of a and r_0. We are entitled, then, to think of these two lengths as experimentally determined quantities.

We may summarize this discussion by saying that the low-energy ($\lesssim 10$ MeV) scattering of the neutron-proton system is determined by two quantities: the scattering length a and the effective range r_0. Consequently, experiments at these energies cannot distinguish between potentials that yield the same values of a and r_0 upon solution of Eqs. 2–37 and 2–42. As the expression $V_0 f(r/r_n)$ makes clear, any potential function contains at least two constants which can, in general, be adjusted to give the experimental values of a and r_0, whatever the form of $f(r/r_n)$ may be. This shows that low-energy scattering experiments cannot determine the shape of the potential; or, alternatively, the low-energy properties of the neutron-proton system are insensitive to the shape of the potential.

2-4 BOUND STATE

We can see that not only scattering, but also the properties of a bound state, may be expressed in terms of a and r_0. For a bound state, the energy is negative, say $- \epsilon$, and

$$k = i\gamma = i \left(\frac{M\epsilon}{\hbar^2} \right)^{1/2}. \tag{2–44}$$

Then Eqs. 2–24 and 2–25 still hold, and the asymptotic form of u given by 2–26 is

$$u \sim \frac{A}{2i} (e^{i\delta} e^{-\gamma r} - e^{-i\delta} e^{\gamma r}). \tag{2–45}$$

For a bound state, the probability of a large separation of the particles must be small, that is, u must behave like $e^{-\gamma r}$. Hence $e^{-i\delta} = 0$, and consequently $i\delta$ is a large positive real number. Also A must be small so that the product $Ae^{i\delta}$ remains finite and is the normalizing factor. These conditions replace the restriction to outgoing waves required in the scattering problem to determine A. However, the argument that leads to $k \cot \delta = -1/a + \frac{1}{2}r_0 k^2$ remains valid. Now $\cot \delta = i \coth i\delta = i$ as $i\delta$ approaches infinity. Thus $k \cot \delta = ik = -\gamma$, and we find

$$\gamma = \frac{1}{a} + \frac{1}{2}r_0\gamma^2. \tag{2–46}$$

This means that the binding energy of the deuteron can be determined from a knowledge of the scattering length and effective range of the force acting between the nucleons.[†] It will be realized that Eq. 2–46 is valid only for sufficiently small γ. Otherwise, higher powers should appear on the right-hand side of the equation. Fortunately, the binding energy of the deuteron turns out to be small enough (2.226 MeV) so that the approximation is good.

2-5 EXPERIMENTAL RESULTS

The preceding discussion of scattering has been carried out without regard to spin dependence of the forces. In a situation in which each neutron can be considered as interacting with just one proton, the encounter will be either a singlet or a triplet one, and if there is no spin polarization (i.e., no preferred direction of spin), the total cross section will be

$$\sigma(E) = 4\pi\lambda^2(\tfrac{1}{4}\sin^2\delta_s + \tfrac{3}{4}\sin^2\delta_t). \tag{2–47}$$

This expression arises since there are three triplet states and one singlet state and hence a three-to-one probability that the nucleons will collide in the triplet state. At very low energies, therefore, we have

$$\sigma(0) = \pi(a_s^2 + 3a_t^2). \tag{2–48}$$

The subscripts refer to the singlet and triplet states.

If, however, the neutron is sufficiently slow so that it is in effective interaction with more than one proton, there will be interference of the waves from the two scatterers. In this case, as in all such wave phenomena, it becomes necessary to add the amplitudes of the waves. This effect is significant for the so-called thermal neutrons, which have energy of a few one-hundredths of an electron volt. Experiments with these neutrons provide information

---

[†] This is true only if the range of the force is much smaller than the extent of the deuteron wave function. For examples of potentials for which Eq. 2–46 is not valid, see K. F. Chong and Y. Nogami, *Am. J. Phys.* **39**, 182 (1971).

about the *coherent-scattering amplitude* a_H for the hydrogen molecule, which is given by

$$a_H = \tfrac{1}{2}(a_s + 3a_t). \tag{2-49}$$

Such experiments are discussed further in Appendix B.

The most accurate determinations of the total cross section for slow neutrons and of the coherent-scattering amplitude are as follows:

$$\sigma(0) = 20.442 \pm 0.23 \text{ b} \quad [9],^{\dagger}$$

$$a_H = -3.707 \pm 0.008 \text{ fm} \quad [10].$$

Substitution of the weighted means of these values into Eqs. 2-48 and 2-49 yields

$$a_s = -23.710 \pm 0.030 \text{ fm},$$

$$a_t = 5.432 \pm 0.005 \text{ fm}.$$

The signs of a_s and a_t are in themselves instructive. A negative scattering length indicates that the corresponding potential does not have a bound state, at least not for any reasonably small value of binding energy. This can be seen immediately from Eq. 2-46, which shows that, if a bound state were assumed and we used this equation to calculate its binding energy ϵ, the essentially positive quantity γ would turn out to be negative. It can also be seen from a consideration of the graphs of the zero-energy wave function. Writing

$$\frac{d^2u}{dr^2} = -V_0 f\left(\frac{r}{r_n}\right)u - \frac{2mE}{\hbar^2}u, \tag{2-50}$$

we notice that at zero energy d^2u/dr^2 is negative until the potential vanishes when it is zero. If the scattering length is negative, we have the result for u_0 shown in Fig. 2-5a. If we now consider the effect of the last term in Eq. 2-50 for small negative E, we see that d^2u/dr^2 is increased so that u slopes upward more rapidly than u_0 and continues to rise after the potential has vanished, giving a wave function that is infinite at large values of r. The case $a > 0$ is illustrated in Fig. 2-5b. Here the effect of the increase in the second derivative is, in general, the same, that is, an unbounded wave function, but it is possible that for certain energies the wave function will decay exponentially as illustrated by the curve u_B, giving a bound state. We deduce, then, that the singlet S-state is unbound. However, the large value of a_s means

---

† The measure 1 barn = 10^{-24} cm^2 is a useful unit for nuclear cross sections.

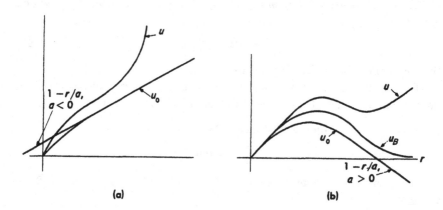

Fig. 2-5 Wave function u_0 for $E = 0$ and u for nonzero E. In (a) the scattering length a is negative, and in (b) it is positive; u_B is the wave function for a bound state.

that it is "only just" unbound, that is, the line $1 - r/a$ is not rising extremely rapidly. This result also suggests very strongly that the deuteron might be in the triplet S-state, an assumption which receives experimental support from measurement of the total angular momentum J, which is 1. Although this would not rule out P- and D-states, as we shall see in Chapter 5 the electromagnetic properties show that the wave function of the deuteron is more than 90% that of the ^3S-state. For many purposes, including those of this section, it is sufficiently accurate to treat the deuteron as a ^3S-state.

Because of these facts, Eq. 2–46 provides us with a good value of the triplet effective range r_{0t}, since ϵ and a are accurately known. The result is

$$r_{0t} = 1.749 \pm 0.008 \text{ fm}.$$

To determine r_{0s} we have to use the results of scattering at small energies in the form of Eq. 2–47, which can be written as

$$\sigma(E) = \pi[(k^2 \cosec^2\delta_s)^{-1} + 3(k^2 \cosec^2\delta_t)^{-1}]$$

$$= \pi\left\{\left[k^2 + \left(\frac{1}{a_s} - \tfrac{1}{2}r_{0s}k^2\right)^2\right]^{-1} + 3\left[k^2 + \left(\frac{1}{a_t} - \tfrac{1}{2}r_{0t}k^2\right)^2\right]^{-1}\right\}. \quad (2\text{–}51)$$

The term in r_{0s} is a small part of the cross section, so that great accuracy in measurements of $\sigma(E)$ is required to obtain a good value of r_{0s}. The available data [11] yield

$$r_{0s} = 2.73 \pm 0.03 \text{ fm}.$$

As we have remarked, the scattering length and effective range are sufficient to fix the constants in any potentials with two parameters. These potentials may be chosen to be of square, Gaussian, exponential, or Yukawa shape.

It is clear that the expressions "range of nuclear forces" and "strength of nuclear potential" are not precise, unless the shape of the potential is specified. Nevertheless, they are frequently used, and it is customary to think of the range of forces as being 1.5–2 fm and the average depth as being 20–35 MeV, depending on the spin state.

2-6 PROTON-PROTON FORCES

Experimentally, the p-p scattering data have been measured with very high precision. The theoretical analysis of the data, however, is not so clear as in the n-p case, because it is difficult to isolate the nuclear and electromagnetic effects. This would be simple if the only electromagnetic effect were a Coulomb potential e^2/r between the two protons. For the present, we naively assume this to be so, in order to understand how the scattering formalism is modified in the presence of a long-range force. At the end of this section we indicate what corrections should be made to this simple analysis.

Irrespective of the electromagnetic effects, some modifications are necessary in the theoretical framework because the two protons are identical particles and are indistinguishable from each other. Experimentally, this means that either one of the two protons is detected at an angle θ (in the center-of-mass frame) from the incident direction. The two protons are scattered back to back in the CM frame. This implies that a proton is detected when the scattering angle is either θ or $\pi - \theta$. Quantum-mechanically, this situation requires the addition of the corresponding scattering amplitudes in order to calculate the proton flux. In Section B–3 the details are given, and it is shown that interference terms appear in the formula for the cross section.

Two protons in the S-state can interact only in the singlet spin state because of the exclusion principle. First, let us suppose that only the Coulomb potential e^2/r is operating between the two protons, and that the nuclear interaction has been switched off. Because the Coulomb interaction has an infinite range, all the partial waves are affected even at low relative energies, unlike the nuclear case in which only the S-state is modified. Let us still concentrate on the S-state. On solving the radial Schrödinger equation with the Coulomb potential in the S-state, we obtain two independent solutions, $F_0(r)$ and $G_0(r)$, which are given in Section B–1. The regular solution is $F_0(r)$, which, for large r, takes the form

$$F_0 \sim \sin(kr - \eta \ln 2kr + \sigma_0), \tag{2-52}$$

where $\eta = \frac{1}{2}kR$, R being the characteristic Coulomb length for protons, given by

$$R = \frac{\hbar^2}{M_p e^2} = 28.8(15) \text{ fm}, \tag{2-53}$$

and σ_0 is a function of k alone. This should be compared with the asymptotic form of the radial wave function in the presence of a short-range nuclear potential $V(r)$ alone (see Eq. 2-26):

$$u \sim \sin(kr + \delta_0). \tag{2-26}$$

In Eq. 2-26, all that the short-range potential did to the unperturbed form $\sin kr$ was to change its phase by $\delta_0(k)$, and this was called the phase shift. In Eq. 2-52, however, we see that the Coulomb potential not only changes the phase by σ_0 but also continually distorts the wave through the factor $\eta \ln 2kr$. This occurs because the Coulomb potential has an infinite range and its presence is felt for all values of r. We still call σ_0 the Coulomb phase shift.

Now consider the case in which both the Coulomb potential and the short-range nuclear potential $V(r)$ are present. Outside the range of $V(r)$, only the Coulomb potential is present, so that for large r the wave function has the same form as Eq. 2-52:

$$u_C \sim \sin(kr - \eta \ln 2kr + \sigma_0 + \delta_0^C), \tag{2-54}$$

where δ_0^C is now the nuclear phase shift with respect to the Coulomb wave function (see Section B-3 for details). If there were no Coulomb potential but only $V(r)$, the asymptotic wave function would be written as

$$u \approx \cos \delta_0 \sin kr + \sin \delta_0 \cos kr. \tag{2-6}$$

Similarly, in the presence of the Coulomb and the nuclear potential,

$$u_C \approx \cos \delta_0^C F_0(kr) + \sin \delta_0^C G_0(kr), \tag{2-55}$$

where the above form is valid for $r > r_n$, the range of the nuclear potential. In this sense, when the Coulomb potential is present in addition to $V(r)$, it is necessary to replace $\sin kr$ by $F_0(kr)$ and $\cos kr$ by $G_0(kr)$. Since the higher partial waves remain unaffected by $V(r)$ at low energies, $\delta_l^C \approx 0$ for $l > 0$. We note that, although $V(r)$ alone gives a phase shift δ_0, and e^2/r gives a phase shift σ_0, the combined form $V(r) + (e^2/r)$ does not yield a phase shift $\sigma_0 + \delta_0$, but gives $\sigma_0 + \delta_0^C$, where $\delta_0^C \neq \delta_0$. This is so because the phase shift is not linearly related to the potential.

In p-p scattering, information about the nuclear phase shifts is obtained from study of the angular distribution of the protons, rather than from measurement of the total cross section. Indeed, the total cross section for pure Coulomb scattering is infinite. The differential scattering cross section[†] of two identical particles of charge e due only to the Coulomb force is (Section B–3)

$$\sigma_{\text{Mott}}(\theta) = \left(\frac{e^2}{4E}\right)^2\left[\operatorname{cosec}^4\frac{\theta}{2} + \sec^4\frac{\theta}{2} - \operatorname{cosec}^2\frac{\theta}{2}\sec^2\frac{\theta}{2}\cos\left(\eta\ln\tan\frac{\theta}{2}\right)\right],$$

$$(2\text{--}56)$$

where all quantities are measured in the CM system. The infinite peak for $\theta = 0$ is associated with the assumption of point changes. The infinity vanishes when the protons are taken to be of finite size. In the presence of $(e^2/r) + V(r)$, the differential cross section has the form (see Section B–3)

$$\sigma(\theta) = \sigma_{\text{Mott}}(\theta) + \sigma_N(\theta) + \sigma_{NC}(\theta).$$

Note that $\sigma_{\text{Mott}}(\theta)$ is calculated by taking all partial waves into account, whereas at sufficiently low energies σ_N involves only the S-wave nuclear-force scattering, that is, $\sigma_N = (1/k^2)\sin^2\delta_0{}^C$ and is isotropic. The third term $\sigma_{NC}(\theta)$ is due to interference and consequently is linearly dependent on $\sin\delta_0{}^C$. This allows us to fix the sign of $\delta_0{}^C$ by observing $\sigma(\theta)$.

Typical low-energy angular distributions are shown in Fig. 2–6. An important feature of these curves is the existence of the minimum due to the nuclear-Coulomb interference term. As the energy increases, the minimum moves to smaller angles. The Coulomb effect produces very little scattering near 90°, and the flat part of $\sigma(\theta)$ in this region is due almost entirely to the nuclear forces. Note that it is only necessary to plot p-p data from 0° to 90°, since the scattering is the same at θ and at 180° $-\theta$, that is, symmetrical about 90°.

Even though the scattering formalism has been modified in the presence of a Coulomb force, it is still possible to write an effective-range expansion like Eq. 2–41. The formula now reads as follows:

$$C^2k\cot\delta_0{}^C + \frac{1}{R}h(\eta) = -\frac{1}{a_p} + \tfrac{1}{2}r_{0p}k^2 - Pr_{0p}{}^3k^4 + Qr_{0p}{}^5k^6 + \cdots. \qquad (2\text{--}57)$$

The quantity C^2, often called the Coulomb penetration factor, is given by

$$C^2 = \frac{2\pi\eta}{e^{2\pi\eta} - 1}. \qquad (2\text{--}58)$$

[†] The differential scattering cross section is defined at the beginning of Chapter 4.

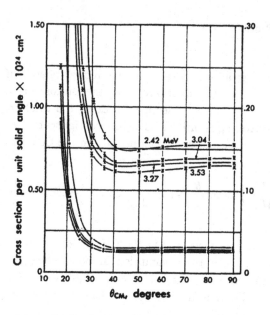

Fig. 2-6 $\sigma(\theta)$ for p-p scattering for energies of 2.42, 3.04, 3.27, and 3.53 MeV in the laboratory system. The lower curves are plotted to the vertical scale on the left; the higher curves, to that on the right. (After Blair et al. [32].)

This is essentially the ratio of the probability of finding the two protons close together to that of finding two uncharged particles there, all other factors being identical. Since the Coulomb force is repulsive, C^2 is always less than 1. It is so small below $E_{\text{lab}} = 200$ keV that there is practically no nuclear scattering, while for $E_{\text{lab}} = 800$ keV, $C^2 \approx 0.5$. The function $h(\eta)$ in Eq. 2–57 is a slowly varying function of η and is defined explicitly by Eq. B–8b. Note that, as $e \to 0$, $R \to \infty$, $C^2 \to 1$, and the left side of Eq. 2–57 becomes simply $k \cot \delta_0$. Also, a_p and r_{0p} are, as before, given by the zero-energy wave function and hence can be calculated for any given potential, just as in the n-p case. Extraction of $\delta_0{}^C$ from the experimental data allows one to plot the left side of Eq. 2–57 against k^2. At sufficiently low energies, this is a straight line, its slope being $\frac{1}{2}r_{0p}$ and its intercept at zero energy giving $-1/a_p$.

We now emphasize that experimentally one measures only various elastic-scattering data such as the differential cross section $\sigma_{pp}(\theta)$, and not the phase shifts $\delta_0{}^C$. In fact, $\delta_0{}^C$ is a model-dependent quantity, in the sense that a simple model has been taken in which the electromagnetic interactions between the two protons are replaced by a simple Coulomb potential. It follows from Eq. 2–57 that a_p, r_{0p}, and P are also, in this sense, model dependent, irrespective of the accuracy of the experiments. Actually, for energies up to about 5 MeV, experimental techniques have been developed such that $\sigma(\theta)$ can be measured to an accuracy of 0.1–0.3%. This precision warrants a more sophisticated model, which we now outline, to isolate the electromagnetic effects from the nuclear ones.

There are several electromagnetic effects that modify the pure Coulomb interaction. A description of these, together with recent compilation of the experimental data, is given by Sher, Signell, and Heller [12]. In addition to the Coulomb potential, which arises from a single photon exchange (Fig. 2–7a), there can be the process illustrated in Fig. 2–7b, in which virtual

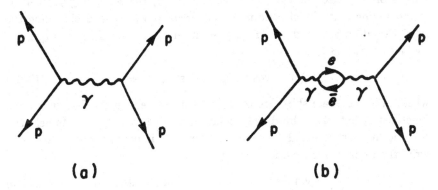

(a)   (b)

Fig. 2-7 The Coulomb force between two protons can be considered to arise from single photon exchange (a). This is modified as a result of vacuum polarization (b), in which a virtual electron-positron pair is created and annihilated during the photon-exchange process.

creation and destruction of an electron-positron pair occur. This process is called vacuum polarization (VP). The potential arising from this, to the lowest order in the fine-structure constant α $(= e^2/\hbar c)$, has been derived by Uehling [13] and is given by

$$V_{VP}(r) = \frac{2\alpha}{3\pi} \frac{e^2}{r} I(2\kappa r), \tag{2-59}$$

where $I(2\kappa r)$ is a long-range smooth function of r, which goes like $e^{-2\kappa r}/(2\kappa r)^{3/2}$ for $\kappa r \gg 1$. Here $(2\kappa)^{-1}$ corresponds to the Compton wavelength of the electron-positron pair,

$$(2\kappa)^{-1} = \frac{\hbar}{2mc} = 193.1 \text{ fm.} \tag{2-60}$$

Hence $V_{VP}(r)$ is a very long-range potential, and this in turn implies that the effective-range formula (Eq. 2–57) should be further modified for VP correction. Had the Coulomb and VP potentials been of short-range, formula 2–41 would have remained unaltered, except that these forces would have modified the nuclear potential slightly and thereby produced different values of a and r_0.

Heller [14] has shown that one can take account correctly of both the Coulomb and the VP potential and still use Eq. 2–57, but with a more complicated left-hand side. The combination of the Coulomb and VP potentials is often called the electric potential. First, the scattering problem is solved taking only the electric potential, $(e^2/r) + V_{VP}(r)$, and obtaining the regular solution $S_0(r)$ and the irregular solution $T_0(r)$. In analogy to Eq. 2–55, we have, for $kr \gg 1$,

$$S_0(r) \sim \cos \tau_0 F_0(kr) + \sin \tau_0 G_0(kr), \tag{2-61}$$

where τ_0 is the phase shift due to $V_{VP}(r)$ with respect to the Coulomb wave functions. Next, the Schrödinger equation is solved taking both the electric and the nuclear potential $V(r)$ into account. It is now most convenient to write the radial wave function for $r > r_n$ as

$$u_E(r) \sim \cos \delta_0{}^E S_0(r) + \sin \delta_0{}^E T_0(r), \tag{2-62}$$

where the subscript E on u denotes that the nuclear problem is being solved in the presence of the electric potential. Here $\delta_0{}^E$ is the nuclear phase shift in the electric potential. When the effective-range formula (Eq. 2–57) is modified to take $V_{VP}(r)$ into account, its left side becomes somewhat more complicated and also $\delta_0{}^C$ is replaced by $\delta_0{}^E$. The values of a_p, r_{0p}, and P will depend on whether one is using Eq. 2–57 and ignoring VP altogether or using the more general effective-range formula involving $\delta_0{}^E$. Heller [14] has shown that a_p and r_{0p} change only very slightly (less than 1% change in a_p) when the VP correction is incorporated. Omission of VP from the analysis, however, results in a value of P, the shape-dependent parameter, that is about 0.02 smaller than the value obtained with VP included in the analysis. This is a big change, since it is of the same order as P itself.

The other main electromagnetic effects are the interaction between the magnetic moments of the protons and the finite-size effect of the protons. The interaction of two magnetic dipoles gives rise to a potential of noncentral character which goes as r^{-3}. Since the potential is fairly long ranged, it should, in principle, be treated on the same footing as the Coulomb and VP potentials, but this has not been done. Rather, its effect is calculated by perturbation theory. Because of the finite size of the proton, both the

Coulomb potential and the magnetic interaction will be modified at short distances. These can be calculated using the charge and magnetic form factors of the proton (see Chapter 3). Details of these corrections are given by Sher et al. [12].

Using the data available for p-p scattering from 1 to 10 MeV, and including all these electromagnetic corrections, Sher et al. find for the singlet S-state

$$a_p = -7.821 \pm 0.004 \text{ fm}, \qquad r_{0p} = 2.830 \pm 0.017 \text{ fm}.$$

It would be interesting to compare these values with the corresponding ones for the singlet-S n-p and n-n interactions. However, this cannot be done directly, since the above numbers are obtained from $\delta_0{}^E$, which is the nuclear phase shift with respect to the electric functions $S_0(r)$ and $T_0(r)$, whereas in the n-p or n-n case the phase shifts are with respect to plane waves, $\sin kr$. In order to make a valid comparison, one should switch off the electric potential, $(e^2/r) + V_{VP}(r)$, retaining only the nuclear part of the p-p interaction, $V(r)$. This $V(r)$ should then be used to calculate $\delta_0(k)$ with respect to plane waves, just as in n-p or n-n case. For potentials that fit the p-p data up to 350 MeV (e.g., the Hamada-Johnston and Yale potentials described in Chapter 5), this leads to $a_N = 16.8$–17.1 fm, where by a_N we denote the scattering length of two protons in the absence of the electric potential [15]. The corresponding effective range r_N changes by only about 1% from the r_{0p} value.

2-7 LOW-ENERGY NEUTRON-NEUTRON (n-n) INTERACTION

Direct n-n scattering experiments by two colliding neutron beams are extremely difficult to perform, although they are becoming feasible [16].

In the absence of direct methods, indirect means are used to get information about the n-n force. The standard method is to study nuclear reactions that produce three particles in the final state, two of which are neutrons. The spectra of the emitted particles are sensitive to the final-state interactions, and in some suitable reactions the scattering length a_n of the n-n interaction can be extracted with reasonable accuracy, although the experimental results are generally rather insensitive to the effective-range parameter.

To illustrate the method. we shall take the following reaction: $\pi^- + d \rightarrow n + n + \gamma$, which was originally suggested by Watson and Stuart [17] to study the n-n scattering length. This reaction will be discussed in some detail because it brings out clearly the physical principles involved and has been the subject of considerable work. Note that in the final state only the two neutrons interact strongly; the third particle is a photon, whose interaction with the two neutrons is very small, being of electromagnetic character. In this sense, the reaction is a "clean" one. The major features of the experimental results are that the γ-ray yield peaks near 131 MeV (the high end of the spectrum) and the neutron yield is peaked around 2.4 MeV. The shapes of the spectra near these peaks are sensitively dependent on the n-n interaction, in particular on the n-n scattering length a_n. The peaks, both in the neutron and in the γ-ray spectrum, have a width of about 2 MeV, which must be measured accurately. We focus our attention on the neutron spectrum, since it is experimentally simpler to study 3-MeV neutrons than 130-MeV γ-rays [18].

In order to understand why these peaks are caused by the n-n interaction, consider what would happen if the two neutrons in the final state formed a bound state, which may be called a dineutron. The π^- is captured by the deuteron in its lowest Bohr orbit, so that it is essentially at rest in the laboratory frame. The π^- rest-mass is 139.58 MeV, of which 136.07 MeV is available for reaction (see Problem 2–8). Although so much energy is released, since the capture takes place at rest the initial total momentum is zero. If the dineutron could exist as a bound state, the final state would involve only two bodies, the dineutron and the photon. By applying the conservation laws of energy and linear momentum, one sees that these particles would be emitted back to back in opposite directions and, furthermore, that their energies would be unique. The dineutron would be emitted with about 4.8 MeV, and the rest of the energy would be carried by the γ-ray. However, if the two neutrons do not form a bound state, there are three particles in the final state, and the available reaction energy can be shared among them in a variety of ways, causing a spectrum in energy in the yield of the detected particle. If these three particles were noninteracting, the spectrum could be obtained from kinematical considerations alone; this is the so-called phase-space factor. However, if the two neutrons interact via a strong attractive force, there will be a tendency for them to recoil with approximately the same velocity in the same direction, causing the γ-ray to be more nearly monochromatic than if there were no n-n force. If the two neutrons were bound, the γ-ray would have been monochromatic at about 131 MeV. Now, instead, there will be a peak in the γ-ray spectrum at this energy, which is also the high-energy end point. Correspondingly, the spectrum of each neutron is peaked around 2.4 MeV. The same principle of enhancement in the cross section due to final-state interaction in three-body decays is used to detect "resonances" in elementary-particle physics [19].

To understand why these peaks should depend sensitively on the low-energy parameter a_n, we note that when the γ-ray has near-maximum energy the relative momentum of the two neutrons is very small. To show this,

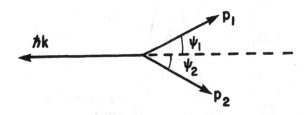

Fig. 2-8 Schematic diagram of the three-body final state in the reaction $\pi^- + d \to n + n + \gamma$. The γ-ray carries a momentum $\hbar k$, while the two neutrons have momenta p_1 and p_2 as shown.

we denote the momenta of the two neutrons by p_1 and p_2 and the photon momentum by $\hbar k$ (Fig. 2–8). By momentum conservation

$$\hbar k = -(p_1 + p_2) = -P,$$

and by energy conservation the total energy in the final state is

$$\hbar ck + \frac{p_1{}^2 + p_2{}^2}{2M} = \hbar ck + \frac{p^2 + \frac{1}{4}\hbar^2 k^2}{M} = 136.07 \text{ MeV},$$

where $p = \frac{1}{2}(p_1 - p_2)$ is the relative momentum of the two neutrons, and $P = (p_1 + p_2)$ is the center-of-mass momentum. Also, M is the neutron mass, and 136.07 MeV is the available reaction energy. This shows that the γ-ray energy completely determines the magnitude of the relative *n-n* momentum p and that k is maximum when $p = 0$. The same statement can be equivalently made for the peak of the neutron spectrum at 2.4 MeV. Consequently the neutrons near the 2.4-MeV peak have $p \approx 0$ and are recoiling in nearly the same direction, with very small scattering angles ψ_1 and ψ_2. We therefore deduce that near the neutron peak the *n-n* interaction can essentially be characterized by a_n alone. To show the sensitivity of the neutron spectrum to a_n, we reproduce in Fig. 2–9 the experimental histograms of Haddock et al. [20], together with the theoretical spectra for $a_n = -16$ fm and $a_n = -27$ fm.

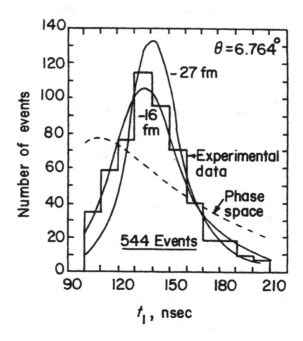

Fig. 2-9 Experimental histogram of the number of events consisting of the detection of a neutron in coincidence with another neutron at an angle of 6.764° and a γ-ray, against the time of flight (in nanoseconds) for a flight path of 118.25 in. The velocity of the neutron, and hence its energy, can be easily deduced from the flight time. Also shown are the theoretical curves for $a_n = -16$ fm and -27 fm, and the spectrum obtained from phase-space considerations alone. (After Haddock et al. [20].)

The main features of the energy dependence of the cross section of the reaction can be deduced very simply. The deuteron is a loosely bound system, the neutron and proton spending a considerable part of the time outside the range of their mutual interaction (see Section 5–1). This suggests that the rate of π^--absorption by the deuteron with subsequent γ-radiation depends essentially on the matrix element of π^--capture by a free proton, modified by the momentum distribution of the proton in the deuteron and by the final-state interaction. The transition cross section is proportional to the square of a matrix element M_s, which involves the initial-state wave functions of the pion and deuteron, the transition matrix T of π^--capture by the proton, and the final-state wave function of the two neutrons. We are interested in finding the p-dependence of the cross section. Because the pion is essentially captured at rest by the proton (which itself has a low kinetic energy), the absorption takes place in the s-state of the π^--p system. For small relative momenta, T, the transition matrix for the capture, can be regarded as a constant. Also, since the Bohr orbit for the pion is about 50 times larger than the deuteron radius, the pionic wave function over the extent of the deuteron can be taken as a constant. Thus the p-dependence in M_s is essentially determined by the overlap of the deuteron wave function and the final-state n-n wave function in the singlet state (because $p \approx 0$, only the s-state is important, and the two neutrons can interact only in the spin singlet state). This we write as

$$M_s = \int \psi_s{}^*(r) r f(r) \, d^3r.$$

Here $\psi_s(r)$ is the S-state n-n wave function, and $rf(r)$ is the deuteron wave function. For simplicity, we assume zero-range n-n force. Then ψ_s is given at all values of r by its asymptotic form:

$$\psi_s(r) = \frac{\sin(pr + \delta)}{pr},$$

where $\delta(p)$ is the S-state n-n phase shift. Thus

$$M_s = \frac{\sin \delta(p)}{p}\left[\int f(r) \cos pr \, d^3r + \cot \delta \int f(r) \sin pr \, d^3r\right].$$

Note that $\cot \delta \rightarrow -(1/pa_n)$ as $p \rightarrow 0$. The spatial extent of the integrands is of the order of the deuteron radius R_d. If $pR_d \ll 1$, we can put $\sin pr \approx pr$ and $\cos pr \approx 1$. Then it is seen that the expression in the brackets is p-independent, and the p-dependence of M_s is given by $M_s = C[\sin \delta(p)/p]$, where C is independent of p. For very small p, the effective-range formula (Eq. 2–41) yields

$$\sin \delta = \frac{pa_n}{(1 + p^2 a_n{}^2)^{1/2}}.$$

Thus the energy dependence of the cross section is given by

$$|M_s|^2 \Omega(p) \propto \Omega(p) \left(\frac{1}{1 + p^2 a_n{}^2}\right),$$

where $\Omega(p)$ is the phase-space factor and is a smooth function of p. The above formula shows that there will be a peak in the cross section for $p = 0$, and that the character of the peak will be determined by $a_n{}^2$. The formula can be generalized to remove the approximation of zero-range n-n force, and this introduces terms linear in a_n as well as the effective range r_{0n} [21]. Actual theoretical analysis has been done in great detail by McVoy [18] and by Bander [22].

We now summarize the experimental situation. A careful study of the reaction $\pi^- + d \rightarrow n + n + \gamma$ has been made by Haddock et al. [20], who detect all three particles in coincidence. Nygren [23] has carefully analyzed their data and gives the value $a_n = -18.42 \pm 1.53$ fm, where a theoretical uncertainty of ± 1 fm is included.

The other promising method of determining a_n is the so-called comparison procedure. The idea is as follows. Reactions involving three-body final states involving two neutrons and two protons are measured and analyzed under equivalent conditions, using the same theoretical models. If one can extract a reasonable value for a_p by this indirect method, confidence can be gained in regard to the corresponding value of a_n. The α-particle spectra have been measured at small scattering angles in the reactions (1) $^3\text{He} + {}^3\text{He} \rightarrow \alpha + 2p$, (2) $^3\text{He} + {}^3\text{He} \rightarrow \alpha + n + p$, and (3) $^3\text{H} + {}^3\text{H} \rightarrow \alpha + 2n$ at a center-of-mass energy of 22.6 MeV in the final states [24]. Using a theoretical model and the known values of low-energy p-p and n-p parameters, it is possible to fit the data of reactions 1 and 2 accurately. Applying the same theoretical formalism to reaction 3, one can find values for the n-n parameters. Reference [24] gives $a_n = -17.4 \pm 1.8$ fm and the n-n effective range $r_{0n} = 2.4 \pm 1.5$ fm.

For a review of the n-n scattering length, the reader is referred to the article by Verondini [33].

2-8 CHARGE INDEPENDENCE OF NUCLEAR FORCES

Table 2–1 lists the experimentally determined low-energy parameters for n-p, p-p, and n-n forces. These data indicate that the singlet S-state nuclear forces for p-p, n-p, and n-n pairs are nearly the same, if the electric potential is removed from the p-p pair. There is a difference of about 6–7 fm between the scattering lengths a_s and a_N (or a_n), and this may seem to be a large difference. However, the scattering length is very sensitively dependent on the potential, particularly when its magnitude is large compared to the range of the nuclear force, r_n. This has been carefully investigated by Moravcsik [25]. If the scattering length changes by Δa because of a change in the potential by ΔV, the relation between $\Delta a/a$ and $\Delta V/V$ is dependent on the shape of the potential. For a singlet-S square well of depth V (see problem 2–4),

$$\frac{\Delta a}{a} \approx -\frac{a}{r_n}\frac{\Delta V}{V}. \tag{2-63}$$

For the $^1S_0$-state, $a/r_n \approx 8$, showing that a 24% change in scattering length can be brought about by only a 3% alteration in the potential. More careful analysis by Henley [21] indicates that the n-p force is only slightly stronger (by about 2%) than the p-p force, while the n-n force is almost identical to the p-p force, all in the $^1S_0$-state. Only the n-p pair can interact in both the $^1S_0$- and the $^3S_1$-state, and no comparison of the n-p force in the $^3S_1$-state can be made.

TABLE 2-1 Low-Energy Parameters for n-p, p-p, and n-n Forces

For n-p system:

$\varepsilon = 2.224644 \pm 0.000034$ MeV
$a_t = 5.425 \pm 0.0014$ fm
$a_s = -23.714 \pm 0.013$ fm   [11]
$r_{0t} = 1.749 \pm 0.008$ fm
$r_{0s} = 2.73 \pm 0.03$ fm

For p-p system:

$a_p = -7.821 \pm 0.004$ fm
$r_{0p} = 2.830 \pm 0.017$ fm   [12]
$a_N \approx -(16.8 - 17.1)$ fm

For n-n system:

$a_n = -17.4 \pm 1.8$ fm; -18.4 ± 1.5 fm [24, 23]
$r_{0n} = 2.4 \pm 1.5$ fm [24]

The above observation suggests a generalized postulate of charge independence, stating that the *nuclear forces* for the p-p, n-n, and n-p pairs are identical in any given state antisymmetric in angular momentum (that is, spin-singlet, l-even, or spin-triplet, l-odd). A less general postulate is that of charge symmetry, which implies the equality of p-p and n-n nuclear forces. Before giving more experimental evidence in support of the above postulates, we shall see what charge independence means within the framework of i-spin (see the last part of Chapter 1).

Since each nucleon has i-spin $\frac{1}{2}$, a two-nucleon system can have total i-spin $T = 0$ or 1. In the notation of Chapter 1, there are four possible states, $^1(\tau)_0$, $^3(\tau)_1$, $^3(\tau)_0$, $^3(\tau)_{-1}$, where the superscript is $2T + 1$ and the subscript is T_3. The charge-independence postulate states that the nuclear force in all three of the $T = 1$ states is the same. It turns out that the nuclear force in the $T = 0$ state is quite different from that of the $T = 1$ state. Actually, the concept of charge independence has been extended to all strong interactions and seems to hold very well. Electromagnetic forces do not obey charge independence, and therefore any test of charge independence should make corrections for this fact.

To bring out some formal aspects of charge independence, we may frame it as follows: strong interactions cannot change the total i-spin T or its third component T_3 of an interacting system, and furthermore are independent of the T_3 quantum number, that is, independent of the total charge of the system. The last part follows because, as Eq. 1–5 shows, the total charge Q of an assemblage of nucleons is given by

$$Q = \frac{A}{2} + T_3, \tag{2–64}$$

where A is the total number of nucleons. Each nucleon of course has $T = \frac{1}{2}$, and their i-spins are added vectorially to give a resultant T, T_3. If the nuclear Hamiltonian is denoted by H_N, this implies that

$$\langle T_b, T_{3b}\alpha_b | H_N | T_a, T_{3a}\alpha_a \rangle = \delta_{T_b T_a} \delta_{T_{3b} T_{3a}} M_{T_a}(\alpha_b, \alpha_a). \tag{2–65}$$

The left-hand side denotes a matrix element between two nuclear states of the various quantum numbers describing the states, T and T_3 have been shown explicitly, and α represents the set of all other quantum numbers. The important point to note is that M_{T_a} is independent of the third component of i-spin, T_3. Since the total i-spin of the system is being conserved, it follows that the total i-spin operator must be the generator of a symmetry operation under which the Hamiltonian H_N remains invariant. This is a general principle in both classical and quantum mechanics (see Goldstein [26]). For example, the conservation of orbital angular momentum implies that the Hamiltonian is invariant under rotations in configuration space, and it is the orbital-angular-momentum operator that generates these rotations. Similarly, the conservation of total i-spin T implies that H_N is invariant under rotations in a space whose axes are given by T_1, T_2, and T_3. In other words, H_N may contain only scalars in i-spin operators. For a two-nucleon system, the only possible scalar in i-spin space is $\tau(1) \cdot \tau(2)$. Thus the most general form of a central two-nucleon potential which is a function of the variable $r = |r_1 - r_2|$ only is

$$V = - [V_1(r) + V_2(r)\sigma_1 \cdot \sigma_2 + V_3(r)\tau_1 \cdot \tau_2 + V_4(r)(\sigma_1 \cdot \sigma_2)(\tau_1 \cdot \tau_2)], \tag{2–66}$$

which is actually completely equivalent to Eq. 2–17. This is so since the potential defined in Eq. 2–17 was written down without regard to the charge of the nucleons. Of course the exclusion principle requires antisymmetry under the exchange of all quantum numbers specifying two fermions, that is, $P_x P_\sigma P_\tau = -1$, and hence, as Eq. 2–15 shows, $P_H = -P_\tau$. Since $\tau_1 \cdot \tau_2 = 2P_\tau - 1$, the formal identity of Eqs. 2–66 and 2–17 is evident.

2-9 EXPERIMENTAL EVIDENCE FOR CHARGE INDEPENDENCE AND CHARGE SYMMETRY

Charge independence manifests its consequences in a variety of areas in nuclear physics—in the nuclear spectra of complex nuclei, in nuclear reactions and their selection rules, and in radioactive transitions. The predictions of charge independence are rather clear cut, so that it is one aspect of the nuclear force which can be tested in a quantitative manner by studying light and complex nuclei. Of course it is true that in nuclei the electromagnetic forces, particularly the Coulomb force between the protons, cannot be ignored, and this violates charge independence. Since nature does not allow us to turn off the electromagnetic forces, we have to subtract their effects theoretically.

The charge-dependent effects due to the electromagnetic field are classified into two types: direct and indirect [27]. The direct effects include the n-p mass difference and the finite size of the charge and magnetic-moment distribution in the system. These involve no meson exchange between the nucleons. The indirect effects include any process in which the electromagnetic interaction modifies the nuclear force. Simple examples of these are the mass splittings between the charged and neutral mesons and the coupling-constant splitting; other examples are given in detailed accounts, for example reference [21]. Often it is possible to estimate theoretically the corrections to charge independence required by the direct electromagnetic effects in a given experiment. However, calculations of the indirect effects, violating charge independence by a few per cent, rest on the less secure theoretical basis of field theory.

The most important direct electromagnetic effect is the presence of the static Coulomb force between protons. Since the Coulomb force is of long range and varies slowly over the dimensions of a nucleus, it does not greatly alter the details of the nuclear wave function, although the Coulomb energy can be appreciable. Fortunately, this energy is not very sensitive to the wave function but is determined primarily by the nuclear radius (see Chapter 3). Thus the Coulomb energy is approximately the same for the ground state and the excited states of a nucleus, so that the relative spacings of the energy levels are primarily determined by nuclear forces alone. This would not be true if some of the excited states were unbound, since then the wave function is so different from a bound-state one that the Coulomb energy alters appreciably.

The speculations concerning charge symmetry are confirmed when we examine nuclear spectra. First, consider a nucleus with Z protons and $Z - 1$ neutrons, and another one with $Z - 1$ protons and Z neutrons. These are called "mirror nuclei," and examples are found in the lighter varieties, for example, (^{11}B, ^{11}C), (^{13}C, ^{13}N), (^{17}O, ^{17}F). The number of n-p bonds is the same in any two mirror nuclei, but the numbers of n-n and p-p bonds in one nucleus differ from those of its mirror. Therefore, if charge symmetry is true, the spectra of mirror nuclei should be very similar. This is confirmed by Fig. 2–10, where the low-lying energy levels of ^{11}C and ^{11}B are displayed.

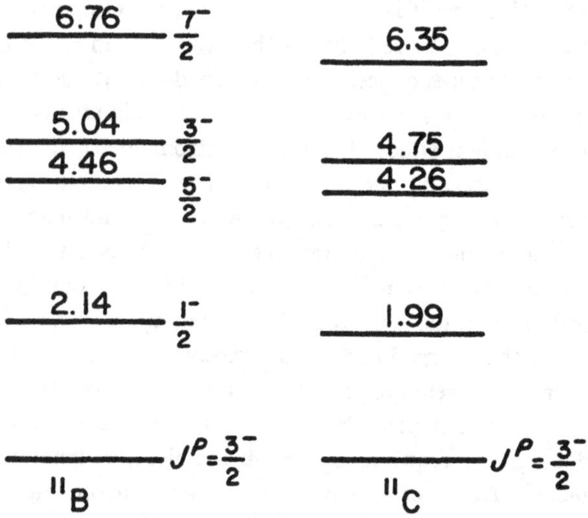

Fig. 2-10 The low-lying energy levels of mirror nuclei ^{11}B and ^{11}C (all energies in MeV). The spin and parity J^P, where known experimentally, are also shown.

A much worse example is given in Fig. 2–11, where the levels of ^{13}C and ^{13}N seem to show a large deviation from charge symmetry. The first excited states differ by more than 700 keV. Actually, the first excited state of ^{13}C is bound with respect to $^{12}C + n$, whereas that of ^{13}N is unstable against proton emission, as shown in the same figure. This results in the

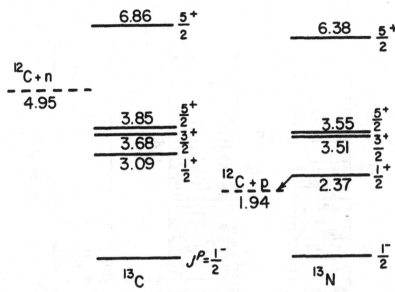

Fig. 2-11 The low-lying energy levels of mirror nuclei ^{13}C and ^{13}N (energies in MeV). The spin-parity assignments are known experimentally. Note that the 2.37-MeV state of ^{13}N is unstable against proton emission.

wave functions of the first excited states of ^{13}C and ^{13}N being vastly different, so that the respective Coulomb energy shifts with respect to the bound ground states are also very different. This type of change in the Coulomb shift due to different boundary conditions on the wave function of the nuclear surface is called the Thomas-Ehrman effect after Thomas [28] and Ehrman [29].

One consequence of i-spin conservation is the existence of the so-called isobaric multiplets. The idea arises in a simple form when strongly interacting elementary particles are considered. A meson or a baryon has a definite T, and the $2T + 1$ possible values of T_3 correspond to different charge states of the particle. These different charge states have nearly the same mass, and all quantum numbers, such as spin, parity, strangeness, and so forth are identical. The small mass differences between the members of a multiplet are believed to be of electromagnetic origin, although a fully quantitative understanding has not yet been achieved.

Similar examples are found in nuclear physics. A given nuclear state has a definite i-spin T. There are $2T + 1$ possible values of T_3 for a given T. Since the nuclear interaction is independent of T_3 (charge independent), all these $2T + 1$ states with differing T_3's but the same T, spin, and parity should be degenerate in energy. Different values of T_3 can occur only in

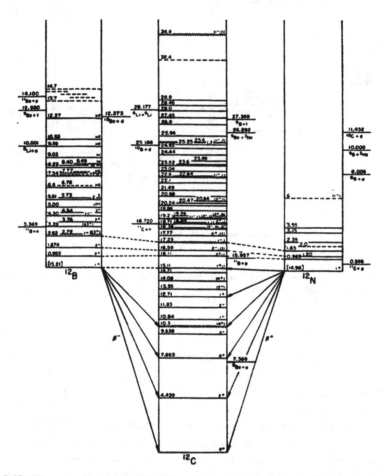

Fig. 2-12 The ground states of ^{12}B, ^{12}N, and the 1$^+$ excited state of ^{12}C form an i-spin triplet, while the ground state of ^{12}C is an i-spin singlet. [From F. Ajzenberg-Selove and T. Lauritsen, *Nucl. Phys.* A114, 1 (1968), Fig. 8.]

different nuclei, since $T_3 = \frac{1}{2}(Z - N)$. Thus each member of the degenerate multiplet comes from a different nucleus. For example, Figure 2-12 shows the ground states of ^{12}B ($T = 1, T_3 = -1$), ^{12}N ($T = 1, T_3 = 1$), and the 1$^+$ excited state of ^{12}C at 15.11 MeV ($T = 1, T_3 = 0$), which form an isobaric triplet. The three levels are not quite degenerate because of Coulomb effects and n-p mass differences. Another well-known example is found in ^{14}C, ^{14}N, and ^{14}O. The ground states of ^{14}C ($T = 1, T_3 = -1$) and ^{14}O ($T = 1, T_3 = 1$)

and the 2.3-MeV excited state of ^{14}N ($T = 1$, $T_3 = 0$) form an isobaric triplet of 0$^+$ states.

In passing, we shall enumerate a few general rules, using the example of Fig. 2-12. For a given nucleus, T_3 is fixed (for instance, $T_3 = -1$ for all states of ^{12}B, and $T_3 = 0$ for all states of ^{12}C). The i-spin for the *ground state* of light nuclei is generally the smallest possible T consistent with T_3, that is, $T = |T_3|$. Thus, for the ground state of ^{12}C, $T = 0$, and for the ground states of ^{12}B and ^{12}N, $T = 1$. In a nucleus, it is energetically more favorable to "pair off" nucleons to spin and i-spin zero, yielding the minimum value of total spin and i-spin. This results in the wave functions of the pairs being symmetrical in space. The nuclear forces are attractive for these symmetrical pairs and produce the state with greatest binding energy in this configuration. For the same reason, even though both ^{12}C and ^{12}B have twelve nucleons each, and even if the forces were perfectly charge independent, their ground states would not have the same energy. The ground state of ^{12}C would be lower because its i-spin, $T = 0$, is less than the i-spin of the ^{12}B ground state, which is $T = 1$.

Isobaric multiplets are found also in heavy nuclei in spite of the Coulomb effects being appreciable. These so-called isobaric analog states are best detected by nuclear reactions, and an exciting new field of research in this area has developed in the last decade.

Charge independence imposes severe selection rules on nuclear reactions, and sometimes the ratio of cross sections of two reactions can be predicted, using only this principle, and without using any dynamic knowledge. To illustrate this, we consider the reactions (1) $p + d \rightarrow {}^3\text{He} + \pi^0$ and (2) $p + d \rightarrow {}^3\text{H} + \pi^+$. The initial state in both reactions has $T = \frac{1}{2}$, $T_3 = \frac{1}{2}$, since the deuteron d has $T = 0$. Consequently, the final states in reactions 1 and 2 should also have $T = \frac{1}{2}$, $T_3 = \frac{1}{2}$. The ^{3}He nucleus has $T = \frac{1}{2}$, $T_3 = \frac{1}{2}$, and ^{3}H has $T = \frac{1}{2}$, $T_3 = -\frac{1}{2}$. The scattering amplitude (the square of which gives the cross section) depends on the overlap of the initial- and final-state wave functions with the interaction in the channel, $T = \frac{1}{2}$. To construct final-state wave functions with $T = \frac{1}{2}$, $T_3 = \frac{1}{2}$, we can use the vector-addition coefficients (see Eq. A–29), which couple two angular momenta. For reaction 1 this coefficient is $(\frac{1}{2}\ 1\ \frac{1}{2}\ 0|\frac{1}{2}\ \frac{1}{2})$, and for reaction 2 it is $(\frac{1}{2}\ 1\ -\frac{1}{2}\ 1|\frac{1}{2}\ \frac{1}{2})$, using the notation of Appendix A. The ratio of the cross sections of reactions 1 and 2 will be determined by the ratio of the squares of these coefficients, all other factors being the same. Thus

$$\frac{\sigma(p + d \rightarrow {}^3\text{He} + \pi^0)}{\sigma(p + d \rightarrow {}^3\text{H} + \pi^+)} = \frac{|(\frac{1}{2}\ 1\ \frac{1}{2}\ 0|\frac{1}{2}\ \frac{1}{2})|^2}{|(\frac{1}{2}\ 1\ -\frac{1}{2}\ 1|\frac{1}{2}\ \frac{1}{2})|^2} = 2.$$

Harting et al. [30] measured this ratio at 591 MeV and found it to be
2.13 ± 0.06. The difference between theory and experiment can be accounted
for if one allows for the differences in the wave functions of ^{3}H and ^{3}He due
to Coulomb force, and the mass difference between the final states, which
causes the phase space factor to differ.

There are nuclear reactions involving more complex nuclei which test
charge symmetry or charge independence. As a simple example, consider
the reaction $^{16}O + d \rightarrow \alpha + {}^{14}N$, leading to various excited state of ^{14}N.
The initial system has $T = 0$; hence finally also we should have $T = 0$.
If the α-particle remains in a $T = 0$ state (more than 20 MeV is required
to excite it to a $T = 1$ state), then only the $T = 0$ states of ^{14}N can be excited.
The low-lying excited state of ^{14}N at 2.3 MeV which has $T = 1$ should not,
therefore, be populated. Experiments by Cerny et al. [31] show this to be
true; for incident deuterons of 24 MeV the yield of this state is only 0.7 ± 0.6%
of the ground-state yield.

Actually, the above observation follows also from the less general postulate
of charge symmetry and hence the experiment is not a real test of charge
independence. To clarify this, we introduce a charge-parity operator P_{ch},
which, operating on an i-spin wave function ϕ of Z protons and N neutrons,
transforms it to ϕ' with Z neutrons and N protons, that is,

$$\phi' = P_{ch}\phi. \tag{2–67}$$

Obviously, $P_{ch}^2 = 1$. Formally, since T_3 is being changed to $-T_3$ by this
operator, it amounts to a rotation of the axes by π about the T_2-axis,

$$P_{ch} = \exp(i\pi T_2). \tag{2–68}$$

For a self-conjugate nucleus ($Z = N$), ϕ' must be the same state as ϕ;
therefore

$$\phi' = P_{ch}\phi = p\phi, \quad \text{where} \quad p = \pm 1. \tag{2–69}$$

In the reaction $^{16}O + d \rightarrow \alpha + {}^{14}N$, the ground states of ^{16}O, d, and α all
have $T = 0$, so that necessarily $p = 1$ for each of these. Parity being a
multiplicative quantum number, it follows that only those states of ^{14}N that
have $p = 1$ can be excited. Therefore the result that the 2.3-MeV state
of ^{14}N is not populated can be explained by assigning to it a negative "charge
parity." Conservation of charge parity only implies charge symmetry here
and does not reflect on charge independence. Other tests of charge symmetry
may be found in β-decay schemes, but we shall not discuss them here.

Conservation of i-spin plays a dominant role in determining the forces between strongly interacting particles. For example, the force between two nucleons can be generated by exchanging various kinds of mesons, and Fig. 2–13a shows two nucleons exchanging a pion (which is the lightest meson) to generate the longest-range interaction, called the one-pion-exchange potential (OPEP). This will be discussed at length in Chapter 5 on the two-nucleon force. The point to note here is that a single pion exchange between two nucleons is allowed by i-spin conservation, since the process $N \to N + \pi$ can go through the channel $T = \frac{1}{2}$. The Λ particle, however, being an i-spin singlet, has to change to a $T = 1$ particle on emitting a pion: $\Lambda \to \Sigma + \pi$, where Σ has $T = 1$, and the pion has also i-spin 1, so that the two can combine to give an initial i-spin of zero. Thus the longest-range force between a Λ and N is of two-pion character. This simple process is shown in Fig. 2–13b.

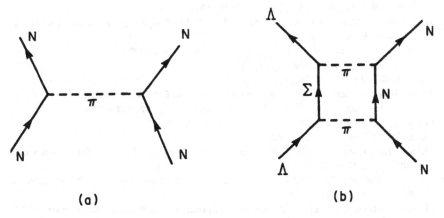

(a) (b)

Fig. 2-13 In (a), N-N scattering is taking place via the exchange of a single pion. In (b), Λ-N scattering is displayed where two pions are exchanged. Note that the i-spin is conserved at each "vertex."

A detailed analysis of charge independence and charge symmetry in nuclear forces is available in the review article by Henley [21]. The best evidence of charge independence and charge symmetry in nuclear forces still comes from the scattering-length parameters of the p-p, n-p, and n-n forces in the $^1S_0$-state. As mentioned before, the data indicate that the n-p force is about 2% stronger than the p-p or n-n force, while the latter two are identical to within 1%, when electromagnetic corrections are made.

REFERENCES

1. D. M. Brink, *Nuclear Forces*, Pergamon Press, New York, 1965.
2. R. P. Feynman and M. Gellman, *Phys. Rev.* **109**, 193 (1958); F. C. Michel, *Phys. Rev.* **133**, B329 (1964). For a review of the subject, see the articles by B. H. McKellar, "*P*- and *T*-Violations in Nuclear Physics," and D. Hamilton, "Experimental Investigations of *P*- and *T*-Violations," in *High-Energy Physics and Nuclear Structure*, edited by S. Devons, Plenum Press, New York-London, 1970.
3. D. P. Boyd, P. F. Donovan, B. Marsh, D. E. Alburger, D. H. Wilkinson, P. Assimakopoulous, and E. Beardsworth, *Bull. Am. Phys. Soc.* **13**, 1424 (1968).
4. K. S. Krane, C. E. Olsen, J. R. Sites, and W. A. Steyert, *Phys. Rev. Lett.* **26**, 1579 (1971).
5. J. H. Bartlett, Jr., *Phys. Rev.* **49**, 102 (1936).
6. E. Majorana, *Z. Phys.* **82**, 137 (1933).
7. W. Heisenberg, *Z. Phys.* **77**, 1 (1932).
8. E. P. Wigner, *Phys. Rev.* **43**, 252; *Z. Phys.* **83**, 253 (1933).
9. T. L. Houk and R. Wilson, *Rev. Mod. Phys.* **39**, 546 (1967); Erratum, *Rev. Mod. Phys.* **40**, 672 (1968).
10. L. Koester, *Z. Phys.* **198**, 187 (1967); **203**, 515(E) (1967); T. L. Houk, D. Shambroom, and R. Wilson, *Phys. Rev. Lett.* **26**, 1581 (1971).
11. R. Wilson, *Comments Nucl. Particle Phys.* **2**, 142 (1968); H. P. Noyes and H. M. Lipinski, *Phys. Rev.* **C4**, 995 (1971).
12. M. S. Sher, P. Signell, and L. Heller, *Ann. Phys.* **58**, 1 (1970).
13. E. A. Uehling, *Phys. Rev.* **48**, 55 (1935).
14. L. Heller, *Phys. Rev.* **120**, 627 (1960).
15. L. Heller, P. Signell, and N. R. Yoder, *Phys. Rev. Lett.* **13**, 577 (1964).
16. C. D. Bowman and W. C. Dickinson, *Rev. Mod. Phys.* **39**, 592 (1967).
17. K. M. Watson and R. N. Stuart, *Phys. Rev.* **82**, 738 (1951).
18. K. W. McVoy, *Phys. Rev.* **121**, 1401 (1961).
19. See, for example, W. R. Fraser, *Elementary Particles*, Prentice-Hall, Englewood Cliffs, N.J., 1966, p. 71.
20. R. P. Haddock, R. M. Slater, Jr., M. Zeller, J. B. Czirr, and D. R. Nygren, *Phys. Rev. Lett.* **14**, 318 (1965).
21. E. M. Henley, in *Isospin in Nuclear Physics*, North-Holland, Amsterdam, 1970.
22. M. Bander, *Phys. Rev.* **134**, B1052 (1964).
23. D. R. Nygren, unpublished thesis, University of Washington, 1968.
24. E. E. Gross, E. V. Hungerford, and J. J. Manify, *Phys. Rev.* **C1**, 1365 (1970).
25. M. J. Moravcsik, *Phys. Rev.* **136**, B624 (1964).
26. H. Goldstein, *Classical Mechanics*, Addison-Wesley, Reading, Mass., 1950.
27. J. S. Leung and Y. Nogami, *Nucl. Phys.* **B7**, 527 (1968).
28. R. G. Thomas, *Phys. Rev.* **80**, 136 (1950).
29. J. B. Ehrman, *Phys. Rev.* **81**, 412 (1951).
30. D. Harting, J. C. Kluyver, A. Kusunegi, R. Rigopoulos, A. M. Sachs, G. Tibell, G. Vanderhaeger, and G. Weber, *Phys. Rev.* **119**, 1716 (1960).
31. J. Cerny, R. H. Pehl, E. Rivet, and B. G. Harvey, *Phys. Lett.* **7**, 67 (1963).
32. J. M. Blair, G. Freier, E. E. Lampi, W. Sleator, and J. H. Williams, *Phys. Rev.* **74**, 553 (1948).
33. E. Verondini, *Riv. Nuovo Cimento* **1**, 33 (1971).

PROBLEMS

2–1. Calculate v/c and λ for neutrons of the following energies: thermal, 500 keV, 5 MeV, 20 MeV, 100 MeV, 400 MeV. Comment on the significance of these values for (a) nucleon-nucleon scattering, (b) study of stable nuclei, and (c) neutron-nucleus reactions.

2–2. Suppose that n-p scattering is studied by scattering neutrons from a nylon target. Explain how you would correct for the presence of carbon and oxygen.

2–3. Consider a square well of radius r_n and depth of potential V. Show that, if two types of particles of equal mass M interact by means of this potential, and if particles of one type are initially at rest and are bombarded by a beam of particles of the other type, with energy $2E$, the scattering phase shift δ for the S-wave is given by

$$k \cot \delta = \frac{\sqrt{k^2 + K^2} + k \tan kr_n \tan \sqrt{k^2 + K^2}\, r_n}{\tan \sqrt{k^2 + K^2}\, r_n - \sqrt{1 + (K^2/k^2)} \tan kr_n}. \tag{1}$$

Also show that the scattering length is

$$a = -\frac{1}{K} \tan Kr_n + r_n,$$

and that the effective range is

$$r_0 = r_n - \frac{1}{3}\frac{r_n^3}{a^2} - \frac{1}{K^2 a}.$$

In these expressions, $k^2 = ME/\hbar^2$, $K^2 = MV/\hbar^2$. For $V = 36.2$ MeV, $r_n = 2.02$ fm, and a bombarding energy of 6 MeV, compare the phase shift for n-p scattering given by Eq. 1 with the effective-range expression and determine the resulting difference in the cross section. How would you know experimentally whether waves of angular momentum > 0 were important?

2–4. In Problem 2–3, it is shown for a square-well potential that the scattering length is

$$a = -\frac{1}{K} \tan Kr_n + r_n.$$

For a potential in the $^1S_0$-state, the quantity Kr_n is only very slightly less than $\pi/2$. Using this fact, show that, if the scattering length a changes by Δa because of a change in the depth of the potential by ΔV, then

$$\frac{\Delta a}{a} \approx -\frac{\pi^2}{8}\left(\frac{a}{r_n}\right)\frac{\Delta V}{V},$$

which is essentially Eq. 2–63 of the text.

2-5. Suppose that the central potential between nucleons in the singlet S-state is known. Denote it by $^1V^s(r)$. Assume similarly that the following potentials are known: $^1V^p$, $^3V^s$, $^3V^p$. Find expressions for the Wigner, Bartlett, Majorana, and Heisenberg potentials in terms of the four V's listed above.

2-6. Imagine two nucleons interacting only in the S-state by a hard-core potential of radius r_c:

$$V(r) = \infty \quad \text{for} \ \ r \leqslant r_c,$$

$$= 0 \quad \text{for} \ \ r > r_c.$$

Show that the s-wave phase shift is given by $\delta(k) = -kr_c$. Hence show that the scattering length a and the effective range r_0 are given by

$$a = r_c, \qquad r_0 = \tfrac{2}{3}r_c.$$

Verify the answer for r_0 by using Eq. 2-43 of the text.

2-7. As is stated in the text, in elastic scattering the outgoing flux of particles must equal the incoming flux, with the result that the phase shift $\delta(k)$ is real. This implies that the scattering "matrix" $S(k) = \exp[2i\, \delta(k)]$ is unitary, that is,

$$S^*(k) = \frac{1}{S(k)} \quad \text{for real } k.$$

One of the simplest forms of the S-matrix obeying unitarity is a rational function of k,

$$S(k) = \left(\frac{k + i\beta}{k - i\beta} \right) \cdot \left(\frac{k - i\alpha}{k + i\alpha} \right),$$

where we assume α and β to be real and positive. Show that for such a case the effective-range expression (Eq. 2-41) is exactly satisfied for all values of k, with

$$a = -\frac{\beta - \alpha}{\alpha\beta}, \qquad r_0 = \frac{2}{\beta - \alpha}.$$

Assuming that $a = -23.7$ fm and $r_0 = 2.7$ fm, find α and β. A potential that exactly reproduces this scattering matrix for all real k is the Bargmann potential [V. Bargmann, *Rev. Mod. Phys.* **21**, 488 (1949)], given by

$$V(r) = 2\beta^2(\alpha^2 - \beta^2)(\beta \cosh \beta r + \alpha \sinh \beta r)^{-2},$$

which asymptotically behaves as $\exp(-2\beta r)$.

2-8. (a) In the reaction

$$\pi^- + d \rightarrow n + n + \gamma,$$

show that the available energy for the reaction is **136.07 MeV**.

(b) Consider the reaction

$$\pi^- + d \to n + n.$$

Assume that the capture of the pion taken place from the s-orbit of the π-mesic atom, and that the spin parity associated with the deuteron is 1^+. Noting that the spin of the pion is zero, prove that the parity associated with it is negative.

2–9. (a) The Λ particle has spin $\frac{1}{2}$, i-spin 0, and charge 0. It decays by weak interactions

$$\text{(A)} \quad \Lambda \to p + \pi^-, \qquad \text{(B)} \quad \Lambda \to n + \pi^0.$$

Charge and angular momentum are conserved in the decay, but the total i-spin T changes by $\frac{1}{2}$. Using the i-spin formalism, find the ratio of protons to neutrons observed in the decay.

(b) Consider the following strong interactions:

$$\text{(C)} \quad \pi^- + p \to K^0 + \Lambda, \qquad \text{(D)} \quad \pi^0 + n \to K^0 + \Lambda.$$

In these reactions, i-spin T and its third component T_3 are conserved. What are the possible values for the i-spin of K^0 for reactions C and D to go? Calculate $\sigma(\pi^- + p \to K^0 + \Lambda)/\sigma(\pi^0 + n \to K^0 + \Lambda)$ for each of these possible values. Knowing the ratio experimentally to be 2, can you predict the i-spin of K^0?

Hint: The i-spin wave functions needed are the following:

$$|\pi^- p\rangle = \sqrt{\tfrac{1}{3}}\,|\tfrac{3}{2}, -\tfrac{1}{2}\rangle - \sqrt{\tfrac{2}{3}}\,|\tfrac{1}{2}, -\tfrac{1}{2}\rangle,$$
$$|\pi^0 n\rangle = \sqrt{\tfrac{2}{3}}\,|\tfrac{3}{2}, -\tfrac{1}{2}\rangle + \sqrt{\tfrac{1}{3}}\,|\tfrac{1}{2}, -\tfrac{1}{2}\rangle.$$

Chapter 3

Nuclear Moments and Nuclear Shapes

3-1 ELECTROMAGNETIC MULTIPOLES

We now examine electromagnetic effects in which the nucleus interacts with electromagnetic systems separated from it by distances that are large on the nuclear scale. In such cases, it is desirable to think of the interaction of fields rather than particles. It is shown in works on electromagnetism that the external effects of any distribution of charges and currents (e.g., a nucleus) can be expressed in terms of a series of "multipole moments." In many instances only a few multipoles of the lower orders need be considered, and it is this feature that makes the use of the series convenient. The simplification depends on two conditions: (1) the velocity of the nuclear particles must be small compared with the speed of light, and (2) the fractional change of the applied field across the nucleus must be small. Clearly, to obtain information about the undisturbed nucleus, a third condition is necessary, namely, the applied field must not be so strong as to change appreciably the arrangement of charges and currents in the nucleus. For example, we must watch for possible polarization of nuclear charges in strong electric fields.

The energy of interaction between an applied electromagnetic field and a charge-current system can be written as

$$H_{el} = q\phi_0 - \mathbf{P} \cdot \mathbf{E}_0 - \mathbf{M} \cdot \mathbf{H}_0 - \frac{1}{6} \sum_{ij} Q_{ij} \left(\frac{\partial E_j}{\partial x_i} \right)_0 - \cdots, \qquad (3\text{-}1)$$

where $\mathbf{E}$ and $\mathbf{H}$ are the applied electric and magnetic fields, and ϕ is the electrostatic potential of the applied field. The subscript 0 indicates that the quantity in question is to be evaluated at the origin of coordinates, which is taken at some convenient point in the charge-current system (say the center of the nucleus in our case). The symbol q denotes the total charge of the system, $\mathbf{P}$ is the electric dipole vector, $\mathbf{M}$ is the magnetic dipole vector, and Q_{ik} is the electric quadrupole tensor. For a charge density ρ and current density $\mathbf{J}$, the explicit expressions are:

$$q = \int \rho \, dV, \tag{3-2a}$$

$$\mathbf{P} = \int \rho \mathbf{r} \, dV, \tag{3-2b}$$

$$\mathbf{M} = \tfrac{1}{2} \int \mathbf{r} \times \mathbf{J} \, dV, \tag{3-2c}$$

$$Q_{ik} = \int \rho(3x_i x_k - \delta_{ik} r^2) \, dV. \tag{3-2d}$$

In quantum mechanics the energy H_{el} becomes an operator, as do the various multipole moments. Expression 3–1 is actually an infinite series of terms of increasing multipole order. It will be observed that the values of the quantities defined in Eqs. 3–2 depend only on the radial and angular distributions of charge and current in the nucleus. Conversely, a knowledge of these multipoles gives information about the size, shape, density, and motion of the matter in nuclei.

Equation 3–1 shows that we may learn experimentally the values of the multipoles either by observing energy shifts due to the interaction of the nucleus with atomic or molecular fields, or by applying an external field of a chosen type. The latter methods are, in general, more accurate since the fields are better known. It is particularly difficult to obtain accurate values of the electric quadrupole from the energies of lines in atomic and molecular spectra, since one is required to know the value, at the nucleus, of the gradient of the electric field of the atomic or molecular electrons, a quantity which is difficult to calculate and is very sensitive to details of the electronic configuration. Experimental methods of measuring nuclear moments are described in a number of books, for example, Ramsey's *Nuclear Moments* [1].

Although the fundamental definition of $\mathbf{M}$ expresses the magnetic dipole vector in terms of the current density $\mathbf{J}$, an alternative form is very important. If $\mathbf{J}$ is produced by an electric charge of density ρ moving with velocity $\mathbf{v}$, then $\mathbf{J} = \rho \mathbf{v}/c$.[†] If the charge density is actually composed of particles of charge e and mass M, the angular-momentum density is $\mathbf{L} = (M\rho/e)(\mathbf{r} \times \mathbf{v})$, and we can write

$$\mathbf{M} = \frac{1}{2} \frac{e}{Mc} \int \mathbf{L} \, dV. \tag{3-3a}$$

For Z protons moving in a nucleus, the quantal analog is

$$\mathbf{M}_{\text{orb}} = \frac{1}{2} \frac{e\hbar}{Mc} \int \psi^* \sum_1^z \mathbf{l}_k \psi \, dV, \tag{3-3b}$$

where $\mathbf{l}_k$ is the angular-momentum operator (divided by $\hbar$) of the kth proton, ψ is the nuclear wave function, and the subscript "orb" is used with $\mathbf{M}$ to stress the fact that this magnetic moment is due to the orbital angular momentum. It will be seen from Eq. 3-3 that, when all particles have the same ratio of charge to mass, the magnetic dipole is parallel to the total angular momentum with which it is associated. The orbital magnetic dipole operator is

$$\mathbf{M}_{\text{orb}} = \mu_0 \sum_{\text{protons}} \mathbf{l}_k, \tag{3-4a}$$

where μ_0 is the nuclear magneton $e\hbar/2m_p c$. In addition to the orbital magnetic moment of the protons, there is the spin magnetic moment of each nucleon,

$$\mathbf{M}_{\text{spin}} = \mu_0 \sum g_k \mathbf{s}_k, \tag{3-4b}$$

where $g = 5.5856$ for protons and -3.8263 for neutrons (see Chapter 1). The close connection of magnetic moment and angular momentum requires that we now survey the angular momenta of nuclei.

3-2 ANGULAR MOMENTUM

The total angular momentum of a nucleus is the vector sum of the angular momenta of the constituent nucleons, which in turn consist of both orbital angular momenta due to the motion about the center of the nucleus and of the intrinsic spin angular momentum of $\frac{1}{2}\hbar$ per nucleon. We shall use capital letters to refer to angular momenta associated with the whole nucleus and lower-case letters to denote those of individual nucleons. We measure all angular momenta in units of $\hbar$. If $\mathbf{l}_k$ is the orbital angular momentum of nucleon k and $\mathbf{s}_k$ is its spin vector, we define

---

[†] The factor c is due to the use of Gaussian units.

$$\mathbf{j}_k = \mathbf{l}_k + \mathbf{s}_k = \mathbf{l}_k + \tfrac{1}{2}\boldsymbol{\sigma}_k,$$

$$\mathbf{L} = \sum_k \mathbf{l}_k, \quad \mathbf{S} = \sum_k \mathbf{s}_k,$$

$$\mathbf{J} = \sum \mathbf{j}_k = \mathbf{L} + \mathbf{S}. \tag{3-5}$$

It should be realized that the only vector mentioned which is necessarily a constant of motion is $\mathbf{J}$, the total angular momentum of the whole nucleus; the eigenvalue of J^2 is $J(J+1)$, with similar notation for the other vectors when applicable. Also, the eigenvalue of J_z is written as M, and corresponding quantities (when they exist) are m_k, m_{l_k}, m_{s_k} $(= \pm \tfrac{1}{2})$, m_{σ_k} $(= \pm 1)$, m_L, m_S. The quantum number J is often referred to rather loosely as the "spin" of the nucleus.

Experimentally, it is found that all nuclei have relatively low spins in their ground states. Although all constituent nucleons have angular momenta of at least $\tfrac{1}{2}\hbar$, the highest confirmed value of J is 6, which occurs in the ground state of ^{50}V. This indicates that the motion of the nucleons and the orientation of their spins are such that most of the angular-momentum vectors cancel each other. Indeed, with no known exception the ground states of nuclides with even Z and even N have zero spin, a fact which suggests the even stronger rule that the nucleon angular momenta cancel in pairs for each type of nucleon. In this case, we might expect that the total angular momentum of an odd-A nuclide would be just that of the last unpaired particle, and that in an odd-odd nuclide only the last proton and last neutron would contribute to $\mathbf{J}$. This provides a very natural explanation of the low values of $\mathbf{J}$ which exist.

Although the facts do not require us to adopt such an extreme degree of pairing of nucleons in order to obtain cancellation of their $\mathbf{j}$-vectors, it is clear that in any reasonable model of nuclear forces cancellation will occur, since any marked difference is unlikely between the energies of nucleons with different orientations of $\mathbf{j}$, that is, different m, but with all other quantum numbers the same. Hence for each set of quantum numbers (including a specific j), we expect to find in the ground state of lowest energy one nucleon with each of the $(2j+1)$-values of m. Of course, only one proton and one neutron can have the same m-value because of the exclusion principle. Now $2j+1$ particles of the same j have a resultant angular momentum of zero. Contributions to $\mathbf{J}$ would come only from the last few particles, which are insufficient in number $(< 2j + 1)$ to provide complete cancellation for the value of j to which they belong. This picture does not necessarily imply that all even-even nuclides have spin zero, since if there are, for example, two protons with the same j, they need not have opposite values of m in order to minimize the energy. Consequently the low actual values of J do suggest more than the minimum amount of angular-momentum cancellation. They also suggest that the j-values of individual nucleons may be constants of motions, an assumption which is implicit in the above argument.

The idea that individual nucleons have a constant angular momentum implies that each nucleon essentially follows a prescribed orbit and does not exchange angular momentum in encounters with other nucleons. In fact, this assumption appears to ignore the nucleon's interaction with other nucleons. The strength and range of the nucleon-nucleon forces might lead one to expect that this is a poor approximation, since collisions should be frequent and a constant value of j most improbable. Nevertheless, it is possible to understand a number of fundamental nuclear properties in terms of a very simple model in which the effects of the nuclear forces are represented only by confining the nucleons in a volume of nuclear dimensions, and all direct nucleon-nucleon interaction is ignored. The reconciliation of this model with the strong nuclear force will be considered in Part II; for the present we shall merely employ the model on a heuristic basis.

3-3 ELECTRIC MOMENTS

The term in the expansion 3–1 of H_{el} which involves the electric multipole of order λ can be expressed in terms of the so-called *electric 2^λ-pole moment operator*, defined by

$$Q_\lambda = e \sum_{k=1}^{z} r_k^\lambda Y_\lambda^0(\theta_k), \tag{3–6}$$

where (r_k, θ_k, ϕ_k) are the coordinates of the kth proton, and Y_λ^0 is a spherical harmonic of order λ. For $\lambda = 2$ it can be shown [1] that the quantal operator for the electric quadrupole tensor is

$$Q_{ij} = \frac{Q}{J(2J-1)} [\tfrac{3}{2}(J_i J_j + J_j J_i) - \delta_{ij} \mathbf{J}^2], \tag{3–7a}$$

where J_i ($i = 1, 2, 3$) are the components of $\mathbf{J}$, and

$$Q = \left(\frac{16\pi}{5}\right)^{1/2} \langle Q_2 \rangle_{M=J} = \left(\frac{16\pi}{5}\right)^{1/2} \int \Psi_{J,J}^* Q_2 \Psi_{J,J} \, dV. \tag{3–7b}$$

Thus the interaction energy H_{el} can be written in terms of Q; similar results hold for higher multipoles.

It will be observed that

$$Q_2 = e \sqrt{\frac{5}{4\pi}} \sum r_k{}^2 P_2(\theta_k)$$

$$= e \sqrt{\frac{5}{16\pi}} \sum (3z_k{}^2 - r_k{}^2) = \sqrt{\frac{5}{16\pi}} Q_{33}, \qquad (3\text{--}8)$$

and that Q is the expectation value of Q_{33} in the state in which the z-component of $\mathbf{J}$ has its maximum value J. The quantity Q is also called the *electric quadrupole moment*, although it differs from $\langle Q_2 \rangle$ by a numerical factor. There is no logical need to introduce Q instead of $\langle Q_2 \rangle$, but this is done since it is the quantity Q which is conventionally quoted in experimental results.[†] Its units are charge times area and are usually written e cm^2 or e b.

We may note that any system with a definite parity cannot have static electric poles of odd order, since the Legendre polynomials of odd order are odd functions of z, whereas $|\psi|^2$ is an even function of (x, y, z) if ψ has definite parity. Hence the integral for the multipole moment $\int |\psi|^2 Q_\lambda \, dV$ has an odd integrand and is zero. In particular, nuclei cannot have electric dipole moments, and, apart from the term $q\phi_0$, the quadrupole is the lowest order of interaction with an external electric field.[‡] There are also restrictions on the angular momentum of a nucleus if it is to have a multipole of given order. Since $P_\lambda(\theta_k)$ is the mathematical form of a wave function of angular momentum λ, and since wave functions of different angular momenta are orthogonal, the factor $\psi^*\psi$ in the integral $\int \psi^*_{J,J} \psi_{J,J} Q_\lambda \, dV$ must have a term corresponding to an angular momentum λ. Also, $\psi_{J,J}$ describes a state of angular momentum J, and the product $\psi^*\psi$ therefore describes states in which two angular-momentum vectors of the same J are added. The possible resultants are all angular momenta $\leqslant 2J$. Consequently, for an electric 2^λ-pole moment to exist, J must be at least $\frac{1}{2}\lambda$. In particular, nuclides of spin 0 or $\frac{1}{2}$ cannot possess static electric quadrupole moments. (The adjective *static* has been introduced to emphasize that we are not discussing the multipole orders which can play a role in γ-radiation from nuclei. These are not so restricted.)

[†] The reader should be warned that even this usage is not without exceptions in the literature.

[‡] It is sometimes argued that the experimental absence of electric dipole moments of nuclei and also of nucleons shows that the states are of definite parity to a very good approximation. However, it can be demonstrated that odd electric moments must vanish if our theories are invariant under time reversal only, without any assumption about parity.

The existence of a quadrupole moment means that the charge distribution is not spherically symmetric. This can be seen from Eq. 3–7b, which is zero unless the density $\psi^*\psi$ has an angular dependent term proportional to $P_2(\theta)$. Less formally, the situation may be illustrated by considering a charge uniformly distributed throughout an ellipsoid while ignoring the fact that charge is actually on the individual protons. Let us call the axis of the ellipsoid the z'-axis, let this axis have length $2a$, and let b be the other semiaxis. Then the charge density is

$$e\rho = \frac{Ze}{V} = \frac{3Ze}{4\pi ab^2}.$$

In analogy with Eqs. 3–8 and 3–2d, we define quadrupole moments with respect to the z'-axis by

$$\langle Q_2' \rangle = \sqrt{\frac{5}{16\pi}} \langle Q_{33}' \rangle = \sqrt{\frac{5}{16\pi}} \int e\rho(3z'^2 - r^2)\, dV$$

$$= \frac{Ze}{\sqrt{20\pi}}(a^2 - b^2) = \frac{3Ze}{4\pi}\sigma a^2, \tag{3-9}$$

where we have introduced as a measure of ellipticity the parameter

$$\sigma = \frac{1}{3}\sqrt{\frac{4\pi}{5}}\frac{a^2 - b^2}{a^2}. \tag{3-10}$$

(The numerical factors will be seen to be convenient later.) If σ is small so that a and b are not very different, it is convenient to define a mean radius by

$$R_0^3 = ab^2, \tag{3-11}$$

so that a sphere of radius R_0 has the same volume as the ellipsoid. Then

$$Q' = \sqrt{\frac{16\pi}{5}} \int \langle Q_2' \rangle = \frac{3Ze}{\sqrt{5\pi}} R_0^2 \sigma \left(1 - 3\sqrt{\frac{5}{4\pi}}\sigma \right)^{-2/3}. \tag{3-12}$$

If we take a z-axis making an angle Θ with a z'-axis (not necessarily a symmetry axis) as in Fig. 3–1, it is not difficult to see that the quadrupole moment with respect to this axis is

$$\langle Q_2 \rangle_\Theta = \sqrt{\frac{5}{16\pi}} \int e\rho(3z^2 - r^2)\, dV = \tfrac{1}{2}(3\cos^2\Theta - 1)\langle Q_2' \rangle. \tag{3-13}$$

Equation 3–13 shows that the largest value of $\langle Q_2 \rangle_\Theta$ occurs when Θ is as

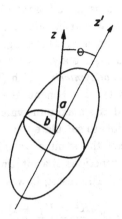

Fig. 3-1 The quadrupole moment of a spheroid.

small as possible. Classically of course it can be zero, but quantally the maximum value of cos Θ is determined by the angular-momentum properties.

If Q' is the intrinsic moment *with respect to the spin direction*, the z'-axis is in the direction of **J**, and this maximum, which occurs when $M = J$, is

$$\cos\Theta = \frac{J}{\sqrt{J(J+1)}}.$$

It follows from Eq. 3–13 that

$$Q = \sqrt{\frac{16\pi}{5}}\,\langle Q_2\rangle_{M=J} = \frac{2J-1}{2(J+1)}\,Q'. \tag{3–14}$$

It is now clear why the quantity Q is defined as the moment in the state $M = J$; it is the maximum observable quadrupole moment. We can think of Q' as the intrinsic quadrupole moment of the system with respect to its spin axis, but since the uncertainty principle prevents the exact location of the angular-momentum vector, only a fraction of Q' is effectively in interaction with an applied field. Indeed we see that, for $J = \frac{1}{2}$, the extent to which the vector **J** can be aligned with the z-axis is so slight that $Q = 0$. This is the "physical" reason behind the general theorem that $Q = 0$ unless $J \geqslant 1$. We note that this implies, not that nuclei with spins 0 and $\frac{1}{2}$ are spherical, but simply that their asymmetry does not show up in static electric quadrupole interactions.

If we think of the nucleus as an ellipsoid of fixed shape, rotating in space, the angular momentum **J** will have a fixed value, called K, for its component along the axis of symmetry. This is a general result in the classical mechanics of the motion of a rigid body. We can connect the quadrupole moment Q', *referred to the axis of symmetry*, with Q, referred to an axis fixed in space, along which **J** has the component $M = J$. In this case, **J** can coincide with neither axis, and the maximum value of K is J. In order to use Eq. 3–13, it must be possible to calculate the expectation value of $\cos^2 \Theta$, where Θ is the angle between the symmetry axis and the axis in space. To do this, we need the wave functions of a rotating ellipsoid, which we shall not encounter until Chapter 8. The result, however, is

$$Q = Q' \frac{3K^2 - J(J+1)}{(J+1)(2J+3)}, \tag{3-15}$$

or, for $K = J$,

$$Q = Q' \frac{J(2J-1)}{(J+1)(2J+3)}. \tag{3-16}$$

Figure 3–2 shows experimental values of quadrupole moments, plotted as

$$\frac{\sqrt{5\pi}}{3} \frac{Q}{ZeR_0^2} \frac{(J+1)(2J+3)}{J(2J-1)}.$$

This ordinate provides a measure of the asymmetry parameter σ, according to Eq. 3–12, in cases where it is reasonable to think of the nucleus as an ellipsoid. For odd proton nuclides the abscissa is Z, for odd neutron nuclides it is N, and for even-even nuclides it is Z. For even-even nuclides and some others, Q' is obtained from Coulomb-excitation cross sections. The curve is drawn in Fig. 3–2 merely to show the trend of the data. Numbers for which nuclei are spherical are listed.

The assumption of an ellipsoidal charge distribution should be contrasted with the simpler picture already suggested in connection with nuclear spins in which particles in the nucleus are paired so that all the particles of a given j constitute a spherically symmetric distribution and the nuclear spin is the angular momentum of the last particle. It can be seen (Problem 3–2) that the quadrupole moment due to a single proton is

$$Q_{sp} = -e \frac{2J-1}{2(J+1)} \langle r^2 \rangle, \tag{3-17a}$$

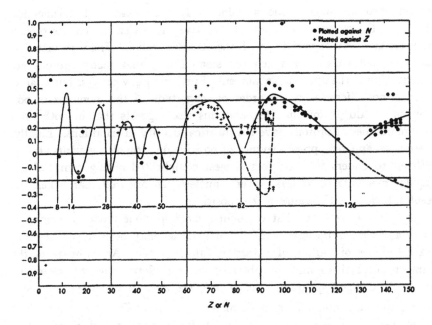

Fig. 3-2 Experimental values of electric quadrupole moments plotted to show the nuclear asymmetry.

where $\langle r^2 \rangle$ is the mean-square distance of the proton from the center of the nucleus. For an odd neutron nucleus, there is a quadrupole moment due to the recoil motion of the rest of the nucleus, which is a charge Z at a distance r_n/A from the center of mass. Hence

$$Q_{sn} = \frac{Z}{A^2} Q_{sp}. \qquad (3\text{–}17b)$$

In these formulas the value $\langle r^2 \rangle$ will be somewhat less than the square of the nuclear radius. Consequently, if this model were valid, values of Q for odd proton nuclides would increase like $A^{2/3}$ and would range from $\sim 10^{-26}$ to $\sim 5 \times 10^{-25}$ e cm^2; for single-neutron nuclei, the values would behave roughly like $A^{-1/3}$ and would lie between 10^{-28} and 10^{-26} e cm^2. These results would imply that the ordinate in Fig. 3–2 would have the value

$$\left(\frac{2J-3}{2J}\right)\left(\frac{\langle r^2 \rangle}{R_0^2}\right)\left(\frac{\sqrt{5\pi}}{3}\right)\left(\frac{1}{Z}\right) \sim \frac{1}{Z}$$

for odd proton nuclides and a value $\sim 1/A^2$ for odd neutron nuclides. Moreover, the predicted signs are all negative. It is clear that these predictions are seriously at variance with the observed values. The very fact that the predicted values are much too small shows that many protons are contributing to the quadrupole moment. Furthermore, when Q is fairly large, there is no difference in the order of magnitude of the moments of odd neutron or odd proton nuclides, and this also suggests that in both cases a large number of protons have nonspherical density distributions. Finally, there is an apparent preponderance of positive quadrupole moments; this is difficult to understand but does suggest complicated arrangements of the angular momenta of at least several nucleons so oriented that, when the resultant $M = J$, the value of Q is positive.

All these facts show that the simple single-particle model is much too naive and that quite large numbers of particles may be involved in cooperative modes of motion of an aspherical character. An estimate of the number of particles may be obtained by considering the model of the uniformly charged spheroid. For $Q > 0$, the amount of charge outside a sphere with the minor axis as diameter is $3\sqrt{5/4\pi}\,\sigma Ze = 1.89\sigma Ze$ to first order in σ. In agreement with the results for spherical nuclei, there will be some reduction of density near the edges of the nucleus, and this means that the actual value of distortion is somewhat greater than the value of σ indicated in Fig. 3–2. These values of σ are already appreciable, and it therefore appears that in many nuclei a large number of nucleons cooperate to produce a spheroidal density distribution.

Another striking feature of the graph of Q is the occurrence of zeros for certain nucleon numbers. From high positive values, Q falls through zero to small negative values not inconsistent with the single-particle value and then rises again rather abruptly to positive values. The zeros on the falling branches of the graph of Q occur for nuclides with values of Z or $N = 8, 16, 20, 28, 38, 50, 82, 126$. Nuclei with these values of Z or N are spherical, and if one or two nucleons are added, the spherical core is apparently not greatly distorted; but as the number of extra nucleons is increased, it seems that the nucleus takes on a spheroidal shape. We shall see that certain other nuclear properties have discontinuities at some of the same nucleon numbers; a fuller discussion and understanding of the behavior of Q will be possible after we have explained some less naive nuclear models. In the meantime let us note that a satisfactory model must include both the apparent single-particle behavior evident in the values of angular momentum and the collective motion implied by the large distortions of nuclei.

3-4 MAGNETIC MOMENTS

The magnetic-moment operator of a nucleus due to the orbital and spin moments of the nucleons can be written as

$$\mathbf{M} = \mu_0 \sum_{k=1}^{A} (g_k^{(l)} \mathbf{l}_k + g_k \mathbf{s}_k), \tag{3–18}$$

where the spin gyromagnetic ratios g_k are as in Eq. 3–4, and the orbital factor $g^{(l)}$ is 1 for protons and 0 for neutrons.

The magnetic moment μ is defined as the expectation value of M_z in nuclear magnetons in the state in which $J_z = J$, that is,

$$\mu = gJ = \frac{1}{\mu_0} \int \Psi_{J,J}^* M_z \Psi_{J,J} \, dV. \tag{3–19}$$

Equation 3–19 also defines the nuclear g-factor. For a nucleus of zero spin and therefore no preferred orientation, $\mu = 0$. This applies in particular to the ground states of even-even nuclides.

We can define the magnetic 2^λ-pole operator by

$$M_\lambda = \mu_0 \sum_{k=1}^{A} [\nabla r_k^\lambda P_\lambda(\theta_k)] \cdot \left[g_k^{(l)} \frac{2}{\lambda+1} \mathbf{l}_k + g_k \mathbf{s}_k \right]. \tag{3–20}$$

The magnetic moment is the special case $\lambda = 1$, namely, $\mu = (1/\mu_0)\langle M_1 \rangle_{M=J}$. The magnetic octupole moment $\langle M_3 \rangle_{M=J}$ has been measured in a few instances [2], but experimental accuracy has not yet proved sufficient to detect any higher magnetic multipoles.

In the extreme single-particle model, the angular momenta and hence the magnetic moments of nucleons cancel in pairs, and the magnetic moment of the nucleus is then simply that of the last odd nucleon. The value depends on whether $J = j = l + \frac{1}{2}$ or $l - \frac{1}{2}$, and it is not difficult (Problem 3–3) to see that the results are

$$j = l + \tfrac{1}{2}: \quad \mu = (j - \tfrac{1}{2})g^{(l)} + \mu_{n,p}, \tag{3–21a}$$

$$j = l - \tfrac{1}{2}: \quad \mu = \frac{j}{j+1}\left[\left(j + \frac{3}{2}\right)g^{(l)} - \mu_{n,p} \right], \tag{3–21b}$$

where $\mu_{n,p}$ is the free magnetic moment of the odd nucleon ($\mu_{n,p} = \frac{1}{2}g_{n,p}$). These values of the magnetic moment for a given j are referred to as the Schmidt limits [3].

Figures 3–3 and 3–4 show the measured values of the magnetic moment as a function of J for odd-A nuclei. The solid lines on the diagrams are

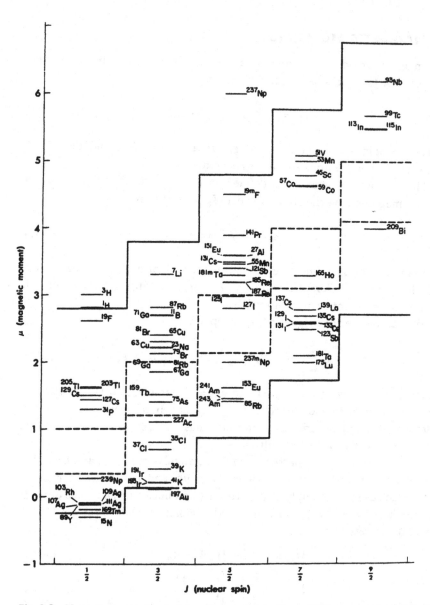

Fig. 3-3 Magnetic moments of odd proton nuclides. The solid lines are Schmidt limits; the dashed lines are single-particle moments for a fully quenched intrinsic magnetic moment.

the Schmidt limits given by Eq. 3–21. Examination of these figures shows certain general features. The moments of most nuclei are not on the Schmidt lines. However, the nuclides in each diagram fall into two rough groups, one near each of the Schmidt lines and about one nuclear magneton away.

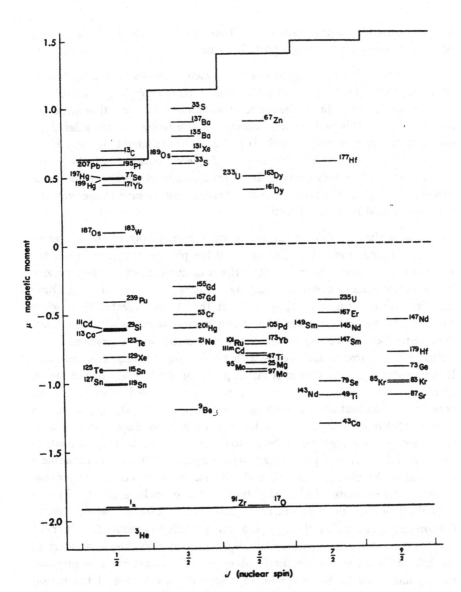

Fig. 3-4 Magnetic moments of odd neutron nuclides. The lines have the same significance as in Fig. 3–3.

It is also remarkable that, except for a few light nuclei and ^{237}Np, all deviations are inward from the Schmidt lines.

These facts stress again that the extreme single-particle model is not in accord with the details of nuclear structure, but that it does seem to explain the general trend of some phenomena. There are two possible causes for the large deviations from the single-particle values:

1. A more realistic description of the nucleus includes very considerable effects of collective motion. Suppose that a reasonably large part of the angular momentum is due to the cooperative orbital motion of other nucleons. Associated with this will be a relatively small g-factor of the order Z/A, since only the protons are charged. It is then to be expected that the μ-value will move in from the Schmidt lines. We need consider, not the whole nucleus involved in this motion, but perhaps only nucleons of the same j-value, although in the latter case it is perhaps not so easy to see why the deviation should be in one direction.

2. The shifts can also be explained by the *quenching* of the nucleon's intrinsic magnetic moments. Let us recall the picture which attributes the anomalous magnetic moment to the effects of currents of charged mesons in the virtual meson cloud surrounding a "bare" nucleon with the Dirac magnetic moment ($\mu_n = 0$, $\mu_p = 1$). It is then to be expected that, when nucleons are brought together, the strong nuclear forces which arise from the intermediary of the meson field will be associated with alterations in the virtual meson currents and hence with changes in the magnetic moment. If this produces a reduction in μ (as a spreading out in space of the meson currents would suggest), we would expect the observed moments to lie between the Schmidt lines and similar lines based on the magnetic moments of bare nucleons. The latter are the dashed lines in Figs. 3-3 and 3-4; they are quite close together in the proton case and coincident in the neutron case, so that it is not a particularly strong confirmation of this hypothesis that most of the moments do lie outside these lines. It is therefore important to obtain an estimate of the magnitude of the quenching effect. This is difficult, however, in the present unsatisfactory state of meson theory. Certain phenomenological theories [4, 5] can be made to produce the observed average magnitude of quenching by adjustment of available parameters in the calculations. To reduce the freedom of the parameters, it is possible, for the mirror nuclei ^{3}H and ^{3}He, to isolate the magnitude of the meson exchange-current contribution with acceptable certainty, since in these simple nuclei the wave functions are reasonably well known [6]. In each case the effect on μ is 0.27 nuclear magneton to the *outside* of the Schmidt limits. With the restriction that the theory used for other nuclei must fit ^{3}He and ^{3}H, it would seem that the quenching effect will not exceed about 0.25 nuclear magneton [7, 8] and may be in the wrong direction. Although

the situation is not clear, it is reasonable to suppose that one important cause of the deviations is the inadequacy of the single-particle model, rather than to attribute them wholly to a quenching phenomenon.

Despite this evidence of the importance of cooperative effects, there are some aspects of the data on magnetic moments that stress the importance of the number of odd particles in the determination of nuclear properties. In approximately twelve odd-Z elements scattered throughout the periodic table, there are two or more isotopes having the same spin but differing by an even number of neutrons. In such cases the magnetic moments are remarkably close; one outstanding exception is ^{151}Eu and ^{153}Eu, and two other less pronounced ones are ^{81,87}Rb and ^{69,71}Ga. There are also three instances of two odd-N isotones for which the magnetic moments have been measured, namely, $N = 65, 85$, and 117; in each case the moments are very close. These facts indicate that, in a nucleus with an odd number of neutrons (protons) and an even number of protons (neutrons), the even-numbered particles play only a small role in determining μ. This can also be seen by plotting the percentage deviation from the Schmidt limit against N or Z, whichever is odd (Fig. 3–5). It will be observed that the two curves are very similar, even though the nuclides being compared are of very different mass. For example, $Z = 29$ corresponds to ^{63}Cu, whereas $N = 29$ is ^{53}Cr. We also note that, as far as magnetic moments are concerned, the departure from the single-particle model does not depend to any extent on whether

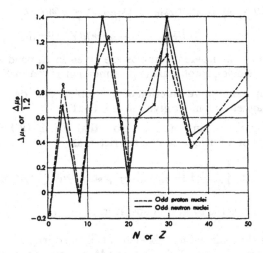

Fig. 3-5 Magnetic-moment deviation from the Schmidt limit as a function of the odd nucleon number. (After Blin-Stoyle [9]).

the odd-numbered nucleons present are neutrons or protons, but does depend rather sensitively on the value of the odd number.

REFERENCES

1. N. F. Ramsey, *Nuclear Moments*, John Wiley, New York, 1953.
2. C. Schwartz, *Phys. Rev.* **97**, 380 (1955).
3. T. Schmidt, *Z. Phys.* **106**, 358 (1937).
4. A. Miyozawa, *Progr. Theor. Phys.* **6**, 801 (1951).
5. A. Russek and L. Spruch, *Phys. Rev.* **87**, 1111 (1952).
6. R. G. Sachs, *Nuclear Theory*, Addison-Wesley, Reading, Mass., 1953.
7. M. Ross, *Phys. Rev.* **88**, 935 (1952).
8. S. D. Drell and J. D. Walecka, *Phys. Rev.* **120**, 1069 (1960).
9. R. J. Blin-Stoyle, *Theories of Nuclear Moments*, Oxford University Press, Oxford, 1957.

PROBLEMS

3–1. (a) Show that the quadrupole moment Q' of a uniformly charged spheroid about the axis of symmetry is $\frac{2}{5}Ze(b^2 - a^2)$, where a, b are the semi-axes, b being along the axis of symmetry.

(b) Show that the quadrupole moment about an axis making an angle β with the axis of symmetry is $(\frac{3}{2}\cos^2\beta - \frac{1}{2})Q'$.

(c) From the values of σ in Fig. 3–2, calculate the percentage difference between the major and minor axes of several typical distorted nuclei.

3–2. Note that in the simple model used in this chapter a nucleon has constant values of l, j, and m. Hence its wave function is

$$\psi(l, j, m) = u(r)[(l\tfrac{1}{2}m - \tfrac{1}{2}\tfrac{1}{2}|jm)\alpha Y_l^{m-(1/2)}(\theta, \phi)$$
$$+ (l\tfrac{1}{2}m + \tfrac{1}{2} - \tfrac{1}{2}|jm)\beta Y_l^{m+(1/2)}(\theta, \phi)].$$

The notation and some needed formulas are given in Appendix A.

Calculate the single-proton quadrupole moment given in Eq. 3–17a.

3–3. Using the same wave function as in Problem 3–2, derive the Schmidt values of the magnetic moment given in Eq. 3–21.

3–4. (a) Using the i-spin notation, show that Eq. 3–18 for the magnetic moment operator of a nucleus can be rewritten as

$$\mathbf{M} = \mu_0\left\{\sum_{k=1}^{A} \tfrac{1}{2}[1 + \tau_3(k)]\mathbf{l}_k + \mu_S\mathbf{S} + \mu_V\sum_{k=1}^{A}\tau_3(k)\mathbf{s}_k\right\}$$

where $\mathbf{S} = \sum_{k=1}^{A}\mathbf{s}_k$ is the total spin operator, $\mu_S = \tfrac{1}{2}(g_p + g_n) = 0.8797$, and $\mu_V = \tfrac{1}{2}(g_p - g_n) = 4.7059$.

(b) Denoting the diagonal matrix element of the operator M_z with respect to the nuclear state $|J, M = J; T, T_3\rangle$ as $\langle M_z\rangle_{T_3}$, show that the sum of the magnetic moments of mirror nuclei is given (in units of μ_0) by

$$\langle M_z \rangle_{T_3} + \langle M_z \rangle_{-T_3} = (\tfrac{1}{2} + \mu_S)J + \frac{1}{J+1}(\tfrac{1}{2} - \mu_S)\langle L^2 - S^2 \rangle.$$

You will need formula A–75a of the appendix. This is known as the mirror theorem of Sachs [R. G. Sachs, *Phys. Rev.* **69**, 611 (1946)].

Chapter 4

Sizes of Nuclei

4-1 INTRODUCTION

We would like to ask these important questions: What are the shapes and sizes of nuclei, and how are the nucleons distributed throughout the nuclear volume?

In Chapter 1 we have already remarked that Rutherford showed in 1911 that nuclei have a radial extent of less than 10^{-12} cm [1]. Rutherford's calculations were based on experiments by Geiger and Marsden [2, 3, 4], in which a beam of α-particles was made to fall on gold atoms in a thin foil. A scintillation detector was used to determine how many α-particles were scattered at each of several angles. If the final direction of motion of a scattered particle is specified by polar angles (θ, ϕ), and if $d\Omega$ is an infinitesimal solid angle located around the ray (θ, ϕ), then the number of particles scattered per second into $d\Omega$ divided by the incident flux is defined as the *differential scattering cross section* and is written as $\sigma(\theta, \phi)\, d\Omega$ (Fig. 4–1). For charged particles scattered by a point charge with potential $Z_2 e/r$, this cross section can be easily calculated classically. Here we have one formula of classical mechanics that carries over unchanged into nonrelativistic quantum theory.[†] The result, which does not depend on ϕ, is

$$\sigma(\theta) = \left(\frac{Z_1 Z_2 e^2}{2\mu v^2}\right)^2 \operatorname{cosec}^4 \tfrac{1}{2}\theta. \tag{4-1}$$

[†] See, for example, K. R. Symon, *Mechanics*, Addison-Wesley, Reading, Mass., p. 120; and L. I. Schiff, *Quantum Mechanics*, McGraw-Hill, New York, p. 114.

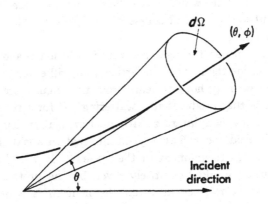

Fig. 4-1   Scattering angles and solid angle.

In this formula, θ is the scattering angle in the center-of-mass system, μ is the reduced mass, v is the initial relative velocity, and $Z_1 e$, $Z_2 e$ are the charges of the two particles. The scattering described by Eq. 4-1 is referred to sometimes as "Coulomb scattering" and sometimes as "Rutherford scattering."

Classically, the distance of closest approach for a given energy is $(2 Z_1 Z_2 e^2 / \mu v^2) = r_C$, say. In Geiger's experiments, if r_C were less than the radius of the gold nucleus, the α-particles would be able to penetrate into the nucleus. In this region the potential would no longer be that of an isolated charge. If, for example, the electric charge $Z_2 e$ were uniformly distributed throughout a sphere of radius R, the potential inside the sphere would be (see Problem 4-11)

$$\frac{Z_2 e}{R}\left[\frac{3}{2} - \frac{1}{2}\left(\frac{r}{R}\right)^2\right] \quad \text{for} \quad r < R. \tag{4-2}$$

Thus, if the α-particles entered the region of positive charge, the angular dependence of their scattering would be quite different from that given by Eq. 4-1. In the experiments that Rutherford analyzed, r_C was as small as 10^{-12} cm, and yet there was no departure from Coulomb scattering. Consequently it was deduced that nuclei have radii of less than 10^{-12} cm.

Since the scattering in question is due to electric forces, one might be tempted to say that all that had been demonstrated was that the protons are inside r_C and that nothing was learned about the location of neutrons.

However, the α-particle would have "felt" nuclear forces if any nuclear matter were outside r_C, and, of course, these forces would also have caused anomalous scattering.

It is clear that, if a scattering experiment is to teach us about the distribution of matter inside the nucleus, a particle that will enter the nucleus must be used. Then we may hope to learn from the angular scattering pattern what the force is that produces this scattering. If for a probe we use very fast electrons, they will interact with nuclear matter only through the electromagnetic field and will thus yield information about the distribution of charge and magnetic moment in the nucleus. It would seem that the charge distribution would be entirely given by the protons. This is not strictly true, however, since although the neutron is a neutral particle, experiments indicate (see Section 4–3) that it has a charge structure, with positive charge mainly in the core and negative charge distributed to a greater degree in the tail. This implies that an incoming high-energy electron may be scattered by the charge distribution of a neutron; consequently, the charge distribution in a nucleus, although governed mainly by the proton distribution, is also influenced by the neutrons. It is reasonable to state that the charge density in a nucleus is generated from the single-particle wave functions of the nucleons, convoluted with the intrinsic charge structure of the nucleons obtained experimentally, with small and spatially smooth corrections for many-body correlations and meson-exchange currents. In the magnetic-moment distribution, of course, both protons and neutrons contribute importantly.

Since the only force of any importance between the electrons and the nucleons is electromagnetic, the electron is an ideal probe particle. Electron beams can be scattered from nuclei, just as α-particles were, and if the electrons are sufficiently energetic, they are not affected by the electrons in the target, but enter the nucleus and provide very good indications of their charge distributions. It will be realized also that any electron which penetrates the nucleus, such as an s-electron in an atom, will have its motion influenced by the nuclear-charge distribution. This will affect the electron's energy, and consequently there is some hope of learning about the nuclear structure from a study of fine structure in X-ray spectra. The difficulty lies in the smallness of these effects, which is due to the fact that only a very small fraction of the motion of a bound electron lies within the nucleus. However, there is another particle which also interacts with nucleons essentially only through electromagnetic forces. This is the muon (μ). Indeed, from the present point of view, it differs importantly from the electron only by its mass, which is $207m_e$, and its instability. It has a half-life at rest of 2.2×10^{-6} sec. However, this is ample time for negative muons (produced, say, in an accelerator) to slow down in matter and to be captured into an orbit about a nucleus. This orbit will have the same properties as an electronic orbit in an atom, except for the mass difference. The radius of the Bohr orbit is inversely proportional to the mass, and thus the μ^- has a much larger fraction of its wave function within the nucleus than has the electron. In heavy elements, the radius of the Bohr orbit for the μ^- is $\sim 10^{-12}$ cm. Consequently the spectra of muonic atoms are more sensitive to nuclear size and structure than are the spectra of ordinary atoms. We shall consider each of these methods in turn, beginning with the most instructive, namely, electron scattering [5, 55].

4-2 ELECTRON SCATTERING

The amount of detail of nuclear structure discovered in an electron-scattering experiment is, of course, governed by the de Broglie wavelength of the electron. Thus, to delineate features having dimensions of the order of 1 fm, it is necessary to use electrons of over 100 MeV. It is essential to employ relativistic wave mechanics to describe such electrons. Also, in the scattering process, electrons of quite high angular momenta are involved, and in calculations for 180 MeV on heavy nuclei as many as ten phase shifts are required to describe the scattering process. In these circumstances, it is not possible to give a closed expression for even point-charge scattering. Analytic forms of $\sigma(\theta)$ have been given for high-energy electrons scattered from a point charge Ze, where $Ze^2/\hbar c = Z/137 \ll 1$ [6, 7, 8]. The first terms of a series in $Z/137$ are

$$\sigma(\theta) = \left(\frac{Ze^2}{2E}\right)^2 \frac{\cos^2 \tfrac{1}{2}\theta}{\sin^4 \tfrac{1}{2}\theta} \left[1 + \frac{\pi Z}{137} \frac{(\sin \tfrac{1}{2}\theta)(1 - \sin \tfrac{1}{2}\theta)}{\cos^2 \tfrac{1}{2}\theta}\right]. \tag{4-3}$$

The factors outside the brackets are referred to as Mott scattering and may be written σ_{Mott}. The quantity E is the center-of-mass energy. For high Z this formula cannot be used, but numerical evaluations are available [9]. It holds, of course, for the center-of-mass system; hence in the laboratory system the apparently energy-independent angular distribution is modified, and the energy of the scattered electrons is a function of the scattering angle.

   Departures from the cross section for a point charge reveal features of the nuclear structure. For heavy nuclei, the effects of this structure also must be handled by numerical computation. However, for light elements with $Z \lesssim 10$, where the second term in the brackets of Eq. 4–2 can be ignored, a valid closed formula can be obtained. The cross section is expressed as the Mott cross section multiplied by a *form factor*:

$$\sigma(\theta) = \sigma_{\text{Mott}}(\theta)[F(q)]^2, \tag{4–4}$$

$$F(q) = \frac{4\pi}{q} \int_0^\infty \rho(r) \sin(qr) r \, dr. \tag{4–5}$$

In this expression, $\mathbf{q}$ is the vector representing the momentum transfer in the collision, and its magnitude q is given by

$$q = \frac{2E}{\hbar c} \sin \tfrac{1}{2}\theta = \frac{2}{\lambda} \sin \tfrac{1}{2}\theta, \tag{4–6}$$

where λ is the reduced de Broglie wavelength of the electron (see Problem 4–1). The derivation of these formulas is valid only for spherically symmetric charge distributions; $\rho(r)$ represents the charge density, but it is normalized to unity, that is, $\int \rho(r) 4\pi r^2 \, dr = 1$, and the charge in the spherical shell of radius r and thickness dr is $Ze\rho(r)4\pi r^2 \, dr$. Experimentally, the form factor is obtained directly for some range of q, determined by the energy and the angular range for which observations are made. It is then desirable to correlate the density $\rho(r)$ with the function F. Let us suppose that the wavelength λ is large compared with nuclear dimensions. Then in Eq. 4–4, since $\rho(r)$ is zero outside the nucleus, we have $qr \ll 1$ and can write

$$F(q) = \frac{4\pi}{q} \int \rho(r)(qr - \tfrac{1}{6}q^3r^3 + \cdots)r \, dr$$

$$= \int \rho(r)4\pi r^2 \, dr - \tfrac{1}{6}q^2 \int r^2 \rho(r)4\pi r^2 \, dr + \cdots$$

$$= 1 - \tfrac{1}{6}(qa)^2 + \cdots. \tag{4–7}$$

We have put $a^2 = \int r^2 \rho(r) 4\pi r^2 \, dr$, and it is seen that a^2 is the mean-square radius of the charge distribution. It is clear that, until λ is decreased to at least nuclear dimensions, it will not be possible to learn anything more than the value of a^2. This is a reflection of the general remark that we must use high energies to learn the finer details of nuclear structure.

It is possible to give an expression for $\rho(r)$, for the Fourier transform of Eq. 4–5 yields

$$\rho(r) = \frac{1}{2\pi^2 r} \int_0^\infty F(q) \sin{(qr)q} \, dq. \tag{4-8}$$

This formula is of limited usefulness until experiments are carried out that give the form factor with considerable accuracy and over a sufficient range of q. The alternative procedure, which is the only one available whenever formula 4–3 does not apply, is to assume many possible forms of the function $\rho(r)$ and, by substitution in Eq. 4–4, discover which are consistent with experiment.

One of the principal results of the electron-scattering experiments is the conclusion that nuclei do not have sharp edges, but rather that there is an appreciable region over which the nucleon density falls away. One form of density dependence that has been studied extensively is

$$\rho_F(r) = \frac{\rho_0}{1 + \exp[(r - c)/a_0]}. \tag{4-9}$$

This distribution has been referred to in the literature as a Fermi density function. It has the property that, for $c \gg a_0$, ρ is essentially ρ_0 until $c - r$ is a few times a_0, and then it falls to negligible values in a distance determined by a_0 and independent of c.

Hence for the same a_0 we obtain density distributions of the type shown in Fig. 4–2, with uniform density in central cores of varying radii, but with the same thickness of boundary layer. For any density distribution with a symmetrical skin, we shall define c as the radius at which the density has fallen to half its maximum value, and shall use the letter t to denote the

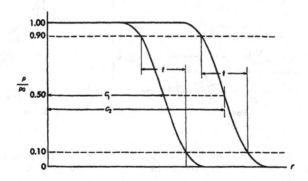

Fig. 4-2 The Fermi-density distributions for two nuclei of different radii.

distance in the surface layer over which the density decreases from 90% to 10% of its maximum.[†] For the Fermi distribution, $t = 4.40a_0$. It will be realized that, as a_0 approaches zero, the distribution approaches that of a uniform sphere of radius c.

The form factors for many shapes of $\rho(r)$ have been calculated. A general result is that the sharper the cutoff of the density, the more pronounced are the oscillations in the cross section, as would be expected, since the cross section is essentially a diffraction pattern for the electron wave. This behavior is also confirmed, by numerical calculation, in the range for which the form-factor approximation is not valid. An example is illustrated in Fig. 4–3, based on calculations carried out by Yennie, Ravenhall, and Wilson [10]. The scattering cross section is shown for 125-MeV electrons incident on gold for three different shapes of charge density. It is seen that the smoothest curve arises from the smoothest (i.e., the exponential) distribution, and the most oscillatory curve is associated with the sharp cutoff of a

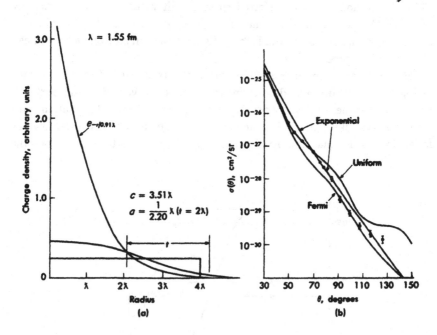

Fig. 4-3 Dependence of the smoothness of the scattering cross section on the smoothness of the density distribution. The densities in (a) give the cross sections in (b). (After Yennie et al. [10].)

[†] Strictly speaking, the definition of c used in the literature is $c\rho(0) = \int_0^\infty \rho(r)\,dr$.

uniform sphere. The intermediate Fermi distribution is nearer to the experimental points than either of these extremes.

The calculations require first evaluating the potential energy inside the nucleus due to the assumed spherically symmetric charge distribution, given by

$$(4\pi)^{-1}V(r) = \left[\frac{1}{r}\int_0^r \rho(r')r'^2\,dr' + \int_r^\infty \rho(r')r'\,dr'\right](-Ze^2). \qquad (4\text{–}10)$$

The wave function describing the electron scattering is determined by solving Dirac equations for all necessary angular momenta, using Eq. 4–10 for the potential function. This is done numerically, and phase shifts are found. The cross section, which is a known function of the phase shifts, is then calculated. This procedure resembles the methods discussed in Chapter 2 for neutron-proton scattering, but is complicated by the fact that the electrostatic forces are not of short range. We shall not discuss the details here; they are given in the original paper [10].

We first discuss an early experimental study on gold made by Hahn, Ravenhall, and Hofstadter [11], using 153- and 183-MeV electrons. This will illustrate the fact that, with electrons in this energy range, it is not possible to determine the details of $\rho(r)$ for $r < 4$ fm. Previous results, such as those at 125 MeV described above, had made it clear that the density should have a "tail" rather like the Fermi shape and be fairly flat inside. Actually the density at small r is very hard to determine, since there is not much charge in a sphere of small radius, say 3 fm. [This geometrical factor is

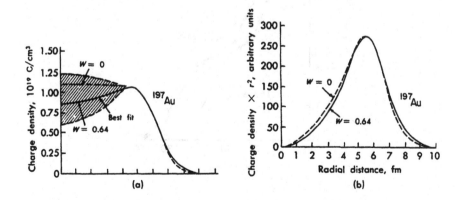

Fig. 4-4 Possible charge densities for gold. (After Hofstadter [5].)

easily taken into account by plotting $4\pi r^2 \rho(r)$, a quantity that is associated with the charge in a spherical shell and that portrays more graphically than $\rho(r)$ itself the distances most important for scattering.] Consequently, in addition to using the Fermi distribution F, Hahn et al. studied shapes with central depressions or elevations, given by

$$\rho(r) = \left(1 + \frac{wr^2}{c^2}\right)\rho_F(r), \qquad (4\text{--}11)$$

which is also called the three-parameter Fermi distribution. The best fit to experiment was obtained with a slight reduction in central density corresponding to $w = 0.64$. However, any curve passing through the shaded area of Fig. 4–4a gives a cross-section curve within experimental error. Figure 4–4b shows clearly the effect of the spherical geometry in reducing the contribution from near the center. On the other hand, the density from about 5 fm out, that is, in the nuclear skin, is quite closely determined. The quality of the agreement with experiment is illustrated in Fig. 4–5.

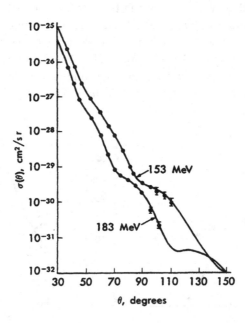

Fig. 4-5 The best-fitting Fermi density for gold gives this excellent agreement with experiment. (After Hofstadter [5].)

Hahn et al. also evaluated the scattering for a trapezoidal density and for the density

$$\rho(r) = \rho_0 \left[1 + \exp\left(\frac{r^2 - c^2}{a^2} \right) \right]^{-1}. \tag{4-12}$$

It was found possible to fit the data equally well with these shapes, but, as a glance at Fig. 4-6 shows, the actual densities are very similar, particularly in the skin. In fact, all shapes that agree with experiment have the

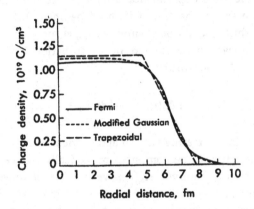

Fig. 4-6 Densities for gold, all of which fit the electron-scattering data. (After Hofstadter [5].)

same values for c and t. Later experiments on muonic X-ray measurements exclude the trapezoidal and the modified Gaussian densities for gold, but cannot differentiate between the Fermi distribution $\rho_F(r)$ and the three-parameter Fermi form of Eq. 4-11 [12].

More information about the charge density $\rho(r)$ can be obtained when electrons of higher energy are used. It should be emphasized, however, that experimental cross sections are obtained over only a limited range of q-values; for example, in ^{208}Pb, which is one of the most thoroughly examined nuclides, data points are available only up to $q \approx 2.7$ fm^{-1}. It is not possible, therefore, even with the present-day experimental data, to specify $\rho(r)$ uniquely. To be more explicit, we assume that there are N data points, $q_1 \cdots q_N$, at which $F(q)$ has been determined experimentally. Let $\rho_0(r)$ be a density distribution that can fit these experimental data within errors. For simplicity, we assume that the Born formula Eq. 4–5 may be used for $F(q)$, although essentially the same conclusions are reached by more rigorous arguments [13]. If we now change $\rho_0(r)$ to $\rho_0(r) + \delta\rho(r)$, with the constraints

$$\int \delta\rho(r)r^2\,dr = 0,$$

$$\int \delta\rho(r)j_0(q_i r)r^2\,dr = 0, \quad i = 1 \text{ to } N, \qquad (4\text{-}13)$$

it follows that the same experimental $F(q)$'s will be reproduced. Friar and Negele [13] have shown that, even if the experimental $\sigma(\theta)$'s are fitted by an exact phase-shift calculation assuming a $\rho_0(r)$, provided that the change $\delta\rho(r)$ is small enough to be treated by perturbation theory, a constraint similar to Eq. 4–13 is obtained:

$$\int \delta\rho(r)K_\alpha(r)r^2\,dr = 0. \qquad (4\text{-}14)$$

Here α stands for the scattering angle θ and incident energy E of the electron, and $K_\alpha(r)$ is a function to be computed numerically. Except at the values of q near the diffraction minima, $K_\alpha(r)$ is very similar in appearance to $j_0(qr)$.

It is most advantageous to combine the electron-scattering data at low and high energies with muonic X-ray data to determine the characteristics of the nuclear-charge distribution. As we saw from Eq. 4–7, at low energies only the rms radius a is obtained from elastic electron-scattering measurements. We shall see in Section 4–4 that the muonic data also determine a accurately. The high-energy measurements, however, scan the regions of the density that have the steepest slope [14, 15]. A careful analysis of ^{208}Pb data on electron scattering and muon X-rays shows [13] that, although the charge distribution $\rho(r)$ can be obtained unambiguously beyond $r > 5$ fm, ambiguities in $\rho(r)$ for the interior region remain, the uncertainty in the central region being the largest, of the order of 4–5%. The charge distribution of ^{208}Pb that yields the best fit to the latest Stanford data on electron scattering and the muonic X-ray data is shown in Fig. 4–7, the ambiguity in $\rho(r)$ being indicated by the shaded envelope [13]. The rms radius a of this distribution is 5.502 ± 0.006 fm. An earlier analysis by Bethe and Elton [15] indicated that there is a 7–10% depression in the central density of ^{208}Pb; the more careful analysis of Friar and Negele [13], on the other hand, shows only a slight depression at $r \approx 1.5$ fm (see Fig. 4–7). The central depression is very

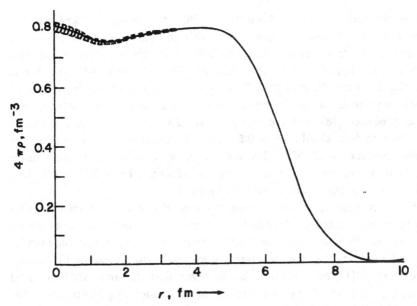

Fig. 4-7 Charge-density distribution for lead, as deduced from electron scattering and muonic X-ray. (After Friar and Negele [13].)

sensitive to the absolute normalization of the incident electron beam, and a 4% change in the normalization can eliminate the depression shown in Fig. 4-7. Furthermore, earlier analysis of high-energy electron-scattering data [16] indicated that minute oscillations in $\rho(r)$ superimposed on a smooth density improve the fit; the analysis of Friar and Negele [13] shows that there is nothing unique about these wiggles and that the data may be fitted with or without them, so long as the constraints of Eq. 4–14 are obeyed. We can state, therefore, that there is no positive evidence for the wiggles in $\rho(r)$ found earlier, but the data do not exclude them either.

Interesting work by electron scattering has also been done on calcium isotopes [18, 19, 20]. It was found that the electron-scattering data at 250 MeV on ^{40}Ca, ^{44}Ca, and ^{48}Ca could be well fitted by the three-parameter Fermi distribution given by Eq. 4–11. The best fit to ^{40}Ca yielded the parameters $c = 3.6685$ fm, $a_0 = 0.5839$ fm, and $w = -0.1017$, indicating a slight elevation at the center. The corresponding values for ^{48}Ca were found to be $c = 3.7369$ fm, $a_0 = 0.5245$ fm, and $w = -0.0300$. Since w is so small, we may still consider c to be the half-density radius, and the surface thickness $t \approx 4.4a_0$. The above parameters then reveal that, although the half-density radius increases from ^{40}Ca to ^{48}Ca, the surface thickness of the heavier isotope is about 10% smaller, with the result that there is more charge in the surface region of ^{40}Ca than of ^{48}Ca. This results in the rms radius of ^{48}Ca being slightly smaller than that of ^{40}Ca. The rms radius of ^{44}Ca is only slightly larger than that of ^{40}Ca—about a quarter of the increase that would be given by a simple $A^{1/3}$ law. On the basis of these distributions, it was predicted [18] that the change in the $2p \rightarrow 1s$ muonic X-ray energy from ^{40}Ca to ^{48}Ca should be $- 0.52$ keV, the negative sign indicating the smaller rms radius of ^{48}Ca. The muonic X-ray measurements were made shortly afterward, and this shift in energy was found to be $- 0.47 \pm 0.12$ keV [21], confirming the electron-scattering results.

For a few years it was erroneously thought that the above results on Ca isotopes demonstrated a reduction in the rms radius of the proton distribution in going from ^{40}Ca to ^{48}Ca. This was very puzzling, since microscopic calculations with realistic two-nucleon forces were unable to reproduce this result. Bertozzi et al. [17] have pointed out, however, that this anomaly is resolved by taking account of the contribution of the added $f_{7/2}$ neutrons to the charge distribution. These neutrons contribute negatively to the charge distribution in the surface, and more than compensate for the increase in the rms radius of the proton distribution. The result is a net decrease in the charge rms radius in going from ^{40}Ca to ^{48}Ca.

In the independent-particle model of the nucleus, each nucleon moves in an average one-body potential well, the latter in general being different for protons and neutrons. The energy levels of the potential well are occupied in accordance with the Pauli exclusion principle. For example, in the lowest s-state, there can be only two protons and two neutrons, since the z-component of the spin of each particle can be up or down. By filling up the orbitals in this manner, one can calculate the resulting density distribution of point nucleons from a knowledge of the single-particle wave functions (see Problem 4-10). Such a "shell-model" calculation of the density distribution shows some wiggles due to shell structure and a somewhat pronounced peaking at the center caused by the filling of s-wave orbitals. In order to obtain the charge distribution, it is necessary to convolute the intrinsic charge structure of the nucleons (Section 4-3) with the appropriate densities. Let $\rho_{SM}^{(p)}(\mathbf{r})$ and $\rho_{SM}^{(n)}(\mathbf{r})$ denote the point distributions of the protons and the neutrons, respectively, as obtained from the single-particle wave functions; and $f_{Ep}(\mathbf{r})$ and $f_{En}(\mathbf{r})$ the spatial charge distributions of the proton and the neutron in the nonrelativistic limit (see Eq. 4-31). Then the charge distribution of the nucleus in this model is

$$\rho(\mathbf{r}) = \int \rho_{SM}^{(p)}(\mathbf{r'}) f_{Ep}(\mathbf{r} - \mathbf{r'}) \, d\mathbf{r'} + \int \rho_{SM}^{(n)}(\mathbf{r'}) f_{En}(\mathbf{r} - \mathbf{r'}) \, d\mathbf{r'}.$$

We have seen that the second term in the right-hand side of the above equation plays an important part in the interpretation of the Ca isotope results. It should be realized that taking an average potential well still leaves out parts of the two-nucleon interaction, which may cause virtual excitation of the nucleons from the lowest occupied orbitals. When the single-particle wave functions are suitably modified to take account of this effect (this is called configuration mixing; see Chapter 7), it is found that the elevation at the center of the calculated $\rho(r)$ is reduced and the wiggles in the density distribution are somewhat smoothed out. It has been shown [22] that agreement with high-energy electron-scattering data may be obtained in this way by using a phenomenological average potential which is smooth and nonlocal.

In the following discussion of light nuclei ($4 \leqslant A \leqslant 16$), we shall generate the proton distributions from assumed one-body potentials, and compare the results with experimentally extracted charge distributions. In other words, we are going to neglect the contribution of the neutrons to the charge distribution by assuming that $f_{En}(r) = 0$, which, as we have pointed out, is not quite right. Detailed elastic scattering data are available for many nuclei, including ^{16}O, ^{12}C, ^{6}Li, and ^{4}He [23]. In the earlier analyses of these results, good agreement was generally obtained by assuming that each proton moves in a harmonic-oscillator shell-model potential, given by

$$V = - V_0 + \tfrac{1}{2} M \omega^2 r^2,$$

where V_0 is a constant, M is the proton mass, ω is a parameter determining the spread of the wave function, and r is the proton coordinate with respect to the center of the well. By putting two protons in the s-shell and $Z - 2$ protons in the p-shell of such a well, the density distribution is given by (see Problem 4–10)

$$\rho_{SM}(r) = \rho_0 \left[1 + \frac{Z - 2}{3} \left(\frac{r}{b} \right)^2 \right] \exp\left(- \frac{r^2}{b^2} \right), \qquad (4\text{--}15)$$

where by ρ_{SM} we denote the "shell-model" density, ρ_0 is the normalization constant, and $b = \sqrt{\hbar/m\omega}$ is called the oscillator parameter. In order to obtain the charge density $\rho(r)$ from this, it is necessary to take account of the finite size of the protons and to correct for the center-of-mass motion. For these details, the reader is referred to Elton's book [23]. When these corrections are made, it is found that good fits to the ^{12}C and ^{16}O data can be obtained for $q \leqslant 400$ MeV/c, with $b = 1.64$ fm for ^{12}C and $b = 1.76$ fm for ^{16}O. However, the experimental form factor of ^{16}O, shown in Fig. 4–8, has two diffraction minima, the first at $q \approx 300$ MeV/c and the second at $q \approx 600$ MeV/c [24]. The form factor generated by a harmonic-oscillator

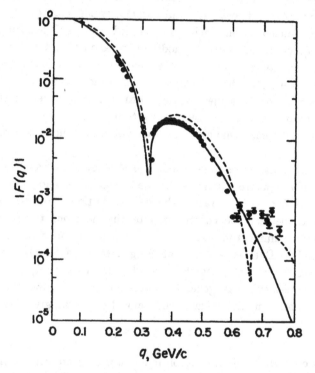

Fig. 4-8 Electron elastic-scattering form factor for ^{16}O. The experimental data of McCarthy and Sick [24] are shown by dots. The full line is the theoretical fit obtained by assuming a closed-shell configuration and using harmonic-oscillator wave functions; the dashed curve is obtained from Woods-Saxon wave functions. (After Donnelly and Walker [25].)

potential yields only one diffraction minimum and therefore cannot fit the high-energy data. A harmonic potential is rather unrealistic in the sense that it keeps increasing with increasing distance, rather than falling off to zero. Any potential with a finite range, on the other hand, gives rise to a form factor that continues to undulate with increasing momentum transfer, giving a series of diffraction minima. Donnelly and Walker [25] have shown that a reasonable fit to the ^{16}O data, including the second diffraction minimum, can be obtained by a suitably chosen potential well of the form

$$V(r) = \frac{V_0}{1 + \exp[(r - c)/a_0]}. \qquad (4\text{-}16)$$

This is known as the Woods-Saxon potential; it falls to half its central value at $r = c$. The form factor of ^{16}O generated by such a well is also shown in Fig. 4-8.

The situation for ^{12}C is shown in Fig. 4-9. The Woods-Saxon potential well predicts a second minimum in the form factor at $q \sim 700$ MeV/c, which is the upper limit of the existing data.

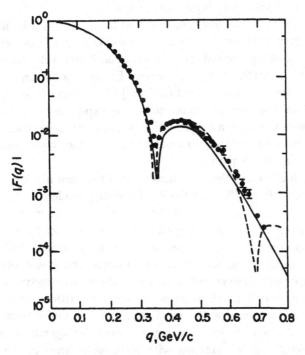

Fig. 4-9 Same as in Fig. 4–8, but for ^{12}C. (After Donnelly and Walker [25].)

The nucleus ^6Li is the lightest stable nucleus with two p-shell nucleons. Its ground-state spin J is 1; consequently it has magnetic dipole and electric quadrupole moments. The simple formula 4–4 should be modified to take account of the fact that these also can contribute to the form factor [26]. The quadrupole moment of ^6Li being very small, $Q = (-8.0 \pm 0.8) \times 10^{-2}\,\text{fm}^2$, the charge part of the form factor may still be assumed to be spherically symmetric, as in Eq. 4–5. There is an additional term in the form factor due to the magnetic interaction, which becomes more important for backward scattering. The charge form factor of ^{6}Li has been extracted experimentally [27] in the range $0.7 < q < 2.62$ fm^{-1}. An excellent fit to the data can be made with

$$|F(q)|^2 = \exp(- a^2 q^2) - c^2 \exp(- b^2 q^2),$$

where $a^2 = 0.87$ fm^2, $b^2 = 1.70$ fm^2, and $c^2 = 0.205$, yielding an rms radius of 2.54 fm. However, the data cannot be fitted by taking a model in which the nucleons are moving in a single-particle well.

The elastic electron-scattering cross section from ^{4}He has been measured and the form factor of ^{4}He extracted for $q^2 = 0.5$–20 fm^{-2} by Frosch et al. [28]. Again the form factor generated by a harmonic-oscillator well cannot fit the data beyond $q^2 = 6$ fm^{-2}. A three-parameter Fermi distribution for $\rho(r)$ fits the data in the entire range. It is found [29] that a form factor produced from a finite potential well is unable to fit the experimental points if the depth of the well is adjusted to yield the experimental separation energies of the proton and neutron. It appears that, as with ^{6}Li, the simple independent-particle picture is not adequate.

A careful calculation of the form factor of ^{6}Li has been made by Bouten, Bouten, and Van Leuven [30]. In the independent-particle model, the wave function of the nucleus is a product of the individual nucleon wave functions, properly antisymmetrized. Such a product-type wave function, however, is not in general an eigenfunction of the total angular momentum. To obtain the form factor of the ^{6}Li ground state, for example, one should project out the $J = 1$ component of the wave function from the antisymmetrized product of the single-particle orbitals. Furthermore, the single-particle orbitals should be chosen so as to minimize the total energy for a given Hamiltonian. Bouten et al. [30] have shown that they can get good agreement with the experimental results of ^{6}Li, starting with reasonable forms of the nuclear Hamiltonian. Such a generalized shell model, therefore, seems to be adequate for light nuclei.

Finally, we mention that the electromagnetic form factors for ^{3}He [31] and the deuteron [32] have been extracted from the elastic scattering data over a wide range of q. For ^{3}He, the charge form factor is known up to $q^2 = 20$ fm^{-2}, while the magnetic form factor is determined up to $q^2 = 12.5$ fm^{-2}. The neutron-proton interaction in the deuteron would be uniquely determined if the deuteron wave function were known for all values of r. The form factor is the most directly related quantity to the wave function that can be determined experimentally, and it is clear that the deuteron form factor can provide some insight into the details of the n-p interaction. To perform the analysis, it is necessary to know the form factors of neutron and proton, which we discuss in the next section. It turns out that most n-p potentials with short-range repulsion which fit the scattering data also fit the deuteron electric form factor reasonably well [32] (see Section 5–2, where we discuss this topic in detail).

For all but these very light nuclei, certain regularities in the behavior of the charge density are important. For each nucleus studied, a rms charge radius a is calculated from

TABLE 4-1 Parameters of Charge Distribution of Some Near-Spherical Nuclei. The distances are given in femtometers. The numbers are taken from a compilation by Überall [55, p. 210]. The last column, "Comments," indicates the place and year in which the analysis was done. The abbreviations are as follows: D: Darmstadt; K: Khartov; O: Orsay; S: Stanford; and Y: Yale. For example, S 59 indicates that the work was done at Stanford in 1959. The accuracy is within 2% for the radii and 10% for the surface thickness.

Nuclide	Skin thickness t	Half-density radius c	Root-mean-square radius a	Equivalent uniform radius R	Comments
⁴He			1.63		[34]
⁶Li			2.54		[27]
¹²C	2.20	2.24	2.50	3.23	S 59
¹⁴N	2.20	2.30	2.45	3.16	S 59
¹⁶O	1.8	2.60	2.65	3.42	
²⁴Mg	2.6	2.85	2.98	3.85	S 56
²⁸Si	2.8	2.95	3.04	3.93	S 56
⁴⁰Ca	2.51	3.60	3.52	4.54	S 65
⁴⁸Ti	2.49	3.74	3.59	4.63	D 66
⁵²Cr	2.33	3.97	3.58	4.62	O 64
⁵⁶Fe	2.5	4.16	3.76	4.85	O 62
⁵⁸Ni	2.46	4.14	3.83	4.94	K 68
⁸⁹Y	2.51	4.76	4.24	5.47	K 67
⁹³Nb	2.52	4.87	4.33	5.59	K 67
¹¹⁶Sn	2.37	5.27	4.55	5.87	O 67
¹²⁰Sn	2.53	5.31	4.64	5.99	O 67
¹³⁹La	2.35	5.71	4.85	6.26	K 67
¹⁴²Nd	1.79	5.83	4.76	6.15	Y 69
¹⁹⁷Au	2.32	6.38	5.33	6.88	S 56
²⁰⁸Pb	2.33	6.54	5.50	7.10	[13]

$$a^2 = \int r^2 \rho(r) 4\pi r^2 \, dr. \tag{4-17}$$

The equivalent uniform distribution is defined as the uniform distribution with the same value of a. Its radius R is given by $R = (5/3)^{1/2}a$. Both these quantities are tabulated in Table 4-1, which lists also the values of the skin thickness t and the half fall-off radius c.

The size and density of a nucleus are determined mainly by the nuclear forces, to which the electrostatic forces are a relatively small correction. If the numbers of neutrons and protons in the nucleus are the same, the nuclear forces being charge independent, we would expect the neutrons and protons to have nearly the same spatial distribution. In heavier nuclei, the neutron number being large, the protons should move in a deeper potential well. However, this will be somewhat offset by the average repulsive Coulomb potential that the proton experiences. There is the additional fact that the single-particle wave function of the proton should decay more rapidly in the asymptotic region than that of the neutron because of the Coulomb barrier. We may expect, therefore, that in the outer edge of the surface there are more neutrons than protons. The densities for protons and neutrons should otherwise be very similar (except for the normalization factor). Experimentally, the absorption of K^--mesons in the nuclear periphery [33] seems to support the view that the distant surface [$r > (c + 1)$ fm] of a heavy nucleus contains more neutrons than protons.

In summary, elastic electron scattering has provided the following information about the spatial distribution of charge in nuclei:

1. For nuclei with $A > 30$, the main features of the charge density may be represented by two- or three-parameter Fermi distributions which have almost uniform central cores.

2. For nuclei with $A \gtrsim 20$, the half-density radius c and the rms radius a are both roughly proportional to $A^{1/3}$. As illustrated in Figs. 4-10 and 4-11, good fits are found to be as follows [34]:

$$c = 1.18A^{1/3} - 0.48 \text{ fm}, \qquad a = 0.82A^{1/3} + 0.58 \text{ fm}. \tag{4-18}$$

3. For lighter nuclei ($4 \leqslant A \leqslant 20$), the central core is not well developed and the density falls steadily with r. The low-q data may be fitted by form factors generated in harmonic-oscillator wells, but refinements are necessary to obtain agreement at higher momentum transfers. The rms radii for these nuclei do not obey Eq. 4-18; they fluctuate a great deal for $A < 6$ and are almost constant for $6 < A < 16$. This is shown in Fig. 4-12.

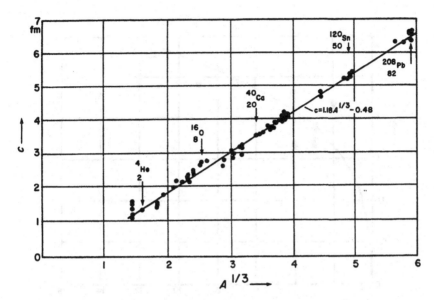

Fig. 4-10 Half-density radius c of the nuclear charge density, plotted as a function of $A^{1/3}$. For many nuclei, more than one experimental point is shown, corresponding to different assumed forms of the distribution. (After Hofstadter [34].)

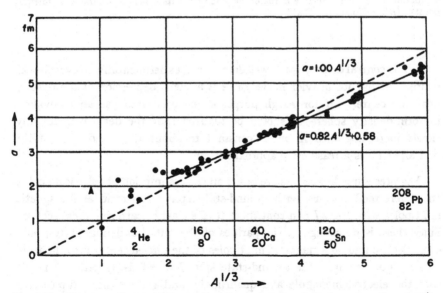

Fig. 4-11 The rms radius, a, of the nuclear charge density, plotted as a function of $A^{1/3}$. Multiplicity of experimental points, for the same value of $A^{1/3}$, corresponds to different assumed charge distributions. (After Hofstadter [34].)

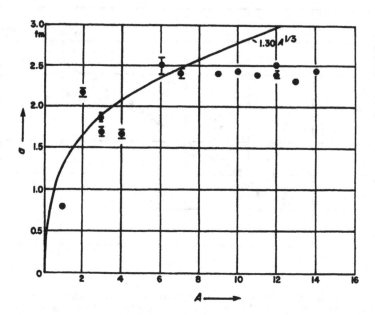

Fig. 4-12 The rms radius, a, of the nuclear charge density for light nuclei ($A \leqslant 14$) as a function of A. The full line is the curve $1.30A^{1/3}$, which fails to fit the experimental points. (After Hofstadter [34].)

4. The central density is not well determined experimentally. Nevertheless, it appears that for heavier nuclei there is a slight depression of the density near the center. If for rough purposes one wishes to use an equivalent uniform-density sphere, it should be realized that the density is actually slowly increasing with A but is constant to about 12%, and $R = 1.2A^{1/3}$ or $1.3A^{1/3}$ fm is a reasonable approximation.

The preceding discussions are based on the assumption of spherical nuclei. In a deformed nucleus with ground-state spin J, electric and magnetic multipoles with $l < 2J$ can contribute to the elastic scattering cross section. Since there is no change in the parities of the initial and final states, only even electric and odd magnetic multipoles of the nucleus enter into the form factor. For example, the ground-state spin J of ^{11}B is $\frac{3}{2}$, with the result that the electric monopole and quadrupole, and the magnetic dipole and octupole, moments of the nucleus contribute to elastic scattering. In the Born approximation (which is reasonable for light nuclei), the charge and magnetic parts of the form factor can again be expressed neatly [35]. The magnetic form factors contribute negligibly for small scattering angles but become important at backward angles. Therefore these can be separated experimentally from the electric part of the form factor, and this has been done, for example, in the case of ^{11}B [36]. Note that deformed nuclei should have smoother scattering cross sections (a) because their random orientations reduce the angular dependence appreciably, and (b) because the extension along the axis is equivalent to a thicker skin, producing less pronounced diffraction minima. These points are confirmed by electron-scattering experiments on Nd isotopes of $A = 142, 144, 146, 148,$ and 150 [53]. The nuclide ^{142}Nd is nearly spherical with a closed neutron shell, while ^{150}Nd is a highly deformed nucleus. A 13% increase in skin thickness is found in going from ^{142}Nd to ^{150}Nd.

4-3 FORM FACTOR OF NUCLEONS

Neutrons and protons are the stable constituents of a nucleus, and their distribution in a localized volume, resulting from the attractive nature of the nucleonic forces, gives rise to the finite size of the nucleus. In Sections 4-1 and 4-2 we inquired in particular about the distribution of the electric charge in various nuclides by electron-scattering experiments. A logical question is whether, by similar experiments, we can gather information about the spatial extent of a single nucleon—in particular, about its distribution of charge and current, which should be responsive to the electromagnetic probe. In relativistic quantum theory, where space and time coordinates are treated on an equal footing, spatial localization of a particle to a point is not possible. In the Dirac theory of a point particle, for example, the particle coordinate cannot be exactly *localized* because of the interference of the positive- and negative-energy components of the wave packet (*Zitterbewegung*), resulting in fluctuation of the particle coordinate of the order of the Compton wavelength of the particle [37]. From this point of view alone, it would be expected that the spatial distribution for the nucleon would spread over about 0.2 fm, its Compton wavelength. However, we also know that the magnetic moments of the nucleons deviate considerably from the Dirac (or "normal") values, indicating additional structure in the current distributions.

We are naturally led to inquire about the origin of this additional structure in the nucleon. In the case of a nucleus, which has stable constituents, it arises from the distribution in space of the constituents themselves. One point of view, to answer the question, would be to assume that a nucleon itself is made up of stable, subnucleonic particles. Indeed, a large body of experimental data for a host of strongly interacting elementary particles can be analyzed by postulating three such basic constituents, called "quarks" [38]. Experimentally, however, these have not been detected as separate entities to date. From a more conventional point of view, the origin of the nucleonic structure can be traced to the Yukawa theory of nuclear force. We noted in Chapter 2 that the force between two nucleons may be considered as due to the exchange of virtual mesons. By the uncertainty principle, it is admissible for a nucleon to create one or more mesons, violating energy conservation by an amount ΔE if these mesons are reabsorbed, either by the nucleon itself or by another nucleon, within a time period $\Delta\tau \sim \hbar/\Delta E$. When the meson is absorbed by a neighboring nucleon, it is the source of the nuclear force; if, on the other hand, the absorption occurs by the same nucleon that emitted the meson, this can be the source of additional structure. The mesons are virtual in the sense that their existence is transitory; a nucleon is continually emitting and absorbing them to form a cloud around it. The electromagnetic composition of a meson-clothed nucleon should therefore reflect the charge and current distributions of this composite system [39]. An electron, since it does not interact strongly, has no meson cloud around it. Nonetheless, it has the same charge as a proton because, in all these virtual processes of emission and absorption of mesons reflecting strong interaction, the charge is conserved.

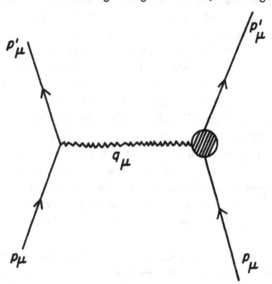

Fig. 4-13 Electron-nucleon scattering by the exchange of a virtual photon of four-momentum q_μ. The structure of the nucleon is schematically shown by the shaded blob.

To probe the structure of a nucleon by elastic electron scattering, very high-energy electrons, whose momenta are of the order of a few GeV/c, are used. It is necessary, therefore, to use a fully relativistic formalism to define the form factors. Furthermore, the recoil of the nucleon cannot be ignored, so that even for elastic scattering the initial and final electron energies are not the same. Figure 4–13 is a schematic diagram of an electron scattering off a nucleon by the exchange of a single photon. In the notation of Sections A–3 and A–4, the initial four-momentum of the electron is denoted by $p_\mu(\mathbf{p}, iE_e)$ and its final four-momentum by $p_\mu'(\mathbf{p}', iE_e')$, and likewise for the nucleon. The shaded blob at the nucleon vertex indicates the composite structure of the proton. The four-momentum transferred by the photon is expressed as $q_\mu = p_\mu - p_\mu'$. With the metric defined in Section A–3, it follows ($c = \hbar = 1$) that

$$p_\mu^2 = p_\mu'^2 = -m^2,$$

$$P_\mu^2 = P_\mu'^2 = -M^2,$$

$$q_\mu^2 = (\mathbf{p} - \mathbf{p}')^2 - (E_e - E_e')^2 \geqslant 0. \tag{4–19}$$

The last equation shows that the exchanged photon is virtual, since for a real photon the mass is zero; hence $q_\mu^2 = 0$. In a relativistic theory, the three components of the electric current and the charge form a four-vector j_μ. This current for the Dirac electron is given by (see Section A–4)

$$j_{e\mu}(x) = ie\bar{\psi}_{p'}(x)\gamma_\mu\psi_p(x), \tag{4–20}$$

where $\psi_p(x)$ is the free four-component Dirac spinor,

$$\psi_p(x) \equiv \psi_p(\mathbf{x}, t) = \psi(\mathbf{p})\exp[i(\mathbf{p}\cdot\mathbf{x} - Et)]. \tag{A–90}$$

It is straightforward to prove that

$$j_{e\mu}(x) = j_{e\mu}^{(1)}(x) + j_{e\mu}^{(2)}(x),$$

with

$$j_{e\mu}^{(1)}(x) = \frac{ie}{2m}\left[-\bar{\psi}_{p'}(x)\frac{\partial\psi_p(x)}{\partial x_\mu} + \frac{\partial\bar{\psi}_{p'}(x)}{\partial x_\mu}\psi_p(x)\right], \tag{4–21}$$

$$j_{e\mu}^{(2)}(x) = \frac{e}{2m}\frac{\partial}{\partial x_\nu}[\bar{\psi}_{p'}(x)\sigma_{\mu\nu}\psi_p(x)], \tag{4–22}$$

where $\sigma_{\mu\nu}$ is defined by Eq. A–134. The quantity $j_\mu^{(1)}$ is analogous to the non-relativistic quantum-mechanical current of a particle without spin, while $j_\mu^{(2)}$ is the current from a magnetic dipole of "normal" moment $e/2m$. When the electron interacts with an electromagnetic field $A_\mu(\mathbf{A}, i\phi)$, the interaction Hamiltonian is given by

$$H_{\text{int}}^e = -\int j_\mu^e(\mathbf{x}) A_\mu(\mathbf{x})\, d^3x. \qquad (4\text{-}23)$$

In the static limit, this gives rise to the usual electrostatic term $e\phi$ and the normal magnetic term $- (e/2m)\,\mathbf{B}\cdot\boldsymbol{\sigma}$.

For the case of a nucleon, we know that the nucleon four-current $j_{N\mu}$ is modified because of the meson clothing around it. Even in the static limit, the magnetic interaction is changed to $- (e/2M)(1 + \kappa)\mathbf{B}\cdot\boldsymbol{\sigma}$, where κ is the anomalous contribution. Furthermore, the structure of the meson cloud must be taken into account. The phenomenological form of the current is now considered to consist of a normal term of the form of Eq. 4–20, and an additional magnetic term of the form of Eq. 4–22:

$$j_{N\mu} = \left[F_1(q_\mu{}^2) ie\bar{\psi}_p\cdot\gamma_\mu\psi_p + F_2(q_\mu{}^2) i\,\frac{\kappa e}{2M} q_\nu\bar{\psi}_p\cdot\sigma_{\mu\nu}\psi_p \right]. \qquad (4\text{-}24)$$

The functions F_1 and F_2 describe the finite electromagnetic structure of the nucleon; for a point particle these should be independent of $q_\mu{}^2$. The normalizations of F_1 and F_2 are so chosen that they yield the correct static values for the charge and magnetic moment, that is,

$$F_1(0) = F_2(0) = 1 \qquad \text{for the proton,}$$

$$F_1(0) = 0, \qquad F_2(0) = 1 \quad \text{for the neutron.} \qquad (4\text{-}25)$$

In order to obtain the correct static magnetic moments, $\kappa = 1.79$ for the proton and $- 1.91$ for the neutron in nuclear magnetons.

The functions $F_1(q_\mu{}^2)$ and $F_2(q_\mu{}^2)$ are called the Dirac and Pauli form factors, respectively. These are still not related directly to the charge and magnetic-moment distributions in the nucleon, since, although the static charge is $eF_1(0)$, the static magnetic moment, in nuclear magnetons, is $F_1(0) + \kappa F_2(0)$. It is customary, therefore, to define the electric and magnetic form factors $G_E(q_\mu{}^2)$ and $G_M(q_\mu{}^2)$:

$$G_E = F_1 - \left(\frac{q_\mu{}^2\kappa}{4M^2}\right)F_2, \qquad G_M = F_1 + \kappa F_2, \qquad (4\text{-}26)$$

which yield the charge and magnetic moment of the physical nucleon in the static limit and also give a simpler form for the cross section.

Experimentally, for protons, G_E and G_M are determined as functions of $q_\mu{}^2$ from a measurement of the electron-proton scattering cross section.

For a point charge with no spin, the differential scattering cross section is given, up to the order $\alpha^2 = 1/137$, by the first term in Eq. 4–3, which is the Mott scattering cross section $\sigma_{\text{Mott}}(\theta)$. For a proton with form factors G_E and G_M, this expression is modified by Rosenbluth [40] to

$$\sigma(\theta) = \sigma_{\text{Mott}}(\theta) \left[\frac{G_E{}^2 + (q_\mu{}^2/4M^2)G_M{}^2}{1 + (q_\mu{}^2/4M^2)} + \frac{q_\mu{}^2}{2M^2} G_M{}^2 \tan^2 \frac{\theta}{2} \right]. \qquad (4\text{–}27)$$

Thus a plot of the ratio of the observed cross section to that of $\sigma_{\text{Mott}}(\theta)$, against $\tan^2 \theta/2$ for fixed $q_\mu{}^2$, should give a straight line, from which values

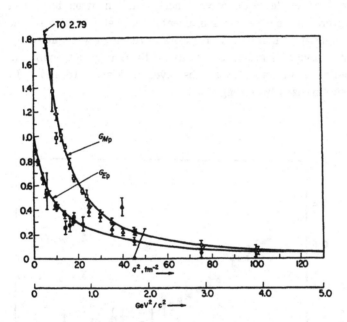

Fig. 4-14 Experimental values of the proton electromagnetic form factors, as determined by elastic electron-proton scattering. The compilation of experimental points is taken from Griffy and Schiff [48]. $\triangle G_{Mp}$, $\blacktriangle G_{Ep}$, Berkelman et al. [41]; $\square G_{Mp}$, $\blacksquare G_{Ep}$, Bumiller et al. [42]; $\otimes G_{Ep}$, Dunning et al. [43]; $\diamondsuit G_{Mp}$, Janssens et al. [44]; $\blacktriangledown G_{Ep}$, Lehmann et al. [45]; $\bigcirc G_{Mp}$, $\bullet G_{Ep}$, Chen et al. [46].

of $G_E{}^2(q_\mu{}^2)$ and $G_M{}^2(q_\mu{}^2)$ can be readily deduced. In Fig. 4–14, we show the experimentally determined form factors of G_{Ep} and G_{Mp} of the proton as functions of the four-momenta squared, the signs having been fixed from the known static values.

Before discussing the physical implication of these results, we present briefly the experimental results for the neutron form factors. Since it is not possible to obtain free neutron targets, the chief source of information is electron-deuteron scattering, where the neutron is loosely bound to the proton. There appear to be some obvious difficulties in the analysis of the experimental data to extract the neutron form factors from this approach. Even if, for simplicity, the deuteron is treated nonrelativistically, the results depend on the form of the deuteron wave function and the meson-exchange currents in the deuteron itself. Both elastic *e-d* scattering and inelastic scattering leading to deuteron disintegration have been performed. If, assuming some form of deuteron wave function, the neutron form factors are deduced from the inelastic measurements, these should also fit the elastic data in a consistent manner. Early work in this area was started by Jankus [47], and a review is given in the article by Griffy and Schiff [48]. The experimental results [49, 54] are displayed in Figs. 4–15 and 4–16, which may be summarized by noting that

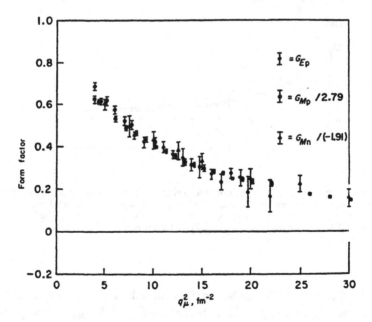

Fig. 4-15 The magnetic form factors of the neutron, as determined from deuteron electrodisintegration (after Hughes et al. [49]). For comparison, the proton form factors G_{E_p} and G_{Mp/μ_p} are also plotted.

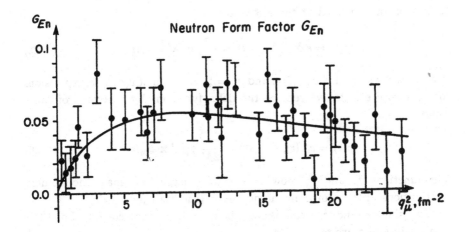

Fig. 4-16 The electric neutron form factor G_{En}, as determined from elastic electron-deuteron scattering, using the Feshbach-Lomon wave function for the deuteron. The solid curve is the fit $G_{En} = - \mu_{Nt} G_{Ep}/(1 + 5.6\tau)$. See Galater et al. [54] for more details.

$$G_{Ep}(q_{\mu}^{2}) \approx \frac{G_{Mp}(q_{\mu}^{2})}{\mu_{p}} \approx \frac{G_{Mn}(q_{\mu}^{2})}{\mu_{n}} \qquad (4\text{-}28\text{a})$$

and

$$G_{En}(q_{\mu}^{2}) \approx - \frac{\mu_{n}\tau}{1 + 5.6\tau} G_{Ep}(q_{\mu}^{2}), \qquad (4\text{-}28\text{b})$$

where $\tau = q_{\mu}^{2}/4M^{2}$.

We now proceed to discuss the physical interpretations of these relativistic form factors. In Sections 4-1 and 4-2 the nonrelativistic charge form factor was simply the Fourier transform of the spatial charge distribution. Can we make similar interpretations in the relativistic case? To answer this question, we note that, from Eq. 4-19, the form factors are defined only for space-like $q_{\mu}^{2} \geqslant 0$. It is possible, therefore, to go to a special Lorentz frame in which q_{μ} has no time-like component, and to define the spatial distributions of charge and magnetic moment as the inverse Fourier transforms of $G_{E}(\mathbf{q}^{2})$ and $G_{M}(\mathbf{q}^{2})$. These spatial distributions, however, are not Lorentz invariant, and, therefore, relativistically it is not possible to define spatial distributions uniquely. The other alternative is to go to the non-relativistic limit. In this limit, the charge and the current-density operators for a *point* proton fixed at the origin are

$$\rho(\mathbf{r}) = e\,\delta(\mathbf{r}), \qquad j(\mathbf{r}) = \frac{e}{2M}\,g_p[\nabla \times \mathbf{s}\,\delta(\mathbf{r})], \tag{4-29}$$

where s is the spin of the proton and $g_p = 2.79$. To modify these expressions for finite-size effects, we replace the delta functions by functions of finite range:

$$\rho(\mathbf{r}) = ef_E(r), \qquad j(\mathbf{r}) = \frac{e}{2M}\,g_p[\nabla \times \mathbf{s}f_M(r)]. \tag{4-30}$$

The functions f_E and f_M now define the charge and magnetic-moment distributions of the proton in the nonrelativistic limit ($\hbar q \ll Mc$).

Two similar functions can likewise be defined for the neutron. For these small values of q, we may write

$$G_E(\mathbf{q}^2) = \int f_E(r)\exp(i\mathbf{q}\cdot\mathbf{r})\,d^3r,$$

$$G_M(\mathbf{q}^2) = \mu \int f_M(r)\exp(i\mathbf{q}\cdot\mathbf{r})\,d^3r. \tag{4-31}$$

Here μ is the static magnetic moment of the nucleon. Small values of q give information for large r, and we can make the following interpretations from the experimental data:

1. For the proton, $f_{Ep}(r)$ and $f_{Mp}(r)$ are very similar in shape for large r, being roughly exponential in character with a rms radius of about 0.8 fm.

2. For the neutron, the magnetic-moment distribution is very similar to that of a proton. The data for the electric part G_{En} are less certain, although the rms radius of the charge distribution of the neutron is precisely known to be 0.36 fm.[†] Recent data [54] indicate that in the nonrelativistic limit the charge distribution of the neutron, $f_{En}(r)$, may be taken to be a super-position of a short-range exponential distribution of positive charge $+ e$ and a longer-range exponential of negative charge $- e$ [17].

Finally, we should remark that on the theoretical side substantial progress has been made in understanding the nucleon structure from the meson cloud point of view. The simplest virtual process in which a charged meson is created and annihilated, with virtual photon coupling to the meson, is not

---

[†] Note that by means of slow neutron scattering from atomic electrons it is known precisely that $(\partial/\partial q^2)G_{En}|_{q^2=0} = 0.021 \pm 0.001$ fm², yielding the neutron rms radius for the electric part to be 0.36 fm [52].

enough to explain the data. In fact, it was this kind of theoretical analysis that first suggested the necessity to include strong two-pion and three-pion correlated states in the meson cloud that can couple to the photon, and these multipion resonances (the ρ-, ω-, and ϕ-mesons) were later experimentally observed (see Section 5–10 for more details on this topic). Details of these calculations are given by Griffy and Schiff [48].

4-4 MUONIC ATOMS

We have already mentioned that muons may form bound orbits about nuclei, just as electrons do. In order to produce muonic atoms, the first step is to produce negative pions in the collisions between a beam of high-energy protons (400–900 MeV) and a target like beryllium or copper. The negative muons are produced in the decay of these pions in flight. The muons thus produced have energies of a few hundred MeV with considerable spread.

TABLE 4-2 Comparison of Characteristics of Muonic and Electronic Atoms. The expressions are for a point-nucleus, Coulomb-field case.

Quantity	Expression	Ratio between muonic and electronic atoms	Typical values for muonic atoms
Orbit radius	$\dfrac{n^2\hbar^2}{Ze^2}\cdot\dfrac{1}{m}$	$\dfrac{m_e}{m_\mu}\approx\dfrac{1}{207}$	5 fm ($Z\approx 50$)
Energy levels E_n	$-\dfrac{(Z\alpha)^2c^2}{2n^2}m$	$\dfrac{m_\mu}{m_e}\approx 207$	$-$ 6 MeV ($Z\approx 50$) ($n=1$)
Fine structure $\left\langle\dfrac{(\mathbf{l}\cdot\boldsymbol{\sigma})Z}{m^2r^3}\right\rangle$	$\alpha Z^4 m$	$\dfrac{m_\mu}{m_e}\approx 207$	550–200 keV in heavy elements
Hyperfine electric quadrupole $\left\langle\dfrac{e^2Q_N}{r^3}\right\rangle$	$\sim Q_N Z^3 m^3$	$\dfrac{m_\mu^{\ 3}}{m_e^{\ 3}}\sim 10^7$	50–500 keV in highly deformed heavy nuclei
Hyperfine magnetic dipole $\left\langle\dfrac{\mu_N\mu_{\text{lepton}}}{r^3}\right\rangle$	$\sim\mu_N Z^3 m^2$	$\dfrac{m_\mu^{\ 2}}{m_e^{\ 2}}\sim 4\times 10^4$	$\sim$ A few keV in heavy nuclei

These muons have to be slowed down to thermal energies by passage through the target material before being captured in atomic orbits. The density of states of high-angular-momentum orbits being large, a muon first enters a bound state in one of these orbits and then cascades down to the ground state.

The muonic atom has characteristics considerably different from those of the electronic one since the muon mass is about 207 times heavier than the electron mass and also because there is no shielding of the nuclear charge in the lower orbits of the muonic atoms. The characteristics of the two types are compared in Table 4–2, following Wu and Wilets [50]. It will be noticed that a muonic orbit with principal quantum number n around 14 has the same radius as the lowest electronic orbit. Therefore, while a muon is in its higher orbits, it interacts strongly with the electrons, and emission of electrons predominates over radiation as a de-excitation mechanism. By the time the principal quantum number n of the muonic orbit reaches a value of 6 or less, its interaction with the electron structure is very small, and one is dealing essentially with a bare, hydrogen-like muonic atom. Transitions involving these low quantum numbers result in the emission of photons, which may have energies up to several MeV. These are still called muonic X-rays, because of the similarity of the processes in which they arise. On reaching the lowest 1s-level, the muon either decays with a half-life of 2.2×10^{-6} s or is captured by a proton in the nucleus in a comparable time, with emission of a neutron and neutrons. It is therefore only possible to study one-muon atoms.

The radius of the lowest muonic orbit in a heavy or medium-heavy nucleus is comparable to the half-density radius of the nucleus. For example, in lead, the orbit with $n = 1$ (the 1s-state) has a radius of 3.1 fm, and the $n = 2$ state radius is about 12 fm. The half-density radius of lead is 6.5 fm, and therefore the muon spends a substantial fraction of its time inside the nucleus, especially in the 1s-orbit. Since it will experience an electrostatic potential of non-Coulomb shape inside the nucleus, its energy levels will be appreciably different from those given by the simple theory for a hydrogen-like atom. This finite-size effect reduces the binding of the 1s-level drastically in a heavy nucleus. For example, the binding of the 1s-level in lead should be about 21 MeV for a point nucleus according to the Dirac theory, whereas the experimental value is approximately 10.6 MeV.

It will also be seen from Table 4–2 that the hyperfine electric quadrupole interaction in a heavy deformed element can be as large as the fine-structure splittings, and these are of the same order as the spacings of the low-lying collective states of a nucleus. With the development of lithium-drifted germanium and silicon detectors, having resolutions of a few keV even for a γ-ray of energy 5 MeV, it has become possible to measure these splittings in heavy nuclei.

As we have seen, measurement of the energies of the X-rays of a muonic atom should provide information about the charge distribution in a nucleus. The first quantitative studies of the X-ray energies, made by Fitch and Rainwater [51], resulted in the determination of the $2p \rightarrow 1s$ transition energy of nine elements with Z ranging from 13 to 83. It is clear that measurement of one energy will provide only one parameter of the nuclear-charge distribution and therefore will not by itself give information about the shape of the radial dependence. Rather, some gross feature of the charge distribution (like the rms radius or some moment of the radius) may be fixed by this form of data.

We have noted that, for a heavy element like lead, the shift in the $1s$-level due to the finite-size effect is very large. For light nuclides, however, the muonic X-ray energies are a rather insensitive measure of nuclear size, since the radii of the Bohr orbits are inversely proportional to Z (see Table 4–2), while the nuclear dimensions are proportional to $A^{1/3}$. Consequently, only a small fraction of the muonic orbit is inside the nucleus. Thus the shift in the $1s$-level for ^{40}Ca is only about 0.07 MeV. For even lighter nuclei, only the s-state energies are much affected, and an estimate of this effect can be made by first-order perturbation calculation. Because of the finite size of the nucleus, the muon experiences a potential $V(r)$, given by Eq. 4–10, instead of the point Coulomb field Ze^2/r, so that the first-order energy shift is

$$\Delta E = \langle \psi_0 | V - \left(-\frac{Ze^2}{r} \right) | \psi_0 \rangle, \tag{4–32}$$

where ψ_0 is the muonic wave function for a point nucleus, obtained by solving the Dirac equation in a Coulomb field. This wave function has a singularity at the origin that goes like $r^{\sigma-1}$, where $\sigma = (1 - \alpha^2 Z^2)^{1/2}$, α being the fine-structure constant $e^2/\hbar c$. The actual wave function for a finite nucleus does not have this singularity, but remains finite. The singularity of ψ_0 at the origin presents no difficulty in the integration indicated in Eq. 4–32, but it is true that ψ_0 is not realistic for very small values of r. This is not significant in lighter nuclei, but we cannot expect Eq. 4–32 to be valid for heavy nuclei. When Eq. 4–32 is simplified, it is found that

$$\Delta E = \frac{2^{2\sigma-1}}{\pi(2\sigma + 1)!} (\alpha Z)^{2\sigma+2} \langle r^{2\sigma} \rangle \left(\frac{\hbar}{\mu c} \right)^{-2\sigma} \mu c^2, \tag{4–33}$$

where

$$\langle r^{2\sigma} \rangle = 4\pi \int \rho(r)^{2\sigma+2} \, dr.$$

For lighter nuclei, σ is nearly unity, and we see that the energy shift measures essentially the mean-square radius. We further note that $\Delta E \propto Z^4$, a result that turns out to be approximately true throughout the periodic table. Thus the experiments of Fitch and Rainwater may be regarded as determining the value of the mean-square radius.

Since these early measurements, great progress has been made in precise measurements of the muonic X-rays, and many transitions can now be accurately determined. In the medium- and heavy-Z regions several energy differences have been measured, but all that is effectively obtained is the shifts in the energy levels of the $1s_{1/2}$-, $2p_{1/2}$-, and $2p_{3/2}$-levels, because the shift in the $3d$-level is not sensitive to charge distribution. For most nuclides, the $3d$-level shift is considerably smaller than the experimental accuracy, and even for $Z \sim 80$ this shift is only about 4 keV, while the accuracy of measurements is of the order of 1 keV. In principle, to analyze the experiments, a nuclear charge distribution $\rho(r; a, b, c, \ldots)$ with appropriate parameters may be assumed and the resulting potential $V(r)$ calculated from Eq. 4–10. The Dirac equations are then solved with this potential inserted, and the energy levels determined. The parameters a, b, c, etc., of the density distribution are adjusted to give agreement between the experimental and theoretically determined energy levels.

At best, the number of nuclear parameters deduced in this way can be only as large as the independent number of levels measured. Actually, when one considers the diminishing sensitivity of the higher energy levels to the nuclear parameters and the limited experimental accuracy, the significant number of parameters deduced is much less. For example, although for most medium and heavy nuclei the $1s_{1/2}$-, $2p_{3/2}$-, and $2p_{1/2}$-levels are accurately determined, information about the $n = 2$ levels becomes sensitive to charge distributions only for $Z \gtrsim 50$. The sensitivity of the fine-structure components $2p_{3/2}$ and $2p_{1/2}$ to the charge distribution is not sufficiently different to yield significant information. For light nuclei, then, only one parameter of the charge distribution (like $\langle r^2 \rangle$) is obtained, whereas for heavier nuclei some additional constraint on $\rho(r)$ can also be found. Typically, a two-parameter Fermi distribution for $\rho(r)$ is assumed. Measurement of the $1s_{1/2}$-level then provides a single relation between the half-density radius c and the surface thickness t, which can be plotted as an isoenergetic contour, as

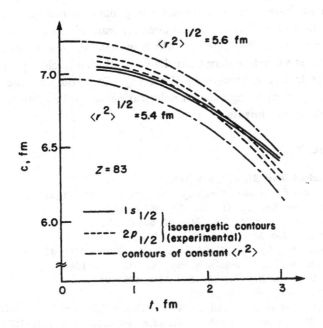

Fig. 4-17 Dependence of the isoenergetic contours of the muonic states on the parameters c and t of the Fermi distribution for the nuclear charge of ^{209}Bi. The contours are for the energies $E(1s_{1/2}) = 5225.5 \pm 3.9$ keV and $E(2p_{1/2}) = 1812.7 \pm 2.7$ keV. The curves for constant $\langle r^2 \rangle^{1/2}$ show that it is determined more accurately than either c or t. [From S. Devons and I. Duerdoth, *Advan. Nucl. Phys.* **2**, 295 (1969).]

shown in Fig. 4–17 for $Z = 83$. Measurement of the $2p$-level gives a different relation between c and t, which is plotted as another contour in the same figure. Intersection of these contours should determine c and t. In practice, however, because of experimental inaccuracies, c and t are determined only within certain limits, with a rather wide variation of t-values possible. In contrast, as is shown in the figure, the value of $\langle r^2 \rangle$ is quite well determined, independently of the $c - t$ combinations permissible by the above procedure. Typically, for the assumed Fermi distribution, c is determined to about $\pm 1\%$ and t to about $\pm 10\%$ for heavy nuclei ($Z \sim 80$). The uncertainties in c and t are strongly correlated, so that a single parameter like $\langle r^2 \rangle^{1/2}$ is found to within about 0.02 fm. In this sense, the accuracy is better than that obtained in the electron-scattering experiments. The results for c and t obtained for heavy nuclei from muonic data are consistent with the electron-scattering results. As already noted, it is often fruitful to analyze the combined muonic X-ray and electron-scattering data to deduce information about the charge distribution, as was done for lead and gold.

Apart from the finite-size effect, which we have discussed, several other effects (such as vacuum polarization, Lamb shift, and nuclear polarization) can alter the transition energies by about 1%. Corrections for these effects (the most important being vacuum polarization) are taken into account in the analyses of experiments.

REFERENCES

1. E. Rutherford, *Phil. Mag.* 21, 669 (1911).
2. H. Geiger and E. Marsden, *Proc. Roy. Soc. (London)* A82, 495 (1909).
3. H. Geiger, *Proc. Roy. Soc. (London)* A83, 482 (1910).
4. H. Geiger and E. Marsden, *Phil. Mag.* 25, 604 (1913).
5. R. Hofstadter, *Rev. Mod. Phys.* 28, 214 (1956).
6. N. F. Mott, *Proc. Roy. Soc. (London)* A135, 429 (1932); A124, 426 (1929).
7. W. A. McKinley and H. Feshbach, *Phys. Rev.* 74, 1759 (1948).
8. R. H. Dalitz, *Proc. Roy. Soc. (London)* A206, 509 (1951).
9. H. Feshbach, *Phys. Rev.* 88, 295 (1952).
10. D. R. Yennie, D. G. Ravenhall, and R. N. Wilson, *Phys. Rev.* 95, 500 (1954).
11. B. Hahn, D. G. Ravenhall, and R. Hofstadter, *Phys. Rev.* 101, 1131 (1956).
12. D. G. Ravenhall, in *Proceedings of the Williamsburg Conference on Intermediate Energy Physics, 1966.* Quoted in "Muonic Atoms," by S. Devons and I. Duerdoth, *Adv. Nucl. Phys.* 2, 363 (1969).
13. J. L. Friar and J. W. Negele, *Nucl. Phys.* A212, 93 (1973).
14. Y. C. Lin, private communication, as reported by H. A. Bethe, *Phys. Rev.* 167, 879 (1968).
15. H. A. Bethe and L. R. B. Elton, *Phys. Rev. Lett.* 20, 745 (1968).
16. J. Heisenberg, R. Hofstadter, J. S. McCarthy, I. Sick, B. C. Clark, R. Herman, and D. G. Ravenhall, *Phys. Rev. Lett.* 23, 1402 (1969).
17. W. Bertozzi, F. Friar, J. Heisenberg, and J. W. Negele, *Phys. Lett.* 41B, 408 (1972).
18. K. J. van Oostrum, R. Hofstadter, G. K. Nöldeke, M. R. Yearian, B. C. Clark, R. Herman, and D. G. Ravenhall, *Phys. Rev. Lett.* 16, 528 (1966).
19. J. B. Bellicard, P. Bounin, R. F. Frosch, R. Hofstadter, J. S. McCarthy, F. J. Uhrhane, M. R. Yearian, B. C. Clark, R. Herman, and D. G. Ravenhall, *Phys. Rev. Lett.* 19, 527 (1967).
20. R. F. Frosch, R. Hofstadter, J. S. McCarthy, G. K. Nöldeke, K. J. van Oostrum, M. R. Yearian, B. C. Clark, R. Herman, and D. G. Ravenhall, *Phys. Rev.* 174, 1380 (1968).
21. R. D. Ehrlich, D. Fryberger, D. A. Jensen, C. Nissim-Sabat, R. J. Powers, V. L. Telegdi, and C. K. Hargrove, *Phys. Rev. Lett.* 18, 959 (1967); 19, 334(E) (1967).
22. L. R. B. Elton and S. J. Webb, *Phys. Rev. Lett.* 24, 145 (1970).
23. L. R. B. Elton, *Nuclear Sizes,* Oxford University Press, Oxford, 1961.
24. J. S. McCarthy and I. Sick: their data are displayed in ref. 25.
25. T. W. Donnelly and G. E. Walker, *Phys. Rev. Lett.* 22, 1121 (1969).

26. R. H. Pratt, J. D. Walecka, and T. A. Griffy, *Nucl. Phys.* **64**, 677 (1965).

27. L. R. Suelzle, M. R. Yearian, and H. Crannel, *Phys. Rev.* **162**, 992 (1967).

28. R. F. Frosch, J. S. McCarthy, R. E. Rand, and M. R. Yearian, *Phys. Rev.* **160**, 874 (1967).

29. B. F. Gibson, A. Goldberg, and M. S. Weiss, *Nucl. Phys.* **A111**, 321 (1968).

30. M. Bouten, M. C. Bouten, and P. van Leuven, *Nucl. Phys.* **A100**, 105 (1967); *Phys. Lett.* **26B**, 191 (1968).

31. J. S. McCarthy, I. Sick, R. R. Whitney, and M. R. Yearian, *Phys. Rev. Lett.* **25**, 884 (1970).

32. J. E. Elias, J. I. Friedman, G. C. Hartmann, H. W. Kendall, P. N. Kirk, M. R. Sogard, L. P. van Speybroeck, and J. K. de Pagter, *Phys. Rev.* **177**, 2075 (1969).

33. C. E. Wiegand, *Phys. Rev. Lett.* **22**, 1235 (1969); H. A. Bethe and P. J. Siemens, *Nucl. Phys.* **B21**, 589 (1970).

34. R. Hofstadter, "Nuclear Radii," in *Nuclear Physics and Technology*, Vol. 2, edited by H. Schopper, Springer-Verlag, Berlin, 1967.

35. See, for example, H. Überall, in *Springer Tracts in Modern Physics*, Vol. 39, edited by G. Höhler, Springer-Verlag, Heidelberg, 1969, p. 18.

36. J. Goldemberg, D. B. Isabelle, D. Vinciguerra, T. Stovall, and A. Bottino, *Phys. Rev. Lett.* **16**, 141 (1965).

37. For a discussion of *Zitterbewegung*, see J. J. Sakurai, *Advanced Quantum Mechanics*, Addison-Wesley, Reading, Mass., 1967, p. 117.

38. M. Gellman and Y. Neeman, *The Eight-Fold Way*, Benjamin, New York, 1964.

39. S. D. Drell and F. Zachariasen, *Electromagnetic Structure of Nucleons*, Oxford University Press, Oxford, 1961, Chap. 1.

40. M. N. Rosenbluth, *Phys. Rev.* **79**, 615 (1960).

41. K. Berkelman, M. Feldman, R. M. Littauer, G. Rouse, and R. R. Wilson, *Phys. Rev.* **130**, 2061 (1963).

42. F. Bumiller, M. Croissiaux, E. Dally, and R. Hofstadter, *Phys. Rev.* **124**, 1623 (1961).

43. J. R. Dunning, K. W. Chen, N. F. Ramsay, J. R. Rees, W. Shlaer, J. K. Walker, and R. Wilson, *Phys. Rev. Lett.* **10**, 500 (1963).

44. T. Janssens, R. Hofstadter, E. Hughes, and M. Yearian, *Phys. Rev.* **142**, 922 (1966).

45. P. Lehmann, R. Taylor, and R. Wilson, *Phys. Rev.* **126**, 1183 (1962).

46. K. W. Chen, A. A. Cone, J. R. Dunning, S. G. T. Frank, N. F. Ramsay, J. K. Walker, and R. Wilson, *Phys. Rev. Lett.* **11**, 561 (1963).

47. V. Z. Jankus, *Phys. Rev.* **102**, 1586 (1956).

48. T. A. Griffy and L. I. Schiff: see the article "Electromagnetic Form Factors," in *High Energy Physics*, Vol. I, edited by E. H. Burhop, Academic Press, New York, 1967.

49. E. B. Hughes, T. A. Griffy, R. Hofstadter, and M. R. Yearian, *Phys. Rev.* **139B**, 458 (1965).

50. C. S. Wu and L. Wilets, *Ann. Rev. Nucl. Sci.* **19**, 527 (1969).

51. V. L. Fitch and J. Rainwater, *Phys. Rev.* **92**, 789 (1953).

52. L. L. Foldy, *Rev. Mod. Phys.* **30**, 471 (1958).

53. J. H. Heisenberg, J. S. McCarthy, I. Sick, and M. R. Yearian, *Nucl. Phys.* **A164**, 340 (1971).

54. S. Galster, H. Klein, J. Mortiz, K. H. Schmidt, D. Wegener, and J. Bleckwenn, *Nucl. Phys.* **B32**, 221 (1971).

55. H. Überall, *Electron Scattering from Complex Nuclei*, Vols. A and B, Academic Press, New York, 1971.

PROBLEMS

4-1. Calculate the magnitude of q (Eq. 4–6) for 60-, 183-, and 550-MeV electrons. For what range of θ would expression 4–7 give cross sections reliable to 10% for 60-MeV electrons on ^{14}N ?

4-2. For the density $\rho(r) = \rho_0 \exp(-r/\alpha)$, show that the rms radius $a = \alpha\sqrt{12}$ and that the form factor is $(1 + \alpha^2 q^2)^{-2}$. Find also a and F for a uniform distribution of radius R. For 183-MeV electrons on ^{14}N, compare the cross sections predicted by the two distributions, using the same a for both.

4-3. (a) Show that, if E ($\gg mc^2$) is the laboratory energy of electrons incident on a nucleus of mass M, the nucleus will acquire kinetic energy

$$E_n = \frac{E^2}{Mc^2} \frac{1 - \cos\theta}{1 + (E/Mc^2)(1 - \cos\theta)}.$$

Evaluate this result for ^{40}Ca and 550-MeV electrons.

(b) Describe some processes other than elastic scattering which electrons of high energy would initiate. Draw a schematic curve representing the number of electrons counted versus energy for a fixed angle of scattering for (i) 180 MeV and (ii) 550 MeV.

4-4. The "elastic peak" in electron scattering is broadened and its intensity is reduced by emission of radiation (both real and virtual) by the scattered electrons. The correction for bremsstrahlung is given by

$$I_1 = I \exp(\delta_B),$$

with

$$\delta_B = \frac{t}{\ln 2} \ln\left(\frac{E}{\Delta E}\right),$$

where I_1 is the corrected intensity at energy E, I is the observed intensity, ΔE is the full width of the peak at half-maximum, and t is the target thickness through which the electrons scattered at the specified angle pass. There is also the correction for radiation processes of energy less than $\Delta'E$, where $\Delta'E$ is the smallest energy resolvable from the main peak. This is given by

$$I_0 = I_1 \exp(\delta_r),$$

where I_0 is the intensity to be compared with a theory of elastic scattering and

$$\delta_r = \frac{4\alpha}{\pi}\left\{K\left[\ln\left(\frac{2E}{mc^2}\sin\tfrac{1}{2}\theta\right) - \frac{1}{2}\right] + \frac{17}{12}\right\},$$

$$K = \ln\frac{E}{\Delta'E} - \frac{13}{12}, \qquad \alpha = \frac{e^2}{\hbar c}.$$

These formulas are discussed by H. A. Bethe and J. Ashkin, *Experimental*

Nuclear Physics, edited by E. Segré (John Wiley, New York, 1953), and by J. Schwinger, *Phys. Rev.* **75**, 898 (1949).

Consider 550-MeV electrons; assume that $\Delta'E = \Delta E = 5$ MeV and calculate δ_r at 30°. Demonstrate that the effect of δ_r on the differential cross section is very slight for angles from $\theta = 30$–150°.

4–5. By assuming a trapezoidal density function

$$\rho(r) = \begin{cases} \rho_0, & r < c - \tfrac{1}{2}t, \\ \rho_0\left(\dfrac{1}{2} - \dfrac{r-c}{t}\right), & c - \tfrac{1}{2}t < r < c + \tfrac{1}{2}t, \\ 0, & r > c + \tfrac{1}{2}t, \end{cases}$$

show that $c \sim A^{1/3}$ implies that the density is independent of A to a good approximation inside the core, that is, for $r < c - \tfrac{1}{2}t$, but that $c - \tfrac{1}{2}t \sim A^{1/3}$ would not have this implication.

4–6. It might be possible to discover by electron-electron scattering whether the apparent structure of the proton is in part a reflection of the breakdown of the Coulomb law. What velocity would be necessary in the center-of-mass system? What energy would be required in the laboratory system?

4–7. Deduce Eq. 4–33, recalling that

$$\int |\psi_0|^2 \, d\Omega \, r^2 \, dr = \left(\frac{2\alpha Z \mu c}{\hbar}\right)^{2\sigma+1} [(2\sigma)!]^{-1} \int_0^\infty r^{2\sigma-2} \exp\left(\frac{-2\alpha Z r \mu c}{\hbar}\right) r^2 \, dr,$$

$$\nabla^2 r^\lambda = \lambda(\lambda+1) r^{\lambda-2},$$

and $\int (U\nabla^2 V - V\nabla^2 U) \, dr = 0$ if V vanishes sufficiently quickly at large r.

The experimental values of the $2p \to 1s\mu^-$ X-rays in titanium and copper are 0.995 and 1.548 MeV, respectively [51]. Show that the values for point nuclei are 1.045 and 1.826 MeV and use the above equation to deduce the value of R, the radius of the equivalent uniform well, in each case.

4–8. Consider a charge density defined by

$$\rho(r) = \rho_0 f(x) = \frac{1}{1 - \tfrac{1}{2}e^{-n}} \begin{cases} 1 - \tfrac{1}{2}e^{-n}e^x, & x \leqslant n, \\ \tfrac{1}{2}e^n e^{-x}, & x \geqslant n; \end{cases}$$

$$x = \frac{nr}{c}.$$

Show that, as n varies from 0 to ∞, this shape varies from exponential through Gaussian and Fermi shapes to a uniform density. Show also that $n = 2\ln 5(c/t)$, where c is the half-density radius and t is the thickness parameter.

This is family II of a set of functions ρ considered by D. L. Hill and K. W. Ford, *Phys. Rev.* **94**, 1617 and 1630 (1954). Use Eq. 34 of their second paper and the integrals given for family II(a) in their appendices to show that, for $A \gtrsim 12$, it is a good approximation to write

$$\frac{Ec}{(Ze)^2} = \frac{3}{5} \frac{1}{c} (1 + 6n^{-2})^{-1} \left[1 - \frac{n^{-2}}{1 + 6n^{-2}} \left(1 - \frac{75}{8n} + \frac{15}{n^2} - \frac{315}{16} \frac{1}{n^3} \right) \right].$$

4–9. (a) Consider electron-proton scattering at an incident electron energy E_e so high that we can write $E_e = p$ (we use units $c = \hbar = 1$ here), where p is the magnitude of its three-momentum. Then show that the quantity q_μ^2, defined in Eq. 4–19, is given by

$$q_\mu^2 = 4E_e E_e' \sin^2 \frac{\theta}{2},$$

where E_e' is the energy of the electron scattered at an angle θ.

(b) Using the result of Problem 4–3a, show that

$$q_\mu^2 = \frac{4E_e^2 \sin^2 \theta/2}{1 + (2E_e/M) \sin^2 \theta/2}.$$

Note that $q_\mu^2 \geqslant 0$.

4–10. In a very simple model of the nucleus, assume that each nucleon is moving in the field of a common central one-body potential. Then the spatial wave function of a nucleon may be written as

$$\phi_{nlm}(\mathbf{r}) = R_{nl}(r) Y_l^m(\hat{r}), \tag{1}$$

where n, l, and m are the principal, orbital, and azimuthal quantum numbers, respectively, of a nucleon; and $R_{nl}(r)$ is the radial and Y_l^m the angular wave function (defined in Eq. A–13a). If, in a given shell (defined by n, l), all the azimuthal states are occupied, the shell is called closed. For such a closed-shell nucleus, show that the single-particle proton density ρ is spherically symmetric:

$$\rho = 2 \sum_{\substack{n,l \\ \text{occupied}}} \frac{2l + 1}{4\pi} [R_{nl}(r)]^2. \tag{2}$$

If one assumes a harmonic-oscillator potential $\frac{1}{2}M\omega^2 r^2$ for the one-body potential, then [$\alpha = (M\omega/\hbar)^{1/2}$]

$$R_{nl}(r) = \sqrt{\frac{2n! \alpha^3}{\Gamma(n + l + \frac{3}{2})}} (\alpha r)^l \exp(-\tfrac{1}{2}\alpha^2 r^2) L_n^{l+1/2}(\alpha^2 r^2), \tag{3}$$

where the associated Laguerre polynomial is defined by

$$L_n^\beta(x) = \sum_{m=0}^n \frac{(n + \beta)!}{(n - m)!(\beta + m)!} \frac{(-x)^m}{m!}, \tag{4}$$

which yields, for example,

$$L_0^{1/2} = 1, \qquad L_0^{3/2} = 1,$$
$$L_1^{1/2} = \tfrac{3}{2} - x, \quad L_1^{3/2} = \tfrac{5}{2} - x,$$
$$L_2^{1/2} = \tfrac{15}{8} - \tfrac{5}{2}x + \tfrac{1}{2}x^2, \quad \text{etc.}$$

Assuming, for ^{16}O, that there are two protons in the $1s$ ($n = 0,\ l = 0$) and six in the $1p$ ($n = 0,\ l = 1$) state, show, using Eqs. 2 and 3, that the proton density is

$$\rho(r) = \frac{2\alpha^3}{\pi\sqrt{\pi}} \exp(-\alpha^2 r^2)(1 + 2\alpha^2 r^2). \tag{5}$$

If the form factor $F(q)$ is defined by

$$F(q) = 4\pi \int_0^\infty \rho(r) j_0(qr) r^2\, dr,$$

show that the form factor $F(q)$ for the ^{16}O-distribution (Eq. 5) is

$$F(q) = \exp\left(-\frac{q^2}{4\alpha^2}\right)\left(8 - \frac{q^2}{\alpha^2}\right), \tag{6}$$

which shows that $F^2(q)$ has only one minimum. Note that this is not true if the nucleons are assumed to move in a finite potential, in which case the tail of the density distribution decays exponentially.

4–11. Isotope shift in atomic X-rays.

From the expression for the orbital radius of an electron given in Table 4–2, show that the lowest electron orbital radius in a heavy atom is about 100 times larger than the nuclear radius R. Even then, the electronic binding energy would undergo a small shift because of the finite size of the nucleus, and the shifts for two different isotopes would not be the same. This is called the isotope shift in X-rays.

A rough estimate of this effect may be made by making the following simplifying assumptions:

1. Screening effects are identical for the two isotopes and cancel out in the isotope shift.

2. The nuclear charge distribution is uniform over a radius R with sharp cutoff, and $R = 1.2A^{1/3}$ fm.

3. The nonrelativistic electronic wave function for a point nucleus is used for a first-order perturbation calculation of the effect. For the lowest electronic state, this wave function is

$$u_0 = \frac{\gamma^{3/2}}{\sqrt{\pi}} \exp(-\gamma r), \qquad \gamma = Z\frac{me^2}{\hbar^2},$$

and the unperturbed energy is

$$E_0 = -\tfrac{1}{2}Z^2 \frac{me^4}{\hbar^2}.$$

(a) Using Eq. 4–10, show that the potential energy of an electron in the field of the nucleus is

$$V(r) = -\frac{Ze^2}{r} \qquad \text{for} \quad r > R$$

$$= \frac{Ze^2}{R}\left(\frac{r^2}{2R^2} - \frac{3}{2}\right) \quad \text{for} \quad r < R.$$

(b) Using perturbation theory, show that the energy shift in the lowest state due to the finite nuclear size is

$$\Delta E = \frac{4Ze^2}{R}\gamma^3 \int_0^R \left(\frac{r^2}{2R^2} - \frac{3}{2} + \frac{R}{r}\right)\exp(-2\gamma r)\, r^2\, dr.$$

(c) Even for heavy atoms $\gamma R \ll 1$, and we may replace the exponential $\exp(-2\gamma r)$ by unity in the above integrand. Show that

$$\Delta E \approx \frac{4}{5} Z^2 |E_0| \left(\frac{R}{a_0}\right)^2,$$

where $a_0 = \hbar^2/me^2$. This shows that $\Delta E \propto Z^4$. For Tl ($Z = 81$), check that this expression yields $\Delta E = 8.25$ eV.

(d) The isotope shift S, in K X-ray between $^{203}_{81}$Tl and $^{205}_{81}$Tl may be found from

$$S = \frac{d}{dR}\Delta E \cdot \frac{dR}{dA}\cdot \Delta A.$$

Show that

$$S = \frac{2}{3}\frac{\Delta A}{A}\cdot \Delta E.$$

The isotope shift S between $^{203}_{81}$Tl and $^{205}_{81}$Tl is thus only of the order of 0.05 eV.

Chapter 5

Internucleon Forces: II

5-1 ELECTROMAGNETIC MOMENTS OF THE DEUTERON AND TENSOR FORCES

The deuteron has been observed to have an electric quadrupole moment $Q = 0.2860 \pm 0.0015$ fm^2 [1, 2] and a magnetic moment $\mu_d = 0.857406 \pm 0.000001$ [3]. Its total angular momentum $J = 1$. We saw in Chapter 2 that the binding energy and angular momentum are consistent with the assumption that the deuteron is in a $^3S_1$-state. However, a system in an S-state can have no quadrupole moment, since its wave function has no angular dependence and the density distribution is therefore isotropic. For two spin-$\frac{1}{2}$ particles whose total angular momentum is 1, the only possible angular-momentum states are $^1P_1$, $^3S_1$, $^3P_1$, $^3D_1$. For central forces, the magnitude of the orbital angular momentum is a constant of motion, but if we consider noncentral forces it is possible to have a combination of states with different values of L but the same J. If parity is to be a good quantum number, we may combine only states of the same parity. Therefore, in addition to the states of definite $\mathbf{L}^2$, we should consider the possible deuteron wave functions,

$$\psi(SD) = \cos \omega \psi(^3S)^1(\tau)_0 + \sin \omega \psi(^3D)^1(\tau)_0, \qquad (5\text{--}1a)$$

$$\psi(PP) = \cos \omega \psi(^1P)^1(\tau)_0 + \sin \omega \psi(^3P)^3(\tau)_0. \qquad (5\text{--}1b)$$

The amount of mixing of the individual states is described by the parameter ω, which is introduced in the form of trigonometric functions, so that $\psi(SD)$ and $\psi(PP)$ are automatically normalized to unity. Both wave functions describe states in which a measurement of J will yield the value 1, but in the state described by Eq. 5–1a a measurement of L will sometimes yield 0 and sometimes 2, the respective probabilities being $\cos^2 \omega$ and $\sin^2 \omega$. Similarly, in the state of Eq. 5–1b, a measurement of S yields 0 or 1 with respective probabilities $\cos^2 \omega$ and $\sin^2 \omega$. The choice among the five possible alternatives for a pure ^3S-state is determined by the value of the magnetic moment.

To calculate the magnetic moment of the deuteron, we make use of the Sachs formula, given in Problem 3–4:

$$\cdot \; [\langle M_z \rangle_{T_3} + \langle M_z \rangle_{-T_3}] = (\tfrac{1}{2} + \mu_s)J + \frac{1}{J+1}(\mu_s - \tfrac{1}{2})\langle S^2 - L^2 \rangle, \quad (5\text{–}2)$$

where $\mu_s = (\mu_p + \mu_n)$ is the isoscalar part of the nucleon magnetic moment. Noting that for the deuteron $T_3 = 0$, $J = 1$, we immediately get its magnetic moment operator in nuclear magnetons as

$$\mu_d = \tfrac{1}{2}[(\tfrac{1}{2} + \mu_s) + \tfrac{1}{2}(\mu_s - \tfrac{1}{2})(S^2 - L^2)]. \quad (5\text{–}3)$$

When μ_d acts on a state of fixed S and L, the operators S^2 and L^2 are to be replaced by the values $S(S+1)$ and $L(L+1)$.

For various possible states the results are as follows:

$^1P_1$: $\mu_d = \tfrac{1}{2}$,

$^3S_1$: $\mu_d = \mu_s = 0.879$,

$^3P_1$: $\mu_d = \tfrac{1}{2}(\mu_s + \tfrac{1}{2}) = 0.689$,

$^3D_1$: $\mu_d = \tfrac{1}{2}(-\mu_s + \tfrac{3}{2}) = 0.310$,

$^3S + {}^3D$: $\mu_d = 0.879 - 0.569 \sin^2 \omega$,

$$^1P + {}^3P: \quad \mu_d = 0.500 + 0.189 \sin^2 \omega + 3.327 \sin 2\omega \int {}^1f\,{}^3f \, dr. \quad (5\text{–}4)$$

In the last of these equations ${}^1f(r)/r$ and ${}^3f(r)/r$ are the radial wave functions in the $^1P_1$- and $^3P_1$-states, respectively.

Only the last two states are capable of yielding the experimental value of 0.857, and the P-state is ruled out by later considerations. At least, this is true if we ignore the possibility of very strong velocity-dependent forces or very considerable meson-exchange magnetic moments. As indicated in Chapter 3, these are not thought to be very large. If all such corrections are ignored, we see from Eq. 5–4 that the deuteron magnetic moment is

$$\mu_d = 0.879 - 0.569 \sin^2 \omega, \qquad (5\text{-}5)$$

and a value of $\sin^2 \omega = 0.04$, that is, a D-state admixture of 4% is needed to give the experimental value of 0.857 nuclear magneton. Note that in Eq. 5-5 only the isoscalar part of the nucleon magnetic moment is involved, and also that an increase in the D-state probability would reduce the magnetic moment. We shall see in the next section that a variety of other experimental data, like those on nucleon-nucleon scattering and the deuteron electromagnetic form factor, as well as results on elastic π-d scattering, indicate that a D-state admixture of about 6–7% is present in the deuteron. This implies that Eq. 5-5 for μ_d must be corrected for meson-exchange currents and other relativistic effects, which must contribute positively to the deuteron magnetic moment to offset the effect of the increase in the D-state probability to more than 4%.

The calculation of the quadrupole moment of the deuteron is straightforward. The wave function (Eq. 5-1a) can be written as

$$\psi(SD) = \frac{1}{r} [\cos \omega \, f_S(r) \mathscr{Y}^1_{1,10} + \sin \omega \, f_D(r) \mathscr{Y}^1_{1,12}], \qquad (5\text{-}6)$$

where $\mathscr{Y}^M_{J,SL}$ is an angular-momentum function, and $f_S(r)$ and $f_D(r)$ are the radial parts of the wave functions, normalized so that $\int_0^\infty f(r) \, dr = 1$. In these expressions, r is the neutron-proton distance, and therefore expression 3-7b yields

$$Q = \left(\frac{16\pi}{5}\right)^{1/2} \int |\psi(SD)|^2 \left(\frac{r}{2}\right)^2 Y_2^0 \, dV, \qquad (5\text{-}7)$$

which leads to

$$Q = \frac{1}{10} \left[\sqrt{2} \cos \omega \sin \omega \int f_S(r) f_D(r) r^2 \, dr - \tfrac{1}{2} \sin^2 \omega \int f_D^2(r) r^2 \, dr \right]. \qquad (5\text{-}8)$$

We note that a pure D-state would have a negative Q, but if the relative phase of the S- and D-wave functions is positive, a positive Q can be obtained. Indeed, for $\sin^2 \omega \sim 0.04$, the leading term of Eq. 5-8 is the more important one, and the observed value of Q requires that $\int f_S f_D r^2 \, dr \sim 10 \times 10^{-26}$ cm^2. Since low values of r will contribute little to this integral, the wave function should have a range of 4 or 5 fm. This is quite consistent with the size of the deuteron suggested in Chapter 2, where we saw that outside an effective range of about 1.7 fm the deuteron wave function decays like $e^{-\gamma r}$, with $1/\gamma = 4.31$ fm, the latter value often being loosely referred to as the "radius of the deuteron." It will be realized that the deuteron is a quite loosely bound nucleus with a large radius comparable to that of a nucleus with as many as 20 nucleons.

Since Q is not subject to uncertain small corrections, as is the magnetic moment, it might be thought that the values of Q would allow a more precise estimate of the percentage of D-state in the wave function. However Eq. 3–29 shows that such a calculation requires knowledge of the radial wave functions, which in turn can be determined only if the nucleon forces are known in some detail. If a specific nucleon force is proposed, both Q and ω can be calculated; the experimental value of the former provides a fairly precise datum which the suggested interaction must fit, but the latter is only roughly fixed by the magnetic moment.

The type of nucleon forces that we have so far considered are essentially central and cannot mix orbital angular momenta. We can generalize the forces by considering velocity-dependent forces or, as we remarked in Section 2–2, by taking a force dependent on the angles between the spin vectors and the join of the two nucleons. A possible noncentral potential which can be formed from the three vectors $\boldsymbol{\sigma}_1$, $\boldsymbol{\sigma}_2$, and $\mathbf{r} = r\hat{\mathbf{r}}$ is

$$V_T = V_T(r)(3\boldsymbol{\sigma}_1 \cdot \hat{\mathbf{r}} \, \boldsymbol{\sigma}_2 \cdot \hat{\mathbf{r}} - \boldsymbol{\sigma}_1 \cdot \boldsymbol{\sigma}_2) = V_T(r)S_{12}. \tag{5-9}$$

The subscript T is used because this potential is called the *tensor* interaction; it is actually a scalar product of two second-rank tensors. It can be demonstrated that this and the exchange potentials considered in Chapter 2 are the only velocity-independent potentials for two nucleons which conserve energy, momentum, total angular momentum, parity, and charge.[†] Since $\mathbf{L}^2$ does not commute with S_{12}, the addition of a potential of this form to the Hamiltonian will mean that $\mathbf{L}^2$ is not a constant of motion and that the states are mixtures of states of different L-values, for example, a mixture of S- and D-states. Consequently the tensor force is capable of explaining the deuteron quadrupole moment. On the other hand, $S^2 = (\boldsymbol{\sigma}_1 + \boldsymbol{\sigma}_2)^2$ does commute with V_T, and S_{12} applied to a singlet wave function gives zero, with the result that a tensor force couples together neither singlet and triplet states nor different singlet states. We may also note that it conserves parity and couples together only states that are both of even L or both of odd L.

The effect on the deuteron quadrupole moment can be understood in a qualitative way by treating the spins classically. In the triplet state they

---

[†] The simplest kind of velocity-dependent force, namely an $\mathbf{L} \cdot \mathbf{S}$-force, does not mix states of different L. Whether or not there are such forces in the nuclear interaction, the deuteron properties show that another type of noncentral force is also present.

are then parallel, and if each makes an angle θ with $\mathbf{r}$, we have

$$V_T = V_T(r)(3\cos^2\theta - 1). \tag{5-10}$$

The corresponding force, $-\operatorname{grad} V$, has a component in the direction of increasing θ equal to $(6/r)V_T(r)\cos\theta\sin\theta$. If $V_T(r)$ is negative, this force will move the vector $\mathbf{r}$ to angles near the z-axis, that is, it will tend to concentrate the motion near $\theta = 0$ and π. Hence, if the other forces by themselves were to produce an S-state, the addition of a small tensor force would modify the density, so that instead of being isotropic it would be peaked along the axis $(Q > 0)$. However, if the system were in a pure D-state under central forces, the density (proportional to $\sin^4\theta$) would be greatest perpendicular to the axis $(Q < 0)$. The addition of the tensor force with $V_T < 0$ would then reduce the magnitude of this negative quadrupole moment. These effects are seen to correspond to the values of Q which are predicted by Eq. 5–8 for a mixed S- and D-state, provided that $\cos\omega\sin\omega > 0$. In other words, a tensor force with $V_T < 0$ produces a state in which the wave functions of the S- and D-states are mixed with coefficients of the same sign, and it is clearly possible, by adjusting the parameters in $V_T(r)$ (of which there must be at least two), to obtain values of Q and ω consistent with experiment.

The quantitative calculation requires the solution of the wave equation:

$$-\nabla^2\psi + (V_{te}(r) + V_T(r)S_{12})\psi = -\gamma^2\psi, \tag{5-11}$$

where $(\hbar^2/M)V_{te}$ is the central potential in triplet even states, $(\hbar^2/M)V_T$ is the tensor potential, and $(\hbar^2/M)\gamma^2$ is the binding energy of the deuteron. Equation 5–11 is simplified by writing ψ in the form of Eq. 5–6, multiplying in turn on the left by $\mathscr{Y}^1_{1,10}$ and $\mathscr{Y}^1_{1,12}$, and using their orthogonality, together with the matrix elements of S_{12} found in Problem 5–2, namely,

$$(\mathscr{Y}^1_{1,10}|S_{12}|\mathscr{Y}^1_{1,10}) = 0, \qquad (\mathscr{Y}^1_{1,12}|S_{12}|\mathscr{Y}^1_{1,12}) = -2,$$

$$(\mathscr{Y}^1_{1,12}|S_{12}|\mathscr{Y}^1_{1,10}) = \sqrt{8}.$$

The results are:

$$\frac{d^2 f_S}{dr^2} - (\gamma^2 + V_{te})f_S = \sqrt{8}\, V_T \tan\omega\, f_D, \tag{5-12a}$$

$$\frac{d^2 f_D}{dr^2} - \left(\gamma^2 + V_{te} + \frac{6}{r^2} - 2V_T\right)f_D = \sqrt{8}\, V_T \cot\omega\, f_S. \tag{5-12b}$$

These radial wave equations are referred to as "coupled", they show that, if a tensor force exists, the system cannot have a fixed L-value, since nonzero values of $\cos \omega f_S$ require nonzero values of $\sin \omega f_D$, and vice versa. At distances greater than the range of nucleon forces, the equations are uncoupled, and the wave functions for a bound state can be written explicitly as

$$f_S = N_S e^{-\gamma r}, \tag{5-13a}$$

$$f_D = N_D e^{-\gamma r}\left(1 + \frac{3}{\gamma r} + \frac{3}{\gamma^2 r^2}\right), \tag{5-13b}$$

for r such that V_T and $V_{te} \ll \gamma^2$. The quantities N_S and N_D are normalizing constants. In addition to satisfying these equations for large r, f_S and f_D must also vanish at the origin or, if one postulates an impenetrable core, at the core radius. For a given set of potentials V_{te} and V_T, these conditions constitute an eigenvalue problem for the binding energy (γ^2) and the proportion of D-state ($\sin^2 \omega$). When these values are found and f_S and f_D are known, the quadrupole moment can be calculated in accordance with Eq. 5-8. The values of γ^2 and Q, and to a lesser degree the value of $\sin^2 \omega$, provide quantities for comparison with experiment, on the basis of which the proposed potentials may be rejected or tentatively accepted. A further test is supplied by the predicted triplet scattering cross section. In a scattering experiment, the wave functions will satisfy equations like 5-12 with $-\gamma^2$ replaced by k^2 and with asymptotic forms of f_S and f_D appropriate to a scattering problem.

The eigenvalue equations 5-12 cannot be solved analytically even for simple square-well potentials. The first numerical calculations were done by Rarita and Schwinger [4], who took V_T and V_{te} to be square-well potentials of the same range but different depths. A large variety of potentials can yield the experimental binding energy, the magnetic dipole moment, and the electric quadrupole moment of the deuteron, provided that the tensor force is strong and has a range comparable to that of the central part. Although the admixture of D-state is only a few per cent, the tensor force makes a substantial contribution to the binding energy. We shall see in a later section that the long-range part of the two-nucleon force is given by the one-pion-exchange potential, which for the triplet-even state is given by

$$-V_0\left[1 + \left(1 + \frac{3}{x} + \frac{3}{x^2}\right)S_{12}\right]\frac{e^{-x}}{x},$$

with $x = r/1.41$ fm, and $V_0 = 11$ MeV. This prediction is modified for $r < 3$ fm by two-meson terms, but it is the outer region which mainly controls a low-energy system such as the deuteron. It will be observed that for $x = 1$ the tensor force is 10 times the central, and even for $x = 3$, at which distance the nuclear force has become quite small, the tensor force is still twice the central force, allowing for the value of S_{12}. Most modern potentials contain the one-pion-exchange tail and fit the deuteron data.

TABLE 5-1 Deuteron Properties of Some Potentials

Potential	Ref.	Per cent D-state	Predicted magnetic moment in nuclear magnetons	Predicted quadrupole moment, fm^2
Hamada-Johnston	36	6.96	0.840	0.281
Feshbach-Lomon	42	4.31	0.854	0.268
Reid hard-core	39	6.50	0.842	0.277
Reid soft-core	39	6.47	0.842	0.280
Ueda-Green	58	5.5	0.856	0.280
Experimental			0.857406 ± 0.000001	0.282 ± 0.002[a]

[a] This was the accepted experimental value of Q before 1972, which the potential models attempted to reproduce. The current value of Q is 0.2860 ± 0.0015 fm^2 [2].

In Table 5–1, the low-energy properties of some of these potentials are listed. These potentials also fit the high-energy phase-shift data and are described in some detail in Sections 5–8 and 5–12. Table 5–1 illustrates the

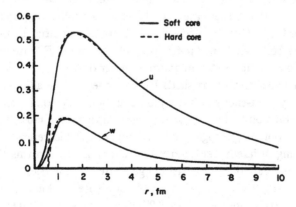

Fig. 5-1 Normalized deuteron wave functions for the Reid soft-core and hard-core potentials [39].

points that the low-energy parameters of the deuteron do not determine a unique potential and that the percentage of D-state admixture (and therefore the central-to-tensor ratio) can vary from potential to potential, even after roughly fitting the deuteron data. We discuss this question in more detail in the next section.

The normalized deuteron wave functions $\cos \omega f_S(r)$ and $\sin \omega f_D(r)$ for the Reid hard-core and soft-core potentials [39] are sketched in Fig. 5–1. Note that the S-state wave function is not negligible even at a distance of 10 fm!

5-2 THE D-STATE ADMIXTURE IN THE DEUTERON- AND MESON-EXCHANGE CURRENTS

We stated in Section 5–1 that a variety of experiments indicate that the D-state probability in the deuteron is in the range of 6–7%. We now briefly list some of these approaches.

1. The most direct method of obtaining information about the nuclear force is to perform nucleon-nucleon scattering, a subject that we shall study in detail in this chapter. The characteristics of the n-p force in the $T = 0$, $J = 1$ channel can be extracted from the scattering data by phase-shift analysis. A measure of the strength of the tensor force is the mixing parameter ε_1 (see Section 5–5), which indicates the proportion of the $^3S_1$- and the $^3D_1$-states in the triplet state. In Fig. 5–18, we show some experimentally determined values of the mixing parameter as a function of the laboratory energy, together with the values computed from various phenomenological potentials. It will be seen that only three potentials give some agreement with the data: of these, the Yale and the Hamada-Johnston potentials also yield $\sin^2 \omega \approx 0.07$, while the Sprung-de Tourreil potential gives ≈ 0.05. Most of the one-boson-exchange potentials have much less D-state probability and also give poorer fit to the mixing-parameter data.

2. Another experiment that sheds light on the D-state admixture in the deuteron is very high-energy elastic pion-deuteron scattering. In Fig. 5–2, the experimental points for the π^+-d elastic differential cross section are shown as a function of $q_\mu{}^2$, together with some of the theoretically predicted results, assuming different deuteron wave functions [5]. In this figure, the Humberston wave function is essentially generated by the Hamada-Johnston potential [36], slightly modified to yield a deuteron binding energy of 2.226 MeV (the H-J potential gives 2.27 MeV) and $\sin^2 \omega = 0.06953$. The wave function GK8 is generated by potential 8 of Glendenning and Kramer [6]; it has a hard core of 0.5 fm and gives $\sin^2 \omega = 0.05622$. Each of the

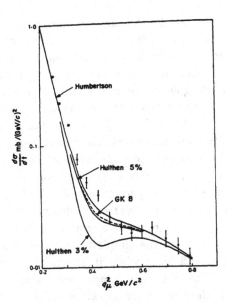

Fig. 5-2 Differential cross section for elastic π^+-d scattering for pions of momentum 3.75 GeV/c, plotted against $q_\mu{}^2$. The theoretical curves are obtained using the Glauber formalism for high-energy scattering [5], employing different deuteron wave functions of varying *D*-state probability (see text).

two Hulthen wave functions have a hard core of 0.43 fm, with $\sin^2 \omega = 0.03$ and 0.05. The radial dependence of $f(r)$ in the Hulthen wave function is simply the difference of two exponentials of differing range, with the longer one corresponding to the binding energy. It will be seen from Fig. 5–2 that the Humberston wave function with about 7% *D*-state probability fits the data best. If the deuteron was completely in the $^3S_1$-state, theory would predict a pronounced dip in the cross section for $q_\mu{}^2 \approx 0.4$ (GeV/c)^2, which is experimentally not seen. In the presence of the *D*-state, there is scattering also from the quadrupole part of the form factor, which fills up this dip. A very similar phenomenon occurs also in the elastic scattering of electrons from deuterons, which we discuss next.

3. Other experiments have examined elastic electron-deuteron scattering. In analyzing the data from these studies, generally the deuteron is treated nonrelativistically; the electron, of course, is relativistic. The ratio of the cross section $\sigma(\theta)$ to the Mott value $\sigma_{\text{Mott}}(\theta)$ can again be expressed as in Eq. 4–27:

$$\frac{\sigma(\theta)}{\sigma_{\text{Mott}}(\theta)} = \left[A(q_\mu{}^2) + B(q_\mu{}^2) \tan^2 \frac{\theta}{2} \right],$$

where $A(q_\mu^2)$ contains dominantly the electric monopole and quadrupole form factors of the deuteron (multiplied by the isoscalar electric form factor of the nucleon to account for its finite size), while $B(q_\mu^2)$ is the contribution of the magnetic part [7]. Even if exchange currents and relativistic corrections for the deuteron are ignored, there is additional uncertainty due to the presence of the isoscalar form factor of the nucleon. Elias et al. [8] have measured $A(q_\mu^2)$ up to $q_\mu^2 = 34$ fm^{-2}. They find that the data can be fitted (ignoring exchange effects) by deuteron wave functions generated by Hamada-Johnston [36] or Reid [39] soft- and hard-core potentials, which have $\sin^2 \omega \approx$ 0.065–0.07 and fit the high-energy N-N scattering data. For other types of deuteron wave functions with lesser percentages of D-state, relatively large meson-exchange effects must be postulated to obtain agreement with the electric data for large q_μ. The situation, of course, is not very clear cut. The contribution to $A(q_\mu^2)$ from the simplest meson-exchange currents is appreciable only for $q_\mu^2 > 12$ fm^{-2}, while it affects $B(q_\mu^2)$ even for very small values of q_μ^2 [7]. That meson-exchange currents must be important for the magnetic part is clear even from the static magnetic moment of the deuteron: a 7% D-state yields a static magnetic moment of 0.840 nm in the absence of exchange effects, compared to the experimental value of 0.857 nm. It is found [7] that this discrepancy for $B(q_\mu^2)$ increases with higher values q_μ^2, the experimental points lying above the ones calculated using deuteron wave functions with 7% D-state and ignoring exchange currents. Summarizing the above statements, we may remark that the electric form factor of the deuteron favors a 6–7% D-state probability, that the behavior of the magnetic form factor clearly points to meson-exchange currents, and finally that the correction due to such currents should be very small for $A(q_\mu^2)$ and should be appreciable only for $B(q_\mu^2)$.

We now explain very briefly the simplest meson-exchange currents that are of importance to the deuteron. As discussed in Section 4–3, an electron can interact with a nucleon by the exchange of a virtual photon—the photon carries an electromagnetic field (or four-potential A_μ) which couples with the nucleonic current. In the deuteron, there are two nucleons, each clothed with its meson cloud. The photon can couple to either one or the other nucleon; it can also couple, as shown in Fig. 5–3a, to a charged pion which is being exchanged between the two nucleons. Since the deuteron has i-spin $T = 0$, it is not difficult to show, however, that this simplest process with the π-π-γ vertex does not contribute to the deuteron form factor. It is desirable to express the pionic exchange current in terms of nucleon variables

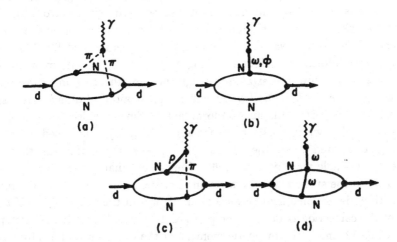

Fig. 5-3 Diagrams for the elastic interaction of a deuteron with an electromagnetic field. Note that (a) is forbidden; (b) is included in the form factor of the nucleon; (c) represents the simplest exchange-current contribution; and (d) is a double-scattering process.

by integrating over the intermediate states. Such a procedure [65] yields for the one-pion exchange current a term proportional to $(\tau_1 \times \tau_2)_3$. This has zero matrix elements between states of the same T. For $T = 0$ states, this is because the current is an isovector, for $T = 1$ states it follows from symmetry under exchange of particles 1 and 2. This exchange current does contribute to electromagnetic transitions between $T = 0$ and $T = 1$ states, such as the photodisintegration of the deuteron. Of course, charged pionic exchange contributes to the nuclear force in $T = 0$ as well as $T = 1$ states; indeed it gives a dependence $\tau_1 \cdot \tau_2$ in the nucleon variables.

The formal result that the one-pion exchange current gives no contribution to matrix elements diagonal in T may appear intuitively reasonable from the following physical picture. Consider each nucleon in a neutron-proton system as spending half its time in each charge state, with the result that positive and negative virtual pions are emitted in such a way that the net charged pionic current is zero. However, the result does, of course, involve a sum over intermediate quantum states and their relative phases are significant, and are such that a contribution arises for matrix elements between two-nucleon states that are not diagonal in T.

The simplest exchange-current contribution is shown in Fig. 5–3c, where a virtual ρ is emitted by one nucleon, it interacts with the electromagnetic field producing a virtual photon and a pion which lands on the second nucleon. This process was first studied in detail by Adler and Drell [9]. In order to estimate the contribution of Fig. 5–3c to the magnetic moment, it is necessary to know the signs and magnitudes of the coupling constants $g_{\rho\pi\gamma}$ and $g_{\rho NN}$. Adler and Drell used a value of $g_{\rho NN}$ obtained from an analysis of the isovector part of the nucleon form factor. Since the decay $\rho \rightarrow \pi + \gamma$ has too small a width to have been observed directly, they used a value for $g_{\rho\pi\gamma}^2$ estimated from the reaction $\gamma + p \rightarrow p + \rho$, assuming that it proceeds by single pion exchange [66]. They found that the contribution of Fig. 5–3c to the static magnetic moment is of the order of $(1–2) \times 10^{-2}$ nm, but with undetermined sign because of ambiguity in the sign of $g_{\rho\pi\gamma}$. If the sign of the correction is taken to be positive, this would explain the discrepancy of 0.017 nm in the magnetic moment of the deuteron that is obtained with 7% D-state.

Although this simple meson-exchange current is also able to improve the fit with $B(q_\mu^2)$, it is possible that many other more complicated processes are also important. For one thing, the simple diagram 5–3c seems to make the fit of $A(q_\mu^2)$ poorer for $q_\mu^2 > 14$ fm^{-2} [7], although its effect on the static quadrupole moment is very small. Blankenbecler and Gunion [10] have recently considered the double-scattering process of Fig. 5–3d, which is not an exchange-current effect. It seems to yield a good fit to the form-factor data of the deuteron (including the magnetic part) with 7% D-state.

Summarizing the present section, it seems reasonable to say that current experimental data favor 6–7% D-state probability in the deuteron wave function, with meson-exchange currents giving important corrections to the magnetic moment and the magnetic form factor.

5-3 THE NUCLEAR FORCE AND GENERAL FEATURES OF HIGH-ENERGY SCATTERING

The nuclear force between two nucleons may be regarded as arising from the exchanges of mesons between them. This basic idea, due to Yukawa [11], is in analogy to the electromagnetic case, where the Coulomb force is regarded as arising from a photon exchange between two charged particles. From the field-theoretical point of view, there is an interaction between the meson and nucleon fields, which gives rise to processes like $N \rightarrow N + \pi$. Since energy cannot be conserved in such a process, it is called "virtual." Since the energy conservation is being violated by $\Delta E \sim m_\pi c^2$, the pion, by the uncertainty principle, has to be reabsorbed within a time Δt, such that $\Delta E \, \Delta t \sim \hbar$. If the virtual pion is absorbed by the same nucleon that emitted it, it contributes to the "self-energy" of the nucleon. If, however, it is absorbed by a neighboring nucleon, there is a contribution to the energy of the two nucleons, which really arises because of the interaction between the meson and nuclear fields. Alternatively, one may eliminate the meson-nucleon interaction and replace it by a N-N potential. The precise definition of a potential from the field-theoretic point of view will be discussed in Section 5–9.

If the pion is assumed to move with the maximum possible speed c, the distance traveled by it in time Δt is $c \, \Delta t = \hbar / m_\pi c = 1.4$ fm, which should be a measure of the range of the nuclear force arising as a result of one-pion exchange. The derivation of this one-pion-exchange potential (OPEP) is given in Section 5–11; it is

$$V(\text{OPEP}) = g^2 \frac{m_\pi{}^3 c^2}{12 M^2} \, (\boldsymbol{\tau}_1 \cdot \boldsymbol{\tau}_2) \left[\boldsymbol{\sigma}_1 \cdot \boldsymbol{\sigma}_2 + S_{12} \left(1 + \frac{3}{x} + \frac{3}{x^2} \right) \frac{e^{-x}}{x} \right], \quad (5\text{–}14)$$

where $g^2 = 15$ is the pseudoscalar coupling constant, and $x = \mu r$, with $\mu = m_\pi c / \hbar$. We shall discuss later the experimental evidence for the fact that the actual nuclear interaction beyond $r > 3$ fm is dominantly $V(\text{OPEP})$. Multiple exchanges of pions would lead to shorter-range forces, since the range is seen to be inversely proportional to the mass of the exchanged particles. For example, the two-pion-exchange potential (TPEP) becomes increasingly dominant as we go to shorter distances, and is much stronger than $V(\text{OPEP})$ for $r \leqslant 1$ fm.

Consider the nucleon that is emitting the virtual pion to be at point $\mathbf{r}_1$. Since the linear momentum is conserved in a virtual process, the nucleon would recoil after emitting the pion to some point $\mathbf{r}_1'$. The second nucleon, in its turn, would also recoil from $\mathbf{r}_2$ to $\mathbf{r}_2'$ while absorbing the pion. Thus, if recoil is taken into account, the interaction potential cannot be only a function of $\mathbf{r}_1 - \mathbf{r}_2$; other relative distances are involved as well. When a potential depends only on $\mathbf{r} = \mathbf{r}_1 - \mathbf{r}_2$, it is called a "local" potential. Quantum-mechanically, the potential V is an operator which, acting on a state vector $|\psi\rangle$, yields a new state $V|\psi\rangle$. In r-space, we can write

$$\langle \mathbf{r} | V | \psi \rangle = \int \langle \mathbf{r} | V | \mathbf{r}' \rangle \langle \mathbf{r}' | \psi \rangle \, d^3 r'$$

$$= \int V(\mathbf{r}, \mathbf{r}') \psi(\mathbf{r}') \, d^3 r'. \quad (5\text{–}15)$$

If $V(\mathbf{r}, \mathbf{r}')$ is a function only of $\mathbf{r}$, then

$$V(\mathbf{r}, \mathbf{r}') = V(\mathbf{r}) \, \delta(\mathbf{r} - \mathbf{r}'); \qquad (5\text{-}16)$$

hence $\langle \mathbf{r}|V|\psi \rangle = V(\mathbf{r})\psi(\mathbf{r})$ for a local potential.

From the meson-theoretical viewpoint, it is natural to assume that the actual N-N potential would be "nonlocal" in nature, obeying Eq. 5-15. The ratio of the spatial extent of the nonlocality to the range of the nuclear force goes roughly as $m_\pi/M \ll 1$. We shall return to the consideration of non-local potentials in a later section.

We stressed previously that, to discover the fine details of a charge distribution or of a force field, it is necessary to use probe particles with de Broglie wavelengths comparable to the dimensions of the details to be investigated. We also pointed out that the impact parameter governs the scattering of partial waves and that the higher-angular-momentum components of an incident wave are not scattered at all by the nuclear force at low energies. Both considerations indicate that it is necessary to resort to high-energy scattering to investigate details of the potential shape and to learn about the two-nucleon force in states other than the S-state.

We can make these ideas more quantitative by asking what regions in the space between two nucleons contribute to the scattering of a given angular-momentum state. The impact parameter $b_l = \sqrt{l(l + 1)}\,\hbar/p$ represents classically the distance of closest approach. Actually the wave function of the lth wave is nonzero at distances smaller than b_l, so that the potential at closer approaches does affect the scattering of the relevant partial wave. By numerical calculations with simple representative potentials, Matsumoto and Watari [12] have shown that changes in the potential for distances of separation less than about $\frac{1}{2}b_l$ have essentially no effect on the lth partial wave. This means, for example, that, whereas the S-wave always senses the whole potential, the P-wave is insensitive to details of the potential for $r < 0.7$ fm, so long as the energy is below 100 MeV. Again, the shape of the potential inside $r < 2.0$ fm is first sensed by the P-wave at about 20 MeV center-of-mass energy, by the D-wave at 60 MeV, and by the F-wave at 120 MeV.

We have thus arrived at the important result that, as the energy is gradually increased, a study of the phase shifts of the higher-angular-momentum states allows us to acquire information about the nuclear potential at steadily decreasing distances of separation. Consequently, it is important to consider higher-energy experiments. There is one restriction to be mentioned; this concerns energies so high that the nucleons have speeds comparable with the speed of light and energies sufficient to create real mesons. In such instances, it is necessary to use relativistic mechanics, in which the very idea of a static potential is an approximation, becoming increasingly poorer as the energy rises. It may be that no potential exists which will precisely fit all data up to 400 MeV, although a potential description for energies up to about 200 MeV should be possible. Indeed, it is to be expected that there is more than one such potential, since any arbitrary energy cutoff precludes a unique determination of the potential at short distances. Fortunately, for a study of atomic nuclei detailed knowledge of relativistic behavior is not necessary, since the momentum distributions of bound nucleons do not extend appreciably above momenta corresponding to about 150–200 MeV.

5-4 POLARIZATION

In addition to differential and total cross-section determinations, experiments on polarization yield important information. A beam of nucleons is polarized if the spins are not randomly distributed but have some preferred orientation. If the number of particles in the beam with spin component parallel to this preferred direction is N_+ and the number with antiparallel spin component is N_-, the polarization is defined[†] (for spin-$\frac{1}{2}$ particles) as

$$P = \frac{N_+ - N_-}{N_+ + N_-}. \tag{5–17}$$

A beam can be polarized if some force acts preferentially on particles with one or the other spin direction. Such a force is provided by an in-homogeneous magnetic field, which deflects in opposite directions particles of opposite magnetic-momentum orientation. This is not the most important polarizing force in nuclear physics, however. The tensor force will produce polarization, for in the two cases in which the spins of incident and target nucleons are parallel or antiparallel, the tensor force will have opposite signs and will produce entirely different angular distributions. Consequently, with a mixture of central and tensor forces acting, the beam of particles

---

[†] It will be noted that Eq. 5–17 is equivalent to $P = \langle \sigma_d \rangle$, the expectation value of the component σ_d of the spin operator σ along the preferred direction d. If this direction is not selected in advance, P can be defined with respect to arbitrary axes and the direction also found. We have

$$P^2 = \langle \sigma_x \rangle^2 + \langle \sigma_y \rangle^2 + \langle \sigma_z \rangle^2,$$

and the preferred direction is specified by the polarization vector

$$\vec{\sigma} = \langle \sigma_x \rangle \mathbf{i} + \langle \sigma_y \rangle \mathbf{j} + \langle \sigma_z \rangle \mathbf{k}.$$

deflected through a given angle will be partially polarized. Other possible noncentral forces depending on the spin vectors would similarly produce differently polarized beams at different scattering angles. Even with an unpolarized beam falling on a target with random spin orientation, partially polarized beams may result, owing to interference between the wave describing particles that are scattered without change of spin direction and the wave describing spin-flipped particles. The phenomenon is an essentially quantum-mechanical effect, and semiclassical reasoning cannot lead to the correct results. Thus, in any scattering process involving a suitable force coupling spin and space coordinates, there is polarization, its sign and magnitude depending on the angle of scattering.

Conversely, one may analyze the polarization of a beam by using the fact that the differential cross section depends on the polarization if the scattering force is spin dependent. If the spins are predominantly up, say, the scattering will be mainly to one or the other side of the beam, in contrast to the scattering of an unpolarized beam, which is always symmetric about the direction of incidence.

To illustrate these points let us consider a double-scattering experiment in which first an unpolarized beam is scattered on an unpolarized target and then a part of the scattered beam undergoes a second scattering, for which the differential scattering cross section is measured (see Fig. 5-4). As a result of the first scattering, the beam incident on the second target is partially polarized, so that the measured scattering has left-right asym-

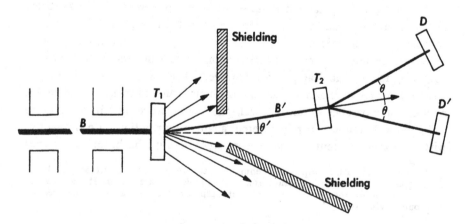

Fig. 5-4   A double-scattering experiment. An unpolarized beam B is scattered from the unpolarized target T_1. The partially polarized beam B' is scattered from a second unpolarized target T_2, and the differential scattering cross section is measured by detector D.

metry about the direction of the intermediate beam. This asymmetry is a simple function of the polarizations of the two scatterings. A number of general results concerning polarization follow from the assumptions of invariance of the forces under coordinate rotations and parity and time reversals. We shall simply state two of these results:[†]

1. For a given scattering angle θ, and for an initial beam with polarization $P_0\mu$, the dependence of the differential cross section on the azimuthal angle ϕ after a single scattering with an unpolarized target is given by a proportionality factor,

$$1 + PP_0 \mathbf{n} \cdot \mathbf{\mu} = 1 + PP_0 \cos\phi, \qquad (5\text{-}18)$$

where P is a number characteristic of the scattering process, and $\mathbf{n}$ is a unit vector perpendicular to the plane of scattering, with an opposite direction for left and right scattering. Precisely,

$$\mathbf{n} = \frac{\mathbf{p}_i \times \mathbf{p}_f}{|\mathbf{p}_i \times \mathbf{p}_f|}, \qquad (5\text{-}19)$$

where $\mathbf{p}_i$ and $\mathbf{p}_f$ are the initial and final center-of-mass momenta. These quantities are shown in Fig. 5-5, where the initial and final center-of-mass momenta $\mathbf{p}_i$ and $\mathbf{p}_f$ define the scattering plane, $\mathbf{p}_f$ is defined as a left scattering, and $\mathbf{p}_f'$ is a right scattering in the same plane. The corresponding perpendiculars $\mathbf{n}$ and $\mathbf{n}'$ are shown. The unit vector $\mathbf{\mu}$ is the direction of the polarization of the incident beam, its component perpendicular to the incident direction is $\mathbf{\mu}_\perp$, and ϕ is the angle between $\mathbf{\mu}_\perp$ and $\mathbf{n}$. The corresponding angle ϕ' for

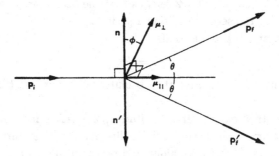

Fig. 5-5 Polarization and left-right asymmetry.

[†] Full details of the theory are given in an article by Wolfenstein [13].

the right scattering is seen to be $\pi - \phi$, since $\mathbf{n}' = -\mathbf{n}$. Note that any component $\mu_\parallel$ is undetected in a single scattering of a polarized beam.

2. After scattering from an unpolarized traget, an initially unpolarized beam has polarization $P\mathbf{n}$, where the polarization P is the quantity that occurs in Eq. 5–18. It depends on the parameters of the specific collision in question, namely, the types of particles, the energy, and the scattering angle θ.

The first result (Eq. 5–18) gives the basic relation between polarization and differential scattering cross section or, equivalently, left-right asymmetry in a scattering plane. The relative difference between the intensities of the two beams $\mathbf{p}_f$ and $\mathbf{p}_f'$ in Fig. 5–5 is

$$e_{lr} = \frac{\sigma(\theta, \phi) - \sigma(\theta, \phi')}{\sigma(\theta, \phi) + \sigma(\theta, \phi')}, \tag{5–20}$$

which, by means of Eq. 5–18, gives

$$e_{lr} = PP_0 \cos \phi. \tag{5–21}$$

With this result we can understand the use of double scattering to determine polarization in a single scattering. Let the two scatterings be in the same plane so that $\mathbf{n}_1$ and $\mathbf{n}_2$ are either parallel or antiparallel, depending on whether the second scattering is to left or right, the first one being to the left as in Fig. 5–4. It is also arranged that the two scattering angles are identical, and the center-of-mass energies are assumed to be the same. This last requirement restricts the usefulness of the method to high energies or heavy targets and even then introduces some error. Within the stated conditions, the collisions are identical and will have the same value of P. After the first collision, result 2 tells us that the beam has polarization $P\mathbf{n}_1$, and therefore for the second collision $P_0\mu$ in Eq. 5–18 is $P\mathbf{n}_1$. Consequently $\cos \phi$ is unity, and Eq. 5–21 yields

$$e_{lr} = P^2. \tag{5–22}$$

A measurement of e_{lr} will give the absolute value, but not the sign of the polarization.

There are other experiments to study polarization in more detail. In general, the polarization is rotated in a collision and may acquire components in the plane of scattering, both along and perpendicular to the final momentum. To study such effects, the beam must first be polarized and then scattered, and its polarization must be analyzed, that is, triple scattering is required. Analysis of polarization by scattering is valid only for the component perpendicular to the scattering plane. Consequently, various different planes must be used in the final scattering. One particular quantity determined is the depolarization D, which enters into a triple-scattering experiment in which all three planes are parallel. Such an experiment determines the polarization after the second scattering, according to Eq. 5–21. The quantity D is so defined that the polarization along n after the second scattering is $(P_2 + DP_1)/(1 + P_1P_2)$; it is seen that, if the first scattering completely polarized the beam $(P_1 = 1)$, $D = 1$ would mean no depolarization in the second scattering. Depolarization is one of the easiest parameters to measure. Determination of three other parameters requires triple scattering in other planes, simultaneous measurement of the polarization of the scattered particle and the recoiling target, and application of magnetic fields to rotate the direction of polarization between collisions [14]. With the availability of polarized proton beams, conventional double- and triple-scattering experiments reduce to single and double scattering; while a polarized target with a polarized beam makes it possible to use single scattering throughout.

5-5 THE SCATTERING MATRIX FOR COUPLED WAVES

In any scattering experiment, the particles are detected only when they have come out of the force range of the target. All the information about the detected particle in these experiments is therefore contained in the asymptotic part of the wave function. For elastic scattering in a central force, we found in Chapter 2 that the asymptotic wave function in the S-state is characterized by phase shifts $\delta_0(k)$. Similar considerations hold also for higher partial waves. The radial wave function in the asymptotic region may be written as

$$u_L \sim A_L(k) \exp[- i(kr - L\pi/2)] - B_L(k) \exp[+ i(kr - L\pi/2)],$$

which is a mixture of outgoing and incoming waves. As we saw in Section 2–3, the scattering matrix $S_L(k)$ is defined as the ratio of the outgoing to the incoming amplitude:

$$B_L = S_L A_L, \tag{5-23}$$

and for this one-channel case $S_L(k) = \exp[2i\, \delta_L(k)]$ is a complex number. Now we consider the more complicated case in which a tensor force is present. As discussed in Section 5–1, such a force mixes states of different L, and a further coupling parameter has to be introduced to describe scattering. For example, consider the triplet case with $J = 2$. With $S = 1$, we can have

$L = 1, 2,$ or 3. Since the tensor force cannot mix parity, the $L = 2$ partial wave remains uncoupled, while the other two, with $L = J - 1$ and $L = J + 1$, are mixed in any state, namely,

$$\Psi_J = \frac{1}{r} [U_J(r) \mathscr{Y}^M_{J,1,J-1} + W_J(r) \mathscr{Y}^M_{J,1,J+1}]. \tag{5-24}$$

In the asymptotic force-free region, each term corresponds to a simple partial wave for $L = J - 1$ or $L = J + 1$. We can then write

$$U_J(r) \sim A_1 \exp\{- i[kr - (J - 1)\pi/2]\} - B_1 \exp\{i[kr - (J - 1)\pi/2]\},$$

$$W_J(r) \sim A_2 \exp\{- i[kr - (J + 1)\pi/2]\} - B_2 \exp\{i[kr - (J + 1)\pi/2]\}.$$

$$\tag{5-25}$$

In this case also, we can define a scattering matrix by the matrix equation

$$\begin{pmatrix} B_1 \\ B_2 \end{pmatrix} = \begin{pmatrix} S_{11} & S_{12} \\ S_{21} & S_{22} \end{pmatrix} \begin{pmatrix} A_1 \\ A_2 \end{pmatrix}, \tag{5-26}$$

where S_{11}, S_{12}, etc., are complex. There are thus eight real parameters in the S-matrix. But conservation of incoming and outgoing flux requires $B^+B = A^+A$, where

$$B = \begin{pmatrix} B_1 \\ B_2 \end{pmatrix} \quad \text{and} \quad B^+ = (B_1{}^*B_2{}^*).$$

Combining this with Eq. 5–26, we obtain

$$S^+S = 1, \tag{5-27}$$

which means that S is unitary. This gives us three equations and reduces the number of parameters to five. If we now invoke time-reversal invariance, which essentially amounts to interchanging the incoming and outgoing waves and taking the complex conjugate of the amplitudes, we get

$$A^* = SB^*. \tag{5-28}$$

When this is combined with $B = SA$, we find that the S-matrix has to be symmetric, that is,

$$S_{12} = S_{21}, \tag{5-29}$$

which provides two more conditions. Thus there are only three independent real parameters in the S-matrix describing scattering in a coupled triplet state of specified J and parity. These are taken to be the two phase shifts and a coupling parameter ε_J.

There are various ways of defining the S-matrix in terms of these parameters, but the one that is most extensively used is due to Stapp, Ypsilantis, and Metropolis [15]. In this definition,[†] the S-matrix for a given J and parity is

$$
S_J = \begin{pmatrix} \exp(i\delta_{J\alpha}) & 0 \\ 0 & \exp(i\delta_{J\beta}) \end{pmatrix} \begin{pmatrix} \cos 2\varepsilon_J & i \sin 2\varepsilon_J \\ i \sin 2\varepsilon_J & \cos 2\varepsilon_J \end{pmatrix} \begin{pmatrix} \exp(i\delta_{J\alpha}) & 0 \\ 0 & \exp(i\delta_{J\beta}) \end{pmatrix}.
$$

(5-30)

Phase shifts so defined are often called "bar" phases, because in early papers they were written as $\bar\delta$. It is incorrect to consider $\delta_{J\alpha}$ and $\delta_{J\beta}$ as the phase shifts for $L = J - 1$ and $L = J + 1$, respectively, since neither of these is an eigenstate of the Hamiltonian, and the complex of three quantities, ε_J, $\delta_{J\alpha}$, $\delta_{J\beta}$, is required to prescribe the possible scattered states. Only in the limit of $E \to 0$ does $\varepsilon_J \to 0$, and then $\delta_{J\alpha} = \delta(L = J - 1, J = J)$, $\delta_{J\beta} = \delta(L = J + 1, J = J)$. However, in the current literature, this loose description is often used. For example, $\delta_{2\alpha}$ and $\delta_{2\beta}$ are denoted by the symbols ${}^3P_2$ and ${}^3F_2$, a notation that we shall also follow. We should mention here that in the above analysis we have not considered inelastic scattering, which may occur for energies above the meson-production threshold, that is, $E_{\text{lab}} > 280$ MeV. At energies up to $E_{\text{lab}} \approx 400$ MeV, meson production is so small that the S-matrix may be approximated as unitary in the elastic channel.

At this point, we stress that the analysis of the N-N interaction is a complicated procedure that advances in several steps. The first important step is to measure the elastic-scattering data. For the range of $E_{\text{lab}} = 0$–400 MeV, there are over 2000 items of experimental data involving p-p and n-p scattering. These are measurements of differential cross section, polarization, triple-scattering parameters, and spin-spin correlation coefficients at various energies and angles. The second step is to find a unique set of phase shifts that. can reproduce the experimental data consistently. Finally, an attempt is made to construct, either theoretically or phenomenologically, models (like a N-N potential) that can yield the same phase shifts.

[†] A different convention sometimes used is due to J. M. Blatt and L. C. Biederharn, *Phys. Rev.* **86**, 399 (1952). Transformation formulas relating the Blatt-Biederharn phases and the "bar" phases are given by Stapp et al. [15]. Unless stated specifically to the contrary, we shall use the bar phases.

In order to appreciate the logic of these steps, it is useful to gain more understanding of the scattering matrix and its relation to experiments. Although we have seen that the S-matrix at a given energy and for specified values of J and parity is determined by three real parameters, it does not follow that only three independent experiments are needed to specify the complete S-matrix at a given energy because, for a beam of high-energy particles, many possible L- (and consequently J-) values are involved, and all these can contribute to the total S-matrix. A particularly simple situation exists for very low-energy scattering where only the $L = 0$ wave contributes and the phases are essentially determined by cross-section measurements alone. But for higher energies, it is more convenient to express the experimental results in terms of a matrix M involving the incident energy k^2 and the scattering angle θ in the CM frame.

In Section 2–3 we considered scattering with a central force in the S-state, and found that (see Eq. 2–29) the scattered wave could be written as $(e^{ikr}/r)[(1/2ik)(e^{2i\,\delta_0(k)} - 1)]$. In this case, we could have alternatively written

$$u_{sc} = \frac{e^{ikr}}{r} M(\mathbf{k}, \mathbf{k}'), \quad k^2 = k'^2, \tag{5-31}$$

with

$$M(\mathbf{k}, \mathbf{k}') = \frac{1}{2ik} \langle \mathbf{k}|(S-1)|\mathbf{k}'\rangle. \tag{5-32}$$

For the S-state in particular, there was no θ-dependence in M. In the more complicated case in which the spin has to be taken into account explicitly, we can generalize this by writing the asymptotic wave function as

$$u \sim e^{ikz}\chi_s{}^{m_s} + \frac{e^{ikr}}{r} \sum_{m_s'} \langle \chi_s{}^{m_s'}|M(k^2,\theta)|\chi_s{}^{m_s}\rangle \chi_s{}^{m_s'}. \tag{5-33}$$

Here $M(k^2, \theta)$ is a matrix in the spin space, related to the total S-matrix as before, and describes the scattering through an angle θ, of an incident particle of spin (s, m_s) with momentum k along the z-axis (J is unspecified), its spin state changing from $\chi_s{}^{m_s}$ to $\chi_s{}^{m_s'}$.

The number of independent parameters in M connected with the change of spin orientations in a collision depends on the intrinsic spin of the colliding particles. For illustration, take the simple case of a nucleon scattering off a spin-zero target like ^{12}C. Then the matrix M can be expanded in the complete set of operators of the incident-particle spin space, which are just the unit matrix and $\boldsymbol{\sigma}$. Thus

$$M = B + \mathbf{C} \cdot \boldsymbol{\sigma},$$

where B and $\mathbf{C}$ are functions of (k^2, θ). Since M has to be invariant under rotation, it must be a scalar. But $\boldsymbol{\sigma}$ is a pseudoscalar; therefore so should be $\mathbf{C}$. The only pseudoscalar other than $\boldsymbol{\sigma}$ in this problem is $\mathbf{k} \times \mathbf{k'}$, where $\mathbf{k}$ and $\mathbf{k'}$ are the initial and final wave vectors in the CM frame. Hence we can finally write

$$M = B(k^2, \theta) + C(k^2, \theta)\boldsymbol{\sigma} \cdot \mathbf{n} \tag{5-34}$$

with $\mathbf{n} = (\mathbf{k} \times \mathbf{k'})/|(\mathbf{k} \times \mathbf{k'})|$ being a unit vector normal to the plane of scattering. In Eq. 5–34 there are four independent real parameters at each energy and angle, since B and C can each be complex. One of these can be regarded as an overall phase factor, leaving three real parameters to be determined by experiments.

In the more complicated case of N-N scattering, similar analysis shows that nine parameters[†] are required to specify M. However, the unitarity requirement on S (and therefore M) reduces the number to five. Thus, at any given energy, five independent experiments must be performed at each angle to determine the scattering matrix completely, for each i-spin state ($T = 0$ or 1).

5-6 EXPERIMENTAL RESULTS AND PRELIMINARY INTERPRETATIONS

Data on nucleon-nucleon scattering are presented in this section in graphical form, based mainly on a summary compiled by Hess [18]. Figure 5–6 shows the differential n-p cross section at various energies from 14 to 580 MeV. Experimental points are not shown, but the curves are drawn through them. Typical errors are of the order of 10%, so that fine details of the curves are not reliable; for example, curves for different energies probably do not cross each other, although there are a few instances in which they appear to do so. The threshold for π-meson production is 290 MeV. Below this energy only elastic scattering is possible, but above it the total and the elastic cross sections differ. The data on total cross section are summarized

---

† More precisely, invariance requirements specify nine parameters for proton-proton scattering. When the nucleons are not identical, an extra term proportional to $(\boldsymbol{\sigma}_i - \boldsymbol{\sigma}_t) \cdot \mathbf{n}$ may occur in the scattering amplitude, where $\boldsymbol{\sigma}_i$ is the spin of the incident particle and $\boldsymbol{\sigma}_t$ that of the target. This term would imply that polarized neutrons would scatter from hydrogen differently than polarized protons from target neutrons. Both experiments have been done [16, 17], and the results are similar. Also, on the basis of charge symmetry, the extra term should not be present since charge symmetry implies no change if all neutrons and protons are interchanged.

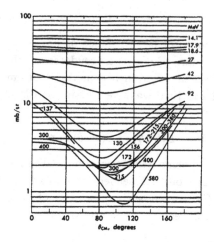

Fig. 5-6 Experimental values of the differential n-p cross section at various energies. (After Hess [18].)

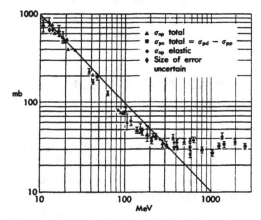

Fig. 5-7 Experimental values of total and elastic n-p scattering. (After Hess [18].)

in Fig. 5-7 [18]. The points labeled σ_{np} are from experiments in which neutrons are incident on hydrogeneous targets. Those labeled σ_{pn} are from an experiment in which protons are incident on deuterium, and the proton-proton cross section is subtracted. This is only an approximation to σ_{np} since it ignores interference terms in the deuteron scattering. The approximation improves, however, as the energy increases, and at the energies shown the correction is about 6 mb. The straight line is proportional to $1/E$.

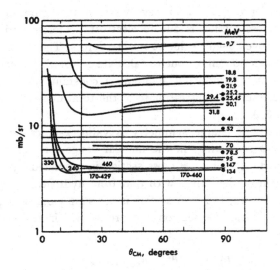

Fig. 5-8 Experimental values of the differential p-p cross section at various energies. (After Hess [18].)

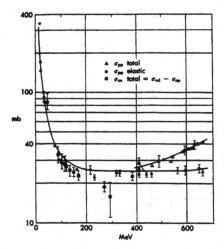

Fig. 5-9 Experimental values of the "total" and "elastic" p-p nuclear cross sections, that is, with Coulomb scattering subtracted, and of the n-n elastic cross section, as determined approximately from $\sigma_{nd} - \sigma_{np}$. (After Hess [18].)

For p-p scattering, the differential cross sections are shown in Fig. 5–8 for energies up to 500 MeV; this is as high as we need go, since we intend to restrict our attention to nuclear structure. We remarked in Chapter 2 that, as the energy increases, the effects of Coulomb scattering are confined to smaller and smaller angles. This is clearly seen in the figure, in which the large small-angle cross section is due to electrostatic forces and the cross section at angles above 20–40° is due to nuclear forces. The marked flatness of the nuclear cross section allows one to calculate a "total" p-p cross section for nuclear forces only, simply by assuming a constant value of $\sigma(\theta)$ for each energy. The results of this estimate of the total nuclear cross section are shown in Fig. 5–9.

If the hypothesis of charge symmetry is valid, n-n scattering should be the same as p-p scattering, in the angular range for which Coulomb effects are unimportant. The n-n cross section is estimated by comparing neutron scattering from deuterium with that from hydrogen. For neutrons scattered from deuterium, the cross section is not simply $\sigma_{nn} + \sigma_{np}$, because the waves scattered from the neutron and proton will interfere coherently and, consequently,

$$\sigma_{nd}(\theta) = \sigma_{nn}(\theta) + \sigma_{np}(\theta) + I(\theta).$$

The interference term $I(\theta)$ is most important at low energies and small scattering angles. At 300 MeV, $I(\theta)$ is estimated to fall rapidly from 0.6 mb/sr at 40° to one-tenth this value at 90°. Figure 5–10 shows $\sigma_{nd} - \sigma_{np}$ for angles above 30°. It is seen that the probable errors exceed the estimated values

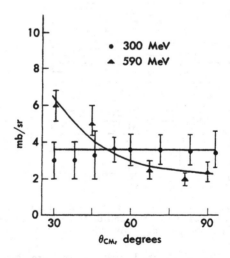

Fig. 5-10   The points show the experimental values of the differential n-n cross section at 300 and 590 MeV, estimated from $\sigma_{nd}(\theta) - \sigma_{np}(\theta)$. The curves show $\sigma_{pp}(\theta)$ at 300 MeV (from Fig. 5–8) and at 590 MeV. Points are available at 300 MeV at angles less than 30°, but these are not shown since at these angles $I(\theta)$ is important for σ_{nd} and the Coulomb scattering is important for σ_{pp}.

of $I(\theta)$. Also, one can estimate that the total cross section may have an error of 6 or 7 mb at about 300 MeV. Some points of σ_{nd} (total) — σ_{np} (total) are shown in Fig. 5–9. Comparison of the neutron-neutron and proton-proton data in Figs. 5–9 and 5–10 gives additional support to the assumption of charge independence, already suggested by the low-energy results.

We shall now consider some of the salient features of these results. An outstanding characteristic is the high probability in neutron-proton collisions that the neutron is scattered backward in the center-of-mass system. For ordinary forces, this is impossible as soon as the bombarding energy becomes large compared with the average potential, since the proton, initially at rest in the laboratory, can pick up energy only through the internucleon force and hence acquires kinetic energy of the same order of magnitude as the nuclear potential energy. It will be recalled that the results at low energy suggested a figure of less than about 50 MeV for the average potential. Consequently, in an elastic collision at energies over 100 MeV, the scattered neutron would have more energy than the recoil proton and, by the conservation laws for a collision, would be scattered forward with a deflection less than 45° in the laboratory or 90° in center-of-mass coordinates.

Reference to Fig. 5–6 indicates that, on the contrary, neutrons are scattered as strongly through 180° as through 0°. Indeed, above about 100 MeV, the back scattering is the more probable. The position of the minimum cross section varies slowly with energy and does not shift from 90° by more than about 10° for energies up to 400 MeV.

This is a clear indication of the presence of exchange forces, for if neutron and proton interchange identity in the collision, backward scattering will be expected rather than forward scattering. Hence an approximately equal mixture of ordinary and exchange forces would produce the observed results. One might expect the exchange force slightly to exceed the ordinary force because $\sigma(\pi) > \sigma(0)$, but in turning from general features to matters of detail, precise calculations become necessary. In any event, it is evident that large exchange forces must exist.

Before the experiments above 100 MeV were done, Robert Serber proposed an equal mixture of ordinary and Majorana forces:

$$V(r)(1 + P_x). \qquad (5\text{–}35)$$

Since $P_x = -1$ in states of odd parity, the Serber hypothesis implies no interaction between nucleons in states of odd L, with consequent symmetry about 90°. We shall see below that forces do act in P- and F-states, but they are weaker than in S-, D-, and G-states. Nevertheless, because of its simplicity, the Serber force has been used in many calculations.

Total n-p and p-p cross sections are markedly different. The n-p cross section follows roughly the $1/E$ law ($\sigma \propto \lambda^2$) up to energies at which inelastic processes are important. The small departures from $1/E$ are attributable to the variation of the phase shifts of the different angular-momentum states as these become important. On the other hand, the p-p elastic cross section is amazingly constant at about 23 mb from 150 to 600 MeV. Above this energy it decreases. Moreover, the p-p cross section has very little angular dependence at any energy except in the small-angle Coulomb scattering region.

This constant flat differential cross section is evidence that we are not dealing with an ordinary attractive force. Of course an S-wave gives an isotropic scattering, and it might be supposed that no forces are evident in the higher-angular-momentum states, even up to 400 MeV. However, this hypothesis is untenable, since the maximum total cross section for an S-wave is $2\pi\lambda^2$ or about 11 mb, which is only half the observed value. The differential p-p scattering cross section is given in Appendix B as

$$\sigma_{pp}(\theta) = |f(\theta)|^2 + |f(\pi - \theta)|^2 - \text{Re}\,[f^*(\theta)f(\pi - \theta)], \qquad (5\text{--}36)$$

where

$$f(\theta) = f_C(\theta) + f_N(\theta), \qquad (5\text{--}37)$$

$f_C(\theta)$ is the Coulomb scattering amplitude, and f_N contains all the effects of nuclear scattering. The formula

$$f_N(\theta) = -\frac{1}{2ik} \sum_l (2l + 1)e^{2i\eta_l}(e^{2i\delta_l} - 1)P_l(\cos\theta), \qquad (5\text{--}38)$$

in which η_l is a Coulomb phase shift and δ_l is a nuclear phase shift, makes clear that the nearly isotropic distribution at high energies must be due to a remarkable compensation among the coefficients of the Legendre polynomials. In the energy region where only S-, P-, and D-waves need be considered, the cross section is made up of an isotropic term proportional to $\sin^2 \delta_0$ due to the ^1S-wave, a term proportional to $\sin^2 \delta_2 \,[P_2(\cos\theta)]^2$ due to the ^1D-wave, a term proportional to $\sin \delta_0 \sin \delta_2$ and $P_2(\cos\theta)$ due to S-D interference, and terms due to the ^3P-wave. For central forces, a P-wave gives a contribution proportional to $\cos^2\theta$, but when tensor forces act, the various substates $^3P_0$, $^3P_1$, and $^3P_2$ have different phase shifts, and a term in $|Y_1^1|^2$ or $\sin^2\theta$ appears. Also, the $J = 2$ state is a mixture of $^3P_2$ and $^3F_2$, and interference terms appear in $\sigma(\theta)$ proportional to the products of first- and third-order spherical harmonics. At energies at which the F-wave separately would have negligible phase shift, the coupling due to the tensor force makes the $^3F_2$-contribution appreciable, and squares of third-order harmonics also appear.

We are now able to see how a flat cross section might arise. The major contributions to $\sigma(\theta)$ come from the ^1S- and ^1D-states, since the force in states of odd L is comparatively weak. The function $P_2(\cos \theta)$ has a zero at 55° and is negative at 90°. Hence the D-wave contributions are as shown in Fig. 5–11. To obtain $\sigma(\theta)$ we combine curves (a) and (b) with the constant S-wave contribution, in proportions determined by δ_0 and δ_2. It is evident that, if curve (a) is added in with a positive coefficient, forward scattering will predominate. However, if curve (a) is added with a negative coefficient, the scattering at 90° can be built up and a curve of the general nature of (c) or (d) can be obtained for $\sigma(\theta)$. This is still not the observed isotropic scattering, since there is a minimum around 45° due to the zero of P_2, but a tensor force acting in the P-states will fill in the minimum. A central-force P-wave merely builds up the forward scattering with a $(\cos^2 \theta)$-term, but the $\sin^2 \theta$ provided by the interference waves between the different P-states

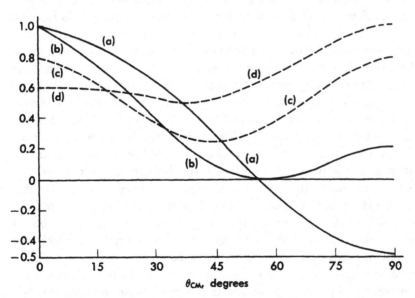

Fig. 5-11 Contributions of S- and D-waves to p-p scattering. Curve (a) is $P_2(\cos \theta)$, curve (b) is $[P_2]^2$, curves (c) and (d) are arbitrary combinations which might represent $\sigma(\theta)$. Curve (c) is $0.3 + P_2(P_2 - 0.5)$, and curve (d) is $0.6 + P_2(P_2 - 1)$.

tends to smooth out this effect, and the P-F interference terms provide angular functions, similar to $\sin^2 \theta \cos^2 \theta$, which maximize around $45°$.

These considerations show that the isotropic proton-proton cross section can be understood on the basis of two essential hypotheses, stipulating (1) a force such that the coefficient of the S-D interference will be negative and (2) tensor forces in the triplet states to build up the central portion of the distribution. It is easy to believe that the relative strengths of the various forces involved might be such as to produce isotropy at one energy; what is not so simple is to achieve it at all energies. As the energy rises, D-scattering should become more important, producing a steady increase in δ_2 and hence a large contribution from curve (b) in Fig. 5–11. In compensation the coefficient of curve (a) must change by the right amount. This coefficient involves the factor $\sin \delta_0 \sin \delta_2$, which consequently must be negative at sufficiently high energies and must decrease as the energy increases.

It was pointed out by Jastrow [19] that δ_0 would have this behavior if the nucleon force had a repulsive core of suitably small radius, surrounded by an attractive potential. For purposes of illustration, let us consider a P-wave in such a potential. The P-wave at 200 MeV senses only the potential outside approximately 0.5 fm. Hence, if the repulsive core is inside this radius, the P-wave will be scattered only by the attractive potential and will have a positive phase shift. As the energy increases and the repulsive potential begins to influence the scattering, the phase shift decreases, and as the energy increases still more, a point will be reached at which the repulsive scattering is predominant and the phase shift actually becomes negative. A very considerable energy is required before this sign change occurs for P-waves or waves of higher L. However, S-waves sense the whole potential at all energies. At low energies, the S-state wave function, which vanishes at the origin, has appreciable amplitude only at distances in the outer attractive region, and hence the S-phase shift is positive. As the energy increases, the wave function is compressed into smaller distances, the effect of the repulsive region becomes more important, and the phase shift, which has been an increasing function of energy, begins to decrease and eventually becomes negative. The energies at which the phase shift reverses its slope and at which it becomes negative are sensitive to the details of the potential shapes, but with potentials that agree with experiment the ^1S-phase shift reaches its maximum between 10 and 20 MeV and becomes negative somewhere around 200 MeV. Thus the uniform differential cross section over the whole region of 150–450 MeV may be understood in terms of the steady changes of phase shift attributable to a potential with a repulsive core.

Other types of forces could introduce angular functions with maxima at 90° and 45°. A force of the form $V(r)\mathbf{L} \cdot \mathbf{S}$ does not mix states of different L, as does the tensor force, but it does yield different scattering for states of the same L and different J. With the phase shifts depending on both L and J, terms in $Y_L^1 Y_{L'}^{1*}$ and $Y_L^2 Y_{L'}^{2*}$ are introduced into $\sigma(\theta)$. Of course, the F-wave is not enhanced at lower energies as it is by the tensor force, and it is reasonable to include the P-scattering while ignoring the P-F interference. The $\mathbf{L} \cdot \mathbf{S}$ operator is zero in singlet states, and for triplet states it is $\frac{1}{2}[J(J+1) - L(L+1) - 2]$. Thus the ratios of the $\mathbf{L} \cdot \mathbf{S}$ forces in the three P-states are $F(^3P_2) : F(^3P_1) : F(^3P_0) = 1 : -1 : -2$. If no other forces act in the P-states, the $\mathbf{L} \cdot \mathbf{S}$ forces produce an angular distribution of the form $a + \sin^2 \theta$, where the constant a is small compared to 1 and vanishes in the Born approximation[†] [20]. Clearly this distribution builds up the 90° cross section. However, it has not been possible to produce an isotropic $\sigma_{pp}(\theta)$ over the whole observed energy range by $\mathbf{L} \cdot \mathbf{S}$ forces alone without the introduction of a repulsive core.

In turning to the results of polarization experiments, we find that an $\mathbf{L} \cdot \mathbf{S}$ force is indicated. This statement is made in the context of an energy-independent potential defined even for very short distances of approach of two nucleons. There are serious questions as to how far this concept should be trusted. These doubts will be discussed after we have explained the results.

We have reached three important conclusions:

1. There is a large exchange force.

2. The singlet force, at least, has a repulsive core.

3. In proton-proton scattering the three P-states have different phase shifts, thus eliminating the minimum in $\sigma(\theta)$ around 45°; we have attributed this P-state splitting to a tensor force, particularly because $^3P_2$-$^3F_2$ coupling will help appreciably to make $\sigma(\theta)$ isotropic.

[†] In Born approximation, the phase shift due to a potential $V(r)$ acting in a state of angular momentum l is

$$\delta = -\left(\frac{Mk}{\hbar^2}\right) \int r^2 V(r) f_l^2(kr) \, dr. \tag{5-39}$$

Although the applicability of this formula is restricted to certain energy regions whose precise limits are determined by l and by the nature of V, the Born approximation provides a useful guide in qualitative discussions. For n-p scattering the function f_l is the spherical Bessel function j_l; for p-p scattering it is the corresponding Coulomb function.

These conclusions have been reached without detailed analysis of the experimental data. To make further progress and, indeed, to confirm our conclusions, exact quantitative fits must be found. For example, is a tensor force in the P-state really indicated, or would the splitting due to an $\mathbf{L} \cdot \mathbf{S}$ force produce sufficient cross section at $45°$? The answers to such questions require exact numerical investigation.

5-7 MORE PRECISE ANALYSIS OF EXPERIMENTAL DATA

The most instructive way of presenting the experimental data is to analyze them for the phase shifts of all possible states in a given energy range. We have mentioned that there is a vast body of experimental measurements in the range $E_{lab} = 0$–450 MeV and that inelastic effects there are small. A considerable number of data have been accumulated also for the energy range 450–750 MeV, and a phase-shift analysis has been attempted [21] after making corrections for inelastic effects. In this book, we shall discuss briefly the results for the energy range 0–450 MeV.

Phase-shift analyses can be done either for the data at a fixed energy or for the data at many different energies. In the single-energy analysis, the aim is to determine a set of phases δ for all states up to a maximum $L = L_{max}$. For $L > L_{max}$, the phases are assumed to be generated by the one-pion-exchange potential. The method of analysis is then to search for sets of phases up to L_{max} which minimize the least-square sum

$$M = \frac{1}{n} \sum_{i=1}^{n} \left[\frac{y_i(\delta) - y_i(\text{exp})}{e_i} \right]^2, \qquad (5\text{--}40)$$

where the y_i are the n quantities observed at that particular energy, $y_i(\delta)$ is the value calculated for a specific set of phase shifts δ, $y_i(\text{exp})$ is the experimental value, and e_i is the probable error in $y_i(\text{exp})$. Such single-energy analyses have been performed at many different energies.

The alternative method is to do an energy-dependent analysis, in which the experimental data in an entire energy range are analyzed simultaneously. In this method, each phase is expanded in terms of some set of energy-dependent functions:

$$\delta(E) = \alpha \delta^{\text{OPEP}}(E) + \sum_{i=1}^{N} \beta_i f_i(E), \qquad (5\text{--}41)$$

where α and β_i are adjustable parameters for states $L < L_{max}$, and $\alpha = 1$, $\beta_i = 0$ for other states. The OPEP value constitutes the leading term, but α remains adjustable because a phase-shift analysis is a method of presenting experimental data and should not prejudge theoretical results.[†] The functions $f_i(E)$ may be chosen with some guidance from theory (the Livermore group [22] does so), but what is of basic importance is that there be enough of them to permit expression 5-41 to represent any smooth function $\delta(E)$. MacGregor, Arndt, and Wright [23] obtain precision fits to 1076 items of p-p data in the range 1–450 MeV by having 26 adjustable parameters in their analysis. They search for all phase parameters up to $L = 5$ and use V_{OPEP} for the higher partial waves. A similar analysis has also been done by the same authors for 990 items of n-p data in the same energy range. Many other groups, notably those of Breit [24], Signell [25], and Kazarinov [27], as well as Hoshizaki [26] and Perring [28], have done phase-shift analyses of the N-N data.

Two main conclusions follow from these extensive analyses: (1) the $T = 1$ scattering matrix is very accurately determined up to $E_{\mathrm{lab}} = 450$ MeV, and (2) the $T = 0$ scattering matrix is also fairly well determined up to $E_{\mathrm{lab}} = 450$ MeV, the accuracy being good at energies between 100 and 200 MeV. Figures 5–12a and 5–12b show the phases up to $L = 2$ that have been extracted by MacGregor et al. [23]. The coupled phases are the nuclear "bar" phase shifts, as defined by Eq. 5–30. It should be noted that single-energy analyses at various energies between 25 and 330 MeV yield results in general agreement with the energy-dependent analysis. Because of deviation from charge independence, the p-p and n-p phases differ slightly. The accuracy of the analyses is such that MacGregor et al. [23] are able to give separate sets of $^1S_0$-phases for p-p and n-p scattering. In Fig. 5–12a, only the p-p $^1S_0$-phase shifts are plotted.

These studies also reveal that in this entire energy range the phases for $L = 4$ and 5 strongly resemble the phases produced by V_{OPEP} alone. Indeed, these phases can be very well fitted by taking the coupling constant g^2 in V_{OPEP} to be a free parameter. The best fit is obtained for a value of g^2 very close to the value of $g^2 = 15$ obtained from π-N scattering data, in accordance with the predictions in Section 5–3.

From Fig. 5–12, it will be seen that both the $^1S_0$- and the $^3S_1$-phases change sign at higher energies, confirming that at short distances the nuclear force is repulsive. The $^1S_0$-phase shift changes sign around $E_{\mathrm{lab}} = 250$ MeV, while the $^3S_1$-phase shift becomes repulsive at about 370 MeV, indicating that the potential is more attractive in this state. This is also manifested in

---

[†] It is important, however, to ensure the correct low-energy behavior of these phases. For example, the parameters of the $L = 0$ phase should be adjusted to give the correct scattering length and effective range at low energies.

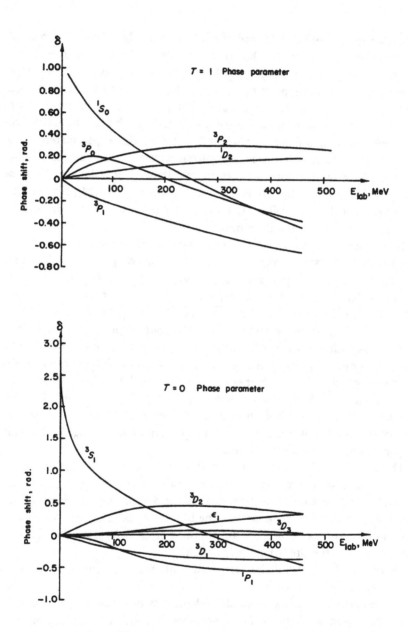

Fig. 5-12 Nuclear phase parameters for nucleon-nucleon scattering for $L \leqslant 2$. In (a) the phase shifts are for the $T = 1$ channel; in (b) the phases are the nuclear "bar" phase shifts for the $T = 0$ channel as defined in Eq. 5–30. The values are taken from the multienergy analysis of MacGregor et al. [23], and continuous curves are drawn through them. The error bars are small and are not shown.

the extreme-low-energy behavior of the phases: as $E_{lab} \to 0$, $\delta(^1S_0) \to 0$, whereas $\delta(^3S_1) \to \pi$, implying that there is a bound state in $^3S_1$.

At low energies ($E_{lab} \leqslant 20$ MeV), it will also be seen that $\delta(^3P_2) \ll |\delta(^3P_1)| < \delta(^3P_0)$. Furthermore, $\delta(^3P_2)$ is positive throughout, $\delta(^3P_1)$ is always negative, and $\delta(^3P_0)$ goes from positive to negative around 200 MeV. We have already mentioned that, for energies up to 20 MeV, the P-wave is essentially scattered by the force outside 2 fm, which is dominantly V_{OPEP}. MacGregor [29] calculated the Born approximation to the ^3P- and ^1D-phase shifts using V_{OPEP} and found that 98% of $\delta(^1D_2)$ comes from the region outside 2 fm; the corresponding figures for the three ^3P-states with $J = 0, 1$, and 2 are 70, 70, and 50%, respectively. The behavior of these phases at low energies can therefore be explained by the long-range part of V_{OPEP}.

This behavior is, of course, representative of a mixture of a fairly weak central force and a dominant tensor force. If we assume a potential $V_C + V_T S_{12}$, there are simple expressions for the uncoupled and the coupled bar phases in the Born approximation [15]. Let δ_C and δ_T denote the Born phases that would be produced by the potentials V_C and V_T, respectively. Then, since the expectation values of S_{12} in the three ^3P-states are -4, 2, and $-\frac{2}{5}$, we have

$$\delta(^3P_0) = \delta_C - 4\delta_T, \qquad \delta(^3P_1) = \delta_C + 2\delta_T, \qquad \delta(^3P_2) = \delta_C - \tfrac{2}{5}\delta_T. \qquad (5\text{-}42)$$

If V_C and V_T are taken to be positive, δ_C and δ_T are negative and it follows that

$$\delta(^3P_0) > \delta(^3P_2) > \delta(^3P_1). \qquad (5\text{-}43)$$

The positive signs for V_C and V_T are predicted by V_{OPEP} in the ^3P-states, since $S = 1$ and $T = 1$, with $(\boldsymbol{\sigma}_1 \cdot \boldsymbol{\sigma}_2) = (\boldsymbol{\tau}_1 \cdot \boldsymbol{\tau}_2) = 1$. We further note that $\delta(^3P_1)$ in Eq. 5–43 is always negative, whereas, so long as δ_C is sufficiently small, $\delta(^3P_0)$ and $\delta(^3P_2)$ can still be positive, in conformity with the experimental findings.

At higher energies, the three ^3P-phases are at direct variance with the behavior shown in Eq. 5–43. Although the $^3P_1$-phase shift remains negative throughout, the $^3P_0$- and $^3P_2$-phases cross over near 100 MeV; indeed, the $^3P_0$-phase shift changes from positive to negative at about 200 MeV. Such behavior cannot be explained by $V_C + V_T S_{12}$ alone, but may be reproduced by invoking a spin-orbit force $V_{LS}\mathbf{L} \cdot \mathbf{S}$, which is of short range and therefore becomes operative at higher energies. With a potential of the form $V_C + V_T S_{12} + V_{LS}\mathbf{L} \cdot \mathbf{S}$, the Born phase shifts are given by

$$\delta(^3P_0) = \delta_C - 4\delta_T - 2\delta_{LS},$$

$$\delta(^3P_1) = \delta_C + 2\delta_T - \delta_{LS},$$

$$\delta(^3P_2) = \delta_C - \tfrac{2}{5}\delta_T + \delta_{LS}. \tag{5-44}$$

We see that, if δ_{LS} is positive, corresponding to a negative V_{LS}, then, when it is sufficiently large, $\delta(^3P_0)$ may change from positive to negative, while $\delta(^3P_2)$ continues to remain positive. The effect of this spin-orbit term is apparent even around 50 MeV, when $\delta(^3P_0)$ changes its slope. Such a spin-orbit force is generated, not by the one-pion-exchange process, but only by the exchange of heavier bosons of spin 1, and is therefore largely restricted within a radius of about 0.7 fm.

Finally, a word of caution is necessary in regard to the interpretation of phase shifts extracted from multienergy analyses. Most of the experimental data are clustered about 10, 25, 50, 95, 140, 210, and 320 MeV. At each of these energies, single-energy analyses have been done, and these are model independent. Multienergy analyses restrict the phase shifts and energy to a narrow band of values, but since the data are at rather widely separated energies, this restriction may be artificial. For example, at 95 MeV, the $^3P_0$-phase shift, according to single-energy analysis, is $11.17 \pm 2.15°$, whereas multienergy analysis gives $10.30 \pm 0.23°$. Thus the error limits in multi-energy analyses should not be taken literally [30].

5-8 PHENOMENOLOGICAL NUCLEON-NUCLEON POTENTIALS

We have seen how a more or less unique set of phase shifts has been extracted in the laboratory energy range 0–450 MeV to correlate a large body of elastic N-N scattering data. This knowledge, however, is insufficient to determine a two-nucleon potential uniquely. Even in the simplest case of a local central potential without a bound state, it is necessary to know the phase shifts of a given partial wave at all positive energies to determine the potential uniquely. In spite of this ambiguity, it is convenient to construct N-N potentials that can reproduce these experimental phases because in the non-relativistic region we have a dynamical model, namely, the Schrödinger equation, which can be applied to any nuclear problem, provided that the potential is known. In atomic physics, for example, the successes of the Coulomb potential are well known. In the nuclear case, although many different potentials can be constructed that fit the experimental phase shifts equally well from 0 to 450 MeV they yield different results in a many-body problem. For two nucleons scattering in the presence of a third nucleon, the scattering need not be elastic, since some energy may be taken away by the third particle. Thus "off-the-energy-shell" matrix elements of the interaction play an important role in determining the properties (like binding energy or equilibrium density) of a many-body system. One motivation for studying nuclear matter (see Chapter 9), for example, is to try to differentiate between potentials that yield the same phase shifts up to about 400 MeV.

Fortunately, some theoretical guide lines restrict the form of the N-N potential. First, there are the invariance principles, which state that the potential should remain invariant under symmetry operations like translation of space or momentum coordinates, rotation and reflection of the space axes, and time reversal. Additionally, using charge independence, the neutrons and protons may be treated on an equal footing for each i-spin state $T = 0$ and 1, so that the nuclear potential must be symmetric with respect to the exchange of the two nucleons. Finally, since we are considering the energy range in which inelastic processes can be ignored, the potential must be unitary. These principles were exploited by Eisenbud and Wigner [31] and later generalized by Okubo and Marshak [32]. The most general functional form of the potential may be written as

$$V = V_C + V_{LS}\mathbf{L} \cdot \mathbf{S} + V_\sigma \boldsymbol{\sigma}_1 \cdot \boldsymbol{\sigma}_2 + V_T S_{12} + V_{\sigma p}(\boldsymbol{\sigma}_1 \cdot \mathbf{p})(\boldsymbol{\sigma}_2 \cdot \mathbf{p}) + V_{\sigma L} Q_{12},$$

$$(5\text{-}45)$$

where each coefficient V_i can itself be written as

$$V_i = V_i{}^0(r, p, L) + V_i{}^\tau(r, p, L)\boldsymbol{\tau}_1 \cdot \boldsymbol{\tau}_2. \tag{5-46}$$

In the above equations, $\mathbf{r} = \mathbf{r}_1 - \mathbf{r}_2$, $\mathbf{p} = \frac{1}{2}(\mathbf{p}_1 - \mathbf{p}_2)$,

$$\mathbf{L} \cdot \mathbf{S} = \frac{1}{2}(\mathbf{r}_1 - \mathbf{r}_2) \times (\mathbf{p}_1 - \mathbf{p}_2) \cdot (\boldsymbol{\sigma}_1 + \boldsymbol{\sigma}_2),$$

and

$$Q_{12} = \frac{1}{2}[(\boldsymbol{\sigma}_1 \cdot \mathbf{L})(\boldsymbol{\sigma}_2 \cdot \mathbf{L}) + (\boldsymbol{\sigma}_2 \cdot \mathbf{L})(\boldsymbol{\sigma}_1 \cdot \mathbf{L})].$$

It turns out that on the energy shell the term containing $V_{\sigma p}$ can be expressed in terms of the other terms in V.

In addition to the above general principles, we have some guidance from meson theory. Early efforts to use field theory to derive the N-N interaction through meson exchanges enjoyed only limited success, since multiple pion exchanges are very important for $r < 2$ fm and are difficult to calculate unambiguously. In the next section, we shall discuss this approach in more detail. We have seen from the phase-shift analyses of higher partial waves, however, that the tail of the *N-N* potential (for $r > 3$ fm) should be of one-pion-exchange character. More progress in this direction has been made with the discovery of multipion resonances, and we shall discuss these later. But even with a completely phenomenological approach, the tail of the potential should be taken as that of OPEP. We now discuss some of the recently developed phenomenological potentials.

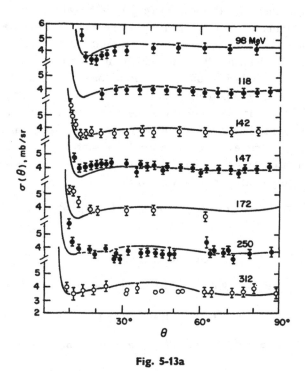

Fig. 5-13a

In Sections 5–6 and 5–7, we made analyses of the scattering data, keeping the potential model in mind. We indicated how the data demand that there be short-range repulsion, exchange force of almost Serber character, and tensor and spin-orbit components in the nucleon-nucleon potential. The earlier attempts to incorporate all these features in a phenomenological potential were due to Brueckner, Gammel, and Thaler [33] and Signell and Marshak [34]. In these attempts, the potential contained only the first four terms of Eq. 5–45, $V_{\sigma p}$ being superfluous for fitting phase shifts, and $V_{\sigma L}$

was ignored. For example, in the Brueckner-Gammel-Thaler potential, each V_i was taken to be a function of the relative distance r alone (which is a special case of Eq. 5–46), and to contain an infinite ("hard") repulsive core of 0.4 fm in each state, followed by an attractive potential of Yukawa

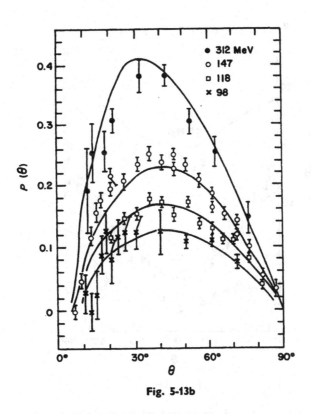

Fig. 5-13b

shape, whose range and strength were adjustable parameters for singlet-even, singlet-odd, triplet-even, and triplet-odd states. There were thus more than twenty adjustable parameters in the potential. In view of the number of data available in those days, the overall fit was reasonable. There were, however, two important shortcomings in this potential. First, it was noted that the singlet-even potential which fitted the $^1S_0$-phases yielded too large a $^1D_2$-phase shift at higher energies. It has been argued [35] that a purely local static potential cannot reproduce the experimental $^1S_0$- and

$^{1}D_{2}$-phases simultaneously. The second defect was that the tail of this potential did not correspond to one-pion-exchange form.

In later potentials given by Hamada and Johnston [36] and by the Yale group [37] both the above shortcomings were rectified. The term

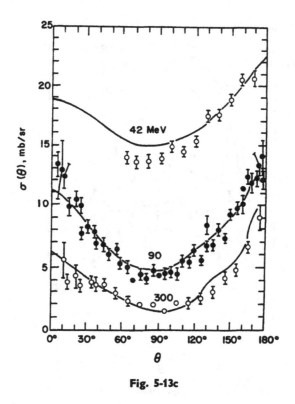

Fig. 5-13c

Fig. 5-13 Comparison of some of the experimentally measured nucleon-nucleon scattering data at various energies with the predictions of the Hamada-Johnston potential [36], shown by the solid lines. The *p-p* differential cross sections are shown in (a); the *p-p* polarization data in (b); and the *n-p* differential cross section in (c).

containing $V_{\sigma L}$ in Eq. 5-45 was included (in slightly different forms) in these potentials, so as to make the interaction weaker in the $^{1}D_{2}$-state than in the $^{1}S_{0}$-state. This L^{2}-dependent potential also helps in depressing the $^{3}D_{2}$-phases, thus making the interaction stronger in the $^{3}S_{1}$-$^{3}D_{1}$ state than in the $^{3}D_{2}$-state. In both the Yale and Hamada-Johnston (H-J) potentials, the hard-core radius was close to 0.5 fm. The overall fit to data was greatly improved. There are minor discrepancies in the H-J potential, such as having some unphysical bound states for very high values of L, which can be removed by altering some parameters slightly [38]. Also, the H-J potential yields too large $^3F_4$-phases at higher energies, resulting in some broad bumps in the calculated p-p differential cross section, which are absent in the experimental data (see Fig. 5–13a). To obtain better agreement with the $^3F_4$-phases, the Yale group arbitrarily take $V_{LS} = 0$ for $J > 2$. Figure 5–13 presents for comparison some of the experimentally determined data at various energies and the theoretical curves given by the H-J potential.

An alternative potential, with different treatment of the repulsive core, has been developed by Reid [39]. For states $J \leqslant 2$, the mathematical form, although equivalent to Eq. 5–45, has a different combination of arbitrary functions—one for each state; for higher J's the one-pion-exchange potential may be used. All states have the OPEP tail. Reid's hard-core potential has a repulsive-core radius of 0.42 fm in all states except $^3S_1$-$^3D_1$, where it is 0.55 fm. Alternatively, he also presents a "soft-core" potential with Yukawa-type repulsion at short distances.

Figure 5–14 illustrates the fact that different potentials can fit the experimental data; the Reid and Hamada-Johnston potentials are shown for the $^1S_0$- (Fig. 5–14a), the $^3P_1$- and $^3P_0$- (Fig. 5–14b) states. For some of the hard-core phenomenological potentials, Signell has computed the least-square sum M defined in Eq. 5–40, using 648 items of 10–330 MeV p-p data taken from ref. [40]. We reproduce his results in Table 5-2:

TABLE 5-2 Goodness of Fit of Some Potentials

Potential	Year	Ref.	M	r_c, fm
Brueckner-Gammel-Thaler	1958	33	106	0.40
Hamada-Johnston	1962	36	2.98	0.4855
Yale	1962	37	3.81	0.5115
Reid hard-core	1968	39	2.72	0.423

In the notation of Eq. 5–40, $y_i(\delta)$ are the values of the observable y_i calculated from the phases generated by the chosen potential. It will be seen that the overall fits of the H-J and Yale potentials are nearly as good as the fit of the more recent Reid potential.

Figure 5–14 shows that the "soft core" of the Reid potential attains a height of more than 800 MeV for $r \sim 0.4$ fm. In this sense, the core is not

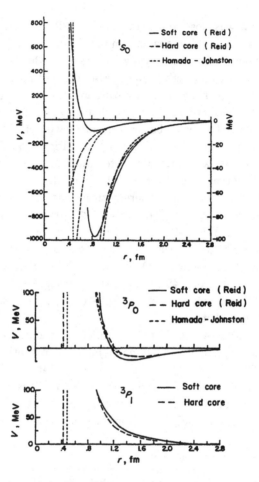

Fig. 5-14 Comparison of the Hamada-Johnston [36] and the Reid hard- and soft-core potentials [39]: (a) in the $^1S_0$-state and (b) in the $^3P_1$- and $^3P_0$-states. (After Reid [39].)

really very soft, but Sprung and Srivastava [41] have shown that it is possible to fit the $^1S_0$-phase shifts accurately with a "supersoft" repulsive core whose maximum height is only about 200 MeV, while still retaining the V_{OPEP} tail. This is made possible by spreading out the repulsive region to about 0.8 fm and at the same time reducing the height of the repulsive core. Sprung and de Tourreil [63] have also constructed supersoft potentials for the other partial waves (including the coupled channels) which fit the experimental phases accurately.

Many alternative and fruitful ways of constructing phenomenological interactions have been reported. One such approach is the so-called boundary-condition model (BCM), mainly developed by Lomon and Feshbach [42]. No attempt is made to specify the potential inside some boundary which delimits the "region of ignorance"; instead, one specifies boundary conditions (usually a separate energy-independent logarithmic derivative of the radial wave function for each partial wave at a common boundary of ~ 0.7 fm) that allow one to carry through a numerical evaluation of the wave function and hence of the phase shifts. The wave function outside the boundary is given by some potential, which may, for example, be generated by various one-boson-exchange processes.

In Section 5–3 we mentioned that the N-N potential is expected, from considerations of the meson theory, to be nonlocal especially for small internucleon distances. A simple form of nonlocality (not given by meson theory) is separable in r and r',

$$V(r, r') = g(r)g(r'), \tag{5-47}$$

where $g(r)$ is a smooth, finite-range function. Such potentials were first introduced by Yamaguchi [43] and later developed by Tabakin[†] [44] and Mongan [45]. The great advantage of this form is its simplicity in calculations. The two-body T-matrix, defined in the next section, can be calculated analytically for such potentials and is itself separable in r and r'. Mitra [46] has shown that the three-body problem can be solved exactly for such potentials. The form given in Eq. 5–47 is called a "rank-one" separable potential, since there is only one term on the right side. In order to fit the S-state phase shifts which change sign, one generally requires a rank-two separable potential with one repulsive and one attractive term. Separable potentials cannot have the local character of OPEP for larger distances and are therefore not very realistic. However, they may be regarded as offering one convenient way of extrapolating the off-shell behavior which is very different from the local approach.

Another class of nonlocal potentials contain momentum dependence explicitly:

$$V(r, p) = V_1(r) + \frac{p^2}{m}V_2(r) + V_2(r)\frac{p^2}{m}, \tag{5-48}$$

where p^2 is the operator $-\hbar^2\nabla^2$. This may be regarded as an expansion

---

† The Tabakin potential yields the value for the quadrupole moment of the deuteron as only 0.128 fm², which is far too low.

of a general nonlocal potential in a power series of p^2, with the leading terms retained. Phenomenologically, $V_1(r)$ is taken as attractive, while $V_2(r)$ is repulsive and of shorter range. Difficulties arise if $V_2(r)$ is attractive for any value of r, because then the interaction becomes more and more attractive for large p^2 and may give rise to bound states with no lower bound [47]. The form of Eq. 5–48 was suggested from a phenomenological viewpoint by Peierls [48], and a preliminary fit to experimental data was obtained by Green [49]. However, further improvements have been carried out within the framework of the one-boson-exchange model, discussed in section 5–12.

5-9 DEFINITION OF A POTENTIAL FROM MESON THEORY

We now proceed to formalize the ideas of Section 5–2 in order to define the concept of a potential arising from meson exchanges, which may be used in the Schrödinger equation. To do this, we first have to define, in terms of a two-body potential V, an energy-dependent operator T whose diagonal matrix elements at the scattering energy contain all the information about elastic scattering. The T-matrix, as it is called, can also be calculated from meson-exchange processes using field theory and hence can be used to make a connection between the potential and meson-theoretical approaches.

In relative coordinates, the Schrödinger equation can be written as ($\hbar^2/M = 1$, $E_k = k^2$)

$$(H_0 + V)|\psi_k\rangle = k^2|\psi_k\rangle,$$

where $H_0 = -\nabla^2$, and we are considering solutions in the continuum, for $E_k \geqslant 0$, which describe scattering. It is necessary to specify the boundary conditions; in particular, the form of $\langle r|\psi_k\rangle$ will depend on whether we want outgoing or incoming spherical waves asymptotically. An alternative way of writing the Schrödinger equation is

$$|\psi_k\rangle = |\phi_k\rangle + G(E_k)V|\psi_k\rangle, \tag{5–49}$$

where $|\phi_k\rangle$ is just the solution of the homogeneous equation

$$(H_0 - k^2)|\phi_k\rangle = 0, \tag{5–50}$$

and $G(E_k)$ is the free Green's function, defined as

$$G(E_k) = \frac{1}{E_k - H_0 + i\varepsilon}, \tag{5–51}$$

where ε is an infinitesimal quantity which is finally put to zero. If we premultiply Eq. 5-49 by $G^{-1}(E_k) = E_k - H_0 + i\varepsilon$, it is immediately seen to be equivalent to the Schrödinger equation. The quantity $i\varepsilon$ is introduced to take care of the singularities of $G(E_k)$ in the k-space and actually incorporates the boundary condition at infinity. It can be shown (see, e.g., Gottfried [50]) that $+ i\varepsilon$ gives rise to an outgoing spherical wave, while $- i\varepsilon$ corresponds to an incoming spherical wave at infinity. In Eq. 5-49, the first term on the right side, $|\phi_k\rangle$, is the incident wave, while the second term, $G(E_k)V|\psi_k\rangle$, is the scattered wave. Details of these interpretations are given by Gottfried. Equation 5-49 is often called the Lipmann-Schwinger equation.

Premultiplying Eq. 5-49 by V, we obtain

$$V|\psi_k\rangle = V|\phi_k\rangle + VG(E_k)V|\psi_k\rangle. \tag{5-52}$$

Now we define an operator $T(E_k)$, such that

$$T(E_k)|\phi_k\rangle = V|\psi_k\rangle. \tag{5-53}$$

It then follows that

$$T(E_k) = V + VG(E_k)T(E_k). \tag{5-54}$$

We can generalize Eq. 5-54 to any E, but $E = E_k = k^2$ corresponds to the physical scattering case. Matrix elements $\langle k'|T(E_k)|k\rangle$ with $k'^2 = k^2$ are "on the energy shell" because, for scattering involving only two bodies, the energy k^2 must be conserved. However, if three or more particles are present, part of the energy may be taken away by a third particle, and the two-body T-matrix may go "off the energy shell."[†] Such a situation may arise, for example, in proton-proton bremsstrahlung, $p + p \rightarrow p + p + \gamma$.

The scattering amplitude $f(k', k)$ is defined as

$$f(\mathbf{k'}, \mathbf{k}) = - 2\pi^2 \int \exp(- i\mathbf{k'} \cdot \mathbf{r'}) \, V(r')\psi_k(\mathbf{r'}) \, d^3r'$$
$$= - 2\pi^2 \langle \phi_{k'}|V|\psi_k\rangle, \tag{5-55}$$

where, for elastic scattering, $k'^2 = k^2$. For this case, it is customary to use the variables k and θ, the angle between $\mathbf{k}$ and $\mathbf{k'}$, and to write $f(\mathbf{k'}, \mathbf{k})$ as $f_k(\theta)$. From the definition of the T-operator, we see that $(k'^2 = k^2)$

$$f_k(\theta) = - 2\pi^2 \langle \mathbf{k'}|T(E = k^2)|\mathbf{k}\rangle. \tag{5-56}$$

[†] More precisely, $\langle \mathbf{k'}|T(E_k)|\mathbf{k}\rangle$ with $k^2 \neq k'^2$ is termed "half off-shell," while $\langle \mathbf{k'}|T(E)|\mathbf{k}\rangle$ with $E \neq k^2 \neq k'^2$ is "completely off-shell." Using unitarity, one can express a completely off-shell matrix element in terms of half-off-shell ones. In the p-p bremsstrahlung process, half-off-shell matrix elements enter into the calculation.

All the information about elastic two-body scattering is contained in $f_k(\theta)$, that is, in the T-matrix on the energy shell. It is shown in most books of quantum mechanics (see, e.g., Gottfried [50]) that

$$f_k(\theta) = \sum_{l=0}^{\infty} \sqrt{4\pi(2l+1)} \frac{1}{k} e^{i\delta_l} \sin \delta_l Y_l^0(\theta), \tag{5-57}$$

where δ_l is the phase shift of the lth partial wave, and Y_l^0 is the standard spherical harmonic function defined in Eq. A–13. We derived the above equation for S-waves in Chapter 2 (see Eq. 2–31).

To summarize, in nonrelativistic scattering theory one defines an operator $T(E)$:

$$T(E) = V + VG(E)T(E), \tag{5-58}$$

with $G(E)$ given by Eq. 5–51, whose diagonal matrix elements on the energy shell, $\langle \mathbf{k}|T(E = k^2)|\mathbf{k}\rangle$, contain all the information about the elastic scattering of two particles. Note that, when calculating T from V by using Eq. 5–58, the intermediate states also contain only two nucleons, because the potential V causes transitions only between two-nucleon states.

From meson theory also, one can calculate the scattering amplitude, which has identical physical content but does not involve the concept of a potential. To do this, one classifies processes in which one, two, three, etc., pions are being exchanged between the two nucleons. With the pion-nucleon coupling constant denoted by g, the meson-theoretical scattering amplitude $\mathscr{T}$ can be expanded in a series

$$\mathscr{T} = g^2 \mathscr{T}^{(1)} + g^4 \mathscr{T}^{(2)} + \cdots, \tag{5-59}$$

where $\mathscr{T}^{(1)}$ is calculated from diagram (a) in Fig. 5–15, $\mathscr{T}^{(2)}$ from diagrams (b) and (c), and so forth. Diagrams (b) and (c) are different, although in both only two pions are exchanged, because the intermediate state of (c) contains two additional virtual pions. Using the techniques of field theory, $\mathscr{T}$ can be calculated completely relativistically. There are prescribed rules, due to Feynman, for calculating the contributions of diagrams like (a), (b), and (c) and hence $\mathscr{T}^{(1)}$, $\mathscr{T}^{(2)}$, etc. [51]. We indicate in Section 5–11 how, for example, $\mathscr{T}^{(1)}$ can be calculated.

To define a potential V, we may now demand that it should be such that the scattering amplitude T calculated from it by using Eq. 5–58 be identical to $\mathscr{T}$ for all k on the energy shell. To see this in more detail, let us expand V in a power series of g^2:

$$V = g^2 V^{(1)} + g^4 V^{(2)} + \cdots. \tag{5-60}$$

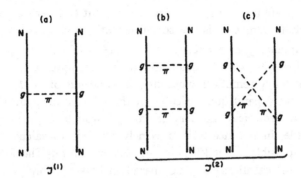

Fig. 5-15 Some meson-exchange processes contributing to the two-nucleon scattering amplitude $\mathscr{T}$. The single-pion-exchange contribution (a) is proportional to g^2 and gives $\mathscr{T}^{(1)}$; the two-pion-exchange processes (b) and (c), proportional to g^4, give $\mathscr{T}^{(2)}$.

Then, using Eq. 5–58, we obtain

$$T = g^2 V^{(1)} + g^4 V^{(2)} + g^4 V^{(1)} G V^{(1)} + \cdots \tag{5-61}$$

$$= g^2 T^{(1)} + g^4 T^{(2)} + \cdots. \tag{5-62}$$

Hence

$$V^{(1)} = T^{(1)}, \tag{5-63}$$

$$V^{(2)} = T^{(2)} - V^{(1)} G V^{(1)}, \tag{5-64}$$

and so forth. Since we demand that $T^{(1)} = \mathscr{T}^{(1)}$, $T^{(2)} = \mathscr{T}^{(2)}$, etc., it follows that

$$V^{(1)} = \mathscr{T}^{(1)},$$

$$V^{(2)} = \mathscr{T}^{(2)} - V^{(1)} G V^{(1)}, \tag{5-65}$$

and so on. Knowing $\mathscr{T}^{(1)}$, $\mathscr{T}^{(2)}$, etc., from field theory, we can, in principle, calculate $V^{(1)}$, $V^{(2)}$, and the higher-order terms. Note that in calculating $\mathscr{T}^{(2)}$ all possible baryonic intermediate states should be taken; that is, the nucleon may have changed to a N^* (spin $\frac{3}{2}$, i-spin $\frac{3}{2}$) in the intermediate state. The two-pion-exchange potential (TPEP) $V^{(2)}$ will therefore depend on what intermediate states have been included in calculating $\mathscr{T}^{(2)}$.

Since the potential V is used in the Schrödinger equation, which is a non-relativistic equation, it makes sense to take the nonrelativistic limits of the field-theoretically calculated amplitudes $\mathscr{T}^{(1)}$, $\mathscr{T}^{(2)}$,.... Unless some extreme approximations are made, the potential V turns out to be energy dependent. An extreme approximation is the so-called static approximation, in which the relativistic energy of a nucleon, $E = \sqrt{p^2c^2 + M^2c^4}$, is simply replaced by Mc^2, putting $p/M = 0$. In this limit, static, local potentials are obtained. Often, one makes systematic expansions of E in powers of $(p/M)^2$ and retains in V terms that depend quadratically on the relative momentum squared.

Although a theoretical basis for calculating V from multiple pion exchange has been made, in practice it is extremely difficult to calculate $V^{(2)}$ and practically impossible to get $V^{(3)}$, etc., by this procedure. The TPEP potential $V^{(2)}$ has been calculated by a number of authors, notably by Cottingham and Vinh Mau [52] and most recently by Partovi and Lomon [53]. A good discussion of TPEP can be found in a review article by Signell [30]. There is experimental evidence that multiple pions are strongly correlated, sometimes forming a metastable state with definite quantum numbers, which may be looked upon as short-lived bosons. The current trend is to replace the multiple-pion-exchange processes between two nucleons by these single boson exchanges, which are also called "resonances." This simplifies things greatly, since only terms like $\mathscr{T}^{(1)}$ have to be evaluated, and the resulting potential can be obtained directly. Before discussing the one-boson-exchange potentials (OBEP) more fully, we present some details about the resonances.

5-10 BOSON RESONANCES

First we discuss the experiments that show that multiple pions can be strongly correlated. In such cases, a two-pion or a three-pion system may be thought of as a metastable state with definite quantum numbers and be regarded as a short-lived meson in its own right. Consider the reaction of monoenergetic pions of high energy (~ 1 GeV) interacting with the protons of liquid hydrogen in a bubble chamber:

$$p + \pi^- \rightarrow \begin{array}{c} \pi^- + \pi^0 + p, \\ \pi^+ + \pi^- + n. \end{array} \tag{5-66}$$

If the two pions in the final state formed a bound state, there would be essentially only two bodies in the final state, and each would have a definite kinetic energy. The situation is analogous to that in Section 2-6, where the final-state interaction between two neutrons was considered. If there is no real bound state of the two pions, but they are strongly correlated, this will show up as a peak when the number of events is plotted against the kinetic energy of the third particle. Relativistically, the energy of a particle is $E^2 = (\mathbf{p})^2 + m^2$, where we have put $c = 1$; $\mathbf{p}$ is the three-momentum and m is the rest-mass of the particle. Thus

$$m^2 = E^2 - (\mathbf{p})^2 = - p_\mu^2, \tag{5-67}$$

where p is the four-momentum with components $(\mathbf{p}, iE)$. It is more convenient, therefore, to plot the number of events in Eq. 5–66 against $- (p_{1\mu} + p_{2\mu})^2$, where $p_{1\mu}$ and $p_{2\mu}$ are the four-momenta of the two pions. The peak in this graph will directly give the mass of the "resonance," and this is shown in Fig. 5–16. The quantity $- (p_{1\mu} + p_{2\mu})^2$ is called the "invariant mass" for reasons explained above. The mass of this "particle," called the ρ-meson, is about 765 MeV with a width Γ of about 125 MeV. This means that this compound state has a lifetime τ of the order of $\hbar/\Gamma \approx 5 \times 10^{-24}$ sec, much too short to be detected experimentally. Note that the mass of a resonance is greater than the combined mass of the two pions, whereas the mass of

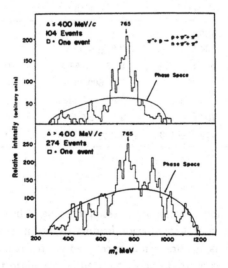

Fig. 5-16 Two-pion resonance ρ in the three-body final state in the reaction

$$\pi^- + p \rightarrow \left. \begin{matrix} p + \pi_T^- + \pi^0 \\ n + \pi^- + \pi^+ \end{matrix} \right\}.$$

The relative intensity of the experimental counts is plotted against the invariant mass of the two pions in the final state. The smooth curve is the phase-space prediction, normalized to the number of events plotted. The upper distribution is for the situation in which the momentum transfer to the nucleon, Δ, < 400 MeV/c, while the lower one is for $\Delta > 400$ MeV/c. The peak at 765 MeV corresponds to the ρ-mass. [After A. R. Erwin, R. March, W. D. Walter, and E. West, *Phys. Rev. Lett.* **6**, 628 (1961).]

a truly bound system is less than the sum of its constituent masses (e.g., the mass of the deuteron is less than the sum of the proton and neutron masses).

We should be able to assign definite quantum numbers to this resonance if it is to be regarded as a separate entity. The resonance is not found in the reaction $\pi^+ + p \to \pi^+ + \pi^+ + n$, hence the i-spin of the ρ-meson is 1. This i-spin of 1 is obtained by combining two pion i-spins of 1 each, and the i-spin part of the wave function is antisymmetric. Since the total wave function of two bosons must be symmetric, and the spin of each pion is zero, it follows that the ρ-meson must have odd-integral total spin, the simplest assignment being 1^-. In this particular case, more direct determination of the quantum numbers is difficult.

TABLE 5-3 Some of the Experimentally Detected Bosons That Contribute to the N-N Interaction. Parameters are taken from "Review of Particle Properties," *Rev. Mod. Phys.* **45**, 51 (1973).

Boson	Mass, MeV	Width	T	J^p
$\pi^\pm$	139.57	0	1	0^-
π^0	134.96	7.8 ± 0.9 eV		
η	548.8 ± 0.6	2.63 ± 0.58 keV	0	0^-
σ_0 [a]	≈ 700	> 600 MeV	0	0^+
ρ	770 ± 5	146 ± 10 MeV	1	1^-
ω	783.8 ± 0.3	9.8 ± 0.5 MeV	0	1^-
ϕ	1019.6 ± 0.3	4.2 ± 0.2 MeV	0	1^-

[a] Also called ε.

By the same methods, many other resonances have been detected. By studying proton-antiproton annihilation leading to multiple-pion production, the ω-meson, which is a three-pion correlated state, has been detected. Table 5–3 lists the bosons that are often used to calculate the OBEP N-N potential. It should be noted that the η-meson is stable against strong interactions and decays only by electromagnetic processes. It should be regarded, not as a multiple-pion resonance, but as a heavy, stable meson in its own right. Bosons with spin parity $(J^p) = 0^+$ are scalars, with $J^p = 0^-$ are pseudoscalars, and with $J^p = 1^-$ and 1^+ are respectively vectors and pseudovectors, because the corresponding fields transform accordingly under proper Lorentz transformation and space inversion (see Section A–4).

5-11 DERIVATION OF THE OBEP

In order to provide a more complete understanding for interested students, we give an outline of the derivation of the OBEP, using the concepts of Section 5-9. The potential will depend, of course, on the type of interaction between the meson field and the nucleons. This, in turn, depends on whether the meson field transforms as a scalar, as a pseudoscalar, or as a vector. We take the example where the exchanged boson is a pion, which has a pseudoscalar field. For simplicity, we ignore the charge of the nucleons and the pion.

Now consider Fig. 5-17, where two nucleons, 1 and 2, are scattered via the exchange of a single pion. The time scale is chosen to run upward, and there are two possible "time orderings" for the process, which are shown in Figs. 5-17a and 5-17b. We concentrate first on Fig. 5-17a. At "vertex" 1, the nucleon with momentum $\mathbf{p}$ emits a pion and goes to state $\mathbf{p}'$. At a later time, the second nucleon with momentum $-\mathbf{p}$ absorbs this virtual pion at vertex 2, itself going over to a state $-\mathbf{p}'$. For convenience, the center-of-mass frame is chosen so that the initial and final total momenta are zero. The initial and final relative momenta between 1 and 2 are $\mathbf{p}$ and $\mathbf{p}'$, respectively. Since the momentum is conserved at each vertex, the pion momentum is $\mathbf{q} = \mathbf{p} - \mathbf{p}'$. The initial total energy of particles 1 and 2 is $2E$, with $E = \sqrt{p^2 + M^2}$, while the final energy is $2E'$, where $E' = \sqrt{p'^2 + M^2}$.

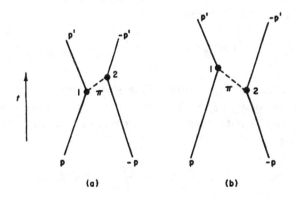

Fig. 5-17 Two-nucleon scattering via the exchange of a single pion.

For elastic scattering, $E = E'$. This is the case that we consider here, since we want to evaluate the scattering amplitude on the energy shell. As shown in Eq. A-136, the matrix element for the process at vertex 1 is

$$i\sqrt{4\pi}\, g \int \bar{\psi}(\mathbf{p}')\, \exp(-\,i\mathbf{p}'\cdot\mathbf{r})\,\gamma_5\phi_\pi(\mathbf{r})\psi(\mathbf{p})\,\exp(i\mathbf{p}\cdot\mathbf{r})\,d^3r, \qquad (5\text{–}68)$$

where $\psi(\mathbf{p})$ is the nucleon spinor and is defined in Eq. A-95. To proceed, we have to quantize the mesic field ϕ_π, giving it the role of creating or annihilating a pion. In this role, it acts as an operator on the meson state. We write the mesic field in the quantized form as

$$\phi_\pi(r) = \sum_{\mathbf{q}'} \frac{1}{\sqrt{2w_{q'}}}\,[a_{\mathbf{q}'}\exp(i\mathbf{q}'\cdot\mathbf{r}) + a_{\mathbf{q}'}{}^{+}\exp(-\,i\mathbf{q}'\cdot\mathbf{r})], \qquad (5\text{–}69)$$

where $w_{q'} = \sqrt{m_\pi^2 + q'^2}$, and $a_{\mathbf{q}'}$ and $a_{\mathbf{q}'}^{+}$ are, respectively, the annihilation and creation operators for a meson of momentum $\mathbf{q}'$. If we denote a state with no meson present by the "vacuum" $|0\rangle$, and a state with one meson of momentum $\mathbf{q}$ by $|1_\mathbf{q}\rangle$, then $a_\mathbf{q}$ and $a_\mathbf{q}^{+}$ are defined by the equations

$$a_\mathbf{q}|0\rangle = 0, \qquad (5\text{–}70)$$

$$a_\mathbf{q}^{+}|0\rangle = |1_\mathbf{q}\rangle, \qquad (5\text{–}71)$$

$$a_\mathbf{q}|1_\mathbf{q}\rangle = |0\rangle. \qquad (5\text{–}72)$$

Note that the initial state, with two nucleons of momenta $\mathbf{p}$ and $-\mathbf{p}$, is a meson vacuum $|0\rangle$, and so also is the final state. The meson field ϕ_π creates a pion at vertex 1, which in turn is annihilated at vertex 2. Let us substitute Eq. 5–69 into Eq. 5–68 to obtain

$$H_{\text{int}} = \sqrt{4\pi}\, ig\bar{\psi}(\mathbf{p}')\gamma_5\psi(\mathbf{p})\,\frac{1}{\sqrt{2w_q}}\,(a_\mathbf{q}^{+} + a_{-\mathbf{q}}). \qquad (5\text{–}73)$$

We call this expression H_{int} because it is still an operator in the meson variable, since the mesic field ϕ_π is an operator in the quantized form. Using Eqs. A-98 and A-110 and taking $E \approx m$, we find the nonrelativistic approximation of this to be

$$H_{\text{int}} = -\,i\sqrt{4\pi}\,\frac{g}{2E}\,(\boldsymbol{\sigma}_1\cdot\mathbf{q})\,\frac{1}{\sqrt{2w_q}}\,(a_\mathbf{q}^{+} + a_{-\mathbf{q}}), \qquad (5\text{–}74)$$

where $\boldsymbol{\sigma}_1$ is to be understood to act on the nonrelativistic spin states of nucleon 1. Note that this is Hermitian, since $\mathbf{p}$ and $\mathbf{p}'$ should be interchanged while taking the Hermitian adjoint. In order to reach the final state from the initial one, two operations of H_{int} are necessary; one for creating a meson and a second for annihilating it. Thus the lowest-order scattering amplitude between relative momenta $\mathbf{p}$ of the initial state and $\mathbf{p}'$ of the final state is given by second-order perturbation theory and is of the order g^2,

$$\langle \mathbf{p}'|\mathcal{T}^{(1)}|\mathbf{p}\rangle = 2\sum_n \langle 0|H_{\text{int}}|n\rangle \frac{1}{\varepsilon_i - \varepsilon_n} \langle n|H_{\text{int}}|0\rangle. \qquad (5\text{-}75)$$

The factor 2 takes account of the diagram in Fig. 5–15b, which gives an identical contribution. For fixed $\mathbf{p}$ and $\mathbf{p}'$, $|n\rangle$ is just a one-meson state with fixed momentum $\mathbf{q}$, that is, $|n\rangle = |1\mathbf{q}\rangle$. The energy denominator is the difference between the initial- and the intermediate-state energies,

$$\varepsilon_i - \varepsilon_n = -\sqrt{m_\pi^2 + q^2} = -w_q. \qquad (5\text{-}76)$$

Using Eqs. 5–70 to 5–73, we get

$$\langle \mathbf{p}'|\mathcal{T}^{(1)}|\mathbf{p}\rangle = -\frac{g^2}{w_q^2}\frac{4\pi}{4E^2}(\boldsymbol{\sigma}_1 \cdot \mathbf{q})(\boldsymbol{\sigma}_2 \cdot \mathbf{q}). \qquad (5\text{-}77)$$

In the static limit, $E^{-2} = (p^2 + M^2)^{-1}$ is simply replaced by M^{-2}, assuming that $(p/M)^2 \ll 1$. According to Section 5–9, we should now equate $\langle \mathbf{p}|\mathcal{T}^{(1)}|\mathbf{p}\rangle$ to $\langle \mathbf{p}|V_{\text{OPEP}}|\mathbf{p}\rangle$, where

$$\langle \mathbf{p}'|V_{\text{OPEP}}|\mathbf{p}\rangle = \int \exp[i(\mathbf{p}' - \mathbf{p})\cdot \mathbf{r}]\, V_{\text{OPEP}}(r)\, d^3r$$

$$= V_{\text{OPEP}}(\mathbf{q}). \qquad (5\text{-}78)$$

In the coordinate space, therefore,

$$V_{\text{OPEP}}(\mathbf{r}) = \frac{1}{(2\pi)^3}\int V_{\text{OPEP}}(\mathbf{q})\exp(i\mathbf{q}\cdot\mathbf{r})\, d^3q$$

$$= -\frac{g^2}{4M^2}\frac{4\pi}{(2\pi)^3}\int d^3q \frac{(\boldsymbol{\sigma}_1 \cdot \mathbf{q})(\boldsymbol{\sigma}_2 \cdot \mathbf{q})}{q^2 + m_\pi^2}\exp(i\mathbf{q}\cdot\mathbf{r}). \qquad '(5\text{-}79)$$

Note that the central part of this potential comes from the integral

$$\frac{1}{(2\pi)^3}\int d^3q \frac{q^2}{q^2 + m_\pi^2}\exp(i\mathbf{q}\cdot\mathbf{r}) = \frac{1}{(2\pi)^3}\int d^3q \left(1 - \frac{m_\pi^2}{q^2 + m_\pi^2}\right)\exp(i\mathbf{q}\cdot\mathbf{r})$$

and therefore contains a delta function $\delta(\mathbf{r})$ at the origin, which should be discarded. One way of doing this is to replace $(\boldsymbol{\sigma}_1 \cdot \mathbf{q})(\boldsymbol{\sigma}_2 \cdot \mathbf{q})$ by $-(\boldsymbol{\sigma}_1 \cdot \boldsymbol{\nabla})(\boldsymbol{\sigma}_2 \cdot \boldsymbol{\nabla})$ and bring it out of the integral. Then

$$V_{\text{OPEP}}(\mathbf{r}) = \frac{g^2}{4M^2} (\boldsymbol{\sigma}_1 \cdot \boldsymbol{\nabla})(\boldsymbol{\sigma}_2 \cdot \boldsymbol{\nabla}) \frac{4\pi}{(2\pi)^3} \int d^3q \frac{\exp(i\mathbf{q} \cdot \mathbf{r})}{q^2 + m_\pi^2}.$$

Since

$$\frac{1}{(2\pi)^3} \int d^3q \frac{\exp(i\mathbf{q} \cdot \mathbf{r})}{q^2 + m_\pi^2} = \frac{1}{4\pi} \frac{\exp(-m_\pi r)}{r},$$

$$V_{\text{OPEP}}(\mathbf{r}) = \frac{q^2}{4M^2} m_\pi(\boldsymbol{\sigma}_1 \cdot \boldsymbol{\nabla})(\boldsymbol{\sigma}_2 \cdot \boldsymbol{\nabla}) \left(\frac{\exp(-m_\pi r)}{m_\pi r} \right). \tag{5-80}$$

It can be shown that

$$(\boldsymbol{\sigma}_1 \cdot \boldsymbol{\nabla})(\boldsymbol{\sigma}_2 \cdot \boldsymbol{\nabla}) \frac{\exp(-m_\pi r)}{m_\pi r}$$

$$= \tfrac{1}{3}m_\pi^2 \left[(\boldsymbol{\sigma}_1 \cdot \boldsymbol{\sigma}_2) + S_{12} \left(1 + \frac{3}{m_\pi r} + \frac{3}{(m_\pi r)^2} \right) \right] \frac{\exp(-m_\pi r)}{m_\pi r}, \tag{5-81}$$

finally giving the form of Eq. 5–14. The extra factor of $(\boldsymbol{\tau}_1 \cdot \boldsymbol{\tau}_2)$ in Eq. 5–14 arises from i-spins of nucleons 1 and 2, which were ignored in this discussion. Since $(\boldsymbol{\sigma}_1 \cdot \boldsymbol{\sigma}_2)(\boldsymbol{\tau}_1 \cdot \boldsymbol{\tau}_2)$ is -3 for all even states, the central part of V_{OPEP} is attractive for these. We find that a strong tensor potential is naturally generated by a pseudoscalar meson exchange. We also note that, even if we did not make the static approximation in Eq. 5–77 of replacing E^{-2} by M^{-2}, the momentum dependence would be very small, of the order of $(p/M)^2$, compared with the static term.

Exchange of a scalar meson also gives rise to an attractive static central potential and, in addition, a two-body spin-orbit potential whose relative strength is proportional to m_s/M, where m_s is the mass of the scalar meson. Also, the momentum-dependent terms, which arise from expanding terms like $(E + M)/2E$, are stronger than those coming from exchange of a pseudoscalar meson.

When a vector meson is exchanged between two nucleons, the static part of the central potential is repulsive. This is analogous to the repulsive Coulomb potential, which arises from the exchange of a vector particle, the photon. It also gives rise to a spin-orbit potential, a tensor potential, and some momentum-dependent terms. The derivations of these potentials arising from scalar and vector meson exchanges are too lengthy to be given here, but they follow essentially the same lines as the pseudoscalar meson case. A detailed account can be found in Moszkowski [54].

5-12 MAIN FEATURES OF THE OBEP

We have seen that the experimental phase-shift data (for $E_{lab} < 450$ MeV) can be reproduced by various forms of phenomenological potentials. Many of these have the OPEP tail, but otherwise contain thirty or more adjustable parameters. In the past few years, however, a better understanding has been gained by appreciating the roles that the various mesons and resonances play in generating the N-N force. These are as follows:

1. The long-range part ($r > 3$ fm) of the N-N potential is given by the one-pion-exchange mechanism.

2. The attractive interaction at intermediate ranges ($0.7 < r < 2$ fm) can be simulated by the exchange of scalar mesons. Most authors introduce two scalar mesons ($J^p = 0^+$): one with $T = 0$, $m \sim 500$ MeV, and the other with $T = 1$, $m > 700$ MeV. Experimentally, at least one broad scalar resonance is seen around 700 MeV with $T = 0$. These scalar resonances may be regarded as substitutes for the two-pion-exchange contribution, which is rather difficult to calculate. The scalar meson also generates a spin-orbit term which is attractive for all states for the isoscalar ($T = 0$) case. In addition, momentum-dependent repulsive potentials are also produced.

3. The central N-N interaction is only weakly attractive in the triplet-odd states. This feature can be explained by exchange of the vector ρ-meson. The resulting potential has the characteristic $(\boldsymbol{\tau}_1 \cdot \boldsymbol{\tau}_2)(\boldsymbol{\sigma}_1 \cdot \boldsymbol{\sigma}_2)$ factor in one of the central terms, which is attractive for even-l and repulsive for odd-l states. It also produces a tensor potential of opposite sign to OPEP. This results in the tensor component of the N-N force being much weaker at intermediate ranges than that of the OPEP, bringing about better agreement with the experimental triplet-D and triplet-P phases. The spin-orbit potential generated by the ρ has a $(\boldsymbol{\tau}_1 \cdot \boldsymbol{\tau}_2)$ factor which makes it repulsive in triplet-even states and attractive in triplet-odd ones. Momentum-dependent terms of the same form as those due to the scalar meson are also produced.

4. The isoscalar vector mesons ω and ϕ contribute a very strong short-range repulsion, which is needed to fit the S-wave phases at higher energies. These mesons also give rise to an attractive spin-orbit potential in all triplet states, which adds to the scalar contribution. In triplet-even states, this nearly cancels out with the repulsive ρ-contribution, leaving only a strongly attractive spin-orbit potential in triplet-odd states. This is needed to explain the experimental splitting of the triplet P-waves above 50 MeV.

In constructing an OBEP, one generally includes all the resonances up to 1 GeV which are listed in Table 5–3 and another scalar meson, σ_1, with

$T = 1$. If all the masses and coupling constants of these bosons were known from other sources, the predictions of such a potential could be compared directly with experiment. Not much information is available, however, about the coupling constants. Before discussing this, we note that two types of coupling, direct and gradient, are possible (see Eqs. A-136 to A-139). For pseudoscalar mesons, these essentially give identical potentials in the lowest order. For each vector meson, however, there are two distinct coupling constants g_v and f_v, as given in Eqs. A-138 and A-139. In the absence of detailed knowledge about all these coupling constants, they can be taken as "search parameters" to obtain the best possible agreement between the experimental and theoretically calculated phase shifts. Additionally, the masses of σ_0 and σ_1 are taken as adjustable, since they parametrize part of the two-pion-exchange contributions. Since the ω and ϕ are so similar, their effects are often included in a single isoscalar vector meson. The tensor coupling f_v for this meson is generally taken to be zero, since this value is indicated from studies of the electromagnetic form factors of nucleons [54]. The pion-nucleon coupling constant g^2 is known to be 15.0 ± 0.3 from π-N scattering data [55] and is generally kept fixed around this value. Even when it is taken as a search parameter, its optimum value comes close to the above number, which puts the OBEP model in a favorable light. After making these simplifying assumptions, there are eight free parameters in the model: m_{σ_0}, m_{σ_1}, g_{σ_0}, g_{σ_1}, g_η, g_ω, g_ρ, and f_ρ.

The most elaborate OBEP models have been constructed by Bryan and Scott [56], Ingber [57], Ueda and Green [58], and Erkelenz, Holinde, and Bleuler [59]. These models have been successful in reproducing the N-N data up to 350-MeV laboratory energy, the fits being comparable to those of purely phenomenological potentials which use many more parameters. These potentials also fit the S-wave data and the deuteron parameters reasonably well, but there are some important discrepancies. Signell [60] and MacGregor, Arndt, and Wright [23] have extracted the (bar) mixing parameter ε_1 between the $^3S_1$- and $^3D_1$-phases from n-p scattering data by single-energy phase-shift analyses at different energies, as shown by the open circles in Fig. 5–18. This parameter[†] is important because it is a measure of the tensor-to-central ratio in the triplet-even $J = 1$, $T = 0$ interaction. For comparison, the

---

[†] In the Blatt-Biedenharn convention, the slope of the mixing parameter ε_1^{BB} at zero energy is directly related to the quadrupole moment Q_d of the deuteron:

$$Q_d \approx \frac{1}{\sqrt{2}} \frac{d\varepsilon_1^{BB}}{dk^2} \bigg|_{k=0} ,$$

where $k^2 = ME_{\text{lab}}/2\hbar^2$. See P. Signell, *Phys. Rev.* C2, 1171 (1970).

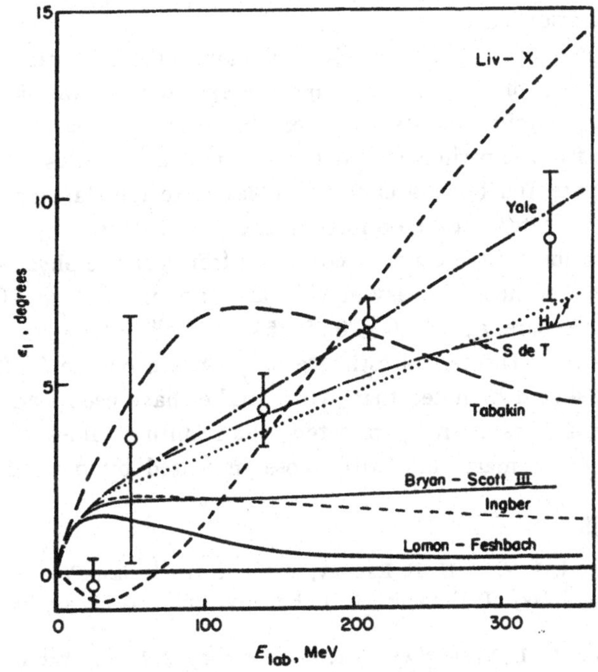

Fig. 5-18 The mixing parameter ε_1 (see Eq. 5–30 in the $T = 0$, $J = 1$ channel in n-p scattering, as extracted by Signell [60] and by MacGregor et al. [23] from single-energy analysis (shown by open circles with error bars) and from multienergy analysis (shown by the dashed curve). Predictions of the Yale [37], H-J [36], Tabakin [44], Lomon-Feshbach [42], Ingber [57], Bryan-Scott [56], and Sprung-de Tourreil [63] potentials are shown for comparison.

results of the multienergy analyses of the Livermore group [23] are also shown, along with the predictions of the different phenomenological potentials. It will be noted that the OBEP potentials generally underestimate ε_1, indicating too strong a central component relative to the tensor strength. This has serious implications in the binding-energy calculation of nuclear matter, which we shall discuss in Chapter 8. We note, however, that Grangé and Preston [64] have proposed an OBEP potential that yields ε_1 in satisfactory agreement with experiment, giving a curve slightly above that labelled S de T in Fig. 5–18.

It should be noted that in the OBEP model there is an ambiguity about how to treat the singularities at the origin in configuration space, which arise if no cutoff is introduced. Bryan and Scott, as well as Ingber, take care of this for $L > 0$ by retaining the p^2-dependent terms. For the S-wave, either an explicit cutoff in the configuration space [56] or pionic form factors [58] must be introduced.

Recently, Partovi and Lomon [53] calculated the potential produced by single exchanges of ρ, ω, η, and π and the exchange of two pions. They did not use the coupling constants as search parameters, but chose values with guidance from experiments and theories in particle physics. Beyond 0.7 fm, their potential bears a close resemblance to the Hamada-Johnston potential, but with too little $\mathbf{L} \cdot \mathbf{S}$ force in the $T = 1$ states.

Obtaining values of the coupling constants from particle physics requires the use of a blend of theoretical assumptions and experimental data. Currently reasonable values are $g_\rho^2 = 0.6 \pm 0.05$ [55], $g_\omega^2/g_\rho^2 = 9.0 \pm 3$ [56], and $f_\rho/g_\rho = 3.7$ [54]. A theoretical estimate of g_η^2 based on the $SU(3)$ model is about 2. It should be noted that, when studies have been made treating the coupling constants as free parameters, the resulting values have agreed in general order of magnitude with those indicated by particle physics.

REFERENCES

1. J. B. Kellogg, I. I. Rabi, N. F. Ramsey, and J. R. Zacharias, *Phys. Rev.* **57**, 677 (1940); H. G. Kolsky, T. E. Phipps, N. F. Ramsey, and H. B. Silibee, *Phys. Rev.* **87**, 395 (1952).
2. R. V. Reid and M. L. Vaida, *Phys. Rev. Lett.* **29**, 494 (1972) and erratum.
3. I. Lindgren, in *Alpha-, Beta-, and Gamma-Ray Spectroscopy*, ed. K. Siegbahn, North-Holland, Amsterdam, 1966.
4. W. Rarita and J. Schwinger, *Phys. Rev.* **59**, 436, 556 (1941).
5. C. Michael and C. Wilkin, *Nucl. Phys.* **B11**, 99 (1969).
6. N. K. Glendenning and G. Kramer, *Phys. Rev.* **126**, 2159 (1962).
7. C. D. Buchanan and M. R. Yearian, *Phys. Rev. Lett.* **15**, 303 (1965).
8. J. E. Elias, J. I. Friedman, G. C. Hartmann, H. W. Kendall, P. N. Kirk, M. R. Sogard, L. P. Van Speybroeck, and J. K. de Pagter, *Phys. Rev.* **177**, 2075 (1969).
9. R. J. Adler and S. D. Drell, *Phys. Rev. Lett.* **13**, 349 (1964).
10. R. Blankenbecler and J. F. Gunion, *Phys. Rev.* **D4**, 718 (1971).
11. H. Yukawa, *Proc. Phys. Math. Soc. Japan* **17**, 48 (1935).
12. M. Matsumato and W. Watari, *Progr. Theor. Phys.* **11**, 63 (1954).
13. L. Wolfenstein, *Ann. Rev. Nucl. Sci.* **6**, 43 (1956).
14. T. V. Kanellopoulos and G. E. Brown, *Proc. Phys. Soc. (London)* **A70**, 690, 703 (1957).
15. H. P. Stapp, T. J. Ypsilantis, and N. Metropolis, *Phys. Rev.* **105**, 302 (1957).
16. O. Chamberlain, R. Donaldson, E. Segre, R. Tripp, C. Wiegand, and T. Ypsilantis, *Phys. Rev.* **95**, 850 (1954).
17. R. T. Siegel, A. J. Hartzler, and W. A. Love, *Phys. Rev.* **101**, 838 (1956).
18. W. N. Hess, *Rev. Mod. Phys.* **30**, 368 (1958).
19. R. Jastrow, *Phys. Rev.* **81**, 165 (1951).
20. K. M. Case and A. Pais, *Phys. Rev.* **80**, 203 (1950).

21. M. H. MacGregor, R. A. Arndt, and R. M. Wright, *Phys. Rev.* **169**, 1149 (1969); N. Hoshizaki, *Rev. Mod. Phys.* **39**, 700 (1967).

22. R. A. Arndt and M. H. MacGregor, *Phys. Rev.* **141**, 873 (1966).

23. M. H. MacGregor, R. A. Arndt, and R. M. Wright, *Phys. Rev.* **182**, 1714 (1969).

24. R. E. Seamon, K. A. Friedman, G. Breit, R. D. Haracz, J. M. Holt, and A. Prakash, *Phys. Rev.* **165**, 1579 (1968).

25. M. D. Miller, P. S. Signell, and N. R. Yoder, *Phys. Rev.* **176**, 1724 (1968).

26. N. Hoshizaki, *Progr. Theor. Phys. Suppl.* **42**, 1 (1968).

27. Yu M. Kazarinov, F. Lehar, and Z. Janout, *Rev. Mod. Phys.* **39**, 571 (1967).

28. J. K. Perring, *Rev. Mod. Phys.* **39**, 550 (1967).

29. M. H. MacGregor, *Phys. Rev.* **113**, 1559 (1959).

30. P. Signell, *Advan. Nucl. Phys.* **2**, 223 (1969).

31. L. Eisenbud and E. Wigner, *Proc. Natl. Acad. Sci. (U.S.)* **27**, 281 (1941).

32. S. Okubo and R. E. Marshak, *Ann. Phys.* **4**, 166 (1958).

33. K. A. Brueckner, J. A. Gammel, and R. M. Thaler, *Phys. Rev.* **109**, 1023 (1958).

34. P. Signell and R. E. Marshak, *Phys. Rev.* **109**, 1229 (1958).

35. D. A. Giltinan and R. M. Thaler, *Phys. Rev.* **131**, 805 (1963).

36. T. Hamada and I. D. Johnston, *Nucl. Phys.* **34**, 382 (1962).

37. K. E. Lassila, M. H. Hull, Jr., H. M. Ruppel, F. A. McDonald and G. Breit, *Phys. Rev.* **126**, 881 (1962).

38. T. Hamada, Y. Nakamura, and R. Tamagaki, *Progr. Theor. Phys.* **33**, 769 (1965).

39. R. V. Reid, *Ann. Phys.* **50**, 411 (1968).

40. H. P. Noyes, P. Signell, N. R. Yoder, and R. M. Wright, *Phys. Rev.* **159**, 789 (1967).

41. D. W. L. Sprung and M. K. Srivastava, *Nucl. Phys.* **A139**, 605 (1969).

42. E. L. Lomon and H. Feshbach, *Rev. Mod. Phys.* **39**, 611 (1967); *Ann. Phys. (N.Y.)* **48**, 94 (1968).

43. Y. Yamaguchi, *Phys. Rev.* **95**, 1635 (1954).

44. F. Tabakin, *Ann. Phys. (N.Y.)* **30**, 51 (1964).

45. T. R. Mongan, *Phys. Rev.* **175**, 1260 (1968); **178**, 1587 (1968).

46. A. N. Mitra, *Nucl. Phys.* **32**, 529 (1962).

47. F. Calogero and Yu A. Simonov, *Phys. Rev. Lett.* **25**, 881 (1970).

48. R. E. Peierls, in *Proceedings of the International Conference on Nuclear Structure, Kingston, Ontario*, Toronto University Press, 1960, p. 7.

49. A. M. Green, *Nucl. Phys.* **33**, 218 (1962).

50. K. Gottfried, *Quantum Mechanics*, Vol. 1, Benjamin, New York, 1966, p. 94.

51. S. S. Schweber, H. A. Bethe, and F. de Hoffman, *Mesons and Fields*, Vol. I, Row, Peterson, 1956, p. 242.

52. W. N. Cottingham and R. Vinh Mau, *Phys. Rev.* **130**, 735 (1960).

53. M. H. Partovi and E. L. Lomon, *Phys. Rev. Lett.* **22**, 438 (1969); *Phys. Rev.* **D2**, 1999 (1970).

54. S. Moszkowski, in *Les Houches Lectures on Nuclear Physics, Grenoble, 1968*, Gordon and Breach, New York, 1969.

55. V. K. Samaranayake and W. W. Woolcock, *Phys. Rev. Lett.* **15**, 936 (1965).

56. R. Bryan and B. L. Scott, *Phys. Rev.* **177**, 1435 (1969).

57. L. Ingber, *Phys. Rev.* **174**, 1250 (1968).

58. T. Ueda and A. E. S. Green, *Phys. Rev.* **174**, 1304 (1968).

59. K. Erkelenz, K. Holinde, and K. Bleuler, *Nucl. Phys.* **A139**, 308 (1969).

60. P. Signell, in *The Two-Body Force in Nuclei*, edited by S. M. Austin and G. M. Crawley, Plenum Press, New York, 1972, p. 9.

61. J. J. Sakurai, *Phys. Rev. Lett.* **17**, 1021 (1966).

62. S. C. C. Ting, in *Proceedings of the Fourth International Conference on High Energy Physics, Vienna, Austria, September 1968*, CERN Scientific Information Service, Geneva, 1968, p. 43.

63. R. de Tourreil and D. W. L. Sprung, *Nucl. Phys.* **A201**, 193 (1973).

64. P. Grangé and M. A. Preston, *Nucl. Phys.* **A204**, 1 (1973).

65. F. Villars, *Helv. Phys. Acta* **20**, 476 (1947).

66. S. M. Berman and S. D. Drell, *Phys. Rev.* **138**, B791 (1964).

PROBLEMS

5-1. Derive Eq. 5-8 from Eq. 5-7.

5-2. Show the following properties of the operator S_{12}.

(a) Applied to any singlet state, S_{12} gives 0.

(b) From parity considerations, S_{12} applied to a triplet function with $L = J \pm 1$ gives a linear combination of the triplet functions for $L = J + 1$ and $J - 1$ and, applied to a triplet function with $L = J$, reproduces the same function multiplied by a constant, that is,

$$S_{12}\mathscr{Y}^M_{J,1L} = a_{L,J+1}\mathscr{Y}^M_{J,1J+1} + a_{L,J-1}\mathscr{Y}^M_{J,1J-1} \tag{1}$$

if $L = J \pm 1$, and

$$S_{12}\mathscr{Y}^M_{J,1J} = a_{J,J}\mathscr{Y}^M_{J,1J}. \tag{2}$$

(c) The operator S_{12} commutes with J_z, and hence the coefficients $a_{L,J+1}$, $a_{L,J-1}$, and $a_{J,J}$ are independent of M.

(d) $a_{J+1,J-1} = a_{J-1,J+1}$.

(e) By considering the particular case in which $\hat{\mathbf{r}}$ is along the z-axis and $M = 1$, and using the fact that $Y_l^m(\theta = 0) = \sqrt{(2l+1)/4\pi}\,\delta_{l0}$, show that

$$a_{J,J} = \langle \mathscr{Y}^M_{J,1J}|S_{12}|\mathscr{Y}^M_{J,1J}\rangle = 2.$$

(f) Similarly, taking the special case $\hat{\mathbf{r}} = \mathbf{k}$, $M = 0$ and $M = 1$, use Eq. 1 above to find two equations between $a_{J-1,J+1}$ and $a_{J-1,J-1}$. Thus show that

$$\langle \mathscr{Y}^M_{J,1J-1}|S_{12}|\mathscr{Y}^M_{J,1J-1}\rangle = \frac{-2(J-1)}{2J+1}, \tag{3}$$

$$\langle \mathscr{Y}^M_{J,1J+1}|S_{12}|\mathscr{Y}^M_{J,1J-1}\rangle = \frac{6\sqrt{J(J+1)}}{2J+1}. \tag{4}$$

Similarly show that

$$\langle \mathscr{Y}^M_{J,1J+1}|S_{12}|\mathscr{Y}^M_{J,1J+1}\rangle = \frac{-2(J+2)}{2J+1}. \tag{5}$$

5-3. Demonstrate either directly or from the formulas in Appendix I of Case and Pais [20], that the P-wave contribution to $\sigma_{pp}(\theta)$ is

$$\pi\lambda^2\{[\tfrac{17}{4}\sin^2\delta_2 + \tfrac{9}{4}\sin^2\delta_1 + \tfrac{1}{2}\sin^2\delta_0$$

$$+ \tfrac{9}{2}\cos(\delta_2 - \delta_1)\sin\delta_2\sin\delta_1 + 2\cos(\delta_2 - \delta_0)\sin\delta_2\sin\delta_0]$$

$$- \sin^2\theta[\tfrac{21}{8}\sin^2\delta_2 + \tfrac{9}{8}\sin^2\delta_1 + \tfrac{27}{4}\cos(\delta_2 - \delta_1)\sin\delta_2\sin\delta_1$$

$$+ 3\cos(\delta_2 - \delta_0)\sin\delta_2\sin\delta_0]\},$$

where δ_0, δ_1, and δ_2 are the ${}^3P_0$-, ${}^3P_1$-, and ${}^3P_2$-phase shifts, respectively. If the phase shifts are small and the Born approximation is valid, show that the contribution from an $\mathbf{L \cdot S}$ force is $9\pi\lambda^2\delta_2^2\sin^2\theta$, where δ_2 is the part of the ${}^3P_2$-phase shift due to the $\mathbf{L \cdot S}$ force.

5-4. (a) For the tensor operators $S_{12}(\mathbf{r})$ defined in the text, prove that

$$S_{12}^2(\mathbf{r}) = 6 + 2\boldsymbol{\sigma}_1 \cdot \boldsymbol{\sigma}_2 - 2S_{12}(\mathbf{r}). \tag{1}$$

Hence show that, acting on a singlet state, S_{12}^2 gives zero, whereas for a triplet state it has a substantial central component.

(b) For plane-wave states in a volume Ω,

$$\psi_{\mathbf{k}_1}(\mathbf{r}_1) = \frac{1}{\sqrt{\Omega}}\exp(i\mathbf{k}_1 \cdot \mathbf{r}_1), \tag{2}$$

prove that the matrix element of V_T defined in Eq. 5–9 is given by the expression

$$\int \psi_{\mathbf{k}_1+\mathbf{q}}^*(\mathbf{r}_1)\psi_{\mathbf{k}_2-\mathbf{q}}^*(\mathbf{r}_2) V_T(r)S_{12}(\mathbf{r})\psi_{\mathbf{k}_1}(\mathbf{r}_1)\psi_{\mathbf{k}_2}(\mathbf{r}_2)\, d^3r_1\, d^3r_2$$

$$= -\left(\frac{4\pi}{\Omega}\right)V_T^{(2)}(q)S_{12}(\mathbf{q}), \tag{3}$$

where

$$V_T^{(2)}(q) = \int_0^\infty j_2(qr)V_T(r)r^2\, dr,$$

j_2 being the spherical Bessel function of order 2, and

$$S_{12}(\mathbf{q}) = \frac{3(\boldsymbol{\sigma}_1 \cdot \mathbf{q})(\boldsymbol{\sigma}_2 \cdot \mathbf{q})}{q^2} - (\boldsymbol{\sigma}_1 \cdot \boldsymbol{\sigma}_2).$$

Comment on the statement that nondiagonal matrix elements (3) are negligible for very small values of q if $V_T(r)$ is of short range. Taking, for simplicity, a Yukawa form for $V_T(r)$ of one-pion-exchange range, estimate the value of q at which $V_T^{(2)}(q)$ is maximum.

5-5. Using the notation of Section 5–9, we can write the Schrödinger equation for energy $E_k = k^2$ as (with kinetic-energy operators $T = H_0$)

$$(k^2 - H_0)|\psi_k\rangle = V|\psi_k\rangle.$$

In r-space this becomes

$$(k^2 + \nabla^2)\psi_k(\mathbf{r}) = \int \langle \mathbf{r}|V|\mathbf{r}'\rangle \psi_k(\mathbf{r}') \, d^3r',$$

and in p-space the same equation is

$$(k^2 - p^2)\psi_k(\mathbf{p}) = \int \langle \mathbf{p}|V|\mathbf{p}'\rangle \psi_k(\mathbf{p}') \, d^3p',$$

where we use the notation $\langle \mathbf{p}|\psi_k\rangle = \psi_k(\mathbf{p})$, etc.

(a) Consider an attractive separable potential acting in the s-state,

$$\langle \mathbf{r}|V|\mathbf{r}'\rangle = -\lambda g(r)g(r'), \tag{1}$$

where λ is a positive constant. Show that the Schrödinger equation in the p-space reduces to

$$(k^2 - p^2)\psi_k(\mathbf{p}) = -\lambda g(p) \int g(p')\psi_k(\mathbf{p}') \, d^3p', \tag{2}$$

where

$$g(p) = \int \langle \mathbf{p}|\mathbf{r}\rangle g(r) \, d^3r$$

$$= \frac{1}{(2\pi)^{3/2}} \int \exp(-i\mathbf{p}\cdot\mathbf{r}) \, g(r) \, d^3r.$$

(b) Noting that Eq. 2 above is of algebraic nature, prove that the outgoing scattering solution is given by

$$\psi_k(\mathbf{p}) = \delta(\mathbf{p} - \mathbf{k}) - \frac{\lambda g(k)}{1 + \lambda \int d^3q [g^2(q)/(k^2 - q^2 + i\varepsilon)]} \frac{g(p)}{k^2 - p^2 + i\varepsilon}. \tag{3}$$

The scattering amplitude for elastic scattering is defined by Eq. 5–55 in the text. Using that definition and Eq. 3, show that

$$f_k = -2\pi^2 \langle \mathbf{k}|V|\psi_k\rangle = \frac{2\pi^2 \lambda g^2(k)}{1 + \lambda \int d^3q [g^2(q)/(k^2 - q^2 + i\varepsilon)]}. \tag{4}$$

(c) Use Eqs. 4 and 5–57 to prove that

$$k \cot \delta = \frac{1 + \lambda I}{2\pi^2 \lambda g^2(k)}, \tag{5}$$

where

$$I = P \int d^3q \, \frac{g^2(q)}{k^2 - q^2}.$$

the symbol P indicating the principal part integral. You will have to make use of the identity

$$\int d^3q \, \frac{g^2(q)}{k^2 - q^2 + i\varepsilon} = P \int d^3q \, \frac{g^2(q)}{k^2 - q^2} - i2\pi^2 k g^2(k).$$

(d) Taking a specific example, let us put

$$V(k, k') = -\frac{\hbar^2}{M} \lambda \frac{1}{\sqrt{k^2 + \eta^2}} \cdot \frac{1}{\sqrt{k'^2 + \eta^2}} \text{ MeV fm}^3, \qquad (6)$$

where we have now written $\hbar^2/M$ explicitly to clarify the units. (Note that λ has units of fm^{-1}.) Using formula 5, show that

$$k \cot \delta = \frac{(k^2 + \eta^2) - 2\pi^2 \lambda \eta}{2\pi^2 \lambda}. \qquad (7)$$

Hence note that the effective-range formula 2–41 is exact for this potential, with scattering length a and effective range r_0 given by

$$\frac{1}{a} = \eta \left(1 - \frac{\eta}{2\pi^2 \lambda} \right), \qquad r_0 = \frac{1}{\pi^2 \lambda}. \qquad (8)$$

Compare these results with those of Problem 2–7. Note that, since the simple Bargmann potential of Problem 2–7 also satisfies the effective-range formula exactly, it can be chosen to be "phase equivalent" to the potential of Eq. 6, that is, the two yield the same phase shift $\delta(k)$ for all positive values of k. This implies that the on-shell values of $\langle \mathbf{k}'|T(k^2)|\mathbf{k}\rangle$ with $k'^2 = k^2$ are the same for the two, but the off-shell matrix elements may be very different.

5–6. The bound-state problem with the potential given by Eq. 1 of Problem 5–5 can be solved exactly. In this case, we put $k^2 = -\alpha^2$, where the binding of the two-body system is $\hbar^2 \alpha^2/M$. Then the Schrödinger equation (Eq. 2) of Problem 5–4 modifies to

$$(\alpha^2 + p^2)\psi_\alpha(\mathbf{p}) = \lambda g(p) \int g(p')\psi_\alpha(\mathbf{p}') \, d^3p'.$$

Show that the bound-state energy α^2 is determined by the equation

$$\frac{1}{\lambda} = \int d^3q \, \frac{g^2(q)}{\alpha^2 + q^2}.$$

Can there be more than one bound state for such a potential?

5–7. Consider a nucleon confined in a very large sphere of radius R, such that its energy eigenvalues are nearly continuous. Let the number of states of a given partial wave l with wave number lying between k and $k + dk$ be denoted by $g_l^{(0)}(k) \, dk$. Next assume the nucleon to be in the same enclosure, but in the presence of a one-body potential $U(r)$ of range $r_n \ll R$. Let the state density in k-space be now denoted by $g_l(k)$.

(a) Prove that

$$g_l(k) - g_l^{(0)}(k) = \frac{2l+1}{\pi} \cdot \frac{\partial \delta_l(k)}{\partial k}, \tag{1}$$

where $\delta_l(k)$ is the phase shift suffered by the nucleon of wave number k in the presence of potential $U(r)$.

Hint: Note that the eigenvalues k are determined in the two cases by the boundary conditions

$$kR + \frac{l\pi}{2} = \pi n \quad \text{[free nucleon]},$$

and

$$kR + \frac{l\pi}{2} + \delta_l(k) = \pi n \quad \text{[nucleon in the field of } U(r)\text{]},$$

where $n = 0, 1, 2$, etc.

(b) Formula 1 shows that, if the phase shift *rises* steeply over a narrow range of k and then flattens, it will give rise to a sharp bump in the level density with a width determined by $\partial \delta_l / \partial k$. Since a discrete bound state may be regarded as a Dirac-delta function in the level density, these sharp bumps in the continuum arising from Eq. 1 may be regarded as "resonances." Invoking the uncertainty principle, note that a particle in a resonant state in the continuum will be trapped there for a finite lifetime.

Look up the energy levels of ^{8}Be in Fig. 12 of T. Lauritsen and F. Ajzenberg-Selove, *Nucl. Phys.* **78**, 1 (1966), where the α-α scattering phase shifts for $l = 0, 2$, and 4 are also shown. Comment on the three lowest-lying levels, 0^+, 2^+, and 4^+, in the light of the above discussion.

A relevant and fascinating account of the synthesis and detection of the ^{8}Be ground state is given by H. Staub, *Adventures in Experimental Physics*, Vol. γ, 1 (1973).

(c) Show that the scattering cross section in the lth partial wave may be written as

$$\sigma_l(k) = \frac{4\pi}{k^2} (2l+1) \frac{1}{\cot^2 \delta_l + 1}. \tag{2}$$

The Born approximation result given by Eq. 5-39 suggests that, for small values of k, $\sin \delta_l \propto k^{2l+1}$, a result that turns out to be correct even in exact analysis. Imagine the phase shift δ_l as a function of k to rise sharply from 0 to π in a short range of k-values and then flatten out. Sketch a plot of σ_l for this situation. Note that σ_l reaches its "unitary limit," $(2l+1)4\pi\lambda^2$, at a value of k for which $\delta_l(k) = \pi/2$.

5-8. Consider the "half-off-shell" scattering amplitude as defined in Eq. 5-55, with $k'^2 \neq k^2$,

$$f(\mathbf{k}', \mathbf{k}) = -2\pi^2 \langle \mathbf{k}'|T(E = k^2)|\mathbf{k}\rangle$$

$$= -2\pi^2 \int \exp(-i\mathbf{k}' \cdot \mathbf{r}) \, V(r)\psi_k(\mathbf{r}) \, d^3r. \tag{1}$$

Show that the S-wave contribution to Eq. 1 can be written as

$$f_0(k', k) = -8\pi^3 \int_0^\infty \frac{\sin k'r}{k'} V(r)u_k(r) \, dr, \tag{2}$$

where $(1/r)u_k(r)$ is the wave function in the S-state. Denoting the asymptotic form of $u_k(r)$ by $v_k(r)$ (see Section 2–3):

$$u_k(r) \sim v_k(r) = \frac{\exp[i \, \delta_0(k)]}{k} \sin(kr + \delta_0), \tag{3}$$

show that

$$f_0(k', k) = f_0(k, k) + (2\pi)^3(k'^2 - k^2) \int_0^\infty \frac{\sin k'r}{k'} [u_k(r) - v_k(r)] \, dr, \tag{4}$$

where $f_0(k, k) = [\exp(i \, \delta_0(k))/k] \sin \delta_0(k)$ is the on-shell scattering amplitude. Note that, for a finite-range potential $V(r)$ with range r_n, the upper limit in the integral of Eq. 4 is just r_n, since, for $r > r_n$, $u_k = v_k$.

The off-shell scattering amplitude $f_0(k', k)$ thus depends sensitively on the Fourier transform of the "wound" $[u_k(r) - v_k(r)]$, while the on-shell part $f_0(k, k)$ is completely determined by the phase shift [T. Fulton and P. Schwed, *Phys. Rev.* **115**, 973 (1959)].

Hint: Note that Eq. 2 may be rewritten as

$$f_0(k', k) = -8\pi^3 \int_0^\infty \frac{\sin k'r}{k'} \left(k^2 + \frac{d^2}{dr^2}\right)[u_k(r) - v_k(r)] \, dr.$$

Chapter 6

Nuclear Binding Energies

6-1 INTRODUCTION

Clearly, a basic property of a nucleus is its total energy or mass in its ground state. We have already remarked that the mass is near $ZM_p + NM_n$; it is its deviation from this value which is of interest. The quantity most easily measured is the mass of the neutral atom, denoted by $M(A, Z)$. We define the binding energy of a nuclide as

$$B(A, Z) = ZM_H + NM_n - M(A, Z), \qquad (6\text{-}1)$$

where M_H is the mass of the hydrogen atom (H^1). The binding energy consists of the binding energy of the nucleus, plus the difference between the binding energy of the electrons and that of Z hydrogen atoms. The electronic contribution is negligible for all usual purposes.[†]

Nuclear masses can be obtained by a number of methods. The most direct technique is mass-spectrometric determination of the mass-to-charge ratio in electric and magnetic fields. Atoms of the desired substance are ionized and introduced into the mass spectrometer, along with a comparison material of known mass and almost the same mass-to-charge ratio. Comparison materials are often various hydrocarbons, and advantage is taken of the presence of multiply charged ions. The procedure usually makes it possible

---

[†] Various other quantities contain the same information as B. One is the *mass excess* $M - A$, where M is expressed in mass units. One mass unit ($= 1$ u $= 931.44$ MeV) is one-twelfth the mass of the neutral atom ^{12}C (1 mu $= 10^{-3}$ u). Another expression that has been used is the *packing fraction* $(A - M)/A$. The authors agree with Wapstra that the latter usage is best avoided.

to arrange a "mass doublet" between the unknown mass and a known mass. In this way, many nuclidic masses have been measured in terms of carefully determined standards [1, 2].[†]

An alternative direct mass measurement may occasionally be made by observing the change in the moment of inertia of a molecule when one isotope of an element is replaced by another. The mass of one of the isotopes must be known. The method is accurate if the molecular spectrum has frequencies in the microwave region; their precise values are related to the rotational energies and hence to the moments of inertia. Only a few values have been determined in this way [3].

The differences between certain nuclear masses can be obtained from the thresholds of various nuclear reactions and radioactive decays. The end-point energies of β-spectra (corrected by two electron masses in positron decay) give the mass difference between parent and daughter atoms. Of course, if the daughter nucleus is left in an excited state, the energy of the subsequent γ-decay must be included. The energy difference between an α-active nuclide (A, Z) and the daughter $(A - 4, Z - 2)$ consists of the kinetic energy of the α-particle, corrected for the recoil of the nucleus, less the binding energy of the α-particle; in this case also, we must be sure that we are dealing with the ground-state transition, or, if we are not, we must include the energy of the de-excitation γ-rays. Nuclear reactions of the types (p, n), (d, p), (n, γ), (d, α), etc., can be used to find mass differences if the energies of the incident and final particles or photons are known. A large number of such reactions have been studied sufficiently accurately to produce mass data. Useful compilations of radioactive and reaction data are published in conference reports and nuclear data tables [4, 5, 6].

By these various methods, many atomic masses have been measured in several ways with results that do not always agree. It is an extensive task to assess the reliability of the various measurements, to reconcile the discrepancies, and to obtain a best value. Wapstra and Gove [7] have prepared an extensive mass table by making a least-squares reduction of the available Q-value and mass-doublet data. The basic data are displayed in Fig. 6–1, which represents the binding energy per nucleon, B/A, plotted against A, for the most stable nuclide for each A.[‡] Determination of the best mass values is a continuing process.

[†] In all work done before 1960 a different mass unit, denoted as amu, was used. It was equal to $\frac{1}{16}$ the mass of the ^{16}O atom.

[‡] Usually there is only one β-stable nuclide for a given A, but in 58 cases there are two such nuclides.

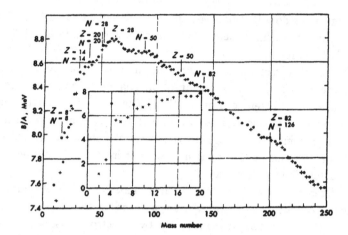

Fig. 6-1 Binding energy per nucleon of the most β-stable isobars. Each point represents an average of a few neighboring nuclides: solid circles for even-even, crosses for odd A. In the insert for low A, oblique crosses refer to odd-odd nuclides. [After Wapstra.]

The most remarkable aspect of the experimental masses revealed in Fig. 6-1 is the approximate constancy of the binding energy per nucleon. Except for the very lightest atoms ($A < 10$), B/A is never less than 7.4 MeV and never greater than 8.8 MeV. This was one of the facts realized in the early 1930's from the mass measurements of Aston [8, 9], and it led Heisenberg [10] to propose exchange forces between nucleons. If each nucleon interacted through an attractive force with all other nucleons, we would expect a potential energy proportional to the number of pairs, that is, roughly proportional to A^2. Indeed, as more particles were added and the attraction became stronger, we might expect the particles to draw closer together, so that the binding would increase faster than A^2. Then B/A would be proportional to at least A. But B/A is nearly constant, and, as more particles are added, nuclear density is also constant. In other words, nuclear forces saturate, as if each nucleon were effectively interacting with only a few others. The saturated value of B/A is reached in nuclei with about 10 or more particles.

We shall see in Chapter 8 that complete understanding of the saturation involves the precise nature of the internucleon forces, especially the contributions of the tensor force, the repulsive core, and the exchange forces, as well as the Pauli exclusion principle. However, the gross features of the variation of the binding energy with mass number A may be understood by considering the nucleus as a zero-temperature Fermi gas of uniform density contained in a volume Ω and by assuming the effective interaction between the nucleons inside the nucleus to be reasonably smooth and its contribution estimable by a perturbation calculation. Before carrying out such a calculation, we try to justify this model of the nucleus in the following section.

6-2 FERMI GAS MODEL OF THE NUCLEUS

A nucleus is a self-bound system consisting of A fermions with a core density which is approximately uniform. In a first approximation, a nucleus can be regarded as a sphere of uniform density having an equivalent radius R (see Chapter 4). The average volume occupied by a nucleon in this picture is $(4\pi/3)r_0^3$, where $r_0 = RA^{-1/3}$ is slightly greater than 1 fm (see Table 4-1). The average distance between two nucleons then is $\sim 2r_0$, which is about 2 fm. The first question to ask is, "Why may the system be regarded as a gas and not, for example, as a solid?" The characteristic feature of a solid is that the constituent atoms in it are localized in space, so that the vibrational energy due to thermal and zero-point motion is small compared to the potential energy between the atoms. This keeps the atoms glued to fixed positions, although they can oscillate about their equilibrium positions. When the temperature is increased so that the thermal energy becomes comparable to the potential energy or larger, the solid turns to a liquid and ultimately a gas. In a nucleus, if it is to be regarded as a solid, a nucleon should be localizable within a distance $\Delta x \sim 1$ fm, since the average interparticle spacing is about 2 fm. This localization would give rise to a zero-point energy $(\hbar^2/2M)[1/(\Delta x)^2] \approx 21$ MeV, which is much larger than the potential energy of two nucleons that are 2 fm apart. This indicates that even in the absence of thermal motion the nucleons may be regarded as a Fermi gas. Had the two-nucleon force been much stronger at the internucleon distance of 2 fm, the Fermi gas picture would cease to be valid.

The next question concerns the interaction between the various nucleons in the gas. Are the nucleons in the gas strongly interacting, or can they be regarded as independent of each other in the first approximation, and the potential energy estimated using perturbation theory with an effective interaction? We saw in Chapter 5 that the interaction between two nucleons is quite strong; hence at first sight the independent-particle picture does not seem plausible. Although we shall give a detailed justification of the independent-particle picture in a later section, two salient points are worth mentioning here. First, although the two-nucleon scattering phase shifts indicate a strong short-range repulsion followed by a deep attractive tail, these could equally well be fitted at low energies by a weak attractive potential which became still weaker as the relative momentum between the pair of nucleons increased. This would result from partial cancellation of the strong repulsive and attractive parts of the potential, leaving a rather weak effective interaction. Second, in the ground state of the Fermi gas all single-particle states up to a certain momentum k_F are fully occupied. A pair of particles of momenta $\mathbf{k}_1$ and $\mathbf{k}_2$ ($k_1, k_2 < k_F$) can only scatter, conserving momentum, to two states $\mathbf{k}_1 + \mathbf{q}$ and $\mathbf{k}_2 - \mathbf{q}$, where these momenta are greater than k_F. This cannot be real scattering, however, since the energy of the pair is not conserved. These virtual scatterings introduce only high-momentum components into the wave function (since k_F is large), modifying it at short relative distances. Since no real scattering can take place, a virtual scattering has to be followed by other energy-nonconserving scatterings to bring the pair back to the configuration in which the states $\mathbf{k}_1$ and $\mathbf{k}_2$ are occupied. Thus no phase shifts can be introduced into the pair wave function inside the nucleus, with the consequence that it "heals" to its unperturbed form in a relatively small distance compared to the interparticle distance. The short-range part of the pair wave function ($r < 1$ fm) may be distorted through the high-momentum components introduced via the virtual scatterings. It is reasonable, therefore, to make a first-order approximation that the nucleons are moving freely within the nuclear volume Ω, with the undistorted spatial wave functions of the form $(1/\sqrt{\Omega}) \exp(i\mathbf{k}_i \cdot \mathbf{r}_i)$. The total wave function of the nucleus is an antisymmetrized product of such single-particle states, taking account, of course, of the spin and i-spin parts also.

A noninteracting Fermi gas has a ground state with all single-particle levels up to the Fermi energy completely filled, and all other levels empty. This corresponds to a Fermi distribution with temperature $\theta = 0°$K. In our simple picture the single-particle levels form a continuum, whereas in a real nucleus they are discrete. A real nucleus cannot absorb an arbitrary amount of energy from an external source because of the discrete nature of its energy levels, but can go to an excited state if the energy supplied equals the gap between the ground and the excited state. Typically, this is of the order of 100 keV–1 MeV, or, in degrees Kelvin[†], 10^9–10^{10} °K. Thus, if a nucleus is in a laboratory of ambient room temperature, or a furnace with ambient temperature of 10^4 °K or so, it is unable to absorb any energy and remains an isolated system of $\theta = 0°$K. If the gas model of the nucleus is taken literally, however, the nucleus can absorb any arbitrary amount of energy, since its energy levels are continuous. Is it justifiable, then, to use the zero-

---

[†] A temperature of $\theta = 1°$K corresponds to an energy $k_B\theta \approx 10^{-4}$ eV, where k_B is Boltzmann's constant.

temperature distribution for a nucleus at room or high ambient temperature? We shall see that for nuclear densities the zero-temperature Fermi distribution yields an average nucleon kinetic energy of ~ 23 MeV or about 10^{11} °K, and even if the individual nucleons are considered to be at room or furnace temperature, their thermal energies are completely negligible in comparison to the kinetic energy of the zero-temperature distribution. Thus, as long as the Fermi energy θ_F (expressed in degrees Kelvin) of the gas is very much greater than the ambient temperature θ, it is valid to use the zero-temperature distribution.

An extreme example of this is the so-called neutron star. As the name implies, it consists mainly of neutrons, and the interior temperature can be as great as 10^9 °K. Neutron stars are believed to be very dense, and the radius may be only 10 km, but the mass of the star is comparable to that of the sun. A simple calculation yields the mass density $\rho \sim 10^{15}$ g/cm³, which is of the same order as the nuclear density. Thus θ_F for a neutron star is $\sim 10^{11}$ °K, which is much higher than the ambient temperature of 10^9 °K. It is proper, therefore, to consider the neutron star to be a zero-temperature Fermi gas.[†] A gas with zero-temperature characteristics is termed completely degenerate, because its behavior is very different from that of a classical gas.

It is now a simple matter to calculate the kinetic energy of a nucleon in this model. Each state of linear momentum **p** can accommodate four nucleons according to the exclusion principle when the spin-isospin degeneracy is taken into account. We assume first that the numbers of protons and neutrons in the nucleus are the same and that the volume Ω is large enough to regard the momentum states as continuous. All states up to a "Fermi momentum" p_F are filled in a zero-temperature gas, so that the total number of nucleons is A. Assuming that there is one quantum state per volume h^3 of the phase space, we obtain

$$A = \frac{4\Omega}{h^3} \int_0^{p_F} d^3p = \frac{16\pi}{3} \Omega \left(\frac{p_F}{h} \right)^3.$$

This yields a relation between $k_F = p_F/\hbar$ and the nucleon density $\rho = A/\Omega$,

$$\rho = \frac{2}{3\pi^2} k_F^3. \tag{6-2}$$

---

† Baym, Bethe, and Pethick [11] have estimated that the uniform neutron gas approximation for neutron stars is valid only for $\rho > 2.4 \times 10^{14}$ g/cm³. For lesser densities, a description in terms of a lattice of neutron-rich nuclei, immersed in a neutron gas, is more appropriate.

Current data on the density of heavy nuclei indicate that $k_F = 1.36$ fm^{-1}, but it is less for lighter nuclei. The total kinetic energy of the gas is

$$\langle T \rangle = 4 \frac{\Omega}{h^3} \int_0^{p_F} \frac{p^2}{2M} d^3p = \frac{3}{5} \frac{\hbar^2 k_F^2}{2M} A, \qquad (6\text{-}3)$$

which shows that the average kinetic energy of a nucleon is 60% of the value for the most energetic particle. It is also seen from Eqs. 6–2 and 6–3 that

$$\langle T \rangle = \frac{3}{5} \frac{\hbar^2}{2M} \left(\frac{3\pi^2}{2} \right)^{2/3} \rho^{2/3} A. \qquad (6\text{-}4)$$

For $k_F = 1.36$ fm^{-1}, the Fermi kinetic energy is 38 MeV and the average kinetic energy of a nucleon is about 23 MeV.

Next, we examine the potential energy of the system. For simplicity, we first take an attractive Wigner type of central two-body effective interaction, which is the same for all states, and consider only the direct term. Since there are $A(A-1)/2 \approx A^2/2$ pairs of nucleons,

$$\langle V \rangle = \frac{A^2}{2} \int \frac{\rho(\mathbf{r}_i)}{A} \frac{\rho(\mathbf{r}_j)}{A} V_W(r_{ij}) \, d^3 r_i \, d^3 r_j, \qquad (6\text{-}5)$$

where $\rho(\mathbf{r})/A$ is the probability of finding a particle at point $\mathbf{r}$. In the gas model ρ is simply A/Ω. Using this and the fact that the interaction depends only on relative coordinates, we obtain

$$\langle V \rangle = \tfrac{1}{2}\rho A \int V_W(r_{ij}) \, d^3 r_{ij} = A\rho \bar{V}_W, \qquad (6\text{-}6)$$

where $\bar{V}_W$ is half the volume integral of the interaction and is independent of A. Combining Eqs. 6–4 and 6–6, we note that for this case the energy per nucleon as a function of density is given by

$$\frac{E}{A} = c_1 \rho^{2/3} - c_2 \rho, \qquad (6\text{-}7)$$

where c_1 and c_2 are positive constants independent of A. The equilibrium density is obtained by minimizing the energy per particle, and we see that a purely attractive force like V_W tends to make $\rho \to \infty$, that is, the system collapses.

As we mentioned before, there are several mechanisms which prevent this collapse, depending on the details of the two-nucleon force. For simplicity, the above analysis was performed by assuming a purely attractive force. If there is a repulsive core in the potential, the particles cannot be packed too closely together, since the repulsion starts to dominate. This has to be reflected in the formalism by noting that, although the probability of finding a particle at any given position is uniform throughout the gas, the probability of finding a second particle at r_i is altered when it is known that there is one at r_j. This "correlation effect" is a function of r_{ij}. We also know that there is a strong tensor component in the two-nucleon force. It does not contribute any binding in the first order to a spherical system, but gives attraction in the second order. Its contribution, however, does not increase linearly with ρ but flattens out with increasing density, ultimately decreasing for very large densities. This helps greatly in producing saturation.

Furthermore, the introduction of exchange forces radically alters the potential contribution. The space-exchange force, for example, is attractive in even states and repulsive in odd states. This certainly reduces the potential energy. For example, the four particles in ^{4}He are all in relative S-states, about the center and in relation to each other, but the fifth particle in ^{5}He must, by the exclusion principle, be in a central p-state, and is in a relative P-state with all the other particles. With a pure space-exchange type of force, the fifth particle would be repelled. One can estimate the effect of exchange forces rather simply in this model. A Majorana-type space-exchange force V_M is attractive for space-symmetric pairs and repulsive for antisymmetric pairs. Let ξ be the fraction of the interacting particles within the nuclear force range which form antisymmetric states in the space coordinates. Then the contribution to the potential energy from the symmetric pairs is, as derived in Eq. 6–6, given by $A\rho(1 - \xi)\bar{V}_M$, and the contribution from the antisymmetric pairs is $- A\rho\xi\bar{V}_M$. The total potential-energy contribution from Wigner- and Majorana-type forces is

$$\langle V \rangle = A\rho[\bar{V}_W + (1 - 2\xi)\bar{V}_M]. \tag{6–8}$$

Note that ξ is a function of the nucleon density ρ, decreasing with decreasing density. This change arises because, for a fixed number of particles, the larger the nucleus, the lower are the kinetic energies, the more the relative momentum of any nucleon pair is reduced, and the greater is the increase in the corresponding de Broglie wavelength. But two particles with an antisymmetric space part of the wave function cannot occur at the same point, for in that case the wave function must vanish, and the probability of their being within a de Broglie wavelength of each other is correspondingly very small. Thus, in Eq. 6–8, $1 - 2\xi$ is a decreasing function of density and helps to bring about saturation.

We see from Eqs. 6–4 and 6–8 that the total energy of the gas is a linear function of the nucleon number A. Furthermore, the saturation density ρ_0, which minimizes the energy per particle, is independent of A because the constants c_1 and c_2 in Eq. 6–7, as well as others that enter with more complicated forces, are determined only by the characteristics of the force. It follows, therefore, that the binding energy per nucleon is a constant independent of A in this simple model, which is the major trend of the accumulated experimental data in Fig. 6–1.

To get a better fit to data we shall have to take account of surface effects due to the finiteness of the nuclear volume, Coulomb forces between protons, and the so-called symmetry and pairing effects. Since these make energy contributions that are not proportional to the number of particles, the binding energy per particle displays some gross A-dependence, which we shall study for each of these effects within the framework of the Fermi gas model.

In evaluating the kinetic energy $\langle T \rangle$ from Eq. 6–4, we assumed the number of momentum states between $\mathbf{p}$ and $\mathbf{p} + d\mathbf{p}$ to be $(4\Omega/h^3)\,d^3p$, which is strictly true only if the surface-to-volume ratio in configuration space is zero. More careful calculations show that this is $(4\Omega/h^3)[1 - (Sh/8\Omega p)]\,d^3p$, where S is the surface area of the nucleus (see Problem 6–6). This has the effect of increasing the average kinetic energy of a nucleon by an additional term which is proportional to the surface-to-volume ratio, that is, $A^{-1/3}$. When this effect is included, the total kinetic energy $\langle T \rangle$ is increased by this surface term, which is proportional to $A^{2/3}$. Surface effects also modify the potential-energy contribution because a nucleon near the surface region will have fewer neighbors. This can be seen in the following manner. The expression $\langle V \rangle$ in Eq. 6–5 may be written as

$$\langle V \rangle = \frac{A^2}{2\Omega} \int V_W(r_{ij})\, d^3r_{ij}\, d^3R_{ij},$$

where $\mathbf{r}_{ij} = \mathbf{r}_i - \mathbf{r}_j$, and $\mathbf{R}_{ij} = (\mathbf{r}_i + \mathbf{r}_j)/2$. In writing Eq. 6–6, we assumed that the integration over d^3R_{ij} is simply Ω. But if the volume Ω is finite (i.e., a sphere of radius R_0), we must impose the conditions

$$|\mathbf{r}_i| = \left| \mathbf{R}_{ij} + \frac{\mathbf{r}_{ij}}{2} \right| < R_0 \qquad \text{and} \qquad |\mathbf{r}_j| = \left| \mathbf{R}_{ij} - \frac{\mathbf{r}_{ij}}{2} \right| < R_0,$$

which restrict the region of integration. For simplicity, if we assume V_W to be an attractive square well of range r_n, a straightforward but tedious calculation shows that Eq. 6–6 is modified to

$$\langle V \rangle = A \rho \bar{V}_w \left[1 - \frac{9}{16} \frac{r_n}{R_0} + \frac{1}{32} \left(\frac{r_n}{R_0} \right)^3 \right]. \tag{6-9}$$

The cubic term in the square bracket is negligible, but the linear term is significant. Assuming that $R_0 = r_0 A^{1/3}$, we find for the potential energy

$$\langle V \rangle = A \rho \bar{V}_w - \tfrac{9}{16} A^{2/3} \rho \bar{V}_w \left(\frac{r_n}{r_0} \right), \tag{6-10}$$

which also shows that the potential energy is reduced by a factor proportional to $A^{2/3}$. The magnitude of the surface term depends on the surface area of the nucleus and as such is shape dependent.

Another important contribution to the nuclear energy comes from the electrostatic field between the protons, which is due to the long-range Coulomb force, and increases quadratically with Z. For a sphere of radius R_0 having constant charge density and total charge Ze, the Coulomb energy is

$$E_{\text{Co}} = \frac{3}{5} \frac{(Ze)^2}{R_0}. \tag{6-11}$$

This is a classical expression.[†] An accurate quantum-mechanical expression requires a good knowledge of the wave function of the protons and includes exchange contributions. In fact, information about the charge distribution and charge radii can be obtained from a knowledge of the Coulomb-energy shift of mirror nuclei and analog states. However, here we are more interested in the general trends, so we simply write, for a spherical nucleus of sharp surface,

$$E_{\text{Co}} = \frac{k_c Z^2}{A^{1/3}}, \tag{6-12}$$

where k_c is a constant.

Meyers and Swiatecki [13] have generalized this expression for the more general case of a deformed nucleus with a diffused surface. If the surface diffuseness is of the Woods-Saxon type (Eq. 4–9), they find the following expression for the Coulomb energy:

$$E_c = E_{\text{Co}} g(\text{shape}) - \left(\frac{\pi^2}{2} \right) \left(\frac{e^2}{r_0} \right) \left(\frac{a_0}{r_0} \right)^2 \left(\frac{Z^2}{A} \right), \tag{6-13}$$

[†] It may be noted that in many older publications the term Z^2 in Eq. 6–11 is replaced by $Z(Z - 1)$ to allow for the granularity of the charge, the argument being that no proton can interact with itself. However, Peaslee [12] has shown that quantally this is incorrect, and the direct Coulomb term should be proportional to Z^2.

where the function $g(\text{shape})$ depends on the deformation parameters and $a_0 = 0.546$ fm is defined in Eq. 4–9. We shall find the form of $g(\text{shape})$ in Chapter 12 for some simple types of deformation.

Adding these energies so far calculated, we may write the binding energy per particle of a spherical nucleus as

$$\frac{B(A)}{A} = -\frac{\langle T \rangle + \langle V \rangle + E_{\text{Co}}}{A}$$

$$= a_V - a_S A^{-1/3} - k_C (A - I_A)^2 A^{-4/3}, \qquad (6\text{–}14)$$

where a_V, a_S, and k_C are, respectively, called the volume, surface, and Coulomb coefficients and are independent of A. In the Coulomb term, we have taken the charge $Z = Z_A$, where Z_A is the atomic number of the most stable isobar of mass number A. Since very roughly $Z_A = \frac{1}{2}A$, we have expressed Z_A as

$$Z_A = \tfrac{1}{2}(A - I_A),$$

where $I_A = N - Z$ and is the neutron excess of the most stable isobar. Although we have obtained the gross A-dependence of the binding energy from the Fermi gas model, it is too much to expect that the numerical values of the various coefficients a_V, a_S, etc., will be predicted accurately. Rather, the point of view is taken that these constants are to be determined by fitting the observed masses, and for this reason the formula is called semiempirical. We shall comment on these numerical coefficients while discussing the com-

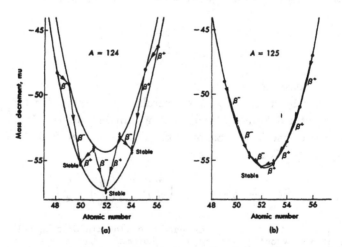

Fig. 6-2 (a) Mass excesses for $A = 124$; the lower curve is for even-even, and the upper one for odd-odd, nuclides. (b) Mass excesses for $A = 125$.

plete mass formula, but we mention here that the predictions of these from our simple model are of the right order of magnitude.

For further enlightenment, we must turn from the general trend of B/A to investigate the details in smaller regions. Let us examine the dependence of the mass decrement on Z for fixed A by studying isobars. It is observed that a plot of mass excess against Z is very accurately parabolic. For odd A, there is one parabola; for even A, the even-even and odd-odd isobars lie on different parabolas. These facts are illustrated in Fig. 6–2 for $A = 124$ and $A = 125$, where the large number of β-emitters known provide several points on each parabola. Note the following points:

1. The binding energy per particle of the most stable even-even nuclide (^{124}Te) is greater than that of the most stable odd-A nuclide (^{125}Te), which in turn is greater than that of any of the odd-odd nuclei.

2. There is only one stable odd-A nuclide, whereas there are more than one stable even-even nuclide with $A = 124$ and no stable odd-odd nuclide. This general pattern is followed throughout the periodic table. In fact, only four very light odd-odd nuclides are β-stable.

It is desirable that we understand these results as well as the parabolic shape. The Coulomb energy is parabolic in Z, but Eq. 6–14 is clearly incomplete, for if we replaced Z_A by a variable Z, then B would be a maximum for $Z = 0$. In other words, the nucleus would be all neutrons, and indeed, if Coulomb energy were the only consideration, this would be true. But the exclusion principle will allow only two neutrons and two protons in a given energy level. If we replaced the 52 protons in ^{124}Te with 52 neutrons, these could not go in the 26 low-energy levels occupied by the protons, but would have to enter the 37th to 62nd energy levels, since the first 36 levels already have their full quotas of neutrons. This change would lead to a tremendous increase in energy. At the other extreme, if there were no Coulomb energy, the exclusion principle would dictate equal numbers of neutrons and protons for stability. If we depart from $Z = N$, the Coulomb energy favors lowering Z, but the exclusion principle leads to an increase of energy. This latter energy is known as the *symmetry* energy, and the balance of Coulomb and symmetry energies determines the most stable charge for a given A.

The symmetry energy may be attributed to changes in both kinetic and potential energy. Although the kinetic-energy correction can be obtained very easily, we shall derive for it a more general expression which we can also use for the potential-energy part. Consider a nucleus with $N - Z = I$. Let the Fermi momenta for neutrons and protons be denoted by $k_F{}^N$ and $k_F{}^Z$, respectively, and let k_F be the Fermi momentum when $N = Z = A/2$. In the gas model, the number of momentum states between k and $k + dk$ is given by $\alpha k^2\, dk$, where α is a constant. We have

$$\alpha \int_0^{k_F^N} k^2\, dk = N, \qquad \alpha \int_0^{k_F^Z} k^2\, dk = Z, \qquad \alpha \int_0^{k_F} k^2\, dk = \frac{A}{2}.$$

It then follows from these equations that

$$k_F^N = k_F\left(1 + \frac{I}{A}\right)^{1/3}, \qquad k_F^Z = k_F\left(1 - \frac{I}{A}\right)^{1/3}. \tag{6-15}$$

Expanding binomially up to the second order, we obtain

$$(k_F^N - k_F) = \left(\frac{1}{3}\frac{I}{A} - \frac{1}{9}\frac{I^2}{A^2}\right)k_F,$$

$$(k_F^Z - k_F) = \left(-\frac{1}{3}\frac{I}{A} - \frac{1}{9}\frac{I^2}{A^2}\right)k_F. \tag{6-16}$$

The total kinetic energy of the system in the gas model is given by

$$\langle T \rangle = \alpha \int_0^{k_F^N} t(k)k^2\, dk + \alpha \int_0^{k_F^Z} t(k)k^2\, dk,$$

where $t(k)$ is the one-body kinetic energy in the momentum space. We define

$$f(y) = \alpha \int_0^y t(k)k^2\, dk.$$

Then, making a Taylor-series expansion, we obtain

$$f(k_F^N) = f(k_F) + \left.\frac{\partial f}{\partial y}\right|_{k_F}(k_F^N - k_F) + \frac{1}{2}\left.\frac{\partial^2 f}{\partial y^2}\right|_{k_F}(k_F^N - k_F)^2 + \cdots$$

$$= f(k_F) + \alpha t(k_F)(k_F)^2(k_F^N - k_F) + \tfrac{1}{2}\alpha\left[\left.\frac{\partial t}{\partial k}\right|_{k_F}(k_F)^2 + 2k_F t(k_F)\right]$$

$$\cdot (k_F^N - k_F)^2 + \cdots,$$

where the first term $f(k_F)$ is the kinetic energy when $N = A/2$, and the other terms are due to neutron excess. We can write a similar expression for $f(k_F^Z)$. The kinetic contribution to the symmetry energy is then

$$\langle T_{\text{sym}} \rangle = \alpha t(k_F)(k_F)^2[(k_F^N - k_F) + (k_F^Z - k_F)]$$

$$+ \tfrac{1}{2}\alpha\left[\left.\frac{\partial t}{\partial k}\right|_{k_F}(k_F)^2 + 2k_F t(k_F)\right][(k_F^N - k_F)^2 + (k_F^Z - k_F)^2].$$

Substituting for $(k_F^N - k_F)$ and $(k_F^Z - k_F)$ from Eq. 6–16 and noting that $\alpha = 3A/2(k_F)^3$, we immediately get

$$\langle T_{\text{sym}} \rangle = \frac{1}{6} \frac{\partial t}{\partial k}\bigg|_{k_F} k_F \frac{I^2}{A}. \tag{6-17}$$

Note that Eq. 6–17 has been derived without specifying the form of $t(k)$. If we put $t(k) = \hbar^2 k^2/2M$, we obtain

$$\langle T_{\text{sym}} \rangle = \tfrac{1}{3} t(k_F) \frac{I^2}{A}. \tag{6-18}$$

Thus the kinetic-energy symmetry term is parabolic in neutron excess, and its coefficient is one-third the kinetic Fermi energy of nuclear matter, or about 13 MeV.

The potential contribution to the symmetry energy comes from two different physical effects. One of these arises because the nucleons inside the nucleus move in an average one-body potential $U(k)$, which is itself momentum dependent, and the other because the $T = 0$ nucleon-nucleon force is more attractive than the $T = 1$ part of the interaction. As we shall see in Chapter 8, the average potential $U(k)$ is most attractive for $k = 0$, and it becomes less attractive with increasing k in an approximately quadratic way. Irrespective of its form, this gives rise to a symmetry energy that equals $\tfrac{1}{6}k_F(\partial U/\partial k)|_{k_F}(I^2/A)$ as before. Calculation of $U(k)$, performed in Chapter 8, indicates that this part of the symmetry-energy coefficient is repulsive and about 7 MeV in magnitude.

To calculate the i-spin part of the symmetry energy, we note that the nuclear force is more attractive for the $T = 0$ n-p (singlet) pairs than for the $T = 1$ n-p, p-p, or n-n (triplet) pairs. For a given A, the number of n-p pairs is a maximum when $N = Z$. With increasing neutron excess, the number of triplet pairs increases, with a corresponding decrease in singlet pairs. This causes a decrease in potential energy. These ideas can be formulated more precisely. In a neutron-rich nucleus, $N = \tfrac{1}{2}(A + I)$ and $Z = \tfrac{1}{2}(A - I)$. The number of triplet pairs is $\tfrac{1}{2}(N^2 + Z^2 + NZ)$, and the number of more attractive singlet pairs is $\tfrac{1}{2}NZ$. Alternatively, we can say that there are $\tfrac{1}{8}(3A^2 + I^2)$ triplet pairs and $\tfrac{1}{8}(A^2 - I^2)$ singlet pairs. If the effective Wigner-type interactions in the $T = 1$ and $T = 0$ states are denoted by V_1 and V_0, respectively, then, following Eq. 6–6, we can write for the direct contribution

$$\langle V \rangle = A\rho(\tfrac{3}{4}\bar{V}_1 + \tfrac{1}{4}\bar{V}_0) + \frac{1}{4}\frac{I^2}{A}\rho(\bar{V}_1 - \bar{V}_0), \tag{6-19}$$

where, as before, $\bar{V}_1$ is half the volume integral of the potential V_1. An estimate in regard to nuclear matter ($I = 0$) indicates that $\langle V \rangle / A \approx -40$ MeV, with the $T = 1$ and $T = 0$ parts contributing about equally, that is, $3V_1\rho \approx V_0\rho \approx -80$ MeV. This shows that the coefficient of the I^2/A term in Eq. 6–19 is about 13 MeV and is repulsive. A more accurate estimate of this term yields about 10 MeV. Thus the symmetry term in the mass formula is repulsive and is given by $a_{sym}(I^2/A)$, where

$$ a_{sym} = \tfrac{1}{3}t(k_F^0) + \tfrac{1}{6}k \left.\frac{\partial U}{\partial k}\right|_{k=k_F} + \tfrac{1}{4}\rho(\bar{V}_1 - \bar{V}_0). \tag{6–20} $$

Each of the terms in the right-hand side of this equation is repulsive, the total being about 30 MeV. In the above analysis of the symmetry energy, we have neglected all surface effects. These reduce the repulsive contribution of the symmetry energy, this contribution being proportional to $(I^2/A) \cdot (1/A^{1/3})$.

Combining all the terms that we have derived, we may write a semi-empirical mass formula for the binding energy $B_{sph}^{LD}(A, Z)$ of a spherical nucleus with a sharp surface as

$$ B_{sph}^{LD}(A, Z) = a_V A - a_S A^{2/3} - k_C Z^2 A^{-1/3} - a_{sym} I^2 A^{-1} $$
$$ + a_{surf\,sym} I^2 A^{-4/3} + \delta\, a_{pair} A^{-\varepsilon}, \tag{6–21} $$

where the last term has been added to account for the different behaviors of even-even, odd-odd, and odd-A nuclei. We have seen that, for a given A, the binding-energy-against-Z parabola for even-even nuclei shows maximum stability, the odd-odd nuclei have the least stability, and the binding energies of the odd-A nuclei are intermediate between the others. This behavior can be reproduced by taking the coefficient a_{pair} to be positive, and assuming that δ is $+1$ for even-even, 0 for odd-A, and -1 for odd-odd nuclei. The origin of this "pairing effect" lies in the special property of the nucleon-nucleon force, which gives extra binding between pairs of like nucleons in similar quantum states. This will be discussed in Chapter 8.

A formula of the type of Eq. 6–21 was first developed by Von Weizsäcker [14] and Bethe and Bacher [15], and the arguments to justify the symmetry term were given by Wigner [16]. In this simple form it neglects the correction to the Coulomb term due to the diffuseness of the surface and, more importantly, deformation and shell effects. A more general form of the mass formula for distorted nuclei has been given by Myers and Swiatecki and is written as

$$ B(A, Z) = C_V A - C_S A^{2/3} \cdot f(\text{shape}) - k_C Z^2 A^{-1/3} g(\text{shape}) $$

$$ + \delta a_{pair} A^{-\varepsilon} + C_d \left(\frac{Z^2}{A}\right) + \delta B $$

$$= B_{\text{dist}}^{LD}(A, Z) + \delta B, \tag{6-22}$$

where

$$C_V = a_V \left[1 - \kappa \left(\frac{N-Z}{A} \right)^2 \right], \qquad C_S = a_S \left[1 - \kappa \left(\frac{N-Z}{A} \right)^2 \right],$$

$$C_d = \frac{\pi^2}{2} \left(\frac{a_0}{r_0} \right)^2 \frac{e^2}{r_0}, \qquad k_C = \frac{3}{5} \frac{e^2}{r_0}, \qquad \varepsilon = 0.5, \tag{6-23}$$

f(shape) and g(shape) are shape-dependent functions dependent also on the deformation parameters of the nucleus, so that $f = g = 1$ for spherical shapes, and δB is the shell correction which arises due to single-particle bunching of levels and cannot be obtained from the liquid-drop part $B_{\text{dist}}^{LD}(A, Z)$. We shall discuss the forms of functions f and g for some simple types of deformation in Chapter 12. The term $C_d(Z^2/A)$ is a correction term to the Coulomb energy to account for the diffuseness of the surface (see Eq. 6–13).

The liquid-drop part of the mass formula $B_{\text{dist}}^{LD}(A, Z)$ has a smooth dependence on A if the binding energies of the most stable isobars are considered. The shell effect gives a superimposed structure on this smooth curve. By assuming a certain theoretical form (and its deformation dependence) for δB, Myers and Swiatecki [13] parametrized it in terms of three adjustable constants. The other adjustable parameters of the formula are a_V, a_S, κ, and r_0 in Eq. 6–22. The two other parameters, a_{pair} and ε, were taken to be 11 MeV and 0.5, respectively, from even-odd mass difference data and not adjusted. The seven free parameters (three for δB and a_V, a_S, κ, r_0) were determined by making a least-squares fit in a multi-dimensional space of about 1200 ground-state experimental binding energies, the deformations being obtained from the experimental quadrupole moments. The best values that Myers and Swiatecki quote are as follows:

$$a_V = 15.68 \text{ MeV}, \qquad a_S = 18.56 \text{ MeV}, \qquad k_C = 0.717 \text{ MeV}, \qquad \kappa = 1.79$$

$$\text{(fixed parameters are } a_{\text{pair}} = 11 \text{ MeV}, \quad \varepsilon = 0.5). \tag{6-24}$$

These values imply $r_0 = 1.2049$ fm, $a_{\text{sym}} = 28.06$ MeV, and $a_{\text{surf sym}} = 33.22$ MeV.

We do not discuss here the form of δB and the associated three parameters taken by Myers and Swiatecki, but it should be realized that the numerical values of the parameters in expression 6–24 are somewhat dependent on the form of δB that is chosen. The importance of the shell effect will be briefly discussed in the next section, and its actual evaluation will be made in Chapter 12.

The above mass formula can be used to find the Z-value of the most stable nucleus for a given A. For simplicity, neglecting the surface-symmetry term in Eq. 6–21, and setting $\partial B/\partial Z|_A = 0$, we find the condition

$$Z = \frac{A/2}{1 + (k_C/2a_{\mathrm{sym}})A^{2/3}}, \qquad (6\text{–}25)$$

which shows that for light nuclei the most stable nucleus has $Z = A/2$, while for heavier nuclei, $Z < N$.

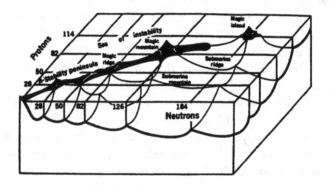

Fig. 6-3 Nuclear stability is illustrated in a scheme that shows a peninsula of known elements and an island of predicted stability (nuclei around $Z = 114$ and $N = 184$) in a sea of instability. Grid lines show magic numbers of protons and neutrons giving rise to exceptional stability. Magic regions of the mainland peninsula are represented by mountains or ridges. [After S. G. Thompson and C. F. Tsang, *Science* **178**, 1047 (1972).]

Using the mass formula, one can also investigate the conditions for spontaneous β-emission, α-decay, or nucleon emission. Figure 6–3 shows a schematic peninsula of stability where elements are known, and a predicted island of stability in the superheavy region which is yet to be found. "Magic" numbers, arising out of shell structure, result in extra stability and are shown schematically as mountains or ridges in this diagram.

6-3 SHELL EFFECTS AND REFINEMENTS IN THE MASS FORMULA

The semiempirical mass formula 6–21 is able to reproduce the main trend in the variation of binding energy of nuclei with mass number A and charge Z. That the finer details are not fitted, however, can be seen clearly when we plot the difference $(M_{exp} - M_{LD})$ versus the neutron number N. Here M_{exp} is the experimental mass of a given nucleus, and M_{LD} is the spherical liquid-drop value given by Eq. 6–21 and the Myers-Swiatecki parameters (6–24). We should mention here that the mass formula derived in Section 6–2 is often called the liquid-drop mass formula because a charged spherical liquid drop has surface tension and Coulomb terms similar to what we have obtained.

The liquid-drop analogy can be extended to calculate the kinetic energy of oscillations of the nucleus about its spherical shape (see Section 9–2). Figure 6–4 shows a plot of $(M_{exp} - M_{LD})$ against N. This quantity is also called the experimental shell effect. It will be seen that the deviations are particularly large for $N = 2$, 8, 20, 28, 50, 82, and 126. Similar large discrepancies show up when the plot is made against proton number Z. Many other nuclear properties show peculiarities at these "magic" numbers. Although the mean deviation over the periodic table is about 3 MeV, in the vicinity of the magic numbers the difference can be as large as 10 MeV. We should not be surprised at this because our derivation did not take account of the "shell effects": the single-particle levels in a real nucleus are discrete with large energy gaps between major shells, while we assumed a smooth, continuous distribution of levels. The local maxima in the binding energies of the heavier nuclei are associated with the filling up of these shells. In Chapter 12 on fission, after we have discussed the shell structure more fully, we shall explain how the semiempirical mass formula can be corrected for these shell effects. Furthermore, corrections have to be applied for deformed nuclei—apart from introducing obvious corrections in the Coulomb energy, deforma-

Fig. 6-4 "Experimental" shell correction $(M_{exp} - M_{LD})$, plotted against the neutron number ▶ number N. Here M_{exp} is the experimental mass of a nuclide (in MeV), while M_{LD} is the mass predicted by the spherical liquid-drop formula 6–21. [After Myers and Swiatecki [13].]

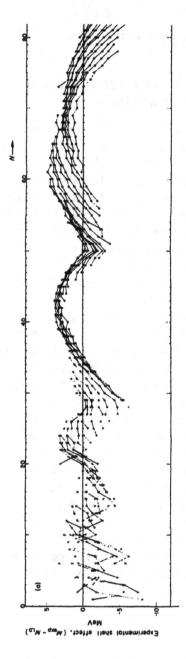

Fig. 6-4 "Experimental" shell correction ($M_{exp} - M_{LD}$), plotted against the neutron number N. Here M_{exp} is the experimental mass of a nuclide (in MeV), while M_{LD} is the mass predicted by the spherical liquid-drop formula 6–21. [After Myers and Swiatecki [13].]

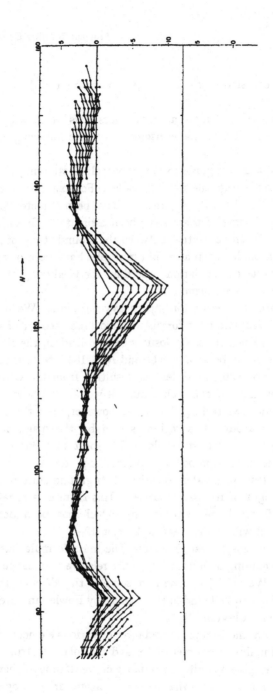

Fig. 6-4 (continued)

tion of the nuclear shape has important effects on the density of single-particle levels.

It is desirable to have more accurate mass laws (necessarily with more parameters than the five we have discussed) for the following applications:

1. To provide a useful guide to experimentalists studying the properties of nuclei far removed from the stability valley. For example, light, neutron-rich nuclei like ^{8}He, ^{11}Li, ^{15}B, ^{19}N, and ^{21}O have been detected [17], and the binding energies of some of these have been estimated. Specific equations relating the mass differences either between the ground states of neighboring nuclei or between the levels in an i-spin multiplet have been developed [18]; these give a reliable extrapolation procedure for predicting the masses of these exotic nuclei (see Problem 6–7).

2. To facilitate the search for superheavy elements. We have already mentioned that a realistic mass formula should take account of shell effects and deformation. The next shell closures beyond lead, in the single-particle model, are predicted to be at $Z = 114$ and $N = 184$. Note that the magic number for Z is not 126, the difference resulting from the Coulomb force. In the absence of shell effects, such a nucleus would have almost no fission barrier. Myers and Swiatecki [13] showed, however, that there may be an island of stability around this region (shown by the magic mountain in Fig. 6–3), where the nuclei are stable with respect to spontaneous fission, α-decay, and β-decay. It is important, in order to predict the lifetime of such a superheavy nucleus, to be able to estimate its ground-state mass and also the shape and height of its fission barrier. This requires a reliable method of calculating δB for large deformations, which has been developed by Strutinsky [19] and will be discussed in Section 12–4.

3. To study nucleosynthesis in stars. The mass formula has played an important role in gaining understanding of the relative abundance of elements in the universe. We briefly outline here some of the ideas on this subject; further information can be found in the booklet by Fowler and Stephens [20], and in the article by Clayton [21].

Figure 6–5 shows the isotopic abundances of various elements in our solar system, obtained mainly from meteorites and the solar spectrum. The data from stellar sources give a similar abundance curve, although there are some differences. The main feature is that 93% of all atoms are hydrogen, followed by about 7% of ^{4}He. All the heavier elements add up to give about 0.1% of the total atoms. Among the heavier elements, there is an overabundance of α-particle nuclei with $A = 16, 20, \ldots, 40$ and also a pronounced peak for

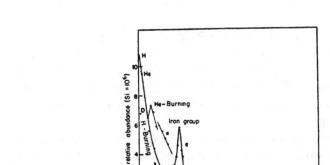

Fig. 6-5 Schematic curve of atomic abundances in our solar system. Note the over-abundances, relative to their neighbors, of the α-particle nuclei $A = 16, 20, \ldots, 40$; the peak at the iron group nuclei, and the twin peaks at $A = 80$ and 90, at 130 and 138, and at 194 and 208. [After E. M. Burbidge, G. R. Burbidge, W. A. Fowler, and F. Hoyle, *Rev. Mod. Phys.* **29**, 547 (1957).]

^{56}Fe. Beyond the iron region there appear double peaks for neutron shell closure at $N = 50$, 82, and 126, which are designated by r (rapid neutron capture) and s (slow neutron capture) in the figure. We shall presently discuss these in some detail.

There are a number of theories of cosmic origins, but all of them include a stage in which protons and electrons have formed hydrogen atoms that are distributed throughout the universe. In this rarefied medium, pockets of gas are formed by the random motion of the atoms. If the local density becomes large enough, gravity causes a growing accumulation of matter, resulting in ever more frequent collisions and a rise in temperature. When the temperature reaches about 10^7 °K, nuclear reactions start and a star is born. The first stage is "hydrogen burning," leading to the formation of helium and the release of more energy. This is the longest stage in the life of a star, occupying about 99% of its lifetime. When the hydrogen is largely depleted, the star core contracts and its temperature keeps rising. At about 10^8 °K "helium burning" takes place, with three He nuclei fusing via a series of nuclear reactions to yield ^{12}C, thus releasing more energy. Further α-

particle additions produce ^{16}O, ^{20}Ne, and ^{24}Mg. As the temperature rises to $6 \cdot 10^8$ °K, the carbon nuclei can fuse with each other and with helium to form heavier nuclei, manufacturing nuclei up to the iron region. The peak in this region appears for ^{56}Fe, which is the most stable nucleus, and results from an approach to equilibrium in a system where a large variety of nuclear reactions are taking place.

Above iron, charged-particle reactions are improbable because of the increasing Coulomb barrier, but it is energetically still favorable to form heavier nuclei by neutron capture. Neutrons are ejected in many of the nuclear reactions taking place, but we shall not go further into the details of the source of neutrons. If there is a low flux of neutrons, the capture process is called slow (s-process), since it occurs on a time scale of 10–1000 years, sufficient to permit β-decay between successive neutron captures. This results in a synthesis pattern near the line of β-stability, and the s-peaks in Fig. 6–5 are for the normal β-stable nuclei. On the other hand, a large neutron flux may be available to the iron group elements for capture on a very short time scale, ~ 0.01–10 s, so that neutron-rich isotopes are generated. The three r-peaks in the figure are thought to represent neutron-rich nuclei like $^{80}_{30}$Zn$_{50}$, $^{130}_{48}$Cd$_{82}$, and $^{195}_{69}$Tm$_{126}$, respectively. In the r-process synthesis, the formation of neutron-excess nuclei is limited by (γ, n) reactions and by β^--decays. One assumes dynamic equilibrium between (n, γ) and (γ, n) processes, and the slow β^--decay causes a leakage from the equilibrium. A detailed knowledge of the energetics (i.e., the mass law) is required to calculate the track of the r-process. In fact, Seeger [22] has shown that both shell effects and deformation energies must be included in the mass formula to obtain the best agreement with experimental data.

In order to examine the predictions of the above theories in a quantitative way, it is necessary, among other things, to have values for the masses of many nuclear species. Many of these are experimentally known, but for the unstable exotic species no longer found on earth, or even readily formed in laboratory reactions, one is forced to rely on a mass formula.

REFERENCES

1. R. C. Barber, R. L. Bishop, L. A. Cambey, H. E. Duckworth, J. D. Macdougall, W. McLatchie, J. H. Ormrod, and P. van Rookhuyzen, in *Nuclidic Masses*, edited by W. H. Johnson, Jr., Springer-Verlag, Berlin, 1964, p. 393.
2. Accounts of mass-spectroscopic measurements of several groups are reported in *Proceedings of the Third International Conference on Atomic Masses, Winnipeg, Manitoba*, edited by R. C. Barber, University of Manitoba Press, 1967, pp. 673–830.

3. S. Geschwind, *Encyclopedia of Physics*, edited by S. Flügge, Vol. 38/1, Springer-Verlag, Berlin, 1958, p. 38.
4. N. C. Rasmussen, V. J. Orphan, and Y. Hukai, in *Proceedings of the Third International Conference on Atomic Masses* (ref. 2), p. 278.
5. V. E. Viola, Jr., J. A. Swant, and J. Graber, *Atomic Data and Nuclear Data Tables* 13, 35 (1974).
6. G. Murray, J. M. Freeman, J. G. Jenkin, and W. E. Burcham, in *Proceedings of the Third International Conference on Atomic Masses* (ref. 2), p. 545.
7. A. H. Wapstra and N. B. Gove, *Nuclear Data Tables* 9, 267 (1971).
8. F. W. Aston, *Proc. Roy. Soc. (London)* A115, 487 (1927).
9. F. W. Aston, *Mass Spectra and Isotopes*, Arnold, London, 1933.
10. W. Heisenberg, *Z. Phys.* 77, 1 (1932).
11. G. Baym, H. A. Bethe, and C. J. Pethick, *Nucl. Phys.* A175, 225 (1971).
12. D. C. Peaslee, *Phys. Rev.* 95, 717 (1954).
13. W. D. Myers and W. J. Swiatecki, *Nucl. Phys.* 81, 1 (1966).
14. C. F. von Weizsäcker, *Z. Phys.* 96, 431 (1935).
15. H. A. Bethe and R. F. Bacher, *Rev. Mod. Phys.* 8, 82 (1936).
16. E. P. Wigner, *Phys. Rev.* 51, 106, 947 (1937).
17. A. M. Poskanzer, S. W. Cosper, E. K. Hyde, and J. Cerny, *Phys. Rev. Lett.* 17, 1271 (1966).
18. G. T. Garvey, W. J. Gerace, R. L. Jaffe, I. Talmi, and I. Kelson, *Rev. Mod. Phys.* 41, S1 (1969).
19. V. M. Strutinsky, *Nucl. Phys.* A95, 420 (1967); A122, 1 (1968).
20. W. A. Fowler and W. E. Stephens, *Resource Letter* OE-1 on "Origin of Elements," American Institute of Physics. This booklet also contains a reprint of the classic paper (often referred to in literature as B²FH) by E. M. Burbidge, G. R. Burbidge, W. A. Fowler, and F. Hoyle, *Rev. Mod. Phys.* 29, 547 (1957).
21. D. Clayton, *Physics Today*, May 1969, p. 28.
22. P. A. Seeger, *Ark. Fysik* 36, 495 (1966).

PROBLEMS

6–1. α-Particles from radon have an energy of 5.486 MeV. By how much do the masses of radon and radium differ?

6–2. When $M(A, Z)$ is greater than $M(A - 4, Z - 2) + M(4, 2)$, the nucleus (A, Z) is unstable against α-emission. Compute the kinetic energy of the emitted α-particle as a function of A, for the range of A for which the energy is positive, on the basis of mass formula 6–21, with the parameters given by Eq. 6–24. Use the most stable charge Z_A for Z. Compare the energies obtained with known α-energies. What are the two lightest naturally occurring α-emitters known? Justify the general lack of known α-emitters for A less than 200, in view of the wide range of A for which the kinetic energy available is positive.

6–3. Using the semiempirical mass formula, discuss the energy parabolas for the ground states of nuclei with $A = 110$, and determine an experimental value for the separation of the parabolas.

6–4. Show, ignoring surface effects, that the states of a Fermi gas in a spherical enclosure of radius R with energy less than $\hbar^2 k^2/2M$ number $(2/9\pi)(kR)^3$.

6–5. Consider a very large number N of noninteracting fermions enclosed in a large volume Ω, and let the surface-to-volume ratio of the enclosure be small enough for surface effects to be negligible. Let the density of the fermions be large enough that the kinetic energy of a particle greatly exceeds its rest mass, so that the particle energy in a state $\mathbf{p}$ is simply $\varepsilon_p = pc$.

(a) Assuming that only two fermions can be accommodated in a given state $\mathbf{p}$, show that the Fermi energy ε_F is given by

$$\varepsilon_F = cp_F = (3\pi^2)^{1/3}\hbar c\rho^{1/3}, \tag{1}$$

where $\rho = N/\Omega$.

(b) Consider a neutron star core with $\rho > 10^{17}$ g/cm³. For such densities, the neutrons are in the extreme relativistic limit, and the above model may be used. In such a gas, a certain fraction of protons, electrons, and other heavier particles will be present. Consider only the protons and electrons, appearing from the reaction

$$n \to p + e^- + \bar{\nu},$$

where the neutrinos escape without interaction. Noting that for the equilibrium condition the Fermi energies must match, that is,

$$\varepsilon_F(n) = \varepsilon_F(p) + \varepsilon_F(e),$$

use Eq. 1 and the fact that the gas is neutral to prove that the proton-to-neutron ratio is $N_p/N_n = \frac{1}{8}$.

6–6. *Surface effects in density of states.* (a) Consider a particle of mass m in an enclosure of volume Ω. Define the number of single-particle states available to the particle between energies ε and $\varepsilon + d\varepsilon$ as $g(\varepsilon)\,d\varepsilon$. Note that the classical partition function at temperature T is given by ($\beta = 1/k_B T$; k_B is the Boltzmann constant)

$$Z_{\text{cl}}(\beta) = \frac{1}{h^3}\int \exp(-\beta p^2/2m)\,d^3p\,d^3r$$

$$= \frac{\Omega}{h^3}(2\pi m)^{3/2}\cdot\frac{1}{\beta^{3/2}}. \tag{1}$$

Alternatively, by writing

$$Z(\beta) = \int_0^\infty g(\varepsilon)\exp(-\beta\varepsilon)\,d\varepsilon, \tag{2}$$

we see that $Z(\beta)$ in general is the Laplace transform of the density of states $g(\varepsilon)$, and the latter can be recovered from $Z(\beta)$ by taking its inverse Laplace transform. Using the classical expression for $Z(\beta)$ given in Eq. 1, and looking up a table of Laplace transforms (e.g., in M. Abramowitz and I. A. Stegun,

Handbook of Mathematical Functions, Dover Publications, New York), show that the classical density of states is

$$g_{cl}(\varepsilon) = \frac{\Omega}{4\pi^2}\left(\frac{2m}{\hbar^2}\right)^{3/2}\sqrt{\varepsilon}. \tag{3}$$

(b) In order to take the surface effects into account, we must recognize that at any finite temperature T the particle has an uncertainty in configuration space characterized by its de Broglie wavelength λ_d:

$$\lambda_d = \frac{\hbar}{p}. \tag{4}$$

In Eq. 4, replace p by its average value $\bar{p}$ at a given temperature T, and, taking Boltzmann weighting, show that

$$\bar{p} = 2\left(\frac{2m}{\pi}\right)^{1/2}\frac{1}{\beta^{1/2}},$$

and hence

$$\lambda_d = \frac{h}{4\sqrt{2\pi m}}\cdot\beta^{1/2}. \tag{5}$$

We now associate a finite radius λ_d with the particle, and modify Eq. 1 to exclude a volume $S\lambda_d$ from Ω (S is the surface area of the enclosure) to obtain

$$Z_{semicl}(\beta) = \frac{\Omega - S\lambda_d}{h^3}(2\pi m)^{3/2}\cdot\frac{1}{\beta^{3/2}}. \tag{6}$$

Taking the Laplace inverse of Eq. 6, show that the surface-corrected density of states is

$$g_{semicl}(\varepsilon) = \frac{\Omega}{4\pi^2}\left(\frac{2m}{\hbar^2}\right)^{3/2}\sqrt{\varepsilon} - \frac{S}{16\pi}\left(\frac{2m}{\hbar^2}\right), \tag{7}$$

which is dependent on the shape of the enclosure.

6–7. Garvey-Kelson mass formula. (a) Although the overall behavior of the binding energy $B(A, Z)$ is well reproduced by the semiempirical mass formula, the local fluctuations in the experimental values due to shell effects make this expression less useful for precise predictions of the bindings of the more exotic nuclides. These local fluctuations can be canceled, however, by constructing mass relations involving adjacent nuclides, such that the *n-n*, *n-p*, and *p-p* bonds cancel off separately. Consider, for example, the simple mass relation

$$M(N, Z) + M(N + 1, Z + 1) = M(N, Z + 1) + M(N + 1, Z), \tag{1a}$$

which can be graphically represented by

$Z+1$

$-$	$+$
$+$	$-$

(1b)

Z

$N\quad N+1$

where each square represents a nuclide with a given N, Z.

Show that the number of n-n and p-p bonds separately and exactly cancel out in Eq. 1, although this is not true for the n-p bonds. Also note that expression 1 is exactly satisfied if the local behavior of the mass can be represented by

$$M(N, Z) = f_1(N) + f_2(Z),$$

where f_1 and f_2 are arbitrary functions.

(b) Since n-p bonds are also important in determining the mass of a nuclide, it is necessary to modify the four-term formula 1 so that these too are canceled. The two six-term formulas which are next in simplicity can be easily constructed graphically by superimposing two square structures corner to corner:

$Z+2$ | $+$ | $-$ |

$Z+1$ | $-$ | $+$ | $+$ |

Z | $+$ | $-$ |

$N\quad N+1\ N+2$

(2)

and

$Z+2$ | $+$ | $-$ |

$Z+1$ | $-$ | $+$ | $+$ |

Z | $+$ | $-$ |

$N\quad N+1\ N+2$

(3)

Formulas 2 and 3, which are known as the Garvey-Kelson mass formulas [18], can be easily written down algebraically.

Show that formula 2 is satisfied exactly if

$$M(A, Z) = f_1(Z) + f_2(N) + f_3(I),\qquad (2')$$

while formula 3 is true if

$$M(A, Z) = g_1(Z) + g_2(N) + g_3(A),\qquad (3')$$

where the f_i's and g_i's are unspecified arbitrary functions, and $I = N - Z$. Equations 2 and 3 have been tested, with certain restrictions, throughout the periodic table and are accurate to within 200 keV. They are very useful, therefore, in predicting the unknown mass of a nuclide if the masses of the other five members of the difference equation are experimentally known.

(c) The Kelson-Garvey approach may also be applied to elementary particles. The semistable baryons with spin-parity $\frac{1}{2}^+$ may be grouped in a multiplet; these are $(n, p, \Lambda, \Sigma^+, \Sigma^0, \Sigma^-, \Xi^0,$ and $\Xi^-)$. Individual members of the multiplet may be specified by i-spin T, its third component T_3, and an additional quantum number S called "strangeness"; the nucleons have $S = 0$, the Λ and Σ's have $S = -1$, and the Ξ's have $S = -2$.

Draw a two-dimensional plot with S as the x-axis and charge C as the y-axis, and place the members of the $\frac{1}{2}^+$ multiplets in the appropriate places. Assume the mass dependence of a baryon to be given by a formula analogous to expression 2', that is,

$$M(C, S) = f_1(C) + f_2(S) + f_3(S - C).$$

Hence derive the mass formula

$$M(\Xi^-) + M(\Sigma^+) + M(n) = M(\Xi^0) + M(\Sigma^-) + M(p). \qquad (4)$$

Look up the masses of the above baryons in the most recent Review of Particle Properties [e. g. *Phys. Lett.* **50B**, 1 (1974)] and confirm that formula 4 is obeyed with high accuracy. This formula, which relates to electromagnetic mass splittings, can also be derived from the SU_3 classification of particles; see, for example, the article on unitary symmetry by P. T. Mathews in *High Energy Physics*, Vol. 1, Academic Press, New York, 1967, p. 392.

6–8. The interaction of a Λ-particle with a nucleon is of a short-range, attractive nature (see Chapter 2). Note that a free Λ has a lifetime of 2.5×10^{-10} s

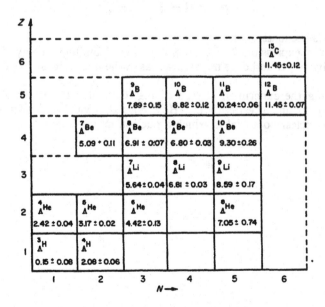

Fig. 6-6 Separation energies of the Λ-particle in light hypernuclei (all energies in MeV).

and decays by the weak interaction

$$\Lambda \rightarrow p + \pi^-$$

with a release of energy $Q = 37.7$ MeV. Because of the attractive interaction of a Λ with nucleons, it can form a bound state in a nucleus, although this interaction is too weak to form a two-body Λ-N bound pair. A nucleus containing a Λ-particle is called a hypernucleus and typically has a lifetime of the order of 10^{-10} s. The separation energy of a Λ-particle in a hypernucleus is denoted by B_Λ, and Fig. 6–6 is a chart of the experimentally known B_Λ's. Although heavier hypernuclei certainly exist, they are very hard to identify experimentally. In Fig. 6–6, the notation is standard: for example, $^5_\Lambda$He denotes that there are five baryons (four nucleons + one Λ) in the system, and that the Λ is trapped in a He nucleus.

(a) In a simple model, the Λ-particle moves in an average potential $U_\Lambda(r_\Lambda)$ which is generated by the two-body Λ-nucleon interaction:

$$U_\Lambda(r_\Lambda) = \int \rho(\mathbf{r}) V(|\mathbf{r} - \mathbf{r}_\Lambda|) \, d^3r, \qquad (1)$$

where $\rho(\mathbf{r})$ is the nuclear density. As the nucleon number A is increased, explain why you expect the depth of U_Λ to remain the same, but its radius R to increase as $A^{1/3}$. Assuming, for simplicity, that U_Λ is a square well of depth D and radius $R = r_0 A^{1/3}$, show that

$$B_\Lambda \approx D - \frac{\pi^2 \hbar^2}{2 m_\Lambda r_0^2 A^{2/3}}, \qquad (2)$$

where m_Λ is the mass of Λ.

(b) From expression 2, we note that the Λ-binding energy B_Λ varies continually with A and does not saturate as in the nuclear case. Explain why this is so.

(c) Using the experimental B_Λ's of Fig. 6–6, plot B_Λ versus $A^{-2/3}$ and extrapolate to find the depth D. How does this value compare with the analogous nuclear potential in which a nucleon moves?

Part II
Nuclear Models

Chapter 7

Single-Particle Model

7-1 INTRODUCTION

One of the main objectives of the study of nuclear physics is understanding the *structure of nuclei*. This expression is used to include all aspects of the motion of intranuclear nucleons: their paths in space, their momenta, the correlations between them, the energies binding them to each other. Mathematically, the complete description of nuclear structure is contained in the correct total wave function of the nucleus. It is clear that there are formidable difficulties in the experimental or theoretical determination of the behavior of all the degrees of freedom of a dynamical system of such complexity as a moderately heavy nucleus. Even if these difficulties could be overcome, it would be necessary to look for some simplified description of nuclei in terms of a number of parameters sufficiently small to permit ready assimilation by the human mind. At the same time, the parameters should be sufficient to give a fairly complete picture of the most important features of a particular nucleus. The problem is an example of the search for a systematic description of nature, by which science has progressed. Both our present inadequate knowledge and the general need for systematization drive us to look for conceptual or mathematical models. From a nuclear model, it must be possible, in a systematic way and without prohibitively lengthy calculation, to predict various observable properties of the nuclides. The test of the usefulness or realism of any model is the extent to which its predictions are confirmed by experiment.

We made some use of certain simple models in the discussions of Part I. Although the agreement of the model with experiment is not precise, we saw that some of the properties of nuclides, such as angular momenta and magnetic moments, suggest the usefulness of the single-particle model. In this model the individual nucleons are considered to move in stationary orbits and are paired off in such a way that the values of many nuclear parameters are determined solely by a single unpaired nucleon. The model contains no correlated or collective motion of several nucleons and no explicit reference to two-body forces between nucleons. These omissions, particularly the latter, suggest that the model must be of strictly limited applicability, and we found, for example, that its predictions of quadrupole moments are completely erroneous. To describe quadrupole moments within the framework of a model, we had recourse to the other extreme of regarding the nucleus as a distorted liquid drop, in which fairly large fractions of the fluid are moving together to produce a nonspherical shape. A simple liquid-drop model contains little reference to the numbers of nucleons present and can hardly predict the discontinuities associated with the "magic" numbers. It is clear that, to incorporate even the main features of nuclear structure, a model must be considerably more complicated than either of these extreme cases.

The more realistic models can be described as generalizations and extensions built on the single-particle model. Despite this model's sweeping simplifications and the neglect of internucleon interactions, it may well be that the single-particle orbits represent fairly well an average of the actual nucleon motions. Certain nuclear quantities are sensitive only to the average motion, but others are strongly affected by the details of the nuclear structure and, in particular, by particle correlations. From this viewpoint, the single-particle model makes a reasonable starting point to which more sophisticated models may be adduced. In this chapter we shall describe in detail the simplest form of shell model, the so-called single-particle model, and in later chapters of Part II we shall examine the embellishments that make it more realistic. In Chapter 8 we shall consider a basic question: "Why does the shell model work at all, even as a first approximation?" In attempting to answer this question, we shall be describing a nuclear theory as distinguished from a model, that is, an effort to build up the description of the nucleus from the fundamental facts concerning nuclear interactions and without introducing any ad hoc assumptions, however plausible they may appear.

7-2 SINGLE PARTICLE ORBITS

If each nucleon moves in a permanent orbit with fixed angular momentum, it is tempting to try to describe the effective force on each particle by means of a central potential $V(r)$. In analogy with atomic electrons, where $V(r) = -Ze^2/r$, one might then expect to find a shell structure in nuclei. This hypothesis was suggested during the 1930's (e.g., ref. 1), but was not widely used because of the theoretical difficulties inherent in the assumption that the strong internucleon forces could average out in such a simple way, and also because of the paucity, at that time, of experimental data that suggested shell structure or quasi-single-particle behavior. However, by 1949, a large number of measurements of nuclear spins and magnetic moments had accumulated, and the discontinuities associated with the magic numbers were recognized. In that year, a description of a single-particle model, which systematically explained these data, was published. Like many other significant developments in physics, the nuclear-shell model was arrived at simultaneously but independently, in this case by Maria Goeppert Mayer [2, 3], acting on a suggestion by E. Fermi, and by O. Haxel, J. H. D. Jensen, and H. E. Suess [4, 5]. Their papers were devoted to describing the postulated single-particle force and showing that there was impressive agreement with experiment. No attention was given to the theoretical basis of the model, and in this book also such discussion is postponed to a later section.

As a starting point, we may confine ourselves to spherical nuclei and remark that the average force on a nucleon at the center is zero. Thus, if the force can be derived from a potential $V(r)$, this potential will be flat at $r = 0$. Moreover, since nuclear matter is of roughly uniform density throughout the nuclear volume, we can expect V to be fairly constant in the nuclear interior and to decrease steadily throughout the boundary region, where the density falls to zero. Indeed, it is reasonable to suppose that $V(r)$ is a function with much the same shape as $\rho(r)$, but with a rather larger range, since the potential will extend beyond the nucleons that produce it.

A difficulty is encountered in connection with the identity of the center of the potential well. The potential is an average over nucleon motions; as such its center has no fundamental physical significance, but coincides *on the average* with the mass center. In a model in which the center of the potential is fixed at the origin, there are oscillations of the mass center about the origin. The energy associated with this mass-center motion is, however, fictitious, since it is in reality the mass center itself that is fixed. This model may thus give rise to spurious states associated with the center-of-mass motion.

In a central potential, the orbital angular momentum of each nucleon is a constant of motion. For each quantum number l there is a series of energy levels, which we shall distinguish by the quantum number n, associated with the number of nodes of the radial wave function. For example, the state of lowest energy with $l = 1$ is called the $0p$-state; the fifth state with $l = 3$ is the $4f$-state, etc.[†] The spacing of the energy levels depends on the form of the potential, but for $V(r)$'s of the general type discussed above, the order of levels is almost fixed. Two extreme cases for which analytic calculations can be made are the harmonic oscillator:

$$V = -V_0 + \tfrac{1}{2}M\omega^2 r^2; \qquad (7\text{-}1)$$

and the infinite square well:

$$V = -V_0, \quad r < R,$$
$$= \infty, \qquad r > R. \qquad (7\text{-}2)$$

Numerical calculations for the finite square well (i.e., $V = 0, r > R$) and for $V(r)$ given by a Fermi distribution like Eq. 4–9 [6] yield level orders intermediate to those of the above two extremes. The levels of the harmonic oscillator are shown on the left of Fig. 7–1; they are evenly spaced and are highly degenerate, since all states with the same value of $2n + l$ have the same energy. On the right of Fig. 7–1 are the levels of the infinite square well. It will be observed that the degeneracies have been removed, the states of

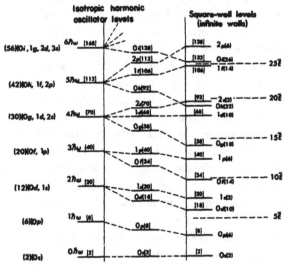

Fig. 7-1 The levels of an harmonic oscillator, a square well, and an intermediate potential shape. The energy units $\tilde{E}$ and $\hbar\omega$ depend on the radii of the potentials.

[†] The spectroscopist's alphabet begins "$s, p, d, f, g, h, i, j, \ldots$" Explicitly, n = number of nodes in the radial wave function, excluding the origin and infinity.

higher angular momentum moving to lower energy. For more realistic potentials intermediate to these extremes, we expect a level spacing like that arbitrarily interpolated in the center of the figure. It should be noted that, although the spacing is different, the order of levels is the same in the two extreme cases, with two exceptions in the range of levels shown, namely, the 0h, 2s states and the 0i, 2p states. In an intermediate potential these pairs of states may be expected to be rather close in energy.

The single-particle model describes a nuclear state by stating how many nucleons are in each of the orbits. For each level, there are 2l + 1 degenerate substates corresponding to different orientations of the angular momentum (i.e., m_l-values), and in accordance with the exclusion principle each of these substates can contain at most two nucleons of each type. Thus a d-level, for example, can contain ten protons and ten neutrons. For the interpolated level sequence, Fig. 7–1 shows, for each level, the number of protons (or neutrons) which it can hold and also the total numbers which can be held in that level and all lower ones. Thus, if these levels were realistic, $^{13}_{6}C_7$ would have two protons and two neutrons in the 0s-level and four protons and five neutrons in the 0p-level. As another example, in $^{87}_{37}Rb_{50}$, all levels below the 1p-level would be completely filled, and this level itself would have its full complement of six neutrons. It would also have three protons, and there would be ten neutrons in the 0g-level.

In a single-particle model, wherever there is a large gap in the spacing of the energy levels, corresponding discontinuities appear in the nuclear properties, such as binding energy, capture cross section, and angular momentum, which mark the magic numbers. Consequently, the magic numbers indicated by Fig. 7–1 are 2, 8, 20, 40, 70, etc. Except for the first three, these are not the magic numbers occurring in the experimental data in Chapter 6. The feature of the shell model that produces energy gaps at the correct magic numbers in a quite natural way is the assumption of a spin-orbit force. Explicitly, it is assumed that, in addition to the static potential already discussed, each nucleon is subject to a potential:

$$V_{ls} = - V(r)\mathbf{l} \cdot \mathbf{s} = - V(r)(\mathbf{r} \times \mathbf{p}) \cdot \mathbf{s}, \tag{7–3}$$

where the vectors refer to the orbital angular momentum, spin, position, and momentum of the individual nucleon. Originally this potential was introduced in 1949 as an hypothesis to explain the magic numbers; but, as we shall see, there is now considerable evidence that it is real. A completely satisfactory theoretical explanation of its presence has not been provided. Its strength shows that it is due to nuclear forces, not to electromagnetic phenomena or small relativistic effects; the tensor force does not seem to account for it wholly, but the two-nucleon $\mathbf{L \cdot S}$ force may provide a source.

Unlike the tensor force, the spin-orbit force does not mix states of different l, since l^2 commutes with $\mathbf{1 \cdot s}$. However, the orientation of $\mathbf{1}$ is not a constant of motion, and only $\mathbf{j}$ has a fixed z-component. There are, of course, two possible j-values for a given l, namely, $j = l \pm \frac{1}{2}$. The factor $\mathbf{1 \cdot s}$ in the potential implies that the energy of a nucleon with parallel orbital and spin angular momenta will differ from that of a nucleon for which these vectors are antiparallel. If $V(r)$ is taken as positive in Eq. 7-3, the potential is attractive when $\mathbf{1}$ and $\mathbf{s}$ are predominantly parallel. Hence each level of the sequence shown in Fig. 7-1 is split into two corresponding to the two j-values, with $j = l + \frac{1}{2}$ being the lower level. It is also clear that the magnitude of the splitting increases with $\mathbf{1}$, and this can be shown to be proportional to $2l + 1$ if the radial dependence of the $\mathbf{1 \cdot s}$ potential is smooth. However, it may, in fact, be largely concentrated near the nuclear boundary, in which case the l-dependence of the splitting is somewhat modified.

With these facts in mind, let us examine Fig. 7-2. An s-state ($l = 0$) is of course not split, and the splitting of p-states is small. Thus the gaps at 2

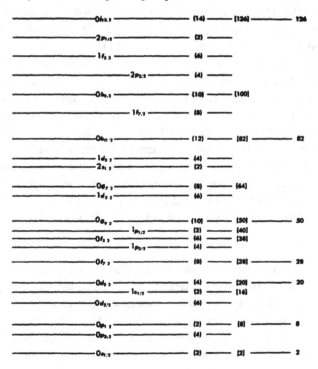

Fig. 7-2 The energy levels of the first 126 neutrons in the simple shell model.

and 8 persist. The $0d_{5/2}$-state is lower and the $0d_{3/2}$-state higher than the original $0d$-state; whether or not they bracket the $1s$-state depends on the magnitude of the $\mathbf{l} \cdot \mathbf{s}$ force, and experimental data indicate that it is sufficiently large to produce the order shown in the figure. Although the $0f_{7/2}$-state is depressed and the $0d_{3/2}$-state is raised by the $\mathbf{l} \cdot \mathbf{s}$ splitting, there is still a gap corresponding to the magic number at 20. On the other hand, the elevation of the $0f_{5/2}$-state produces an energy gap at nucleon number 28.

The substitution of 50 for 40 as a magic number is achieved by the splitting of the $0g$-level, whose l-value is sufficiently high so that a pronounced gap appears above the $0g_{9/2}$-level (50 nucleons) which has been depressed so close to the $1p_{1/2}$-state that there is no large gap at 40 nucleons. Similarly, the large splitting of the $0h$-level has removed any large gaps at numbers 70 and 92 and introduced the correct magic number 82. The i-level splitting has introduced the magic number 126.

Within each shell the level order will depend on the specific properties of the potentials assumed. The level order in Fig. 7-2 is that which, for neutrons, is found to agree with experiment. For $Z > 50$, the proton level scheme is different in that higher angular momenta tend to have lower energies. The level order is $0g_{7/2}$, $1d_{5/2}$, $1d_{3/2}$, $2s_{1/2}$, and $0h_{11/2}$.

7-3 EXTREME SINGLE-PARTICLE MODEL AND SPIN

In general, the extent of the agreement of the various nuclear models with experiment will be discussed in later chapters, individually devoted to the different observable quantities. However, it is convenient to consider, in this part of the book, the predictions concerning nuclear spin, one of the simpler nuclear properties, which served as a fundamental guide in the development of the models. In the simplest form of the shell model, the extreme single-particle model, the nucleons are supposed, in the ground state, to have dynamically paired motions so that many of the nuclear properties are due only to the last unpaired nucleon. It is also assumed that the neutron and proton states fill independently, that is, that the state into which a given proton goes is independent of the number of neutrons in the nucleus, and vice versa. In such a system, every even-even nuclide has zero spin, every odd-A nuclide has the angular momentum of the unpaired particle, and the angular momentum of odd-odd nuclides cannot be predicted, since there is nothing to indicate which of the various possible resultants of the j-vectors of the two unpaired particles has lowest energy. This is a very unsophisticated model, ignoring as it does all nucleon correlation. A more realistic approach would consider only the nucleons in closed shells to form an inert core and take into account the internucleon forces between particles in unfilled orbitals.

TABLE 7-1 Shell-Model Configurations for Odd-A Nuclides.
(a) Odd-N Nuclides

Nuclide	N	Z	J	μ nm	Q barns	Configuration of Neutrons		
						$0s_{1/2}$	$0p_{3/2}$	$0p_{1/2}$
n	1	0	1/2	− 1.9131		1		
^{3}He	1	2	1/2	− 2.1276		1		
^{7}Be	3	4	3/2			2	1	
^{9}Be	5	4	3/2	− 1.1776	0.05	2	1	
^{11}C	5	6	3/2	± 1.03	0.031	2	3	
^{13}C	7	6	1/2	0.7024		2	4	1
^{15}O	7	8	1/2	0.7189		2	4	1

| | 8 | | | | | | | |

Nuclide	N	Z	J	μ nm	Q barns	$0d_{5/2}$	$1s_{1/2}$	$0d_{3/2}$
^{17}O	9	8	5/2	− 1.8937	− 0.026	1		
^{19}Ne	9	10	1/2	− 1.887		0	1	
^{19}O	11	8	5/2			3		
^{21}Ne	11	10	3/2	− 0.6618	0.09	3		
^{23}Mg	11	12	3/2			3		
^{23}Ne	13	10	5/2	− 1.08		5	deformed	
^{25}Mg	13	12	5/2	− 0.8551	0.22	5		
^{27}Si	13	14	5/2			5		
^{27}Mg	15	12	1/2			6	1	
^{29}Si	15	14	1/2	− 0.5553	$|Q| < 10^{-4}$	6	1	
^{31}S	15	16	1/2			6	1	
^{33}S	17	16	3/2	0.6433	− 0.055	6	2	1
^{35}Ar	17	18	3/2	0.63		6	2	1
^{35}S	19	16	3/2	$\begin{cases}+ 1.00 \text{ or} \\ - 1.07\end{cases}$	0.04	6	2	3
^{37}Ar	19	18	3/2	0.95		6	2	3
^{39}Ca	19	20	3/2			6	2	3

| | 20 | | | | | | | |

Nuclide	N	Z	J	μ nm	Q barns	$0f_{7/2}$		
^{37}S	21	16	7/2			1		
^{39}Ar	21	18	7/2	− 1.3		1		
^{41}Ca	21	20	7/2	− 1.595		1		
^{41}Ar	23	18	7/2			3		
^{43}Ca	23	20	7/2	− 1.317		3		
^{45}Ti	23	22	7/2	0.095	0.015	3		
^{47}Ti	25	22	5/2	− 0.7881	0.29	5		
^{49}Cr	25	24	5/2	0.476		5		
^{49}Ti	27	22	7/2	− 1.1039	0.24	7		
^{51}Cr	27	24	7/2	0.933		7		
^{53}Fe	27	26	7/2			7		

Nuclide	N	Z	J	μ nm	Q barns	Configuration of Neutrons			
	28								
						$1p_{3/2}$	$0f_{5/2}$	$1p_{1/2}$	$0g_{9/2}$
^{51}Ti	29	22	3/2			1			
^{53}Cr	29	24	3/2	− 0.4744	0.026	1			
^{55}Fe	29	26	3/2			1			
^{57}Ni	29	28	3/2			1			
^{55}Cr	31	24	3/2			3			
^{57}Fe	31	26	1/2	0.0906		3			
^{59}Ni	31	28	3/2			3			
^{61}Zn	31	30	3/2			3			
^{59}Fe	33	26	3/2			3	2		
^{61}Ni	33	28	3/2	− 0.7487	0.128	3	2		
^{63}Zn	33	30	3/2	− 0.282	0.29	3	2		
^{63}Ni	35	28	3/2			3	4		
^{65}Zn	35	30	5/2	0.769	− 0.026	4	3		
^{65}Ni	37	28	5/2			4	5		
^{67}Zn	37	30	5/2	0.8754	0.17	4	5		
^{69}Ge	37	32	5/2	0.735	0.028	4	5		
^{69}Zn	39	30	1/2			4	6	1	
^{71}Ge	39	32	1/2	0.546		4	6	1	
^{73}Ge	41	32	9/2	− 0.8792	− 0.173	4	6	2	1
^{75}Se	41	34	5/2		1.0	} deformed			
^{77}Kr	41	36	5/2						
^{75}Ge	43	32	1/2	0.508		4	6	1	4
^{77}Se	43	34	1/2	0.534		4	6	1	4
^{79}Kr	43	36	1/2			4	6	1	4
^{79}Se	45	34	7/2	− 1.02	0.9	4	6		7 ?
^{81}Kr	45	36	7/2			4	6		7 ?
^{81}Se	47	34	1/2			4	6	1	8
^{83}Kr	47	36	9/2	± 0.970	0.26	4	6	2	7
^{83}Se	49	34	9/2			4	6	2	9
^{85}Kr	49	36	9/2	± 1.005	0.43	4	6	2	9
^{87}Sr	49	38	9/2	− 1.093	0.30	4	6	2	9
^{89}Zr	49	40	9/2			4	6	2	9
	50								
						$1d_{5/2}$	$0g_{7/2}$	$2s_{1/2}$	$1d_{3/2}$	$0h_{11/2}$
^{87}Kr	51	36	7/2			0	1		
^{89}Sr	51	38	5/2			1			
^{91}Zr	51	40	5/2	− 1.303		1			
^{93}Mo	51	42	5/2			1			
^{91}Sr	53	38	5/2			3			
^{93}Zr	53	40	5/2			3			

Nuclide	N	Z	J	μ nm	Q barns	Configuration of Neutrons				
						$1d_{5/2}$	$0g_{7/2}$	$2s_{1/2}$	$1d_{3/2}$	$0h_{11/2}$
^{95}Mo	53	42	5/2	− 0.9133	± 0.12	3				
^{97}Ru	53	44	5/2			3				
^{95}Zr	55	40	5/2			5				
^{97}Mo	55	42	5/2	− 0.9325	± 1.1	5				
^{99}Ru	55	44	5/2	0.623	⩾ 0.05	5				
^{101}Pd	55	46	7/2			5				
^{97}Zr	57	40	1/2			6	1			
^{99}Mo	57	42	1/2			6	1			
^{101}Ru	57	44	5/2	− 0.69		5	2			
^{105}Cd	57	48	5/2	− 0.7385	0.43	5	2			
^{103}Ru	59	44	5/2			5	4			
^{105}Pd	59	46	5/2	− 0.616	0.8	5	4			
^{107}Cd	59	48	5/2	− 0.6163	0.68	5	4			
^{109}Cd	61	48	5/2	− 0.8270	0.69	5	6			
^{109}Pd	63	46	5/2			5	8			
^{111}Cd	63	48	1/2	− 0.5943		6	6	1		
^{113}Sn	63	50	1/2	0.879		6	6	1		
^{113}Cd	65	48	1/2	− 0.6217		6	8	1		
^{115}Sn	65	50	1/2	− 0.918		6	8	1		
^{117}Te	65	52	1/2			6	8	1		
^{115}Cd	67	48	1/2	− 0.6484		6	8	1	0	2
^{117}Sn	67	50	1/2	− 1.00		6	8	1	0	2
^{119}Te	67	52	1/2	0.25		6	8	1	0	2
^{119}Sn	69	50	1/2	− 1.046		6	8	1	0	4
^{121}Sn	71	50	3/2	0.70	0.08	6	8	2	1	4
^{123}Te	71	52	1/2	− 0.7359		6	8	1	0	6
^{125}Te	73	52	1/2	− 0.8871		6	8	1	0	8
^{125}Sn	75	50	11/2			6	8	2	0	9
^{127}Te	75	52	3/2	0.66		6	8	2	1	8
^{129}Xe	75	54	1/2	− 0.7768		6	8	1	0	10
^{131}Ba	75	56	1/2			6	8	1	0	10
^{129}Te	77	52	3/2	0.67		6	8	2	1	10
^{131}Xe	77	54	3/2	0.6908	− 0.12	6	8	2	1	10
^{133}Ba	77	56	1/2			6	8	1	0	12
^{131}Te	79	52	3/2			6	8	2	1	12
^{133}Xe	79	54	3/2			6	8	2	1	12
^{135}Ba	79	56	3/2	0.8365	0.18	6	8	2	1	12
^{137}Ce	79	58	3/2	0.90		6	8	2	1	12
^{135}Xe	81	54	3/2			6	8	2	3	12
^{137}Ba	81	56	3/2	0.9357	0.28	6	8	2	3	12
^{139}Ce	81	58	3/2	0.90		6	8	2	3	12
^{141}Nd	81	60	3/2			6	8	2	3	12
^{143}Sm	81	62	3/2			6	8	2	3	12

Nuclide	N	Z	J	μ nm	Q barns	Configuration of Neutrons					
	82										
						$1f_{7/2}$	$0h_{9/2}$	$2p_{3/2}$	$1f_{5/2}$	$2p_{1/2}$	$0i_{13/2}$
^{141}Ce	83	58	7/2	0.9		1					
^{143}Nd	83	60	7/2	− 1.08	− 0.48	1					
^{143}Ce	85	58	3/2	1.0		3 ?					
^{145}Nd	85	60	7/2	− 0.66	− 0.25	3					
^{147}Sm	85	62	7/2	− 0.813	− 0.20	3					
^{147}Nd	87	60	5/2	0.577		5 ?					
^{149}Sm	87	62	7/2	− 0.670	0.058	5					
^{149}Nd	89	60	5/2			7 ?					
^{151}Sm	89	62	7/2			7					
^{153}Gd	89	64	3/2			} deformed					
^{155}Dy	89	66	3/2	0.21							

$N = 91\text{--}116$: Nonspherical Nuclei

Nuclide	N	Z	J	μ nm	Q barns	$1f_{7/2}$	$0h_{9/2}$	$2p_{3/2}$	$1f_{5/2}$	$2p_{1/2}$	$0i_{13/2}$
^{195}Pt	117	78	1/2	0.6060		8	10	4	6	1	6
^{197}Hg	117	80	1/2	0.524		8	10	4	6	1	6
^{199}Pb	117	82	5/2			8	10	4	5	0	8
^{201}Po	117	84	3/2			8	10	3	6	0	8?
^{197}Pt	119	78	1/2			8	10	4	6	1	8
^{199}Hg	119	80	1/2	0.5027		8	10	4	6	1	8
^{201}Pb	119	82	5/2			8	10	4	5	0	10
^{203}Po	119	84	5/2			8	10	4	5	0	10
^{199}Pt	121	78	5/2			8	10	4	5	0	12
^{201}Hg	121	80	3/2	− 0.5567	0.45	8	10	3	6	0	12?
^{203}Pb	121	82	5/2			8	10	4	5	0	12
^{205}Po	121	84	5/2	0.26	0.17	8	10	4	5	0	12
^{203}Hg	123	80	5/2	0.84	$\pm < 13$	8	10	4	5	0	14
^{205}Pb	123	82	5/2			8	10	4	5	0	14
^{207}Po	123	84	5/2	0.27	0.28	8.	10	4	5	0	14
^{205}Hg	125	80	1/2	0.5911		8	10	4	6	1	14
^{207}Pb	125	82	1/2	0.5895		8	10	4	6	1	14
^{209}Po	125	84	1/2	0.76		8	10	4	6	1	14

Nuclide	N	Z	J	μ nm	Q barns	$1g_{9/2}$
	126					
^{209}Pb	127	82	9/2			1

$N > 130$: Nonspherical Nuclei

(b) Odd-Z Nuclides

Nuclide	Z	N	J	μ nm	Q barns	Configuration of Protons		
						$0s_{1/2}$	$0p_{3/2}$	$0p_{1/2}$
^{1}H	1	0	1/2	2.7928		1		
^{3}H	1	2	1/2	2.9789		1		
^{7}Li	3	4	3/2	3.2564	$-$ 0.04	2	1	
^{11}B	5	6	3/2	2.6885	0.04	2	3	
^{13}B	5	8	3/2			2	3	
^{13}N	7	6	1/2	$-$ 0.3221		2	4	1
^{15}N	7	8	1/2	$-$ 0.2831		2	4	1

| | 8 | | | | | | | |

Nuclide	Z	N	J	μ nm	Q barns	$0d_{5/2}$	$1s_{1/2}$	$0d_{3/2}$
^{17}F	9	8	5/2	4.7224		1		
^{19}F	9	10	1/2	2.6288		0	1	
^{21}Na	11	10	3/2	2.386				
^{23}Na	11	12	3/2	2.2175	0.14	(3)		
^{25}Na	11	14	5/2			(3)	deformed	
^{25}Al	13	12	5/2			(5)		
^{27}Al	13	14	5/2	3.6414	0.15	(5)		
^{29}P	15	14	1/2	1.2349		6	1	
^{31}P	15	16	1/2	1.1317		6	1	
^{33}Cl	17	16	3/2			6	2	1
^{35}Cl	17	18	3/2	0.8218	$-$ 0.079	6	2	1
^{37}Cl	17	20	3/2	0.6841	$-$ 0.062	6	2	1
^{39}Cl	17	22	3/2			6	2	1
^{37}K	19	18	3/2	0.203		6	2	3
^{39}K	19	20	3/2	0.3914	0.057	6	2	3
^{41}K	19	22	3/2	0.2149	0.076	6	2	3
^{43}K	19	24	3/2	0.163		6	2	3
^{45}K	19	26	3/2	0,173		6	2	3

| | 20 | | | | | | | |

Nuclide	Z	N	J	μ nm	Q barns	$0f_{7/2}$		
^{41}Sc	21	20	7/2	5.43		1		
^{43}Sc	21	22	7/2	4.62	$-$ 0.26	1		
^{45}Sc	21	24	7/2	4.7564	$-$ 0.22	1		
^{47}Sc	21	26	7/2	5.34	$-$ 0.22	1		
^{49}Sc	21	28	7/2			1		
^{47}V	23	24	3/2			(3) ?		
^{49}V	23	26	7/2	4.5		3		
^{51}V	23	28	7/2	5.149	$-$ 0.05	3		
^{53}V	23	30	7/2			(3) deformed ?		
^{51}Mn	25	26	5/2	3.583		(5) deformed ?		

Nuclide	Z	N	J	μ nm	Q barns	Configuration of Protons			
						$0f_{7/2}$			
^{53}Mn	25	28	7/2	5.046		5			
^{55}Mn	25	30	5/2	3.468	0.40	(5)			
^{55}Co	27	28	7/2	4.3		7			
^{57}Co	27	30	7/2	4.58		7			
^{59}Co	27	32	7/2	4.62	0.40	7			
	28								
						$1p_{3/2}$	$0f_{5/2}$	$1p_{1/2}$	$0g_{9/2}$
^{61}Cu	29	32	3/2	2.13		1			
^{63}Cu	29	34	3/2	2.223	− 0.180	1			
^{65}Cu	29	36	3/2	2.382	− 0.195	1			
^{67}Cu	29	38	3/2			1			
^{67}Ga	31	36	3/2	1.850	0.22	3			
^{69}Ga	31	38	3/2	2.016	0.19	3			
^{71}Ga	31	40	3/2	2.562	0.12	3			
^{71}As	33	38	5/2			4	1		
^{73}As	33	40	3/2			3	2		
^{75}As	33	42	3/2	1.439	0.29	3	2		
^{77}Br	35	42	3/2			3	4		
^{79}Br	35	44	3/2	2.106	0.31	3	4		
^{81}Br	35	46	3/2	2.270	0.26	3	4		
^{83}Br	35	48	3/2			3	4		
^{85}Br	35	50	3/2			3	4		
^{81}Rb	37	44	3/2	2.05		3	6		
^{83}Rb	37	46	5/2	1.40		4	5		
^{85}Rb	37	48	5/2	1.3524	0.29	4	5		
^{87}Rb	37	50	3/2	2.7500	0.14	3	6		
^{89}Y	39	50	1/2	− 0.1373		4	6	1	
^{91}Y	39	52	1/2	0.164		4	6	1	
^{93}Nb	41	52	9/2	6.167	− 0.22	4	6	2	1
^{95}Nb	41	54	9/2	6.3		4	6	2	1
^{99}Tc	43	56	9/2	5.68	0.3	4	6	2	3
^{103}Rh	45	58	1/2	− 0.0883		4	6	1	6
^{103}Ag	47	56	7/2	4.4		4	6	2	7 ?
^{105}Ag	47	58	1/2	0.101		4	6	1	8
^{107}Ag	47	60	1/2	− 0.1135		4	6	1	8
^{109}Ag	47	62	1/2	− 0.1305		4	6	1	8
^{111}Ag	47	64	1/2	− 0.145		4	6	1	8
^{113}Ag	47	66	1/2	0.159		4	6	1	8
^{109}In	49	60	9/2	5.53	1.20	4	6	2	9
^{111}In	49	62	9/2	5.53	1.18	4	6	2	9
^{113}In	49	64	9/2	5.523	1.14	4	6	2	9

Nuclide	Z	N	J	μ nm	Q barns	Configuration of Protons			
						$1p_{3/2}$	$0f_{5/2}$	$1p_{1/2}$	$0g_{9/2}$
^{115}In	49	66	1/2	5.5357	1.16	4	6	1	10
^{117}In	49	68	9/2			4	6	2	9

	50								

Nuclide	Z	N	J	μ nm	Q barns	$0g_{7/2}$	$1d_{5/2}$	$0h_{11/2}$	$1d_{3/2}$	$2s_{1/2}$
^{115}Sb	51	64	5/2	3.46	-0.20	0	1			
^{117}Sb	51	66	5/2	2.67	-0.30	0	1			
^{119}Sb	51	68	5/2	3.45	-0.21	0	1			
^{121}Sb	51	70	5/2	3.359	-0.26	0	1			
^{123}Sb	51	72	7/2	2.547	-0.37	1				
^{125}Sb	51	74	7/2	2.6		1				
^{123}I	53	70	5/2			2	1			
^{125}I	53	72	5/2	3.0	-0.89	2	1			
^{127}I	53	74	5/2	2.808	-0.79	2	1			
^{129}I	53	76	7/2	2.617	-0.55	3				
^{131}I	53	78	7/2	2.74	-0.40	3				
^{133}I	53	80	7/2	2.84	-0.26	3				
^{135}I	53	82	7/2			3				
^{125}Cs	55	70	1/2	1.41		\} deformed ?				
^{127}Cs	55	72	1/2	1.46						
^{129}Cs	55	74	1/2	1.479						
^{131}Cs	55	76	5/2	3.54	-0.70	4	1			
^{133}Cs	55	78	7/2	2.578	-0.003	5				
^{135}Cs	55	80	7/2	2.729	0.044	5				
^{137}Cs	55	82	7/2	2.838	0.045	5				
^{139}La	57	82	7/2	2.778	0.22	7				
^{141}Pr	59	82	5/2	4.3	-0.07	8	1			
^{143}Pr	59	84	7/2			7	2			
^{143}Pm	61	82	7/2	3.9		7	2			
^{147}Pm	61	86	7/2	2.7	0.7	(7)	(2)			
^{149}Pm	61	88	7/2	3.3		(7)	(2) \} deformed			
^{151}Pm	61	90	5/2	1.8	1.9	(8)	(1)			
$Z = 63$–78: Nonspherical Nuclei										
^{191}Au	79	112	3/2	0.137		8	6	12	3	
^{193}Au	79	114	3/2	0.139		8	6	12	3	
^{195}Au	79	116	3/2	0.147		8	6	12	3	
^{197}Au	79	118	3/2	0.1449	0.58	8	6	12	3	
^{199}Au	79	120	3/2	0.270		8	6	12	3	
^{195}Tl	81	114	1/2	1.56		8	6	12	4	1
^{197}Tl	81	116	1/2	1.55		8	6	12	4	1
^{199}Tl	81	118	1/2	1.59		8	6	12	4	1
^{201}Tl	81	120	1/2	1.60		8	6	12	4	1

Nuclide	Z	N	J	μ nm	Q barns	Configuration of Protons					
						$0g_{7/2}$	$1d_{5/2}$	$0h_{11/2}$	$1d_{3/2}$	$2s_{1/2}$	
^{203}Tl	81	122	1/2	1.611		8	6	12	4	1	
^{205}Tl	81	124	1/2	1.627		8	6	12	4	1	
	82										
						$0h_{9/2}$	$1f_{7/2}$	$2p_{3/2}$	$1f_{5/2}$	$2p_{1/2}$	$0i_{13/2}$
^{199}Bi	83	116	9/2			1					
^{201}Bi	83	118	9/2			1					
^{203}Bi	83	120	9/2	4.59	-0.64	1					
^{205}Bi	83	122	9/2	5.5		1					
^{209}Bi	83	126	9/2	4.24	-0.35	1					
^{211}At	85	126	9/2			3					
				$Z > 86$: Nonspherical Nuclei							

We find that this more realistic single-particle model makes the same predictions about spin, that is, it predicts that the lowest state of an even number of neutrons or protons in the same j-shell is $J = 0$ and that the lowest state of an odd number is $J = j$. Consequently, the experimental values of J for odd-A nuclides should allow us to determine the level order of the shell model.

Table 7–1 shows the experimentally known spin values and the expected configurations of nucleons. It is seen that the successive states of the level sequence are filled in order until N or $Z = 33$ is reached. There is one exception, however, to this statement, namely, the ninth proton is in the $0d_{5/2}$-level in ^{17}F, but is in a $1s_{1/2}$-state in ^{19}F. The two levels are close together, and this behavior demonstrates that small variations in the effective central potential occur from nuclide to nuclide. In particular, the neutron number is observed to have some effect on the method of filling the proton shells, making it clear that the assumed independence of the neutron and proton shells is only approximate. Attention should also be directed to nucleon number 11 (^{21}Ne, ^{23}Na), in which three nucleons in $d_{5/2}$-orbitals are believed to couple their angular momenta to a total J of $\frac{3}{2}$, and to number 25 (^{47}Ti, ^{55}Mn), in which $f_{7/2}$-nucleons are coupled to $J = \frac{5}{2}$. These anomalous cases will be considered below.

At nucleon number 33 a new principle becomes evident. The $1p_{3/2}$-shell is filled by the 32nd neutron or proton, and we would expect to find the 33rd in the $0f_{5/2}$-state. However, the nuclides with 33 neutrons or protons have spin $\frac{3}{2}$, as do many of the nuclides with 35 or 37 nucleons of one kind. Similarly, we would expect the $1d_{5/2}$-state to fill at $N = 56$, but in fact spins of $\frac{5}{2}$ are still found for nuclides with 59 and 61 neutrons, and in ^{101}Ru, which has 57 neutrons. Evidently, the higher-angular-momentum states (e.g., the $0g_{7/2}$-state in this case) fill in pairs, so that when the 59th neutron is added, it does not enter as the third $g_{7/2}$-neutron; instead the closed $d_{5/2}$-shell is broken up, and the configuration is $(1d_{5/2})^5(0g_{7/2})^4$.

The probable explanation of this behavior is the pairing energy. Consider a nuclide with an even number N of neutrons having binding energy $B(N)$. When two neutrons are added, it is found that $B(N + 2) - B(N)$ is always greater than $2[B(N + 1) - B(N)]$. In other words, there exists extra pairing energy between two neutrons in the same shell over and above the "individual" neutron binding energies. The same phenomenon is observed with protons. The source of this extra energy is not accounted for in the extreme single-particle model, but when we take internucleon interactions into consideration, we shall see that the pairing energy increases with the j-value of the particles (see Problem 7–5). Thus, although the $0g_{7/2}$-level is above the $1d_{5/2}$-level, binding energy can be gained by breaking the $d_{5/2}$-shell and adding an extra pair in the $g_{7/2}$-state. Allowing for this principle of the filling by pairs of the $0f_{5/2}$-shells, the $0g_{9/2}$-proton shell, the $0g_{7/2}$-neutron shell, the $0h_{11/2}$-shells, and the $0i_{13/2}$-neutron shell, and remembering that certain levels (such as the $0g_{7/2}$- and $1d_{5/2}$-proton states) are very close in energy, we see that the single-particle model accounts in a very natural way for the J-values of almost all the odd-A nuclides. Except for two or three isolated cases such as ^{79}Se and ^{201}Hg, the exceptional J-values all occur for $155 \leqslant A \leqslant 180$ and $N \geqslant 140$. In these regions, the nuclei are considerably deformed from sphericity, and we cannot expect the level sequence of a spherical potential well to apply. There are also deformed nuclei in other parts of the periodic table, and the present level scheme cannot be applied in these cases; the correct scheme for such nuclides is discussed in Chapter 9.

The unforced prediction of such a large number of experimental J-values was historically one of the most important indications that the single-particle model might have a considerable range of validity. There were others, which we shall mention later, but we may also recall that the single-particle model predicts magnetic moments on the Schmidt lines. The nature of the experimental agreement is such that an underlying significance of the model is confirmed, while it is also made clear that refinements of the simple model are necessary. We shall refer to the model already described as the *extreme single-particle model*; when we allow for the interactions between particles in unfilled shells and consider filled shells to form an inert core, we shall say that we are using the *single-particle model*.

7-4 SINGLE-PARTICLE MODEL

In this model, we attribute all the low-energy properties of a nuclide to its "loose" particles outside closed shells. We assume that the interactions of these loose particles with each other do not perturb them appreciably from the single-particle orbits described by the quantum numbers (n, l, j). This is equivalent to ignoring any correlations in the motion of the particles (such as formation of clusters); the model will not be expected to be any improvement over the extreme single-particle model for treating topics (such as total binding energy) for which correlations are important.

In the extreme single-particle model, all the different states that can be formed by k particles with the same (n, l, j) have the same energy. This degeneracy is removed by considering the interaction between the particles. The calculations on which the extended single-particle model is based require that the interaction be strong enough to remove the degeneracies but not so strong, compared with the spin-orbit force, that j ceases to be a good quantum number for each nucleon. This assumption is part of the single-particle model. It is sometimes possible to relax this approximation and, at the same time, maintain the language of the single-particle model by taking for the particle state a superposition of the wave functions of two or three (n, l, j)-states whose energies are close to one another; such a procedure is called *configuration mixing*.

It is not implied that the interaction potential v_{ij} between particles in a shell is the same as that in the two-particle system. A large part of the actual interaction is accounted for in the single-particle shell-model potential, which represents the average effect on one nucleon of all the other nucleons. The force that removes the degeneracy of shell-model states is consequently often referred to as a remanent effective interaction. Its exact form is hard to calculate; for many purposes, it is satisfactory to take simple analytic forms with adjustable parameters.

In a discussion of the light nuclei ($A \lesssim 50$), the i-spin plays an important role. We define the total i-spin of a nucleus as

$$\mathbf{T} = \sum_{i=1}^{A} \mathbf{t}_i, \tag{7-4}$$

where $\mathbf{t}_i$ is the i-spin vector of the ith nucleon. For any nuclear state, $T_3 = \frac{1}{2}(Z - N)$ and is therefore a good quantum number. To the extent to which the internucleon forces are charge independent, T^2 is also a constant of motion. We have seen that the two-nucleon nuclear force is very probably charge independent to quite a good approximation, but, of course, the Coulomb force is not. Hence, in the lighter nuclei where the Coulomb energy is a small part of the total, it is reasonable to treat $T^2 = T(T + 1)$ as a quantum number of the system, but, as the value of Z^2/A increases, this approximation breaks down.[†] The dividing line between the two regions is also characterized by increasing neutron excess and by the fact that, when T is no longer a good quantum number, the neutrons and protons fill different single-particle shells.

The exclusion principle requires that the wave function be antisymmetric under interchange of all coordinates of any pair of nucleons. If we are dealing with a shell containing only protons, say, this requirement means that the angular-momentum wave function must be antisymmetric in all pairs; if we are dealing with a shell containing both neutrons and protons and if T is a good quantum number, the angular-momentum function must be symmetric or antisymmetric for exchange of a given pair, depending on the symmetry of the i-spin function for the same exchange.

These stipulations imply certain restrictions on the states that can be formed. For example, it is not possible to construct an antisymmetric state with J odd from two particles with the same value of j, as can be seen by noting that the appropriate wave function for two (n, l, j)-particles combining to give angular momentum (J, M) is[‡]

$$u_n(r_1)u_n(r_2) \sum_m (jjm\, M - m|JM)\chi_j{}^m(1)\chi_j{}^{M-m}(2). \qquad (7\text{–}5)$$

When particles 1 and 2 are interchanged, expression 7–5 becomes

$$u_n(r_1)u_n(r_2) \sum_m (jjm\, M - m|JM)\chi_j{}^m(2)\chi_j{}^{M-m}(1);$$

or, changing the summation index by introducing $m' = M - m$, we have

$$u_n(r_1)u_n(r_2) \sum_{m'} (jj\, M - m'\, m'|JM)\chi_j{}^{m'}(1)\chi_j{}^{M-m'}(2)$$

$$= u_n(r_1)u_n(r_2) \sum_{m'} (-)^{2j-J}(jjm'\, M - m'|JM)\chi_j{}^{m'}(1)\chi_j{}^{M-m'}(2), \qquad (7\text{–}6)$$

where we have used the property

$$(j_1 j_2 m_1 m_2|JM) = (-)^{j_1+j_2-J}(j_2 j_1 m_2 m_1|JM).$$

Since j is half an odd integer, 7–6 and 7–5 differ only by the factor $(-)^{J+1}$, and we observe that antisymmetric functions exist only for even J.

[†] Formally, the Coulomb term in the Hamiltonian is $\sum (e^2/r_{ij})(\tfrac{1}{2} + t_{3,i})(\tfrac{1}{2} + t_{3,j})$. This operator does not commute with T^2, and therefore T is not a good quantum number.

[‡] The Clebsch-Gordan symbols are defined in Appendix A.

There are similar restrictions for states of more than two particles. For example, it can be shown that the three-particle configuration $(n, l, j)^3$ has no state $J = \frac{1}{2}$ antisymmetric for exchange of all three pairs of nucleons. A discussion of these cases is considerably more complicated and may be found in Chapter XII of Rose [7] or in Edmonds and Flowers [8]. Table 7–2 shows the possible antisymmetric states for k particles, for values of $j \leqslant \frac{11}{2}$. The possible states for $2j + 1 - k$ particles, that is, k holes in a shell, are the same as for k particles.

TABLE 7-2 Possible Total Spins J for Various Configurations $(j)^k$. From Mayer and Jensen [9].

$j = \frac{3}{2}$	$j = \frac{5}{2}$	$j = \frac{7}{2}$
$k = 1: \frac{3}{2}$	$k = 1: \frac{5}{2}$	$k = 1: \frac{7}{2}$
$= 2: 0, 2$	$= 2: 0, 2, 4$	$= 2: 0, 2, 4, 6$
	$= 3: \frac{3}{2}, \frac{5}{2}, \frac{9}{2}$	$= 3: \frac{3}{2}, \frac{5}{2}, \frac{7}{2}, \frac{9}{2}, \frac{11}{2}, \frac{15}{2}$
		$= 4: 0, 2 \text{ (twice)}, 4 \text{ (twice)}, 5, 6, 8$

$$j = \frac{9}{2}$$

$k = 1: \frac{9}{2}$

$= 2: 0, 2, 4, 6, 8$

$= 3: \frac{3}{2}, \frac{5}{2}, \frac{7}{2}, \frac{9}{2} \text{ (twice)}, \frac{11}{2}, \frac{15}{2}, \frac{17}{2}, \frac{21}{2}$

$= 4: 0 \text{ (twice)}, 2 \text{ (twice)}, 3, 4 \text{ (3 times)}, 5, 6 \text{ (3 times)}, 7, 8, 9, 10, 12$

$= 5: \frac{1}{2}, \frac{3}{2}, \frac{5}{2} \text{ (twice)}, \frac{7}{2} \text{ (twice)}, \frac{9}{2} \text{ (3 times)}, \frac{11}{2} \text{ (twice)}, \frac{13}{2} \text{ (twice)}, \frac{15}{2} \text{ (twice)}, \frac{17}{2} \text{ (twice)}, \frac{19}{2}, \frac{21}{2}, \frac{25}{2}$

$$j = \frac{11}{2}$$

$k = 1: \frac{11}{2}$

$= 2: 0, 2, 4, 6, 8, 10$

$= 3: \frac{3}{2}, \frac{5}{2}, \frac{7}{2}, \frac{9}{2} \text{ (twice)}, \frac{11}{2} \text{ (twice)}, \frac{13}{2}, \frac{15}{2} \text{ (twice)}, \frac{17}{2}, \frac{19}{2}, \frac{21}{2}, \frac{23}{2}, \frac{27}{2}$

$= 4: 0 \text{ (twice)}, 2 \text{ (3 times)}, 3, 4 \text{ (4 times)}, 5 \text{ (twice)}, 6 \text{ (4 times)}, 8 \text{ (4 times)}, 9 \text{ (twice)}, 10 \text{ (3 times)}, 11, 12 \text{ (twice)}, 13, 14, 16$

$= 5: \frac{1}{2}, \frac{3}{2} \text{ (twice)}, \frac{5}{2} \text{ (3 times)}, \frac{7}{2} \text{ (4 times)}, \frac{9}{2} \text{ (4 times)}, \frac{11}{2} \text{ (5 times)}, \frac{13}{2} \text{ (4 times)}, \frac{15}{2} \text{ (5 times)}, \frac{17}{2} \text{ (4 times)}, \frac{19}{2} \text{ (4 times)}, \frac{21}{2} \text{ (3 times)}, \frac{23}{2} \text{ (3 times)}, \frac{25}{2} \text{ (twice)}, \frac{27}{2} \text{ (twice)}, \frac{29}{2}, \frac{31}{2}, \frac{35}{2}$

$= 6: 0 \text{ (3 times)}, 2 \text{ (4 times)}, 3 \text{ (3 times)}, 4 \text{ (6 times)}, 5 \text{ (3 times)}, 6 \text{ (7 times)}, 7 \text{ (4 times)}, 8 \text{ (6 times)}, 9 \text{ (4 times)}, 10 \text{ (5 times)}, 11 \text{ (twice)}, 12 \text{ (twice)}, 13 \text{ (twice)}, 14 \text{ (twice)}, 15, 16, 18$

The quantum numbers J and T are not sufficient to fix the state of a system of k nucleons. In this case two other quantum numbers, the seniority, s, and the reduced i-spin, t, may be used. To define these we consider the symmetry of the system. Each particle has available only two i-spin states, and hence at most two particles can be in a given space state, with the same (n, l, j, m). Thus there may be space-symmetric pairs, which will automatically be antisymmetric in i-spin. In addition to particles that are so paired, there may be particles whose space states are different from those of all other particles, and which form space-exchange antisymmetric pairs with other nucleons. These pairs are i-spin symmetric, that is, $T = 1$. Of these space-antisymmetric pairs, some may have resultant $J = 0$. The seniority quantum number s is defined as the number of particles left in the shell after removal of any such antisymmetric pairs for which $T = 1$, $J = 0$. The reduced i-spin t is defined as the i-spin of these remaining s nucleons.[†]

If the interaction potential v_{ij} between two particles is sufficiently small compared with the single-particle forces, its contribution to the energy is simply the expectation value $\langle \sum_{i<j=1}^{k} v_{ij} \rangle$ evaluated for the single-particle state in question. If v_{ij} is a Wigner potential (i.e., a central, nonexchange force), it is possible to give a closed expression for the energy contribution in terms of J, T, s, t, and certain integrals of the radial function [8, 10]. However, the nuclear potential contains spin-exchange forces. In this case, a general formula for $\langle \sum v_{ij} \rangle$ can be given only under the further assumptions that (a) the force range is much less than the radius of the nucleus, and (b) all particles in the shell have the same charge. If these conditions are satisfied, only the singlet (antisymmetric) states contribute to v_{ij}. The result for a potential of form of Eq. 2–17 is

$$\langle \sum v_{ij} \rangle = (W + M - H - B)V_0(2j + 1)f_{nl}\left[k - \frac{2j + 3 - s}{2(j + 1)}s\right], \quad (7\text{–}7)$$

where

$$V_0 = 4\pi \int V_0(r)r^2 \, dr \ [‡]$$

and

$$f_{nl} = \frac{1}{4} \int [ru_{nl}(r)]^4 \frac{dr}{4\pi r^2}.$$

The expression $(W + M - H - B)V_0(r)$ is the potential in the $^1S(n, p)$-system, which is attractive. Since s is necessarily less than $2j$ and f_{nl} is positive, we see that the lowest energy is obtained with the smallest value of s. From the definition of s, it is clear that, for an even number of nucleons, k, the smallest possible value of s is zero, whereas for odd k the lowest s is 1. But

---

[†] Further discussions of these concepts may be found in Elliott and Lane [10], Section 20, and Edmonds and Flowers [8].

[‡] Actually Assumption (a) is used in the calculation by taking $V_0(r) = V_0 \, \delta(r)$.

$s = 0$ means that all nucleons are coupled in pairs to $J = 0$, while $s = 1$ means that all but one nucleon are so coupled. Hence the ground-state spin of an odd number of like nucleons with the same (n, l, j) is $J = j$.

Before we can use the preceding results to predict the spin of a nucleus, we must make the further assumption that the effective interparticle forces between nucleons in different orbits l and l' are weak compared to those between two particles in the same orbit. A possible basis for such an assumption lies in the short-range nature of nuclear forces and in the small overlap of the l- and l'-wave functions. This simplification means that the angular-momentum coupling within a shell, whether filled or not, is not upset by internucleon interaction with other shells. Hence, in an even-Z, odd-N nucleus, the protons still couple to $J = 0$, and the neutrons to $J = j_n$, the j-value of the neutron shell which is filling. Then the resultant angular momentum of the whole nucleus is $J = j_n$. Similarly, in an odd-Z, even-N nucleus, $J = j_p$, and in an even-even nucleus, J remains zero. These results show why the simple shell model predicts correct angular-momentum values in spite of its complete disregard of specific internucleon forces in unfilled shells.

For odd-odd nuclides the ambiguity inherent in the extreme shell model is not removed. We still have a neutron shell $(n_n l_n j_n)^{k_n}$ with total angular momentum j_n, and a proton shell $(n_p l_p j_p)^{k_p}$ with angular momenta coupled to j_p. The resultant nuclear J may be any value from $|j_n - j_p|$ to $j_n + j_p$. To remove the degeneracy of these J-values, we must evaluate the energy of the forces between individual neutrons and protons in the two unfilled shells, assuming that these forces do not change the states calculated for the individual shells. In symbols we evaluate

$$\sum_{i=1}^{k_n} \sum_{j=1}^{k_p} ((j_n)_{j_n}^{k_n}(j_p)_{j_p}^{k_p}, J|V_{ij}|(j_n)_{j_n}^{k_n}(j_p)_{j_p}^{k_p}, J), \qquad (7\text{--}8)$$

where $|(j_n)_{j_n}^{k_n}(j_p)_{j_p}^{k_p}, J)$ is the wave function for the states in which the k_n neutrons are coupled to yield j_n, the k_p protons are coupled to give j_p, and then the two resulting states are coupled so that their vector sum $\mathbf{j}_n + \mathbf{j}_p$ has magnitude J. To evaluate expression 7–8 in a simple, general form, it is again necessary to assume very short-range forces, that is, to express the potential as a delta function, $V_0(\mathbf{r}_{ij}) = V_0\delta(\mathbf{r}_{ij})$. In this limit, space exchange, P_x, is meaningless, and both Heisenberg and Bartlett forces are spin-exchange forces, while Majorana and Wigner forces have no exchange character. Thus we write:

$$v_{ij} = -(1 - \alpha + \alpha\boldsymbol{\sigma}_i \cdot \boldsymbol{\sigma}_j)V_0\,\delta(\mathbf{r}_{ij}), \qquad (7\text{--}9)$$

where

$$\alpha = \tfrac{1}{2}(H + B) = \tfrac{1}{2}(1 - W - M).$$

Then the energy (Eq. 7–8) is found to be [11, 12]

$$E[(j_n)_{j_n}^{k_n}(j_p)_{j_p}^{k_p}, J] = E_0(k_n, k_p) + E_\sigma$$
$$= \lambda_n \lambda_p E_0(1, 1) + (k_n k_p - \lambda_n \lambda_p)(1 - \alpha)F_0(l_n, l_p) + E_\sigma,$$

$$(7\text{--}10)$$

where

$$\lambda_n = \frac{2j_n + 1 - 2k_n}{2j_n - 1},$$

$$F_0(l_n, l_p) = \frac{V_0}{4\pi} \int (r u_{l_n})^2 (r u_{l_p})^2 \frac{1}{r^2}\, dr,$$

$$E_0(1, 1) = \tfrac{1}{2}(1 - \alpha)F_0(l_n, l_p)(2j_n + 1)(2j_p + 1)(j_n j_p \tfrac{1}{2} - \tfrac{1}{2}|J0)^2$$

$$\times (2J + 1)^{-1}\left\{\frac{[(j_n + \tfrac{1}{2}) + (-)^{j_n + j_p + J}(j_p + \tfrac{1}{2})]^2}{J(J + 1)} + 1\right\},$$

$$E_\sigma = \tfrac{1}{2}\alpha F_0(l_n, l_p)(2j_n + 1)(2j_p + 1)(j_n j_p \tfrac{1}{2} - \tfrac{1}{2}|J0)^2(2J + 1)^{-1}$$

$$\times \left\{\frac{[(j_n + \tfrac{1}{2}) + (-)^{j_n + j_p + J}(j_p + \tfrac{1}{2})]^2}{J(J + 1)} - [1 + 2(-)^{l_n + l_p + J}]\right\}.$$

The energy E_σ is due to the forces that involve exchange and is independent of the number of nucleons in the shells.

Even for a single neutron and proton in each shell, the energy, which is given by

$$E[(j_n)^1(j_p)^1, J] = E_0(1, 1) + E_\sigma,$$

is seen to be a complicated function of J. We can derive some insight into the situation by noting that the potential (Eq. 7–9) implies that the triplet interaction is V_0 and the singlet is $(1 - 4\alpha)V_0$. If the net spin-exchange force is attractive, α is positive and the triplet interaction is stronger than the singlet. Of course, this is the situation for two free nucleons. We therefore expect the orientation of the two nucleons to favor the triplet spin state, that is, σ_n and σ_p parallel. If $j_p - l_p$ and $j_n - l_n$, each of which is $\pm \tfrac{1}{2}$, are of opposite sign, parallel σ's are obtained for opposite orientations of $\mathbf{j}_n$ and $\mathbf{j}_p$. Thus, in this case, we expect $J = |j_n - j_p|$. On the other hand, if $j_p - l_p$ and $j_n - l_n$ are of the same sign, the tendency toward parallel spins makes $J = j_n + j_p$ the state of lowest energy. However, even the state with $J = |j_n - j_p|$ is not a pure triplet state but contains a singlet component; for a numerical example, see Problem 7-3. Hence, for the above argument to hold, it is necessary that the preference for the triplet state be fairly pronounced. This means that α cannot be too small. De Shalit [11] and Brink [13] have shown that even for forces of finite range, provided $\alpha > \frac{1}{10}$, the state $J = |j_n - j_p|$ is energetically well below the other possible states if $j_p - l_p$ and $j_n - l_n$ are of opposite sign. When these quantities are of the same sign, the energy of the state $J = j_n + j_p$ is lowered, and if α is large enough, it may cross the state $|j_n - j_p|$. However, the spacing between the states is fairly small, so that reasonable perturbations of our idealized picture might well alter the order.

These results are summarized in *Nordheim's rules* [14], as modified by Brennan and Bernstein [15]. If we define Nordheim's number $N = j_p - l_p + j_n - l_n$, the spin J of an odd-odd nuclide is given by

strong rule: $\quad N = 0, \qquad J = |j_n - j_p|$;

weak rule: $\quad N = \pm 1, \quad J$ is either $|j_n - j_p|$ or $j_n + j_p$.

The same rules hold for k_n and $k_p > 1$, with the understanding that j_n and j_p are replaced by J_n and J_p, the total angular momenta of the configurations $j_n^{k_n}$ and $j_p^{k_p}$ in neighboring odd-A nuclides. In exceptional cases, J_n and J_p may not be equal to j_n and j_p. These rules have been theoretically established by De Shalit and Walecka [16] on the basis of a spin-dependent residual interaction; they also clarify the nature of some of the effects which can invert the expected order of the closely spaced states when the weak rule applies.

When confronted with experimental data, Nordheim's rules are found to work quite well, most of the exceptions being for light nuclei. A number of examples are listed in Problem 7-4; an examination of the observed spins and parities will show that, when two orbits are close together in energy, an odd nucleon is not always in the same orbit in the odd-odd nuclide as in the corresponding odd-A one.

7-5 CONFIGURATION MIXING

As we have remarked before, it is strictly correct to consider particles to move in pure (n, l, j)-orbits only if there are no interparticle interactions in addition to the effective single-particle potential. But, in the absence of such interactions, all states of k particles in the same configuration will be degenerate; we have removed this degeneracy by introducing interactions and calculating the energy splitting between different states by first-order perturbation theory. However, although we have in this way taken account of the energy effects of the interaction, we have assumed that it is not sufficiently strong to alter the wave functions of the states appreciably. This may well be a good approximation, but its validity should at least be investigated.

Let us consider a nucleus with the extreme single-particle wave function $\psi_0(JM)$, corresponding to the configuration

$$(n_1, l_1, l_1 + \tfrac{1}{2})^{k_1}(n_1, l_1, l_1 - \tfrac{1}{2})^{K_1}(n_2, l_2, l_2 + \tfrac{1}{2})^{k_2}(n_2, l_2, l_2 - \tfrac{1}{2})^{K_2}\cdots . \quad (7\text{--}11)$$

Usually only one or two of the shells will be unfilled, that is, the majority of the $k_i = 2l_i + 2$, and the majority of the $K_i = 2l_i$. The various ways of coupling the vectors $\mathbf{j}_i$ in the unfilled shells lead to various states of different J, and in the absence of nucleon-nucleon interactions all these states are degenerate with energy E_0. The state $\psi_0(JM)$ is a particular one, which we have chosen to study. If we now introduce an internucleon potential v_{ij}, this state will alter its energy, to first order, by an amount $(\psi_0|\sum_{i<j} v_{ij}|\psi_0)$, thus removing the degeneracy. But, at the same time, the wave function is altered by the interaction and, in first-order perturbation theory, can be represented as a sum of various states as follows:

$$\psi_{JM} = \psi_0(JM) + \sum_p \alpha_p \psi_p(JM) + \sum_q \alpha_q \psi_q(JM), \quad (7\text{--}12)$$

where

$$\alpha_p = \frac{(\psi_0|\sum_{i<j} v_{ij}|\psi_p)}{E_0 - E_p}, \quad (7\text{--}13)$$

with a similar expression for α_q. The states p and q are single-particle states, with unperturbed energies E_p and E_q, and the sums include all such states of the same J and M for which α_p or α_q is nonzero. From the form of Eq. 7–13 it is seen that an admixed state p or q can differ from the unperturbed state ψ_0 by having at most two particles in different states. The states p are distinguished from the states q in that, in the states p, all nucleons have the same n- and l-values as in ψ_0. Hence the configuration of a state ψ_p is

$$(n_1, l_1, l_1 + \tfrac{1}{2})^{k_1 - a_1}(n_1, l_1, l_1 - \tfrac{1}{2})^{K_1 + a_1}(n_2, l_2, l_2 + \tfrac{1}{2})^{k_2 - a_2}(n_2, l_2, l_2 - \tfrac{1}{2})^{K_2 + a_2}\cdots ,$$

$$(7\text{--}14)$$

with $a_i \doteq -2, -1, 0, 1,$ or 2 and $\sum_i |a_i| = 1$ or 2.

A typical state of type q has the configuration

$$(n_1, l_1, l_1 + \tfrac{1}{2})^{k_1 + b_1}(n_1, l_1, l_1 - \tfrac{1}{2})^{K_1 + \beta_1}(n_2, l_2, l_2 + \tfrac{1}{2})^{k_2 + b_2}(n_2, l_2, l_2 - \tfrac{1}{2})^{K_2 + \beta_2}\cdots ,$$

$$(7\text{--}15)$$

with b_i and $\beta_i = -2, -1, 0, 1,$ or $2, \sum_i (|b_i| + |\beta_i|) = 2$ or 4, and $b_i + \beta_i \neq 0$ for at least one value of i. This ensures that in a state q at least one nucleon has changed its n and/or l.

It will be observed that, if states of type p only were admixed, each nucleon could still be assigned a specific l-value, although j would no longer be a good quantum number, but that for states of type q neither l nor j is, in general, a constant of motion. Note also that in both types of admixture we envisage exciting to unfilled levels nucleons initially in both filled and unfilled shells.

The coefficients α_p and α_q measure the amount of admixture. For first-order perturbation theory to hold and, *a fortiori*, to ignore configuration mixing, it is necessary that all the $|\alpha|^2 \ll 1$. Usually only a few states are appreciably mixed into the wave function, and Eq. 7–13 shows that these will be the states whose unperturbed energies E_p or E_q are near E_0 or those for which the matrix element $(\psi_0 |v_{ij}| \psi_p)$ is extraordinarily large. Generally speaking, the single-particle states that are quite close in energy have different l-values (see Fig. 7–2). This fact seems to suggest that certain states of type q may be important in configuration mixing. On the other hand, two nucleons with the same n and l have very similar wave functions, so that the matrix element may be large, even though the energy difference is not particularly small. Thus states of type p are also important.

A reasonable amount of configuration mixing does not change the conclusions that we have reached about the energies of the various states, since our calculations have allowed already for the first-order energy correction. Thus none of the conclusions about ground-state spins will be changed, although we can perhaps appreciate more clearly now how the order of neighboring states is occasionally inverted because of higher-order effects— for example, the inversions that make Nordheim's weak rule a tendency rather than a precise result. However, other properties of nuclides may be more strongly modified. For example, magnetic moments may be substantially changed by the admixture of a state with an appreciably different single-particle moment. We shall study such effects in later chapters.

REFERENCES

1. W. Elsasser, *J. Phys. Rad.* **5**, 625 (1934).
2. M. G. Mayer, *Phys. Rev.* **75**, 1969 (1949).
3. M. G. Mayer, *Phys. Rev.* **78**, 16 (1950).
4. O. Haxel, J. H. D. Jensen, and H. E. Suess, *Phys. Rev.* **75**, 1766 (1949).
5. O. Haxel, J. H. D. Jensen, and H. E. Suess, *Z. Physik* **128**, 295 (1950).

6. A. Ross, R. D. Lawson, and H. Mark, *Phys. Rev.* **102**, 1613; **104**, 401 (1956).

7. M. E. Rose, *Elementary Theory of Angular Momentum*, John Wiley, New York, 1957.

8. A. R. Edmonds and B. H. Flowers, *Proc. Roy. Soc. (London)* **A214**, 515 (1952).

9. M. G. Mayer and J. H. D. Jensen, *Elementary Theory of Nuclear Shell Structure*, John Wiley, New York, 1955.

10. J. P. Elliott and A. M. Lane, "The Nuclear Shell Model," *Encyclopedia of Physics*, Vol. XXXIX, Springer-Verlag, Berlin, 1957.

11. A. De Shalit, *Phys. Rev.* **91**, 1479 (1953).

12. C. Schwartz, *Phys. Rev.* **94**, 95 (1954).

13. D. M. Brink, *Proc. Phys. Soc.* **A67**, 757 (1954).

14. L. Nordheim, *Phys. Rev.* **78**, 294 (1950).

15. M. H. Brennan and A. M. Bernstein, *Phys. Rev.* **120**, 927 (1960).

16. A. De Shalit and J. D. Walecka, *Nucl. Phys.* **22**, 184 (1961).

PROBLEMS

7-1. Show that $(E_{l-1/2} - E_{l+1/2}) \propto (2l + 1)$ if $V(r)$ in V_{ls} is constant. How does this formula change if $V(r)$ has a maximum near the nuclear surface?

7-2. It has been suggested that the Coulomb field may be responsible for the different level orders for neutrons and protons for N or $Z > 50$. Calculate the difference in electrostatic energy between a $0g_{7/2}$- and a $1d_{5/2}$-proton, assuming that the potential is that of a uniform, spherically symmetric charge distribution and using harmonic-oscillator wave functions. For a numerical estimate, use a suitable typical nucleus.

7-3. Show that the wave function $\langle (d_{5/2})^1 (p_{1/2})^1, 2|$ with $M = 2$ can be written as

$$\frac{1}{3\sqrt{5}} \{ {}^3(\sigma)_1 [\sqrt{2}\, Y_2{}^1(1) Y_1{}^0(2) - 5 Y_2{}^2(1) Y_1{}^{-1}(2)] - {}^3(\sigma)_{-1} Y_2{}^2(1) Y_1{}^1(2)$$

$$+ {}^3(\sigma)_0 [-\sqrt{2}\, Y_2{}^1(1) Y_1{}^1(2) + 3 Y_2{}^2(1) Y_1{}^0(2)]$$

$$+ {}^1(\sigma)_0 [-\sqrt{2}\, Y_2{}^1(1) Y_1{}^1(2) + 2 Y_2{}^2(1) Y_1{}^0(2)] \},$$

and show that this wave function is $\frac{2}{13}$ singlet state and $\frac{11}{13}$ triplet. Similarly, demonstrate that $\langle (d_{5/2})^1 (p_{1/2})^1, 3|$ is $\frac{1}{3}$ singlet and $\frac{2}{3}$ triplet. Does this confirm Nordheim's rule?

7-4. For the following nuclides, the most likely experimental ground-state spins and parities are given in parentheses. State the probable single-particle configurations in each case and comment on any unusual circumstances:

^{3}He($\frac{1}{2}^+$), ^{9}Be($\frac{3}{2}^-$), ^{14}N(1^+), ^{15}N($\frac{1}{2}^-$), ^{17}F($\frac{5}{2}^+$), ^{21}Ne($\frac{3}{2}^+$), ^{27}Al($\frac{5}{2}^+$), ^{30}P(1^+), ^{33}S($\frac{3}{2}^+$),

^{34}Cl(0^+), ^{35}Cl($\frac{3}{2}^+$), ^{38}K(3^+), ^{40}K(4^-), ^{42}K(2^-), ^{43}Ca($\frac{7}{2}^-$), ^{50}V(6^+), ^{53}Mn($\frac{7}{2}^-$),

^{53}Cr($\frac{3}{2}^-$), ^{54}Mn(2^+), ^{65}Cu($\frac{3}{2}^-$), ^{66}Cu(1^+), ^{66}Ga(1^+), ^{69}Ga($\frac{3}{2}^-$), ^{97}Mo($\frac{5}{2}^+$), ^{99}Tc($\frac{9}{2}^+$),

^{103}Rh($\frac{1}{2}^-$), ^{105}Pd($\frac{5}{2}^+$), ^{106}Rh(1^+), ^{121}Sb($\frac{5}{2}^+$), ^{122}Sb(2^-), ^{123}Sb($\frac{7}{2}^+$), ^{123}Te($\frac{1}{2}^+$),

^{130}Cs(1+), ^{131}Cs($\frac{5}{2}$+), ^{133}Cs($\frac{7}{2}$+), ^{131}Xe($\frac{3}{2}$+), ^{134}Cs(4+), ^{138}La(5+), ^{199}Hg($\frac{1}{2}$−), ^{203}Tl($\frac{1}{2}$+), ^{206}Tl(0−), ^{209}Bi($\frac{9}{2}$−), ^{210}Bi(1−).

7-5. (a) From formula 7-7, calculate the pairing energy $v = [B(N + 2) - B(N)] - 2[B(N + 1) - B(N)]$, when N is even and is the number of particles in a given shell (n, l, j), and B is the binding energy of the ground state. Show that v increases with j, provided f_{nl} does not decrease as l increases.

(b) If an (n, l, j)-nucleon is bound by an energy E in the effective single-particle potential $V_c + V_{ls}$, compare the binding energy of the lowest state with $2j$ (n, l, j)-nucleons and N (n', l', j')-nucleons with that of the lowest state with $(2j + 1)$ (n, l, j)-nucleons and $(N - 1)$ (n', l', j')-nucleons, and thus explain the filling in pairs of high-angular-momentum states in the single-particle model.

(c) Give qualitative arguments to show that f_{nl} tends to increase with l.

7-6. Show that, if Nordheim's rules were strictly true, there would be no odd-odd nuclides with ground state 0+ or 1−. Find some exceptions to these results.

7-7. Show that the mean-square radius of a particle in the oscillator state with quantum numbers n, l, m is

$$\alpha^{-2}(2n + l + \tfrac{3}{2}),$$

where $\alpha = (m\omega/\hbar)^{1/2}$.

Find the mean-square radius of the whole nucleus for ^{40}Ca, ^{41}K, and ^{60}Ni.

Chapter 8

Correlations in Nuclear Matter

8-1 INTRODUCTION

In Chapter 7 we saw how satisfactorily the shell model "explains" a substantial number of nuclear properties. And yet at first sight one might not expect a shell model to be at all appropriate for nuclei. After all, it implies that each nucleon moves in a single-particle orbital and, therefore, that the wave function of a nucleon and hence its behavior at a given position are independent of whether or not another nucleon is close to it. Since there is a strong short-range nucleon-nucleon force, this approximation would seem to be unreasonable. The explanation lies to a great extent in the operation of the Pauli exclusion principle, which modifies the classical behavior of interacting fermions, as well as in certain properties of the interaction between nucleons. In this chapter we shall examine the derivation of properties of nuclei, starting from known interactions between nucleons *in vacuo*, that is, free nucleons.

In order to understand the main features, it is useful to introduce the concept of *nuclear matter*. Nuclear matter is a medium of infinite extent, of uniform density, consisting of equal numbers of neutrons and protons, and in which the Coulomb force between protons is ignored. In such a system, each particle moves with constant momentum except when it interacts with another. This plane-wave motion is the analog of the closed orbital of a nucleon in the shell model of a finite nucleus. In each case the nucleon "feels" an average field of some kind, which we shall seek to understand. In the finite case the geometry is such that the average field permits a set of discrete bound states; in the infinite case the geometrical symmetry insists that uncorrelated states be states of constant momentum, but the average field will alter the relationship between the momentum p and the energy e_p of a particle so that $p^2 \neq 2Me_p$.

It is important to realize some of the consequences of a momentum-dependent "potential." A potential in this sense is defined as the difference between the total energy of the particle and its kinetic energy, $p^2/2M$. In infinite, uniform nuclear matter, the properties are the same at all points of space, since the system is translationally invariant. Thus the energy of a particle cannot depend on its location but may depend on its momentum. Alternatively, it may be said that the energy of a particle can depend on the location of other particles in its vicinity, but that this local structure cannot be at variance with the homogeneity of nuclear matter; this is possible since the particles are not, in fact, localized and the overall density is constant. Although in the neighborhood of a particle center there may be considerable fluctuations in the density of other particles, the distribution of these centers can be considered to be uniform and continuous, leading to a homogeneous density distribution.

On the other hand, in a finite nucleus, density is a function of position, there is no translational invariance, and the energy of a particle may depend not only on the correlation of other particles in its immediate position, but also on its position with respect to the center of the nucleus. It could also be said that the potential energy can depend on both momentum and position.

We shall now demonstrate the equivalence of the concepts of a nonlocal potential and a momentum-dependent one. If the vectors $\mathbf{r}$ are measured from a fixed center, the energy at a point $\mathbf{r}$ can depend, in the infinite homogeneous medium, only on vectors $\mathbf{r}' - \mathbf{r}$ connecting the point with other points in its immediate vicinity, and the resultant total density cannot depend on $\mathbf{r}$. In a finite nucleus, not only is there this possibility but also the potential may depend directly on $\mathbf{r}$. Thus, in general, the Schrödinger equation becomes

$$\frac{-\hbar^2}{2M} \nabla_r^2 \psi(\mathbf{r}) + \int d\mathbf{r}' \, U(\mathbf{r}, \mathbf{r}' - \mathbf{r})\psi(\mathbf{r}') = e\psi(\mathbf{r}). \qquad (8\text{--}1)$$

This form of energy operator (which is not diagonal in coordinate space) can be understood as reflecting correlations, whereby the presence of a particle at position $\mathbf{r}$ influences the probability of finding another at a point $\mathbf{r}'$, and this in turn affects the energy of the particle at $\mathbf{r}$ and leads to a potential energy in the form of the integral shown.

If the wave functions are plane waves of constant momentum, appropriate as the first approximation for an infinite medium in which U does not depend on its first argument $\mathbf{r}$, Eq. 8–1 becomes

$$e_p = \frac{p^2}{2M} + \int d\mathbf{r}'\, U(\mathbf{r}' - \mathbf{r}) \exp\left[\frac{i\mathbf{p} \cdot (\mathbf{r}' - \mathbf{r})}{\hbar}\right].$$

The integral is not a function of $\mathbf{r}$ but depends only on $\mathbf{p}$, being, in fact, the Fourier transform of $U(\mathbf{r}' - \mathbf{r})$. If we call this $U(\mathbf{p})$, we see that

$$e_p = \frac{p^2}{2M} + U(\mathbf{p}). \tag{8–2}$$

This demonstrates the equivalence between the nonlocal and momentum-dependent formalisms. It is readily generalized to the case in which the wave functions are not plane waves. Two cases will interest us, the correlated wave functions of nuclear matter and those of finite nuclei, but in any case the wave function can be written as a superposition of plane waves, namely,

$$\psi(\mathbf{r}) = \int \phi(\mathbf{p}) \exp(i\mathbf{p} \cdot \mathbf{r}/\hbar)\, d\mathbf{p}.$$

The potential term in Eq. 8–1 becomes

$$\int d\mathbf{r}'\, U(\mathbf{r}, \mathbf{r}' - \mathbf{r})\psi(\mathbf{r}')$$

$$= \iint d\mathbf{r}'\, \phi(\mathbf{p}) U(\mathbf{r}, \mathbf{r}' - \mathbf{r}) \exp(i\mathbf{p} \cdot \mathbf{r}'/\hbar)\, d\mathbf{p}$$

$$= \int \exp(i\mathbf{p} \cdot \mathbf{r}/\hbar)\, d\mathbf{p}\, \phi(\mathbf{p}) \int d\mathbf{r}'\, U(\mathbf{r}, \mathbf{r}' - \mathbf{r}) \exp\left[\frac{i\mathbf{p} \cdot (\mathbf{r}' - \mathbf{r})}{\hbar}\right]$$

$$= \int \exp(i\mathbf{p} \cdot \mathbf{r}/\hbar)\, d\mathbf{p}\, \phi(\mathbf{p}) U(\mathbf{r}, \mathbf{p})$$

$$= \int d\mathbf{p}\, U(\mathbf{r}, \mathbf{p})\phi(\mathbf{p}) \exp(i\mathbf{p} \cdot \mathbf{r}/\hbar)$$

$$= U(\mathbf{r}, \mathbf{p}_{0p}) \int d\mathbf{p}\, \phi(\mathbf{p}) \exp(i\mathbf{p} \cdot \mathbf{r}/\hbar)$$

$$= U(\mathbf{r}, \mathbf{p}_{0p})\psi(\mathbf{r}) \tag{8–3}$$

where $\mathbf{p}_{0p} = -i\hbar\nabla_r$. Of course, in the infinite case U does not depend on $\mathbf{r}$.

To some approximation, nuclear matter is an idealization of the constant-density central region of a large nucleus, where the nucleons may be "unaware" of the boundaries of the nucleus. Since, in fact, neither the average internucleon distance in real nuclei nor the nuclear force range is exceedingly small compared with the dimensions of even large nuclei, it is more accurate to say that the nuclear-matter idealization needs to be corrected for surface effects, effects of neutron excess, and Coulomb effects in order to make comparisons with experiment. However, the semiempirical mass formula contains precisely such terms, and it is reasonable, therefore, to demand that the calculations of nuclear matter predict correctly a density equal to that in the region of constant density of heavy nuclei and a binding energy per particle equal to the volume term of the semiempirical formula. These quantities are [1, 2]

$$\rho_0 = 0.170 \text{ nucleons fm}^{-3} \tag{8-4a}$$

and

$$E/A = -15.68 \text{ MeV per particle.} \tag{8-4b}$$

It might be thought that the actual central density would be less than the value appropriate for hypothetical nuclear matter, since real protons are repelled by electrostatic force; however, Brandow [3] found that the surface tension compensates for this Coulomb pressure.

For fermions, only one particle may occupy a given quantum state, and, as we saw in Section 6-2, a good first approximation to nuclear matter is a zero-temperature gas enclosed in a large volume and with all states occupied up to the Fermi momentum $p_F = \hbar k_F$. If the volume is denoted by Ω, the wave function of a particle with momentum[†] $\mathbf{k}_\alpha$, normalized to one particle in the volume, is

$$\phi_\alpha \equiv \phi(\mathbf{k}_\alpha) = \Omega^{-1/2} \cdot \exp(i\mathbf{k}_\alpha \cdot \mathbf{r}). \tag{8-5}$$

Similarly, a finite system will have its n lowest energy levels occupied, if there are n particles.

The density (in particles per unit volume) is given, in either finite or infinite systems, by summing the square of the wave function over all occupied states, that is,

$$\rho = \sum_\alpha |\phi_\alpha|^2. \tag{8-6a}$$

For the infinite system, this expression becomes

[†] We shall frequently speak of "momentum" k, even though k is, of course, actually the associated wave number. Note also, as indicated in Section 6-2, that $\sum_\alpha \rightarrow \Omega \int d^3k/(2\pi)^3$ for one particle per state.

$$\rho = \frac{k_F{}^3}{6\pi^2}.$$

(8-6b)

Of course, in nuclear matter there are four particles per momentum state, and, as in Eq. 6-2,

$$\rho_0 = \frac{2}{3\pi^2} k_F{}^3$$

(8-7)

with $k_F = 1.36 \text{ fm}^{-1}$.

8-2 EXCLUSION PRINCIPLE CORRELATIONS

In the first paragraph of this chapter, we noted that it is difficult to believe that interacting particles can move in an approximately uncorrelated way, because we have the classical picture that for noninteracting uncorrelated particles the probability of finding a particle in a prescribed volume element is independent of the location of a second particle and, therefore, the particles have reasonable likelihood of being close together, a situation which the actual repulsive nuclear force would prevent. In fact, the probability of finding a particle at $\mathbf{r}_1$ and one at $\mathbf{r}_2$ is, classically,

$$P_{cl} \, d\mathbf{r}_1 \, d\mathbf{r}_2 = \rho(1)\rho(2) \, d\mathbf{r}_1 \, d\mathbf{r}_2 = \sum_m |\phi_m(1)|^2 \, d\mathbf{r}_1 \sum_n |\phi_n(2)|^2 \, d\mathbf{r}_2, \quad (8\text{-}8)$$

where $\rho(1)$ represents the density at $\mathbf{r}_1$, and the sums on m and n extend over all the occupied states. We shall consistently use this summation convention.

However, quantum statistics alter this situation drastically, and for identical fermions, that is, those with the same charge and spin state, the probability of finding two particles, one at $\mathbf{r}_1$ and one at $\mathbf{r}_2$, one in state α and one in state β, is

$$\tfrac{1}{2}|\phi_\alpha(1)\phi_\beta(2) - \phi_\alpha(2)\phi_\beta(1)|^2 \, d\mathbf{r}_1 \, d\mathbf{r}_2.$$

We note immediately the well-known fact that two identical fermions cannot be at the same point, but we can be more precise than this. The probability of finding one such fermion of the system at $\mathbf{r}_1$ and another at $\mathbf{r}_2$ is clearly the sum of the above expression over occupied states, namely,

$$P_F \, d\mathbf{r}_1 \, d\mathbf{r}_2 = \tfrac{1}{2} \sum_m \sum_n |\phi_m(1)\phi_n(2) - \phi_m(2)\phi_n(1)|^2 \, d\mathbf{r}_1 \, d\mathbf{r}_2,$$

$$P_F = \sum_m \sum_n |\phi_m(1)|^2|\phi_n(2)|^2 - \text{Re} \sum_m \sum_n \phi_m(1)\phi_n(2)\phi_m{}^*(2)\phi_n{}^*(1).$$

(8-9a)

The first term of P_F is simply $\rho(1)\rho(2)$, and we see that the fermion and classical probability densities differ by the correlation

$$P_F - P_{cl} = - C^2(1, 2) = - \text{Re} \sum_m \sum_n \phi_m(1)\phi_n^*(1)\phi_m^*(2)\phi_n(2)$$

$$= - |\sum_m \phi_m(1)\phi_m^*(2)|^2. \tag{8-9b}$$

For the infinite medium, we note that, where $\mathbf{r} = \mathbf{r}_1 - \mathbf{r}_2$,

$$\sum_m \phi_m(1)\phi_m^*(2) = \frac{4\pi}{(2\pi)^3} \int_0^{k_F} j_0(kr)k^2\, dk$$

$$= \frac{1}{2\pi^2} k_F^3 \frac{j_1(k_F r)}{k_F r}.$$

The first line follows simply from Eq. 8–5 and the partial wave expansion of a plane wave. The result is

$$C^2(1, 2) = \rho^2 \left[\frac{3j_1(k_F r)}{k_F r}\right]^2 = P_{cl}F_1(k_F r), \tag{8-10a}$$

where we have introduced a correlation function F_1. In this case $P_{cl} = \rho^2$, and, using Eq. 8–9, we note that

$$P_F(1, 2) = \rho^2[1 - F_1(k_F r)]. \tag{8-10b}$$

The quantity $F_1(x)$ is plotted in Fig. 8–1. It has the value unity at $r = 0$, remains large up to $x \sim 2$, and is small (less than 0.1) for $x > 3$. With $k_F = 1.36$ fm^{-1} corresponding to nuclear-matter density, these two values of x correspond to $r = 1.5$ and 2.2 fm. The short-range part of the two-nucleon force between identical nucleons will therefore be suppressed because

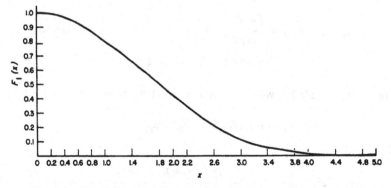

Fig. 8-1 The Pauli correlation function F_1 for an infinite medium.

$P_F(1, 2)$ is small where the force is large. It is apparent that the exclusion principle alone accounts for the possibility of using an effective force in nuclear matter, much weaker than the free nucleon-nucleon force, for collisions between pairs of identical particles. Such states are spin triplet and have odd angular momentum. The exclusion principle creates a "wound" or "hole" in the wave function for other states as well, including neutron-proton pairs, but this phenomenon occurs only when a force is acting to scatter the particles. We discuss this in the next section.

It is instructive to perform the same calculation for a finite nucleus. Consider ^{16}O with the 1s- and 1p-states filled in accordance with the shell model, and assume that the states can be described by harmonic oscillator functions. These are the uncorrelated states ϕ_m. To evaluate the correlation, it is convenient to take it as Re $\sum_{m,n} \phi_m(1)\phi_n^*(1)\phi_m^*(2)\phi_n(2)$, to use the expressions in Problem 4–10 for the wave functions, together with the spherical harmonic addition theorem (Eq. A–70), and to replace r_1 and r_2 by relative and center-of-mass coordinates. The result is found to be

$$C^2(\mathbf{r}, \tilde{\mathbf{R}}) = \left(\frac{\alpha^2}{\pi}\right)^3 [1 + \tfrac{1}{2}\alpha^2(\tilde{R}^2 - r^2)]^2 \exp\left[-\tfrac{1}{2}\alpha^2(r^2 + \tilde{R}^2)\right], \quad (8\text{–}11)$$

where α is the oscillator constant $(M\omega/\hbar)^{1/2}$, $\tilde{\mathbf{R}} = \mathbf{r}_1 + \mathbf{r}_2$, and $\mathbf{r} = \mathbf{r}_1 - \mathbf{r}_2$. This is the correlation in the probability of finding two identical nucleons, for example, neutrons with spin up, one at $\mathbf{r}_1$ and the other at $\mathbf{r}_2$. The density is

$$\rho(1) = \sum_m |\phi_m(1)|^2 = \left(\frac{\alpha^2}{\pi}\right)^{3/2} (1 + 2\alpha^2 r_1^2) \exp(-\alpha^2 r_1^2). \quad (8\text{–}12)$$

(Of course, the nuclear density is four times this value.) The classical joint probability is found to be

$$P_{cl} = \rho(1)\rho(2) = \left(\frac{\alpha^2}{\pi}\right)^3 \{[1 + \tfrac{1}{2}\alpha^2(r^2 + \tilde{R}^2)]^2 - \alpha^4 r^2 \tilde{R}^2 \cos^2\theta\}$$
$$\times \exp\left[-\tfrac{1}{2}\alpha^2(r^2 + \tilde{R}^2)\right],$$

where θ is the angle between the vectors $\mathbf{r}$ and $\tilde{\mathbf{R}}$. Consequently

$$P_F(1, 2) = \rho(1)\rho(2)[1 - F_2(\mathbf{r}, \tilde{\mathbf{R}})]$$

with

$$F_2(\mathbf{r}, \tilde{\mathbf{R}}) = \frac{[1 + \tfrac{1}{2}\alpha^2(\tilde{R}^2 - r^2)]^2}{[1 + \tfrac{1}{2}\alpha^2(r^2 + \tilde{R}^2)]^2 - \alpha^4 r^2 \tilde{R}^2 \cos^2\theta}. \quad (8\text{–}13)$$

The function F_2 is shown in Fig. 8–2 for three special cases:

1. The center of mass of the two particles is at the center of the nucleus ($\tilde{R} = 0$).

2. The center of mass is at the position of maximum density ($\tilde{R} = \sqrt{2}/\alpha$) and r is parallel to $\tilde{R}$, that is, the particles lie on the same nuclear diameter and at points of different density.

3. The mass center is at $\tilde{R} = \sqrt{2}/\alpha$ but r is perpendicular to $\tilde{R}$, that is, the particles are at points of equal density.

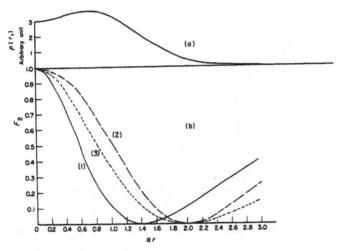

Fig. 8-2 (a) The density of ^{16}O in the harmonic-oscillator shell model. (b) The correlation function F_2 defined in Eq. 8–13 for three cases described in the text.

It is seen that in all these cases the correlation does not become very small until r reaches values somewhat greater than α^{-1}. It is small for a range of r near this value, and although F_2 again becomes appreciable, this does not happen until one of the particles (or both of them) is in a region of low density. Thus we see that for ^{16}O the exclusion principle correlation for identical fermions will suppress the effects of the interaction at internucleon distances up to about α^{-1} and also to some extent at distances larger than $2\alpha^{-1}$. In this model, the rms radius of ^{16}O happens to be $1.5\alpha^{-1}$; it is 2.6 fm. Consequently the force is reduced for $r \lesssim 1.6$ fm, and there may be a good possibility of understanding the real nucleus in terms of small perturbations of the shell model.

Although the applicability of the nuclear-matter approximation to any part of a nucleus as small as ^{16}O might be questioned, it is tempting nevertheless to compare the correlation calculated for the finite case with that obtained for nuclear matter of the same density. Combining Eqs. 8–6b and 8–12 and using the above value for α, we find that the density at the center of ^{16}O corresponds to $k_F = 1.26$ fm^{-1} and that the maximum density (at $r_1 = 1/\alpha\sqrt{2}$) corresponds to $k_F = 1.34$ fm^{-1}. Hence in Fig. 8–3 we show the nuclear-matter correlation $F_1(k_F r)$ for $k_F = 1.30$ fm^{-1} and compare it with $F_2(r, 0)$,

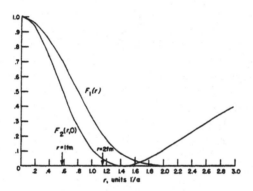

Fig. 8-3   Comparison of the Pauli correlation functions for an infinite medium and for like particles in ^{16}O.

which is the function for two particles with mass center at the center of the nucleus, and each a distance $r/2$ from the center. It is seen that, at least as far as estimates of the Pauli correlation effect are concerned, calculations with the infinite medium would give a very good approximation to the finite case. Of course, the questions of what value of k_F to use, and of how to take account of angular dependence, become more acute for particles that are not equidistant from the origin.

8-3 BETHE-GOLDSTONE EQUATION AND G-MATRIX

There is a further reason why the wave function of two nucleons is small at distances of close approach—the repulsive core of the nucleon-nucleon potential v. Although this core may be quite strong, the product $v\psi$ of the core potential multiplied by the true wave function is finite. Of course, if v is infinite, ψ is zero. We might, therefore, consider it useful to write

$$G\phi = v\psi,$$

where ϕ is the model two-body wave function (e.g., oscillator or plane wave), and the equation defines G. In this way we replace the actual potential v by a quantity G so that G acting on the model function is the same as v applied to the actual wave function; it might be that for many purposes G could serve as an effective potential. Moreover, G might be reasonably smooth compared with the actual potential. These suggestions are confirmed by the theory that follows in this chapter.

We have still to justify the use of uncorrelated waves to describe strongly interacting particles. Such a wave function is clearly wrong within the distance around each nucleon where correlations are significant. We shall assume that, during the collision of two of the nucleons, all others continue to move undisturbed with constant momentum in nuclear matter or in shell-model orbitals in finite nuclei. This assumption amounts essentially to neglect of clusters of three or more nucleons, an assumption which is justified *a posteriori* if the calculations give a small "healing" distance for the wave function of the two that collide. If we confine ourselves to two-nucleon clusters, we are using the *independent-pair model* (originally so called by Gomes, Walecka, and Weisskopf [4]). Although it is necessary to allow for larger clusters, the independent-pair model provides an understanding of much of the physics of nuclear matter; and, as we shall see, it is a well-justified low-density approximation.

Let the model wave functions be denoted by $\phi_\alpha(\mathbf{r})$. We shall consider A particles and denote their coordinates by $\mathbf{r}_1, \mathbf{r}_2, \ldots, \mathbf{r}_A$ with respect to a fixed center. We shall suppress spin and i-spin coordinates for the moment. Suppose that the two particles moving freely in states k and l collide and that all other particles in states $m, \ldots, n$ continue undisturbed. The wave function of the colliding particles is modified; let us call it ψ_{kl}. The wave function of the whole system must be antisymmetric in coordinates of all the nucleons, and it is, therefore,

$$\Psi_{kl} = \mathscr{A}\{\psi_{kl}(\mathbf{r}_1, \mathbf{r}_2)\phi_m(\mathbf{r}_3) \cdots \phi_n(\mathbf{r}_A)\}, \qquad (8\text{–}14)$$

where the $\mathscr{A}$ indicates that all terms formed by antisymmetric exchange of $\mathbf{r}_1, \ldots, \mathbf{r}_A$ are to be taken.

The wave functions $\phi_\alpha(\mathbf{r})$ are eigenfunctions of a single-particle Hamiltonian. For nuclear matter, they are plane waves, each with its momentum $\mathbf{p}_\alpha$; for finite nuclei, they are some sort of orbitals. Thus

$$H_0(i)\phi_\alpha(\mathbf{r}_i) \equiv [T(i) + U(i)]\phi_\alpha(i) = e_\alpha\phi_\alpha(i). \tag{8-15}$$

In nuclear matter the single-particle field U is a function of momentum only; in the finite nucleus it depends on position (e.g., oscillator or Woods-Saxon potential) but can also be momentum dependent. Temporarily we shall not specify $U(i)$ further.

The colliding pair feel the same overall potential as do all the particles and in addition experience the free nucleon-nucleon potential; this is the essence of the independent-pair approximation. Hence the Schrödinger equation is

$$[H_0(1) + H_0(2) + v_{12}]\psi_{kl}(1, 2) = e_{kl}\psi_{kl}(1, 2). \tag{8-16}$$

The energy e_{kl} is not the unperturbed energy $e_k + e_l$. We shall examine the physical significance of this later.

The antisymmetry demanded in the wave function (Eq. 8–14) of the whole system and the assumption that each state, $m,\ldots, n$, contains precisely one nucleon mean that ψ_{kl} is orthogonal to all the filled states, except of course states k and l. In other words, if ψ_{kl} is expanded in eigenfunctions of H_0, it contains no component of states $m,\ldots, n$, since these are filled by other nucleons. With the normalization usual in the literature,

$$\psi_{kl}(1, 2) = \phi_k(1)\phi_l(2) + \sum_{a,b} a_{ab}\phi_a(1)\phi_b(2)$$
$$+ \sum_a a_{ka}\phi_k(1)\phi_a(2) + \sum_a a_{al}\phi_a(1)\phi_l(2). \tag{8-17}$$

We use $a, b, c,\ldots$ for states unoccupied in the ground state and $k, l, m,\ldots$ for filled states. Note that

$$\langle\phi_k\phi_l|\psi_{kl}\rangle = 1. \tag{8-18}$$

We can write Eq. 8–16 as

$$(e_k + e_l)\phi_k(1)\phi_l(2) + \sum_{\alpha,\beta}{}' a_{\alpha\beta}(e_\alpha + e_\beta)\phi_\alpha(1)\phi_\beta(2) + v_{12}\psi_{kl} = e_{kl}\psi_{kl},$$

where the summation $\sum'$ over the single-particle states α, β is such that $\alpha = k, a, b, c,\ldots$; $\beta = l, a, b, c,\ldots$; $\alpha \neq \beta$; and $(\alpha, \beta) \neq (k, l)$. If we take the scalar product of the above expression with $\langle\phi_\alpha(1)\phi_\beta(2)|$, we obtain

$$a_{\alpha\beta}(e_{kl} - e_\alpha - e_\beta) = \langle\phi_\alpha(1)\phi_\beta(2)|v_{12}|\psi_{kl}(1, 2)\rangle$$
$$= \langle\alpha\beta|v|\psi_{kl}\rangle. \tag{8-19}$$

We shall usually abbreviate $\langle\phi_\alpha(1)\phi_\beta(2)|$ to the form $\langle\alpha\beta|$. Substitution of Eq. 8-19 into 8-17 yields

$$|\psi_{kl}(1,2)\rangle = |\phi_k(1)\phi_l(2)\rangle + \sum_{\alpha,\beta}{}' \frac{\langle\alpha\beta|v|\psi_{kl}\rangle}{e_{kl} - e_\alpha - e_\beta}\, |\phi_\alpha(1)\phi_\beta(2)\rangle,$$

which may be written as

$$|\psi_{kl}(1,2)\rangle = |\phi_k(1)\phi_l(2)\rangle + \sum_{\alpha,\beta}{}' \frac{1}{e_{kl} - H_0}\, |\alpha\beta\rangle\langle\alpha\beta|v|\psi_{kl}\rangle, \qquad (8\text{-}20)$$

where $H_0 = H_0(1) + H_0(2)$. It is usual to introduce the projection operator Q, that takes care of the Pauli principle:

$$Q_{kl} = \sum_{ab} |ab\rangle\langle ab| + \sum_a |ka\rangle\langle ka| + \sum_b |bl\rangle\langle bl|. \qquad (8\text{-}21)$$

Then Eq. 8-20 may be rewritten as

$$|\psi_{kl}\rangle = |\phi_k\phi_l\rangle + \frac{Q_{kl}}{e_{kl} - H_0}\, v|\psi_{kl}\rangle. \qquad (8\text{-}22)$$

This is the form of the Bethe-Goldstone equation [5], and a number of important results follow from it.

Equation 8-22 suggests a perturbation procedure obtained by continually substituting expression 8-22 itself for ψ whenever it appears at the end of the equation, that is,

$$|\psi_{kl}\rangle = |\phi_k\phi_l\rangle + \sum_{\alpha\beta}{}' \frac{|\alpha\beta\rangle}{e_{kl} - (e_\alpha + e_\beta)} \langle\alpha\beta|v|kl\rangle$$

$$+ \sum_{\alpha\beta}{}' \sum_{\gamma\delta}{}' \frac{|\alpha\beta\rangle}{e_{kl} - (e_\alpha + e_\beta)} \langle\alpha\beta|v|\gamma\delta\rangle \frac{1}{e_{kl} - (e_\gamma + e_\delta)} \langle\gamma\delta|v|kl\rangle$$

$$+ \cdots, \qquad (8\text{-}23)$$

where, as before, the Pauli principle restricts the intermediate states. This expression is of little value for computation unless it converges rapidly, and, unfortunately, nuclear potentials do not produce rapid convergence since the repulsive core, even when finite, contributes largely to the matrix elements $\langle\alpha\beta|v|\gamma\delta\rangle$ between uncorrelated states.

Expansion 8-23 provides the formal justification for the arguments concerning intermediate states in Section 6-2. For nuclear matter, $|\alpha\beta\rangle$ are states in which the total momentum $\mathbf{P} = \mathbf{p}_\alpha + \mathbf{p}_\beta$ is a constant of motion. Moreover, the nuclear potential conserves momentum. Hence $\langle\alpha\beta|v|kl\rangle$

vanishes, unless $\mathbf{p}_\alpha + \mathbf{p}_\beta = \mathbf{p}_k + \mathbf{p}_l$. Consequently, no matter to how high an order we go in the series of Eq. 8–23, all the intermediate states $|\alpha\beta\rangle$, $|\gamma\delta\rangle,\ldots$ have the same total momentum. Furthermore, k and l are inside the Fermi sea, and momentum conservation does not allow only one of the particles to be scattered above the Fermi sea. This implies that in this case there are no components of ψ_{kl} like the last two in Eq. 8–17, where only one of the particles is excited. Consequently the operator Q_{kl} may be written in nuclear matter as

$$Q = \sum_{ab} |ab\rangle\langle ab|. \qquad (8\text{–}24)$$

Clearly this simplification does not apply to the finite nucleus. The result is valid only in the independent-pair approximation; once we consider the interaction of three particles, we can conserve momentum in the chain $|klm\rangle \rightarrow |abm\rangle \rightarrow |kbc\rangle$ and the last of these is an intermediate state (in the second-order term) where one of the particles is in a level below k_F.

One of the major consequences of the Bethe-Goldstone equation is the "healing" of the wave function, that is, the fact that correlations which constitute a "wound" in the model wave function $\phi_k\phi_l$ are confined to short relative distances. Note that the Bethe-Goldstone equation 8–22 is very similar in structure to the Lipmann-Schwinger equation 5–49 for two-nucleon scattering in free space. In the latter case, however, there is a singularity in the energy denominator, and a term $+ i\varepsilon$ is required for an outgoing wave of the same momentum, as in Eq. 5–51. This singularity introduces a phase shift in the scattered wave that modifies the asymptotic form. In the Bethe-Goldstone equation 8–22, on the other hand, the operator Q_{kl} ensures that no singularity is present in the energy denominator and the wound contains no asymptotic modification of $\phi_k\phi_l$. Consequently there is no real scattering in nuclear matter, and the wave function ψ_{kl} must heal to $\phi_k\phi_l$ for large relative distances. That there is no real scattering in nuclear matter is physically understandable, since the Pauli principle allows only scatterings that defy the conservation of energy. These virtual scatterings above the Fermi sea serve to distort the relative wave function at short distances. A more detailed discussion of healing will be found in ref. [6].

It is useful to define a defect wave function

$$\zeta_{kl} = \phi_k\phi_l - \psi_{kl}, \qquad (8\text{–}25)$$

which tends to zero for large values of $|\mathbf{r}_1 - \mathbf{r}_2|$ as the wave function heals. Indeed, the integral

$$\kappa_{kl} = \int |\zeta_{kl}|^2 \, d\mathbf{r}_1 \, d\mathbf{r}_2 \qquad (8\text{-}26)$$

is often called the "wound" in the wave function. It is easy to see from Eq. 8–17 that

$$\kappa_{kl} = \sum_{ab} |a_{ab}|^2 + \sum_{a} [|a_{ka}|^2 + |a_{al}|^2]; \qquad (8\text{-}27)$$

in words, κ_{kl} is the probability of finding the actual system excited out of the noninteracting Fermi sea, relative to the probability of finding it in the state $|kl\rangle$.

As we have seen, the wound ζ contains as Fourier components no states below the Fermi level. Hence the range of ζ is restricted to distances of the order of k_F^{-1} or shorter, that is, less than about 1.5 fm. In other words, the exclusion principle has the effect of restricting the correlations between particles to short distances. In Section 8–2 we considered noninteracting identical fermions and saw that the exclusion principle introduced correlations on the length scale of k_F^{-1}. We now have seen that, even when there is interaction, and even when a pair of nonidentical fermions in nuclear matter is considered, the fact that all the low-momentum fermion states are filled means that only short-range correlations are possible. This is the basis for our earlier statement that the operation of the exclusion principle is one of the reasons why shell-model wave functions are valid first approximations for many purposes: they are modified only at relatively short distances. Of course, the nature of the modification does depend on the potential v or, more specifically, on the matrix elements $\langle ab|v|kl \rangle$.

We may also introduce a quantity G, mentioned earlier as an effective interaction in the sense that

$$G\phi_k\phi_l = v\psi_{kl}. \qquad (8\text{-}28)$$

Multiplying Eq. 8–22 by v and using this definition G, we obtain

$$G\phi_k\phi_l = v\phi_k\phi_l + v \frac{Q_{kl}}{e_{kl} - H_0} G\phi_k\phi_l$$

or

$$G = v + v \frac{Q_{kl}}{e_{kl} - H_0} G. \qquad (8\text{-}29a)$$

Now in this equation for the infinite medium the only reference to states k and l is in the energy e_{kl}, and therefore it is usual to define $G(\omega)$ as

$$G(\omega) = v + v \frac{Q}{\omega - H_0} G, \qquad (8\text{--}29\text{b})$$

with Q from Eq. 8–24. The quantity G is also a function of the Fermi level k_F, through Q, and of course it depends on the assumed Hamiltonian H_0.

Let us now consider once again the complete system of A particles. Its Hamiltonian is

$$H = \sum_{i=1}^{A} T(i) + \tfrac{1}{2} \sum_{ij=1}^{A} v(i, j), \qquad (8\text{--}30)$$

and its ground state corresponds to the lowest eigenvalue of

$$(H - E)\Psi = 0. \qquad (8\text{--}31)$$

We also consider the Hamiltonian for the single-particle model where particles are in the states $k, l, m, \ldots$. This is

$$H_0 = \sum_{i=1}^{A} H_0(i), \qquad (8\text{--}32)$$

and it defines an orthogonal set of states Φ_i of the A-particle system by

$$(H_0 - W_i)\Phi_i = 0. \qquad (8\text{--}33)$$

The lowest model state Φ_0 is that in which all levels are filled up to the Fermi level; its energy is

$$W_0 = \sum_k e_k,$$

using the notation of Eq. 8–15. The wave function Φ_0 is simply the Slater determinant of the occupied states ϕ_k.

To obtain a Bethe-Goldstone equation for the A-body system we may generalize the techniques already used. We first write

$$H = H_0 + V, \qquad (8\text{--}34)$$

where $V = \tfrac{1}{2} \sum v(i, j) - \sum U(i)$. Then, by writing

$$\Psi = \Phi_0 + \sum a_i \Phi_i$$

and noting that

$$(H_0 - E)(\Phi_0 + \sum a_i \Phi_i) + V\Psi = 0,$$

it is easy to see, taking the scalar product with Φ_0, that

$$E = W_0 + (\Phi_0|V|\Psi). \qquad (8\text{--}35)$$

Also, generalizing the method used to obtain Eq. 8–22, we find

$$\Psi = \Phi_0 + \frac{Q'}{E - H_0} V\Psi, \tag{8-36a}$$

$$G'\Phi_0 = V\Psi, \tag{8-36b}$$

$$G' = V + V\frac{Q'}{E - H_0}G', \tag{8-36c}$$

and

$$Q' = 1 - |\Phi_0\rangle\langle\Phi_0|. \tag{8-36d}$$

These equations form the basis for a rigorous approach to the many-body problem. Primes are used to emphasize the many-body nature of G' and Q'.

For the moment we wish to use only the equation for the energy of the system, E (Eq. 8–35). We work within the independent-pair approximation, so that Ψ is of the form

$$\Psi = N \sum_{kl} \Psi_{kl},$$

where Ψ_{kl}, as given in Eq. 8–14, represents a state in which only the k-l pair interacts and N is a normalization constant $2/A(A-1)$.

Of the terms in $(\Phi_0|V|\Psi)$, we first consider $(\Phi_0|\sum_i U(i)|\Psi_{kl})$. Since $U(i)$ is a single-particle operator, it cannot link states in which more than one particle changes state; and since the filled states other than k, l all remain filled in Ψ_{kl}, only diagonal terms survive for these other filled states. Thus the expression is

$$\sum_{m \neq k,l} \langle m|U|m\rangle + \tfrac{1}{2}\langle\phi_k(1)\phi_l(2) - \phi_l(1)\phi_k(2)|U(1) + U(2)|\psi_{kl}(1, 2) - \psi_{kl}(2, 1)\rangle,$$

where the factor $\tfrac{1}{2}$ comes from the antisymmetric normalization of Ψ_{kl}.

In finite nuclei the terms in ψ_{kl} that have a single-particle excitation cause an extra complication, but for nuclear matter the single-particle potentials U link the term $\phi_k\phi_l$ only to the term $\phi_k\phi_l$ in ψ_{kl}, and the result for the above expression is simply $\sum \langle m|U|m\rangle$ summed over all occupied states m, including k and l.

For the two-particle potential in V, we have

$$\langle\Phi_0|\sum v(i, j)|\Psi\rangle = \langle\Phi_0|v(1, 2)|\sum_{kl} \Psi_{kl}\rangle$$

$$= \sum_{kl} \tfrac{1}{2}\langle\phi_k(1)\phi_l(2) - \phi_k(2)\phi_l(1)|v(1, 2)|\psi_{kl}(1, 2) - \psi_{kl}(2, 1)\rangle$$

$$= \sum_{kl} [\langle kl|v|\psi_{kl}(1, 2)\rangle - \langle kl|v|\psi_{kl}(2, 1)\rangle],$$

where we have used the facts that $v(1, 2)$ is symmetric and that $\psi_{kl}(2, 1) = \psi_{lk}(1, 2)$.

Hence in nuclear matter Eq. 8–35 becomes

$$E = W_0 + \tfrac{1}{2} \sum_{kl} (\langle kl|G|kl\rangle - \langle kl|G|lk\rangle) - \sum \langle m|U|m\rangle, \qquad (8\text{–}37a)$$

$$= \sum_m \langle m|T|m\rangle + \tfrac{1}{2} \sum_{kl} (\langle kl|G|kl\rangle - \langle kl|G|lk\rangle). \qquad (8\text{–}37b)$$

Equation 8–37b demonstrates one reason for thinking of G as an effective interaction, for its matrix elements between uncorrelated states appear in the total energy precisely where one would see the matrix elements of the potential energy between actual wave functions.

For the finite nucleus it is necessary to add other terms to account for the permissible one-particle excitations, namely, the terms

$$\tfrac{1}{2}N \sum_{kla} (a_{al}\langle \phi_k|U|\phi_a\rangle + a_{ka}\langle \phi_l|U|\phi_a\rangle),$$

which, by Eq. 8–19, can be reduced to

$$N \sum_{kla} \frac{\langle ka|G|kl\rangle}{e_{kl} - e_a - e_k} \langle l|U|a\rangle. \qquad (8\text{–}38)$$

Although this sum runs over an infinity of unoccupied states a, convergence arises from the energy denominator as e_a increases, and also because reasonable U and G give small matrix elements between occupied states and highly excited states.

The results 8–37 and 8–38 are important in that they show how to compute the energy of the system within the limitations of the independent-pair approximation. One postulates any convenient single-particle Hamiltonian H_0, finds its states ϕ_α, uses Eq. 8–29 to calculate $G(e_{kl})$, and substitutes the result in Eqs. 8–37 and 8–38. It will be seen that there appears to be nothing unique about the choice of the single-particle potential U, except that the independent-pair approximation will presumably not be valid unless the *occupied* eigenstates ϕ_m bear a reasonable resemblance to the actual wave functions outside the region of correlation. This is necessary not only because of the basic idea of the independent-pair approximation, but also because the treatment of the exclusion principle would be incorrect if the occupied ϕ_m's did not span the space of the real occupied states. We shall say more later about the choice of U, and we shall see that for a certain choice the term 8–38 will be canceled by terms from clusters of more than two particles.

With this choice of potential, Eq. 8-37 applies to both finite and infinite cases, and we note that to determine the energy of the system we need only the diagonal matrix elements of G for occupied states although of course both the direct and the exchange elements are required. But one gap remains before we can carry out the prescription of the last paragraph in order to find E, namely, before we can indicate methods of solution for the G-matrix in Eq. 8-29 we need a value for e_{kl}.

Let us examine e_{kl}. We take the scalar product of Eq. 8-16 with $\langle \phi_k \phi_l |$ to obtain

$$\langle kl|H_0(1) + H_0(2)|\psi_{kl}\rangle + \langle kl|v|\psi_{kl}\rangle = e_{kl}\langle kl|\psi_{kl}\rangle$$

and use Eqs. 8-15 and 8-18 to get

$$\varepsilon_{kl} = e_{kl} - e_k - e_l = \langle kl|v|\psi_{kl}\rangle = \langle kl|G(e_{kl})|kl\rangle. \qquad (8\text{-}39)$$

This suggests a self-consistency technique, in that we could assume a value for e_{kl}, calculate G, use Eq. 8-39 to find a new value for e_{kl}, and continue such cycles until convergence was reached. However, this is not necessary for nuclear matter, because in that case ε_{kl} is infinitesimal.

This is perhaps best seen by noting that v conserves total momentum and depends only on the relative coordinate of the two interacting particles. We have already noticed that all the terms of ψ have the same total momentum. For plane waves

$$\phi_\alpha(\mathbf{r}_1)\phi_\beta(\mathbf{r}_2) = \frac{1}{\Omega}\exp[i(\mathbf{p}_\alpha \cdot \mathbf{r}_1 + \mathbf{p}_\beta \cdot \mathbf{r}_2)] = \frac{1}{\Omega}\exp[i(\mathbf{p}_{\alpha\beta} \cdot \mathbf{r} + \mathbf{P}_{\alpha\beta} \cdot \mathbf{R})]$$

with

$$\mathbf{p}_{\alpha\beta} = \tfrac{1}{2}(\mathbf{p}_\alpha - \mathbf{p}_\beta), \qquad \mathbf{r} = \mathbf{r}_2 - \mathbf{r}_1;$$

$$\mathbf{P}_{\alpha\beta} = \mathbf{p}_\alpha + \mathbf{p}_\beta, \qquad \mathbf{R} = \tfrac{1}{2}(\mathbf{r}_1 + \mathbf{r}_2).$$

Consider the matrix element $\langle \alpha\beta|v|\psi_{kl}\rangle$, where α and β are any two momentum states above or below k_F. We write

$$\psi_{kl} = |kl\rangle + \sum a_{ab}|ab\rangle, \qquad (8\text{-}40)$$

where, as in Eq. 8-19,

$$a_{ab} = \langle ab|v|\psi_{kl}\rangle(e_{kl} - e_a - e_b)^{-1}. \qquad (8\text{-}41)$$

We expand the potential in its Fourier components:

$$v(\mathbf{r}) = \frac{1}{(2\pi)^3}\int v(\mathbf{\kappa})\exp(i\mathbf{\kappa} \cdot \mathbf{r})\, d^3\kappa. \qquad (8\text{-}42)$$

Then

$$\langle\alpha\beta|v|\psi_{kl}\rangle = \frac{1}{\Omega^2}\int dr\, dR\, d\kappa\, \exp(-i p_{\alpha\beta}\cdot r - i P_{\alpha\beta}\cdot R)$$

$$\times \frac{1}{(2\pi)^3} v(\kappa)\exp(i\kappa\cdot r)\{\exp(i p_{kl}\cdot r + i P_{kl}\cdot R)$$

$$+ \int da\, a_{ab}\exp(i p_{ab}\cdot r + i P_{ab}\cdot R)\}.$$

The last term is written as a single sum because b is determined when a is given, since $P_{kl} = P_{ab} = p_a + p_b$. Since $P_{\alpha\beta}$ is also the same as P_{kl}, the integrand is independent of R and the integral over R yields the volume Ω. Hence, performing integrations on r also, we obtain

$$\langle\alpha\beta|v|\psi_{kl}\rangle = \frac{1}{\Omega}\int d\kappa\, v(\kappa)\left[\delta^{(3)}(\kappa - p_{\alpha\beta} + p_{kl}) + \int da\, a_{ab}\delta^{(3)}(\kappa - p_{\alpha\beta} + p_{ab})\right]$$

$$= \frac{1}{\Omega}\left[v(p_{\alpha\beta} - p_{kl}) + \int da\, a_{ab}v(p_{\alpha\beta} - p_{ab})\right]$$

$$= \frac{1}{\Omega}\left[v(p_\alpha - p_k) + \int da\, a_{ab}v(p_\alpha - p_a)\right]. \tag{8-43}$$

In the last step we used again the fact that the total momentum is conserved.

Let us now consider $\varepsilon_{kl} = \langle kl|v|\psi_{kl}\rangle$. We see that

$$\varepsilon_{kl} = \frac{1}{\Omega}\left[v(0) + \int da\, a_{ab}v(p_k - p_a)\right].$$

The first term $v(0)$ is the zero-frequency Fourier transform of the potential. For a potential of finite range, this is finite, and since it is divided by the large volume Ω, this contribution to ε_{kl} is infinitesimal. Can the second term give a finite contribution to ε_{kl}? It could do so only if the integral were to be very large, and this in turn could happen only if a_{ab} became large. We see from Eq. 8-41 that this can happen if e_{kl} is near $e_a + e_b$, which is, in turn, a possibility if k and l are just below the Fermi surface and a and b just above it, with e_{kl} being near $e_k + e_l$. Moreover, since the energy spacing of momentum states is inversely related to the size of the enclosure, this is just the right kind of behavior to compensate for the $1/\Omega$ factor. But the precise result depends also on the nature of the potential. If we trace back through the derivation, we see that the integral in ε_{kl} is essentially the Fourier transform of $v\zeta$, where

ζ is the defect wave function.[†] We have already seen that ζ is of very short range, and hence all these Fourier components within the Fermi sea also are finite. Thus for nuclear matter ε_{kl} is infinitesimal.

We note that this conclusion does not follow if the force is of longer range, or if the spectrum is discrete as in finite nuclei. We remark in passing that, when there is a finite value of ε_{kl} for k and l near the Fermi level, the extra binding arises for like particle pairs, and we find the phenomenon of super-conductivity both in the electron gas in metals and in real nuclei (as opposed to nuclear matter), in the first case because the electromagnetic force is of long range and in the second case because the bound states have discrete spectra.

The solution of the Bethe-Goldstone equations (Eqs. 8–36) of the general many-body problem is an extensive topic in theoretical physics, about which a great deal of literature exists. We do not intend to describe the subject in this book, but we shall find it convenient to quote some results. We have already suggested that the independent-pair approximation is valid only if the density is sufficiently low so that clusters of larger numbers of particles can be ignored. Particles are in interaction only when their relative wave function is correlated; if we write c for an average healing distance of the wave-function defect, the probability that the motion of one particle is correlated with that of some one other particle is of order $(c/r_0)^3$, where r_0 is the radius of the sphere containing, on the average, one particle, that is, its volume is $1/\rho_0$. But if κ is the probability that the motion of one particle is correlated with that of another, the probability that a particle has two other particles within correlation distance is κ^2. Hence, if we write $\bar{v}$ for an average interaction energy for two correlated particles, we would expect the contributions to the binding energy per particle from two-, three-, and four-particle clusters to be, respectively, of order $(1/2!)\kappa\bar{v}$, $(3/3!)\kappa^2\bar{v}$, and $(6/4!)\kappa^3\bar{v}$, that is, $\frac{1}{2}\kappa\bar{v}$, $\frac{1}{2}\kappa^2\bar{v}$, and $\frac{1}{4}\kappa^3\bar{v}$.

Estimation of κ requires knowledge of the wave-function defect, and it is possible to be rather more accurate than simply to replace κ by $(c/r_0)^3$. In fact, if one particle is interacting with another, its motion is modified, that is, it is excited out of the Fermi sea. Hence the probability of interaction for a particle

---

[†] Explicitly,

$$\int d\mathbf{r}\, \exp(-i\mathbf{p}_{kl} \cdot \mathbf{r}) v(r) \zeta_{kl}(r) = \int d\mathbf{r} \sum_a \exp(-i\mathbf{p}_{kl} \cdot \mathbf{r}) v(r) a_{ab} \exp(i\mathbf{p}_{ab} \cdot \mathbf{r})$$

$$= (2\pi)^{3/2} \sum_a a_{ab} v(\mathbf{p}_k - \mathbf{p}_a),$$

where b is fixed by $\mathbf{P}_{ab} = \mathbf{P}_{kl}$.

is the same as the probability of excitation, and this is simply related to the wound κ_{kl}, defined in Eq. 8–26. In fact, the probability per particle of such excitation is

$$\kappa = \frac{1}{A} \sum_{k<l} \kappa_{kl} = \frac{1}{A} \sum_{k<l} \langle \zeta_{kl} | \zeta_{kl} \rangle = \frac{A}{2} \langle \zeta | \zeta \rangle_{\text{av}}, \tag{8-44}$$

where the average is over all pairs k-l. For nuclear matter, if we write

$$\phi_{kl} = \left(\frac{1}{\Omega}\right) \exp(i\mathbf{k}_1 \cdot \mathbf{r}_1) \exp(i\mathbf{k}_2 \cdot \mathbf{r}_2) |SM_S\rangle |TM_T\rangle$$

and

$$\zeta_{kl} = \left(\frac{1}{\Omega}\right) \exp(i\mathbf{P} \cdot \mathbf{R}) \sum_{JLM} 4\pi(i)^L \zeta_{JLS}(kr)$$
$$\times (LSM_L M_S|JM) Y_L^{M_L}(\mathbf{r}) Y_L^{M_L*}(\mathbf{k}) |SM_S\rangle |TM_T\rangle, \tag{8-45}$$

we find

$$\kappa = \tfrac{1}{8}\rho \sum_{JLS} (2J + 1)(2T + 1)\langle \zeta_{JLS} | \zeta_{JLS} \rangle_{\text{av}(k)}, \tag{8-46}$$

where, as indicated, the average is taken over the relative momentum k. We see that, if the defect wave function ζ is nonzero in a sphere of radius c where it has an average value θ/Ω, κ is indeed of order $\theta^2(c/r_0)^3$, as our qualitative reasoning suggested.

We shall see below that for nuclear matter κ is of the order 0.15; its exact value depends, of course, on density and on the internucleon force. Now the average kinetic energy per particle $(0.6\epsilon_F)$ is 23 MeV at $k_F = 1.36$ fm^{-1}. Therefore, to give the observed binding of 16 MeV, taking two-, three-, and four-body clusters into account, we must have

$$\tfrac{1}{2}\kappa \bar{v}(1 + \kappa + \tfrac{1}{2}\kappa^2) = 39 \text{ MeV}.$$

Taking $\kappa = 0.15$, we find that the two-body clusters contribute about 33.6 MeV to the binding energy, while the three- and four-body clusters give 5.0 and 0.4 MeV, respectively.

We should note that the small parameter of the expansion is κ, the wound, and not G, the interaction.

The three-body clusters are contributing about 5 MeV of the binding energy (although only one-eighth of the potential energy) and hence can hardly be ignored. If the many-body Bethe-Goldstone equations 8–36 are treated analogously to the two-body equations 8–16 to 8–20, a series for the

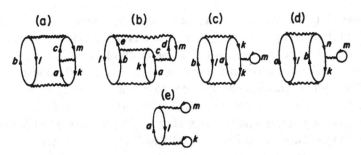

Fig. 8-4 Three-body cluster diagrams. Downward-pointing arrows refer to states in the Fermi sea; upward-pointing ones, to states above the sea. Arrows pointing into a vertex indicate the states occupied before the interaction; arrows pointing away from a vertex, the states occupied afterward. For example, the bottom wavy line of diagram (a) indicates the interaction in which the states k and l $(e_k, e_l < e_F)$ scatter to states a and b, and the next wavy line refers to the interaction in which particles in states a and m scatter to states k and c. Diagrams of this kind are extensively used for actual calculation in many-body theory, but we do not give the details here. Each diagram represents a term in the total energy. Summation on all state labels is implied.

ground-state energy can be obtained. When only two-nucleon forces are taken, all the expressions are related to a succession of virtual excitations involving only two particles and all can be given in terms of the *two-body* G-matrix. We have described a two-body cluster as a correlation ζ_{kl} in the wave function, which we have pictured as a collision causing the virtual excitation of two particles above the Fermi sea, immediately followed by a second interaction refilling their original states in the Fermi sea. A three-body cluster involves simultaneous interaction of three particles, which, with a two-body force, must be pictured as a collision of two particles, producing excitation out of the Fermi sea, a subsequent interaction of one of these with a third particle, exciting it out of the Fermi sea, and then interactions refilling the three original states. The expressions for the energies of three-body clusters are of third or higher order in the G-matrix. For example, a third-order term is

$$- (ml|G|bc)(ck|G|am)(ab|G|kl)(e_a + e_b - e_{kl})^{-1}(e_c + e_b - e_m + e_k - e_{kl})^{-1}.$$

This may be described as due to the interaction of particles in states k and l which go to states a and b, with the particle so placed in state a interacting with the particle in state m to reoccupy state k and place a particle in state c, where it interacts with the particle in state b to re-establish the undisturbed Fermi sea. As an aid to visualizing this term, we show the appropriate Goldstone diagram in Fig. 8-4a. This term is an energy properly attributed to a three-body cluster since it involves the simultaneous interaction of particles in three occupied states k, l, m. The diagram is said to contain three "hole lines," those with downward-pointing arrows labeled k, l, m.

A three-body cluster contribution of a higher order in G appears in Fig. 8–4b. We see that states k and l scatter to a and b, a interacts with m to produce two states above the Fermi sea (c and d), and c and b interact to reoccupy state k and put a particle in state e, which interacts with the particle in state d to reoccupy states l and m. Although this gives a term

$$- \frac{(ml|G|de)(ke|G|cb)(cd|G|am)(ab|G|kl)}{(e_a + e_b - e_{kl})(e_b + e_c + e_d - e_m - e_{kl})(e_d + e_e - e_m + e_k - e_{kl})}$$

of fourth order in G, it is part of the three-body-cluster energy since only three hole states appear. Because the small parameter for a series expansion is κ, not G, it is important to sum all the terms, of whatever order, which constitute the three-body cluster. Bethe [7] showed how to perform this sum, using the methods developed by Faddeev [7] for the three-body problem.

There is an important class of three-body-cluster terms which require separate treatment. These are the ones that contain a diagonal matrix element $\langle mn|G|mn \rangle$ involving two hole states. Such factors are said to produce the "hole self-energy," since they describe the change in energy of a particle in an occupied state due to its interaction with the other particles of the system. The simplest of these diagrams appears in Fig. 8–4c; its contribution to the energy is

$$\langle kl|G|ab \rangle \langle ab|G|kl \rangle \langle km|G|km \rangle (e_a + e_b - e_{kl})^{-2}.$$

It will be recalled (Eq. 8–34) that the perturbation to H_0 in the exact theory is the nucleon-nucleon potential minus the assumed single-particle U, and hence that there are terms in the energy expansion which involve U. We saw in Eq. 8–37 how U contributes in the independent-pair approximation. The next terms to involve U are of the type illustrated by Fig. 8–5a. This diagram represents an energy

$$- \langle kl|G|ab \rangle \langle ab|G|kl \rangle \langle k|U|k \rangle (e_a + e_b - e_{kl})^{-2}.$$

We see that, if we choose U so that

$$\langle k|U|k \rangle = \sum_m \langle km|G|km \rangle,$$

all the terms represented by Fig. 8–4c would cancel those represented by Fig. 8–5a. But there are also exchange diagrams giving the same contributions as Fig. 8–4c except that $\langle km|G|km \rangle$ is replaced by $- \langle km|G|mk \rangle$. It is a

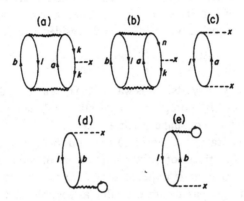

Fig. 8-5 Some diagrams involving U. The single-particle potential U is indicated by a dashed line with a cross at the end. Diagram (a) represents an energy contribution $\langle kl|G|ab\rangle\langle ab|G|kl\rangle\langle k|U|k\rangle(e_a + e_b - e_{kl})^{-2}$.

very important and nontrivial result of the general theory that all self-energy terms to all orders cancel against the terms involving U, provided U satisfies

$$\langle k|U|k\rangle = \sum_m \langle km|G(e_k + e_m)|km - mk\rangle, \qquad (8\text{--}47)$$

where $|km - mk\rangle = |k(1)m(2)\rangle - |m(1)k(2)\rangle$, and we have explicitly shown the starting energy $\omega = e_k + e_m$, for which G is to be evaluated (see Eq. 8-29b). This result is referred to as the BBP theorem after the initials of its authors: Bethe, Brandow, and Petschek [8].

It will be recalled that the selection of a single-particle Hamiltonian H_0 has until now been arbitrary. The choice of Eq. 8–47 for the potential energy of occupied states simplifies the calculation of three-particle clusters and gives more rapid convergence to the cluster expansion. Although this property is its rigorous justification, the choice is also suggested by intuitive reasoning. We have noted that, if the independent-pair approximation is to be valid (in the sense that we can ignore clusters of more than two particles), the eigenfunctions ϕ_k of H_0 must be good approximations to the actual wave functions outside the correlation region. In other words, U should reflect the "average" or "background" effect that a particle experiences from being embedded in the real medium. Since G is in a very real sense the effective interaction between particles in the medium, and since the potential energy of any particle is simply due to its interaction with the other particles, we would reason that the potential energy for a particle in a state m would be the interaction energy of a pair of particles in states m and n, summed over all occupied states n. In other words, we would consider it reasonable to choose

$$\langle m|U|m \rangle = \sum_n \langle mn|G|mn - nm \rangle, \qquad (8\text{-}47)$$

just as the full theory suggests. This way of looking at U makes it clear why the term "self-energy" is appropriate for the class of three-body contributions subsumed by this choice of single-particle potential.

We shall demonstrate in Chapter 10 that Eq. 8-47 is also the condition defining the single-particle potential for a system of fermions with a two-particle interaction G, if it is required that a Slater determinant of single-particle states $\phi_k(\mathbf{r}_i)$ give as accurate a value as possible for the total energy. This is the Hartree-Fock condition.

This condition does not completely define U, since it specifies only its diagonal matrix elements between hole states. Because of momentum conservation these are the only matrix elements that arise in the case of a uniform medium. In finite nuclei, however, if there is a hole k, a nucleon in another normally occupied state n may fill state k by interacting with a nucleon in state m; this is portrayed in Fig. 8-4d. Also nucleons in states k and l may interact, leaving one nucleon in state k and the other in a particle state a; a filling of state l may then occur by interaction with any occupied state (Fig. 8-4e). But analogous processes may also occur through the intermediary of the single-particle interaction, as indicated in Figs. 8-5b to 8-5e. The general theory also shows that all such terms cancel if

$$\langle k|U|l \rangle = \sum_m{}' \langle km|\tfrac{1}{2}[G(e_k + e_m) + G(e_l + e_m)]|lm - ml \rangle, \qquad (8\text{-}48\text{a})$$

$$\langle k|U|a \rangle = \sum_m \langle km|G(e_k + e_m)|am - ma \rangle. \qquad (8\text{-}48\text{b})$$

Notice that the term 8-38 in the independent-pair energy corresponds to Fig. 8-5d, and is removed by this choice of U.

The convergence of the cluster expansion for the energy depends on the value of U. If the independent-pair approximation is to be valid, higher cluster contributions must be small. We see that many of them are canceled if U is chosen so that Eqs. 8-48 hold. This, then, we must do. The specification is not complete, since we have not listed a condition for the value of U in normally unoccupied states $\langle a|U|a \rangle$. It might be hoped that this could be chosen so that some other higher clusters were canceled, as well as the self-energy terms. To do this, however, proves much more difficult, and in most calculations up to 1972 the particle energies e_a have been taken arbitrarily. If third- and fourth-order clusters are then small, we have made a reasonable choice. Alternatively, we could test the sensitivity of the independent-pair result to the values taken for e_a; if the sensitivity is great, higher-order clusters are important, since the sum of the cluster series to all orders is independent of U. The results for nuclear matter have been found to be insensitive to the choice of the particle potential, provided it is small, and higher-order clusters also make small contributions in this case. Hence, for nuclear matter, it has been usual to take $\langle a|U|a \rangle = 0$. At the time of writing an equally satisfactory treatment of the particle spectrum has not been found for finite nuclei.

For the "real" three-body clusters (i.e., for terms that do not involve self-energies) and for realistic internucleon forces, the contribution to the binding energy per particle in nuclear matter is found to be about 2 MeV attraction [9]. This is less than the approximate value of 5 MeV obtained above for $\frac{1}{2}\kappa^2\bar{v}$, but the result is reasonable, because κ^2 is the estimated probability of all three-body clusters, including those producing self-energy contributions, whereas the latter have been included in the calculated two-body contribution by the correct choice of U.

The contribution of four-body correlations has been estimated by Day [10]. There are about 100 different terms, and the computation is rather uncertain; it seems reasonable to quote a potential energy of -1.1 ± 0.5 MeV. To this can be compared the order-of-magnitude estimate $\frac{1}{4}\kappa^3\bar{v}$, which is 0.4 MeV.

It seems completely reasonable to expect the energy due to five-body correlations to be quite small, of the order of $\frac{1}{8}\kappa^4\bar{v}$ or 0.02 MeV. It may be even less, because two particles have only a small probability of being as close together as the healing distance unless they are in a relative S-state, and it is impossible to put more than four nucleons simultaneously in relative S-states.

The following equations summarize the calculation of the binding energy of nuclear matter or of a finite nucleus in the independent-pair approximation:

$$H_0 = T + U, \tag{8-49a}$$

$$H_0\phi_\alpha = e_\alpha\phi_\alpha, \tag{8-49b}$$

$$G(\omega) = v + v\frac{Q}{\omega - H_0}G(\omega), \tag{8-49c}$$

$$Q = \sum_{ab}|ab\rangle\langle ab| + \sum_{a}|ka\rangle\langle ka| + \sum_{b}|bl\rangle\langle bl|, \tag{8-49d}$$

$$E = \sum_k \langle k|T|k\rangle + \tfrac{1}{2} \sum_{kl} \langle kl|G(e_k + e_l)|kl - lk\rangle, \qquad (8\text{-}49\text{e})$$

$$\langle k|U|k\rangle = \sum_l \langle kl|G(e_k + e_l)|kl - lk\rangle, \qquad (8\text{-}49\text{f})$$

$$E_k = \langle k|T|k\rangle + \langle k|U|k\rangle, \qquad (8\text{-}49\text{g})$$

$$e_a = \,?, \qquad (8\text{-}49\text{h})$$

$$\langle k|U|a\rangle = \sum_m \langle km|G(e_k + e_m)|am - ma\rangle, \qquad (8\text{-}49\text{i})$$

$$\langle k|U|l\rangle = \sum_m \langle km|\tfrac{1}{2}[G(e_k + e_m) + G(e_l + e_m)]|lm - ml\rangle, \qquad (8\text{-}49\text{j})$$

$$\langle a|U|b\rangle = \,? \qquad (8\text{-}49\text{k})$$

The choice of U is restricted by Eqs. 8–49i and 8–49j so as to ensure reasonable convergence of the cluster expansion. The spectrum of the unoccupied states and the matrix elements of U between these states are not similarly indicated by the considerations already mentioned, and this fact is indicated by the question marks in Eqs. 8–49h and 8–49k; these values remain arbitrary.

This set of equations can be dealt with only by an iterative approach. The process for nuclear matter is indicated in Fig. 8–6. Since the model wave functions are plane waves, a given value of density determines the Pauli operator Q from Eq. 8–49d, and one can also compute $\langle kl|v|\alpha\beta\rangle$. One assumes a spectrum, that is, the values $e_\alpha = e(k_\alpha)$. These three elements— Q, $\langle kl|v|\alpha\beta\rangle$, $\{e_\alpha\}$—permit the calculation of G-matrix elements from Eq. 8–49c. Equations 8–49f and 8–49g are then used to find $\langle k|U|k\rangle$ and e_k for various

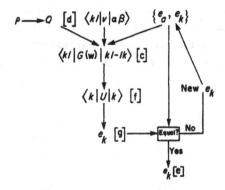

Fig. 8-6 Brueckner self-consistency for nuclear matter.

values of k. If the values found for the energies e_k of the occupied levels are not equal to the starting values, one replaces the old values with the new and repeats the cycle until convergence is obtained. When the input and output values of e_k do agree to within some prescribed limit, one then calculates E, the binding energy of the system, using Eq. 8–49e. By following this procedure, one is said to have established "Brueckner self-consistency," since this is the method of solution of the infinite-medium problem developed by Brueckner in his pioneering work on nuclear matter [11]. Note that the unoccupied-state spectrum, e_a, is not found self-consistently but is fixed arbitrarily.

In dealing with a finite nucleus, there is a second self-consistency problem, which arises because the single-particle wave functions ϕ_α are not known *a priori*. As Fig. 8–7 indicates, one starts with an assumed potential U which generates a set $\{\phi_\alpha\}$ and the corresponding energies e_α. Together with the mass and charge numbers A and Z of the nucleus being studied, the set $\{\phi_\alpha\}$ determines Q through Eq. 8–49d. If $N \neq Z$, there are two different Fermi levels, or, in other language, Q is a function of i-spin. Next, G is determined by Eq. 8–49c, and the Brueckner self-consistency cycle is followed, as shown by the double lines in Fig. 8–7. When self-consistency is reached on e_k, Eqs. 8–49i and 8–49j are used to compute $\langle k|U|\alpha\rangle$. One adds these quantities to the matrix elements of kinetic energy formed from the original wave functions $\{\phi_\alpha\}$ in order to obtain $\langle k|H_0|\alpha\rangle$. If the initial assumption for U were the self-consistent value, this last quantity would be diagonal; when it is not, one diagonalizes the matrix found and obtains new eigenfunctions

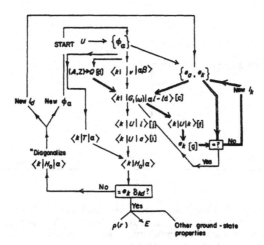

Fig. 8-7 Brueckner-Hartree-Fock self-consistency for a finite nucleus.

$\{\phi_\alpha\}$ and new energies e_α. In principle, this process is repeated until convergence is obtained. One then may calculate the energy of the nucleus and various other properties.

In practice, this is a very complicated procedure. The finding of a self-consistent single-particle U, even one permitted to be nonlocal, is at best an approximate process. Indeed, what we are doing is attempting to find the best single-particle description of the nucleus. This in itself is a matter to which we devote a later chapter. It is the Hartree-Fock problem, and the statement that "one diagonalizes" a trial $(k|H_0|\alpha)$ must be regarded as a schematic description of one of a number of possible approximations to obtain good $\{\phi_\alpha\}$. Although it would be beyond the scope of this book to discuss the details of methods of solution, we shall quote below some results for finite nuclei.

There is another practical difficulty, connected with the Pauli operator. As Fig. 8–7 emphasizes, each successive set of functions $\{\phi_\alpha\}$ implies a different Q. Even worse, Q is defined in terms of single-particle wave functions. But so long as there exist correlations, this is only an approximation of the Pauli principle. We have remarked that the wound integral κ_{kl} is a measure of the depletion of the state $|kl\rangle$. The independent-pair model ignores this effect for all states but those of the two interacting particles. In fact, however, all the "occupied" states are somewhat depleted, and some of the "unoccupied" states are partially filled. We sometimes say that the Fermi level is not sharp,

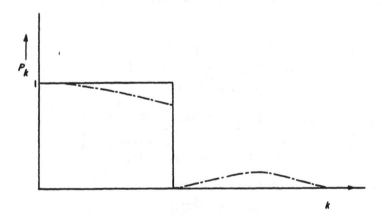

Fig. 8-8 A schematic representation of the state occupation number. The solid line is a sharp Fermi surface, $P_k = 1$, $k < k_F$, and zero otherwise. The dot-dash line shows the effect of correlations in particle occupancy. The shape of this curve depends on the nature of the two-body interaction. For $k > k_F$, P_k will be largest when k is near the maximum of the Fourier transform of $v(r)\zeta(r)$, as can be deduced from Eq. 8–41.

even though, as Fig. 8–8 suggests, the probability of occupation is discontinuous.

To be precise, Eq. 8–27 shows that κ_{kl} is the relative probability of the excitation out of the Fermi sea of the pair of particles that are asymptotically in states k and l, with the probability normalized so that there is one particle in each of these model states. Now let us write P_k for the actual occupation of each state k. Then, when the particle which asymptotically is in state ϕ_k is interacting with that in state ϕ_l, the normalization of the joint occupation is $P_k P_l$, and $\kappa_{kl} P_k P_l$ is the probability that the interaction of a particle in state k with one in state l will lift particles out of the Fermi sea. Hence $\frac{1}{2} \sum_l \kappa_{kl} P_k P_l$ is the probability that a particle asymptotically in state k will leave the Fermi sea due to all its interactions. The factor $\frac{1}{2}$ arises because half of the particles excited in the interaction of k with l can be considered to come from each of the states. But the particle in state k either leaves the sea or remains in state k, and therefore

$$P_k + \tfrac{1}{2} \sum_l \kappa_{kl} P_k P_l = 1, \tag{8–50}$$

that is,

$$P_k = (1 + \tfrac{1}{2} \sum_l \kappa_{kl} P_l)^{-1}.$$

To orient ourselves concerning the values of P_k, we might assume them all to be equal to an average value $\bar{P}$. Then, recalling Eq. 8–44, we see that, if we also take an average for κ_{kl}, we have

$$\bar{P} = (1 + \kappa \bar{P})^{-1},$$

which implies

$$\bar{P} \approx 1 - \kappa + 2\kappa^2. \tag{8–51}$$

Since κ is about 0.15, $\bar{P} \approx 0.9$. When higher accuracy is required, one finds individual P_k's from

$$P_k = (1 + \tfrac{1}{2}\bar{P} \sum_l \kappa_{kl})^{-1} \tag{8–52}$$

and could then iterate again with these values of P_k.

The significant point is that the occupation probabilities are very near unity. Nevertheless, it is more accurate to "renormalize" the energy and the potential by taking

$$\langle k|U|k \rangle = \sum_l \langle kl|G(e_k + e_l)|kl - lk \rangle P_l, \tag{8–53a}$$

$$e_k = \langle k|T|k \rangle + \langle k|U|k \rangle, \qquad (8\text{–}53\text{b})$$

$$E = \sum_k \langle k|T|k \rangle + \tfrac{1}{2} \sum_{kl} \langle kl|G|kl - lk \rangle P_k P_l + \sum_k (1 - P_k) \langle k|U|k \rangle$$
$$\qquad (8\text{–}53\text{c})$$

$$= \sum_k e_k - \tfrac{1}{2} \sum_k \langle k|U|k \rangle P_k. \qquad (8\text{–}53\text{d})$$

Of course, with $P_k = 1$ these equations reduce to our previous results.

The equations can be given some plausibility. If we think of $\langle k|U|k \rangle$ as the potential on a particle in state k due to its interaction with other occupied states, Eq. 8–53a can be seen as allowing for the fact that the other states are occupied only partially. Rigorous theory, in fact, shows that these expressions are the correct ones to allow for this effect [12, 13].

In terms of computation, it will be realized that in principle P_k can be evaluated once $G(\omega)$ is known (see Problem 8–5), and the Brueckner self-consistency cycle can then be suitably modified.

In practice, the introduction of occupation probabilities is important for the value of e_k, but not for the calculation of the total energy. If one puts $P_k = 1$, expression 8–53b for e_k will not be close to the actual removal energy of a particle from state k; if one uses the renormalized expression, the result is near the observed separation energy. From Eq. 8–53a it is evident that the effect is to reduce the attractive potential in each state by about 10%; since the kinetic and potential energies in e_k are both large and of opposite sign, the change is appreciable.

But in the total energy E this loss of attraction is compensated for by the attraction from the correlated parts of the wave functions—in other words, by the interaction between particles excited out of the Fermi sea. It is the last term of Eq. 8–53c that represents this contribution.

We can see these facts explicitly as follows. In Eq. 8–53a, the renormalized value for $\langle k|U|k \rangle$ differs from its unrenormalized value because $P_l \neq 1$ and because the matrix element of G is evaluated at a different energy since $e_k + e_l$ has changed. But the derivative of G with respect to starting energy, $\partial G/\partial \omega$, is of the order of the wound κ_{kl} (see Problem 8–5), and since the change in e_k is also of this order, the change in $G = (\partial G/\partial \omega)(\delta e_k + \delta e_l)$ is of second order. Hence, if we write $P_k = 1 - d_k$, we have for the changes due to renormalization to second order (where A indicates "antisymmetrized")

$$\delta e_k = \delta \langle k|U|k \rangle$$

$$= - \sum_l \langle kl|G|kl \rangle_A d_l - 2 \sum_{lm} \left(\frac{\partial G}{\partial \omega} \right)_{kl,A} \langle km|G|km \rangle_A d_m,$$

$$\delta E = \sum_k \delta e_k - \tfrac{1}{2}\sum_k [(\langle k|V|k\rangle + \delta e_k)(1 - d_k) - \langle k|V|k\rangle]$$

$$= \tfrac{1}{2}\sum_k \delta e_k(1 + d_k) + \tfrac{1}{2}\sum_k d_k\langle k|U|k\rangle$$

$$= \tfrac{1}{2}\sum_{kl} \langle kl|G|kl\rangle_A(d_k - d_l - d_l d_k) - \sum_{klm} \left(\frac{\partial G}{\partial \omega}\right)_{kl,A} \langle k_m|G|k_m\rangle_A d_m.$$

Since all the d_k are roughly equal to κ, $\delta e_k \approx -\kappa\langle k|U|k\rangle$. On the other hand, each term of δE is of order $\kappa^2 G$; and, since the sum $\sum_{k,l} G$ is the total potential energy of pairs (i.e., $\kappa\bar{v}$), $\delta E \approx \kappa^3\bar{v}$, which is of the order of magnitude of the energy of four-body clusters: about 1 MeV per particle.

Indeed, it is not necessary to renormalize in a nuclear-matter calculation, since the effect will arise as a four-body contribution and the single-particle potential U is not compared directly with experiment. It is vital to do so, however, for a finite nucleus in order to have a realistic evaluation of the measurable energies of the discrete single-particle states.

8-4 RESULTS IN NUCLEAR MATTER AND FINITE NUCLEI

It will be apparent to the reader that the crucial step in finding G is the solution of the Bethe-Goldstone equation. There are basically two approaches to this: one is a matrix inversion, and the other involves the solution of an integrodifferential equation. In either case it is necessary to deal with the fact that our basic states are single-particle states best suited to expression in terms of the coordinates of the separate particles, whereas the interaction is a function of relative coordinates. Fortunately, this is not a particularly difficult problem for plane-wave states. Indeed, as is apparent from Eq. 8-43, the matrix elements of v are independent of the total momentum $\mathbf{P}$ and are

$$\langle \phi_\alpha \phi_\beta|v|\phi_\gamma\phi_\delta\rangle = (2\pi)^3\delta^{(3)}(\mathbf{P}_{\alpha\beta} - \mathbf{P}_{\gamma\delta})\Omega^{-1}v(\mathbf{p}_{\alpha\beta} - \bar{\mathbf{p}}_{\gamma\delta}),$$

$$v(\kappa) = \int v(\mathbf{r}) \exp(-i\kappa \cdot \mathbf{r}) \, d^3r. \tag{8-54}$$

In the equation

$$G = v - v\frac{Q}{H_0 - e_{kl}}G$$

the energy denominator will, in general, depend on P. It is true, of course, that P cancels from the kinetic-energy parts of H_0 and of $e_k + e_l$, but the potential is a function of momentum. We have remarked that the unoccupied states are usually taken to have zero potential energy, but even so the hole states have e_k as a function of $\mathbf{p}_k$. If e_k is a function of $p_k{}^2$, then e_{kl} will be a function of $p_k{}^2 + p_l{}^2 = \tfrac{1}{2}P^2 + 2p_{kl}^2$, and the dependence of G on P may be quite simple in this respect.

However, the projection operator Q is a function of P, for it is easy to describe the exclusion of the Fermi sea in terms of $\mathbf{p}_a$ and $\mathbf{p}_b$ but not so simple in terms of $\mathbf{P}_{ab}$ and $\mathbf{p}_{ab}$. The condition, of course, is that $|\tfrac{1}{2}\mathbf{P} \pm \mathbf{p}| \geqslant k_F$. Both inequalities are satisfied if $p \geqslant k_F + \tfrac{1}{2}P$, and both fail if $p^2 < k_F{}^2 - \tfrac{1}{4}P^2$. (Of course $\tfrac{1}{2}P \leqslant k_F$.) When p lies between these two limits, the two momenta $\mathbf{p}_a$ and $\mathbf{p}_b$ both are outside k_F, provided $\pm\,\mathbf{P}\cdot\mathbf{p} \geqslant k_F{}^2 - \tfrac{1}{4}P^2 - p^2$. In other words, in the expression $Q|\mathbf{P}, \mathbf{p}\rangle$,

$$Q = 0 \quad \text{if} \quad p^2 < k_F{}^2 - \tfrac{1}{4}P^2,$$

$$Q = 1 \quad \text{if} \quad p \geqslant k_F + \tfrac{1}{2}P, \tag{8-55}$$

$$Q = 1 \quad \text{if} \quad k_F{}^2 - \tfrac{1}{4}P^2 < p^2 \leqslant (k_F + \tfrac{1}{2}P)^2,$$

and

$$|\cos\theta| \leqslant \frac{\tfrac{1}{4}P^2 + p^2 - k_F{}^2}{Pp},$$

where θ is the angle between $\mathbf{p}$ and $\mathbf{P}$.

The states of fixed p are distributed in angle with a density proportional to solid angle $d\Omega$, that is, $d\phi\,|d\cos\theta|$. Consequently, the average value of Q for a given value of p in the third case is

$$Q_{\text{av}} = \frac{\tfrac{1}{4}P^2 + p^2 - k_F{}^2}{Pp} \tag{8-56}$$

in the sense that

$$\langle \mathbf{P}', \mathbf{p}'|\mathfrak{O}Q|\mathbf{P}, \mathbf{p}\rangle = Q_{\text{av}}\langle \mathbf{P}', \mathbf{p}'|\mathfrak{O}|\mathbf{P}, \mathbf{p}\rangle, \tag{8-57}$$

which is exact if the matrix element of the operator $\mathfrak{O}$ does not depend on the angle between $\mathbf{P}$ and $\mathbf{p}$, but otherwise is an approximation. Of the matrix elements entering the Bethe-Goldstone equation, those of v depend on the angle between $\mathbf{p}'$ and $\mathbf{p}$ (Eq. 8–54) and those of $H_0 - e_{kl}$ do not depend on $\mathbf{P}\cdot\mathbf{p}$ until we consider terms of the order $p_k{}^4$. Consequently, the use of the angle-average Q in Eq. 8–56 is quite accurate in nuclear-matter calculations.

Therefore, although the nuclear-matter G-matrix is diagonal in $\mathbf{P}$, it does depend on it parametrically through Q and H_0. We can replace $\langle\alpha\beta|G|kl\rangle$ with $\langle\mathbf{p}_{\alpha\beta}|G(\mathbf{P})|\mathbf{p}_{kl}\rangle$, where only relative coordinates now occur in the wave functions. A further step is usually taken before any computation is done. Since the nucleon-nucleon potential is different in different spin-isospin states, it is desirable to expand the plane waves in angular-momentum components. It is usual to write k instead of $\mathbf{p}_{\alpha\beta}$. A complete set of quantum numbers specifying a two-nucleon collision consists of the momenta, spin components, and charges or, equivalently, $\mathbf{P}$, $\mathbf{k}$, M_S, and M_T. To expand such a state we write

$$|\mathbf{P}, \mathbf{k}, M_S, M_T\rangle = \sum_{JLMS} c_{JLMS} |\mathbf{P}, \mathbf{k}, J, L, M, S, M_S, (T), M_T\rangle. \quad (8\text{--}58)$$

The total i-spin T is shown in parentheses since its value is determined when L and S are fixed. Explicitly, the well-known expansion for a plane wave in spherical harmonics can be rewritten to yield

$$|\alpha\beta - \beta\alpha\rangle = \phi_A(\mathbf{p}_\alpha, \mathbf{p}_\beta, M_S, M_T)$$

$$= \Omega^{-1} \exp(i\mathbf{P}\cdot\mathbf{R}) \sum_{JLMS} 4\pi i^L (LSM_L M_S|JM)$$

$$\times Y_L^{M_L*}(\hat{\mathbf{k}}) j_L(kr) |JLMS(T)M_T\rangle. \quad (8\text{--}59a)$$

Similarly, we write

$$|\psi_{\alpha\beta} - \psi_{\beta\alpha}\rangle = \psi_A(\mathbf{p}_\alpha, \mathbf{p}_\beta, M_S, M_T)$$

$$= \Omega^{-1} \exp(i\mathbf{P}\cdot\mathbf{R}) \sum_{JLMS} 4\pi i^L (LSM_L M_S|JM) Y_L^{M_L*}(\hat{\mathbf{k}})(kr)^{-1}$$

$$\times u_{JLS}^M(r) |JLMS(T)M_T\rangle, \quad (8\text{--}59b)$$

$$= \Omega^{-1} \exp(i\mathbf{P}\cdot\mathbf{R}) \sum_{JLMS} 4\pi i^L (LSM_L M_S|JM) \sum_{L'} Y_{L'}^{M_L*}(\hat{\mathbf{k}})(kr)^{-1}$$

$$\times u_{L'L}^{JS}(r) |JL'MS(T)M_T\rangle, \quad (8\text{--}59c)$$

where

$$|JLMS(T)M_T\rangle = \sum_m (LSmM_S|JM) Y_L^m(\hat{\mathbf{r}}) |S\, M-m\rangle |TM_T\rangle$$

and $M_L = M - M_S$. The equations are somewhat simplified when one can choose the z-axis in the direction of the relative momentum, for then $M_L = 0$, the spherical harmonic becomes $\sqrt{(2L+1)/4\pi}$, and the sum over M reduces to one term, $M = M_S$. Since each term of these sums is a state of specified parity, spin, and i-spin, one can readily form matrix elements of v between states expanded in this way.

Equation 8–59b for ψ is a general expression for the wave function of two nucleons and is written in this way since it goes asymptotically to the free motion ϕ if the radial function u_{JLS}^M approaches $krj_L(kr)$. It is helpful also to write the defect function $\zeta = \phi - \psi$ in a similar expansion, replacing u by a radial defect function $\chi(r)$, so that

$$\chi_{JLS}^M = krj_L(kr) - u_{JLS}^M, \tag{8-60a}$$

$$\chi_{L'L}^{JS} = \delta_{LL'}krj_L(kr) - u_{L'L}^{JS}. \tag{8-60b}$$

The two forms of ψ need some explanation. They differ in the radial functions u, a difference which arises only in the presence of tensor forces. Let us first explain why the radial function in Eq. 8–59b depends on M. If the axis is taken as the direction of $\mathbf{k}$, M is simply the spin component along $\mathbf{k}$. But the tensor force contains the term proportional to $\boldsymbol{\sigma}_1 \cdot \hat{\mathbf{r}} \boldsymbol{\sigma}_2 \cdot \hat{\mathbf{r}}$, which depends on the spin orientations. If $M = 1$, both spins are along $\mathbf{k}$; if $M = 0$, the resultant is perpendicular. Hence the effective force is different in states of different M when L is fixed. [For example, $\langle JLM|v(r)\boldsymbol{\sigma}_1 \cdot \hat{\mathbf{r}} \boldsymbol{\sigma}_2 \cdot \hat{\mathbf{r}}| JLM\rangle$ is $(11/210)v(r)$ for $J = 1, L = 2, M = 1$ and is $(4/21)v(r)$ for $J = 1, L = 2$, $M = 0$.] This dependence of the radial function on M is inconvenient; and, as we already know from the chapters on the deuteron and nucleon scattering, it is desirable to adopt a different description, using states in which L is not a constant of motion, but in which both values, $L = J \pm 1$, are present. There are two orthogonal states, in each of which one of these L values is dominant (i.e., is the incoming wave in a scattering experiment). In our notation $u_{L'L}^J$, L is the dominant value specifying which solution is being used and L' is the angular momentum actually present with amplitude u. It is only in coupled triplet states that we shall employ two suffixes on the radial functions; otherwise one will do, and L' is put equal to L in Eq. 8–59c.

At this point we wish to define a G-matrix for each angular-momentum state. The basic idea of the G-matrix is that when the G-operator acts on the unperturbed state it reproduces the effect of the actual potential applied to the actual state. Now, for each uncoupled angular-momentum state defined by JLS, there is a potential function $v_L{}^{JS}(r) = \langle JLS|v|JLS\rangle$, and we define

$$G_L{}^{JS}krj_L(kr) = v_L{}^{JS}u_L{}^{JS}(r; k). \tag{8-61a}$$

When there are coupled states, we have to recall that we are dealing with a true wave function of the form

$$\psi = \sum_{L''} u_{L''L}^{JS}(r)\Phi(JL''S),$$

where Φ stands for the angular-momentum part of ψ, and that v couples two states so that

$$v\Phi(JLS) = \sum_{L'} v_{L'L}^{JS}(r)\Phi(JL'S).$$

Hence

$$v\psi = \sum_{L'L''} v_{L'L''}^{JS} u_{L''L}^{JS}\Phi(JL'S),$$

and we require that G applied to the free wave give this result. Therefore, we define angular-momentum components of G by

$$G|k, JLS\rangle = \sum_{L'} G_{L'L}^{JS}|k, JL'S\rangle = v\psi.$$

It follows that

$$G_{L'L}^{JS} krj_L(kr) = \sum_{L''} v_{L'L''}^{JS} u_{L''L}^{JS}. \tag{8-61b}$$

The G-matrix also depends on P and ω, and for uncoupled states it satisfies the equation

$$(k|G_L{}^{JS}|k_0) = (k|v_L{}^{JS}|k_0) - \sum_{k'} \frac{(k|v_L{}^{JS}|k')(k'|G_L{}^{JS}|k_0)}{e(\frac{1}{2}P + \mathbf{k}') + e(\frac{1}{2}P - \mathbf{k}') - \omega} Q(k', P), \tag{8-62}$$

where $(k|v_L{}^{JS}|k_0)$ and $(k|G_L{}^{JS}|k_0)$ are Bessel transforms, namely,

$$(k|v_L{}^{JS}|k_0) = \int j_L(kr)v_L{}^{JS}(r)j_L(k_0r)r^2\,dr,$$

and where $Q(k', P)$ is given by Eqs. 8–55 and 8–56. Equation 8–62 may be solved by matrix inversion, for if we suppress the various suffixes and write

$$(k|A|k') = \frac{(k|v|k')Q(k', P)}{e(\frac{1}{2}P + \mathbf{k}') + e(\frac{1}{2}P - \mathbf{k}') - \omega},$$

we have

$$(k|G|k_0) = (k|v|k_0) - \sum_{k'} (k|A|k')(k'|G|k_0). \tag{8-63}$$

For a numerical computation, we replace the continuous variable k by a set of selected values k_s. These need not be distributed uniformly, but should be chosen so that they are more closely packed in regions where the matrix elements are large. Then Eq. 8–63 is simply

$$\sum_t (\delta_{rt} + A_{rt})G_{ts} = v_{rs}, \tag{8-64}$$

which is solved by inverting the matrix $1 + A$. In practice, this operation is readily handled by modern computers and the number of k-values needed is not excessive [14].

In dealing with coupled states, the principle is the same but the two matrices G_{LL}^{JS} and $G_{L'L}^{JS}$ satisfy coupled equations, namely,

$$
(k|G_{LL}|k_0) = (k|v_{LL}|k_0) - \sum_{k'} [(k|A_{LL}|k')(k'|G_{LL}|k_0)
$$

$$
+ (k|A_{LL'}|k')(k'|G_{L'L}|k_0)],
$$

$$
(k|G_{L'L}|k_0) = (k|v_{L'L}|k_0) - \sum_{k'} [(k|A_{L'L}|k')(k'|G_{LL}|k_0)
$$

$$
+ (k|A_{L'L'}|k')(k'|G_{L'L}|k_0)],
$$

where

$$
(k_1|v_{L_1 L_2}|k_2) = (k_1 L_1 JS|v|k_2 L_2 JS);
$$

with a similar definition for A. In matrix notation analogous to Eq. 8–64, these two equations are

$$
(1 + A_{LL})G_{LL} + A_{LL'}G_{L'L} = v_{LL},
$$

$$
A_{L'L}G_{LL} + (1 + A_{L'L'})G_{L'L} = v_{L'L}, \qquad (8\text{–}65)
$$

which give

$$
[1 + A_{LL} - A_{LL'}(1 + A_{L'L'})^{-1}A_{L'L}]G_{LL} = v_{LL} - A_{LL'}(1 + A_{L'L'})^{-1}v_{L'L},
$$

$$
[1 + A_{L'L'} - A_{L'L}(1 + A_{LL})^{-1}A_{LL'}]G_{L'L} = v_{L'L} - A_{L'L}(1 + A_{LL})^{-1}v_{LL}.
$$

$$
(8\text{–}66)
$$

This technique is in many ways the most straightforward and useful procedure for finding G and hence the energy of the system and the wave functions ψ. It will not work if v has an infinite hard core, for then the matrix elements of v are infinite. However, it works with the realistic potentials with finite repulsion, even though v reaches quite high values in some cases.

The differential-equation technique arises from writing Eq. 8–22 in terms of the wave-function defect as

$$
|\zeta_{kl}\rangle = \frac{Q_{kl}}{H_0 - e_{kl}} v|\psi_{kl}\rangle
$$

or

$$
(H_0 - e_{kl})\zeta_{kl} = Q_{kl}v(\phi_k\phi_l - \zeta_{kl}). \qquad (8\text{–}67)
$$

By writing Q_{kl} in the angle-average approximation (Eq. 8–56), it becomes a straightforward but tedious matter to rewrite this equation in terms of the relative coordinate r for each state of the angular-momentum decomposition. The Hamiltonian H_0 acts on ζ_{kl}, which contains only momenta above the Fermi sea. Hence, in order to reduce the equation to one in the radial coordinate, some assumption is needed about the effect of the operator H_0 on states above the Fermi level. Of course, this is also true in the matrix-inversion method, since the two methods are identical in physical assumptions. Equation 8–62 contains these energies, $e(\frac{1}{2}\mathbf{P} \pm \mathbf{k}')$; the advantage of the matrix-inversion method is that the form of e can be taken arbitrarily. In differential equations, on the other hand, the potential-energy function $U(\mathbf{p})$ in H_0 would introduce several difficulties if it were more complicated than a quadratic. When it is quadratic, H_0 (acting on states with $p > p_F$) can be written as

$$H_0(i) = A + \frac{p_i{}^2}{2Mm^*},$$

where A is a constant and m^* is an "effective mass." If $m^* = 1$, the potential energy of particle states is constant and the last term of H_0 is just the kinetic energy. With this assumption,

$$H_0 - e_{kl} = -\left(\frac{\hbar^2}{Mm^*}\right)\nabla^2 + \frac{P^2}{4Mm^*} + 2A - e_{kl}$$

$$= \frac{\hbar^2}{Mm^*}(-\nabla^2 + \gamma^2), \tag{8–68}$$

where the quantity γ^2 is positive, since, if the system is bound, $e_{kl} < 0$.

For numerical work, it is more convenient to use a projection operator $R = 1 - Q$, since it spans a finite region of momentum space, namely, the Fermi sphere. We may write Eq. 8–67 as

$$\frac{\hbar^2}{Mm^*}(-\nabla^2 + \gamma^2)\zeta = v\psi - Rv\psi. \tag{8–69}$$

The projection operator is

$$R = \int d^3k' \, d^3P' |\mathbf{k}', \mathbf{P}'\rangle\langle\mathbf{k}', \mathbf{P}'|,$$

where $\mathbf{k}'$ and $\mathbf{P}'$ define vectors $\mathbf{p}_1$ and $\mathbf{p}_2$ inside the Fermi sea. Using the angle-average approximation, we can make the angular-momentum reduction and find

$$\left[-\frac{d^2}{dr^2}+\frac{L'(L'+1)}{r^2}+\gamma^2\right]\chi^{JS}_{L'L}(r)$$

$$=\frac{Mm^*}{\hbar^2}\sum_{L''}\left[v^{JS}_{L'L''}(r)u^{JS}_{L''L}(r)-rj_{L'}(kr)\frac{2}{\pi}\int_0^\infty k'^2\,dk'\int_0^\infty r'\,dr'\,R_0(k',P)\right.$$

$$\left.\times r'j_{L'}(k'r')v^{JS}_{L'L''}(r')u^{JS}_{L''L}(r')\right] \tag{8–70}$$

where

$$R_0(k',P)=\begin{cases}1 & \text{if }\;k'^2<k_F{}^2-\tfrac14 P^2,\\[4pt] 0 & \text{if }\;k'\geqslant k_F+\tfrac12 P,\\[4pt] \dfrac{k_F{}^2-(\tfrac12 P-k')^2}{Pk'} & \text{otherwise.}\end{cases}$$

This set of integrodifferential equations has been solved by a number of techniques of approximation. One of particular interest is to ignore the Pauli principle in the first approximation, that is, to put $R_0 = 0$. Then the equation is simply a differential equation. The G-matrix so obtained is corrected for the Pauli principle by a perturbation procedure. This is an essential step in the "reference spectrum method" [8].

The reader may also notice that Eq. 8–70 makes explicit the healing of the wound in the wave function. At distances of separation where the potential has vanished, the equation is just

$$\left[-\frac{d^2}{dr^2}+\frac{L'(L'+1)}{r^2}+\gamma^2\right]\chi=0,$$

the solution to which is a Hankel function with the exponential decay factor $\exp(-\gamma r)$. The expression given for γ^2 in Eq. 8–68 directly connects the size of the wound with the energies involved—those of the interacting particles, e_{kl}, and those of two particles of the same total momentum but above the Fermi sea.

We have given this detail of the methods of solution because a discussion of the physics of the results of calculations relies heavily on the different behaviors of the several angular-momentum states. Many calculations of the properties of nuclear matter have been made since the pioneering work of Brueckner and his collaborators [11] in 1955. During the same years knowledge of the free two-nucleon force also improved steadily, and, naturally, successive nuclear-matter calculations were carried out with the contemporary forces. It was never entirely clear, however, which features of the results were attributable to the force used and which to improvements in calculating techniques. We shall discuss a variety of illustrative results in the literature, with a variety of forces.

In principle, if one has good control of the accuracy of the calculation, the resulting binding energy and density can be used as criteria for choice among different postulated nucleon-nucleon forces, all of which fit the two-nucleon data. The free nucleon-nucleon scattering involves matrix elements of the potential only between states of the same energy; but, as a glance at almost any of the equations of Section 8–3 (e.g., Eq. 8–20) makes clear, a calculation in a many body system involves matrix elements between two-nucleon states of different energy. Hence it is the "off-energy-shell" behavior that distinguishes which of a class of otherwise identical potentials are physically acceptable (see Section 5–9).

As long as at least one other entity is involved and can absorb some energy, a process will involve off-shell matrix elements. For example, the properties of the triton and of nucleon-deuteron reactions depend on off-shell behavior, and so do proton-proton bremsstrahlung ($p + p \rightarrow p + p + \gamma$) and inelastic electron scattering from the deuteron. However, at reasonable energies these processes are not sensitive tests, since most potentials that have the same two-body phase shifts do not differ much for relatively small departures from the energy shell.

The matrix elements of G—and hence the properties of nuclear matter— can in a good approximation be related directly to the wave-function defect. The approximation is to ignore the Pauli operator ($R_0 = 0$) in Eq. 8–70 for $\chi(r)$. We then can start with the definition of $G_{L'L}^{JS}$ in Eq. 8–61b and write

$$\langle k'|G_{L'L}^{JS}|k\rangle = \int r^2\, dr\, j_{L'}(k'r) G_{L'Lj}^{JS} j_L(kr)$$

$$= \int r\, dr\, j_{L'}(k'r) k^{-1} \sum_{L''} v_{L'L''}^{JS} u_{L''L}^{JS}$$

$$= \int r\, dr\, j_{L'}(k'r) \frac{\hbar^2}{Mm^*k} \left[-\frac{d^2}{dr^2} + \frac{L'(L'+1)}{r^2} + \gamma^2 \right] \chi_{L'L}^{JS}$$

using Eq. 8–70. We then apply the differential operator to the Bessel function to get

$$\langle k'|G_{L'L}^{JS}|k\rangle = \frac{\hbar^2}{Mm^*kk'} (k'^2 + \gamma^2) \int k'r j_L(k'r) \chi_{L'L}^{JS}(r)\, dr. \qquad (8\text{--}71)$$

The important result is that, except for Pauli corrections, the off-shell G-

matrix elements are essentially Bessel transforms of the wave defect. This implies that two potentials will have similar nuclear-matter properties only if they have similar defects (i.e., similar short-range correlations) [15].

We may also note that, although G and χ are functions of P, the total momentum of the colliding pair, their dependence on P is quite weak. Whereas the effective interaction of two particles is a sensitive function of their relative momentum, just as their free interaction is, the dependence on P arises only through the energy denominator and the Pauli operator Q. In both these factors, the process of averaging over the angle between $\mathbf{k}$ and $\mathbf{P}$ reduces markedly the extent of the dependence on the total momentum. Table 8–1 shows the $^1S_0$ G-matrix for various values of P and k and makes clear that for many purposes it is possible to ignore the dependence on P and merely evaluate G for an average P. This is done in most calculations, and we shall not refer to P in most of the remaining discussion.

TABLE 8-1 $^1S_0$ G-Matrix Elements (MeV fm^3) for Reid Soft-Core Potential for $k_F = 1.4$ fm^{-1}. From Siemens [16].

k/k_F	P/k_F			
	0.1	0.35	0.65	0.9
0.1	− 884	− 885	− 891	− 903
0.35	− 634	− 636	− 642	
0.65	− 322	− 324	− 329	
0.9	− 121	− 123		

The matrix elements important for the properties of nuclear matter and of reasonably heavy finite nuclei involve the scattering of two particles within the Fermi sea to states with relative momenta two or three times k_F; this corresponds to moving about 200 MeV off shell. Unfortunately, there are only a limited number of experimentally determined numbers for nuclear matter, and in finite nuclei, where more parameters may be compared, the calculations are not yet quite so well under control. There is another regime in which some checks are now available, namely, neutron stars. The best approach for improving understanding seems to be the development of greater accuracy in calculations of medium-mass nuclei.

Turning to the results for nuclear matter, let us first consider some typical wave functions. Figure 8–9 shows the defect function $\chi_{L'L}^{JS}(r)$ for several partial waves and for a typical value of the relative momentum k of the colliding

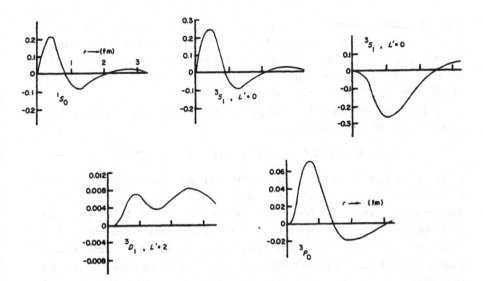

Fig. 8-9 The defect function χ for some typical partial waves. The potential is the Reid soft core, $k_F = 1.36$ fm^{-1}, $k = 0.91$ fm^{-1}, and an appropriate average value of P is used.

nucleons: $0.65k_F$. These calculations of Siemens [16] are for normal nuclear density $k_F = 1.36$ fm^{-1} and for the Reid soft-core potential, described in Chapter 5 (Fig. 5–14). Since this is a "realistic" potential, the results are physically instructive. It is seen at once that the wave function heals rapidly, and that the wound is most pronounced in the S-states. The defect χ is a measure of how the true wave function u differs from the free wave function, which is $krj_L(kr)$. Asymptotically, the free wave function is a sine wave of amplitude unity. Of course, the free wave function for $L' = 2$ is zero for a collision asymptotically in the triplet S-channel, as is also the $L' = 0$ function for a collision asymptotically in the triplet D-state. We see that the tensor force introduces quite an appreciable component of $L = 2$ into the short-range correlation of a pair of nucleons otherwise interacting in the triplet S-state.

We noted in Section 8–3 (Fig. 8–8) that the most highly populated states above the Fermi sea are those for which the Fourier transform of $v\zeta$ is a maximum. It can be seen that for the $L' = 0$ states ζ is mostly inside about 0.8 fm; and if one multiplies in the Reid soft-core potential of Fig. 5–14, one sees that $v\zeta$ also lies mostly inside this distance. Consequently, the Fourier transform will maximize for relative momenta about $\pi/(0.8 \text{ fm}) \approx 4 \text{ fm}^{-1} \approx 3k_F$. This corresponds to kinetic energy of relative motion of about 300 MeV. More exact calculations indicate a slightly lower energy, but this is a very simple confirmation of the statement that we are concerned with matrix elements 100 or 200 MeV off the energy shell. What is more, the matrix elements of G are also related to the transform of ζ (Eq. 8–71) and hence also maximize in this same region of intermediate states.

In Eq. 8–46 we defined κ, the probability of finding a particle excited outside the Fermi sea. The quantity

$$\kappa_{L'L}^{JS}(k) = \tfrac{1}{8}\rho(2J + 1)(2T + 1)\langle \zeta_{L'L}^{JS}(\mathbf{r}, k)|\zeta_{L'L}^{JS}(\mathbf{r}, k)\rangle \tag{8-72}$$

is the contribution of the indicated angular-momentum state to the probability of excitation due to collisions of relative momentum k. Summing these quantities for fixed k, we find the probability $\kappa(k)$ of excitation when two particles of relative momentum k interact. (The calculation should be done for an average value of P.) Knowledge of χ permits us to examine these quantities, some of which are listed in Table 8–2. Although the values in the table do not make possible a precise comparison of the values for different k, the values for the S-states are not, in fact, very sensitive to the relative momentum [16]. The table shows the dependence on density for a given force, the Reid soft-core potential. It will be seen that, as the density increases, the effective interaction in states of higher angular momentum also increases, although even for twice normal nuclear density ($k_F = 1.7$ fm^{-1}) the incident S-states still account for 83% of the excitation. For the two $L' = 0$ states the defect lies mostly in the region of the repulsive core (Fig. 8–9), and it is therefore not surprising that, while the amount of the wounds in these states increases with density, the relative contribution to the total wound changes only slowly. On the other hand, the "defect" in the $L' = 2$ component of the incident $^3S_1$-state is the whole wave function generated in this substate by the tensor force and therefore is of much longer range. Notice that the fractional contribution of this state to the total wound decreases rapidly with increasing density.

This difference can be understood when it is realized that the shorter-range χ's contain mainly states of relatively large k, that is, the correlations are admixtures into the unperturbed ϕ of intermediate states substantially above the Fermi sea. On the other hand, a defect of longer range, about 3 or 4 fm, contains a significant amount of excitation of small k, and hence of states not far from the Fermi sea.

In Fig. 8–10, particles with momenta m, n inside the Fermi sphere interact. If their virtual excitation involves fairly large relative momenta such as k', the intermediate-state points are a and b. As the sphere grows, there is no

TABLE 8-2 Excitation Probabilities from Wound Integrals.

| | Reid Potential | | | | | | | | OBEP | |
| | $k_F = 1.0$ fm^{-1} [a] $k = 0.65k_F = 0.65$ fm^{-1} | | $k_F = 1.4$ fm^{-1} [a] $k = 0.65k_F = 0.91$ fm^{-1} | | $k_F = 1.5$ fm^{-1} [b] $k = 0.57k_F = 0.86$ fm^{-1} | | $k_F = 1.7$ fm^{-1} [a] $k = 0.65k_F = 1.10$ fm^{-1} | | $k_F = 1.5$ fm^{-1} [b] $k = 0.86$ fm^{-1} | |
	$\kappa^{JS}_{L'L}$	$\kappa^{JS}_{L'L}/\kappa$	$\kappa^{JS}_{L'L}$	$\kappa^{JS}_{L'L}/\kappa$	$\kappa^{JS}_{L'L}$	$\kappa^{JS}_{L'L}/\kappa$	$\kappa^{JS}_{L'L}$	$\kappa^{JS}_{L'L}/\kappa$	$\kappa^{JS}_{L'L}$	$\kappa^{JS}_{L'L}/\kappa$
$^3S_1$ κ^{11}_{20}	0.0573	0.58	0.0644	0.45	0.0699	0.45	0.0735	0.36	0.0370	0.35
$^3S_1$ κ^{11}_{00}	0.0188	0.19	0.0321	0.23	0.0388	0.25	0.0542	0.27	0.0282	0.26
$^1S_0$	0.0126	0.13	0.0237	0.17	0.0285	0.18	0.0396	0.20	0.0265	0.25
All S states	0.0887	0.90	0.1202	0.85	0.1372	0.88	0.1673	0.83	0.0917	0.86
$^3P_0$	0.0009	0.01	0.0036	0.03	0.0038	0.02	0.0085	0.04	0.0037	0.03
$^3P_1$	0.0028	0.03	0.0056	0.04	0.0051	0.03	0.0086	0.04	0.0048	0.05
$^3P_2$ κ^{21}_{31}	0.0017	0.02	0.0028	0.02	0.0027	0.02	0.0037	0.02	0.0025	0.02
$^3P_2$ κ^{21}_{11}	0.0017	0.02	0.0024	0.02	0.0021	0.01	0.0025	0.01	0.0018	0.02
$^1P_1$	0.0020	0.02	0.0057	0.04	0.0055	0.04	0.0099	0.05	0.0013	0.01
$^3D_1$	0.0001	0.00	0.0002	0.00	0.0003	0.00	0.0004	0.00		
$^3D_2$	0.0003	0.00	0.0005	0.00	0.0003	0.00	0.0004	0.00	0.0003	0.00
$^1D_2$	0.0001	0.00	0.0003	0.00	0.0002	0.00	0.0006	0.00	0.0001	0.00
Total $\kappa(k)$	0.0983		0.1413		0.1569		0.2019		0.1062	

[a] Ref. 16.
[b] Ref. 17.

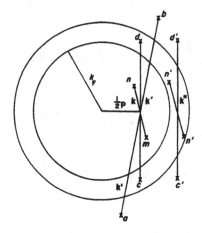

Fig. 8-10 Momentum-space diagram showing some intermediate-state vectors and Fermi spheres for two densities.

change in the number of such states a and b, but the amount of excitation of this kind will increase because there are more states (e.g., m' and n') that can have the same relative momentum k and the same excitation k'. In contrast, if there is a relatively small relative momentum (k'') in the intermediate state, the nucleons in that state have momenta such as c and d. When the density is increased, contributions to the correlation from states c and d are removed, since they are now within the Fermi sea. It is true that the relative momentum k' is still available from scattering to c' and d' from states m' and n', but as the size of the sphere increases the relative importance of the smaller components decreases. Thus, although all the wounds increase with density, the relative importance of a wound due to a long-range force decreases.

   This behavior of the wound due to the tensor force is one of the two main reasons for the so-called saturation of nuclear binding energy, the fact that, as more nucleons are added to a nucleus, the binding energy per particle remains roughly constant. With ordinary attractive forces the potential energy per particle would increase steadily with the density, and since the system would seek the state of lowest energy, stable nuclei would be indefinitely dense and tightly bound. Real nuclear matter saturates, implying minimum energy per particle for the actual density ρ_0 and increasing energy if the density varies from ρ_0. Consequently, there must be some parts of the effective internucleon force which become less attractive as the density increases. The long-range and strong tensor force plays this role for densities near values appropriate to nuclei.

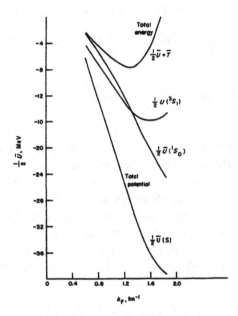

Fig. 8-11 Certain contributions to binding at various densities; $\bar{U}$ is the average potential energy per particle due to pair interactions in the indicated states. Also shown on the same scale is the sum of the potential total and the kinetic energy per particle. (Data from [18] and D. W. L. Sprung as quoted in [1].) The potential is Reid soft core.

This is illustrated in Fig. 8–11, which shows the binding due to the ^1S-, ^3S-, and $^3D_1$-states. The potential in a state m is given by

$$U(m) = \sum_n \langle mn|G|mn - nm \rangle$$

$$= \frac{8}{\pi} \frac{\hbar^2}{M} \sum_{JSL} (2J+1)(2T+1) \int_0^{k_{max}} (k|G_{LL}^{JS}|k)F(k)k^2\,dk, \quad (8\text{–}73)$$

where $k_{max} = \frac{1}{2}(k_F + k_m)$ and is the largest value of relative momentum k possible for m fixed and n extending over all occupied states. Strictly speaking, G should be evaluated at each P and the permitted range of P should be summed over, but it is adequate to evaluate G at an average value of P, as this formula suggests. The quantity $F(k)$ takes account of the effect of the exclusion principle on the angular integration and is

$$F(k) = 1, \qquad\qquad 0 < k < \tfrac{1}{2}(k_F - k_m)$$

$$= \frac{1}{2k_m k}(\tfrac{1}{4}k_F{}^2 + k_m k - \tfrac{1}{4}k_m{}^2 - k^2), \quad \tfrac{1}{2}(k_F - k_m) < k < \tfrac{1}{2}(k_F + k_m).$$

If this is now summed over m, we find the average potential energy per particle as

$$\bar{U} = \frac{8}{\pi} \frac{\hbar^2}{M} \sum_{JSL} (2J + 1)(2T + 1) \int_0^{k_F} (k|G_{LL}^{JS}|k)\left(1 - \frac{3}{2}\frac{k}{k_F} + \frac{1}{2}\frac{k^3}{k_F{}^3}\right) k^2\, dk.$$

$$(8\text{–}74)$$

Each term in this sum is described as the binding due to a specific angular-momentum state. The total energy per particle is $E/A = \bar{T} + \tfrac{1}{2}\bar{U}$. In Fig. 8–11 we show the contributions for the S-states. Of course, the "initial" states ϕ of colliding pairs are all states of fixed L; and when the force is central, there are direct contributions to the diagonal G-matrix from the term $(\phi|v|\phi)$, since a central force links states of the same angular momentum. But the tensor force has zero matrix elements between two S-states and hence can contribute only through $(\phi|v|\zeta)$, since the intermediate states in ζ contain D-states. Recall that

$$|\zeta\rangle = \frac{Q}{e} G|\phi\rangle = \sum_a \frac{|a\rangle\langle a|G|\phi\rangle}{e}.$$

When G contains a tensor operator, and $|\phi\rangle$ is a ^3S-state, some of the states $|a\rangle$ will be $^3D_1$. This means that the tensor force can contribute only through excitations outside the Fermi sphere and that the considerations associated with Fig. 8–10 are important. Of course, the central force contributes also in this way, but since there is a direct contribution as well, the central-force contribution to the G-matrix does not fall off with increasing density. Thus G_0^{00} is monotonic, but G_{00}^{11} results from an interplay of central and tensor contributions and behaves as shown by the curve labeled $^3S_1$ in Fig. 8–11. For low density, where the exclusion principle is not as important, the triplet state has greater binding than the singlet, as we know from the free two-body scattering at low energies. But as the tensor force is progressively more inhibited, the effective singlet strength becomes greater. In the figure the sum of the two potential-energy contributions is shown, and so is the result of adding the average kinetic energy. It is seen that this total energy has a minimum at k_F of about 1.3 fm^{-1} and an energy per particle of -7.8 MeV.

These numbers are surprisingly close to the experimental results (1.36 fm^{-1}, — 15.7 MeV), considering that we have ignored interactions in all states with $L > 0$.

This brings us to the second main cause of saturation: the exchange character of the nuclear force, which, as we saw in Sections 5–6 and 5–7, keeps the interaction in P-states fairly weak, with the $^3P_2$-force attractive, the $^3P_1$ repulsive, and the $^3P_0$ attractive for energies below about 200 MeV. At low densities, the relative momenta in nuclear matter are of course small and states of positive angular momentum are very little affected; however, as the value of k_F increases, so do the distortions in the wave functions of states having $L \geqslant 1$. This is very noticeable in Table 8–2. Consequently, if it were not for the weak and partly repulsive nature of the P-states, we would expect increasing negative potential energy as the density rose, and there would be no saturation.† Actually, the contributions for various states are as shown in Table 8–3. Although this table is for the Reid soft-core potential, the behavior of other potentials is similar. To illustrate this fact, the table also shows some results for a typical OBEP, which can be compared at two densities.

The increasing importance of higher-angular-momentum states is evident, but the triplet P-states consistently cancel each other, so that the whole P-contribution is almost entirely that of the singlet state, which is relatively weak and is repulsive. Hence this particular balance of the $\boldsymbol{\sigma}_1 \cdot \boldsymbol{\sigma}_2$, $\boldsymbol{\tau}_1 \cdot \boldsymbol{\tau}_2$, and $\mathbf{L} \cdot \mathbf{S}$ forces is very important for saturation. The D-states do not exhibit such cancellation and their contribution has reached 6 MeV in the region of nuclear density; if there were no cancellations in the P-states, the latter would contribute an attraction somewhere between that of the S- and that of the D-states, perhaps 15 MeV, and saturation would not be in sight.

Of course, at still higher densities the states with $L \geqslant 2$ will be increasingly important and might well override the effects in the S- and P-states which have produced the minimum in the energy curve. This might then be a local minimum, and there might be states of much lower energy at much higher density. This is prevented by a third fact: the repulsive core plays an important role when the average separation of nucleons is of the order of twice the core radius, or about eight times the normal nuclear density. Calculations are not reliable for densities higher than about $3\rho_0$, or k_F much greater than 2 fm^{-1}, because we do not know the details of the force at short distances. It is possible to prescribe necessary conditions on analytic expres-

---

† Recall that, although the tensor force would be somewhat inhibited, it has nonzero *direct* matrix elements in all triplet states except the S-state.

TABLE 8-3 Contributions to the Energy of Nuclear Matter of Various States, Taking Two-Body Clusters Only. The calculations are by D. W. L. Sprung as quoted in ref. 1 except for those labeled with superscript a, which are by P. Grangé. The OBEP force is that described in ref. 17. The contributions of higher states are estimated in different ways: those labeled b are estimated by Sprung using the phase-shift approximation, and those labeled c are based on the "OPEP $+ \sigma$" model and include only $J = 3, 4, 5$ (see ref. 19).

k_F, fm^{-1}:	1.0	1.0 (OBEP)^a	1.2	1.4	1.5^a	1.5 (OBEP)^a	1.6	1.7
ρ, fm^{-3}:	0.068	0.068	0.117	0.185	0.228	0.228	0.276	0.332
$^3S_1$	-10.15	-11.46	-13.08	-15.19	-15.77	-19.96	-15.89	-15.67
$^1S_0$	-8.36	-8.54	-12.24	-16.35	-18.43	-18.58	-20.30	-22.10
$^3P_0$	-1.34	-1.34	-2.36	-3.56	-4.18	-4.17	-4.77	-6.54
$^3P_1$	3.02	3.03	6.15	11.08	14.40	14.42	18.30	22.90
$^3P_2$	-1.93	-1.93	-4.21	-7.91	-10.38	-10.37	-13.36	-16.84
$^1P_1$	0.55	1.37	1.30	2.76	3.90	5.43	5.34	7.17
$^3D_1$	0.40	0.36	0.87	1.62	2.12	1.97	2.67	3.31
$^3D_2$	-1.17	-1.09	-2.58	-4.85	-6.36	-6.02	-8.13	-10.19
$^1D_2$	-0.65	-0.69	-1.48	-2.89	-3.88	-4.15	-5.08	-6.54
$^3F_2$	-0.11	-0.11	-0.30	-0.64	-0.88	-0.89	-1.20	-1.58
$J \geqslant 3$	0.11^b	0.30^c	0.22^b	0.33^b	1.15^c	0.88^c	0.35^b	0.26^b
Sum	-19.63	-20.08	-27.71	-35.58	-38.33	-41.42	-42.05	-44.50
Kinetic	12.44	12.44	17.92	24.38	28.00	28.00	31.85	35.95
E/A	-7.19	-7.64	-9.79	-11.20	-10.34^c	-13.43	-10.20	-8.55
ΣS	-18.51	-20.00	-25.32	-31.54	-34.20	-38.54	-36.19	-37.77
ΣP	0.30	1.13	0.88	2.37	3.74	5.31	5.51	6.69
ΣD	-1.42	-1.53	-3.19	-6.12	-8.12	-8.20	-10.54	-13.42

sions for nuclear forces to ensure that there is not a regime of collapsed nuclei of very high density and binding [20]. Most of the current forces do not satisfy the conditions, but since the expressions for the forces are not to be taken seriously for relative momenta greater than, say, 5 fm^{-1}, corresponding to a kinetic energy of several hundred MeV, there is no problem. We are, of course, required to assume that strong repulsion arises at higher momenta, but it is also true that in a high-density regime excited nucleon states will be present, both Δ's and N's and the strange baryons Λ's and Σ's.

Having seen the mechanism of saturation, we may now ask how well the quantitative results agree with experiment. Table 8–3 shows that for a given density two potentials, both acceptable on the energy shell, may give binding energies differing by 2 or 3 MeV. Comparison of the Reid soft core and the OBEP at $k_F = 1.5$ fm^{-1} shows two main differences, the $^3S_1$- and $^1P_1$-states.

The OBEP singlet P-potential is that predicted by the OBEP theory; it gives $^1P_1$-phase shifts noticeably different from those of the Reid potential, both above 200 MeV (where it yields the better agreement with experiment) and below 50 MeV. In the latter region there is need for improved experimental information, since current data have been subject to considerable uncertainty; some accurate $^1P_1$-phase shifts would probably distinguish between the more and the less repulsive forces now possible.

The different results for the $^3S_1$-state can be attributed to the different values of the ratio of central force to tensor force in the two potentials. With the OBEP more of the binding is due to the central force, and although the

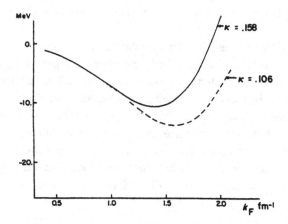

Fig. 8-12 Binding energy of nuclear matter as a function of Fermi momentum. Upper curve: Reid soft core, lower curve: OBEP of Grangé and Preston [17]. Two-body correlations only. The quantity κ is the wound at $k_F = 1.5$ fm^{-1}, $k = 0.86$ fm^{-1}.

quadrupole moment of the deuteron is fitted, the percentage of D-state is lower. This is reflected in a lower value of the scattering parameter ε_1, as discussed in Chapter 5, but one which nevertheless may lie in the experimentally determined range. The result in nuclear matter is that the Pauli-effect inhibition of the tensor force is not quite so effective in preventing the growth of the $^3S_1$-contribution. A considerable range of central-tensor mixtures is permitted by the two-nucleon data [21], and rather different results ensue for nuclear matter throughout the range, as has been shown by Afnan, Clement, and Serduke [22]. Since the tensor suppression mechanism is relatively less significant for OBEP, one would expect that the density at which the energy saturates would be higher. This is borne out by computation, and Fig. 8–12 gives the results. The Reid potential[†] saturates at $k_F = 1.42$ fm^{-1} with $E/A = -10.59$ MeV, and the OBEP at $k_F = 1.58$ fm^{-1} with $E/A = -13.58$ MeV.

Both results are at densities greater than the experimental number $k_F = 1.36$ fm^{-1}, but so far we have considered only two-body clusters. We must now consider additional effects. We shall discuss three- and four-body clusters and relativistic kinematics.

We have already indicated that the work of Dahlblom [9] gives an attractive contribution of three-body clusters of about 2 MeV. A more precise calculation by Grangé for the Reid soft-core potential gives a value of -1.59 MeV at $k_F = 1.36$ fm^{-1}. The three-body contribution becomes less attractive at high density. These numbers are uncertain by ± 0.5 MeV. Also we have already given the result of a calculation of the four-body-cluster energy; it is -1.1 ± 0.5 MeV and also becomes more attractive with increasing density.

Three-body forces must be distinguished from the effects of three-body clusters. We have defined a three-body cluster as a situation in which, after two nucleons have interacted, and while they are still in a virtual excited state, one of them interacts with a third nucleon, and then further interaction returns the three particles to their initial states. All the interactions are two-nucleon ones. A three-body force arises when two nucleons interact to produce a virtual excited state which contains some entity other than nucleons, and while this state exists one of its constituent parts interacts with a third nucleon. The effect cannot be attributed to a succession of two-nucleon interactions.

[†] This result differs slightly from Sprung's [2], quoted as $k_F = 1.44$ fm^{-1}, $E/A = -11.25$ MeV. The discrepancy is due chiefly to the different treatments of the states with $J > 2$.

Some diagrams may help to make the difference clear. Figure 8–13a represents a typical three-body cluster. Nucleons in states k and l interact and are excited to momenta a and b, a and m interact and receive momenta c and k, and c and b interact, returning to momenta l and m. This is a three-body cluster and, in fact, is the same as the cluster represented in Fig. 8–4a. The wavy lines represent the complete two-nucleon interaction due to all possible exchanges of pions and other bosons with appropriate intermediate states. For example, if we use a dotted line to represent a pion exchange and a heavy line to represent a Δ(1236), diagram (a) would contain diagrams (b) and (c) and a host of others. In diagram (b) we deal only with a succession of single-pion exchanges; this is a portion of what occurs in a three-body cluster. In diagram (c) we show another portion, where the result of the first pion exchange is a virtual state containing a nucleon and a Δ(1236), which then exchange a second pion, resulting in a state with two nucleons in states a and b above the Fermi sea. Then a interchanges two pions with m before c

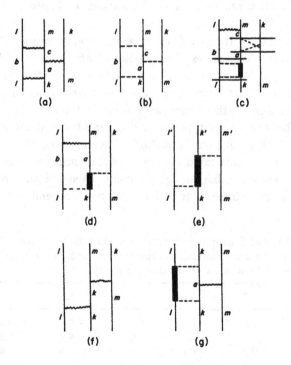

Fig. 8-13 Diagrams to illustrate three-body forces in nuclear matter.

and k result. But all processes between states l, k and a, b and between a, m and c, k are just parts of the two-nucleon interaction. In contrast, the bottom part of diagram (d) represents a portion of the three-body forces, because the virtual state with which the third nucleon interacts is not a nucleonic state; there is no way in which this process reduces to a succession of applications of the two-nucleon force. Diagram (d) then is a three-body contribution to the energy of nuclear matter, involving a three-body force followed by a two-nucleon force.

The diagonal three-body force can also contribute to nuclear matter. It is represented by diagram (e), in which k', l', m' are any permutation of k, l, m. Note that diagram (f), on the other hand, does not represent even a three-body cluster. It merely represents two completely distinct two-body clusters (the states that we would describe as ψ_{kl} and ψ_{km}), and it reminds us to include in the binding energy the quantities $(\phi_{kl}|v|\psi_{kl})$ and $(\phi_{km}|v|\psi_{mk})$. This remark should delineate clearly the nature of the three-body force: the interaction in nuclei is not the superposition of the free interaction for all nucleon pairs; rather, the virtual meson and baryon fields are modified by the presence of the other nucleons, as we already noticed in discussing magnetic moments.

The calculation of the contribution of three-body forces entails all the complexities and uncertainties of calculations both of boson exchanges and of nuclear matter. Nevertheless, the contributions due to the $\Delta(1236)$ should be by far the most important, and they have been estimated [23, 24, 25]. They represent a part—the longer-range part—of the force due to two-pion exchange. The interactions typified by diagram (d) are attractive and contribute about 2 MeV per particle binding. Those represented by diagram (e) are repulsive and become rapidly more so as the density increases. Finally, a third process shown in diagram (g) is found to be significant, though small. Table 8–4 shows these three contributions. The result depends on the cancella-

TABLE 8–4 Potential Energy per Particle from Three-Body Forces. Each number is in MeV and represents the contribution from the indicated class of diagram, including exchange among the final states. From Green et al. [23].

k_F, fm^{-1}:	1.2	1.3	1.4	1.5	1.6	1.7
Fig. 8–13d	− 1.6	− 1.9	− 2.3	− 2.7	− 3.2	− 3.8
Fig. 8–13e	1.0	1.7	2.6	4.0	5.8	8.4
Fig. 8–13g	− 0.3	− 0.1	0.1	0.3	0.5	0.6
Total	− 0.9	− 0.3	0.4	1.6	3.1	5.2

tion between the two larger terms, each of which is sensitive to some assumptions in a lengthy calculation; in particular, the two-nucleon correlation is important. The repulsion of diagram (e) may be substantially overestimated.

It will be realized that all calculations have used nonrelativistic quantum mechanics. This can be changed only when a fully relativistic many-body theory, including the meson fields, is developed. However, it is desirable to make allowance for the most simple relativistic effect, the altered relationship between energy and momentum. This "minimal" allowance for relativity has been estimated by Brown, Jackson, and Kuo [26] and results in additional binding of about 0.5 MeV at normal density.

The various components of the binding energy are collected in Tables 8-5 and 8-6. Since the four-nucleon correlations and some of the three-nucleon correlations have been calculated only for $k_F = 1.36$ fm^{-1}, their dependence on k_F has to be estimated. It is difficult, therefore, to be very precise about

TABLE 8-5 Nuclear-Matter Energy (MeV per Particle). The first three contributions are based on the Reid soft-core potential. The bracketed numbers are estimates by Bethe [1].

k_F, fm^{-1}:	1.2	1.36	1.6
Two-body correlation	− 27.71	− 34.02	− 42.05
Three-body correlation	(− 1.6)	− 1.59	(− 1.1)
Four-body correlation	(− 0.9)	− 1.1	(− 1.5)
Three-body forces	− 0.9	0.1	3.1(?)
Minimal relativity	(− 0.35)	− 0.5	(− 0.7)
Kinetic energy	17.92	22.97	31.85
Total E/A	− 13.5	− 14.1	− 10.4

TABLE 8-6 Nuclear-Matter Energy (MeV per Particle). The two-body and part of the three-body correlations are for OBEP [17]; the rest of the three-body correlations and the four-body correlations are for the Reid soft-core potential. The bracketed numbers are estimates similar to those of Bethe [1].

k_F, fm^{-1}:	1.36	1.5	1.7
Two-body correlation	− 35.42	− 41.43	− 49.14
Three-body correlation	− 1.49	(− 1.9)	(− 2.6)
Four-body correlation	− 1.09	(− 1.4)	(− 1.8)
Three-body forces	0.1	1.6	5.2
Minimal relativity	− 0.5	(− 0.6)	(− 0.8)
Kinetic energy	22.97	28.00	35.95
Total E/A	− 15.4	− 15.7	− 13.2

the saturation density and energy. The values of Table 8–5 suggest $k_F = 1.32 \, \text{fm}^{-1}$ and $E/A = -14.2$ MeV per particle for the Reid soft-core potential, whereas those of Table 8–6 give $k_F = 1.46 \, \text{fm}^{-1}$ and $E/A = -15.8$ MeV per particle for the OBEP. Since some components of the energy are in doubt by as much as 0.5 MeV, both results are consistent with the experimental binding energy. Also, the equilibrium density is very sensitive to the k_F-dependence of the higher-order corrections, even though these may be relatively small near equilibrium. In other words, the energy-density curve for pair correlations alone is quite flat in the region of nuclear densities, and the small higher-order terms have a marked effect on the curvature and the minimum of the total energy-density function.

Since the two forces that we have considered are typical of realistic nucleon-nucleon forces, we may conclude that the energy and the density of nuclear matter are understood and are consistent with the properties of the actual forces. It should be stressed, however, that this result depends on the inclusion of the higher-order terms. We have already noted that the equilibrium values for pair correlations only occur at excessively high densities and at binding energies 2–4 MeV too small. Other potentials with equivalent phase shifts can be devised which will alter these numbers, but a potential that would account for the total binding by the two-body correlation alone should probably be ruled out on the grounds that the higher-order terms do contribute a few MeV.

Nuclear matter, being in its lowest energy state, is at zero temperature. Hence the connection between the pressure, the density, and the energy per particle is

$$P = -\frac{d(E/A)}{dv} = \rho^2 \frac{d(E/A)}{d\rho}, \qquad (8\text{–}75)$$

where $v = 1/\rho$ is the volume per particle. In other words, the pressure is the rate at which the energy per particle decreases as the volume per particle increases. At equilibrium the pressure is zero. If, because of external agencies, the system is in a region of positive pressure, it remains stable in the sense that in losing energy it expands, decreasing in density and reaching equilibrium, where it can exist in isolation. If, however, the pressure is negative, the system breaks up since it loses energy by forming isolated lumps of matter at equilibrium density. In the case of nuclear matter, the pressure is of course negative for $\rho < \rho_0$, and therefore an isolated uniform assembly of protons and neutrons at $\rho < \rho_0$ will coalesce into regions of normal density ρ_0 with "empty" space between them. Of course, there can be a stable situation with $\rho < \rho_0$ in completely different geometries such as a surface layer on a finite nucleus.

It is important to know whether nuclei are compressible. If they are readily so, a common form of excitation of a nucleus would be density oscillations, or the so-called breathing mode, in which the nucleus would alternately contract and expand without change in shape. A measure of the energy needed to produce a change from the equilibrium density ρ_0 is clearly the value of the second derivative of E/A at equilibrium. In fact, it is usual to define the "incompressibility" as

$$K = k_F^2 \left[\frac{d^2(E/A)}{dk_F^2}\right]_0 = 9\rho_0^2 \left[\frac{d^2(E/A)}{d\rho^2}\right]_0. \qquad (8\text{--}76)$$

Excited states corresponding to density changes are not observed, at least in the low-lying regions of nuclear spectra. We deduce that K must be appreciable, and a value of at least 120 MeV is indicated. (See Problem 8-8.) The value of K is of course sensitive to the higher-order corrections, since these affect the shape of the energy-density curve. For the calculations in Tables 8-5 and 8-6 the results are, respectively, 170 and 180 MeV. The two-body correlation alone provides only 130 and 120 MeV for the same two potentials.

Nuclear-matter calculations can be made with unequal numbers of neutrons and protons. There are two distinct Fermi momenta, and some additional complications arise both in handling the two exclusion-principle operators Q which now arise and in weighting the various spin-isospin states. Nevertheless the computations are feasible. For small neutron excess, it is possible in this way to find the coefficient of the symmetry-energy term in the semiempirical mass formula. We write

$$\frac{E(\rho, \alpha)}{A} = \frac{E(\rho, 0)}{A} + \varepsilon_1(\rho)\alpha^2 + \varepsilon_2(\rho)\alpha^4 + \cdots, \qquad (8\text{--}77)$$

where

$$\rho = \rho_n + \rho_p, \qquad \alpha = \frac{\rho_n - \rho_p}{\rho} = \frac{N - Z}{A}.$$

The quantity $\varepsilon_1(\rho_0)$ should be the coefficient $-a_{\text{sym}}$ of formulas 6-21 and 6-24. The results are satisfactory [16], giving

$$\varepsilon_1(\rho) = 32\left(\frac{\rho}{\rho_0}\right)^{2/3} \text{ MeV}.$$

It may be noticed that with the number of protons fixed the binding of the whole system increases with each neutron added by an amount $[\partial E(A,Z)/\partial A]_Z$. Using the constants of the semiempirical formula, one finds that this derivative vanishes for Z/A about $\frac{1}{3}$. Hence, when there are twice as many neutrons as protons, any additional neutrons are unbound. In other words, a system with α greater than about $\frac{1}{3}$ is unstable in isolation. But of course it could exist if additional attraction, other than nuclear forces, could hold the neutrons. Such an attraction is the force of gravity, and there is little doubt that there are dense stars which consist primarily of neutrons with a few protons and other charged baryons accompanied by electrons and negative muons. The structure of such stars is a fascinating subject for study, ranging from the low-density regime where there are nuclear clusters, that is, very large "nuclei," with free neutrons and electrons in the relatively empty space between them [27], through the regime of neutron matter, which is nuclear matter with α quite close to unity (a few protons) [28, 29, 30], and then into the very-high-density regime [31, 32]. Unfortunately, however, this topic is somewhat outside the scope of this book.

At the end of Section 8–3 we described the procedure for a renormalized Brueckner-Hartree-Fock calculation (RBHF) for a finite nucleus and stressed its extreme complication. One of the computational problems is the necessity to calculate matrix elements like $\langle kl|v|\alpha\beta\rangle$, where the single-particle wave functions are determined self-consistently by the Hartree-Fock process and are functions of the coordinates of the nucleons, while v depends on the relative coordinate. We saw that in nuclear matter there are two simplifications: the wave functions must be plane waves, and plane waves can easily be expressed in relative and center-of-mass coordinates. There is a set of finite range functions such that $\phi_\alpha(\mathbf{r}_1)\phi_\beta(\mathbf{r}_2)$ can be written as a finite sum of terms $\phi_\gamma(\mathbf{r})\phi_\delta(\mathbf{R})$; these are the harmonic-oscillator functions. (See Problem 8–11.) But no other set has this property, and, of course, the self-consistent potential U does not turn out to be an harmonic oscillator. The only practical approach to the computation of matrix elements is to express the actual wave functions in terms of a complete set of oscillator functions. If the actual functions are fairly close to oscillator functions, only a few terms are needed in such a series. Fortunately this is often the case; indeed, for many purposes it is not an entirely misleading approximation to take the wave function of a deeply bound nucleon to be a single oscillator function.

The second major technical task is to deal with the exclusion principle. The operator Q must span the complete Hilbert space except for the A occupied states, which are the self-consistent $\{\phi_\alpha\}$. Since the solution of the

Bethe-Goldstone equation requires that terms like $\langle 1|v|i\rangle\langle i|G|2\rangle(E_i - \omega)^{-1}$ be summed over unoccupied states i, there must be a convenient expression also for the states outside the Fermi sea.

A third, more basic problem is to decide on the spectrum E_i for unoccupied states. As was emphasized in Section 8–3, the results are quite sensitive to this choice, unless one proceeds to calculate three- and four-body clusters. Otherwise, one should choose the excited-state spectrum to cancel important higher-order terms. In nuclear matter, sufficient work has been done on three-body clusters to suggest a reasonable excited-state spectrum; there $U = $ constant is reasonable for states above the Fermi level. But there are no calculations of three-body clusters in finite nuclei. Consequently, in a RBHF calculation one should study the effects of different assumptions concerning the excited spectrum.

In these circumstances, the early calculations were not complete. Some did not obtain a self-consistent set $\{\phi_\alpha\}$, some applied Q only to the states of the oscillator basis, not to the "real" states, some assumed the excited states to be plane waves, some took them as oscillators, some took the excited states to have zero potential energy, some assumed oscillator spectra or modifications of these.

It is most important to "renormalize," that is, to allow for the partial occupation of states in the Fermi sea. In some papers, this is done by means of a "saturation potential" or "rearrangement energy," introduced into Eq. 10–13. There are now much improved calculations, but it remains true that in 1973 it was not clear to what extent the inadequacies of the results obtained were to be attributed to approximations in the calculations and to what extent they were due to the assumed properties of the nucleon-nucleon force.

The most complete calculations have difficulty principally with the spectrum of the unoccupied states. Two prescriptions have been commonly used. One is to take $(a|U|a) = 0$, since in nuclear matter this leads to small higher-order cluster contributions. This argument is not particularly strong, since higher-order clusters have not been calculated in finite nuclei. Also in nuclear matter the results for pair correlations are insensitive to $(a|U|a)$ as long as it is small, whereas, as we shall see, the results for finite nuclei are sensitive. This prescription means that the energy of such states is the kinetic energy only. The Hamiltonian is written as QTQ; this is not the free-particle Hamiltonian, since the projection operators prevent it from having matrix elements between particle states and hole states, that is, the occupied and unoccupied states remain orthogonal.

The other common approach is to take the spectrum to be that of a harmonic oscillator, but shifted by an arbitrary amount:

$$e_a = \hbar\omega(2n_a + l_a + \tfrac{3}{2}) - C - B_0\theta_a, \tag{8-78}$$

where ω is the oscillator frequency used for the basis states, C and B_0 are arbitrary constants, and θ_a is unity if a is a state in a certain arbitrarily chosen set near the Fermi level and is zero for states above that set. If $B_0 = 0$ and $C \neq 0$ the whole spectrum is lowered in energy, whereas if $C = 0$ and $B_0 \neq 0$ only a certain number of states have their energies lowered. It is argued that the lowest unoccupied single-particle level of a finite nucleus may be a bound state and therefore should have negative energy, or it may be unbound by a small amount. In any case its energy should be near zero; this is achieved by putting

$$C + B_0 = \hbar\omega(2n' + l' + \tfrac{3}{2}), \tag{8-79}$$

where n' and l' are the quantum numbers of the first unoccupied state. For example, if one is studying ^{16}O, the relevant states are the $1s$ and $0d$, which give $C + B_0 = 7\hbar\omega/2$. Alternatively, to come close to the free-particle prescription, one could take $C + B_0$ at half the above value, since then the energy of the lowest state would be just its kinetic energy; or, by using a somewhat greater value, one could subtract the average potential energy of the first few low-lying unoccupied states.

Tables 8–7 and 8–8 display some results for ^{16}O and ^{40}Ca. The first point to note is that columns A, B, and C in Table 8–7 show that the results are insensitive to the oscillator frequency of the basis states. They should, in principle, be independent of ω since the oscillator states serve only as a complete set in which the Hartree-Fock states are expanded for computational purposes. The actual dependence reflects the fact that the calculations are not completely self-consistent [34]. Next we examine the sensitivity to the intermediate-state energies by comparing columns B and D of Table 8–7. In column B, the lowest intermediate-state energy is zero and the two-particle intermediate states are reduced in energy by $7\hbar\omega$, provided that

$$2n_\alpha + l_\alpha + 2n_\beta + l_\beta \leqslant \bar{N} - 4 \tag{8-80}$$

with $\bar{N} = 18$. This is a substantial number of shifted states. In column D, all states are shifted but the energy of the two-particle intermediate states falls by only $2\hbar\omega$. It will be seen that the differences in the energies and the occupation probabilities are very marked. This can be understood as follows. Lowering the intermediate-state spectrum reduces the denominators in the

TABLE 8-7 Renormalized Brueckner-Hartree-Fock Calculations for ^{16}O. A: $\hbar\omega = 12.5$, $B_0/\hbar\omega = 3.5$, $C = 0$, $\bar{N} = 18$; B: $\hbar\omega = 14.0$, $B_0/\hbar\omega = 3.5$, $C = 0$, $\bar{N} = 18$; C: $\hbar\omega = 15.5$, $B_0/\hbar\omega = 3.5$, $C = 0$, $\bar{N} = 18$; D: $\hbar\omega = 14.0$, $B_0 = 0$, $C = \hbar\omega$. A, B, C, D: Reid soft-core potential; E: OBEP potential, neutron and proton states not distinguished, $\hbar\omega = 13.3$, for spectrum see text. The total energy E includes Coulomb energy. Energies are in MeV, the mean-square mass and charge radii are in fm.

	Experiment[a]	A[b]	B[b]	C[b]	D[c]	E[d]
Neutrons						
$e(0s_{1/2})$	-40 ± 8	-36.9	-37.0	-37.4	-38.4	-40.8
$e(0p_{3/2})$	-21.8	-18.3	-18.2	-18.3	-17.0	-19.9
$e(0p_{1/2})$	-15.7	-15.3	-15.2	-15.1	-13.3	-16.5
$e(0d_{5/2})$	-4.14					-2.59
$e(1s_{1/2})$	-3.27					-0.15
$P(0s_{1/2})$		0.82	0.82	0.82	0.88	
$P(0p_{3/2})$		0.78	0.78	0.78	0.90	
$P(0p_{1/2})$		0.78	0.78	0.78	0.90	
Protons						
$e(0s_{1/2})$		-34.2	-34.3	-34.7	-35.2	
$e(0p_{3/2})$		-15.8	-15.7	-15.7	-14.0	
$e(0p_{1/2})$		-12.9	-12.7	-12.6	-10.4	
$P(0s_{1/2})$		0.81	0.82	0.82	0.88	
$P(0p_{3/2})$		0.78	0.78	0.78	0.90	
$P(0p_{1/2})$		0.79	0.77	0.78	0.90	
$-E/A$	7.98	6.57	6.49	6.47	4.19	7.57
r_{mass}		2.44	2.43	2.41		2.64
r_c	2.73	2.57	2.57	2.55	2.61	

[a] As quoted in ref. 33.
[b] Ref. 34.
[c] Ref. 35.
[d] Ref. 33.

Bethe-Goldstone equation and makes the G-matrix more attractive. This, in turn, gives more binding, both in single-particle states and overall. The shifts in the occupied state are smaller than those in the intermediate states which caused them, and therefore the energy gap is reduced self-consistently. Moreover, this implies more excitation, that is, a larger wound κ, and a smaller occupation probability P. These effects are also apparent in comparing columns A and C of Table 8–8.

The effect of lowering more states can be seen in columns C, D, and E of Table 8–8. In each case the states are lowered by $4.5\hbar\omega$, but the number

TABLE 8-8 Renormalized Brueckner-Hartree-Fock Calculations for ^{40}Ca. Proton states are not tabulated, but are included in the energy E. A: $\hbar\omega = 12.5$, $B_0/\hbar\omega = 4$, $C = 0$, $\bar{N} = 20$; B: same as A, but not renormalized; C: $\hbar\omega = 12.5$, $B_0/\hbar\omega = 4.5$, $C = 0$, $\bar{N} = 20$; D: $\hbar\omega = 12.5$, $B_0/\hbar\omega = 4.5$, $C = 0$, $\bar{N} = 26$; E: $\hbar\omega = 12.5$, $B_0 = 0$, $C/\hbar\omega = 4.5$. A, B, C, D, E: Reid soft-core potential; F: OBEP, neutron and proton states not distinguished, $\hbar\omega = 15$ MeV, spectrum described in text. Energies are in MeV, radii in fm.

	Experiment[a]	A[b]	B[b]	C[b]	D[b]	E[b]	F[c]
Neutrons							
$e(0s_{1/2})$	-50 ± 11	-54.8	-68.2	-54.7	-58.2	-70.7	-55.2
$e(0p_{3/2})$		-35.1	-45.3	-34.9	-37.5	-46.6	-36.4
$e(0p_{1/2})$	-34 ± 6	-32.4	-41.4	-32.2	-34.5	-42.6	-33.7
$e(0d_{5/2})$		-17.5	-24.7	-18.0	-19.7	-25.6	-19.7
$e(1s_{1/2})$	-18.1	-15.8	-21.8	-16.4	-17.9	-23.0	-19.6
$e(0d_{3/2})$	-15.6	-13.4	-18.8	-14.1	-15.5	-20.1	-15.8
$e(0f_{7/2})$	-8.36	3.1	0.54				-6.6
$e(1p_{3/2})$	-6.2	1.9	1.1				0.1
$P(0s_{1/2})$		0.83	1	0.82	0.81	0.80	
$P(0p_{3/2})$		0.84	1	0.81	0.80	0.80	
$P(0p_{1/2})$		0.83	1	0.81	0.79	0.79	
$P(0d_{5/2})$		0.84	1	0.79	0.77	0.73	
$P(1s_{1/2})$		0.84	1	0.79	0.77	0.72	
$P(0d_{3/2})$		0.83	1	0.78	0.76	0.71	
$-E/A$	8.55	4.99	4.21	5.83	6.98	10.7	8.49
r_{mass}							3.38
r_c	3.49	3.19	3.05	3.14	3.06	2.85	

[a] As quoted in ref. 33.
[b] Ref. 34.
[c] Ref. 33.

of states progressively increases, $\bar{N}$ being 20, 26, and infinity. It is seen that this has a marked effect of the kind just described, since more states are at energies capable of participating in the correlation wound. The multiplicity of states of higher spin helps in this process.

A comparison of columns A and B of Table 8–8 demonstrates the importance of renormalization. As we pointed out at the end of Section 8–3, the single-particle energies are very sensitive to renormalization, but the total energy is affected less drastically.

It has been suggested that the intermediate spectrum should be chosen to give a good fit to the observed single-particle bound-state energies. It is safer, however, to consider the choice an open question.

The columns so far discussed are for the Reid soft-core potential. The last column of each table reports a calculation with the same OBEP as we used for nuclear-matter comparisons. In this case the intermediate-state spectrum was taken as discrete through a "valence" band—a few low-lying unoccupied states—and as continuous and free above a specified cutoff energy. This appears to be a good way to treat the Pauli principle. The discrete states were shifted by the average potential energy in the valence band; this would be similar to a B_0 of about $2\hbar\omega$ for ^{16}O and perhaps $3\hbar\omega$ for ^{40}Ca.

It will be observed that with the Reid soft-core potential a substantial spectral shift is required to produce total binding near the experimental value, and that, when this is made, the nuclear radius decreases to values substantially lower than experimental ones. Indeed, when the calculation is adjusted to give radii near experiment, the binding energy is considerably too small. The OBEP, on the other hand, does better at fitting both quantities simultaneously. We should be cautious, however, about what this means physically. The OBEP gives more binding in nuclear matter also, but in nuclear matter we know that there is 3 MeV or so of binding per particle from many-body clusters, many-body forces, and relativistic effects. All of these will occur in finite nuclei but have not been calculated. It is difficult, therefore, to know what answer to desire for a calculation (such as RBHF) based on the independent-pair approximation.

Because of this uncertainty, and also because of increased simplicity in computational aspects and perhaps also greater richness in physical insight, another approach to finite nuclei has also been pursued. This approach emphasizes the idea of thinking of G as an effective interaction to be used with the wave functions ϕ, instead of using v with the real wave functions ψ, and then it exploits the fact that outside a relatively short distance ψ heals to ϕ. Since $G\phi = v\psi$, it follows that, if G is expressed as an interaction in coordinate space, it is identical to v for separations greater than the healing distance. The correlations ζ are of short range compared to the radii of nuclei. Nuclei, except in a relatively small surface region, have slowly changing density. In a nucleus the states ϕ are finite in range; but just as the plane waves of nuclear matter with $k < k_F$ change slowly within a distance of the order of the healing distance, so also do the ϕ's of the finite nucleus. Would it not then seem reasonable that the correlation ζ in a two-nucleon collision taking place in the central region of a nucleus should be very similar to the correlation in nuclear matter of the same density? Since two colliding nucleons have their interaction effectively modified only by others nearby, is it not reasonable to expect to obtain good results for finite nuclei simply by taking G over from nuclear matter and using it as the nucleon-nucleon interaction in finite nuclei? In this way, the short-range correlations are built into the interaction and the wave functions themselves can be the simple single-particle ϕ's of a shell model or of a Hartree-Fock calculation.

Of course, G depends on density, and therefore we adopt a local density approximation. It is assumed that the effective interaction of two nucleons in a finite nucleus is equal to their effective interaction in nuclear matter with constant density equal to the density in the region where the nucleons interact, that is, at a density which is some mean value of the densities at nucleon 1 and nucleon 2. Both arithmetic and geometric means have been used in actual calculations. This basic approach was taken first by Brueckner, Gammel, and Weitzner [36]. In their work, they preserved the nonlocal character of the G-matrix. It was later suggested by Brandow that the theory would be not only simpler, but also more amenable to the use of physical insight, if a local effective interaction could be substituted without much loss of accuracy. This suggestion has been the basis of a considerable amount of more recent work.

We shall write $V_{L'L}^{JS}(r)$ for the effective two-nucleon interaction. At a given density in nuclear matter, this quantity is, in fact, the G-matrix. Using Eq. 8–61b, we can write

$$V_{L'L}^{JS}(r, k, P, k_F; v) = \sum_{L''} \frac{v_{L'L''}^{JS} u_{L''L}^{JS}(r)}{kr j_L(kr)}. \tag{8–81}$$

The basic requirement of an effective interaction is that its matrix elements $\langle k'|V|k\rangle$ be as close as possible to those of G. The V defined in Eq. 8–81 of course satisfies this exactly if u is taken as the result of a self-consistent Brueckner nuclear-matter calculation. There will still be some freedom in regard to the value of V, because of some arbitrariness in the choice of the energy spectrum of the intermediate states in the calculation. The various arguments attached to V emphasize that it depends on the free nucleon-nucleon interaction v, on the density or k_F, and on k and P. We can immediately assume that an average value of P has been used for each k. But the dependence on k is inescapable, and it is this momentum dependence which makes the interaction nonlocal. (The reader may recall Eq. 8–3.)

In order to simplify this expression and obtain a local effective interaction, one may average appropriately over k. A natural thing to do is simply to weight the contribution for each k by the probability of finding relative momentum k in matter having Fermi momentum k_F [37]. The quantities that enter into a calculation of binding energy are the *diagonal* matrix elements of the effective interaction for $k < k_F$. These, then, are the quantities averaged, leading to

$$V_{L'L}^{JS}(k, k_F) = \frac{\int_0^{k_F} \omega(k) j_{L'}(kr) V_{L'L}^{JS}(r, k, k_F) j_L(kr) \, dk}{\int_0^{k_F} \omega(k) j_{L'}(kr) j_L(kr) \, dk}, \qquad (8\text{--}82)$$

where

$$\omega(k) = 12k^2(k_F - k)^2(k + 2k_F).$$

The result is that, on the average, the diagonal matrix elements of V are the same as those of G, but no particular matrix element $(k'|V|k)$ necessarily agrees with $(k'|G|k)$.

Once one has fixed a local potential, one has fixed both its diagonal and its off-diagonal matrix elements. Conversely, once one has fixed the diagonal elements *for all* k, one has fixed the potential. The off-diagonal elements are important in that they give the excitation of intermediate states, and they do play a role in Hartree-Fock calculations. The use of the nuclear-matter G, averaged in this way, should lead to reasonable off-diagonal elements.

Sprung [38] suggested that it would be more reasonable to use an effective interaction which reproduces exactly the diagonal elements for $k < k_F$, and which also has the same general appearance as the averaged V when regarded as a function of r. This latter criterion should ensure reasonable behavior for the matrix elements not fixed by Sprung's first requirement. To do this, it is of course necessary to resort to phenomenology. Sprung and Banerjee obtain a satisfactory effective interaction which is a sum of five Gaussian functions of r with appropriate constants.

Of course, if one fixes the interaction for all states J, S, L', L, one can then express the effective potential as a linear combination of central, tensor, $L \cdot S$, L^2, and exchange forces. Since the L^2-force is rather awkward, Negele [37] in his work simplified V further by averaging V_{LL}^{JS} over J and L for a given S and T to obtain an effective central force.

When $N \neq Z$, it is desirable to have a different effective interaction for each kind of nucleon. It is customary to use the local proton density for proton-proton interactions, the local neutron density for neutron-neutron collisions, and an average of the two densities for neutron-proton interactions.

It must be emphasized that the density dependence of the effective interaction is a vital feature. As will be seen in Chapter 10, the rearrangement energy due to this is an essential component of any understanding of nuclear energies. Equally important for explaining other effects is the change in the single-particle wave function that follows from the density dependence.

If the reader recalls that the density dependence of the total binding energy of nuclear matter is quite different from that of the contribution from two-body clusters, he may well question the usefulness of using the two-body G as the effective interaction in the finite case. There is also the fact that the two-body clusters alone underbind nuclear matter by several MeV per particle and give an incorrect equilibrium density. There is no reason to expect significantly less from higher-order effects in finite nuclei, although they would be very difficult to compute. Yet the goal of Hartree-Fock calculations with effective interactions is to study the nucleus in terms of effective two-body correlations. Consequently, it is important to use an effective interaction that gives the right binding energy at actual nuclear densities. To achieve this, the workers in this field [39] arbitrarily increase the strength of the attractive interaction at short range. It is desirable to use the actual strength for $r \gtrsim 2$ fm, however, since the purpose of the increased attraction is to allow for higher-order effects, which are mostly of short range. Of course, this adjustment does not necessarily produce the right density dependence, and one might think of adjusting the effective interaction to improve this aspect. But the more one adjusts, the more phenomenological the interaction becomes.

In order to give some notion of the nature of the effective interaction, Figure 8–14 shows the tensor components of V plotted against the Reid soft-core potential from which they were derived. The remarkable smoothing and weakening of the potential at distances less than 1 fm are apparent, as is the agreement of V and v at larger distances.

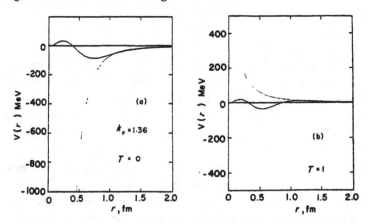

Fig. 8-14 Tensor components of the effective interaction at $k_F = 1.36$ fm^{-1} for $T = 0$ and 1. For comparison, we also show, as a dotted curve, corresponding components of the free potential, from which the effective interaction was derived. (From ref. 38.)

We have not discussed to any extent the changes in the theory that arise if energy levels are degenerate and are not completely filled. This situation occurs in any nucleus where some shell-model states are not fully occupied, that is, in all but magic nuclei. The basic theory of such "valence shells" has been supplied by Brandow [40], but most calculations have been performed by assuming effective interactions. Some cases are discussed in Chapter 10.

Just as an effective interaction is required if ϕ's are used instead of ψ's, so one needs effective operators for other observables such as electromagnetic moments. The following schematic equation is required:

$$(\phi_{mn}|\theta_{\text{eff}}|\phi_{kl}) = (\psi_{mn}|\theta|\psi_{kl}). \qquad (8-83)$$

One can write

$$\psi = \Omega\phi, \qquad (8-84)$$

where Ω, sometimes called the "wave operator," is clearly a formal way of representing the correlations. In this notation

$$\theta_{\text{eff}} = \Omega\theta\Omega^{-1}. \qquad (8-85)$$

This rather formal approach leads to complicated calculations, and more phenomenological approaches, such as "effective charges," are usually employed. However, the above presentation shows that such quantities can, in principle, be computed.

8-5 PAIRING CORRELATIONS AND THE GROUND STATE

In the preceding sections of this chapter we considered a theory of nuclear structure, potentially of very considerable accuracy, but based on the underlying picture of particles moving freely except for short-range correlations arising during the transient existence of clusters of two or more particles. This picture was shown to be possible because of the exclusion principle and the specific nature of the two-nucleon force; nevertheless the model is essentially that of a gas of fermions with very-short-range spatial correlations. These remarks apply to the main features of both nuclear matter and of real nuclei.

The question arises whether or not this quasi-gas state, which we call the *normal* state, is the nuclear ground state. To be more specific, we are asking this question: "Are there other types of states with long-range spatial correlations with lower energy?" In respect to correlations, such states would resemble liquids rather than gases, with wave functions so different that they would not be obtainable from a perturbative calculation starting with quasi-gas wave functions. A similar situation arises in the superconductivity of metals. Following Cooper [41], let us first consider the simple situation of a pair of nucleons outside a quiescent Fermi sphere and interacting with an attractive force. It will be shown that under these circumstances the pair would form a bound state with a ground-state energy lower than that in the quasi-gas model, and the wave function of the pair would extend over a relatively large spatial distance. Furthermore, on considering the density of the two-nucleon states, we would find an energy gap between the ground state and the excited states. This bound pair, which is formed as long as the interaction between the two nucleons is attractive, no matter how weak, is popularly called the Cooper pair.

In this model, we emphasize that the Fermi sea in the presence of which the two nucleons are interacting is considered quiescent, that is, the nucleons inside the sea are inert and are not influenced in any way; the only effect they have is to block all momentum states up to k_F. Consider two identical nucleons of momenta $\mathbf{k}_1, \mathbf{k}_2$ $(k_1, k_2 > k_F)$, with center-of-mass momentum $\mathbf{K} = \mathbf{k}_1 + \mathbf{k}_2$. Since the interaction potential V_{12} is a function of relative coordinates only, $\mathbf{K}$ must be conserved. Furthermore, since we are interested in the ground state, it is reasonable to take the two identical nucleons to be of opposite spins. For a complete set of two-nucleon wave functions we take the plane-wave product $(1/\Omega) \exp[i(\mathbf{k}_1 \cdot \mathbf{r}_1 + \mathbf{k}_2 \cdot \mathbf{r}_2)]$, which in relative coordinates is $(1/\Omega) \exp[i(\mathbf{k} \cdot \mathbf{r} + \mathbf{K} \cdot \mathbf{R})]$, where $\mathbf{k} = \frac{1}{2}(\mathbf{k}_1 - \mathbf{k}_2)$, $\mathbf{R} = \frac{1}{2}(\mathbf{r}_1 + \mathbf{r}_2)$, and $\mathbf{r} = \mathbf{r}_1 - \mathbf{r}_2$. For a given $\mathbf{K}$, the center-of-mass wave function of the pair remains unaffected by the interactions and is simply $(1/\sqrt{\Omega}) \exp(i\mathbf{K} \cdot \mathbf{R})$. For the relative wave function, we can attempt to take a linear combination of plane-wave states to minimize the energy. Thus we write the pair wave function as

$$\psi_K(\mathbf{R}, \mathbf{r}) = \frac{1}{\sqrt{\Omega}} \exp(i\mathbf{K} \cdot \mathbf{R}) \sum_k c_k \frac{1}{\sqrt{\Omega}} \exp(i\mathbf{k} \cdot \mathbf{r}). \qquad (8\text{--}86)$$

Substituting this in the two-nucleon Schrödinger equation

$$\left[-\frac{\hbar^2}{2M} (\nabla_1^2 + \nabla_2^2) + V_{12} \right] \Psi_K = E \Psi_K,$$

we obtain

$$(e_{k_1} + e_{k_2} - E)c_k + \sum_{k'} c_{k'} \langle \mathbf{k} | V_{12} | \mathbf{k}' \rangle = 0, \qquad (8\text{--}87)$$

where $e_{k_i} = (\hbar^2/2M)k_i^2$; $\mathbf{k}_i = (\mathbf{K}/2) \pm \mathbf{k}$, with $\mathbf{k}$ a variable; and

$$\langle \mathbf{k}|V_{12}|\mathbf{k'}\rangle = \frac{1}{\Omega}\int d^3r \exp[i(\mathbf{k'}-\mathbf{k})\cdot\mathbf{r}]\,V_{12}(r).$$

In Eq. 8–87, it is understood that only pairs with fixed center-of-mass momentum $\mathbf{K}$ are being considered. Note from Eq. 8–87 that, in the absence of V_{12}, $E = e_{k_1} + e_{k_2}$, whose lowest limit is simply $2e_F$, e_F being the Fermi energy.

In order to solve Eq. 8–87 analytically, we now make some simplifying assumptions about the matrix elements $\langle \mathbf{k}|V_{12}|\mathbf{k'}\rangle$. We assume that

$$\langle \mathbf{k}|V_{12}|\mathbf{k'}\rangle = -G, \quad k_F < k_i, k_i' \leqslant k_F + \delta k_F,$$

$$= 0, \qquad \text{otherwise,}$$

where G is a positive constant.[†] We shall discuss the rationale of this approximation later. With this V_{12}, it follows from Eq. 8–87 that, when $\mathbf{k}$ is such that $\mathbf{k}_1$ and $\mathbf{k}_2$ lie in the region where G is nonzero,

$$c_{\mathbf{k}} = \frac{CG}{\varepsilon_K + e_k - E},$$

where $C = \sum_{\mathbf{k'}} c_{\mathbf{k'}}$, the sum in $\mathbf{k'}$ is over the allowed region where V_{12} is finite, and

$$\varepsilon_K + e_k = \frac{\hbar^2}{M}\left(\frac{K^2}{4} + k^2\right) = \frac{\hbar^2}{2M}(k_1{}^2 + k_2{}^2).$$

A little algebraic manipulation now yields the eigenvalue equation

$$\frac{1}{G} = \sum_{\mathbf{k}} \frac{1}{\varepsilon_K + e_k - E}. \tag{8-88}$$

The graphical solutions of this equation are easily found from Fig. 8–15, where the right-hand side of Eq. 8–88 is plotted as a function of E. The unperturbed energies of the pair, for a given center-of-mass energy ε_K, are shown by the dotted vertical lines. The actual eigensolutions are shown by crosses where the right-hand side equals $1/G$. It will be seen from this figure that only the lowest eigenvalue is shifted appreciably downward in energy from its unperturbed value of $2e_F$, the others remaining almost unaffected. In the case that we are considering, the unperturbed spectrum actually forms a continuum; nevertheless our analysis shows that the lowest-energy state of the pair will

---

[†] Not, of course, the G-Matrix.

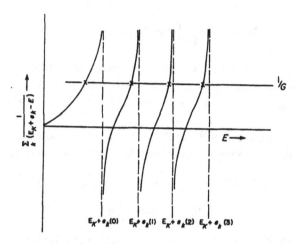

Fig. 8-15 The graphical solutions of Eq. 8–88 are denoted by crosses, showing that the lowest eigenvalue, designated by the cross at the extreme left, is lowered appreciably from the unperturbed position, whereas the others are little affected.

be shifted downward from the other states, giving rise to an energy gap in the spectrum.

The binding energy of the ground state of the pair may be found analytically by rewriting Eq. 8–88 as

$$\frac{1}{G} = \int_{e(0)}^{e(\text{max})} \frac{N(K, e)\, de}{\varepsilon_K + e - E},$$

where we have replaced the sum over $\mathbf{k}$ by an integration, $N(K, e)$ being the unperturbed density of two-nucleon states of total momentum K and energy $e = (\hbar^2/M)k^2$. The energy limits $e(0)$ and $e(\text{max})$ arise from the restriction $k_F < k_1, k_2 \leqslant k_F + \delta k_F$. Replacing $N(K, e)$ over this narrow range of e by the density of states at the Fermi surface, $N[K, e(0)]$, we can take N out of the integral sign and perform the integration analytically. A little algebra shows that the lowest eigenvalue E_0 of the pair is given by

$$E_0 = e(0) + \varepsilon_K - E_B, \tag{8–89a}$$

where E_B is the binding energy of the pair:

$$E_B = \frac{e(\text{max}) - e(0)}{\exp\left\{\dfrac{1}{N[K, e(0)] \cdot G}\right\} - 1}. \tag{8–89b}$$

Note that E_B is strongly dependent on the product $N[K, e(0)] \cdot G$. The quantity $N(K, e)$ in turn is strongly dependent on K, being maximum for $K = 0$ and going rapidly to zero. Thus Eq. 8–89b shows that E_B will be largest for $K = 0$, that is, for two nucleons of momentum, spin $(k_i, \uparrow)$ and $(- k_i, \downarrow)$. It is also possible to show that the bound-state pair wave function has considerable spatial extent [41].

In this idealized model, some important physical points have emerged. The inert Fermi sea plays a crucial role in providing a large number of nearly degenerate states at the Fermi surface, without which the bound pair cannot be formed. Essentially, the wave function of the pair is formed by a coherent superposition of these nearly degenerate states, at practically no cost in kinetic energy, to take maximum advantage of the *attractive* two-nucleon potential. It is essential to keep in mind that the interaction must be attractive for two nucleons colliding head on ($\mathbf{K} = 0$) at the Fermi surface to form the bound pair. For nuclear matter this corresponds to a single nucleon of energy $(\hbar^2/2M)(2k_F)^2 \approx 160$ MeV colliding with a nucleon at rest. As we know, the two-nucleon interaction in the S-state becomes progressively less attractive at higher energies, turning repulsive at about $E_{\text{lab}} \approx 200$ MeV. Thus, at nuclear-matter density, the S-state interaction at the Fermi surface is weak but still attractive, giving rise to a small gap. For finite nuclei, the local Fermi momentum is much smaller in the surface region, and consequently a larger gap is formed [42].

Although the idealized model of the Cooper pair is very instructive, it is not very realistic, since in the many-body problem we expect a large number of such pairs to be present, with strong spatial overlap between the wave functions of the different pairs and numerous pair-pair collisions [43]. The basic many-body problem here is to construct a superconducting ground state $|\Psi_0\rangle$, formed as a coherent superposition of normal (quasi-gas) configurations $|\Phi_n\rangle$,

$$|\Psi_0\rangle = \sum_n c_n |\Phi_n\rangle, \tag{8-90}$$

so that the energy will be as low as possible and lower than in the normal ground state. In terms of the Hamiltonian H (which we will presently specify), the energy of the superconducting ground state is

$$E_0 = \langle \Psi_0 | H | \Psi_0 \rangle = \sum_{n,n'} c_n c_{n'}^* \langle \Phi_{n'} | H | \Phi_n \rangle. \tag{8-91}$$

A straightforward diagonalization procedure would be very complicated and not too instructive. In the following discussion, we shall isolate, for a simplified interaction, the configurations $|\Phi_n\rangle$ whose superposition may lead to a value of E_0 lower than that of the lowest lying "shell-model" state $|\Phi_n\rangle$.

Obviously, if H is such that all its off-diagonal matrix elements are negative, then, with the proper choice of c_n's, all the terms in the sum of Eq. 8-90 can be in phase, yielding a low E_0. Generally, however, the matrix elements $\langle\Phi_{n'}|H|\Phi_n\rangle$ are expected to vary randomly in sign with random variations of normal-state configurations. This random behavior cannot be corrected by adjusting the signs of the coefficients c_n (say N in number), since there are N^2 matrix elements with random signs. Note that the normal configurations $|\Phi_n\rangle$ are determinantal states, although the superconducting state $|\Psi_0\rangle$ is not. If H contains only one- and two-body operators, $\langle\Phi_{n'}|H|\Phi_n\rangle$ is just a sum of one- and two-body matrix elements. Let us restrict the sum in Eq. 8-90 by choosing nonzero c_n's for only those normal-state configurations $|\Phi_n\rangle$ in which the single-particle orbitals are filled pairwise; that is, if an orbital $(\mathbf{k}_i, \uparrow)$ is occupied, its time-reversed state $(-\mathbf{k}_i, \downarrow)$ is also occupied. The total momentum for each of these $|\Phi_n\rangle$'s is thus zero.

The two-body matrix elements arising from $\langle\Phi_{n'}|H|\Phi_n\rangle$ will be of the form $\langle\mathbf{k}_i(1), -\mathbf{k}_i(2)|V_{12}|\mathbf{k}_i'(1), -\mathbf{k}_i'(2)\rangle$, where we have suppressed the spin indices. If the interaction is local and of very short range, all these matrix elements will have the same sign. To see this, observe that we can write, using the notation developed earlier in this section,

$$\langle\mathbf{k}_i(1), -\mathbf{k}_i(2)|V_{12}|\mathbf{k}_i'(1), -\mathbf{k}_i'(2)\rangle = \frac{1}{\Omega}\int \exp(i\mathbf{q}\cdot\mathbf{r})\, V_{12}(\mathbf{r})\, d^3r,$$

where $q = \mathbf{k}_i' - \mathbf{k}_i$ and $\mathbf{r} = \mathbf{r}_1 - \mathbf{r}_2$. If we take for $V_{12}(\mathbf{r})$ the extreme example of a delta function,

$$V_{12}(\mathbf{r}) = -G\Omega\, \delta(\mathbf{r}), \tag{8-92}$$

we obtain the simple result

$$\langle\mathbf{k}_i(1), -\mathbf{k}_i(2)|V_{12}|\mathbf{k}_i'(1), -\mathbf{k}_i'(2)\rangle = -G. \tag{8-93}$$

Such a simplified pairing potential, effective in a narrow band about the Fermi surface, should give rise to a superconducting solution with energy less than that of the normal ground state. It may be verified that for nuclei, where the single-particle wave functions are not plane waves, the corresponding pairing matrix elements (Eq. 8-93) are not quite equivalent to a delta-function potential, although strong similarities remain.

Before considering the details of the superconducting wave function, we inquire whether any part of the two-nucleon interaction may be effectively regarded as a delta-function potential of type 8–92. It is true that the free two-nucleon potential is of short range compared to the dimensions of a heavy nucleus, but we also know that the extreme-short-range part of the potential is repulsive rather than attractive. We must remember, however, that the pairing matrix elements 8–93 are taken in a narrow band of momenta near the Fermi surface, which, for nuclear-matter density, corresponds to $E_{lab} \approx 160$ MeV, as we have already discussed. At such energies, the *effective* two-nucleon interaction (i.e., the G-matrix) is still weakly attractive. What we want to see is whether a part of this effective two-nucleon interaction, operative on the nucleons near the Fermi surface, may have some characteristics of a delta-function type of potential. Assuming, for simplicity, the effective interaction V_{12} to be local, we may decompose it in the following convenient manner:

$$V_{12}(|\mathbf{r}_1 - \mathbf{r}_2|) = \sum_l \frac{2l+1}{4\pi} v_l(r_1, r_2) P_l(\cos \theta_{12}), \qquad (8\text{–}94\text{a})$$

where θ_{12} is the angle between the vectors $\mathbf{r}_1$ and $\mathbf{r}_2$. Using Eq. A–70, we may rewrite Eq. 8–94a in the form

$$V_{12}(|\mathbf{r}_1 - \mathbf{r}_2|) = \sum_{l,m} v_l(r_1, r_2) Y_l{}^m(\mathbf{r}_1) Y_l{}^{m*}(\mathbf{r}_2), \qquad (8\text{–}94\text{b})$$

where a little algebra will show that

$$v_l(r_1, r_2) = \frac{2}{\pi} \int V(q) j_l(qr_1) j_l(qr_2) q^2 \, dq,$$

with

$$V(q) = \int \exp(i\mathbf{q} \cdot \mathbf{r}) \, V_{12}(r) \, d^3r.$$

It should be realized, from Eq. 8–47, that the effective two-nucleon interaction gives rise to a self-consistent one-body average potential U, which, neglecting exchange, may be written in terms of the single-particle density ρ as

$$U(\mathbf{r}_1) = \int \rho(\mathbf{r}_2) V_{12}(|\mathbf{r}_1 - \mathbf{r}_2|) \, d^3r_2$$

$$= \sum_{l,m} Y_l{}^m(\mathbf{r}_1) \int v_l(r_1, r_2)\rho(\mathbf{r}_2) Y_l{}^{m*}(\mathbf{r}_2) \, d^3r_2$$

$$\equiv \sum_{l,m} Y_l{}^m(\mathbf{r}_1) U_{lm}(r_1). \qquad (8\text{–}95)$$

In phenomenological nuclear calculations, the one-body potential U is taken to consist of a spherically symmetric part U_0 and a quadrupole part U_2, which provides the deformed field (see Sections 9–6 and 9–8). The dipole $l = 1$ term can connect only states of opposite parity (belonging thus in different major shells), and as such is not important. Occasionally, even some higher multipole terms like $l = 4$ are included in the deformed field U (see Eq. 9–170). Even then, it is clear that the average deformed one-body potential U does not take any account of the higher multipoles of the effective interaction $V(|\mathbf{r}_1 - \mathbf{r}_2|)$ in Eq. 8–94, which are thus left as part of a residual interaction rich in high-multipole components. Now a delta function is also rich in high multipoles, as can be seen from the expansion

$$\delta(\mathbf{r}_{12}) = \sum_l \frac{2l+1}{4\pi r_{12}^2}\, \delta(r_1 - r_2) P_l(\cos\theta_{12}). \qquad (8\text{–}96)$$

Because of this common property, the residual interaction containing the higher multipoles may be thought of as a delta-function-like potential with an attractive strength, thus giving rise to pairing. It should be emphasized that the short-range repulsive part of the free two-body interaction has already been accounted for in the wound of the relative wave function and has nothing to do with pairing. It is the residual attractive effective interaction, which is rich in high multipoles and in this respect is analogous to a delta-function type of potential, which gives rise to pairing.

In a many-body problem with the schematic residual interaction 8–93 responsible for pairing, we may write the Hamiltonian as

$$H = \sum_\nu e_\nu a_\nu{}^+ a_\nu - G \sum_{\mu,\nu>0} a_\mu{}^+ a_{\bar{\mu}}{}^+ a_{\bar{\nu}} a_\nu \qquad (8\text{–}97)$$

in the second-quantized notation; here the e_ν's are the single-particle energies of the nucleons in an average potential U, and the two-body residual interaction term containing G acts in a narrow band of states around the Fermi surface. In nuclear matter, μ stands for the state $(\mathbf{k}_i, \uparrow)$ and $\bar{\mu}$ for its conjugate $(-\mathbf{k}_i, \downarrow)$. In calculations of practical concern, however, we deal with finite nuclei; in a spherical-potential well μ stands for the quantum numbers n, l, j, m, with $\mu > 0$ meaning[†] $m > 0$. For convenience, we suppress the quantum numbers n and l, and examine under what approximations a general Hamiltonian of the form (see Eq. C–29)

[†] For the quantum numbers in a deformed well, see Eq. 9–162; in this case $\mu > 0$ means $\Omega > 0$.

$$H = \sum_{j,m} e_j a_{jm}^+ a_{jm} + \tfrac{1}{2} \sum_{\substack{j_1,j_2,j_1',j_2' \\ m_1,m_2,m_1',m_2'}} \langle j_1 m_1, j_2 m_2 | V_{12} | j_1' m_1', j_2' m_2' \rangle a_{j_1 m_1}^+ a_{j_2 m_2}^+ a_{j_2' m_2'} a_{j_1' m_1'}$$

$$(8\text{-}98)$$

reduces to the schematic form 8-97. We can write, using Eq. A-33,

$$|j_1' m_1', j_2' m_2'\rangle = \sum_J (j_1' j_2' m_1' m_2' | JM) |(j_1' j_2') JM\rangle.$$

We saw that in nuclear matter pairing is important for nucleons coupled to $K = 0$; we now assume that in finite nuclei the pairing matrix elements are finite only for nucleons coupled to $J = 0$. The experimental evidence for this will be discussed presently. Thus, taking only the $J = 0$ term in the above sum, we obtain

$$|j_1' m_1', j_2' m_2'\rangle_{J=0} = \frac{(-)^{j'-m'}}{\sqrt{2j'+1}} \delta_{j_1' j_2'} \delta_{m_1', -m_2'} |(j'j')00\rangle.$$

Making a similar decomposition for $\langle j_1 m_1, j_2 m_2 |$, we may write for Eq. 8-98

$$H = \sum_j e_j a_{jm}^+ a_{jm} + \tfrac{1}{2} \sum_{\substack{jm, \\ j'm'}} \frac{\langle (jj)00 | V_{12} | (j'j')00 \rangle}{\sqrt{(2j+1)(2j'+1)}}$$

$$\cdot\, a_{jm}^+ (-)^{j-m} a_{j,-m}^+ (-)^{j'-m'} a_{j',-m'} a_{j'm'}.$$

If we now put

$$-G = \frac{\langle (jj)00 | V_{12} | (j'j')00 \rangle}{\sqrt{(j+\tfrac{1}{2})(j'+\tfrac{1}{2})}}, \qquad\qquad (8\text{-}99)$$

$$a_\mu^+ \equiv a_{jm}^+, \qquad a_{\bar\mu}^+ \equiv (-)^{j-m} a_{j,-m}^+,$$

$$a_\nu \equiv a_{j'm'}, \qquad a_{\bar\nu} \equiv (-)^{j'-m'} a_{j',-m'},$$

we obtain Eq. 8-97, where $\mu, \nu > 0$ indicates that the sum over m and m' are taken over positive values only. It is clear from Eq. 8-99 that actually G should depend on the orbitals j, j'; for simplicity in calculation, however, an average constant value of G over a band of states near the Fermi surface is chosen, so that it may be taken outside the summation sign. Note that the Hamiltonian 8-97 is a schematic representation of a residual interaction which is rich in high multipoles and is rather weak; it therefore can scatter nucleon pairs only in a narrow band around the unperturbed Fermi sea. Thus the sum over the states μ and ν in Eq. 8-97 includes only a limited number of states over the Fermi sea for which G is nonzero. In typical calculations, the value of G is adjusted to restrict the sum over an energy band of about 1.5 $\hbar\omega$ on either side of the Fermi energy, where $\hbar\omega$ is the shell spacing.

We shall now try to construct a trial wave function for the superconducting ground state $|\Psi_0\rangle$ of the schematic Hamiltonian 8–97. Note that the residual interaction is such as to cause scattering of pairs $(\nu, \bar{\nu})$ coupled to $K = 0$ (or $J = 0$ in finite geometry) to other conjugate states $(\mu, \bar{\mu})$. Consequently, we should not use a trial wave function as one in which each pair state is definitely occupied or definitely empty, since the pairs could not scatter and lower the energy in that case. Rather, we should introduce an amplitude, say V_μ, that the orbitals $(\mu, \bar{\mu})$ are occupied in Ψ_0, and consequently an amplitude $U_\mu = (1 - V_\mu{}^2)^{1/2}$ that the orbitals $(\mu, \bar{\mu})$ are empty. Since a large number of scatterings in and out of a given pair state $(\mu, \bar{\mu})$ are taking place, the amplitudes V_μ and U_μ describe the average rather than the instantaneous behavior.

Consider the state $(U_\mu + V_\mu a_\mu{}^+ a_{\bar{\mu}}{}^+)|0\rangle$, where $|0\rangle$ denotes the vacuum containing no nucleons. This describes the situation in which the pair state $(\mu, \bar{\mu})$ is occupied with probability $V_\mu{}^2$, and is empty with probability $U_\mu{}^2$. Bardeen, Cooper, and Schrieffer [44] approximated the ground state $|\Psi_0\rangle$ by a trial wave function $|\text{BCS}\rangle$, in which they assumed that the occupancies of the different pair states are uncorrelated, that is,

$$|\text{BCS}\rangle = \prod_{\mu>0} (U_\mu + V_\mu a_\mu{}^+ a_{\bar{\mu}}{}^+)|0\rangle. \qquad (8\text{--}100)$$

In order to find the lowest-energy state with this form of wave function, we must take U_μ and V_μ to minimize $\langle \text{BCS}|H|\text{BCS}\rangle$. But note that the wave function $|\text{BCS}\rangle$ does not describe a state with a fixed number of nucleons; rather it is an admixture of states with different numbers of nucleons, a price that must be paid for its simplicity. We should therefore find the energy minimum subject to the condition that the mean particle number has the desired value N, that is, $\langle \text{BCS}|\hat{N}|\text{BCS}\rangle = N$, where $\hat{N} = \sum_\nu a_\nu{}^+ a_\nu$. This means that we must introduce a Lagrange multiplier λ and adjust the coefficients U_μ and V_μ in the trial wave function (Eq. 8–100) to minimize the expectation value of $\langle \text{BCS}|H'|\text{BCS}\rangle$, where

$$H' = \sum_\nu (e_\nu - \lambda) a_\nu{}^+ a_\nu - G \sum_{\mu,\nu>0} a_\mu{}^+ a_{\bar{\mu}}{}^+ a_{\bar{\nu}} a_\nu. \qquad (8\text{--}101)$$

In order to evaluate this expectation value, it is most convenient to introduce the so-called Bogolyubov-Valatin transformation [45, 46]:

$$\alpha_\mu{}^+ = U_\mu a_\mu{}^+ - V_\mu a_{\bar{\mu}}, \qquad \alpha_\mu = U_\mu a_\mu - V_\mu a_{\bar{\mu}}{}^+. \qquad (8\text{--}102a)$$

These are unitary transformations, so that the anticommutation relations between the new operators α_μ, α_ν^+ are the same as those of fermions (not that $U_\mu^2 + V_\mu^2 = 1$),

$$\{\alpha_\mu, \alpha_\nu^+\} = \delta_{\mu\nu}, \qquad \{\alpha_\mu, \alpha_\nu\} = \{\alpha_\mu^+, \alpha_\nu^+\} = 0.$$

Note that it follows from Eq. 8–102a that

$$\alpha_{\bar\mu}^+ = U_\mu a_{\bar\mu}^+ + V_\mu a_\mu, \qquad \alpha_{\bar\mu}^- = U_\mu a_{\bar\mu}^- + V_\mu a_\mu^+, \qquad (8\text{–}102b)$$

as may be easily verified in the spherical representation, where $\alpha_\mu \equiv \alpha_{jm}$, $\alpha_{\bar\mu} \equiv (-)^{j-m}\alpha_{j,-m}$. It is much simpler to work with the transformed creation and annihilation operators α_μ^+, α_μ (called quasiparticle operators), since the trial wave function $|\text{BCS}\rangle$ is a quasiparticle vacuum, that is,

$$\alpha_\mu|\text{BCS}\rangle = 0 \quad \text{for all } \mu. \qquad (8\text{–}103)$$

It is clear in Appendix C (see Eq. C–33) that in the independent-particle model the sharp Fermi surface behaves like a vacuum to particle and hole operators; from Eq. 8–103 we note that this simplicity of the independent-particle model is regained by working with quasi-particle operators α_μ on the state $|\text{BCS}\rangle$, although this is not a determinantal state. Obviously, the wave function $\alpha_\mu^+|\text{BCS}\rangle$ represents a state with a single quasiparticle in orbital μ. Such a quasiparticle, quite literally, is a mixture of a particle state μ, with amplitude U_μ, and a hole state $\bar\mu$, with amplitude V_μ. It follows that, deep inside the Fermi sphere, a quasiparticle is simply a hole, while far above the Fermi surface it is a particle.

In order to work out explicitly the expectation value $\langle\text{BCS}|H'|\text{BCS}\rangle$, we need the inverse relations of Eq. 8–102a, which are

$$a_\mu = U_\mu\alpha_\mu + V_\mu\alpha_{\bar\mu}^+, \qquad a_{\bar\mu} = U_\mu\alpha_{\bar\mu} - V_\mu\alpha_\mu^+, \qquad (8\text{–}102c)$$

and the corresponding Hermitian relations for the creation operators a_μ^+, $a_{\bar\mu}^+$. Using Eqs. 8–102 and 8–103, we can now easily calculate the expectation value of the first term in H', that is,

$$\langle\text{BCS}|\sum_\nu (e_\nu - \lambda)a_\nu^+ a_\nu|\text{BCS}\rangle = \sum_{\nu>0}(e_\nu - \lambda)\langle\text{BCS}|a_\nu^+ a_\nu + a_{\bar\nu}^+ a_{\bar\nu}^+|\text{BCS}\rangle$$

$$= 2\sum_{\nu>0}(e_\nu - \lambda)V_\nu^2. \qquad (8\text{–}104)$$

In evaluating the expectation value of the two-body interaction term in H' with respect to $|\text{BCS}\rangle$, we can use Wick's theorem (Appendix C), which states that in such a case only the fully contracted terms may contribute to the expectation value. Thus

$$\sum_{\mu,\nu>0} \langle \text{BCS}|a_\mu{}^+ a_{\bar\mu}{}^+ a_{\bar\nu} a_\nu|\text{BCS}\rangle$$

$$= \sum_{\mu,\nu>0} \{\langle a_\mu{}^+ a_{\bar\mu}{}^+\rangle\langle a_{\bar\nu} a_\nu\rangle + \langle a_\mu{}^+ a_\nu\rangle\langle a_{\bar\mu}{}^+ a_{\bar\nu}\rangle - \langle a_\mu{}^+ a_{\bar\nu}\rangle\langle a_{\bar\mu}{}^+ a_\nu\rangle\}, \quad (8\text{--}105)$$

where the $\langle\ \rangle$ denote the expectation value with respect to the state $|\text{BCS}\rangle$. A little algebra, making use of Eq. 8–103 and the anticommutation relations, would show that the last term on the right-hand side of Eq. 8–105 vanishes, and that

$$\sum_{\mu,\nu>0} \langle \text{BCS}|a_\mu{}^+ a_{\bar\mu}{}^+ a_\nu a_\nu|\text{BCS}\rangle = (\sum_{\nu>0} U_\nu V_\nu)^2 + \sum_{\nu>0} V_\nu^4.$$

Combining this result with Eq. 8–104, we finally obtain

$$\langle \text{BCS}|H'|\text{BCS}\rangle = \sum_{\nu>0} [2(e_\nu - \lambda)V_\nu^2 - GV_\nu^4] - G(\sum_{\nu>0} U_\nu V_\nu)^2. \quad (8\text{--}106)$$

We now impose the condition

$$\frac{\partial}{\partial V_\nu} \langle \text{BCS}|H'|\text{BCS}\rangle = 0,$$

which immediately yields

$$4(e_\nu - \lambda)V_\nu - 4GV_\nu^3 - 2G(\sum_{\mu>0} U_\mu V_\mu)\left(U_\nu + V_\nu \frac{\partial U_\nu}{\partial V_\nu}\right) = 0.$$

Noting, from the normalization condition $U_\nu^2 + V_\nu^2 = 1$, that $V_\nu(\partial U_\nu/\partial V_\nu) = -V_\nu^2/U_\nu$, and simplifying, we obtain the equation

$$2(e_\nu' - \lambda)U_\nu V_\nu = \Delta(U_\nu^2 - V_\nu^2), \quad (8\text{--}107)$$

where

$$e_\nu' = e_\nu - GV_\nu^2, \qquad \Delta = G\sum_{\mu>0} U_\mu V_\mu. \quad (8\text{--}108)$$

Note that e_ν' is the renormalized single-particle energy in the orbital ν, which includes the contribution $-GV_\nu^2$ of the pairing interaction to the one-body field. By squaring Eq. 8–107 and making use of the relation $U_\nu^4 + V_\nu^4 = 1 - 2U_\nu^2 V_\nu^2$, we obtain

$$U_\nu^2 V_\nu^2 = \frac{\Delta^2}{4[(e_\nu' - \lambda)^2 + \Delta^2]} = \frac{1}{4}\left[1 - \frac{(e_\nu' - \lambda)^2}{(e_\nu' - \lambda)^2 + \Delta^2}\right]. \quad (8\text{--}109)$$

A little algebra will now yield the important relations

$$U_\nu^2 = \frac{1}{2}\left[1 + \frac{e_\nu' - \lambda}{\sqrt{(e_\nu' - \lambda)^2 + \Delta^2}}\right]$$

and (8–110)

$$V_\nu^2 = \frac{1}{2}\left[1 - \frac{e_\nu' - \lambda}{\sqrt{(e_\nu' - \lambda)^2 + \Delta^2}}\right].$$

Two more equations may be written down in a neat form to solve for the parameters Δ and λ, starting from the Hamiltonian 8–97. The first can be obtained by substituting for U_ν, V_ν from Eq. 8–109 into Eq. 8–108, giving

$$\frac{1}{G} = \frac{1}{2}\sum_{\nu>0}\frac{1}{\sqrt{(e_\nu' - \lambda)^2 + \Delta^2}}. \tag{8–111}$$

The other equation is simply a statement of the condition that the mean number of particles is N:

$$N = 2\sum_{\nu>0}V_\nu^2 = 2\sum_{\nu>0}\left[1 - \frac{e_\nu' - \lambda}{\sqrt{(e' - \lambda)^2 + \Delta^2}}\right]. \tag{8–112}$$

Note the similarity of Eq. 8–111 and the Cooper formula 8–88; as before, we can expect an energy gap in the excitation spectrum of the nucleus. For a given choice of G and the single-particle spectrum, the coupled equations 8–111 and 8–112 have to be solved for λ and Δ.

Note, from Eq. 8–106, that the ground-state energy is given by

$$\langle BCS|H|BCS\rangle = \sum_{\nu>0}2e_\nu V_\nu^2 - \frac{\Delta^2}{G} - G\sum_\nu V_\nu^4, \tag{8–113}$$

which reduces, for $G = 0$, to a sum over occupied single-particle energies, which is different from the self-consistent Hartree-Fock result (Eq. 10–21).

It is clear from Eq. 8–100 that the wave function $|BCS\rangle$ represents an admixture of states with an even number of particles of one kind; the constraint is therefore made that the mean number $N = 2\sum_{\nu>0}V_\nu^2$ is even. The $|BCS\rangle$ state may be regarded as an average of the wave functions of a few neighboring nuclides with particle numbers $N, N - 2, N + 2, \ldots$. However, for large N, the spread $\langle BCS|(\hat{N} - N)^2|BCS\rangle$ is small, and the contribution of the desired nuclide N is predominant. With this limitation, the state $|BCS\rangle$ represents the ground state of an even-even nucleus, and it is a quasi-particle vacuum.

To see the structure of excited states, it is useful to express the particle-number operator $\hat{N}$ in terms of quasiparticle operators by using Eq. 8–103. A little algebra yields

$$\hat{N} = \sum_{\nu>0} [2V_\nu^2 + \hat{\mathscr{N}}_\nu(U_\nu^2 - V_\nu^2) + 2U_\nu V_\nu(\alpha_\nu{}^+\alpha_{\bar\nu}{}^+ + \alpha_{\bar\nu}\alpha_\nu)], \quad (8\text{--}114)$$

where $\hat{\mathscr{N}}_\nu = (\alpha_\nu{}^+\alpha_\nu + \alpha_{\bar\nu}{}^+\alpha_{\bar\nu})$. Consider now the one quasiparticle state $\alpha_\mu{}^+|\text{BCS}\rangle$. Since $\langle\text{BCS}|\hat{N}|\text{BCS}\rangle = N$, it follows from Eq. 8–114 that

$$\langle\text{BCS}|\alpha_\mu\hat{N}\alpha_\mu{}^+|\text{BCS}\rangle = N + (U_\mu^2 - V_\mu^2) = N + 1 - 2V_\mu^2.$$

If the state μ is far above the Fermi sea, V_μ^2 is negligible and the expectation value is just $N + 1$. In this situation, the state $\alpha_\mu{}^+|\text{BCS}\rangle$ represents an odd nucleus. A state with two quasiparticle excitations, $\alpha_\mu{}^+\alpha_{\bar\mu}{}^+|\text{BCS}\rangle$, yields the expectation value of $\hat{N}$ as

$$N + 2(U_\mu^2 - V_\mu^2) = N + 2\left[\frac{e_\mu' - \lambda}{\sqrt{(e_\mu' - \lambda)^2 + \Delta^2}}\right].$$

Only if e_μ is close to λ is this N; at the other extreme, with $(e_\mu - \lambda) \gg \Delta$, it is $N + 2$. Consequently, the low-lying intrinsic states of even-even nuclei are double quasiparticle excitations. Triple quasiparticle excitations are rather high excited states of odd-A nuclei.

The energy of a quasiparticle can be calculated by evaluating the quantity $\langle\text{BCS}|(\alpha_\nu H\alpha_\nu{}^+) - H|\text{BCS}\rangle$, which turns out to be

$$E_\nu = \sqrt{(e_\nu' - \lambda)^2 + \Delta^2}. \quad (8\text{--}115)$$

As we have seen, for an even-even nucleus, the ground state is a quasiparticle vacuum, while the first excited state must have at least two quasiparticles present, say in orbitals ν and ν'. Thus the excitation energy of the first excited state in an even-even nucleus is $E_\nu + E_{\nu'} \geq 2\Delta$. This is so because in the BCS approximation the quasiparticles are noninteracting. There is thus an energy gap of at least 2Δ between the ground state and the first excited state. We should point out that we are here considering the first excited state of an intrinsic (as opposed to collective) nature; a coherent combination of these two-quasiparticle intrinsic states may, of course, give rise to a collective excitation (like the first 2^+ state in a near-spherical nucleus) at an energy lower than 2Δ. In a nucleus containing an odd number of particles, however, the ground state already contains a quasiparticle in an orbital ν (say), and an excited state may be obtained by changing this orbital to ν', so that the excitation energy is simply $E_\nu - E_{\nu'}$, with no gap in the energy spectrum.

In the absence of pairing, it is clear from Eq. 8–108 that Δ vanishes, with U_ν and V_ν becoming 0 and 1, respectively, if $e_\nu < \lambda$, and 1 and 0 if $e_\nu > \lambda$. In other words, in the absence of pairing, λ is the sharp Fermi surface of the independent-particle model. The presence of the pairing force, however, "smears" the Fermi level; particle states ν such that $|e_\nu - \lambda| \gg \varDelta$ are still almost completely full or completely empty, but states with $|e_\nu - \lambda| \sim \varDelta$ are only partially occupied. In this sense, λ is an average Fermi level. More precisely, since it is the Lagrange multiplier of the number operator, it is the chemical potential of the system.

From Eq. 8–111 it is clear that, if the pairing strength G is so small that

$$\frac{G}{2} \sum_{\nu>0} \frac{1}{|e_\nu' - \lambda|} < 1, \tag{8–116}$$

no real solution for $\varDelta$ can be found from the gap equation. One would suspect from the Cooper pair problem, however, that a gap should exist no matter how weak G is (as long as the pairing interaction is attractive), provided there are pairs in an unfilled shell. The accuracy of the BCS solution may be tested in an exactly solvable two-level model [47], in which the two levels, each of degeneracy η, are separated by an energy $\hbar\omega$. In the symmetric model, the number of particles is η, so that in the absence of interaction the lower level is completely filled. The chemical potential λ is halfway between the two levels, so that $|e_\nu' - \lambda| = \tfrac{1}{2}\hbar\omega$ for all ν. It is clear from Eq. 8–116, then, that no gap would be found in the BCS solution for $G \leqslant G_c$, where

$$\frac{G_c}{2} \cdot 2\eta \frac{2}{\hbar\omega} = 1, \quad \text{or} \quad G_c = \frac{\hbar\omega}{2\eta}. \tag{8–117}$$

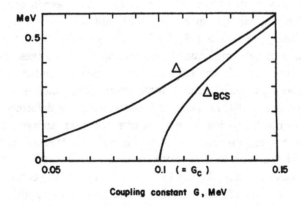

Coupling constant G, MeV

Fig. 8-16 Behavior of the BCS and the "exact" solutions for the gap parameter as a function of the coupling constant G in the symmetric two-level model. The degeneracy factor $\eta = 5$, and the level spacing $\hbar\omega = 1$ MeV here. (After Nogami [49].)

The calculated gap in the BCS approximation, Δ_{BCS}, for $\eta = 5$ and $\hbar\omega = 1$ MeV is shown in Fig. 8–16; it will be seen that this vanishes at $G_c = 0.1$ MeV, in agreement with Eq. 8–117. The inaccuracy in the BCS solution is due to the particle number fluctuation in the wave vector |BCS⟩. Ideally, one should project out from this state the component with N particles and then solve the variational problem with respect to $V_\nu{}^2$, a very complicated procedure [48]. Nogami and his coworkers [49] have devised a simpler technique of suppressing the number fluctuations in |BCS⟩, and the Δ calculated by their method (which is very close to the exact value) is also shown in Fig. 8–16. It will be seen that Δ approaches zero smoothly as $G \rightarrow 0$, in contrast to Δ_{BCS}.

We should emphasize that our discussion in this section has been based on the physical picture of pairs of nucleons, coupled to $J = 0$, being scattered by the pairing interaction, resulting in a depression of the $J = 0$ ground state from all the other excited states. This is true for pairs interacting in the i-spin $T = 1$ state; for example, the two valence neutrons in the $d_{5/2}$-shell in ^{18}O form a 0^+, $T = 1$ ground state (see Chapter 10). However, for a n-p pair, there is a strong attractive interaction in the $T = 0$ state too. The situation is complicated by the fact that two nucleons in a j-shell with $T = 0$ may form only angular-momentum states $1^+, 3^+, 5^+, \ldots$ (Problem 10–1), and several of these states may occur in energy near the $J = 0$, $T = 1$ paired state. The formalism developed in this section is thus applicable directly to pairing between n-n and p-p pairs, but not n-p pairs. This is a serious limitation of the method. Various attempts have been made to incorporate n-p pairing in a more general formalism, and Lane [50] and Rowe [51] review these methods. Despite this limitation of the BCS method, it has been highly successful, particularly in the calculation of the wave functions of low-lying states of nuclei with one closed shell [52], and in the calculation of the potential-energy surfaces of deformed nuclei (see Section 9–6). In these calculations, the residual interaction in the Hamiltonian is taken to consist of the pairing interaction and a quadrupole-quadrupole interaction, which is the $l = 2$ part of V_{12} in expansion 8–94, responsible for producing the deformed field.

Finally, we must mention some of the directly observable effects of pairing on nuclear properties, on the basis of which Bohr, Mottelson, and Pines [53] first suggested that the BCS theory of metallic superconductivity could be applied to nuclear physics. Figure 8–17 shows a plot of the energy of the first excited *intrinsic* state of stable nuclei against the mass number A. In this plot, we have not included the low-lying collective excited states of rotational or vibrational nature, and that is why the term intrinsic is used. It will be seen that there is an energy gap of at least 900 keV in the intrinsic excitation

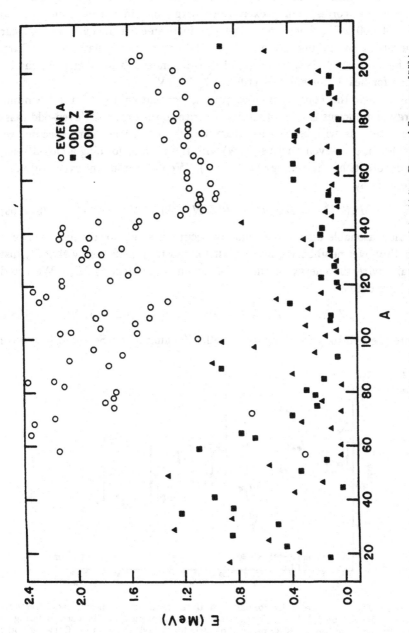

Fig. 8-17 The energy of the first excited *intrinsic* state of stable nuclei. (After Bes and Sorensen [55].)

of even-even nuclei, and the gap is in fact much larger for lighter nuclei. In contrast, the corresponding excitation energy in odd nuclei is only of the order of 100 keV. Such an energy gap in even-even nuclei would appear naturally if the ground state is superconducting, as we have seen. From Fig. 8–17 we conclude that the gap Δ is A-dependent, being larger for smaller A and for heavy nuclei ($A > 140$), $2\Delta \gtrsim 1$ MeV.

Although the contribution of pairing correlation to the total binding energy of a nucleus is very small, it gives rise to the observed "even-odd mass difference," as will be recalled from Eq. 6–21. To be more specific, we consider an even-even nucleus (Z, N) with the protons forming a closed shell, and denote its binding energy by $B(Z, N)$. We may define the even-odd mass difference by

$$P_n(Z, N) = B(Z, N) + B(Z, N - 2) - 2B(Z, N - 1). \quad (8\text{–}118a)$$

On the reasonable assumption that the occupancies $V_\nu{}^2$ for the neutrons vary smoothly over neighboring nuclides, it follows from BCS theory that P_n just equals twice the energy of the odd-neutron quasiparticle, $2E_\nu$. We could similarly define

$$P_p(Z, N) = B(Z, N) + B(Z - 2, N) - 2B(Z - 1, N), \quad (8\text{–}118b)$$

where the neutrons form a closed shell. Kisslinger and Sorensen [52] have

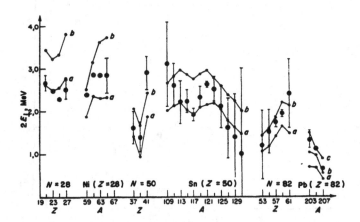

Fig. 8-18 The heavy dots with error bars show the experimental even-odd mass differences P_n or P_p defined by Eq. 8–118. The little dots are theoretical points, corresponding to twice the energy of the lowest-lying quasiparticle for the odd-A isotopes. Curves a and b correspond to $G = 19/A$ and $23/A$, respectively, while curve c, for lead only, corresponds to $G = 30/A$. (After Kisslinger and Sorensen [52].)

compared the experimental quantities P_n and P_p with the corresponding predictions of the BCS calculation for various choices of G, and this comparison is shown in Fig. 8–18.

The partial filling of states near the Fermi level is one of the most important consequences of the pairing force. It is similar in many ways to interconfiguration mixing, and will have the same type of effects on the energy levels in odd nuclides and also on electromagnetic moments, transition rates, and nuclear reactions. Here we shall mention an experiment which directly measures U_ν^2 and V_ν^2. The (d, p) and (d, t) reactions are, respectively, stripping and pick-up reactions in which a deuteron loses and picks up a neutron when it strikes a heavier nucleus. Consider, for example, the deuteron stripping process with an even-even target of spin parity 0^+. The conservation of parity and of angular momentum requires a unique value, l, for the orbital angular momentum of the stripped neutron for a transition to a final state of angular momentum and parity j, π. The proton angular distribution fixes the l-value and thus helps to determine the j, π of the final state. Also, the magnitude of the cross section determines the so-called spectroscopic factor, which is the

TABLE 8-9 Experimental and Theoretical Spectroscopic Factors Observed in (d, p) Reactions Leading to the 0^+ Ground State of Near-Spherical Nuclei. After Sorensen et al. [54].

Target	j	l	$S_{\exp}$	S_{theor}
^{53}Cr	$\frac{3}{2}$	1	0.91	0.97
^{57}Fe	$\frac{1}{2}$	1	0.072 ± 0.01	0.067
^{61}Ni	$\frac{3}{2}$	1	2.0 ± 0.3	1.7
^{67}Zn	$\frac{5}{2}$	3	2.2 ± 0.3	2.9
^{77}Se	$\frac{1}{2}$	1	0.68 ± 0.01	0.83
^{91}Zr	$\frac{5}{2}$	2	1.44 ± 0.21	1.6
^{95}Mo	$\frac{5}{2}$	2	2.48 ± 0.37	2.7
^{99,101}Ru	$\frac{5}{2}$	2	2.74 ± 0.4	3.3
^{105}Pd	$\frac{5}{2}$	2	1.74 ± 0.25	2.8
^{115}Sn	$\frac{1}{2}$	0	1.08 ± 0.16	1.03
^{117}Sn	$\frac{1}{2}$	0	1.4 ± 0.2	1.2
^{119}Sn	$\frac{1}{2}$	0	1.3 ± 0.2	1.4
^{125}Te	$\frac{1}{2}$	0	1.2 ± 0.3	0.99

fraction of the "even 0^+ state plus neutron" in the final-state wave function. In this example, the cross section should be proportional to the probability of an even nucleus picking up a neutron in state j, m, which is determined by the probability that this state is vacant, that is, by $U_j{}^2$. As another example, if the (d, p) reaction from the j-state of an odd target (Z even, N odd) is studied, for example, $^{53}\mathrm{Cr}(d, p)^{54}\mathrm{Cr}$, the cross section is proportional to $V_j{}^2$.

In Table 8-9, we compare the theoretical and experimental occupation factors $S = (2j + 1)V_j{}^2$ observed in deuteron stripping leading to the 0^+ ground state of near-spherical nuclei [54]. The theoretical numbers are determined from a BCS calculation, using the pairing plus quadrupole model of Kisslinger and Sorensen [52], and a small correction in S is included because of the small admixture of collective components in the odd-A ground state. More examples of nuclear reactions (including two-nucleon transfer) in which pairing is important are given by Bes and Sorensen [55].

Finally, we shall see in Section 10–8 that in microscopic calculations of the collective inertial parameters of a deformed nucleus the pairing interaction plays a very important role.

REFERENCES

1. H. A. Bethe, *Ann. Rev. Nucl. Sci.* **21**, 93 (1971).
2. D. W. L. Sprung, *Advan. Nucl. Phys.* **5**, 225 (1972).

References 1 and 2 are comprehensive reviews of the theory both of nuclear matter and of finite nuclei.

3. B. H. Brandow, unpublished thesis, Cornell University, 1964.
4. L. C. Gomes, J. D. Walęcka, and V. F. Weisskopf, *Ann. Phys. (N.Y.)* **3**, 241 (1958): An *anschaulich* presentation of the basic principles of nuclear-matter calculations.
5. H. A. Bethe and J. Goldstone, *Proc. Roy. Soc. (London)* **A238**, 551 (1957); J. Goldstone, *Proc. Roy. Soc. (London)* **A239**, 267 (1957).
6. B. D. Day, *Rev. Mod. Phys.* **39**, 719 (1967).
7. H. A. Bethe, *Phys. Rev.* **B138**, 804 (1965); L. D. Faddeev, *Zh. Eksp. Teor. Fiz.* **39**, 1459 (1960), and *Dokl. Acad. Nauk S.S.S.R.* **138**, 565 (1961): English translation, *Sov. Phys. JETP* **12**, 1014 (1961), and *Dokl.* **6**, 384 (1962); R. Rajaraman and H. A. Bethe, *Rev. Mod. Phys.* **39**, 745 (1967).
8. H. A. Bethe, B. H. Brandow, and A. G. Petschek, *Phys. Rev.* **129**, 225 (1963).
9. T. K. Dahlblom, *Acta Acad. Aboensis* **B29**, 6 (1969); T. K. Dahlblom, K. G. Fogel, B. Qvist, and A. Törn, *Nucl. Phys.* **56**, 177 (1964); B. D. Day, *Phys. Rev.* **187**, 1269 (1969).
10. B. D. Day, *Phys. Rev.* **151**, 826 (1966).
11. K. A. Brueckner and C. A. Levinson, *Phys. Rev.* **97**, 1344 (1955); K. A. Brueckner, *Phys. Rev.* **97**, 1353 (1955), and **100**, 36 (1955); K. A. Brueckner and J. L. Gammel, *Phys. Rev.* **109**, 1023 (1958).

12. B. H. Brandow, *Phys. Rev. Lett.* **22**, 173 (1969); *Ann. Phys. (N.Y.)* **57**, 214 (1970).
13. R. L. Becker, *Phys. Rev. Lett.* **24**, 400 (1970).
14. M. I. Haftel and F. Tabakin, *Nucl. Phys.* **A158**, 1 (1970).
15. M. A. Preston and R. K. Bhaduri, *Phys. Lett.* **6**, 193 (1963); *Can. J. Phys.* **42**, 696 (1964).
16. P. J. Siemens, *Nucl. Phys.* **A141**, 225 (1970).
17. P. Grangé and M. A. Preston, *Nucl. Phys.* **A204**, 1 (1973).
18. P. K. Banerjee and D. W. L. Sprung, *Can. J. Phys.* **49**, 1899 (1971).
19. P. Grangé and M. A. Preston, *Phys. Lett.* **42B**, 35 (1972).
20. F. Calogero and Yu. A. Simonov, *Nuovo Cimento* **B64**, 337 (1969); *Phys. Rev. Lett.* **25**, 881 (1970); *Lett. Nuovo Cimento* **4**, 219 (1970).
21. C. Michael and C. Wilkin, *Nucl. Phys.* **B11**, 99 (1969).
22. I. R. Afnan, D. M. Clement, and F. J. D. Serduke, *Nucl. Phys.* **A170**, 625 (1971).
23. A. M. Green, T. K. Dahlblom, and T. Kouki, *Nucl. Phys.* **A209**, 52 (1973).
24. B. H. J. McKellar and R. Rajaraman, *Phys. Rev.* **C3**, 1877 (1971).
25. B. A. Loiseau, Y. Nogami, and C. K. Ross, *Nucl. Phys.* **A165**, 601 (1971).
26. G. E. Brown, A. D. Jackson, and T. T. S. Kuo, *Nucl. Phys.* **A133**, 481 (1969).
27. J. W. Negele and D. Vautherin, *Nucl. Phys.* **A207**, 321 (1973).
28. J. Nemeth and D. W. L. Sprung, *Phys. Rev.* **176**, 1496 (1968).
29. J. R. Buchler and L. Ingber, *Nucl. Phys.* **A170**, 1 (1971).
30. S. L. Schlenker and E. L. Lomon, *Phys. Rev.* **C3**, 1487 (1971).
31. V. R. Pandharipande, *Nucl. Phys.* **A174**, 641 (1971).
32. M. Miller, C. W. Woo, J. W. Clark, and W. J. TerLouw, *Nucl. Phys.* **A184**, 1 (1972).
33. P. Grangé and M. A. Preston, *Nucl. Phys.* **A219**, 266 (1974).
34. K. T. R. Davies and R. J. McCarthy, *Phys. Rev.* **C4**, 81 (1971).
35. K. T. R. Davies, R. J. McCarthy, and P. U. Sauer, *Phys. Rev.* **C6**, 1461 (1972).
36. K. A. Brueckner, J. L. Gammel, and H. Weitzner, *Phys. Rev.* **110**, 431 (1958).
37. I. J. Donnelly, *Nucl. Phys.* **A111**, 201 (1968); R. K. Bhaduri and C. S. Warke, *Phys. Rev. Lett.* **24**, 1379 (1968); J. W. Negele, *Phys. Rev.* **C1**, 1260 (1970); P. Siemens, *Nucl. Phys.* **A141**, 225 (1970).
38. D. W. L. Sprung and P. K. Banerjee, *Nucl. Phys.* **A168**, 273 (1971); P. K. Banerjee and D. W. L. Sprung, *Can. J. Phys.* **49**, 1899 (1971); D. W. L. Sprung, *Nucl. Phys.* **A182**, 97 (1972).
39. X. Campi and D. W. L. Sprung, *Nucl. Phys.* **A194**, 401 (1972); J. W. Negele, ref. 37.
40. B. H. Brandow, *Rev. Mod. Phys.* **39**, 771 (1967).
41. L. N. Cooper, *Phys. Rev.* **104**, 1189 (1956).
42. E. Jakeman and S. A. Moszkowski, *Phys. Rev.* **141**, 933 (1966); S. Nagata and H. Bando, *Phys. Lett.* **11**, 153 (1964).
43. J. R. Schrieffer, 1972 Nobel lecture, reprinted in *Phys. Today*, **26**, 23 (1973).
44. J. Bardeen, L. N. Cooper, and J. R. Schrieffer, *Phys. Rev.* **108**, 1175 (1957).
45. N. N. Bogolyubov, *Nuovo Cimento* **7** (ser. 10), 794 (1958).
46. J. G. Valatin, *Nuovo Cimento* **7** (ser. 10), 843 (1958).
47. M. Rho and J. O. Rasmussen, *Phys. Rev.* **135**, B1295 (1964).
48. A. K. Kerman, R. D. Lawson, and M. H. Macfarlane, *Phys. Rev.* **124**, 162 (1961).
49. Y. Nogami, *Phys. Lett.* **15**, 335 (1965); H. C. Pradhan, Y. Nogami, and J. Law, *Nucl. Phys.* **A201**, 357 (1973).

50. A. M. Lane, *Nuclear Theory*, Benjamin, New York, 1964, p. 67.
51. D. J. Rowe, *Nuclear Collective Motion*, Methuen, London, 1970, p. 202.
52. L. S. Kisslinger and R. A. Sorensen, *K. Danske Vidensk. Selsk. Mat.-Fys. Medd.* **32**, No. 9 (1960); *Rev. Mod. Phys.* **35**, 853 (1963).
53. A. Bohr, B. R. Mottelson, and D. Pines, *Phys. Rev.* **110**, 936 (1958).
54. R. A. Sorensen, E. D. Lin, and B. L. Cohen, *Phys. Rev.* **142**, 729 (1966).
55. D. R. Bes and R. A. Sorensen, *Advan. Nucl. Phys.* **2**, 129 (1969).

PROBLEMS

8–1. In the text, in considering exclusion-principle correlations in ^{16}O, we defined a function F_2 as the ratio of the actual joint probability $P_F(1, 2)$ to the classical joint probability $P_{cl} = \rho(1)\rho(2)$. However, it would be useful to define the ratio of $P_F(1, 2)$ to the classical probability of finding both fermions at the same radius as their mass center, that is, $[\rho(\frac{1}{2}\mathbf{R})]^2$. Show that this quantity is

$$F_3(\mathbf{r}, \mathbf{R}) = \left(1 - \frac{\alpha^2 r^2}{2 + \alpha^2 R^2}\right)^2 \exp(-\tfrac{1}{2}\alpha^2 r^2).$$

8–2. There is a formula connecting the spherical Bessel function j_l with the harmonic-oscillator radial function R_{nl}. It is given in Problem 8–11b.

Consider like particles in ^{16}O, separated by a distance r with mass center at the position of maximum density in the nucleus and with (i) $\mathbf{r}$ perpendicular to $\mathbf{R}$ and (ii) $\mathbf{r}$ at 45° to $\mathbf{R}$. Obtain the exclusion-principle correlations and consider how best to represent them by a careful choice of k_F in a function $F_1(k_F r)$, coming from nuclear matter.

8–3. Considering the derivation of Eq. 8–37, show that with

$$\Psi = \binom{A}{2}^{-1} \sum_{kl} \Psi_{kl}$$

it follows that

$$\langle \Psi | \Psi \rangle = \binom{A}{2}^{-1} \sum_{kl} \langle \psi_{kl} | \psi_{kl} \rangle.$$

Show further that in nuclear matter $\langle \psi_{kl} | \psi_{kl} \rangle$ is independent of k and l, and hence

$$\langle \Psi | \Psi \rangle = \langle \psi_{kl} | \psi_{kl} \rangle.$$

Note also that $\langle \Psi | \Phi_0 \rangle = 1$.
Establish that

$$\langle \Phi_0 | \sum v(i, j) | \Psi \rangle = (\Phi_0 | v(1, 2) | \sum \Psi_{kl}).$$

8–4. It is asserted in the caption to Fig. 8–8 that, by considering Eq. 8–41, one can show that the most frequently occupied intermediate state in nuclear matter lies near the maximum of the Fourier transform of $v(r)\zeta(r)$. Show that this is true.

8–5. Consider two G-matrices, G_A and G_B, one associated with potential v_A, Pauli operator Q_A, and energy denominator $e_A = H_{0A} - \omega_A$, and the other with v_B, Q_B, e_B. Define $\Omega = 1 - (Q/e)G$.

(a) Note that $\psi = \Omega\phi$.

(b)
$$G_A(\omega_A) = v_A - \frac{Q_A}{H_{0A} - \omega_A} G_A.$$

Show that G is Hermitian, given that v and H_0 are.

(c) Show that

$$G_A - G_B = \Omega_B^\dagger (v_A - v_B)\Omega_A - G_B \left(\frac{Q_A}{e_A} - \frac{Q_B}{e_B}\right) G_A.$$

(This is shown in Appendix A of ref. 8.)

(d) If $v_A = v_B$ and $Q_A = Q_B = Q = Q^2$, note that

$$G_A - G_B = G_B \frac{Q}{e_B} (e_A - e_B) \frac{Q}{e_A} G_A = (1 - \Omega_B^\dagger)(e_A - e_B)(1 - \Omega_A)$$

and show that

$$\left\langle \phi_{kl} \left| \frac{\partial G(\omega)}{\partial \omega} \right| \phi_{kl} \right\rangle = - \langle \zeta_{kl} | \zeta_{kl} \rangle.$$

(e) Indicate how this leads to the formula

$$P_k = \left(1 - \sum_l \left\langle \phi_{kl} \left| \frac{\partial G}{\partial \omega} \right| \phi_{kl} \right\rangle_A P_l \right)^{-1}$$

and discuss the usefulness of this formula in a Brueckner calculation. [See refs. 34 and 35 and R. J. McCarthy and K. T. R. Davies, *Phys. Rev.* **C1**, 1640 (1970).]

8–6. Deduce Eq. 8–59a; that is, starting with the antisymmetrized state ϕ_A in which two nucleons have momenta $\mathbf{p}_\alpha$ and $\mathbf{p}_\beta$ and specified third components M_S and M_T of spin and i-spin, write it as in the text. You will want to use the angular-momentum expansion of the plane wave and also to introduce suitable complete sets of angular-momentum states.

8–7. In Eq. 8–76 we defined the incompressibility K for nuclear matter. In thermodynamics the isothermal compressibility is defined as

$$K_T^{\text{th}} = -\frac{1}{V} \left(\frac{\partial V}{\partial P}\right)_T,$$

where P, V, and T are the pressure, volume, and temperature, respectively, of the system. Also, the free energy is defined as $F = E - TS$, where E is the energy of the system and S is the entropy. The pressure is the rate of change of free energy with volume at constant temperature:

$$P = -\left(\frac{\partial F}{\partial V}\right)_T.$$

(a) Prove the equivalence of the two forms of K in Eq. 8–76 and show that still another form is

$$K = \left[r_0^2 \frac{\partial^2}{\partial r_0^2}\left(\frac{E}{A}\right)\right]_{\text{equm}},$$

where r_0 is a length (e.g., the radius of a sphere) specifying the volume per particle.

(b) For a zero-temperature system like the nucleus, prove that

$$P = \rho^2 \frac{\partial}{\partial \rho}\left(\frac{E}{A}\right)_{T=0},$$

which shows that the net pressure vanishes at equilibrium density. Hence show from the definition of isothermal compressibility that at equilibrium of a zero-temperature system

$$K_{T=0}^{\text{th}} = \left[\rho^3 \frac{\partial^2}{\partial \rho^2}\left(\frac{E}{A}\right)\right]_{\rho=\rho_{\text{sat}}}^{-1}.$$

Note that the nuclear physics definition of incompressibility is not very different from the reciprocal of the thermodynamic definition of compressibility; in fact,

$$\tfrac{1}{9}\rho K = \frac{1}{K_{T=0}^{\text{th}}}.$$

In thermodynamics, the name compressibility for K_T^{th} is a reasonable one; if K_T^{th} is numerically large, it signifies that the system is easily compressible. On the other hand, in nuclear physics, it is reasonable to call K the incompressibility. Nevertheless one does find K referred to in the literature as the compression modulus.

8–8. Consider a spherical nucleus of uniform density, and suppose that the density oscillates, each nucleon moving radially in and out. When the density is $\rho = \rho_0(1 + x)$, a particle that is a distance r from the center when the density is ρ_0 will have radial velocity $-\tfrac{1}{3}r\dot{x}$. Deduce that the kinetic energy associated with these oscillations is $\tfrac{1}{30}Mr_0^2A^{5/3}\dot{x}^2$ and that the potential energy is $\tfrac{1}{18}AKx^2$, where M is the mass of a nucleon, r_0 is the constant in the $A^{1/3}$-law for nuclear radius, and K is the incompressibility. Find the frequency of oscillation and show that, unless K is at least 120 MeV, there will be such oscillations at energies as low as 15 MeV in medium-mass nuclei.

8-9. Consider Eq. 8-29 for the motion of two nucleons in states k_1, k_2 in nuclear matter. In this case, as discussed in the text, we may replace $e_{k_1 k_2}$ by

$$e_{k_1} + e_{k_2} = \frac{\hbar^2}{2M}(k_1^2 + k_2^2) + U(k_1) + U(k_2).$$

In short-hand notation, we can write Eq. 8-29 as

$$G = v + v\frac{Q}{e}G,\tag{1}$$

where it is understood that the operator Q/e is diagonal, and

$$\left\langle k_1', k_2' \left| \frac{Q}{e} \right| k_1', k_2' \right\rangle$$

$$= \frac{Q(k_1', k_2')}{(\hbar^2/2M)(k_1^2 + k_2^2 - k_1'^2 - k_2'^2) + U(k_1) + U(k_2) - U(k_1') - U(k_2')}.$$

Let us study the effect of splitting the two-nucleon potential v into two parts,

$$v = v_s + v_l,\tag{2}$$

where the short-range part v_s contains the strong repulsive core plus most of the attraction, while v_l consists of the weak long-range tail.

(a) Define the following G-matrices for the short-range potential v_s:

$$G_s = v_s + v_s\frac{1}{e}G_s\tag{3}$$

and

$$G_s^F = v_s + v_s\frac{1}{e_0}G_s^F.\tag{4}$$

Note that in both these definitions the Pauli operator Q is absent. In Eq. 3 the energy denominator is the same as in Eq. 1, while in Eq. 4 e_0 is the free-energy denominator, without the one-body potentials U. Assuming the Hermiticity of G_s and G_s^F, prove that

$$G_s = G_s^F + G_s^F\left(\frac{1}{e} - \frac{1}{e_0}\right)G_s.\tag{5}$$

(b) Noting the relation

$$v_s = \left(1 + G_s\frac{1}{e}\right)^{-1}G_s,$$

prove that, to second order in v_l and G_s,

$$G = v_l + G_s + G_s \left(\frac{Q-1}{e} \right) G_s + v_l \frac{Q}{e} v_l + v_l \frac{Q}{e} G_s + G_s \frac{Q}{e} v_l, \qquad (6)$$

and, to the same order,

$$G = v_l + G_s{}^F + G_s{}^F \left(\frac{1}{e} - \frac{1}{e_0} \right) G_s{}^F + G_s{}^F \left(\frac{Q-1}{e} \right) G_s{}^F$$

$$+ v_l \frac{Q}{e} v_l + v_l \frac{Q}{e} G_s{}^F + G_s{}^F \frac{Q}{e} v_l. \qquad (7)$$

This equation was first obtained by Moszkowski and Scott [*Ann. Phys.* **11**, 65 (1960)]. Note that, since the short-range part v_s is very strong, it scatters nucleons high above the Fermi sea in the intermediate states, and the Pauli principle has, therefore, only a small effect on the two-particle wave function at short distances. Thus $G_s{}^F$, the free G-matrix for the short-range part, should be adequate for v_s. The long-range part v_l is rather weak, and the Pauli principle drastically suppresses its effects in higher orders; in Eq. 7, the second-order term $v_l(Q/e)v_l$ has been taken into account. The term

$$G_s{}^F \left(\frac{1}{e} - \frac{1}{e_0} \right) G_s{}^F$$

is called the dispersion term; it takes into account the fact that the nucleons are moving in the presence of an average field U. The other terms in Eq. 7 are the Pauli correction term and the two interference terms. With the choice of v_s to be described below, the convergence of expansion 7 is very good for central potentials.

(c) In binding-energy calculations, one is interested in the diagonal matrix elements $\langle k|G|k \rangle$, summed over the relative momenta **k** of the occupied states (Eq. 8–37). Moszkowski and Scott chose the separation distance so that, for a given k,

$$\langle k|G_s{}^F|k \rangle = 0. \qquad (8)$$

This makes v_s dependent on k, and hence it becomes non-Hermitian, introducing some correction terms into expansion 7 in third order. Note that

$$\langle k|G_s{}^F|k \rangle = \int_0^{r=d} \Phi_k{}^* v_s \Psi_k{}^F \, d^3r, \qquad (9)$$

where Φ_k is the relative unperturbed two-nucleon wave function, and $\Psi_k{}^F$ is given by

$$- \frac{\hbar^2}{M} \nabla^2 \Psi_k{}^F + v_s \Psi_k{}^F = \frac{\hbar^2}{M} k^2 \Psi_k{}^F.$$

Prove that Eq. 8 will be satisfied if

$$\left.\frac{\nabla \Psi_F}{\Psi_F}\right|_{r=d} = \left.\frac{\nabla \Phi^*}{\Phi^*}\right|_{r=d}. \qquad (10)$$

With this choice of d, the net effect of the second-order terms in expansion 7 is found to be rather small, and in the first approximation G is just the long-range potential v_l. For example, for a central potential with a hard core and a spin-averaged strength acting in the S-state, v_l contributed -37.29 MeV to the potential energy per particle in nuclear matter at $k_F = 1.4$ fm^{-1}, while all the second-order terms contributed only 1.65 MeV. In the more realistic situation in which tensor force is present, there are some technical complications in the method because different partial waves are coupled and Eq. 8 cannot be simultaneously satisfied for two partial waves with the same d. Also, since the first-order contribution of the tensor component of v_l to nuclear-matter binding vanishes, and its second-order contributions are large, one is less sure of the convergence of Eq. 7.

8–10. Consider the equation

$$|\Psi_{k,\omega}\rangle = |k\rangle + \frac{P}{\omega - t_{rel}}\, v|\Psi_{k,\omega}\rangle, \qquad (1)$$

where v is the two-body potential, P denotes the principal value, and t_{rel} is the relative kinetic-energy operator,

$$t_{rel}|k\rangle = \frac{\hbar^2}{M}\, k^2|k\rangle.$$

This equation, for the choice of $\omega = (\hbar^2/M)k^2$, is identical to the Lippmann-Schwinger equation 5–49, and describes the scattering of two nucleons. It is also of the same structure as Eq. 8–29, except that the propagator here is for free scattering, without the Pauli operator Q.

(a) Consider the case when $\omega < 0$. In such a situation, there can be no singularity in the propagator, and the principal value sign P is superfluous. Show in this case that the coordinate space representation of Eq. 1 is

$$\langle r|\Psi_{k,\omega}\rangle = \exp(ik \cdot r) - \frac{M}{4\pi\hbar^2}\int \frac{\exp(-\gamma|r-r'|)}{|r-r'|}\, v(r')\langle r'|\Psi_{k,\omega}\rangle\, d^3r', \qquad (2)$$

where $|\omega| = \hbar^2\gamma^2/M$. We have taken unit normalization volume for plane waves, that is, $\langle r|k\rangle = \exp(ik \cdot r)$, and used the relation

$$\sum_{k'} = \frac{1}{(2\pi)^3}\int d^3k'.$$

(b) Making the expansion

$$\langle r|\Psi_{k,\omega}\rangle = \sum_{l=0}^{\infty}[4\pi(2l+1)]^{1/2}i^l u_l(k,\gamma;r)Y_l^0(\hat{k},\hat{r}), \qquad (3)$$

show that the asymptotic form for the radial function u_l is given by

$$u_l(k, \gamma; r) \sim j_l(kr) + b_l(k, \gamma) \frac{\exp(-\gamma r)}{r}, \qquad (4)$$

where b_l is real. Equation 4 shows that for negative ω the two-body relative wave function "heals" to its unperturbed value $j_l(kr)$, the healing distance being governed by the parameter γ. For nuclear matter, the healing is due to the Pauli operator Q in Eq. 8–29; however, its effect may be simulated by making the parameter ω negative in the free-scattering equation. This is the essence of the so-called reference spectrum method due to Bethe, Brandow, and Petschek [8].

8–11. (a) *Harmonic-oscillator wave functions in relative coordinates and the Brody-Moshinsky transformation brackets.* Consider a particle of mass M moving in a harmonic-oscillator potential $\frac{1}{2}M\omega^2 r_1^2$, r_1 being the position coordinate of the particle with respect to the center of the oscillator well. The eigenfunction of the particle, in an orbital (n_1, l_1, m_1), is denoted by

$$\langle r_1 | n_1 l_1 m_1 \rangle = \phi_{n_1 l_1 m_1}(r_1)$$

and is defined by Eqs. 1 and 3 of Problem 4–10, the corresponding eigenenergy being $(2n_1 + l_1 + \frac{3}{2})\hbar\omega$. For two particles of identical mass M, moving in a common harmonic-oscillator well without interaction, the Hamiltonian is

$$H = \frac{p_1^2}{2M} + \frac{1}{2}M\omega^2 r_1^2 + \frac{p_2^2}{2M} + \frac{1}{2}M\omega^2 r_2^2. \qquad (1)$$

Transform this Hamiltonian to a relative frame by making the transformation

$$\mathbf{r} = \frac{\mathbf{r}_1 - \mathbf{r}_2}{\sqrt{2}}, \qquad \mathbf{R} = \frac{\mathbf{r}_1 + \mathbf{r}_2}{\sqrt{2}}, \qquad (2)$$

and likewise for the momenta. Show that Eq. 1 is exactly separable into relative and center-of-mass variables, and remains identical in form. If we denote the quantum numbers for relative and CM motion by n, l and N, L, it follows from conservation of angular momentum that

$$l_1 + l_2 = \lambda = l + L, \qquad (3a)$$

and from conservation of energy that

$$2n_1 + l_1 + 2n_2 + l_2 = 2n + l + 2N + L. \qquad (3b)$$

Denoting the eigenvector of the two particles coupled to angular momentum λ and z-component μ by $|n_1 l_1, n_2 l_2; \lambda\mu\rangle$, we may make the transformation

$$|n_1 l_1, n_2 l_2; \lambda\mu\rangle = \sum_{nlNL} |nl, NL; \lambda\mu\rangle\langle nl, NL; \lambda\mu|n_1 l_1, n_2 l_2; \lambda\mu\rangle. \qquad (4)$$

It can be shown that the brackets in the extreme right of Eq. 4 are independent of the magnetic quantum number μ, and they vanish if conditions 3 are not obeyed. These transformation brackets, commonly known as the Brody-Moshinsky coefficients, are tabulated by T. A. Brody and M. Moshinsky "Table of transformation brackets for nuclear shell model calculations." (Gordon and Breach, New York, 1967) and are extensively used in finite-nucleus calculations.

(b) *The Kallio approximation for the relative radial wave function.* As seen in (a), making the transformation to relative and CM coordinates leaves the Hamiltonian of Eq. 1 identical in form, and consequently we may write for the relative radial wave function in the relative coordinate an expression identical in form to Eq. 3 of Problem 4–10:

$$R_{nl}(r) = \sqrt{\frac{2n!\alpha^3}{\Gamma(n+l+\frac{3}{2})}}\,(\alpha r)^l \exp(-\tfrac{1}{2}\alpha^2 r^2)\, L_n^{l+1/2}(\alpha^2 r^2), \qquad (5)$$

where $\alpha = (M\omega/\hbar)^{1/2}$; n, l are now the quantum numbers for relative motion, and the Laguerre polynomial $L_n^{l+1/2}$ was defined in Problem 4–10.

Consider the special case of $l = 0$. Using the asymptotic expansion, valid for large n,

$$L_n^{1/2}(\alpha^2 r^2) \sim \exp(\alpha^2 r^2/2)\,\frac{1}{\sqrt{n}}\,\frac{\sin(\sqrt{n}\,2\alpha r)}{\alpha r}\,,$$

and the limit

$$\lim_{n\to\infty} \frac{n!}{\Gamma(n+\frac{3}{2})} = \frac{1}{\sqrt{n}}\,,$$

show that, for $n \to \infty$,

$$R_{n0}(r) = \sqrt{\frac{2\Gamma(n+\frac{3}{2})\alpha^3}{\pi n!}}\,2j_0(2\sqrt{n}\,\alpha r). \qquad (6)$$

Note that here

$$\mathbf{r} = \frac{1}{\sqrt{2}}\,(\mathbf{r}_1 - \mathbf{r}_2) \equiv \frac{\mathbf{r}_{12}}{\sqrt{2}}\,.$$

Expression 6 can be generalized for any l, as was shown by A. Kallio [*Phys. Lett.* 18, 53 (1965)]. He gave the approximate relation

$$R_{nl}(r_{12}) \approx \sqrt{\frac{2\Gamma(n+l+\frac{3}{2})\alpha^3}{\pi \cdot n! \nu^l}} \cdot 2^{l+1} j_l(\sqrt{2n+l+\tfrac{3}{2}}\,\alpha r_{12}), \qquad (7)$$

with[†] $\nu = 4n + 2l + 3$. Note that this equation is consistent with Eq. 6. Equation 7 is surprisingly accurate for low values of n, including $n = 0$, provided $r_{12} \lesssim 2$ fm. The accuracy increases for large n.

[†] There is a misprint in Kallio's paper in the definition of ν.

(c) Recall that the lth partial wave of the plane wave is proportional to $j_l(k_{12}r_{12})$, where we are using the definitions $\mathbf{r}_{12} = \mathbf{r}_1 - \mathbf{r}_2$, $\mathbf{k}_{12} = \frac{1}{2}(\mathbf{k}_1 - \mathbf{k}_2)$. The relative kinetic energy of two free particles is of course $\hbar^2 k_{12}^2/M$, whereas for two particles moving in a harmonic well the relative energy is $e_{nl} = (2n + l + \frac{3}{2})\hbar\omega$. However, only half of this, that is,

$$\tfrac{1}{2}e_{nl} = \left(n + \frac{l}{2} + \frac{3}{4}\right)\hbar\omega$$

is the relative kinetic energy. We may think, then, that $R_{nl}(r_{12})$ may be proportional to $j_l(k_{\mathrm{eff}}r_{12})$, where

$$\frac{\hbar^2 k_{\mathrm{eff}}^2}{M} = \tfrac{1}{2}e_{nl}.$$

We see from Eq. 7, however, that

$$k_{\mathrm{eff}} = \sqrt{2n + l + \tfrac{3}{2}}\,\alpha,$$

so that

$$\frac{\hbar^2 k_{\mathrm{eff}}^2}{M} = e_{nl}. \tag{8}$$

Can you explain this result qualitatively, remembering that Eq. 7 is valid only for small values of r_{12}?

8–12. Consider the model of a single shell in which there are η equally spaced single-particle levels in an energy interval $\hbar\omega$. Since each level is doubly degenerate, the shell can accommodate a maximum of up to 2η fermions of one kind. Assume that a pairing interaction of constant strength G acts in this shell, and that the total number of particles in the shell is η.

(a) Using Eq. 8–112, show that the chemical potential λ is in the middle of the shell, symmetrically placed with respect to the single-particle levels.

(b) Noting that the average level density in the shell is $\eta/\hbar\omega$, and replacing the sum over the levels by an integral in the gap equation 8–111, show that

$$\Delta = \frac{\hbar\omega}{2 \sinh\,(\hbar\omega/G\eta)}. \tag{1}$$

In the limit $G\eta \gg \hbar\omega$, this gives

$$\Delta \approx \tfrac{1}{2}G\eta, \tag{2}$$

while for $G\eta \ll \hbar\omega$ we obtain

$$\Delta \approx \hbar\omega \exp\left(-\frac{\hbar\omega}{G\eta}\right). \tag{3}$$

This model may roughly represent the situation in the half-filled shell of a well-deformed nucleus, where, in the first approximations, the single-particle levels (each doubly degenerate) may be taken to be uniformly spaced. Taking the "experimental" values of Δ and $\hbar\omega$ in the deformed rare-earth region, which of the above limiting equations (2 or 3) would you use to estimate $G\eta$? What is your numerical estimate for $G\eta$?

8-13. The pairing strength G, defined by Eq. 8–93, is not a constant for the realistic case of a finite-range interaction V_{12}. A rough estimate of the numerical value of G and its A-dependence may be obtained using the plane-wave approximation and assuming that V_{12} is the S-wave, $T = 1$ nucleon-nucleon potential. Using these assumptions and the definitions $\mathbf{r} = \mathbf{r}_1 - \mathbf{r}_2$, $\mathbf{k} = \frac{1}{2}(\mathbf{k}_1 - \mathbf{k}_2)$, show from Eq. 8–93 that the diagonal matrix element is given by

$$G(k, k) = -\frac{1}{\Omega} \int V_{12}(r) j_0^2(kr)\, d^3r.$$

Using Eq. 5–39, show that this may be written as

$$G(k, k) \approx -\frac{4\pi\rho}{A} \frac{\hbar^2}{Mk} \delta_0^{\text{Born}}(k), \tag{1}$$

where $\delta_0^{\text{Born}}(k)$ is the S-wave, $T = 1$ Born phase shift of the two-nucleon potential, and ρ is the nuclear density.

For numerical estimate, we may take for k the value corresponding to the head-on collision of two nucleons at the Fermi surface, that is, $k = k_F$. Also, for such large values of k, we may assume $\delta_0^{\text{Born}}(k) \approx \delta_0(k)$. For large nuclei, taking the nuclear-matter value of $k_F = 1.36$ fm^{-1}, and using Fig. 5–12a to estimate δ_0, show that Eq. 1 gives $G \approx 14/A$ MeV, whereas using an average value of $k_F = 1.25$ fm^{-1} for a nucleus gives $G \approx 21/A$ MeV. The latter number is close to the value of G actually chosen in calculations.

Chapter 9

Collective Nuclear Motion

9-1 INTRODUCTION

We have pointed out that the shell and independent-particle models explain many nuclear properties, but fail to account for the large nuclear quadrupole moments and spheroidal shapes which many nuclei have. It is clear that such effects cannot be obtained from any model that considers the pairwise filling of the individual orbits of a spherical potential to be a good approximation to nuclear structure. Such large effects can arise only from coordinated motion of many nucleons. Let us visualize the nucleons as moving in essentially single-particle paths, such as the properties of nuclear matter would lead us to expect in a finite nucleus; at each instant the nucleons form a pattern in space. Let us suppose that at some moment this pattern is, say, ellipsoidal. The motion is collective, or cooperative, if this pattern changes very slowly, or, in the extreme case, is permanent. The particles move through many circuits of their orbits, but in a sufficiently correlated manner, so that at all times the overall space pattern is essentially the same or slowly changing. This type of correlation is very different from the short-range correlations in nuclear matter; it is a long-range correlation and may be thought of as a surface effect. Also, the specific values of the parameters involved depend on the exact number of nucleons in the nucleus.

We may characterize such motion by saying that the particle motion and surface motion are coupled. Because the surface is distorted at some moment, the potential "felt" by a particle is not spherically symmetric. If this distortion lasts for a sufficient time—a few nucleon periods—the particles will move in orbits appropriate to an aspherical shell-model potential. This particle motion in turn preserves the nuclear shape. In nearly spherical nuclei, there is only very weak coupling of this type, the particles have no directional correlation, and the resultant average shape is spherical.

To express the particle-surface coupling mathematically, it is necessary to introduce some *collective variables* to describe the cooperative modes of motion. It is also convenient to take parts of the energy, angular momentum, and other dynamical variables and to ascribe these parts to collective motion. For example, the correlated particle motion may be such that the pattern of particles changes slowly, maintaining a fixed shape, but with the orientation in space altering. Looked at from outside, the nucleus of constant shape is rotating. It is natural to attribute kinetic energy and angular momentum to this rotating mass and to consider the rest of the nuclear kinetic energy and angular momentum to be internal. Also, we may remove from the potential energy of the system a term to express the work done in deforming it from spherical shape; such a term will depend only on collective variables if they are suitably chosen, and we may call it the deformation energy. It will include the change in Coulomb energy of a continuous distribution of charge, as well as the change in surface energy. If this potential energy is a minimum for the smallest surface of a given volume, the first terms in a Taylor-series expansion will be quadratic in the parameters describing the departure from sphericity. Because of the great resistance of nuclear matter to compression, it is usual to consider deformations of fixed volume only and to treat the collective potential energy as quadratic in the simplest approximation.

Collective variables are useful only if the energy can be separated in this way into terms that depend on the collective variables and terms that do not. The other variables may be called intrinsic. If there is only a weak collective-intrinsic coupling, the system may be studied by perturbation theory:

$$H = H_c \text{ (collective variables)} + H_i \text{ (intrinsic variables)} + H_{coup}.$$

An important question concerns the identity of the intrinsic variables; this is, in general, a difficult problem. A simple example of a collective variable is the mass center; for a two-body system, the corresponding intrinsic variable is the particle separation $r_1 - r_2$, and there is no coupling. But, as we mentioned in connection with the shell model, there are no such simple intrinsic variables for the N-body system. If we continue to use as intrinsic variables the coordinates of the particles with respect to a fixed origin, we have too many variables, since we also have the collective coordinates. On the other hand, if we use the correct number of variables, the intrinsic variables are complicated. We shall say more about the separation of collective and intrinsic motions and their interactions after we have formalized the concepts.

A nonspherical equilibrium shape might be thought of as arising as a result of the opposing tendencies of the "outer" nucleons to polarize the nuclear core and of the core to resist this polarization and maintain a spherical shape; this possibility was first suggested by Rainwater [1] in 1950. The model was worked out in detail by Aage Bohr and extended and generalized principally by Bohr and Ben R. Mottelson and various coauthors in a comprehensive series of papers during the decade 1950–1960. The most instructive of these papers are probably refs. 2, 3, 4, and 5.

Two different nuclear models are considered in these works. The earlier (and for many purposes the simpler) separates the nucleus into a core and extracore nucleons. The core is treated macroscopically as a deformable drop of nuclear "liquid," in interaction with the few extra nucleons in an unfilled shell. The second model is really an extended shell model; the shell-model potential is assumed to be nonspherical, the energies of the single particles in the nonspherical potential are calculated, and the distortion that gives the minimum energy is taken as the distortion actually found. One is faced here with the problem inherent in any nuclear shell model: the need to understand the model in terms of more fundamental interactions. The long-range correlations are replaced by the assumption of a permanently distorted potential.

In the early development of a collective model, all nuclei were broadly classified into two categories: spherical and permanently deformed. In spherical nuclei the nuclear excitations were assumed to be small-amplitude vibrations of the harmonic type about the equilibrium spherical shape, while in deformed nuclei the low-energy modes were rotations and small-amplitude vibrations about the deformed shape. The experimental data accumulated in the last decade or so have revealed that, while there do exist many well-deformed nuclei with rotational spectra, the simple vibrational picture for spherical nuclei is not a realistic one. Recent experimental techniques make it possible to measure the quadrupole moments of excited states, and it is found [6] that even the first excited 2^+ states of most of the so-called spherical nuclei (e.g., the Sn and Te isotopes) are appreciably deformed. Furthermore, when we examine the level structures of many of these so-called vibrational nuclei up to an excitation of about 5 MeV, well-defined rotation-like bands appear, just as in permanently deformed nuclei. We shall give examples of these quasi-rotational bands in a later section.

These experimental findings make us believe that the concept of a spherical nucleus making small harmonic vibrations about its spherical shape is too naive and does not conform with reality. The concept of a nucleus having a spherical shape in its *ground state*, however, is a valid one. If the potential energy for deformation from this shape goes up steeply, the nucleus is stiff, while if the potential-energy curve against deformation is very flat, the nucleus is soft and may contain many modes of deformation. Nuclei of the latter type are called transitional and occur in regions where either permanent deformation or stiff sphericity is about to set in (e.g., some of the Nd isotopes, where deformation is setting in, and others in the Pt-Os region, where stiff sphericity is about to set in). An important advance in recent times has been a gain in understanding these transitional nuclei. Before going into these more complicated cases, we give, in the next few sections, accounts of the simplest vibrational and rotational models for stiff nuclei.

9-2 THE VIBRATIONAL MODES OF A SPHERICAL NUCLEUS

A spherical nucleus, in the simplest model, may be assumed to be a charged liquid drop, its excitation modes arising from small oscillations about the equilibrium spherical shape. Although, as mentioned in Section 9-1, this is not too realistic, this model introduces the concept of phonon excitations in an elegant manner and is useful for more sophisticated theories. We therefore first consider the modes of small oscillations of a charged liquid drop. The case of an incompressible liquid was first worked out by Lord Rayleigh [7]. The assumption of incompressibility is reasonable for the low-lying states of a nucleus; it means that, for a given mass of nuclear matter, shape changes can take place only with the restriction of volume conservation. For a nucleus, Lord Rayleigh's classical equations of motion must be quantized, and we should allow for a type of motion more general than the irrotational[†] fluid flow considered by him.

Consider a spherical drop of radius R_0. When distorted, its surface is conveniently parametrized by the expression

$$R(\theta, \phi) = R_\alpha \left[1 + \sum_{\lambda=0}^{\infty} \sum_{\mu=-\lambda}^{\lambda} \alpha_{\lambda\mu} Y_{\lambda\mu}(\theta, \phi) \right], \tag{9-1}$$

[†] *Irrotational* is used here in the technical hydrodynamic sense; it does *not* imply that elementary volumes of the fluid may not move in circular paths, but it does imply that each elementary volume is not rotating about an axis moving with its own center. In other words, in successive positions of the volume element, the orientations of its sides in space are unaltered. The general motion of a rigid body is *not* irrotational, but fluids may move in this way.

where θ and ϕ are polar angles with respect to an arbitrary space-fixed axis, and R_α is chosen to preserve volume conservation. For reality, $\alpha_{\lambda\mu} = (-)^\mu \alpha^*_{\lambda-\mu}$. Volume conservation implies that

$$\frac{4\pi}{3} R_0{}^3 = \int_0^{2\pi} d\phi \int_0^\pi \sin\theta\, d\theta \int_0^{R(\theta,\phi)} r^2\, dr$$

$$= \frac{4\pi}{3} R_\alpha{}^3 \left[1 + \frac{3\alpha_{00}}{\sqrt{4\pi}} + \frac{3}{4\pi} \sum_{\lambda,\mu} |\alpha_{\lambda\mu}|^2 + 0(\alpha^3) + \cdots \right], \qquad (9\text{--}2)$$

where the higher-order terms are neglected, assuming the deformation to be small. Deformation of the type α_{00} in the first order causes a volume change costing too much deformation energy and is therefore discarded in the treatment of low-lying modes. Furthermore, the potential energy of the drop should not depend on the location of the center of mass, which, to first order, is given by $\alpha_{1\mu}$ ($\mu = -1, 0, 1$). By constraining the center of mass at the origin, we obtain $\alpha_{1\mu} = 0$. Hence we can write

$$R(\theta, \phi) = \frac{R_0}{\xi} \left[1 + \sum_{\lambda=2}^\infty \sum_{\mu=-\lambda}^\lambda \alpha_{\lambda\mu} Y_{\lambda\mu}(\theta, \phi)\right], \qquad (9\text{--}3)$$

where ξ is a scale factor required by the condition of constant volume and is given by

$$\xi \approx 1 + \frac{1}{4\pi} \sum_{\lambda=2}^\infty \sum_\mu |\alpha_{\lambda\mu}|^2. \qquad (9\text{--}4)$$

If we ignore the volume changes to second order in the α's, then ξ is unity, and we obtain

$$R(\theta, \phi) = R_0 \left[1 + \sum_{\lambda=2}^\infty \sum_{\mu=-\lambda}^\lambda \alpha_{\lambda\mu} Y_{\lambda\mu}(\theta, \phi)\right], \qquad (9\text{--}5)$$

an expression which we use in the following treatment for simplicity. Of course, the nucleus does not have a sharp surface like that of a liquid drop, but appropriate modifications can be made in the calculation of deformation energy for the case of a diffuse surface. Any collective motions are expressed by letting $\alpha_{\lambda\mu}$ vary with time. In the quadratic approximation, the kinetic energy is of the form

$$T = \tfrac{1}{2} \sum_{\lambda,\mu} B_\lambda |\dot{\alpha}_{\lambda\mu}|^2. \qquad (9\text{--}6)$$

In the particular case of the irrotational flow of a constant-density fluid, Rayleigh's calculations show that

$$B_\lambda = \frac{\rho R_0^5}{\lambda}, \tag{9-7}$$

where ρ is the density. This expression for the inertial parameter B_λ turns out to be quite unrealistic in the nuclear case, and a microscopic model is necessary for a more realistic estimate. Because of distortion from sphericity, the potential energy changes: the surface energy increases as a result of an increase in the surface area of the drop, and the Coulomb energy decreases with increasing distortion. This change in the potential energy from its spherical value is denoted by V and is given in the quadratic approximation by

$$V = \tfrac{1}{2} \sum_{\lambda,\mu} C_\lambda |\alpha_{\lambda\mu}|^2. \tag{9-8}$$

For the classical liquid,

$$C_\lambda = C_\lambda^{(1)} - C_\lambda^{(2)}, \tag{9-9}$$

with

$$C_\lambda^{(1)} = SR_0^2(\lambda - 1)(\lambda + 2), \tag{9-10}$$

and

$$C_\lambda^{(2)} = \frac{3}{2\pi} \frac{Z^2 e^2}{R_0} \frac{\lambda - 1}{2\lambda + 1}, \tag{9-11}$$

where S in Eq. 9-9 is the surface tension. For a charged drop to perform oscillations about the spherical shape, the C_λ's should be positive. This classical estimate of C_λ is useful for nuclei, as we shall see in Chapter 12 on fission, provided we use appropriate values of R_0 and S based on the semiempirical mass formula.

With expressions 9-6 and 9-8 for T and V, the Lagrangian $L = T - V$ is seen to be a sum of separate terms for each of the generalized coordinates $\alpha_{\lambda\mu}$, and each of these terms is the Lagrangian of a simple harmonic oscillator. The frequency associated with the variable $\alpha_{\lambda\mu}$ is

$$\omega_\lambda = \left(\frac{C_\lambda}{B_\lambda}\right)^{1/2}. \tag{9-12}$$

As mentioned earlier, we have excluded from our considerations the $\lambda = 0$ mode, whose excitation should be at a much higher energy than the incompressible shape vibrations, as well as the $\lambda = 1$ modes, which merely describe the vibrations of the center of mass of the drop.

In order to quantize the collective modes of the drop and to introduce the concept of phonon excitations [2], we first define the momentum conjugate to the variable $\alpha_{\lambda\mu}$ as

$$\pi_{\lambda\mu} = \frac{\partial T}{\partial \dot{\alpha}_{\lambda\mu}} = B_\lambda \dot{\alpha}^*_{\lambda\mu}.$$

The Hamiltonian for the drop then becomes

$$H = T + V = \sum_{\lambda,\mu} \left(\frac{1}{2B_\lambda} |\pi_{\lambda\mu}|^2 + \tfrac{1}{2} C_\lambda |\alpha_{\lambda\mu}|^2 \right). \tag{9-13}$$

As the different λ-modes of oscillation are completely uncoupled in this approximation, we consider, for quantization, only the case of $\lambda = 2$, and drop the $\lambda = 2$ suffix from B_2 and $\alpha_{2\mu}$ for simplicity. In the quantization procedure, the dependence of the inertial parameter B on the coordinates α_μ is ignored, and the variables α_μ and π_μ are observables with the commutator $[\alpha_\mu, \pi_\mu] = i\hbar$. In the usual way, we define

$$\alpha_\mu = \sqrt{\frac{\hbar}{2B\omega}} (b_\mu + (-)^\mu b^+_{-\mu}),$$

$$\pi_\mu = i \sqrt{\frac{\hbar B\omega}{2}} (b_\mu^+ - (-)^\mu b_{-\mu}), \tag{9-14}$$

where $\omega = \omega_2$ is given by Eq. 9-12, and the operator b_μ and its Hermitian conjugate b_μ^+ will turn out to be the annihilation and creation operators for phonons. When the number operator $\hat{n}_\mu$ is introduced by the definition

$$\hat{n}_\mu = b_\mu^+ b_\mu, \tag{9-15}$$

the Hamiltonian of Eq. 9-13 is diagonal in the n_μ-representation,[†] so that any eigenstate of the Hamiltonian can be specified by the number of phonons present in the state. The energy eigenvalues are given by

$$E_N = \hbar\omega \sum_\mu (\tfrac{1}{2} + n_\mu) = \hbar\omega(\tfrac{5}{2} + N), \tag{9-16}$$

where $N = \sum n_\mu = 0, 1, 2$, etc., and $\lambda = 2$. Thus the ground state is a state with no phonons present, while the first excited state has only one phonon excitation and is fivefold degenerate, since the azimuthal quantum number μ can take any of the integral values $-2, -1$, to $+2$. A similar analysis can be carried through for any λ, except that the phonon excitation energies

---

[†] Indeed, $H = \sum_{\lambda,\mu} \hbar\omega_\lambda (b^+_{\lambda\mu} b_{\lambda\mu} + \tfrac{1}{2})$.

are higher. In general, a state with $n_\lambda = 1$ is $(2\lambda + 1)$-fold degenerate and, therefore, presumably has angular momentum λ. It can be shown for irrotational motion that a phonon of type $\lambda\mu$ carries angular-momentum quantum number λ, with z-component μ and parity $(-)^\lambda$.

The energy $\hbar\omega_\lambda$ is a fairly rapidly increasing function of λ. If we use the classical hydrodynamic expressions 9–7 and 9–9, we readily see that $\omega_3 \approx 2\omega_2$ and $\omega_4 = 3\omega_2$. Consequently, we need consider only low values of λ in order to examine low-lying nuclear states. If there are nuclei that can oscillate collectively about a spherical shape, the first excited vibrational state would have one $\lambda = 2$ phonon present and would therefore be a 2^+ state. One $\lambda = 3$ phonon has about the same energy as two $\lambda = 2$ phonons, so the second vibrational state would be either a 3^- state or one of the states formed from coupling the two angular momenta of two units, namely, 0^+, 2^+ or 4^+. The degeneracy of the last three states will normally be removed by perturbations which have not been considered, but at least the center of gravity of the three levels should be at an energy about twice the first

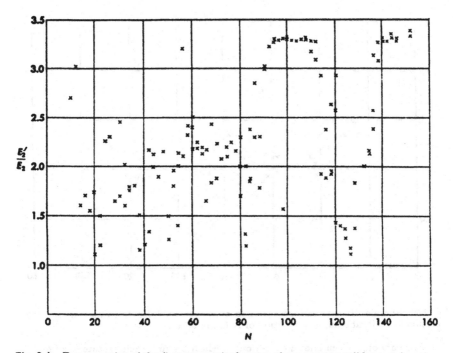

Fig. 9-1 Energy ratios of the first two excited states of even-even nuclides as a function of N.

2+ level. For a rough indication of the excitation energy to be expected we could use the irrotational fluid approximations 9–7, 9–10, and 9–11; for A near 100, $\hbar\omega_2$ is slightly over 2 MeV, falling to about 1 MeV for A near 200. These energies are thus somewhat smaller than particle excitations for even-even nuclei, especially near closed shells, and we might very well expect to find that the lowest levels of even-even spherical nuclei are collective vibrations.

One test of this hypothesis would be to examine the ratio E_2'/E_2 of the energies of the first and second 2+ states; it should be about 2. Figure 9–1 shows the ratio for the first and second excited states[†] of even-even nuclides; we see that there are regions where the ratio is slightly over 2, with abrupt transitions where it is 3.3 and the second state is 4+, not 2+. For nuclides with E_2'/E_2 near 2, the energies E_2 are smaller than the hydrodynamic estimate by a factor of about 2, but this is not inconsistent with the assumption that they are vibrational levels. It does mean that the hydrodynamic model for C_λ and B_λ is not very good. Figure 9–2 shows a plot of the energies

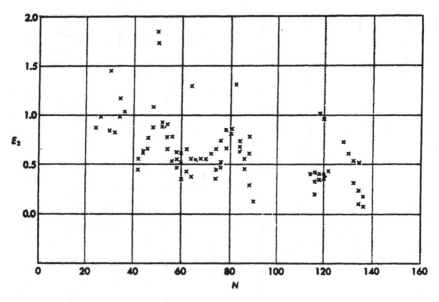

Fig. 9-2 Energies of first excited states in even-even nuclides, for which the energy is presumably vibrational, as a function of N (E_2 in MeV).

---

† There are only seven known exceptions in which the first excited state of an even-even nucleus is not a 2+. These are ^{4}He (0+), ^{14}C (1−), ^{16}O (0+), ^{40}Ca (0+), ^{72}Ge (0+), ^{90}Zr (0+), and ^{208}Pb (3−).

E_2 against N, and it is seen that there is a general decrease of E_2, as forecast by the theory.

Many items of information, both from nuclear reactions and from γ-ray spectra, indicate the presence of 3^- levels at an energy of about 2–3 MeV, decreasing with A. This result gives further support to the collective picture. If the energy of the collective mode is close to that of some single-particle excitations, these may be mixed by interaction between the particle degrees of freedom and the collective degrees of freedom. The wave functions of the actual nuclear states are then only partially collective, and the collective state is distributed over several actual states. Specificially collective properties are not so pronounced. For 3^- states this mixing is important in some cases, and we should expect even higher vibrational states to be so distributed among the nuclear states that the collective description loses its validity almost entirely.

Negative-parity states can also be obtained in spherical even-even nuclei by the combination of one quadrupole and one octupole phonon. There are five states, 1^-, 2^-,..., 5^-, which, though degenerate in first approximation, actually split just as the triad 0^+, 2^+, 4^+ does. Such states lie above the 3^- state, and are almost certainly strongly perturbed and mixed with single-particle excitation.

The wave functions of these vibrational states are simply the harmonic-oscillator functions with $\alpha_{\lambda\mu}$ as variables. The ground-state wave function is

$$\psi_0 \propto \exp\left(-\sum_{\lambda,\mu} \frac{B_\lambda \omega_\lambda |\alpha_{\lambda\mu}|^2}{2\hbar}\right), \tag{9–17}$$

and a state with one phonon of type $\lambda\mu$ present is

$$|\psi(n_{\lambda\mu} = 1)\rangle = b^+_{\lambda\mu}|\psi(n_{\lambda\mu} = 0)\rangle. \tag{9–18}$$

We see from Fig. 9–1 that there are many nuclei, particularly in the region $40 < N < 80$, for which the ratio $E_2'/E_2 \approx 2$, in agreement with the harmonic vibrational model. Does this mean that this simple model is valid for these nuclei? Experimentally, there are not many cases in which the near-degenerate triplet of 0^+, 2^+, and 4^+ is found for the two-quadrupole phonon states, casting some doubt on the model. Let us, however, take two particularly favorable examples in which all three members of the triplet have been detected, and examine the experimental data more closely. In Fig. 9–3, we show the low-lying spectra of $^{114}_{48}\mathrm{Cd}$ and $^{122}_{52}\mathrm{Te}$ for this purpose. In $^{114}_{48}\mathrm{Cd}$, we find the first excited 2^+ state at 0.558 MeV, followed by five,

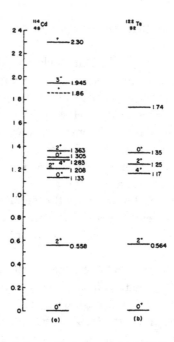

Fig. 9-3 Low-lying excited states of (a) $^{114}_{48}$Cd and (b) $^{122}_{52}$Te.

not three, excited states at about twice this energy. There is also a 3⁻ state
at 1.945 MeV. In order to salvage the vibrational model, we could assume
that the lowest three of the five excited states starting at 1.133 MeV are
two-phonon states, while the other two, the 0⁺ at 1.305 MeV and the 2⁺
at 1.363 MeV, are three-quadrupole phonon states somehow brought down
in energy. Actually, it can be shown that three bosons of spin 2⁺ each can
combine symmetrically to form $J = 0^+, 2^+, 3^+, 4^+,$ and 6^+; this would mean
that the 3⁺, 4⁺, and 6⁺ states have been pushed up in energy, with the
centroid of the quintuplet lying around the three-phonon energy of about
1.8 MeV. The spectrum of the other nucleus, $^{122}_{52}$Te, is much simpler, and we
find the expected triplet at the two-phonon energy. These energy considerations alone, then, may convey the impression that the vibrational model is
reasonable for these nuclei. We shall now see, however, that a consideration
of the electromagnetic properties of these states will lead us to believe that
something is seriously wrong with the model even in the case of $^{122}_{52}$Te, and
will point to the pitfalls of judging the validity of a model from energy
considerations alone.

To consider the electromagnetic properties of these states, we modify the definition of the electric 2^λ-pole moment operator, given by Eq. 3–6, to

$$Q_{\lambda\mu} = \int r^\lambda Y_\lambda{}^\mu(\Omega)\rho(r)\,d^3r, \qquad (9\text{–}19)$$

for a continuous charge distribution with charge density ρ. In our model of the oscillating drop, the charge density is constant up to a radius $R(\theta, \phi)$, given by Eq. 9–5, with a value $3Ze/4\pi R_0{}^3$, and is zero outside. Thus we obtain, for a charged drop,

$$Q_{\lambda\mu} = \frac{3Ze}{4\pi R_0{}^3} \int Y_\lambda{}^\mu(\theta, \phi)\sin\theta\,d\theta\,d\phi \int_0^{R(\theta,\phi)} r^{\lambda+2}\,dr.$$

A straightforward evaluation of this integral yields

$$Q_{\lambda\mu} = \frac{3Ze}{4\pi} R_0{}^\lambda \alpha_{\lambda\mu}^* + 0(\alpha^2). \qquad (9\text{–}20)$$

If we now use the second-quantized expression 9–14 for $\alpha_{\lambda\mu}$, we note that the multipole operator $Q_{\lambda\mu}$ contains only a single creation and a single annihilation operator for a phonon of the type λ-μ; therefore $Q_{\lambda\mu}$ can connect states differing by only one phonon of multipole λ. Applying this result to the low-lying quadrupole states, we find that

(i) there should be strong $E2$-transitions from the first excited 2^+ state to the ground state and from the members of the two-phonon triplet to the first excited 2^+ state;

(ii) there should be no crossover $E2$-transition from the second 2^+ state to the ground state; and

(iii) since the diagonal matrix element $\langle 1\text{ phonon}|Q_{2\mu}|1\text{ phonon}\rangle = 0$ in this model, the quadrupole moment of the first excited 2^+ state should be zero.

Experimentally, information about electromagnetic transitions can be obtained by Coulomb excitation of a nucleus from its ground state by scattering with ions, or by measuring the electromagnetic decay widths of excited states. To be more quantitative, let us define the transition probability for emission of a photon of energy $\hbar\omega = \hbar kc$, angular momentum λ, μ, and electric or magnetic type, with the nucleus going from an initial state $|J_i M_i\rangle$ to a final state $|J_f M_f\rangle$, as in Appendix E, viz.

$$T_{if}(\sigma\lambda\mu) = \frac{8\pi(\lambda + 1)}{\lambda[(2\lambda + 1)!!]^2} \frac{k^{2\lambda+1}}{\hbar} |\langle J_f M_f|O_{\lambda\mu}|J_i M_i\rangle|^2, \qquad (9\text{–}21)$$

where $O_{\lambda\mu}$ stands for the electric or magnetic multipole operator, and the states $|JM\rangle$ are normalized to unity. Since we are not usually interested in the orientation of either the initial or the final nucleus, we sum over M_f and average over M_i, defining the so-called reduced matrix element, which contains information about the nuclear wave function,

$$B(\sigma\lambda, J_i \to J_f) = (2J_i + 1)^{-1} \sum_{M_i, M_f} |\langle J_f M_f | O_{\lambda\mu} | J_i M_i \rangle|^2, \qquad (9\text{-}22)$$

and obtain the relation

$$T(\sigma\lambda\mu, J_i \to J_f) = \frac{8\pi(\lambda + 1)}{\lambda[(2\lambda + 1)!!]^2} \frac{k^{2\lambda+1}}{\hbar} B(\sigma\lambda, J_i \to J_f). \qquad (9\text{-}23)$$

For electric transitions, the multipole operator $Q_{\lambda\mu}$ for a continuous charge distribution is given by Eq. 9-20, and the reduced matrix element from a one-phonon state $(n_\lambda = 1)$ to a no-phonon state $(n_\lambda = 0)$ is found easily:

$$B(E\lambda, n_\lambda = 1 \to n_\lambda = 0) = \left(\frac{3}{4\pi} ZeR_0^\lambda\right)^2 \frac{\hbar}{2\sqrt{B_\lambda C_\lambda}}, \qquad (9\text{-}24)$$

where we have made use of expansion 9-14. Characteristics (i) and (ii) listed above are thus subject to the selection rule

$$\Delta n_\lambda = \pm 1. \qquad (9\text{-}25)$$

Furthermore, using the properties of b_μ, we can obtain the following sum rule:

$$\sum_f B(E\lambda, n_\lambda J_i \to (n_\lambda - 1), J_f) = n_\lambda B(E\lambda, n_\lambda = 1 \to n_\lambda = 0), \qquad (9\text{-}26)$$

where the summation f is over all states with a given J_f and $n_\lambda - 1$ phonons. If we take the particular case of $n_\lambda = 2$, there is only one state with $n_\lambda = 1$, and we obtain

$$B(E2, 2' \to 2) = 2B(E2, 2 \to 0), \qquad (9\text{-}27a)$$

$$B(E2, 4 \to 2) = 2B(E2, 2 \to 0), \qquad (9\text{-}27b)$$

where $2'$ denotes the second excited 2^+ state, and 4 the first excited 4^+ state.

Experimental analyses confirm point (i) above—that the $E2$-transition from the ground state to the one-phonon 2^+ state is several times stronger than the single-particle estimate. In ^{114}Cd and ^{122}Te, for example, this transition is about 35 times stronger than the noncollective value. It is also found that, although the crossover transition is not zero, it is very small

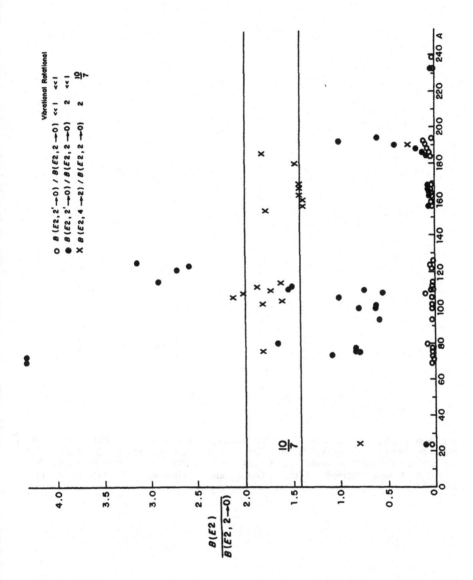

Fig. 9-4a The reduced $E2$-transition probabilities for the decay of the second excited 2^+ state ($2'^+$) and the first excited 4^+ state, relative to the probability for the first 2^+ state. (After Nathan and Nilsson [8].)

Fig. 9-4b Adopted values of the static quadrupole moments of the first excited 2^+ states of even-even nuclei as determined from reorientation experiments. The solid lines are through the rotation-model values obtained from $B(E2, 0^+ \rightarrow 2^+)$ data, where the sign remains undetermined. [After A. Christy and O. Hausser, *Nucl. Data Tables*, **11**, 281 (1973).]

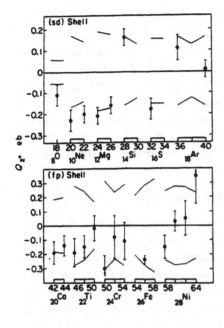

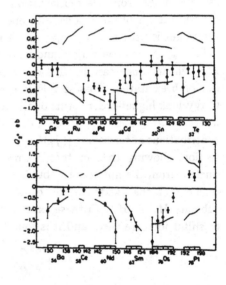

Fig. 9-4b

when compared with the probability of direct transition between adjacent 2^+ and 0^+, confirming conclusion (ii). It is generally found, however, that Eq. 9-27a is not quantitatively satisfied in most cases, typical deviations being about 30–50%. Following Nathan and Nilsson [8], we show in Fig. 9-4a the crossover to ground-state ratio $B(E2, 2' \to 0)/B(E2, 2 \to 0)$, as well as the quantities $B(E2, 2' \to 2)/B(E2, 2 \to 0)$ and $B(E2, 4 \to 2)/B(E2, 2 \to 0)$ as found experimentally as functions of mass number A. Also included in this figure are well-deformed rotational nuclei, which we shall discuss in Section 9–5. For most vibrational nuclei, it will be seen from the figure that Eq. 9–27b is well satisfied, but not Eq. 9–27a.

The quadrupole moments of excited states are extracted from Coulomb excitation experiments with heavy-ion projectiles through the reorientation effect [6]. The static quadrupole moment of a nuclear state $|JM\rangle$ is defined by

$$Q_{J^\pi} = e\langle JJ|\sqrt{16\pi/5}\, r^2 Y_{20}(\theta\phi)|JJ\rangle,$$

where (r, θ, ϕ) are the nuclear coordinates. Figure 9–4b shows the static quadrupole moment Q_{2^+} of the first excited 2^+ states of even-even nuclei, as determined from the reorientation experiments. In Coulomb excitation experiments, a beam of ions with energy less than the Coulomb barrier of the target is used as the projectile. These induce a time-dependent electric field gradient at the target nucleus, causing transitions to excited states. For light ions one can use first-order perturbation theory to show that the Coulomb excitation cross section is directly proportional to $B(E\lambda)$, but for heavy ions higher-order terms of the interaction, involving various possible intermediate states, should be included. For example, one should include contributions of terms like $|\langle J_f M_f|Q_{\lambda\mu}|J_f M_f'\rangle\langle J_f M_f'|Q_{\lambda\mu}|J_i M_i\rangle|^2$ to the cross section, showing that in these terms the projectile not only excites the nucleus from the initial state, but also continues to interact with the λ-pole moment of the excited state, causing a "reorientation" of the magnetic substate $M_f' \to M_f$. In these experiments, commonly α-particles or oxygen or sulfur ions are used, and it is found that the ratio (see Eq. A–83)

$$\frac{|\langle 2||Q_2||2\rangle|}{|\langle 2||Q_2||0\rangle|} \approx 1$$

for many of the vibrational nuclei. This implies a substantial quadrupole moment for the first 2^+ excited states of these nuclei, contradicting conclusion (iii) of the harmonic model, and pointing to the fact that modifications should be made for possibly large anharmonic effects in the model. It is expected that the quadrupole moment of an excited state will depend sensitively on the density distribution of the nucleus in the excited state. Anharmonic effects will introduce into the first 2^+ state some admixture of two-phonon contribution, thus changing its wave function and the quadrupole moment [9, 10]. In a later section, we shall outline a theory, due to Baranger and Kumar [11, 12], which can handle spherical, deformed, and transitional nuclei within a single semiclassical framework.

9-3 COLLECTIVE MODES OF A DEFORMED EVEN-EVEN NUCLEUS

In this section we shall study the collective modes of a liquid drop with a spheroidal equilibrium shape. As was emphasized in Chapter 6, the equilibrium shape of a classical liquid drop is always spherical. However, in making the liquid-drop analogy to a nucleus, only the part of the energy that is smoothly varying with deformation and mass number is considered, and this minimizes at zero deformation. By making the adiabatic approximation, the individual nucleon motion may be decoupled from the collective modes, and a nucleon may be considered to be moving in an average potential well whose shape and orientation change only slowly in time, compared with the orbital periods of intrinsic motion. However, the individual nucleonic motion in a potential well gives rise to a nonsmooth shell effect in energy, which is deformation dependent, and this may favor equilibrium deformation in certain regions of mass number. This point is discussed further in Section 12–4. In the rare-earth region, nuclei with $60 \leqslant Z \leqslant 72$ and $90 \leqslant N \leqslant 110$ are found to have deformed ground states. For even-even nuclei the spins of individual nucleons pair off to zero, and the low-lying states may be considered to be arising from the rotational and vibrational modes of a deformed liquid drop. In this section, therefore, we ignore the intrinsic nucleonic part of the wave function.

The general equation of shape, Eq. 9–5, involving the $\alpha_{\lambda\mu}$ is suitable for studying vibrations of nearly spherical nuclei, but for deformed nuclei a different set of coordinates is more convenient. The $\alpha_{\lambda\mu}$ describe changes of shape with respect to a set of axes fixed in space. If a nucleus has a permanent nonspherical shape which is rotating, the $\alpha_{\lambda\mu}$ will change in time, although we would not say that the nuclear shape is altering. Consequently, we prefer to specify the orientation of the nucleus by giving the Euler angles $(\theta_1, \theta_2, \theta_3)$ of the principal axes of the nucleus with respect to the space-fixed axes,[†] and by introducing other parameters that specify the shape of the nucleus with respect to its own principal (body-fixed) axes.

We shall confine our attention to quadrupole shapes $\lambda = 2$. In the body-fixed system, we define the nuclear surface by

---

† For definitions of Euler angles and $\mathscr{D}$-functions, see Section A-2.

$$R = R_0 \left[1 + \sum_{\mu} a_{2\mu} Y_2{}^{\mu}(\theta', \phi')\right]. \tag{9-28}$$

Comparing this with Eq. 9-5, we have

$$\sum_{\mu} \alpha_{2\mu} Y_2{}^{\mu}(\theta, \phi) = \sum_{\mu} a_{2\mu} Y_2{}^{\mu}(\theta', \phi'). \tag{9-29}$$

In the rotation $(\theta_1, \theta_2, \theta_3)$ from the space-fixed to the body-fixed axes, the spherical harmonics transform according to Eq. A-73:

$$Y_\lambda{}^{\mu}(\theta, \phi) = \sum_{\rho} Y_\lambda{}^{\rho}(\theta', \phi') \mathscr{D}^{\lambda}_{\mu\rho}(\theta_1, \theta_2, \theta_3), \tag{9-30}$$

where the rotation matrix $\mathscr{D}^{\lambda}_{\mu\rho}(\theta_i)$ is defined in Eq. A-53. Substituting the above expression for $Y_\lambda{}^{\mu}(\theta, \phi)$ into Eq. 9-29, we obtain

$$a_{2\rho} = \sum_{\mu} \alpha_{2\mu} \mathscr{D}^2_{\mu\rho}(\theta_i) \tag{9-31a}$$

and

$$\alpha_{2\mu} = \sum_{\rho} a_{2\rho} (\mathscr{D}^2_{\mu\rho})^*. \tag{9-31b}$$

The body-fixed frame is chosen to be the principal axes, such that $a_{21} = a_{2,-1} = 0$ and $a_{22} = a_{2,-2}$ (see Problem 9-3). Thus a_{20} and a_{22}, together with the three Euler angles $(\theta_1, \theta_2, \theta_3)$, completely describe the system, replacing the five $\alpha_{2\mu}$. It is more convenient to use the variables β and γ instead of a_{20} and a_{22}:

$$a_{20} = \beta \cos \gamma, \qquad a_{22} = \frac{1}{\sqrt{2}} \beta \sin \gamma. \tag{9-32}$$

Using these expressions in Eq. 9-28 and looking up the relevant $Y_2{}^{\mu}$'s in Eq. A-14, we find

$$R - R_0 = \sqrt{\frac{5}{16\pi}} R_0 \beta [\cos \gamma (3 \cos^2\theta' - 1) + \sqrt{3} \sin \gamma \sin^2 \theta' \cos 2\phi']. \tag{9-33}$$

The quantity β is a measure of the total deformation of the nucleus from sphericity, since

$$\beta^2 = \sum_{\mu} a_{2\mu}^2 = \sum_{\mu} |\alpha_{2\mu}|^2. \tag{9-34}$$

The significance of γ is seen by writing the increments in lengths along the body-fixed axes x', y', and z':

$$\delta R_{z'} = R(0, \phi') - R_0 = \sqrt{\frac{5}{4\pi}} R_0 \beta \cos \gamma,$$

$$\delta R_{x'} = R\left(\frac{\pi}{2}, 0\right) - R_0 = \sqrt{\frac{5}{4\pi}} R_0 \beta \cos\left(\gamma - \frac{2\pi}{3}\right),$$

$$\delta R_{y'} = R\left(\frac{\pi}{2}, \frac{\pi}{2}\right) - R_0 = \sqrt{\frac{5}{4\pi}} \beta R_0 \cos\left(\gamma - \frac{4\pi}{3}\right).$$

If we denote, for convenience in notation, the body-fixed axes x', y', z' by 1, 2, 3, the above expressions may be summarized as follows:

$$\delta R_\kappa = \sqrt{\frac{5}{4\pi}} \beta R_0 \cos\left(\gamma - \frac{\kappa 2\pi}{3}\right), \quad \kappa = 1, 2, 3. \qquad (9\text{--}35)$$

The value $\gamma = 0$ yields a prolate (cigar-shaped) ellipsoid with the 3-axis as its symmetry axis, while $\gamma = 2\pi/3$ and $\gamma = 4\pi/3$ are prolate ellipsoids with the 1- and 2-axes as symmetry axes. Similarly $\gamma = \pi$, $\gamma = \pi/3$, and $\gamma = 5\pi/3$ are oblate ellipsoids about the 3, 2, and 1 symmetry axes, respectively.

Note that, if we make a cyclic permutation of the axes 1, 2, 3, the nuclear surface remains unaltered. It is easy to see from Eq. 9–35 that this cyclic permutation could be effected by changing γ to $(\gamma - 2\pi/3)$. Thus γ, $(\gamma - 2\pi/3)$, and $(\gamma - 4\pi/3)$ all describe the same nuclear surface. Furthermore, from Eq. 9–35 we see that γ and $-\gamma$ define the same nuclear surface. Therefore

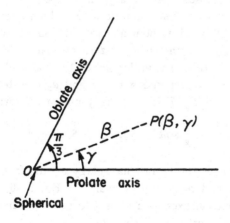

Fig. 9-5 The point P in the β-γ polar coordinates represents a definite asymmetric quadrupole shape. The length OP represents the magnitude of deformation β, and the angle that it makes with the horizontal axis measures the asymmetry parameter γ.

the γ-dependence of any physical quantity which is only a function of nuclear shape should be through the variable $\cos 3\gamma$. It is clear that the nuclear shape can be specified in a two-dimensional β-γ plane in polar coordinates, with the length of the vector representing β and its angular coordinate denoting γ, as shown in Fig. 9–5. Moreover, since only those functions of γ can come for which $f(\gamma) = f(\gamma - 2\pi/3)$, that is, $f(\gamma + \pi/3) = f(\gamma - \pi/3)$, and, in addition, $f(\gamma) = f(-\gamma)$, it follows that $f(\gamma)$ is symmetrical about the line $\gamma = \pi/3$. It is thus sufficient to take a 60° wedge in the β-γ plane, as in Fig. 9–5, to specify all possible quadrupole shapes. Any point on the $\gamma = 0$ line represents a prolate shape with the axis of symmetry along the 3-axis, while a point on the $\gamma = \pi/3$ line is an oblate shape with the 2-axis as the axis of symmetry. Any point within the wedge not lying on either of these lines is a deformed shape with no axis of symmetry.

We must now investigate the transformation of the kinetic energy

$$T = \frac{1}{2} \sum_{\mu} B |\dot{\alpha}_{2\mu}|^2$$

to the body-fixed system. From Eq. 9–32, we find

$$\dot{\alpha}_{2\mu} = \sum_{\rho} \dot{a}_{2\rho} \mathscr{D}_{\mu\rho}^{2*}(\theta_i) + a_{2\rho} \sum_{\theta_i} \dot{\theta}_i \frac{\partial}{\partial \theta_i} \mathscr{D}_{\mu\rho}^{2*}(\theta_i).$$

Substituting this in the expression for T, we find that there are three types of terms in the kinetic energy: (1) terms quadratic in $\dot{a}_{2\rho}$, which represent vibrations of the ellipsoid with change in shape, but the orientation θ_i unaltered in time; (2) terms quadratic in $\dot{\theta}_i$, which denote change in the orientation of the ellipsoid without a change in shape; and (3) cross terms like $\dot{a}_{2\rho}\dot{\theta}_i$, which may be shown to be zero by using the properties of the $\mathscr{D}$-matrices. Note, further, that in this picture there are only two independent modes of vibrations, corresponding to $\dot{a}_{20}$ and $\dot{a}_{22}$. After some nontrivial algebra, the expression for T can be finally put in the form [2]

$$T = \tfrac{1}{2}B(\dot{\beta}^2 + \beta^2\dot{\gamma}^2) + \frac{1}{2} \sum_{\kappa=1}^{3} \mathscr{I}_\kappa \omega_\kappa^2, \tag{9–36}$$

where the first term is the vibrational energy arising from terms quadratic in $\dot{a}_{2\rho}$, and the second term is rotational. In Eq. 9–36, $\boldsymbol{\omega}$ is the angular velocity of the principal axes with respect to the space-fixed axes,[†] and the effective moment of inertia $\mathscr{I}_\kappa$ is given by

---

[†] Expressions for $\boldsymbol{\omega}$ are given, for example, in Eq. 4–103 of H. Goldstein, *Classical Mechanics*, Addison-Wesley, Reading, Mass., 1950.

$$\mathscr{I}_\kappa = 4B\beta^2 \sin^2\left(\gamma - \frac{\kappa 2\pi}{3}\right). \tag{9-37}$$

Note that, if there is an axis of symmetry, the moment of inertia for collective rotations about this axis vanishes, that is, if $\gamma = 0$ or π, $\mathscr{I}_3 = 0$. In this case, the moments about the other axes are equal:

$$\mathscr{I}_1 = \mathscr{I}_2 = \mathscr{I} = 3B\beta^2. \tag{9-38}$$

We note that to obtain the moment of inertia we must first obtain the inertial parameter B from a more basic model. If we use the value of B given by Eq. 9–7 for the irrotational fluid, $\mathscr{I}$ can easily be expressed in terms of the moment of inertia of a rigid sphere of the same radius and density:

$$\mathscr{I} = \left(\frac{45}{16\pi}\right)\mathscr{I}_{\text{rig}}\beta^2, \tag{9-39}$$

with

$$\mathscr{I}_{\text{rig}} = \tfrac{2}{5}AMR_0^2, \tag{9-40}$$

where M, as usual, is the nucleon mass. We see that, since β is small, very little of the nuclear matter is actually taking part in the effective rotation. Although β and γ are fixed so that a given shape is rotating, it is more realistic to think of the rotation as the motion of a wave around the nuclear surface, carrying relatively little matter with it. These conclusions, including in particular the belief that the moment of inertia is zero about the symmetry axis, are based on the hydrodynamical model. We shall have more to say on this point later in the section.

We should now find the potential-energy term of the deformed drop. For the spherical drop, Eq. 9–8 immediately gives

$$V(\beta) = \tfrac{1}{2}C\beta^2, \tag{9-41}$$

where we have dropped the subscript 2 from C with the understanding that only quadrupole modes are being considered. Such a potential could be represented in the β-γ plane as shown in Fig. 9–6a. If we were to quantize the motion of a liquid drop in the present formalism, we would replace the velocities in the kinetic-energy expression by the conjugate momenta, and these in turn by differential operators in the Schrödinger representation. For the spherical case, the Hamiltonian takes the form

$$H = T_\beta + T_\gamma + \sum_{\kappa=1}^{3} \frac{L_\kappa^2}{2\mathscr{I}_\kappa} + \tfrac{1}{2}C\beta^2, \tag{9-42}$$

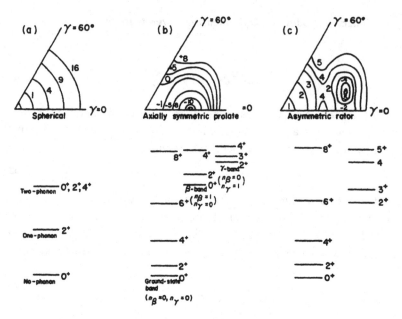

Fig. 9-6 Schematic plots of $V(\beta, \gamma)$ in arbitrary units for (a) spherical, (b) axially symmetric prolate, and (c) asymmetric shapes, with the corresponding low-lying energy levels.

where T_β and T_γ together constitute the vibrational kinetic energy, the third term is rotational, and L_κ is the component of angular momentum along the κ-axis. The quantities L_κ are referred to the moving axes and do not have the same commutation properties as angular-momentum components along fixed axes. The difference is only in the sign of the commutators:[†] $[L_1, L_2] = -iL_3$, but $[L_x, L_y] = iL_z$; consequently only slight changes are involved in handling them. In particular, although L^2 and L_z are constants of motion, in general no component of L_κ is. If we solve the Schrödinger equation with the Hamiltonian 9-42 and values of B and C given as before by Eqs. 9-7 and 9-11, we obtain the vibrational spectrum shown in Fig. 9-6a, which is identical to that of Eq. 9-16. This is not surprising, since we are solving essentially the same problem in the body-fixed frame, and actually making the analysis much more difficult by doing so.

This approach of describing the motion in terms of β- and γ-vibrations coupled with rotation is not of value if the equilibrium shape of the nucleus is strictly spherical; but, as we saw in the earlier section, this is too unrealistic a model. The vast majority of nuclei either seem to have a well-deformed equilibrium shape ("stiff" nuclei) or are transitional, being very soft to

---

[†] See, for example, Section 101 of L. D. Landau and E. M. Lifshitz, *Quantum Mechanics (Non-Relativistic Theory)*, Addison-Wesley, Reading, Mass., 1959.

nuclear deformation. For these "soft" nuclei, it is not meaningful to say that the equilibrium shape is characterized by a single point in the β-γ plane, since $V(\beta, \gamma)$ has no well-localized minimum in this plane. For such nuclei in particular, the method of solving the Schrödinger equation with the Bohr Hamiltonian

$$H = T_\beta + T_\gamma + \sum_{\kappa=1}^{3} \frac{L_\kappa^2}{2\mathcal{J}_\kappa} + V(\beta, \gamma) \tag{9-43}$$

is of great value. We shall return to transitional nuclei after discussing the simpler case of a stiff deformed nucleus.

If the inertial parameters and the potential-energy function $V(\beta, \gamma)$ are known (let us say from some basic microscopic calculation), the energy spectrum and the wave function can, in principle, be obtained by solving the Schrödinger equation with the Bohr Hamiltonian (Eq. 9-43). Much physical insight, however, can often be gained by examining the potential function $V(\beta, \gamma)$, which mainly governs the type of spectrum to be expected. In Fig. 9-6b, for example, a potential function which has a well-defined minimum for $\gamma = 0$ and $\beta = \beta_0$ is plotted. Although kinetic-energy effects are important, we can qualitatively argue that the nucleus in its ground state will tend to have the shape with $\gamma = 0$, $\beta = \beta_0$ in this case, so that the nuclear density has maximum overlap with $V(\beta, \gamma)$. As we have already seen, such a shape corresponds to an axially symmetric prolate ellipsoid with the 3-axis as its axis of symmetry. We can thus expect a rotational ground-state band of an axially symmetric rotor, where no vibrational quanta have been excited. We noted before that in the body-fixed axes two modes of vibrations are possible, $\dot{a}_{20}$ and $\dot{a}_{22}$. In a spheroidal shape, the multipole order λ of phonons is not actually a good quantum number, but its projection along the symmetry axis, ν ($= 0$ or 2 here), is. The vibrations are expected to be about the equilibrium values $\beta = \beta_0$ and $\gamma = 0$. For small vibrations, at any instant of time t,

$$\beta(t) = \beta_0 + \delta\beta(t) \qquad \text{and} \qquad \gamma(t) = \delta\gamma(t). \tag{9-44}$$

Using Eq. 9-32, we then obtain, to first order,

$$\dot{a}_{20} = \delta\dot{\beta}, \qquad \dot{a}_{22} = \frac{1}{\sqrt{2}} \beta_0 \, \delta\dot{\gamma}. \tag{9-45}$$

There are thus two types of vibrational quanta: the β-type, with the phonon angular-momentum projection zero along the symmetry axis, and the γ-type, with projection 2 along the original symmetry axis. We may picture β-vibrations as oscillations of the nuclear shape with its axis of symmetry preserved, while in γ-vibrations the nucleus loses its axial symmetry. If we denote the number of β-type phonons by n_β and the number of γ-type phonons by n_γ, in the ground-state rotational band $n_\beta = n_\gamma = 0$. We should then expect two more low-lying rotational bands, built on vibrational states $n_\beta = 0$, $n_\gamma = 1$ and $n_\beta = 1$, $n_\gamma = 0$. This kind of spectrum for a prolate axially symmetric nucleus is shown in Fig. 9–6b. Note that for the spheroid both the inertial parameter B and the force constant C can be quite different for β- and γ-vibrations, so that $\hbar\omega_\beta \neq \hbar\omega_\gamma$.

Another example of $V(\beta, \gamma)$ for a stiff nucleus is shown in Fig. 9–6c; in this case there is a well-defined minimum at a nonzero value of $\gamma = \gamma_0$ and $\beta = \beta_0$. From the preceding considerations, we can expect the low-lying spectrum to be that of an asymmetric rotor, and this is shown in Fig. 9–6c. In the following section, we shall consider both cases, Fig. 9–6b and Fig. 9–6c, in more detail.

9-4 SYMMETRIES OF THE COLLECTIVE WAVE FUNCTION FOR WELL-DEFORMED, EVEN-EVEN NUCLEI

In this section, we shall discuss in some detail the wave function of the so-called stiff, deformed nuclei, for which the potential function $V(\beta, \gamma)$ has a well-localized minimum. As before, we consider the low-lying states of even-even nuclei and disregard the intrinsic nucleonic part of the wave function. The potential function $V(\beta, \gamma)$ is such that there is a minimum for $\beta = \beta_0$, $\gamma = \gamma_0$; γ_0 may or may not be zero. In addition to rotation, the nucleus will vibrate around the equilibrium shape; for small oscillations the potential can be expanded about (β_0, γ_0) and written in the form

$$V(\beta, \gamma) \approx V_0 + \tfrac{1}{2}C_\beta(\beta - \beta_0)^2 + \tfrac{1}{2}C_\gamma(\gamma - \gamma_0)^2. \qquad (9\text{–}46)$$

We also note that rotations and vibrations are actually coupled because the moments of inertia $\mathscr{I}_\kappa$ are functions of β, γ. To extract the rotation-vibration coupling term in an approximate manner, we can expand $1/\mathscr{I}_\kappa(\beta, \gamma)$ in Eq. 9–43 about the equilibrium (β_0, γ_0):

$$\frac{1}{\mathscr{I}_\kappa(\beta, \gamma)} = \frac{1}{\mathscr{I}_\kappa(\beta_0, \gamma_0)} + \left[\frac{\partial}{\partial\beta}\frac{1}{\mathscr{I}_\kappa(\beta, \gamma_0)}\right]_{\beta_0}(\beta - \beta_0)$$

$$+ \left[\frac{\partial}{\partial\gamma}\frac{1}{\mathscr{I}_\kappa(\beta_0, \gamma)}\right]_{\gamma_0}(\gamma - \gamma_0) + \cdots. \qquad (9\text{–}47)$$

Using expressions 9–46 and 9–47, we can now write, for a well-deformed nucleus, the Hamiltonian 9-43 as,

$$H = T_\beta + \tfrac{1}{2}C_\beta(\beta - \beta_0)^2 + T_\gamma + \tfrac{1}{2}C_\gamma(\gamma - \gamma_0)^2$$

$$+ \sum_{\kappa=1}^{3} \frac{L_\kappa{}^2}{2\mathscr{I}_\kappa(\beta_0, \gamma_0)} + (U_1 + U_2), \qquad (9\text{-}48)$$

where U_1, U_2 are the rotation-vibration interaction terms. If we take the hydrodynamic form 9-37 for the moment of inertia and $\gamma_0 = 0$, then, to first order in $\beta - \beta_0$ and γ, we obtain

$$U_1 = -\frac{\hbar^2}{\mathscr{I}_0}(L^2 - L_3{}^2)\,\frac{\beta - \beta_0}{\beta_0},$$

$$U_2 = -\frac{\hbar^2}{2\mathscr{I}_0}(L_+{}^2 + L_-{}^2)\,\frac{\gamma}{\sqrt{3}}, \qquad (9\text{-}49)$$

where the operators L_+ and L_- are the raising and lowering operators defined in Eq. A-1. Whereas U_1 is diagonal in K, U_2 can mix different bands with $\Delta K = \pm 2$. The effect of $U_1 + U_2$ may be treated in perturbation theory in an approximate manner; in a more careful treatment [13], expansion 9-47 is carried up to second order, and the total Hamiltonian is diagonalized in the basis generated by neglecting the coupling.

We shall soon examine the effects of such terms on the rotational spectrum. If this coupling is neglected, the Bohr Hamiltonian (Eq. 9-48) is separable in vibrational and rotational parts. In the general case of three unequal $\mathscr{I}_\kappa$, the projection of the angular momentum about the 3-axis, K, is not a good quantum number. The wave function, however, can still be expressed as a linear combination of the $(2J + 1)$ $\mathscr{D}^J$'s. We can write

$$\Psi = f_{J,\tau,\nu}(\beta) \sum_{K=-J}^{J} g_K{}^{J,\tau}(\gamma)\mathscr{D}^J_{MK}(\theta_1, \theta_2, \theta_3). \qquad (9\text{-}50)$$

The rotational wave functions $\mathscr{D}^J_{MK}$ are simultaneous eigenfunctions of J^2, J_z, and J_3 as given in Eq. A-71. Besides J and M, three other quantum numbers are required to specify the state completely. For the axially symmetric case, for example, these could be K and the phonon numbers n_β and n_γ, as mentioned in the preceding section. In the general case of Eq. 9-50, τ stands for two of the quantum numbers, associated with γ-vibrations and rotations, and ν stands for the third, associated with β-vibrations.

Certain symmetries restrict the generalities of the function $g(\gamma)$ and the allowed values of J and K in this model. Using Eq. 9-32, we see that the space-fixed $\alpha_{2\mu}$'s are expressible in terms of the body-fixed parameters as

$$\alpha_{2\mu} = \beta \cos \gamma \mathscr{D}^{2*}_{\mu 0}(\theta_i) + \frac{1}{\sqrt{2}} \beta \sin \gamma [\mathscr{D}^{2*}_{\mu 2}(\theta_i) + \mathscr{D}^{2*}_{\mu,-2}(\theta_i)], \qquad (9-51)$$

where we have chosen the body-fixed frame to be right-handed. The wave function should be single valued in the quantities α_μ (we drop the suffix 2), which specify the shape of the nucleus uniquely, but various choices of the body-fixed variables† (β, γ, and the Euler angles $\theta_1, \theta_2, \theta_3$) are possible which all give the same α_μ's. The wave function 9–50 must be invariant under any change of the body axes which does not alter the α_μ's. It is important to note that the body-fixed frame (1, 2, 3) is specified by a set of Euler angles $(\theta_1, \theta_2, \theta_3)$, and we have chosen a right-handed system. By making changes in these values of Euler angles (i.e., by performing a set of rotations), we can never go from the right-handed frame (1, 2, 3) to the left-handed frame $(-1, -2, -3)$ in this three-dimensional system. It is possible, as we shall see, to go from (1, 2, 3) to $(-3, -2, -1)$ by a series of rotations, but this is still a right-handed system and does *not* mean an inversion. Since we describe the $\mathscr{D}$-functions in terms of the Euler angles, and these variables cannot take a right-handed system to a left-handed one, we cannot attach a parity quantum number to the rotational function $\mathscr{D}^J_{MK}(\theta_i)$ in the general case (when M and K are non-zero). However, in the special case when $K = 0$, it may be seen from Eq. A–66 that $\mathscr{D}^J_{M0}(\theta_1, \theta_2, \theta_3) = [4\pi/(2J + 1)]^{1/2} \cdot Y_J^M(\theta_2, \theta_1)$, and θ_3 is redundant. The variables θ_2, θ_1 can now be considered to specify the orientation of a single vector $\mathbf{r}$. By changing θ_2 to $\pi - \theta_2$, θ_1 to $\pi + \theta_1$, we can transform $\mathbf{r}$ to $-\mathbf{r}$, that is, a parity operation. Thus the spherical harmonics have parity, $(-)^J$, in this sense. In the general case, the total wave function of the rotating nucleus will still have a parity, but it will be the parity of the intrinsic nucleonic wave function $\chi(\mathbf{r}_i)$, which we have ignored in this discussion.

We now consider the group of transformations on the body axes (1, 2, 3) which retain the right-handedness, but change the variables γ, θ_i in such a manner as to keep the α_μ's invariant. Any such transformation can be described by a group of rotations, and these should leave the wave function 9–50 invariant. These transformations describe all possible ways in which the *right-handed* body axes can be relabeled: (1) without changing any of the signs of the axes 1, 2, or 3; (2) changing only one of the signs; (3) changing any two of the signs; and finally (4) changing all three signs, but in all cases retaining the right-handedness. We consider each of these cases, 1–4, in turn.

---

† The variable β, by Eq. 9–34, is a scalar and remains invariant under rotation.

1. These are simply the cyclic permutations of the axes $(1, 2, 3)$, and we define the operator R_3 so that $R_3(1, 2, 3) = (3, 1, 2)$. Obviously, there are three such possibilities, and $R_3{}^3 = 1$.

2. The sign of one axis can be reversed, while maintaining the right-handedness, by making a rotation of $\pi/2$ about any one of the axes. For example, if we rotate $(1, 2, 3)$ about the 3-axis by $\pi/2$, we obtain $(1', 2', 3')$, where $1' = 2$, $2' = -1$, and $3' = 3$. This operator is denoted by $R_2{}^{(3)}$, where the parenthetical superscript denotes the axis about which the rotation has been performed:

$$R_2{}^{(3)}(1, 2, 3) = (2, -1, 3). \tag{9-52}$$

These indices can be cyclically permuted in three possible ways. Furthermore, we could have applied any *one* of the operators $R_2{}^{(1)}$, $R_2{}^{(2)}$, or $R_2{}^{(3)}$ to $(1, 2, 3)$; thus there are nine possible combinations $(1', 2', 3')$ where only one sign is negative and the system is right-handed.

3. Similarly, two of the signs in the axes could be reversed in $(1, 2, 3)$ by making a rotation of π about any one of the axes. Defining this rotation by $R_1{}^{(\kappa)}$, we see, for example, that

$$R_1{}^{(1)}(1, 2, 3) = (1, -2, -3). \tag{9-53}$$

As in case (2), there are again nine such transformations in the right-handed system with two of the signs negative.

4. Finally, we note that, by taking suitable combinations of R_1 and R_2, we can reverse all the signs in $(1, 2, 3)$. For example,

$$R_2{}^{(3)}R_1{}^{(1)}(1, 2, 3) = R_2{}^{(3)}(1, -2, -3) = (-2, -1, -3), \tag{9-54}$$

where we have used Eqs. 9–52 and 9–53. Note that we cannot obtain the left-handed system $(-1, -2, -3)$ from $(1, 2, 3)$ by any set of rotations. There are, clearly, three different possibilities, corresponding to the three cyclic permutations of $(-2, -1, -3)$.

In all, therefore, there are 24 different sets of axes $(1', 2', 3')$ in the right-handed frame which can alter the body-fixed variables in such a way as to keep the α_μ's invariant. Since these are all obtainable through various combinations of the operators R_1, R_2, and R_3, we should demand that the wave function 9–50 be invariant under these operations. An alternative, but less transparent, way of noting that there are 24 such combinations is to realize that $(R_3)^3 = 1$, $[R_2{}^{(3)}]^4 = 1$, and $[R_1{}^{(1)}]^2 = 1$, that all possibilities are covered by the operators $[R_1{}^{(1)}]^a \cdot [R_2{}^{(3)}]^b \cdot (R_3)^c$, and that there are

$3 \times 4 \times 2 = 24$ such distinct operators. We can now examine how the $\mathscr{D}$-functions and γ transform under the transformations R_1, R_2, and R_3. The rotation operator $R_1^{(1)}$ is a rotation of π around the 1-axis, changing 2 to -2 and 3 to -3. From Eq. 9–35, this does not change γ, but geometrical considerations show that $\theta_1 \to (\pi + \theta_1)$, $\theta_2 \to (\pi - \theta_2)$, and $\theta_3 \to (2\pi - \theta_3)$. Thus

$$R_1^{(1)}\mathscr{D}^J_{MK}(\theta_1, \theta_2, \theta_3)$$

$$= \mathscr{D}^J_{MK}(\pi + \theta_1, \pi - \theta_2, 2\pi - \theta_3)$$

$$= \exp[iM(\pi + \theta_1)] \, d^J_{MK}(\pi - \theta_2) \exp(-iK\theta_3), \quad \text{using Eq. A–53,}$$

$$= \exp(iM\theta_1) \, (-)^J d^J_{M,-K}(\theta_2) \exp(-iK\theta_3), \quad \text{using Eq. A–59,}$$

$$= (-)^J \mathscr{D}^J_{M,-K}(\theta_1, \theta_2, \theta_3). \tag{9–55}$$

The rotation operator $R_2^{(3)}$ is a rotation by $\pi/2$ about the 3-axis; it changes γ to $-\gamma$ and the Euler angles transform as $\theta_1 \to \theta_1$, $\theta_2 \to \theta_2$, $\theta_3 \to \theta_3 + \pi/2$. Thus

$$R_2^{(3)}\mathscr{D}^J_{MK}(\theta_1, \theta_2, \theta_3) = \mathscr{D}^J_{MK}\left(\theta_1, \theta_2, \theta_3 + \frac{\pi}{2}\right)$$

$$= \exp(iK\pi/2)\,\mathscr{D}^J_{MK}(\theta_1, \theta_2, \theta_3). \tag{9–56}$$

Finally, R_3, a cyclic permutation of the axes, is equivalent to the rotation $\theta_1 = 0$, $\theta_2 = \pi/2$, $\theta_3 = \pi/2$, and changes γ to $(\gamma - 2\pi/3)$. We can therefore write

$$R_3\mathscr{D}^J_{MK}(\theta_1, \theta_2, \theta_3) = \sum_{K'} \mathscr{D}^{J*}_{K'K}(0, \pi/2, \pi/2)\mathscr{D}^J_{MK'}(\theta_i). \tag{9–57}$$

By using Eqs. 9–55 to 9–57, it can be verified that under the transformations R_1, R_2, and R_3 the expression for $\alpha_{2\mu}$ as given in Eq. 9–51 remains invariant. We now apply these transformations to the wave function 9–50 to find its symmetry properties. It should be realized that the intrinsic part of the wave function, χ, is not included in Eq. 9–50, and this part will also transform under the symmetry operations. For even-even nuclei, the low-lying states are given, however, by the rotational and vibrational modes with the intrinsic spin zero, and we can ignore the intrinsic part. For odd nuclei, this is not the case, and it is necessary to consider the transformation of χ explicitly under the symmetry operations.

Consider first the effect of $R_2^{(3)}$ on the wave function 9–50. If we demand that $R_2^{(3)}\Psi = \Psi$, then, using Eq. 9–56 and recalling that $\gamma \to -\gamma$ under the operation of $R_2^{(3)}$, we obtain

$$g_K{}^J(-\gamma)\exp(iK\pi/2) = g_K{}^J(\gamma).\tag{9-58}$$

Applying then the condition $[R_2^{(3)}]^2\Psi = \Psi$, we see that $g_K{}^J(\gamma) = g_K{}^J(\gamma) \cdot \exp(iK\pi)$; hence K must be an even integer for $g_K{}^J$ to be nonzero.

Next, we consider the operator $R_1^{(1)}$ on Ψ, noting that γ remains unchanged. Since $[R_1^{(1)}]^2 = 1$, it follows that, if we take a wave function $[1 + R_1^{(1)}]\Psi$, it will be invariant under the operation $R_1^{(1)}$. Recalling Eq. 9-55, we can therefore write an invariant wave function:

$$\Phi = f_{J,\tau,\nu}(\beta)\sum_{\substack{K=0\\(\text{even})}}^{J} g_K{}^{J,\tau}(\gamma)|JMK\rangle,\tag{9-59}$$

where $|JMK\rangle$ is a normalized rotational wave function given by

$$|JMK\rangle = \left[\frac{2J+1}{16\pi^2(1+\delta_{K0})}\right]^{1/2}(\mathscr{D}^J_{MK} + (-)^J\mathscr{D}^J_{M,-K}).\tag{9-60}$$

If there is axial symmetry, K is a good quantum number, and the collective wave function can simply be written as

$$\Phi = f_{J,K,n_\beta}(\beta)g_{K,n_\gamma}^J(\gamma)|JMK\rangle.\tag{9-61}$$

Note that, for $K = 0$, Eq. 9-60 implies that only even values of J are allowed. In this analysis, however, only the quadrupole oscillations were considered. More generally, it is possible to have rotational bands built on octupole vibrations, in which case, for $K = 0$, only an odd sequence of J values appears. Consider, for simplicity, an extra vibrational degree of freedom $a_{30}(t)Y_3^0(\theta', \phi')$ of the nuclear shape. The length of a vector in the body-fixed frame, given by Eq. 9-28, now generalizes to

$$R(\theta', \phi') = R_0[1 + \sum_\mu a_{2\mu}Y_2^\mu(\theta', \phi') + a_{30}Y_3^0(\theta', \phi')].$$

If we apply the rotation operator $R_1^{(1)}$ to the body-fixed frame, the same vector length is given in a new frame with $\theta' \to (\pi - \theta')$, $\phi' \to (2\pi - \phi')$. Under this operation, then, the Y_2^μ's do not change sign, but $Y_3^0(\pi - \theta', 2\pi - \phi') = -Y_3^0(\theta', \phi')$. In order that the length remain invariant under this transformation, the $a_{2\mu}$'s remain the same (i.e., β and γ do not change under $R_1^{(1)}$), but a_{30} must change sign. But a one-phonon harmonic-wave function always changes sign if the dynamic variable changes sign; hence the vibrational octupole one-phonon state $\phi_{\text{vib}}^{\lambda=3}$ changes sign under the operation $R_1^{(1)}$. Thus an unsymmetrized wave function $\phi_{\text{vib}}^{\lambda=3}(a_{30}) \cdot \mathscr{D}^J_{MK}(\theta_i)$ transforms as

$$R_1^{(1)}[\phi_{\text{vib}}^{\lambda=3}(a_{30})\mathcal{D}_{MK}^J(\theta_i)] = - \phi_{\text{vib}}^{\lambda=3}(a_{30})(-)^J\mathcal{D}_{M,-K}^J(\theta_i). \qquad (9\text{-}62)$$

In order to make the total wave function Φ invariant under $R_1^{(1)}$, we again take the combination

$$(1 + R_1^{(1)})\phi_{\text{vib}}^{\lambda=3}(a_{30})\mathcal{D}_{MK}^J(\theta_i) = (\mathcal{D}_{MK}^J - (-)^J\mathcal{D}_{M,-K}^J)\phi_{\text{vib}}^{\lambda=3}(a_{30}),$$

which shows that for $K = 0$ only odd-J sequences are allowed.

We have thus generated the normal collective $K = 0$ bands in even-even nuclei:

$$K\pi = 0^+, \quad J = 0^+, 2^+, 4^+, \ldots, \qquad (9\text{-}63a)$$

$$K\pi = 0^-, \quad J = 1^-, 3^-, 5^-, \ldots . \qquad (9\text{-}63b)$$

Sequence 9–63a is seen in the ground-state and β-vibrational rotational bands, while 9–63b is observed as the rotational band built on the $K = 0$ octupole vibration. It is worth mentioning that Das Gupta and Volkov [14] have pointed out the possibility of observing anomalous excited-state bands:

$$K\pi = 0^+, \quad J = 1^+, 3^+, 5^+, \qquad (9\text{-}64a)$$

$$K\pi = 0^-, \quad J = 0^-, 2^-, 4^-, \ldots, \qquad (9\text{-}64b)$$

which may arise if the vibrational $\lambda = 2$ state changes sign under $R_1^{(1)}$ whereas the vibrational $\lambda = 3$ state does not. This is not the case for the normal surface vibrations that we have considered, but may be realized, for example, if surface vibrations of the Y_{20}-type are coupled to the intrinsic spins of the nucleons. The 0^- and 2^- states at 1819 and 1856.9 keV in ^{176}Hf have been identified [15] as members of the anomalous sequence 9–64b.

9-5 COLLECTIVE SPECTRA OF EVEN-EVEN NUCLEI

We first consider the simple case of a nucleus with a well-localized deformation in the β-γ plane and further simplify the situation by assuming that it possesses an axis of symmetry, which we choose to be the 3-axis. The wave function of such a system is then given by Eq. 9–61, which separates into a vibrational part $f(\beta)g(\gamma)$ and a rotational part, provided the rotation-vibration coupling terms are neglected. The rotational energy is simply given by

$$\langle JMK|\sum_\kappa \frac{L_\kappa^2}{2\mathcal{I}_\kappa}|JMK\rangle = \hbar^2\left\langle\frac{L^2 - L_3^2}{2\mathcal{I}} + \frac{L_3^2}{2\mathcal{I}_3}\right\rangle$$

$$= \hbar^2\frac{J(J+1) - K^2}{2\mathcal{I}} + \frac{\hbar^2 K^2}{2\mathcal{I}_3}, \qquad (9\text{-}65)$$

where we have put $\mathscr{I}_1 = \mathscr{I}_2 = \mathscr{I}$. We consider first the ground-state rotational band of such a system, where $n_\beta = n_\gamma = 0$. The angular momentum $\mathbf{J}$ in this band is arising solely from rotational motion ($\mathbf{J} = \mathbf{L}$), and K is its projection along the symmetry axis. Any rotation of the frame about the symmetry axis, 3, should leave the rotational wave function invariant or, at most, introduce a phase factor. If we apply such a rotation about the 3-axis by an arbitrary angle θ, then, by Eq. A-46, the wave function $|JMK\rangle$ transforms to

$$\exp(-i\theta J_3)|JMK\rangle = \left[\frac{2J+1}{16\pi^2(1+\delta_{K0})}\right]^{1/2}$$

$$\cdot [\exp(-i\theta K)\,\mathscr{D}^J_{MK} + (-)^J \exp(i\theta K)\,\mathscr{D}^J_{M,-K}]$$

and hence axial symmetry demands that $K = 0$. This means that the rotational angular momentum $\mathbf{L}$ is perpendicular to the 3-axis, with the result that the nucleus is actually rotating about *an axis perpendicular to the axis of symmetry*. We see from Eq. 9–60 that for $K = 0$ only even values of J are allowed; hence the ground-state rotational band is given by

$$E_J = \frac{\hbar^2}{2\mathscr{I}} J(J+1), \tag{9–66}$$

with

$$K = 0; \qquad J = 0, 2, 4, \ldots . \tag{9–67}$$

We should remember that Eq. 9–67 is valid only for the quadrupole distortions that we have considered. If it is assumed that the intrinsic part of the wave function has positive parity, the parities of the rotational levels Eq. 9–67 are also positive. Note that for excited-state bands n_β or $n_\gamma \neq 0$, and it is possible to have nonzero values of K. For nonzero K, J takes all values $K, K + 1, K + 2$, etc.

In this section we have been considering only the collective modes of motion, with no reference to their coupling to intrinsic coordinates. In even-even nuclei where the intrinsic angular momentum is zero, we may expect a ground-state rotational band as given by Eq. 9–66. For the low-lying states of this rotational band, we may ignore the coupling due to vibrational and intrinsic modes of excitations of higher energy, and assume the moment of inertia to be a constant in the simplest approximation. Using Eq. 9–66, we then find that $E_4/E_2 = 10/3$, $E_6/E_2 = 7$, and $E_8/E_2 = 12$. The ground-state rotational band of ^{160}Dy, as found experimentally by Johnson, Ryde,

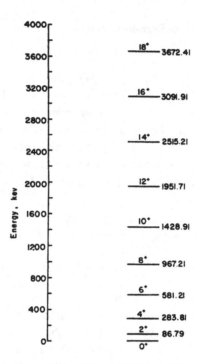

Fig. 9-7 The ground-state rotational band of ^{160}Dy. (After Johnson et al. [16].)

and Hjorth [16], is shown in Fig. 9–7. In this experiment, a beam of α-particles of energy near 43 MeV impinges on a target of ^{160}Gd. These α-particles impart a large amount of angular momentum to the compound nucleus, and a final state of ^{160}Dy in an angular-momentum state of $J \approx 20$ is formed with the ejection of four neutrons. The nucleus de-excites predominantly by cascading down the rotational band through $E2$-transitions, and the associated γ-ray cascades are detected with Ge(Li) detectors. From the experimental spectrum of ^{160}Dy, we note that the ratios of E_4, E_6, and E_8 to E_2 are 3.27, 6.70, and 11.14, respectively, these being somewhat smaller than the values 3.33, 7, and 12 given by $J(J + 1)$. This decrease of a few per cent may be understood in terms of a weak coupling between these low-lying rotational modes and the vibrational modes. In general, such couplings depend on $\mathbf{J}^2$ (see Eq. 9–49) and contribute to a change of energy in the second order of perturbation theory. Consequently, they introduce a correction term $-B[J(J + 1)]^2$. The first three or four excited states may be well fitted by the formula

$$E_J = \frac{\hbar^2}{2\mathscr{I}} J(J + 1) - BJ^2(J + 1)^2, \qquad (9\text{–}68)$$

with $\hbar^2/2\mathscr{I} = 14.53$ keV and $B = 18$ eV. This situation is fairly typical for even-even nuclei in the region $A = 150$–190 and for $A > 220$. Such a perturbative approach for the rotation-vibration coupling is not adequate for the higher members ($J > 10$) of the rotational band or even for the low-lying states of transition nuclei like the Os isotopes, where the γ-vibrational levels are appreciably lowered in energy and the rotation-vibration interaction is substantial. Faessler, Greiner, and Sheline [13] have isolated the rotation-vibration interaction terms U_1 and U_2 by carrying expansion 9–47 up to the second order, assuming the hydrodynamical expression 9–37 for the moment of inertia. They diagonalize the Hamiltonian 9–48 with the basis states 9–50 to obtain the collective states.

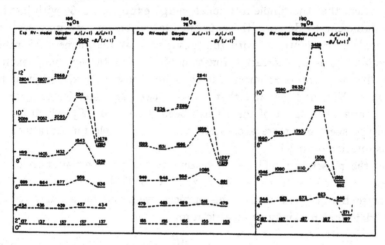

Fig. 9-8 Comparison of the various models for the energy spectra of Os isotopes (energies in keV). (After Faessler et al. [13].)

Figure 9–8 shows the results of their calculation for the Os isotopes, where their values (column 2) are compared with the experimental spectrum (column 1), as well as the predictions of Eqs. 9–66 (column 4) and 9–68 (column 5). In the Davydov model [17, 18, 19], whose predictions are shown in column 3, the levels are generated by the rotations of an asymmetric rotor. We shall discuss this model briefly later in this section; in general, its predictions are very similar to those of the axially symmetric model with rotation-vibration coupling included in the manner prescribed by Faessler et al. [13]. Although the success of this rotation-vibration model is impressive, as seen in Fig. 9–8, we should remember that the form of the rotation-vibration interaction term was extracted from the hydrodynamical formula 9–37 for $\mathscr{I}_\kappa$, which cannot be realistic.

The moment of inertia $\mathscr{I}$ obtained by empirical fitting of the first few low-lying levels is about half the rigid value $\mathscr{I}_{\rm rig} = \frac{2}{5}AMR_0{}^2$ in the case of ^{160}Dy. As we shall see, the distortion β may be determined in the rotational model from the experimental determination of the reduced $E2$-transition probability from the 0^+ to 2^+ state; using the hydrodynamical formula 9–39 with this value of β yields a moment of inertia that is about five times less than the observed $\mathscr{I}$. This poor agreement is typical of the irrotational fluid model and demonstrates the necessity of a microscopic model for the evaluation of the moment of inertia, as discussed in Section 10–8.

It is clear that the simple rotational-energy expression 9–66 with fixed $\mathscr{I}$ is not sufficient to describe the experimental spectrum; we may still use this simple form, however, with the provision that the moment of inertia is variable and is an increasing function of J. For the low-lying excited levels, the two-parameter formula 9–68 is an equivalent expression of the same fact. We shall see later that a better two-parameter formula for E_J with a variable moment of inertia has been developed [20], which can fit the energy spectrum more accurately and is applicable for vibrational as well as rotational nuclei.

For the present, it will be interesting to extract the variation of the moment of inertia with increasing angular velocity from the experimental data presented in Fig. 9–7. For this purpose, we define the angular velocity appropriate for an axially symmetric rotor:

$$\hbar\omega_{\rm rot} = \frac{dE}{d\sqrt{J(J+1)}}, \tag{9–69}$$

where we are identifying the angular momentum with $\hbar\sqrt{J(J+1)}$. Equation 9–69 can be rewritten as

$$\hbar^2\omega_{\rm rot}^2 = 4J(J+1)\left[\frac{dE}{dJ(J+1)}\right]^2. \tag{9–70}$$

The moment of inertia is defined by the relation

$$\mathscr{I}\omega_{\rm rot} = \hbar\sqrt{J(J+1)}. \tag{9–71}$$

Using Eqs. 9–70 and 9–71, we obtain

$$\frac{2\mathscr{I}}{\hbar^2} = \left[\frac{dE}{dJ(J+1)}\right]^{-1}. \tag{9–72}$$

Note that, by extracting the energy derivative $dE/dJ(J+1)$ from the experimental spectrum, we can calculate ω_{rot} and $\mathscr{I}$ from Eqs. 9–70 and 9–72, and study the variation of $\mathscr{I}$ with ω_{rot}. To evaluate the energy derivative from the experimental data, we may write

$$\frac{dE}{dJ(J+1)} = \frac{\Delta E}{\Delta J(J+1)} = \frac{E(J) - E(J-2)}{4J - 2}. \qquad (9\text{–}73)$$

Also, the factor $J(J+1)$ in Eq. 9–70 is replaced by its mean value between J and $J-2$, which is $(J^2 - J + 1)$. It is now possible, using the experimental spectrum of Fig. 9–7, to calculate the variation of $2\mathscr{I}/\hbar^2$ with $(\hbar\omega_{\text{rot}})^2$, and the resulting plot is shown in Fig. 9–9. This figure also shows† the curves for ^{168}Yb, ^{158}Dy, and ^{162}Er [16]. Actually, only discrete points corresponding to discrete even J-values are obtained in this figure, and these are joined by a smooth line. It will be seen that, whereas in ^{168}Yb the increase in $\mathscr{I}$ is smooth throughout, in the other nuclei there is an abrupt increase in the moment of inertia at $J \sim 14$–16, beyond

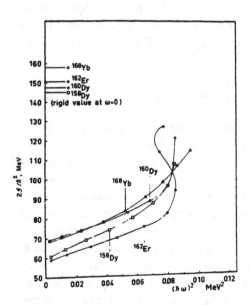

Fig. 9-9 The observed moments of inertia as functions of the square of the angular velocity. (After Johnson et al. [16].)

† Many other nuclides show the back-bending behavior. A detailed listing of experimental data is given by O. Saethre et al., *Nucl. Phys.* **A207**, 486 (1973).

which it tends to approach the rigid value. At least a partial explanation of this behaviour may be given in terms of the "Coriolis antipairing effect," predicted earlier by Mottelson and Valatin [21]. The following discussion is restricted to energy considerations.

Because of pairing interaction, as we have seen, the ground state of an even-even nucleus is lowered by the pairing energy $\langle E_{\text{pair}} \rangle$, which is typically only about -2 MeV in the rare-earth region. The pairing interaction, as we shall see, has a rather drastic effect on the moment of inertia, $\mathscr{I}$. Without the coherence of the pairing correlations, the moment of inertia of a system of fermions moving in an average potential is just the rigid value $\mathscr{I}_{\text{rig}}$ [22]; with the pairing interaction it is about halved in ^{160}Dy. Let us write the energy of the rotating system as

$$E = \frac{\hbar^2}{2\mathscr{I}} J(J+1) + \langle E_{\text{pair}} \rangle, \qquad (9\text{-}74)$$

where $\langle E_{\text{pair}} \rangle \approx -2$ MeV. For lower values of J, the nucleus prefers the superfluid state, because although the rotational energy has increased because of a drastic reduction in $\mathscr{I}$, this is more than compensated for by the gain in pairing energy. However, if J is large enough, it would be energetically more favorable for the nucleus to go over to the normal (nonsuperfluid) state, where $\mathscr{I} = \mathscr{I}_{\text{rig}}$ and $\langle E_{\text{pair}} \rangle = 0$. This should occur, in this simple picture, for a critical value of $J = J'$ such that[†]

$$\frac{\hbar^2}{2}\left(\frac{1}{\mathscr{I}} - \frac{1}{\mathscr{I}_{\text{rig}}}\right) J'(J'+1) = -\langle E_{\text{pair}} \rangle. \qquad (9\text{-}75)$$

For a rough estimate in ^{160}Dy, we assume that $\mathscr{I}_{\text{rig}} \approx 2\mathscr{I}$ and $\hbar^2/2\mathscr{I} \approx 15$ keV; then, using Eq. 9-75, we obtain $J' \approx 16$.

From Fig. 9-9 it is clear that, even well before this drastic change takes place in the moment of inertia $\mathscr{I}$, there is an almost linear increase in its value with increasing ω^2, reflecting the coupling of the rotational motion with vibrational and intrinsic modes of motion. For well-deformed nuclei, this rise in $\mathscr{I}$ with increasing J could be interpreted as due to an increase in the deformation parameter β, which adjusts itself to minimize the total energy of the nucleus. With increasing angular frequency, the energy of a state of given J is determined not only by the rotational kinetic energy, but also by a potential-energy term that represents the energy of other modes of motion. Furthermore, the two may be strongly coupled, the intrinsic wave functions being dependent on the angular velocity. We may

---

† See Section 10-8.

thus generalize Eq. 9–66 to

$$E_J = \frac{\hbar^2 J(J+1)}{2\mathscr{I}(\omega)} + V(\omega), \tag{9-76}$$

where ω is the rotational frequency (the subscript "rot" has been dropped to make the notation less cumbersome), and $V(\omega)$ is the potential energy. The moment of inertia $\mathscr{I}$ at any ω would adjust itself to minimize E_J, so that

$$\frac{dE_J}{d\mathscr{I}} = -\frac{\hbar^2 J(J+1)}{2\mathscr{I}^2(\omega)} + \frac{dV(\omega)}{d\mathscr{I}} = 0.$$

The above relation can be written, using Eq. 9–71, as

$$\frac{dV(\omega)}{d\mathscr{I}} = \tfrac{1}{2}\omega^2. \tag{9-77}$$

Since the experimental data displayed in Fig. 9–9 suggest that ω^2 is linearly related, up to a certain value of J, to $\mathscr{I} - \mathscr{I}_0$, $\mathscr{I}_0$ being the moment of inertia for $J = 0$, we can integrate Eq. 9–77 to obtain

$$V(\omega) = \tfrac{1}{2}C(\mathscr{I} - \mathscr{I}_0)^2,$$

where C is a constant. The energy of the state with angular momentum J can thus be written, from Eq. 9–76, using two parameters $\mathscr{I}_0$ and C:

$$E_J = \frac{\hbar^2 J(J+1)}{2\mathscr{I}} + \tfrac{1}{2}C(\mathscr{I} - \mathscr{I}_0)^2, \tag{9-78}$$

where $\mathscr{I}$ for each J is determined by the condition

$$\frac{dE_J}{d\mathscr{I}} = 0. \tag{9-79}$$

Expression 9–78 for E_J was proposed by Mariscotti, Scharff-Goldhaber, and Buck [20] and is known as the variable-moment-of-inertia (VMI) model. It has been very successful not only in the rotational but also in the vibrational and transitional regions. As an example of this, in Fig. 9–10 the experimental ground-state bands of a few even-even nuclei, together with the calculated values in the VMI model, are shown.

To appreciate that the above form of variable moment of inertia may arise from stretching of the effective radius during rotation, it is instructive to consider a very simple classical problem [23]: a diatomic rotor with an interaction potential, shown in Fig. 9–11. When the system is not rotating,

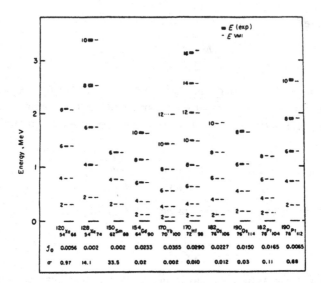

Fig. 9-10 Experimental and calculated ground-state bands of some even-even nuclei. For each nucleus, the experimental energies are shown on the left, and the calculated values, in the VMI model (Eq. 9–78), are on the right. The values of the parameters $\mathscr{I}_0$ and $\sigma = [\mathscr{I}^{-1}(d\mathscr{I}/dJ)]_{J=0} = \frac{1}{2}C\mathscr{I}_0^3$ are listed at the bottom of the figure. (After Mariscotti et al. [20].)

the masses rest at the equilibrium separation r_0. With rotation, there is some stretching in this separation, and the total energy is given by

$$E = \tfrac{1}{2}m\dot{\mathbf{r}}_1^2 + \tfrac{1}{2}m\dot{\mathbf{r}}_2^2 + V(r), \qquad (9\text{--}80)$$

where $\mathbf{r}_1$ and $\mathbf{r}_2$ are the coordinates of the two atoms with respect to some

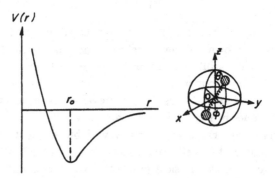

Fig. 9-11 A diatomic rotor with an interaction potential $V(r)$.

arbitrary origin, and $r = r_1 - r_2$. We choose the center of mass $R = \frac{1}{2}(r_1 + r_2)$ to be at rest and transform Eq. 9-80 to polar coordinates. The motion takes place in a plane, and, defining the angular momentum L by

$$L = \mu r^2 \omega = \mathscr{I}\omega, \qquad \mu = \frac{m}{2} \qquad (9\text{-}81)$$

we obtain from Eq. 9-80

$$E = \tfrac{1}{2}\mu \dot{r}^2 + \frac{L^2}{2\mu r^2} + V(r). \qquad (9\text{-}82)$$

Now $\mathscr{I} = \mu r^2$ and $\mathscr{I}_0 = \mu r_0^2$; and, assuming that r does not deviate excessively from r_0, we get

$$\mathscr{I} \approx \mathscr{I}_0 + 2\mu r_0(r - r_0)$$

or

$$r - r_0 = (\mathscr{I} - \mathscr{I}_0)C_1, \qquad (9\text{-}83)$$

where $C_1 = 2\mu r_0$. Similarly, the potential may be expanded about the equilibrium position r_0 to give

$$V(r) = V(r_0) + \tfrac{1}{2}C_2(r - r_0)^2$$

$$= V(r_0) + \tfrac{1}{2}C(\mathscr{I} - \mathscr{I}_0)^2,$$

where $C = C_2/C_1^2$. Substituting this into Eq. 9-82, we obtain

$$E = [\tfrac{1}{2}\mu \dot{r}^2 + V(r_0)] + \frac{L^2}{2\mathscr{I}} + \tfrac{1}{2}C(\mathscr{I} - \mathscr{I}_0)^2. \qquad (9\text{-}84)$$

In a given rotational band, the quantity in the brackets remains unchanged. Furthermore, the minimization condition $dE/d\mathscr{I} = 0$ immediately yields

$$\mathscr{I} - \mathscr{I}_0 = \frac{1}{2C}\,\omega^2,$$

showing that the increase in the moment of inertia is linearly proportional to ω^2.

The preceding exercise is by no means a derivation of the VMI formula; rather, it is meant as an illustration of the stretching effect on the moment of inertia. An analogous quantum-mechanical treatment has been made by Volkov [24], and a many-body problem has been treated in a similar spirit by Das, Dreizler, and Klein [25]. It is not well understood why the VMI equation 9–78 works so well for near-spherical and transitional nuclei.[†]

Coming back to axially symmetric deformed nuclei, we show the energy spectra of $^{164}_{68}$Er and $^{232}_{92}$U in Figs. 9–12 and 9–13, respectively, as typical examples. As mentioned before, in addition to the ground-state band,

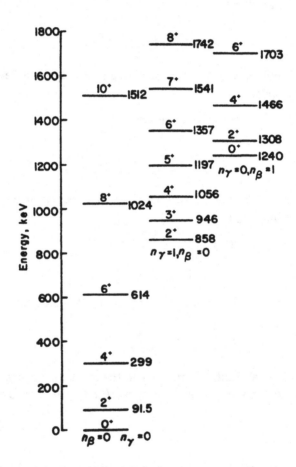

Fig. 9-12 The ground-state, γ-, and β-vibrational bands of $^{164}_{68}$Er.

[†] A more complicated and interesting behavior, termed "forking," is found in $^{100}_{46}$Pd and $^{102}_{46}$Pd, where the ground-state band forks into two branches at $J = 6^+$ and 8^+, with one branch exhibiting back-bending at this low value of J. See S. Scharff-Goldhaber, M. McKeown, A. H. Lumpkin, and W. F. Piel, Jr., *Phys. Lett.* **44B**, 416 (1973).

rotational bands are also built on collective vibrational excitations. The β-vibrational band head in ^{168}Er is at 1.24 MeV; while the γ-vibrational band head is somewhat lower in energy, at 858 keV. The β-band, on the other hand, is lower than the γ-band in ^{232}U, as shown in Fig. 9–13. Although small departures from axial symmetry take place in γ-vibrations, K can

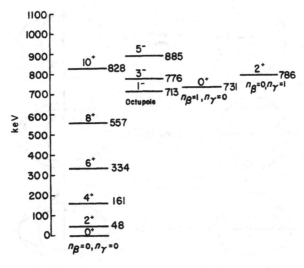

Fig. 9-13 The ground-state and the octupole bands in $^{232}_{92}$U. Also shown are the β- and γ-vibrational band heads. [After W. Elze,and J. R. Huizenga, *Nucl. Phys.* A187, 545 (1972).]

still be considered to be approximately a good quantum number. In many deformed nuclei, there is a negative-parity rotational band, with its lowest level 1$^-$, as, for example, the 713-keV level of ^{232}U in Fig. 9–13. This band is interpreted as based on octupole vibrations of the prolate equilibrium shape. Such vibrations can carry 0–3 units of angular momentum parallel to the 3-axis. This vibrational angular momentum with 3-component ν is coupled to the angular momentum of a rotator with 3-component zero to give a resultant $K = \nu$. In general, the states have $J = K, K + 1$, etc., but we have seen that for the negative-parity $K = 0$ band the states are restricted to be 1$^-$, 3$^-$, etc. These and other octupole states, as well as the rotational bands built on the quadrupole vibrations, have been recognized in a large number of deformed nuclei [26]. Because of the form of the Legendre polynomial $P_3(\cos \theta)$, octupole vibrations of the $K = 0^-$ type may be described as pear-shaped, with the bulge alternating between one end of the nucleus and the other. The various simple forms of shape oscillations are schematically illustrated in Fig. 9–14.

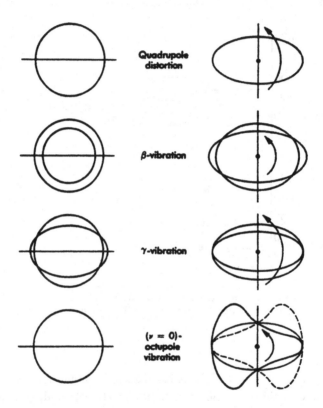

Fig. 9-14 The simple modes of vibrations of a distorted nucleus. The diagrams on the left represent cross sections perpendicular to the 3-axis; the lines are the 1-axis. In the diagrams on the right, the lines are the 3-axis and the cross sections are in the 2-3 plane; the arrows indicate the direction of rotation in a possible rotational state.

It will now be instructive to examine some of the electromagnetic transition properties in this model, one of the interesting quantities being the reduced transition probability $B(E2)$, defined in Eq. 9–22. We shall calculate this quantity for the ground-state rotational band with $K = 0$; more general results (e.g., interband transitions) can similarly be worked out. Adapting

Eq. 9-22 for the rotational case, quite generally, we obtain

$$B(E2, J_i, K \to J_f K') = (2J_i + 1)^{-1} \sum_{M_i, M_f} |\langle J_f M_f K'|Q_{2\mu}|J_i M_i K\rangle|^2, \quad (9\text{-}85)$$

where the operator $Q_{2\mu}$ in the space-fixed coordinates is defined by Eq. 3-6. This operator can be expressed in terms of $Q'_{2\mu}$ in the intrinsic (body-fixed) frame as

$$Q_{2\mu} = \sum_{\nu} \mathscr{D}^2_{\mu\nu} Q'_{2\nu}. \quad (9\text{-}86)$$

In writing the wave function $|JMK\rangle$ of the nuclear state, we have been suppressing until now the intrinsic part χ of the wave function; we now put it in explicitly and write, for the $K = 0$ band, from Eq. 9-60,

$$|JM0\rangle = \left(\frac{2J+1}{8\pi^2}\right)^{1/2} \mathscr{D}^J_{M0}(\theta_i)\chi_0(\mathbf{r}_1 \cdots \mathbf{r}_A), \quad (9\text{-}87)$$

where χ_0 is the intrinsic wave function in the body-fixed frame with $K = 0$. We then obtain

$$\langle J_f M_f 0|Q_{2\mu}|J_i M_i 0\rangle$$

$$= \frac{1}{8\pi^2} \sqrt{(2J_f + 1)(2J_i + 1)} \sum_{\nu} \int \mathscr{D}^{J_f*}_{M_f 0}(\theta_i) \mathscr{D}^2_{\mu\nu}(\theta_i) \mathscr{D}^{J_i}_{M_i 0}(\theta_i) \, d\Omega \langle \chi_0|Q'_{2\nu}|\chi_0\rangle.$$
$$(9\text{-}88a)$$

In this expression, the integral over the $\mathscr{D}$'s is nonzero only if $\nu = 0$; hence only this term contributes in the sum over ν. The integrations over the three $\mathscr{D}$'s can be done using Eq. A-78, and we get

$$\langle J_f M_f 0|Q_{2\mu}|J_i M_i 0\rangle$$

$$= \langle \chi_0|Q'_{20}|\chi_0\rangle \sqrt{\frac{2J_i + 1}{2J_f + 1}} (J_i 2M_i\mu|J_f M_f)(J_i 200|J_f 0). \quad (9\text{-}88b)$$

We therefore obtain

$$\frac{1}{2J_i + 1} \sum_{M_i, M_f} |\langle J_f M_f 0|Q_{2\mu}|J_i M_i 0\rangle|^2$$

$$= |\langle \chi_0|Q'_{20}|\chi_0\rangle|^2 |(J_i 200|J_f 0)|^2 \left[\frac{1}{2J_f + 1} \sum_{M_f} \sum_{M_i} |(J_i 2M_i\mu|J_f M_f)|^2\right].$$

For a given M_f, $\sum_{M_i} (J_i 2M_i\mu|J_f M_f)^2 = 1$; hence the quantity in brackets is unity. We can therefore write

$$B(E2, J_i 0 \to J_f 0) = (J_i 200|J_f 0)^2 |\langle \chi_0|Q'_{20}|\chi_0\rangle|^2. \quad (9\text{-}89)$$

The intrinsic quadrupole moment Q' in a given band K is defined by

$$Q_K' = \left(\frac{16\pi}{5}\right)^{1/2} \langle \chi_K | Q_{20}' | \chi_K \rangle, \tag{9-90}$$

which is related to the quadrupole moment Q in the space-fixed axes by Eq. 3-15. We can then write

$$B(E2, J_i 0 \to J_f 0) = \left(\frac{5}{16\pi}\right) Q_0'^2 \cdot (J_i 200 | J_f 0)^2. \tag{9-91}$$

For the reduced transition probability from the ground state 0^+ to the first excited 2^+ state in the $K = 0$ band, we simply obtain, from Eq. 9-91,

$$B(E2, 0^+ \to 2^+) = \left(\frac{5}{16\pi}\right) Q_0'^2 = 5B(E2, 2^+ \to 0^+). \tag{9-92}$$

Since the quadrupole moment in deformed nuclei is large, the $E2$-transition rates will be enhanced greatly from the single-particle estimate. A rough single-particle estimate of the quadrupole moment was given in Eq. 3-17a; it varies as $A^{2/3}$. It is customary to take the single-particle (SP) unit for $B(E2)$ as ([27] and Appendix E)

$$B(E2, 0^+ \to 2^+)_{\text{SP}} = 2.95 \times 10^{-5} A^{4/3} e^2 \, b^2, \tag{9-93}$$

where the letter b stands for 1 barn = 10^{-24} cm^2. A typical experimental $B(E2, 0^+ \to 2^+)$ for well-deformed rare-earth nuclei is about 5 e^2 b^2, while in the actinides it is about twice this value. We thus see that there is an enhancement in the experimental value of $B(E2)$ by a factor of 100–200 from the single-particle estimates.

We can easily connect $B(E2, 0^+ \to 2^+)$ to the deformation parameter β, defined by Eq. 9-34. For a prolate nucleus, Eq. 9-33 reduces to

$$R(\theta') = R_0 \left[1 + \sqrt{\frac{5}{16\pi}} \beta(3\mu^2 - 1) \right], \quad \mu = \cos\theta'. \tag{9-33}$$

The intrinsic quadrupole moment of a prolate shape with a sharp surface defined by $R(\theta')$ and a density $\rho = 3Ze/4\pi R_0^3$ is then

$$Q_0' = \int \rho r^2 Y_{20}(\theta', \phi') \, d\mathbf{r} = \frac{3Ze}{4\pi R_0^3} \cdot 2\pi \sqrt{\frac{5}{16\pi}} \int_{-1}^{+1} (3\mu^2 - 1) \, d\mu \int_0^{R(\theta')} r^4 \, dr,$$

where we have used the relation

$$Y_{20} = \sqrt{\frac{5}{16\pi}} (3\mu^2 - 1).$$

A straightforward calculation then yields,[†] to second order in β,

$$Q_0' = \frac{3}{\sqrt{5\pi}} ZeR_0^2 \beta (1 + 0.36\beta + \cdots). \qquad (9.94)$$

Substituting this value of Q_0' in Eq. 9–92, we find, to the leading order in β,

$$B(E2, 0^+ \rightarrow 2^+) = \left(\frac{3ZeR_0^2}{4\pi}\right)^2 \beta^2, \qquad (9\text{--}95)$$

so that the deformation β can be deduced from the experimental $B(E2)$-values. These values are shown for the even-even rare-earth and actinide nuclei in Fig. 9–15.

Another interesting result that can immediately be deduced from Eq. 9–91 is that the ratio

$$\frac{B(E2, 4^+ \rightarrow 2^+)}{B(E2, 2^+ \rightarrow 0^+)} = \frac{(4200|20)^2}{(2200|00)^2} = \frac{10}{7}; \qquad (9\text{--}96)$$

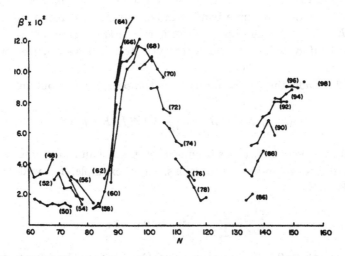

Fig. 9-15 Plot of β^2 as a function of the neutron number N as deduced from reduced E2-transition probabilities to the first excited 2^+ state. The points of same Z are joined together, with Z shown within brackets on the figure.

---

† Note that the shape defined by Eq. 9–33 is *not* an ellipsoid. For an ellipsoid, one can define a parameter $\beta = (4/3)\sqrt{\pi/5}(\Delta R/R_0)$, ΔR being the difference between the major and minor semiaxes. For such a shape,

$$Q_0' = \frac{3}{\sqrt{5\pi}} ZeR_0^2 \beta (1 + 0.16\beta + \cdots).$$

this ratio was 2 for vibrational nuclei (cf. Eq. 9–27b). Experimentally this is found to be well satisfied for well-deformed nuclei, as shown in Fig. 9–4.

We derived Eq. 9–89 for $B(E2)$-transitions within a $K = 0$ band; the result can be easily generalized for a nonzero K. In this case, the nuclear wave function $|JMK\rangle$ must be properly symmetrized (see Eq. 9–60), including the intrinsic state χ. There would, in principle, be cross terms in the matrix element 9–89 involving K and $-K$. Consider, for greater generality, transitions between different K-bands for a multipole operator $Q_{\lambda\mu}$; then the matrix element $\langle J_fM_fK_f|Q_{\lambda\mu}|J_iM_iK_i\rangle$ has to be evaluated. Consequently an integral of the type

$$\int \mathscr{D}^{J_f}_{M_fK_f}(\theta_i)\mathscr{D}^{\lambda}_{\mu\nu}(\theta_i)\mathscr{D}^{J_i}_{M_iK_i}(\theta_i)\, d\Omega \tag{9-97}$$

is involved; this is nonzero only if

$$\Delta K = |K_f - K_i| \leqslant \lambda, \tag{9-98}$$

since $|\nu| \leqslant \lambda$. This is called the K-selection rule in electromagnetic transitions and is broken to the extent that K is not a good quantum number. Furthermore, for transitions within a band with nonzero K, the cross terms involve $\Delta K = 2K \leqslant \lambda$. For quadrupole transitions, therefore, cross terms can contribute only if $K = \frac{1}{2}$ or 1. For even-even nuclei, if we consider only the in-band transitions in the ground-state ($K = 0$) or the β- ($K = 0$) and γ- ($K = 2$) bands, these cross terms do not appear, and we obtain

$$B(E2, J_iK \to J_fK) = \frac{5}{16\pi} Q_K'^2(J_i2K0|J_fK)^2. \tag{9-99}$$

Electromagnetic transitions from one rotational band to another are very weak compared to in-band transitions. For example, for well-deformed nuclei, the ratio

$$\frac{B(E2, 2_\gamma{}^+ \to 0_g{}^+)}{B(E2, 2_g{}^+ \to 0_g{}^+)} \approx 0.02\text{--}0.04, \tag{9-100}$$

increasing to about 0.07 for the transitional Os isotopes. The other interesting ratio which is also experimentally extracted is

$$\frac{B(E2, 2_\gamma{}^+ \to 2_g{}^+)}{B(E2, 2_g{}^+ \to 0_g{}^+)} \approx 0.10 \tag{9-101}$$

for well-deformed nuclei, but it increases rapidly for transition nuclei, reaching a value of about 0.7 for ^{192}Os and becoming larger than unity in some of the Pt isotopes. Excluding for the moment these transitional nuclei, we have the concept of a "band" as one in which there are strong electromagnetic transitions within the states of the band, subject, of course, to the selection rules imposed by angular momentum, parity, and K-quantum number, with weak interband transitions. It should be emphasized that in a theoretical calculation of ratios 9–100 and 9–101, account must be taken of the rotation-vibration interaction. In its simplest form, we have seen, for example, that U_2 (defined in Eq. 9–49) can connect a state (J, K, n_γ) with $(J, K \pm 2, n_\gamma \pm 1)$. Preston and Kiang [28] have given explicit expressions for interband $B(E2)$'s, using a perturbative approach in mixing the states, while a more complete treatment has been made by Faessler et al. [13]. The latter authors compare their theoretical values with experimental data, and the agreement is reasonable even in the transition region. As has been already pointed out, this success should be viewed with caution [29], however, since the rotation-vibration interaction term has been extracted from the hydrodynamical formula for the moment of inertia (see Eq. 9–37).

The hydrodynamical form of the moment of inertia (9–37) and the vibrational kinetic energy (Eq. 9–36) were derived for nuclei deviating slightly from the spherical shape. For well-deformed nuclei, different mass parameters B can be expected for the ground-state, the β-, and the γ-vibrational bands [30]. Retaining only up to quadratic terms, we may generalize the vibrational kinetic energy to [11]

$$T_{\text{vib}} = \tfrac{1}{2}B_\beta\dot{\beta}^2 + \tfrac{1}{2}B_\gamma\beta^2\dot{\gamma}^2 + B_{\beta\gamma}\beta\dot{\gamma}\dot{\beta}. \tag{9–102}$$

For spherical nuclei, $B_\beta = B_\gamma = B_0$ (the ground-state inertial parameter) and $B_{\beta\gamma} = 0$. Using this approach, we can generalize the $B(E2)$-expression 9–24 of vibrational nuclei to interband transitions [29, 31]

$$B(E2, 0_g{}^+ \to 2_\beta{}^+) = \left(\frac{3}{4\pi} ZeR_0{}^2\right)^2 \tfrac{1}{2}\hbar(B_\beta\omega_\beta)^{-1}\left(1 - \frac{6\hbar^2}{\mathscr{I}\hbar\omega_\beta}\right)^2 \tag{9–103a}$$

and

$$B(E2, 0_g{}^+ \to 2_\gamma{}^+) = \left(\frac{3}{4\pi} ZeR_0{}^2\right)^2 \beta_0{}^2\hbar(B_\gamma\omega_\gamma)^{-1}\left(1 - \frac{2\hbar^2}{\mathscr{I}\hbar\omega_\gamma}\right)^2, \tag{9–103b}$$

where the effects of rotation-vibration interactions have been included to first order. Using Eq. 9–103, one can extract the values of B_β and B_γ from the measured interband $B(E2)$-values and the energies of the β- and γ-vibrational states. It is found that the values of B_β and B_γ thus obtained exceed the ground-state mass parameter B_0 $(= \mathscr{I}/3\beta_0{}^2)$ by a factor of 3–4.

Until now, the low-lying collective levels of an even-even nucleus with permanent quadrupole deformation have been described as rotations of an axially symmetric deformed shape, coupled with β- and γ-vibrational modes which are treated as dynamical variables. We should mention that there is an alternative description of these low-lying levels, due to Davydov and Filippov [17] and Davydov and Chaban [19], in which the nuclear shape is considered to be asymmetric, with all three components of the moment of inertia $\mathscr{I}_\kappa$ unequal. In the simplest form of this model [17], β and γ are considered to be fixed parameters for a given nucleus (and not as dynamic variables) with $\gamma \neq 0$. In terms of the Bohr Hamiltonian 9-43, the potential function $V(\beta, \gamma)$ is assumed to have a very deep minimum for a well-defined value of β and $\gamma \neq 0$, so that the stiffness parameters C_β and C_γ of Eq. 9-46 are very large. This implies that the β- and γ-vibrational levels, although present in principle, come at very high excitations, and the low-lying collective levels are simply given by the eigenvalues of the asymmetric rotor

$$H_{AR} = \sum_{\kappa=1}^{3} \frac{L_\kappa{}^2}{2\mathscr{I}_\kappa}, \qquad (9\text{-}104)$$

where $\mathscr{I}_\kappa$, as before, is given by Eq. 9-37 in the hydrodynamical model, with β and γ as fixed parameters. To obtain the eigenvalues of H_{AR}, we can diagonalize it in the basis states (Eq. 9-60), that are eigenfunctions of an axially symmetric rotor,

$$|JMK\rangle = \left[\frac{2J+1}{16\pi^2(1+\delta_{K0})} \right]^{1/2} [\mathscr{D}^J_{MK}(\theta_i) + (-)^J \mathscr{D}^J_{M,-K}(\theta_i)], \qquad (9\text{-}60)$$

with $K = 0, 2, 4$, etc., and

$$J = \begin{cases} K, K+1, K+2, \dots & \text{if } K \neq 0, \\ 0, 2, 4, \dots & \text{if } K = 0. \end{cases} \qquad (9\text{-}105)$$

The eigenfunctions of H_{AR} may then be expressed as

$$|\Psi_{JM}\rangle = \sum_{K \leq J} g^J_{K_i}(\gamma) |JMK\rangle, \qquad (9\text{-}106)$$

where the coefficients $g^J_{K_i}$ are to be determined through diagonalization, and the index i indicates that several states of the same J may occur. An examination of Eqs. 9-105 and 9-106 immediately shows that in expansion 9-106 we get

(i) only one term with $J = 0$ ($K = 0$),
(ii) no possible state for $J = 1$,

   (iii) two terms with $J = 2$ $(K = 0, 2)$,
   (iv) one term with $J = 3$ $(K = 2)$,
   (v) three terms with $J = 4$ $(K = 0, 2, 4)$,

and so on. Diagonalization of H_{AR} in the basis 9–60 would thus yield one 0^+ state, two 2^+ states, one 3^+ state, three 4^+ states, and so on. To determine these levels we simply form for each value of K the equation

$$\langle JMK| \sum_{\kappa=1}^{3} \frac{L_\kappa^2}{2\mathscr{I}_\kappa} - E|\Psi_{JM}\rangle = 0.$$

Using the matrix elements[†] of L_κ^2, we obtain for these equations

$$g_{K+2}^J(\mathscr{I}_2 - \mathscr{I}_1)\mathscr{I}_3[(J - K)(J - K - 1)(J + K + 1)(J + K + 2)]^{1/2}$$

$$+ g_K^J\left[2(\mathscr{I}_1 + \mathscr{I}_2)\mathscr{I}_3(J^2 + J - K^2) + 4\mathscr{I}_1\mathscr{I}_2 K^2 - 8\left(\frac{E}{\hbar^2}\right)\mathscr{I}_1\mathscr{I}_2\mathscr{I}_3\right]$$

$$+ g_{K-2}^J(\mathscr{I}_2 - \mathscr{I}_1)\mathscr{I}_3[(J + K)(J + K - 1)(J - K + 1)(J - K + 2)]^{1/2} = 0.$$

$$(9\text{–}107)$$

The determinant of the coefficients of the g's must be zero for a solution. This yields an equation for E of the degree of the number of equations. For example, for $J = 2$, the equations are

$$\sqrt{6}\,g_2(\mathscr{I}_2 - \mathscr{I}_1)\mathscr{I}_3 + g_0\left[6(\mathscr{I}_1 + \mathscr{I}_2)\mathscr{I}_3 - 4\left(\frac{E}{\hbar^2}\right)\mathscr{I}_1\mathscr{I}_2\mathscr{I}_3\right] = 0,$$

$$g_2\left[2(\mathscr{I}_1 + \mathscr{I}_2)\mathscr{I}_3 + 8\mathscr{I}_1\mathscr{I}_2 - 4\left(\frac{E}{\hbar^2}\right)\mathscr{I}_1\mathscr{I}_2\mathscr{I}_3\right] + \sqrt{6}\,g_0(\mathscr{I}_2 - \mathscr{I}_1)\mathscr{I}_3 = 0,$$

and we obtain two states whose energies are the roots of

$$\left(\frac{E}{\hbar^2}\right)^2 - 2\left(\frac{E}{\hbar^2}\right)(\mathscr{I}_1^{-1} + \mathscr{I}_2^{-1} + \mathscr{I}_3^{-1})$$

$$+ \tfrac{3}{8}[\mathscr{I}_1^{-2} + \mathscr{I}_2^{-2} + 6\mathscr{I}_1^{-1}\mathscr{I}_2^{-1} + 8(\mathscr{I}_1^{-1} + \mathscr{I}_2^{-1})\mathscr{I}_3^{-1}] = 0. \quad (9\text{–}108)$$

[†] The relevant matrix elements are

$$\langle JK|L_3^2|JK\rangle = \hbar^2 K^2, \qquad \langle JK|L_1^2|JK\rangle = \langle JK|L_2^2|JK\rangle = \frac{\hbar^2}{2}[J(J + 1) - K^2],$$

and

$$\langle JK|L_1^2|J, K + 2\rangle = -\langle J, K|L_2^2|J, K + 2\rangle$$

$$= \frac{\hbar^2}{4}[(J - K)(J - K - 1)(J + K + 1)(J + K + 2)]^{1/2}.$$

See A. S. Davydov, *Quantum Mechanics*, §44, Pergamon Press, 1965.

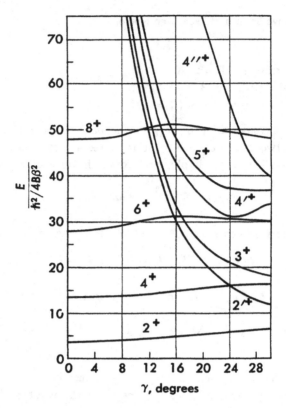

Fig. 9-16 Rotational levels of asymmetric distorted nuclei. [From G. R. De Mille et al., *Can. J. Phys.* **37**, 1036 (1959).]

If we consider the axially symmetric case $\mathscr{I}_1 = \mathscr{I}_2$, $\mathscr{I}_3 \to 0$, we note that the two roots are $3\hbar^2/\mathscr{I}_1$ and $2\hbar^2/\mathscr{I}_3$; the first of these is the solution $(\hbar^2/2\mathscr{I})J(J+1)$ for axial symmetry, and the second is very large. As the asymmetry increases, the values become more nearly comparable. The energies of the lowest levels are shown in Fig. 9–16 as functions of the asymmetry, expressed in terms of the variable γ through the hydrodynamic equations 9–37 for $\mathscr{I}_\kappa$. The levels present at $\gamma = 0$ are only slightly affected by an increase in γ, but the levels that are infinite at this separation are brought down to comparable energies.

A slight departure of the energies of the normal 2^+, 4^+, etc., states from the $J(J+1)$-ratios is predicted by the asymmetric model, but it would be difficult to disentangle this effect from the rotation-vibration coupling. There is a more significant test. Equations 9–108 and 9–107 show that the energies E_2 and E_2' of the two 2^+ states and the energy of the 3^+ state are related by

$$E_2 + E_2' = 2\hbar^2(\mathscr{I}_1^{-1} + \mathscr{I}_2^{-1} + \mathscr{I}_3^{-1}) = E_3. \qquad (9\text{-}109)$$

This prediction is confirmed with an error of approximately 2% in about a dozen cases, and, in addition, the higher rotational levels also fit the model fairly well. Values of γ that fit experimental energies may therefore be deduced.

A rough plot of γ as a function of A is shown in Fig. 9–17. Although K is not a good quantum number, there is only small admixture for small γ-values, and even for larger ones the lowest energy levels for each J are predominantly $K = 0$. The values of some of the coefficients g_{Ki}^J are shown in Table 9–1, demonstrating the extent of mixing of different K-values in the states of nuclei without axial symmetry.

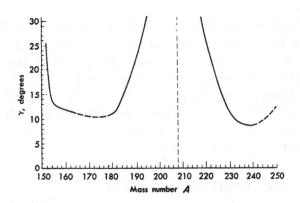

Fig. 9-17 Values of the asymmetry parameter γ, as a function of A. The smooth curve is based on values calculated from the values of E_2 and E_2' in ref. 18.

This simple model predicts only one 0^+ state, which is, in fact, the ground state. We saw, however, in Figs. 9–12 and 9–13, that there is another low-lying 0^+ state in many deformed nuclei which is interpreted as the β-vibrational band head in the axially symmetric model. Davydov and Chaban [19] have modified the asymmetric rotor model to incorporate β-vibrations; in this modified version β is considered to be a dynamic variable but γ is a fixed nonzero parameter. The results of this model, both in energy levels and in electromagnetic transition probabilities, are very similar to those of the axially symmetric model with rotation-vibration coupling included [13].

TABLE 9-1   Coefficients g_{Ki}^J. After Davydov et al. [17, 18].

γ	0	10°	20°	25°	30°
g_{01}^2	1	1.000	0.996	0.974	0.866
g_{21}^2	0	7.8×10^{-3}	0.0872	0.227	0.500
g_{02}^2	0	-7.5×10^{-3}	-0.0867	-0.226	-0.500
g_{22}^2	1	1.000	0.996	0.975	0.866
g_{01}^4	1	0.999	0.955	0.852	0.739
g_{21}^4	0	0.030	0.296	0.522	0.661
g_{41}^4	0	10^{-4}	0.010	0.043	0.125
g_{02}^4	0	-0.030	-0.296	-0.523	-0.559
g_{22}^4	1	0.999	0.954	0.842	0.500
g_{42}^4	0	0.004	0.043	0.128	0.661

Microscopic equilibrium calculations [32, 33, 34], which we shall discuss in Section 9–9 indicate, however, that all rare-earth nuclei, with the possible exception of some transition nuclei, energetically prefer a prolate axially symmetric equilibrium shape. Therefore the asymmetric-rotor model should not be taken literally, but is to be regarded as a different description of the same physical behavior. If a nucleus is oscillating about axial symmetry with an rms value γ_0, one would expect its collective levels to be very like those of a rotator with fixed asymmetry γ_0.

9-6 EVEN-EVEN TRANSITION NUCLEI

In the preceding sections, we considered the collective spectra of nuclei by making the simple assumption that the deformation potential $V(\beta, \gamma)$ in the Bohr Hamiltonian 9–43 has a well-defined minimum in β-γ space, so that an expansion like Eq. 9–46 is meaningful about the equilibrium deformation (β_0, γ_0). For the particular case of $V(\beta, \gamma)$ being γ-independent and $\beta_0 = 0$, we obtained the vibrational spectrum of spherical nuclei. In these simple models, the nonlinearity in the Bohr Hamiltonian 9–43 that is introduced due to the β-γ dependence of the inertial parameter B and the moments of inertia $\mathscr{I}_\kappa$ either was ignored by replacing β, γ in the inertial parameters by their equilibrium values, or at best was treated in a perturbative manner, using the hydrodynamical expression 9–37 for $\mathscr{I}_\kappa$. We found that this simple approach worked very well for permanently deformed rare-earth and actinide nuclei and fared rather poorly for spherical nuclei. We also mentioned that there are nuclei in the transition region (e.g., the lighter isotopes of Sm, Gd, and Nd with $N < 90$ and the isotopes of Os and Pt) for which neither the vibrational nor the rotational model seems to apply in its simple form. We shall find in this section that the simple assumption of a well-localized minimum in $V(\beta, \gamma)$ is no longer applicable for these transitional nuclei, the minimum being very flat and spread out in the β-γ space. For this reason, the inertial parameters can no longer be even approximately replaced by their values at a single point (β_0, γ_0) in the β-γ space, but the nonlinearity of the Schrödinger equation with the Bohr Hamiltonian must be taken into account. Before outlining a theory due to Baranger and Kumar which tackles these points, let us examine some experimental data in the transition region.

It is found experimentally that permanent deformation sets in rather abruptly in the rare earths at neutron number $N = 90$. Thus, although the low-lying spectrum of $^{150}_{62}Sm_{88}$ is vibrational in appearance, $^{152}_{62}Sm_{90}$ is clearly rotational (Figs. 9–18a and 9–18b). On the other hand, the transition from rotational to apparently vibrational characteristics is more gradual in the region of neutron number $N = 110$–120 for the W, Os, and Pt isotopes.

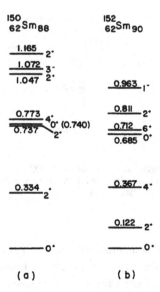

Fig. 9-18 Low-lying energy levels of (a) $^{150}_{62}Sm$ and (b) $^{152}_{62}Sm$ (energies in MeV).

These have been studied in great detail, both experimentally and theoretically.

Figure 9–19 shows a few positive-parity low-lying excited states of ^{148}Sm [35]. Vibrational characteristics are apparent in the spectrum, with $E(4^+)/E(2^+) \approx 2$ and the ratio $B(E2, 4^+ \to 2^+)/B(E2, 2^+ \to 0^+)$ being closer to 2 than to 10/7. However, a closer examination of the higher excited states suggests that these can be grouped into rotation-like bands with spin sequences $(0^+, 2^+, 4^+, 6^+, \ldots)$, $(0^+, 2^+, 4^+, \ldots)$, and $(2^+, 3^+, 4^+, \ldots)$, which tend to go over, with increasing neutron number, to the rotational bands. These quasi-bands are present not only in transition nuclei but also throughout the vibrational region and have been systematically classified by Sakai [35].

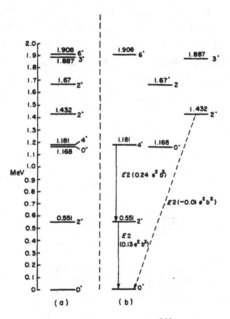

Fig. 9-19 A few positive-parity excited states of $^{148}_{62}$Sm. In (a), the experimental sequence is shown; in (b), this is classified in various quasi-bands. A few measured $B(E2)$-values are also shown (all energies in MeV).

The systematics of these quasi-rotational bands in spherical and transitional nuclei demonstrate that vibrational and rotational characteristics can coexist naturally, and indicates the need for a comprehensive theory that can treat vibrational, rotational, and transitional characteristics on an equal footing.

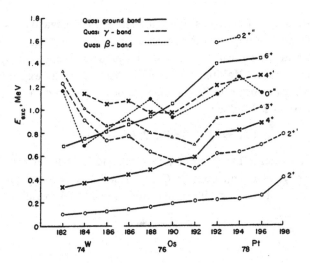

Fig. 9-20 The quasi-ground, quasi-β- and quasi-γ-bands for W, Os, and Pt isotopes. (After D. Cline, in *Colloquium on Intermediate Nuclei*, Institut de Physique Nucleaire d'Orsay, 1971.)

These quasi bands can be examined in detail in the W-Os-Pt region, as shown in Fig. 9-20. From this figure, it will be seen that $^{182}_{74}W_{108}$ has the rotational spectrum of a well-deformed nucleus, while the heavier isotopes of Pt appear to be vibrational. A closer look at the spectra of ^{190}Os and ^{192}Os and a comparison of these with Fig. 9-16, which shows the rotational levels of asymmetric distorted nuclei, suggest that both these nuclei may be asymmetric rotors with $\gamma \sim 24\text{--}28°$.

In general, it is not possible to determine whether the nuclear shape is prolate or oblate from a knowledge of the energy spectrum or the $B(E2)$-values. What is required is an experimental measurement of the static quadrupole moment Q_J of a definite nuclear state of angular momentum J. In the rotational model, the static quadrupole moment can be easily related to the intrinsic quadrupole moment Q_K', defined in Eq. 9-90. The static quadrupole moment of a state with angular momentum J in the $K = 0$ band is given by

$$Q_J = \langle J, M = J, K = 0|Q_{20}|J, M = J, K = 0\rangle$$

$$= (J2J0|JJ)(J200|J0)Q_0', \quad \text{using Eq. 9-88b,}$$

$$= -\frac{J}{2J + 3}Q_0'. \tag{9-110}$$

For a prolate shape, Q_0' is positive and hence Q_J is negative, whereas for an oblate shape the opposite condition holds. Reorientation experiments yield the sign and magnitude of the static quadrupole moment Q_{2+} of the first excited 2^+ state. Making use of the rotational model expressions 9–92 and 9–110, we can obtain only the magnitude of Q_{2+} from a measurement of the $B(E2, 0^+ \rightarrow 2^+)$ value:

$$Q_{2+} = \pm \tfrac{8}{7}[\tfrac{1}{5}\pi B(E2, 0^+ \rightarrow 2^+)]^{1/2}. \qquad (9\text{–}111)$$

Although expression 9–111 is strictly true only for a well-deformed nucleus, it is found to yield magnitudes of Q_{2+} in qualitative agreement with the measured values of Q_{2+} for vibrational and transitional nuclei as well. As

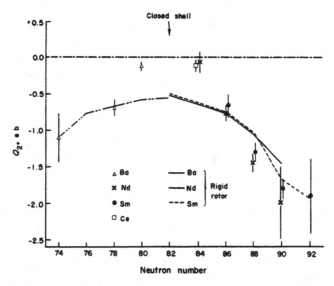

Fig. 9-21 The measured static quadrupole moments of the first 2^+ state of even-even isotopes in the $N = 82$ region as a function of N. The lines in this figure show the static quadrupole moments derived from the $B(E2, 0^+ \rightarrow 2^+)$ values, assuming a prolate rotor model. (From Gertzman et al. [36].)

an illustration of this, we show in Fig. 9–21 the measured Q_{2+} values and the corresponding predictions of Eq. 9–111, using the prolate rotor model for some near-spherical and transitional nuclei in the $N = 82$ region [36].

The static quadrupole moments Q_{2+} have been measured for some of the Os and Pt isotopes by Pryor and Saladin [37]. Their experimental results, displayed in Fig. 9–22, clearly indicate that, whereas the Os isotopes are prolate in shape in the 2+ state, the Pt isotopes are oblate. The dotted lines in the figure join the theoretical values of the calculations of Kumar and

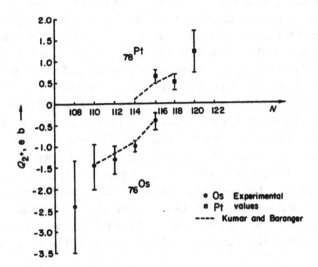

Fig. 9-22 The measured static quadrupole moments of the first 2+ state, Q_{2^+}, for Pt and Os isotopes. For comparison, we also show the theoretical values of Kumar and Baranger [12], through which dotted lines are drawn.

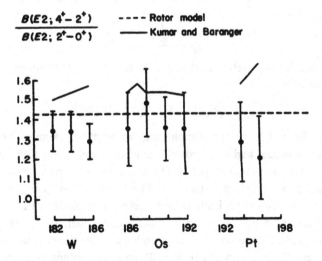

Fig. 9-23 The measured ratio $B(E2, 4^+ \rightarrow 2^+)/B(E2, 2^+ \rightarrow 0^+)$ for W, Os, and Pt isotopes. Also shown, by the dotted line, is the rotational model estimate $(= \frac{10}{7})$, and, by the solid line, the results of the calculation of Kumar and Baranger [12].

Baranger [12], which predicted successfully this prolate-to-oblate transition in these nuclei.

Detailed $B(E2)$-measurements have been made for the W, Os, and Pt isotopes [38]. Figure 9–23 shows the ratio $B(E2, 4^+ \rightarrow 2^+)/B(E2, 2^+ \rightarrow 0)$ for the ground-state band of some of these isotopes. The simple rotational-model prediction is 10/7, whereas the vibrational-model value is 2. It is clear that the vibrational model is not applicable for even the Pt isotopes, in which the spacings of the energy levels appear to be vibrational.

We shall now briefly outline the theory of Kumar and Baranger, which treats spherical, deformed, and transitional nuclei within a single framework. These authors have written a series of papers on this topic, the most important ones being refs. 11, 12, and 39. In ref 11, a dynamic theory of quadrupole collective motion for even-even nuclei is proposed, which is applied to the transitional nuclei in the W-Os-Pt region in ref. 12. In ref. 39, the symmetries and the numerical solution of the Bohr Hamiltonian are discussed in detail. Here we shall only touch on the main points of the Kumar-Baranger theory and discuss some of the important results of the calculations in the transition region.

Baranger and Kumar, following Bohr, assume the collective quadrupole motion of a nucleus to be adiabatic and uncoupled to higher-lying octupole modes and single-particle excitations. As such, their theory is applicable only to doubly even nuclei. They consider the generalized Bohr Hamiltonian

$$H = T_{\text{vib}} + \sum_{\kappa=1}^{3} \frac{L_\kappa^2}{2\mathscr{I}_\kappa} + V(\beta, \gamma), \qquad (9\text{–}112)$$

with the vibrational kinetic energy (Eq. 9–102) containing up to quadratic terms in the velocities:

$$T_{\text{vib}} = \tfrac{1}{2} B_\beta \dot{\beta}^2 + \tfrac{1}{2} B_\gamma \beta^2 \dot{\gamma}^2 + B_{\beta\gamma} \beta \dot{\gamma} \dot{\beta}. \qquad (9\text{–}102)$$

Unlike the Bohr theory, no simplifying assumptions about the inertial parameters are made, and the hydrodynamical model is not invoked. Most importantly, the nonlinear dependences of the inertial parameters on β, γ are retained, and harmonic expansions like 9–46 or 9–47 are *not* made to simplify the solution of the Bohr Hamiltonian. The theory can thus handle anharmonic effects in both spherical and deformed nuclei, as well as couplings between the ground-state and β- and γ-modes of vibrations. The six inertial parameters $B_\beta, B_\gamma, B_{\beta\gamma},$ and $\mathscr{I}_\kappa$ ($\kappa = 1$–3) and the potential energy V are calculated self-consistently as functions of β, γ, starting from a Hamiltonian that contains a spherical one-body potential, a quadrupole-quadrupole two-body residual interaction, and the two-body pairing potential, in a manner to be described briefly. The full collective wave function with angular momentum J, z-component M, and other quantum numbers α is expanded as

$$\Phi_{\alpha JM} = \sum_{K \geqslant 0} A_{\alpha JK}(\beta, \gamma) \phi^J_{MK}(\theta_1, \theta_2, \theta_3), \qquad (9\text{–}113)$$

where $\phi^J_{MK}(\theta_i)$ is the rotational wave function given by Eq. 9–60. As before, for $K = 0$, only even J's are allowed, and for a nonzero K, J takes all values K, $K + 1$, etc. When expression 9–113 is substituted in the Schrödinger equation with the Bohr Hamiltonian 9–112, a set of coupled partial differential equations for the functions $A_{\alpha JK}(\beta, \gamma)$ is obtained. These equations are solved numerically in a triangular β-γ space, since $0 \leqslant \gamma \leqslant \pi/3$. The equilateral triangular area in the β-γ space is divided into a large number of small triangular meshes, and the six inertial parameters and $V(\beta, \gamma)$ are calculated in each mesh. The Schrödinger equation is then solved by a variational method, the variational parameters being the values of the component A_K of the wave function at each mesh.

The most important part of the Kumar-Baranger theory is the calculation of the inertial parameters and $V(\beta, \gamma)$; this should, ideally, be done starting from realistic nucleon-nucleon forces. It is crucial to identify the collective variables β and γ in terms of the particle coordinates in such a calculation, and this is an intractable problem with a two-body force of general form. To surmount this difficulty, Kumar and Baranger use an effective phenomenological Hamiltonian for the nucleons, which contains the two important ingredients of the realistic interaction: the quadrupole-quadrupole force, responsible for collective quadrupole modes of motion, and the pairing force, which gives rise to an energy gap at the Fermi surface and drastically affects the values of the inertial parameters. Such a pairing-plus-quadrupole model of the residual interaction has proved fairly successful in describing the properties of single closed-shell nuclei for low excitations [40]. The effective Hamiltonian is thus written as

$$\mathcal{H} = H_S + V_Q + V_P, \qquad (9\text{–}114)$$

where H_S is a one-body spherically symmetric Hamiltonian generated by the core (e.g., a harmonic-oscillator potential with spin-orbit force), V_Q is the quadrupole interaction

$$V_Q = -\chi \sum_{i<j} r_i^2 r_j^2 \sum_\mu (-)^\mu Y_2^\mu(\hat{r}_i) Y_2^{-\mu}(\hat{r}_j) \qquad (9\text{–}115)$$

acting between the nucleons in the partly filled major oscillator shell, and V_P is the pairing potential with matrix elements

$$\langle i\bar{i}|V_P|j\bar{j}\rangle = -G, \tag{9-116}$$

where i, j are two single-particle states, and $\bar{i}, \bar{j}$ their time-reversed states. It is important to realize that this form of the pairing force with constant matrix element should be used only in a limited subspace in the neighborhood of the Fermi surface, and that the strength of the parameter G is dependent on the number of states $i, j,\ldots$ included around the Fermi surface. The pairing potential is assumed to act only between identical fermions, the neutron-proton pairing being neglected. Similarly, in doing calculations with V_Q, the oscillator space is restricted to the partly filled major shell and the adjacent major shell of opposite parity.

With a realistic effective force (such as the nuclear G-matrix), certain matrix elements are important for producing quadrupole deformation. It is found that with appropriate choices of G and χ these matrix elements can be reproduced approximately in the limited oscillator space defined above, but not outside it. It is therefore important to restrict the oscillator space while working with the phenomenological Hamiltonian 9-114.

It has been assumed that the nucleons in the filled shells give rise to a harmonic-oscillator type of spherical one-body potential in H_S (Eq. 9-114), whose central part is simply $\frac{1}{2}M\omega_0^2 r_i^2$ for the ith nucleon. If the contribution of the pairing force to the average potential is neglected, the noncentral quadrupole interaction V_Q should give rise to an average one-body potential which is noncentral. For simplicity, we may assume that the equipotential surface in a deformed nucleus has the same shape as the nuclear density. Consider, then, a point r_i on a distorted equipotential surface, given by

$$r_i = r_0[1 + \sum_\mu a_{2\mu}Y_2^\mu(\theta', \phi')] \tag{9-117}$$

in analogy with Eq. 9-28. For a different equipotential surface, r_0 would change. The value of the potential for the particular equipotential specified by r_0 is $V(r_0)$. We see that, if we take the harmonic form of the potential, denoted by V_h,

$$V_h(r_0) = \frac{1}{2}M\omega_0^2 r_0^2$$

$$= \frac{1}{2}M\omega_0^2 r_i^2[1 + \sum_\mu a_{2\mu}Y_2^\mu(\theta', \phi')]^{-2}$$

or

$$V_h(\beta, \gamma, r_i) \approx \tfrac{1}{2} M\omega_0^2 r_i^2 \left\{ 1 - 2\beta \left[\cos\gamma Y_2^0(\theta', \phi') + \frac{1}{\sqrt{2}} \sin\gamma \, (Y_2^2 + Y_2^{-2}) \right] \right\}$$

$$= V_{\text{spherical}}(r_i) + V_D(\beta, \gamma, r_i). \tag{9–118}$$

Note that the deformation parameters (β, γ) now refer to the equipotential surfaces. Recall that the noncentral one-body potential V_D is arising because of the interaction of the ith particle with all the other extracore particles j through the two-body quadrupole interaction V_Q. If the exchange contribution is neglected (as is a reasonable approximation for a long-range potential), V_D should be given by

$$V_D(\mathbf{r}_i) \approx - \sum_\mu r_i^2 Y_2^\mu(\hat{r}_i) D_\mu,$$

with

$$D_\mu = \chi \int \rho(r_j) r_j^2 (-)^\mu Y_2^\mu(\hat{r}_j) \, d^3 r_j$$

$$= \chi \langle Q_\mu' \rangle. \tag{9–119}$$

Here ρ is the density of the intrinsic shape. For a nucleus with shape oscillations, ρ is a function of time, and the self-consistency condition is a dynamic one, to be satisfied at all times. The β-γ dependence in V_D arises through the density $\rho(\beta, \gamma)$, and we can put

$$D_0 = D \cos\gamma, \qquad D_2 = \frac{1}{\sqrt{2}} D \sin\gamma,$$

obtaining the identical expression (9–118) for V_D, if we identify $D = M\omega_0^2\beta$.

We now take the deformed one-body potential $H_S + V_D(\beta, \gamma)$, and, for each point of the β-γ mesh mentioned earlier, calculate the single-particle energy levels and the wave functions. Next, for each mesh point, the pairing potential V_P is used to do a BCS calculation on these single-particle levels, obtaining the neutron and proton pairing gaps and the occupation probabilities. Finally, the expectation value of the original Hamiltonian (Eq. 9–114) is taken with the BCS wave function to determine the collective potential energy $V(\beta, \gamma)$, to be used in the Bohr Hamiltonian. The inertial parameters B are calculated microscopically within this framework, using the time-dependent Hartree-Fock theory, with the dynamic self-consistency (Eq. 9–119) maintained. We shall not give any details of this calculation here, but a simpler formulation will be outlined in Section 10–8.

We now proceed to discuss some of the main results of the above calculation for the W-Os-Pt transition nuclei. Figure 9–24a shows $V(\beta, \gamma)$ for ^{184}W,

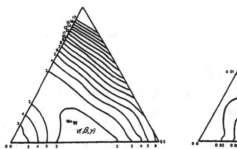 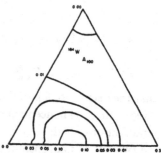

Fig. 9-24 The calculated contour maps of (a) $V(\beta, \gamma)$ and (b) the ground-state wave function A_{100} for the nucleus ^{184}W. The potential minimum occurs for a prolate shape. (From Kumar and Baranger [12].) In this, as well as Figs. 9–25 and 9–26, the units for $V(\beta, \gamma)$ are arbitrary.

as calculated by Kumar and Baranger [12] by the above method, as a set of equipotential lines. The lowest potential minimum occurs for a prolate ($\gamma = 0$) deformed shape in this nucleus. It is also interesting to examine the coefficients $A_{\alpha JK}(\beta, \gamma)$ of the wave function defined in Eq. 9–113. The nuclear state of a given J may occur several times; $\alpha = 1$ denotes the particular J-state coming with the lowest energy, $\alpha = 2$ the state with the same J of next higher energy, and so on. Furthermore, since K need not be a good quantum number, a state of given α and J may have several K-values mixed. The K-probability of a particular state with given α and J may be defined as

$$N_K^{\alpha J} = \int |A_{\alpha JK}|^2 \, d\tau, \qquad \sum_K N_K^{\alpha J} = 1, \qquad (9\text{–}120)$$

where the integration is over the β-γ space. Thus, for a prolate well-deformed nucleus, we would expect for the first 2^+ state ($\alpha = 1$, $J = 2$) $N_0^{12} = 1$, and all other N_K's to be zero.

Figure 9–24b depicts the contour plots of the ground-state A_{100} for ^{184}W in the β-γ plane, showing that the wave function is concentrated mostly in the prolate side around the potential minimum. For this nucleus, the first 2^+ state has $N_0^{12} = 0.995$ and $N_2^{12} = 0.005$, showing that this is a nearly pure $K = 0$ state, in agreement with the prolate rotor model (see Table 9–2). The calculated $V(\beta, \gamma)$ and A_{100} for ^{190}Os and ^{196}Pt are also shown in Figs. 9–25 and 9–26, respectively. In ^{190}Os, the lowest minimum in $V(\beta, \gamma)$ occurs for an asymmetric shape ($\beta > 0$, $\gamma = 30°$), but the potential is very shallow as a function of γ. This results in the wave function A_{100}

TABLE 9-2 The K-Probability of the Nuclear Wave Functions. The probability measures the contribution of a K-component to the normalization integral. The $K = 0$ contribution can be deduced by subtracting from unity the $K = 2$ contribution (if $J = 2$) or the sum of the $K = 2, 4$ contributions (if $J = 4$). After Kumar and Baranger [12].

State[a]	K	^{182}W	^{184}W	^{186}W	^{186}Os	^{188}Os	^{190}Os	^{192}Os	^{192}Pt	^{194}Pt	^{196}Pt
2^+	2	0.002	0.005	0.013	0.013	0.033	0.066	0.174	0.275	0.414	0.512
$2'^+$	2	0.449	0.642	0.791	0.919	0.923	0.899	0.775	0.683	0.533	0.415
$2''^+$	2	0.528	0.328	0.170	0.093	0.110	0.190	0.236	0.159	0.188	0.373
4^+	2	0.005	0.011	0.027	0.027	0.071	0.132	0.218	0.302	0.365	0.380
4^+	4	0.000	0.001	0.002	0.002	0.006	0.015	0.049	0.099	0.185	0.279

State[b]		Prolate rotor		Oblate rotor		Harmonic vibrator
2^+	2	0.0		0.75		0.275
$2'^+$	2	1.0 (γ-band)		0.25 (γ-band)		0.661
$2''^+$	2	0.0 (β-band)		0.75 (β-band)		0.275
4^+	2	0.0		0.313		0.275
4^+	4	0.0		0.547		0.107

[a] W-Os-Pt region.
[b] Some model cases.

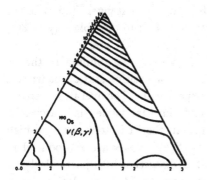
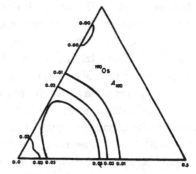

Fig. 9-25 (a) The calculated contour map $V(\beta, \gamma)$ for ^{190}Os. The potential is nearly γ-independent, although a very shallow minimum occurs for an asymmetric shape ($\gamma = 30°$). (b) The corresponding ground-state wave function A_{100}. (From Kumar and Baranger [12].)

being spread over a large β-γ region instead of being concentrated around the minimum. Finally, in ^{196}Pt, although the minimum of $V(\beta, \gamma)$ occurs for an oblate shape, the wave function is spread out to an even greater degree over various β-γ values.

 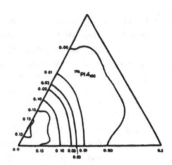

Fig. 9-26 (a) The calculated $V(\beta, \gamma)$ for ^{196}Pt. The minimum is on the oblate side.
(b) The ground-state wave function A_{100} is spread over a large region of the β-γ plane.

Thus the various simple models for stiff nuclei—prolate or oblate rotor or harmonic vibrator—are no longer very meaningful for such cases. This is more evident in Table 9-2, where the calculated K-probabilities N_K (Eq. 9-120) are tabulated for these nuclei and compared with the model cases. Although the W isotopes are of prolate shape, there is strong coupling between the β- and γ-bands, resulting in large deviations from the simple rotational model. In the Os isotopes, Kumar and Baranger find that the potential minimum shifts from the prolate ($\gamma < 30°$) to the oblate ($\gamma > 30°$) side in going from ^{188}Os to ^{192}Os, and the nuclei ^{190}Os and ^{192}Os may be considered asymmetric in shape. Finally, the potential minima in the Pt isotopes occur for oblate shapes, giving rise to a positive static quadrupole moment Q_{2+} for these in accordance with experiments. The overall agreement of the calculated energy spectra with experimental values is reasonable, although the calculated quasi-γ $2,^+$-energies in the Pt isotopes are too high by 100–200 keV. The main trends in the $B(E2)$-values are reproduced in the calculation, but there are quantitative disagreements, particularly in interband transitions. Perhaps the major triumph of the calculation lies in its correct prediction of the prolate-to-oblate transition in these nuclei, and in the applicability of the scheme to soft as well as stiff nuclei, irrespective of shape.

Finally, we should mention that the Davydov-Chaban model outlined in Section 9–5, which implicitly assumes a very steep γ-dependence of $V(\beta, \gamma)$ and a potential minimum for a fixed nonzero γ, also works reasonably well in this transition region, although its basic assumption seems to be in apparent contradiction with the microscopic calculation of $V(\beta, \gamma)$. The point to remember is that in the Os and Pt isotopes, as we have seen, the microscopically calculated $V(\beta, \gamma)$ is nearly γ-independent, and this results in the nucleus sweeping a large range of γ-values, giving an appreciable rms value of the asymmetry parameter γ. The alternative phenomenological description of the Davydov model with a fixed $\gamma = \gamma_{rms}$ thus works reasonably well. Even before the detailed calculations of Kumar and Baranger were performed, these points were realized by Gunye, Das Gupta, and Preston [41] in a simple microscopic calculation of $V(\beta, \gamma)$ for these transition nuclei. They did not, however, carry out any quantitative calculation of the wave functions by solving the Bohr Hamiltonian.

9-7 STRONG COUPLING OF PARTICLE AND COLLECTIVE MOTIONS

The preceding discussions of collective modes of even-even nuclei were carried out without considering the coupling of these modes to single-particle excitations, which occur at higher energies in these nuclei because of the energy gap at the Fermi surface. For even-odd or odd-odd nuclei, however, the single-particle excitations are lowered in energy, and for a proper description of the low-lying levels of these nuclei it is necessary to take into account the coupling of nucleonic and collective modes of motion. In these nuclei, there are rotational levels built on various single-particle excitations. In a phenomenological treatment of the problem, one makes the usual adiabatic assumption that the collective Hamiltonian H_c, describing the slow surface oscillations of the nuclear shape, can be separated from the much faster particle motion in an average distorted one-body potential V_h:

$$H = H_c + H_p, \tag{9-121}$$

with

$$H_p = \sum_{i=1}^{k} [T_i + V_h(\beta, \gamma; \mathbf{r}_i, \mathbf{l}_i, \mathbf{s}_i)], \tag{9-122}$$

where H_p is the particle Hamiltonian. The shape of the distorted potential, V_h, is related to the nuclear shape and is described by the parameters β, γ in the body-fixed frame. The number of particles, k, summed over in the particle Hamiltonian, depends on the sophistication of our model. In the extreme single-particle model, we may consider only one particle in an odd nucleus and lump all others together in the core; or, with a little more sophistication, we may let i extend over all the particles in the last unfilled shell. Since the change in the shape of the nucleus in this model must be small during the time that a particle takes to complete several cycles of its orbit, β and γ occurring in V_h are just average values and not dynamical variables. As before, we consider only the quadrupole deformations, and solve the single-particle Schrödinger equation

$$H_p \psi_\alpha = [T + V_h(\beta, \gamma; \mathbf{r}, \mathbf{l}, \mathbf{s})] \psi_\alpha = e_\alpha(\beta, \gamma) \psi_\alpha. \qquad (9\text{--}123)$$

This calculation yields a set of single-particle states which constitutes a generalization of the shell model. In this simple model, for given β and γ, the k nucleons fill up the lowest orbitals beyond the filled core, and their total energy is simply the deformation potential $V(\beta, \gamma)$ of the Bohr Hamiltonian. In this static approach, the equilibrium shape of the nucleus is determined by the values of β, γ that minimize the total particle energy.

Recalling that the one-body potential V_h is generated by the two-body nucleon-nucleon interaction, we have thus traced the potential $V(\beta, \gamma)$ to the interparticle energy. However, the rotational energy, associated solely with the orientation of the distorted system, remains as the collective Hamiltonian H_c, for Eq. 9–123 yields particle wave functions ψ_α referred to a body-fixed coordinate system. This is a reasonable picture, since the nuclear orientation changes much more rapidly than does its shape; whereas β and γ may be considered constant at their average values, the same obviously does not apply to the Euler angles. Then the complete wave equation for the nucleus is taken as

$$(T_{\text{rot}} + \sum_p H_p)\Phi = E\Phi, \qquad (9\text{--}124)$$

where

$$T_{\text{rot}} = \sum_{\kappa=1}^{3} \frac{L_\kappa{}^2}{2\mathscr{I}_\kappa} = \sum_\kappa \frac{(J_\kappa - j_\kappa)^2}{2\mathscr{I}_\kappa}, \qquad (9\text{--}125)$$

and $\mathbf{J}$, $\mathbf{L}$, and $\mathbf{j}$ are the total, core, and particle angular momenta, respectively. The last, $\mathbf{j}$, is not necessarily for only one particle, but is the angular momenta of all the particles not included in the core.

Before examining the intrinsic particle wave functions, we shall look at the structure of the total wave function Φ, considering first the axially symmetric case, for which

$$T_{\text{rot}} = \frac{\hbar^2}{2\mathscr{I}} [(\mathbf{J} - \mathbf{j})^2 - (J_3 - j_3)^2] + \frac{\hbar^2}{2\mathscr{I}_3} (J_3 - j_3)^2. \qquad (9\text{--}126)$$

For convenience, we have removed the factor $\hbar$ from each angular-momentum operator. Because V_h is assumed to be axially symmetric, the azimuthal angles of the particles do not appear in the Hamiltonian, and therefore

j_3 is a constant of motion. We shall denote it by Ω. Moreover, J^2 is, of course, a constant of motion, and so is J_3. We obtain

$$
\begin{aligned}
T_{\rm rot} &= \frac{\hbar^2}{2\mathcal{I}}[J(J+1) + \mathbf{j}^2 - 2\mathbf{J}\cdot\mathbf{j} - (K-\Omega)^2] + \frac{\hbar^2}{2\mathcal{I}_3}(K-\Omega)^2 \\
&= \frac{\hbar^2}{2\mathcal{I}}[J(J+1) - K^2 - \Omega^2] + \frac{\hbar^2}{2\mathcal{I}_3}(K-\Omega)^2 \\
&\quad - \frac{\hbar^2}{2\mathcal{I}}(J_+ j_- + J_- j_+) + \frac{\hbar^2}{2\mathcal{I}}\mathbf{j}^2.
\end{aligned} \tag{9-127}
$$

The first two terms of Eq. 9–127 are now constants, whereas the last terms still involve operators. The quantities $J_\pm$ and $j_\pm$ are the operators $J_1 \pm iJ_2$ and $j_1 \pm ij_2$. The term containing these operators couples rotational and particle angular momenta and will be abbreviated as RPC. Since in classical mechanics the "potential" energy of the Coriolis force on a particle may be written as $\boldsymbol{\omega}\cdot\mathbf{L}$, we see that the RPC-term is indeed of the same origin as a Coriolis force. The last term, $\hbar^2\mathbf{j}^2/2$, contains only particle coordinates and can be absorbed in the particle Hamiltonian. We have

$$
\left\{\frac{\hbar^2}{2\mathcal{I}}[J(J+1) - K^2 - \Omega^2] + \frac{\hbar^2}{2\mathcal{I}_3}(K-\Omega)^2 + H_0 + RPC\right\}\Phi = E\Phi,
$$

$$\tag{9-128a}$$

$$
H_0 = \sum_p H_p + \frac{\hbar^2}{2\mathcal{I}}\left(\sum_p \mathbf{j}_p\right)^2. \tag{9-128b}
$$

We shall presently prove that in the axially symmetric case $K = \Omega$. In this important special case, the Coriolis term may be treated as a perturbation. The operators $J_\pm$ have matrix elements only for states that differ by one unit in K; hence our assumption that K is a good quantum number is not valid in the presence of RPC. If we take a wave function with definite K as a zero-order approximation to Φ, RPC mixes in states of other K. But in a perturbation calculation, the admixture coefficient contains in the numerator the matrix elements of RPC and in the denominator the energy difference between the two states. Since for axial symmetry $K = \Omega$, states of different K are actually different particle states. The matrix element of RPC, on the other hand, is of the same order as rotational level spacing, proportional to $\hbar^2/2\mathcal{I}$. Consequently (ignoring the special case of $K = \frac{1}{2}$), for nuclei with well-developed axial symmetry, such that the rotational spacing is small compared with particle-excitation energies, the Coriolis term may be ignored, *except when states of the same J and parity but different K come very close in energy*. In the latter situation, although the matrix element of RPC between states of K and $K \pm 1$ is small, the small energy denominators in the perturbation terms make the higher-order terms important, and appreciable mixing is possible even between states differing in K by more than 1. For example, it is experimentally inferred that in $^{163}_{67}\text{Ho}_{96}$ the state of $J = \frac{7}{2}^+$, $K = \frac{1}{2}$ at 431.2 keV is mixed through RPC with the particle state $J = \frac{7}{2}^+$, $K = \frac{7}{2}$ at 439.9 keV [42], leading to appreciable $E1$-transition from the 431.2-keV level to the $J = \frac{7}{2}^-$, $K = \frac{7}{2}$ ground state, which should otherwise be K-forbidden $[\Delta K > \lambda \, (= 1)]$.

The effect of RPC for the special case of $K = \frac{1}{2}$ is especially large, since in the symmetrized zero-order wave function there are components of $K \, (= \frac{1}{2})$ and $- K \, (= - \frac{1}{2})$, which can become coupled by the RPC interaction.

Let the normalized intrinsic wave function χ_Ω be defined by

$$H_0 \chi_\Omega = e_\Omega \chi_\Omega, \tag{9-129}$$

where we have shown explicitly only the quantum number Ω, although others are required to specify a state completely. Then a solution of Eq. 9-128a ignoring RPC is $\chi_\Omega \mathscr{D}^J_{MK}(\theta_i)$ in the body-fixed frame. This solution, however, is not invariant under the symmetry operation $R_1^{'(\kappa)}$, which is a rotation through π about the nuclear axis and leaves the space-fixed α_μ's the same. We recall, from Eq. 9-55, that the rotation operator $R_1^{(1)}$ transforms $\mathscr{D}^J_{MK}$ into $(-)^J \mathscr{D}^J_{M,-K}$, and we see that the properly symmetrized and normalized solution is

$$\Phi(JMK\Omega) = \left(\frac{2J+1}{16\pi^2} \right)^{1/2} [\mathscr{D}^J_{MK}(\theta_i)\chi_\Omega + (-)^J \mathscr{D}^J_{M,-K}(\theta_i) R_1^{(1)} \chi_\Omega]. \tag{9-130}$$

Because the potential V_h is not spherically symmetric, the average force on the particle is not central and $\mathbf{j}^2$ is not a constant of motion. Since in practice it is simpler to deal with states of constant angular momentum, we introduce the expansion

$$\chi_\Omega = \sum_j c_{j\Omega} \chi_{j\Omega}. \tag{9-131}$$

It is not difficult to determine the value of $R_1^{(1)} \chi_{j\Omega}$. The first step is to employ the very general property of the $\mathscr{D}$-matrix which allows the conversion of any angular function or any spherical tensor from the nuclear to the space-fixed axes. In this case, using Eq. A-72, we obtain

$$\chi_{j\Omega} = \sum_{\mu} \mathscr{D}_{\mu\Omega}^{j*}\chi'_{j\mu}, \tag{9-132}$$

where $\chi'_{j\mu}$ is the wave function in the space-fixed axes, and μ is the component of j on the z-axis. The operator $R_1^{(1)}$, of course, has no effect on χ', and

$$R_1^{(1)}\chi_{j\Omega} = \sum_{\mu} \chi'_{j\mu} R_1^{(1)} \mathscr{D}_{\mu\Omega}^{j*}(\theta_i).$$

From Eqs. A–63 and A–64, we have

$$\mathscr{D}_{\mu\Omega}^{j*}(\theta_i) = (-)^{\mu-\Omega}\mathscr{D}_{-\mu,-\Omega}^{j}(\theta_i) \tag{9-133}$$

and

$$R_1^{(1)}\mathscr{D}_{\mu\Omega}^{j*}(\theta_i) = (-)^{\mu-\Omega}(-)^j\mathscr{D}_{-\mu,\Omega}^{j}(\theta_i).$$

Using these relations, it is straightforward to show that

$$R_1^{(1)}\chi_{\Omega} = \sum_{j} (-)^{j-2\Omega}c_{j\Omega}\chi_{j,-\Omega}, \tag{9-134}$$

with the result that the symmetrized wave function becomes

$$\Phi(JMK\Omega) = \left(\frac{2J+1}{16\pi^2}\right)^{1/2} [\mathscr{D}_{MK}^{J}\chi_{\Omega} + (-)^{J-2\Omega}\mathscr{D}_{M,-K}^{J}\sum_{j}(-)^{j}c_{j\Omega}\chi_{j,-\Omega}]. \tag{9-135}$$

This equation applies to the most general case, even with three unequal moments of inertia, for which the wave function is a linear superposition $\sum_{K,\Omega} A_{K\Omega}\Phi(JMK\Omega)$.

It is now easy to prove, using Eq. 9–135, that, for the axially symmetric case, $K = \Omega$. Rotation by an arbitrary angle θ about symmetry axis 3 should leave the wave function $\Phi(JMK\Omega)$ invariant or, at most, introduce a phase factor. Using Eqs. 9–131 to 9–133, we see that

$$\exp(-i\theta J_3) \mathscr{D}_{MK}^{J}\chi_{\Omega} = \exp[-i\theta(K-\Omega)] \mathscr{D}_{MK}^{J}\chi_{\Omega}, \tag{9-136}$$

so that the rotation operator $\exp(-i\theta J_3)$ acting on the two terms on the right-hand side of Eq. 9–135 would yield different phase factors unless $K = \Omega$. This implies that in the axially symmetric case the rotational angular momentum of the core, **L**, is perpendicular to the symmetry axis, with the nucleus rotating about an axis perpendicular to the symmetry axis.

There are a few important cases in which Eq. 9–135 simplifies. For axial symmetry, $K = \Omega$, and the coefficient of the second term is $(-)^{J-1}$, since $2K$ is an odd number. Another important case is that of only one extracore particle—the extreme single-particle model. For this there is an alternative representation for χ_Ω in terms of single-particle functions for which the constants of motion are the orbital angular momentum l, its component Λ on the 3-axis, and Σ, the component of spin on the 3-axis. Of course, $\Sigma = \Omega - \Lambda$. We have

$$\chi_\Omega = \sum_{l,\Lambda} a_{l\Lambda\Omega}\chi_{l\Lambda\Omega}. \qquad (9\text{--}137)$$

Since parity is a good quantum number, the values of l are either all even or all odd. The coefficients $c_{j\Omega}$ are connected with the coefficients $a_{l\Lambda\Omega}$ simply by angular-momentum addition relations, and

$$c_{j\Omega} = \sum_{l,\Lambda} (l\tfrac{1}{2}\Lambda\Sigma|j\Omega)a_{l\Lambda\Omega}. \qquad (9\text{--}138)$$

By considering inversion of the 3-axis, we can show that $a_{l\Lambda\Omega}$ must equal $a_{l,-\Lambda,-\Omega}$. Using these results and the fact that

$$(l_1 l_2 m_1 m_2|jm) = (-)^{l_1+l_2-j}(l_1 l_2 - m_1 - m_2|j-m),$$

we find

$$c_{j\Omega} = (-)^{j-(1/2)}\Pi_\chi c_{j,-\Omega},$$

where $\Pi_\chi = (-)^l$ is the parity of χ. In this case, the last sum in Eq. 9–135 is $(-)^{2j-(1/2)}\Pi_\chi\chi_{-\Omega} = (-)^{1/2}\Pi_\chi\chi_{-\Omega}$; and if, furthermore, we have axial symmetry, then

$$\Phi(JMK) = \left(\frac{2J+1}{16\pi^2}\right)^{1/2} [\mathscr{D}^J_{MK}\chi_K + (-)^{J-(1/2)}\Pi_\chi\chi_{-K}\mathscr{D}^J_{M-K}]. \qquad (9\text{--}139)$$

Let us now return to a consideration of the Coriolis term, which connects only collective states for which the K-values differ by 1. Since the nuclear state $\Phi(JMK)$ contains collective wave functions with both $\pm K$, the Coriolis term contributes directly to the energies of states with $K = \tfrac{1}{2}$, but in no other case. We see readily from Eq. 9–135 with $K = \Omega = \tfrac{1}{2}$ that

$$\langle \text{RPC} \rangle = \frac{\hbar^2}{2\mathscr{J}} a(-)^{J+(1/2)}(J + \tfrac{1}{2}), \qquad (9\text{--}140)$$

where a, called the *decoupling parameter*, is given by

$$a = -\sum_j (-)^{j+(1/2)}(j + \tfrac{1}{2})|c_{j(1/2)}|^2. \qquad (9\text{--}141)$$

The energy of a nuclear state is given by Eqs. 9–128, 9–129, and 9–140 for the axially symmetric nucleus, as

$$E_{J,K} = e_K + \left(\frac{\hbar^2}{2\mathscr{J}}\right)[J(J+1) - 2K^2 + \delta_{K,(1/2)}a(-)^{J+(1/2)}(J+\tfrac{1}{2})]. \quad (9\text{-}142)$$

We see that a rotational band is built on each particle state. The collective angular momentum has no component on the 3-axis, and K is due entirely to the particle angular momentum and is constant throughout the band. All higher values of $J \geqslant K$ are permitted, and, when $K \neq \tfrac{1}{2}$, the lowest state has $K = J$, and successive states have $J = K + 1$, $K + 2, \ldots$. The excitation above the ground state of a given band is

$$\Delta E_J = \left(\frac{\hbar^2}{2\mathscr{J}}\right)[J(J+1) - K(K+1)]. \quad (9\text{-}143)$$

When $K = \tfrac{1}{2}$, the level order is determined by the value of a in the manner shown in Fig. 9-27. For example, for a between -2 and -3, the level order is $\tfrac{3}{2}, \tfrac{1}{2}, \tfrac{7}{2}, \tfrac{5}{2}, \tfrac{11}{2}, \tfrac{9}{2}$, etc.

Just as with even nuclei, the vibration-rotation interaction complicates these results. When this small correction is incorporated, the energies become

$$E_{J,K} = e_K - 2\left(\frac{\hbar^2}{2\mathscr{J}}\right)K^2$$

$$+ \left(\frac{\hbar^2}{2\mathscr{J}}\right)[J(J+1) + \delta_{K,(1/2)}a(-)^{J+(1/2)}(J+\tfrac{1}{2})]$$

$$- B[J(J+1) + \delta_{K,(1/2)}a(-)^{J+(1/2)}(J+\tfrac{1}{2})]^2. \quad (9\text{-}144)$$

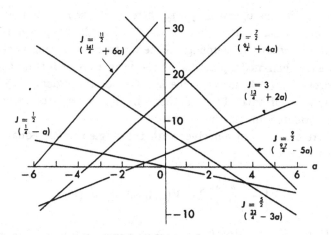

Fig. 9-27 The dependence of the rotational-energy levels on the decoupling parameter a in a $K = \tfrac{1}{2}$ band. The ordinate is $E_J - e$ in the notation of Eq. 9-142.

As we have already remarked, departures from Eq. 9–142 may also be produced by RPC in exceptional cases. This situation arises in ^{183}W, where a $K = \frac{1}{2}$ ground state and a $K = \frac{3}{2}$ excited state are fairly close in energy, with rotational bands built on both of these states. In these rotational bands, certain levels have very similar energies, and the band structure does not follow the level spacings of Eq. 9–142. Kerman [43] has shown that the perturbation due to the Coriolis term can explain the discrepancies. More interesting is the mixing of states with $\Delta K > 1$ because of RPC when the energies are very close, as was mentioned earlier in the case of ^{163}Ho. In such cases it is necessary to diagonalize RPC in the basis of the uncoupled states (Eq. 9–135) to determine the extent of K-mixing.

9-8 PARTICLE STATES IN DISTORTED NUCLEI

For further progress in studying odd nuclei, we must obtain an understanding of the states of motion of the particles outside the core. The fundamental paper on this subject was written by Nilsson [4] in 1955; he dealt only with an axially symmetric nucleus, and his work was extended to the nonaxial shape by Newton [44].

The intrinsic wave function is given by Eq. 9–129:

$$\left[\sum_p H_p + \left(\frac{\hbar^2}{2\mathscr{I}} \right) \left(\sum_p \mathbf{j}_p \right)^2 \right] \chi = e\chi.$$

If it were not for the operators $\mathbf{j}_p \cdot \mathbf{j}_q$, there would be no coupling whatever between the nucleons. If we are to keep our model simple, it is essential to eliminate interparticle coupling. Since we treat the strong nuclear forces by an average shell-model potential, it is clearly a minor further approximation to drop the $\mathbf{j}_p \cdot \mathbf{j}_q$ terms, since $\hbar^2/2\mathscr{I}$ is of the order of 15 keV and $\sum_{p,q} \mathbf{j}_p \cdot \mathbf{j}_q$ is itself small because there are many pairs of nucleons with oppositely oriented angular momenta.

We therefore concern ourselves with the single-particle equation in the body-fixed frame,

$$\left[-\left(\frac{\hbar^2}{2M} \right) \nabla^2 + V(\mathbf{r}_i, \mathbf{l}_i, \mathbf{s}_i) \right] \psi_\alpha = e_\alpha \psi_\alpha. \tag{9-145}$$

The one-body potential $V(\mathbf{r}_i, \mathbf{l}_i, \mathbf{s}_i)$ contains a spin-orbit term, a term proportional to $\mathbf{l}_i^2$, and a distorted harmonic-oscillator potential of the form 9–118,

$$V_h(\beta, \gamma, r_i) = \tfrac{1}{2} M \omega_0^2 r_i^2 \left\{ 1 - 2\beta \left[\cos \gamma \, Y_2^0(\theta', \phi') + \frac{1}{\sqrt{2}} \sin \gamma (Y_2^2 + Y_2^{-2}) \right] \right\},$$

$$(9\text{--}146)$$

where in V_h we include the spherically symmetric part coming from the core. Just as in the spherical-shell model, the harmonic-oscillator form is used for convenience in calculation; a more realistic radial dependence like that of the Woods-Saxon form would involve more numerical work. For $\gamma = 0$, we obtain axial symmetry, with only a $P_2(\cos \theta')$ deformation remaining in Eq. 9–146. An alternative form for V_h is an anisotropic harmonic oscillator:

$$V_h = \tfrac{1}{2} M (\omega_1^2 x_1^2 + \omega_2^2 x_2^2 + \omega_3^2 x_3^2), \qquad (9\text{--}147)$$

where (x_1, x_2, x_3) refer to the intrinsic body-fixed axes. For axial symmetry about x_3, $\omega_1 = \omega_2$; and a little algebra shows that for this case the form 9–147 contains only P_2-type deformation. It is assumed further that the volumes enclosed by the equipotential surfaces do not change with deformation, implying that

$$\omega_1 \omega_2 \omega_3 = \text{constant}. \qquad (9\text{--}148)$$

The single-particle equation to be solved is therefore

$$\left(-\frac{\hbar^2}{2M} \nabla^2 + V_h + V_{1,s} \right) \psi_\alpha = e_\alpha \psi_\alpha, \qquad (9\text{--}149)$$

where $V_{1,s}$ contains the spin-orbit term and a term proportional to l^2.

There are two obvious alternatives in solving this problem. The first is to take a basis generated by the spherically symmetric harmonic oscillator, $- (\hbar^2/2M)\nabla^2 + \tfrac{1}{2} M \omega_0^2 r_i^2$, and to diagonalize the Hamiltonian occurring in Eq. 9–149 in this basis. This has the disadvantage that the term $r_i^2 Y_2^0$ coming in V_h will couple a state in a major oscillator shell N $(= 2n + l)$ with others of $N \pm 2$. The other possibility is to take the Cartesian form 9–147 for V_h and to generate the Cartesian basis of eigenstates $|n_1, n_2, n_3\rangle$ of the Hamiltonian

$$\sum_{\kappa=1}^{3} \left(-\frac{\hbar^2}{2M} \frac{\partial^2}{\partial x_\kappa^2} + \tfrac{1}{2} M \omega_\kappa^2 x_\kappa^2 \right).$$

It is then necessary to diagonalize the total Hamiltonian of Eq. 9–149 in this basis. It is very difficult, however, to work with $l \cdot s$ and l^2-type potentials in a Cartesian basis.

The first alternative, therefore, is the more practical one, and Nilsson has shown that by using the following scale transformation one can eliminate the coupling between different major shells. This transformation is given by

$$\xi_\kappa = \left(\frac{M\omega_\kappa}{\hbar}\right)^{1/2} x_\kappa, \qquad (9\text{-}150)$$

which corresponds to the adoption of different units of length along the three axes. When H_h is written for the anisotropic part of the Hamiltonian, it is transformed by Eq. 9-150 to

$$H_h = -\frac{\hbar^2}{2M} \nabla^2 + V_h = \sum_\kappa \tfrac{1}{2}\hbar\omega_\kappa \left(-\frac{\partial^2}{\partial \xi_\kappa{}^2} + \xi_\kappa{}^2\right). \qquad (9\text{-}151)$$

As before, this is simply the sum of the Hamiltonians of three linear oscillators; the corresponding quantum states have energies $\Sigma(n_\kappa + \tfrac{1}{2})\hbar\omega_\kappa$, where n_1, n_2, n_3 are the numbers of excitations in the respective linear oscillators. As we have mentioned, however, it is more convenient to use spherical coordinates; hence we define

$$\rho^2 = \sum_\kappa \xi_\kappa{}^2 \quad \text{and} \quad \nabla_\xi{}^2 = \sum_\kappa \frac{\partial^2}{\partial \xi_\kappa{}^2}.$$

Then it is easy to rewrite the Hamiltonian 9-151 as

$$H_h = H_0 + H_\varepsilon \qquad (9\text{-}152)$$

with

$$H_0 = \tfrac{1}{2}\hbar\omega_0(-\nabla_\xi{}^2 + \rho^2),$$

$$H_\varepsilon = \tfrac{1}{2}\hbar\omega_0 \sum_\kappa \varepsilon_\kappa \left(-\frac{\partial^2}{\partial \xi_\kappa{}^2} + \xi_\kappa{}^2\right), \qquad (9\text{-}153)$$

and $\omega_\kappa = \omega_0(1 + \varepsilon_\kappa)$. The ε_κ's specify the deformation; for the axially symmetric case, we define[†]

$$\varepsilon_1 = \varepsilon_2 = \tfrac{1}{3}\varepsilon, \qquad \varepsilon_3 = -\tfrac{2}{3}\varepsilon, \qquad (9\text{-}154)$$

yielding

$$\omega_1 = \omega_2 = \omega_\perp = \omega_0\left(1 + \frac{\varepsilon}{3}\right) \quad \text{and} \quad \omega_3 = \omega_0\left(1 - \frac{2\varepsilon}{3}\right).$$

[†] Our ε is the same as the one defined by Nilsson in the Appendix of ref. 4. In current literature, ε is taken to be the quadrupole deformation parameter, while deformations of higher multipoles are introduced as $\varepsilon_l P_l(\cos\theta)$, $l > 2$.

The volume-conservation condition (Eq. 9–148) immediately shows that ω_0 is deformation dependent:

$$\omega_0(\varepsilon) = \mathring{\omega}_0(1 - \tfrac{1}{3}\varepsilon^2 - \tfrac{2}{27}\varepsilon^3)^{-1/3}, \tag{9-155}$$

with $\mathring{\omega}_0 = \omega_0(0)$.

To construct a basis from the eigenfunctions of H_0, we define a pseudo angular momentum

$$\boldsymbol{\lambda} = -i\hbar\boldsymbol{\xi} \times \boldsymbol{\nabla}_\xi. \tag{9-156}$$

Of course, $\boldsymbol{\lambda}$ differs from the true angular momentum $-i\hbar\mathbf{r} \times \boldsymbol{\nabla}$ by terms of first order in ε. Nevertheless, the ξ-system is preferable, because, as is seen from the form of Eq. 9–153, the remaining term H_ε is diagonal in n_1, n_2, and n_3 and hence in $N = n_1 + n_2 + n_3$. Thus the major-shell mixing has been eliminated from the ξ-system. We now define the basis $|N, l, \Lambda, \Sigma\rangle$ by

$$H_0|N, l, \Lambda, \Sigma\rangle = (N + \tfrac{3}{2})\hbar\omega_0|N, l, \Lambda, \Sigma\rangle,$$

$$\boldsymbol{\lambda}^2|N, l, \Lambda, \Sigma\rangle = l(l + 1)\hbar|N, l, \Lambda, \Sigma\rangle,$$

$$\lambda_3|N, l, \Lambda, \Sigma\rangle = \Lambda\hbar|N, l, \Lambda, \Sigma\rangle,$$

$$s_3|N, l, \Lambda, \Sigma\rangle = \Sigma\hbar|N, l, \Lambda, \Sigma\rangle. \tag{9-157}$$

It is more convenient now to take in the single-particle Hamiltonian[†] a spin-orbit potential of the form $C\boldsymbol{\lambda}\cdot\mathbf{s}$ and a quadratic potential $D\boldsymbol{\lambda}^2$ in terms of the pseudo angular momentum rather than the true angular momentum. These terms are chosen phenomenologically to make the energy levels for spherical nuclei agree with the shell model. The $\boldsymbol{\lambda}^2$-dependent term decreases the energy of the higher angular-momentum states from their oscillator values, as observed empirically. Nilsson has shown [4, Appendix A] that there is very little difference in the energies found whether one uses the true or the pseudo angular momenta, and consequently we may now attempt to solve the single-particle wave equation

$$(H_0 + H_\varepsilon + C\boldsymbol{\lambda}\cdot\mathbf{s} + D\boldsymbol{\lambda}^2)\psi_\alpha = e_\alpha\psi_\alpha. \tag{9-158}$$

We can diagonalize the above Hamiltonian in the basis $|N, l, \Lambda, \Sigma\rangle$ defined in Eq. 9–157, so that ψ_α may be written as a linear combination of states $|N, l, \Lambda, \Sigma\rangle$. Note that C and D are constants in a given shell and that H_ε, $C\boldsymbol{\lambda}\cdot\mathbf{s}$, or $\boldsymbol{\lambda}^2$ cannot mix N; hence it is enough to diagonalize within

---

† The coefficients C and D are both negative.

a single major shell. However, even in the absence of the $\lambda \cdot s$ potential, the pseudo angular momentum l is not a good quantum number because of H_ε, which can mix l with $l \pm 2$. This is readily seen by considering the matrix element $\langle N', l', \Lambda', \Sigma' | H_\varepsilon | N, l, \Lambda, \Sigma \rangle$. By expanding the states $|N, l, \Lambda, \Sigma \rangle$ in the Cartesian basis $|n_1, n_2, n_3, \Sigma \rangle$ and noting the form of H_ε in Eq. 9–153, we find that $N' = N$ and, further, that the kinetic and potential parts in H_ε contribute equally. We then obtain

$$\langle N', l', \Lambda', \Sigma' | H_\varepsilon | N, l, \Lambda, \Sigma \rangle$$

$$= \delta_{NN'} \cdot \delta_{\Sigma\Sigma'} \langle N', l', \Lambda', \Sigma' | \hbar\omega_0 \sum_\kappa \varepsilon_\kappa \xi_\kappa^2 | N, l, \Lambda, \Sigma \rangle$$

$$= \delta_{NN'} \cdot \delta_{\Sigma\Sigma'} \langle N', l', \Lambda', \Sigma' | \tfrac{1}{3}\hbar\omega_0(\xi_1^2 + \xi_2^2 - 2\xi_3^2) | N, l, \Lambda, \Sigma \rangle.$$

But $(\xi_1^2 + \xi_2^2 - 2\xi_3^2)$ is proportional to Y_2^0, showing by the Wigner-Eckart theorem that l and l' may differ by ± 2, whereas Λ cannot change because of H_ε.

We could also write V_h as given in Eq. 9–147 in terms of the deformation parameter ε in the axially symmetric case. A little algebra shows that

$$V_h = \tfrac{1}{2} M \omega_0^2 r^2 \left[\left(1 + \frac{2}{9}\varepsilon^2 \right) - \frac{4}{3}\varepsilon\left(1 - \frac{\varepsilon}{6} \right) P_2(\cos\theta') \right], \quad (9\text{–}159a)$$

$$= \tfrac{1}{2}\hbar\omega_0\rho^2[1 - \tfrac{2}{3}\varepsilon P_2(\cos\theta_t)], \quad (9\text{–}159b)$$

where $\cos^2\theta' = x_3/r$, and $\cos^2\theta_t = x_3/\rho$. If we neglect second-order terms in ε, the form 9–159a is equivalent to Eq. 9–146 for $\gamma = 0$, with the relation

$$\varepsilon = \frac{3}{2}\sqrt{\frac{5}{4\pi}}\,\beta = 0.95\beta. \quad (9\text{–}160)$$

We have also seen in Eq. 9–155 that ω_0 is a function of the deformation parameter ε. It is sometimes more convenient to use the parameter

$$\eta = \frac{2\varepsilon\hbar\omega_0(\varepsilon)}{-C}, \quad (9\text{–}161)$$

where C is the coefficient of the $\lambda \cdot s$ term. The quantities C and D are chosen to yield the shell model for spherical nuclei; D is varied from shell to shell. In the literature the notations $\kappa = -C/2\hbar\mathring{\omega}_0$ and $\mu = 2D/C$ are used. Reasonable values are $\hbar\mathring{\omega}_0 = 41A^{-1/3}$ MeV and $\kappa = 0.05$ for $N > 50$, increasing to $\kappa = 0.08$ for $N < 20$. Values of μ vary from 0 to 0.70, depending on the shell.

Nilsson's results are given in extensive tables and graphs in refs. 4 and 5. We reproduce these graphs as Figs. 9-28 to 9-32, for the regions of the periodic table in which there are large distortions. These graphs show the energy of each state as a function of the distortion ε. For Z and $N > 20$, only positive distortions are shown. The line representing each state is labeled

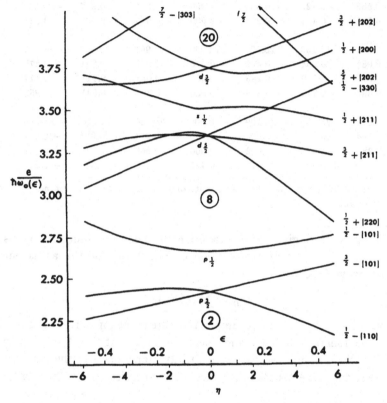

Fig. 9-28 Single-particle levels for $8 < Z < 20$, $8 < N < 20$. (After Mottelson and Nilsson [5].) The corresponding wave functions are given in ref. 4, Tables I and I(a).

by the value of Ω, the parity, and the triad $[Nn_3\Lambda]$. Although n_3 is not a constant of motion in the spherical representation, it becomes very nearly so for very large distortions. This would be expected physically as the energy difference $\hbar\omega_1 - \hbar\omega_3$ increases.

An example of Nilsson's wave-function tables is shown in Table 9-3. The energies in the first row of the table for each state are in units $\hbar\omega_0$.

TABLE 9-3 An Example of Nilsson's Wave Functions[a]. $N = 5$, $\Omega = \frac{7}{2}$. Base vectors: $|553 +\rangle$, $|533 +\rangle$, $|554 -\rangle$.

$\eta = -6$	-4	-2	0	2	4	6	$+\infty$
-16.090	-15.493	-14.007	-11.400	-8.393	-5.255	-2.055	
-0.615	-0.485	-0.214	0.000	0.094	0.144	0.177	
-0.583	-0.723	-0.954	1.000	-0.994	-0.987	-0.982	[503]
0.531	0.493	0.210	0.000	-0.055	-0.067	-0.068	
-20.498	-18.153	-16.194	-15.000	-13.961	-12.885	-11.757	
0.082	0.254	0.432	-0.426	-0.372	-0.321	-0.279	
-0.717	-0.655	-0.285	0.000	-0.086	-0.111	-0.117	[514]
-0.692	-0.712	-0.856	0.905	0.924	0.941	0.953	
-27.812	-26.755	-26.199	-26.000	-26.046	-26.260	-26.588	
0.784	-0.837	-0.876	0.905	-0.924	-0.936	-0.944	
-0.382	0.220	0.092	0.000	-0.066	-0.114	-0.149	[523]
0.489	-0.501	-0.473	0.426	-0.378	-0.333	-0.295	

[a] It may be noted that the coefficients of the base states are not normalized in Nilsson's paper [4].

The other entries give the composition of the wave function by listing quantities $a_{l\Lambda}$ such that the solution of Eq. 9–158 for the actual single-particle wave function is

$$\psi_{\alpha\Omega} = \sum a_{l\Lambda} |Nl\Lambda\Sigma\rangle. \tag{9–162}$$

Of course, $\Lambda + \Sigma = \Omega$. For example, the state in the top portion of Table 9-3 has wave function $|533 +\rangle$ and energy -11.400 at $\eta = 0$, the spherical shape. This is simply an $f_{7/2}$ ($m = \frac{7}{2}$) wave function. As the deformation increases, the wave function becomes more complicated, so that for $\eta = 2$ it is

$$0.094|553 +\rangle - 0.994|533 +\rangle - 0.055|554 -\rangle.$$

At this stage, the only valid quantum numbers are N and Ω. The wave functions could also be represented as a combination of states with fixed values of n_1, n_2, and n_3; instead we have used states of fixed l and Λ, and these states are themselves linear combinations of the (n_1, n_2, n_3)-states. As $\eta \to \infty$, the contribution of the state with $\Lambda = 4$, which is never very large, tends to zero, and we are left with the linear combination of $|553 +\rangle$ and $|533 +\rangle$ that describes the $n_3 = 0$ state. The entry [503] is therefore made for $[Nn_3\Lambda]$ in the final column of the table.

The second state in Table 9–3 is, for a spherical nucleus,

$$- 0.426|553 +\rangle + 0.905|554 -\rangle.$$

The coefficients are simply specific values of Clebsch-Gordan coefficients, and this state is actually $h_{9/2}$ $(m = \frac{7}{2})$. The third state in the table is $h_{11/2}$ $(m = \frac{7}{2})$ at zero deformation. The two upper states can be seen in Fig. 9–31, and the third state in Fig. 9–29. Similar states for neutrons can be found

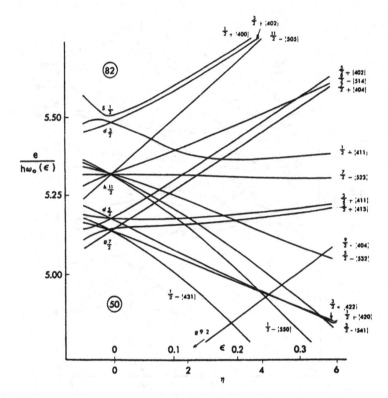

Fig. 9-29 Single-proton levels for $50 < Z < 82$. (After Mottelson and Nilsson [5].) The corresponding wave functions are given in ref. 4, Tables I and I(a).

in Fig. 9–30; the neutron states have rather different wave functions, because comparison with experiment suggests different values for the empirical constants in the theory.

The rule for determining the asymptotic quantum numbers n_3 and Λ is straightforward. For positive deformation, ω_3 is less than ω_1 and therefore, for a fixed N, the state of highest energy has the lowest value of n_3: $n_3 = 0$. The next highest state is $n_3 = 1$, etc. In our example, the three states for $N = 5$, $\Omega = \frac{7}{2}$ are assigned successively the asymptotic numbers $n_3 = 0, 1, 2$. The oscillator motion is such that Λ is even or odd as $N - n_3$ is even or odd; this determines Λ. In our example, the highest state has $N - n_3 = 5$, and therefore $\Lambda = 3$; the next state has $n_3 = 1$ and $\Lambda = 4$; and the lowest state has again $\Lambda = 3$.

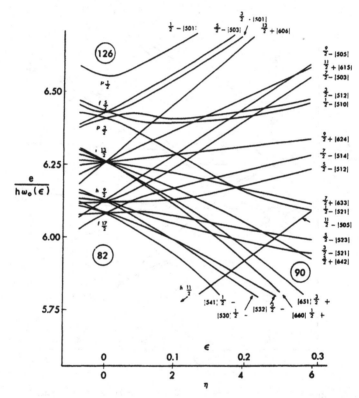

Fig. 9-30 Single-neutron levels for $82 < N < 126$. (After Mottelson and Nilsson [5].) The corresponding wave functions are given in ref. 4, Tables I and I(a).

In some processes selection rules (for β- and γ-radiation) are associated with changes in n_3 and Λ. Although these asymptotic quantum numbers are not strictly valid at the usual nuclear deformation, the rules are observed to hold. This implies that the actual wave function contains only a small component corresponding to other than the asymptotic values of n_3 and Λ. To study this point, it is more convenient to express the wave functions as combinations of $|Nn_3\Lambda\Sigma\rangle$ than as Nilsson's base vectors $|N, l, \Lambda, \Sigma\rangle$. Such

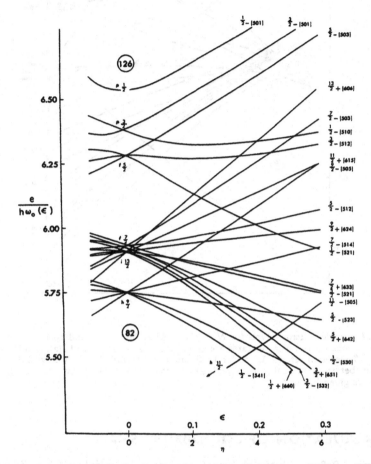

Fig. 9-31 Single-proton levels for $Z > 82$. (After Mottelson and Nilsson [5].) The wave functions of the $N = 4$ states are given in ref. 4, Tables I and I(a), and those for the $N = 5$ states in ref. 5, Appendix I.

a translation may be found in a paper by Rassey [45]. His Table II shows that the three states in our example are, respectively, 95% $|503\rangle$, 85% $|514\rangle$, and 86% $|523\rangle$ for $\eta = 4$; the corresponding figures for $\eta = 6$ are 97%, 89%, and 89%.

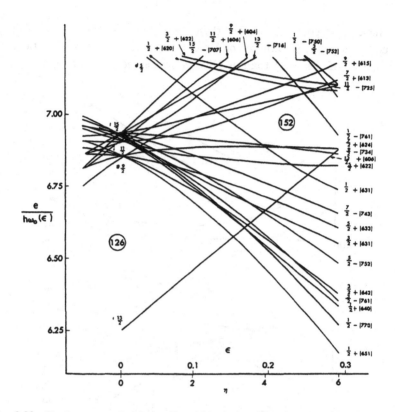

Fig. 9-32 Single-neutron levels for $N > 126$. (After Mottelson and Nilsson [5].) The wave functions for the $N = 6$ states are given in ref. 4, Tables I and I(a), and those for the $N = 7$ states in ref. 5, Appendix I. Some adjustments of energy have been made to obtain better agreement with experiment in this figure and the two preceding ones. Details are given in ref. 5.

9-9 CALCULATION OF EQUILIBRIUM SHAPE

In Section 9-6, we outlined how the nuclear potential energy $V(\beta, \gamma)$ is determined self-consistently from a phenomenological two-body Hamiltonian consisting of quadrupole-quadrupole and pairing interactions. It is possible to start with an even simpler approach and to assume that the nucleons are moving in a deformed one-body potential of the Nilsson or Newton type, with the single-particle energies e_α determined by Eq. 9-158 or a slightly generalized form to include axial asymmetry. Although it is implicitly assumed that this average potential is generated by two-body interactions, no attempt is made in this phenomenological approach to calculate it self-consistently. All residual interactions are neglected in the simplest version, and the effect of Coulomb interaction is approximately taken into account by choosing different values of the oscillator parameter $\overset{\circ}{\omega}_0$ for neutrons and protons. The total intrinsic energy of the nucleus is then

$$E = \left\langle \sum_i t_i + \sum_{i<j} V_{ij} \right\rangle = \left\langle \sum_i \frac{p_i^2}{2M} \right\rangle + \frac{1}{2} \left\langle \sum_i V_i \right\rangle,$$

where the sum i runs over occupied states only, and

$$V_i = V_h + V(\boldsymbol{\lambda}, \mathbf{s})$$

is the distorted one-body potential of Section 9-8.

Writing $H_i = t_i + V_i$, we can easily express E as

$$E = \frac{3}{4} \left\langle \sum_i H_i \right\rangle - \frac{1}{4} \left\langle \sum_i (V_i - t_i) \right\rangle. \tag{9-163}$$

Since a harmonic form (Eq. 9-147) has been assumed for V_h, for which the kinetic and potential energies of V_h are equal, we immediately obtain

$$E = \tfrac{3}{4} \sum_i e_i - \tfrac{1}{4} \sum_i \langle C\boldsymbol{\lambda}_i \cdot \mathbf{s}_i + D\lambda^2 \rangle. \tag{9-164}$$

Formula 9-164 is particularly simple, since it involves only single-particle energies e_i and an easily calculated correction due to the angular-momentum terms. It is then possible to calculate E as a function of the deformation parameter ε (or, in the axially asymmetric case, ε_κ) and minimize it for equilibrium. Such a calculation was done by Mottelson and Nilsson [5] for the axially symmetric V_h defined in Eq. 9-159 for rare-earth and actinide nuclei, imposing, of course, the volume-conservation condition 9-148. Their

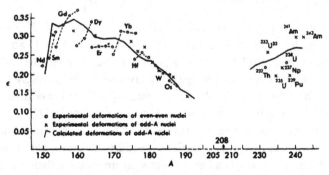

Fig. 9-33 Nuclear equilibrium deformation ε. The solid line is the calculated value. Circles denote experimental points for even-even nuclides; crosses, those for odd A. (After Mottelson and Nilsson [5].)

results, shown in Fig. 9–33, are in good agreement with experimentally extracted ε-values from $B(E2)$-measurements. Bes and Szymanski [46] refined this method of calculation by the inclusion of two-body pairing correlations. In the presence of a pairing force, the occupation probability $V_i{}^2$ of a single-particle state i varies smoothly with energy, instead of being only 1 or 0. Thus pairing effects decrease the importance of any special ordering of single-particle levels, making the variation of E with ε gradual. Das Gupta and Preston [32] essentially used the same method for even-even rare-earth nuclei ^{156}Gd-^{172}Yb, but with an anisotropic harmonic-oscillator potential with λ^2 and $\lambda \cdot s$ interactions, to demonstrate that an axially symmetric prolate shape is favored energetically in all cases. The preponderance of axially symmetric shapes over asymmetric ones is attributed by Kumar and Baranger [33] to the pairing force. Their calculations clearly show that the pairing force always energetically favors keeping the nucleus as symmetric as possible. A very strong pairing produces complete symmetry (i.e., the spherical shape), while a more reasonable weaker value yields axial symmetry. If the pairing interaction strength were reduced to half, for example, in their calculation, many nuclei (e.g., the W isotopes) would prefer an asymmetric rather than a symmetric shape.

The next question to be answered is, "Granted an axially symmetric shape, why do most nuclei prefer the prolate side $(\varepsilon > 0)$?" We consider first whether the above statement is still true in an axially symmetric anisotropic harmonic-oscillator potential without any l^2 or $l \cdot s$ types of interactions. The energy of a nucleon in the quantum state (n_1, n_2, n_3) is given by

$$
\begin{aligned}
e_{n_1, n_2, n_3} &= (\tfrac{1}{2} + n_1)\hbar\omega_1 + (\tfrac{1}{2} + n_2)\hbar\omega_2 + (\tfrac{1}{2} + n_3)\hbar\omega_3 \\
&= \hbar\omega_0(\varepsilon)\left[\frac{3}{2} + N + \frac{\varepsilon}{3}(N - 3n_3)\right],
\end{aligned} \tag{9-165}
$$

with $N = n_1 + n_2 + n_3$. It follows that, as the major shell N is being filled by nucleons, for large values of n_3 such that $3n_3 > N$ it is advantageous to put nucleons in prolate orbits $(\varepsilon > 0)$, whereas for smaller values of n_3 $(3n_3 < N)$ oblate orbits are energetically favored. It is also easy to see from Eq. 9–165 that the lowest prolate orbit with $n_3 = N$ is lower in energy than the lowest oblate orbit with $n_3 = 0$. Thus, in this model, the nuclei in the first half of the shell would be prolate and those in the latter half oblate.

In the presence of the $l \cdot s$ and l^2 potentials, however, this conclusion is no longer true. First, the $l \cdot s$ force would split up a major shell into subshells. Second, it is no longer true that at the beginning of the shell the orbit on the prolate side is always the lowest. For example, in Fig. 9–31 for the proton orbits for $Z > 82$, the $\frac{9}{2}^-$ [505] level on the oblate side is lower in energy than the $\frac{1}{2}^-$[541] level on the prolate side for very small deformation. This results in a negative quadrupole moment for ^{209}Bi (Table 7–1), and a similar, more pronounced effect is found just beyond the shell closure of $Z = 50$. For larger values of ε, there is frequent crossing of levels between different subshells, and the l^2-dependent term can bring down in energy many levels with large angular-momentum components. Numerical calculations show that it is the l^2-dependent potential that favors the prolate shapes more than the oblate.

Until now, we have considered in this section only the Y_2^0-type of deformation for axially symmetric well-deformed nuclei. It would be of

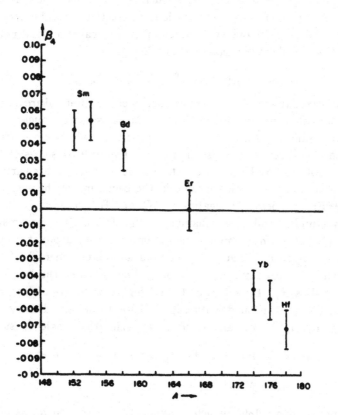

Fig. 9-34 Experimentally extracted β_4-values of some doubly even rare-earth nuclei. (After Hendrie et al. [47].)

interest to investigate deformation modes involving higher multipoles. To
find whether permanent deformations of $Y_4{}^0$- and $Y_6{}^0$-types are present in
rare-earth nuclei, Hendrie et al. [47] have scattered α-particles off these
targets, measuring the angular distribution of the scattered particles. These
α-particles at an incident energy of 50 MeV carry enough angular momentum
to transfer large units of angular momentum directly to members of the
ground-state rotational band. The inelastically scattered α-particles interact
mostly at the surface of the target nuclei because of strong absorption in
the interior. For direct excitation of the target to a state J, the angular
distribution of the inelastically scattered α's is determined by the Jth multipole
component of the nuclear field. Hendrie et al. [47] measured the differential
cross sections to the various rotational levels up to the 6^+ state, and analyses
of the data yield the shape of the nuclear field up to the Y_6-mode. In this
analysis, they assume that the target nucleus is a perfect rotor at least up
to the 8^+ state, and that the α-nucleus interaction may be represented
by a deformed complex one-body potential (the so-called optical potential).
The shape of this deformed potential is taken as

$$R = R_0(1 + \beta_2 Y_2{}^0 + \beta_4 Y_4{}^0 + \beta_6 Y_6{}^0) \tag{9--166}$$

in the body-fixed frame. The other parameters of the potential are determined
from an optical-model analysis of the neighboring spherical nuclei. The deformation parameters β_2, β_4, etc., are adjusted to yield the best fit to the
experimental differential cross-section data. Figure 9–34 shows the results
for β_4 thus obtained by Hendrie et al. for various even-even rare-earth nuclei.
We note that β_4 has a positive value in the beginning of the shell, goes to
zero for ^{166}Er, and becomes negative for Yb and Hf isotopes.

Can we understand this behavior of the deformation parameter β_4?
Bertsch [48] has given a simple explanation, assuming a prolate spheroidal
shape of the nucleus. Figure 9–35 shows a prolate spheroid of uniform
density with a sharp surface. The dotted line indicates the largest inside
sphere of radius R_0; the quadrupole and higher moments of this spheroid
arise from the outer polar density caps. If we denote the density of these
polar caps by $\Delta\rho(\mathbf{r})$, we can introduce an azimuthal angle θ_0 such that

$$\Delta\rho(\mathbf{r}) = C \quad \text{for} \quad 0 < \theta < \theta_0 \quad \text{and} \quad \pi - \theta_0 < \theta < \pi,$$

$$= 0 \quad \text{otherwise.}$$

We can now expand $\Delta\rho(\mathbf{r})$, which is axially symmetric, in terms of $P_L(\mu)$,
where $\mu = \cos\theta$. Let

$$\Delta\rho(\mathbf{r}) = \sum_L C_L P_L(\mu), \tag{9-167}$$

the coefficients C_L being directly related to the β_L's defined in Eq. 9-166. Then

$$C_L = C \left[\int_{\mu_0}^1 P_L(\mu)\, d\mu + \int_{-1}^{-\mu_0} P_L(\mu)\, d\mu \right],$$

giving

$$C_L = 0 \qquad \text{for } L \text{ odd},$$

$$= 2C \int_{\mu_0}^1 P_L(\mu)\, d\mu \quad \text{for } L \text{ even}. \tag{9-168}$$

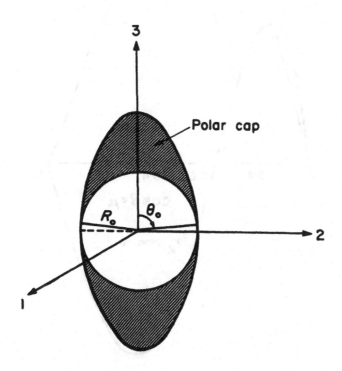

Fig. 9-35 A prolate spheroid of uniform density with a sharp surface. The dotted line shows the largest inside sphere of radius R_0.

We can also calculate the Lth moment of these caps, which is

$$Q_L = 2\pi \sqrt{\frac{2L+1}{4\pi}} \int_{-1}^{1} \Delta\rho(\mu) R_0^L P_L(\mu) R_0^2 \, d\mu$$

$$= 2\pi \sqrt{\frac{2L+1}{4\pi}} R_0^{L+2} C_L. \tag{9-169}$$

We thus see from Eqs. 9–168 and 9–169 that both C_L (i.e., β_L) and Q_L are

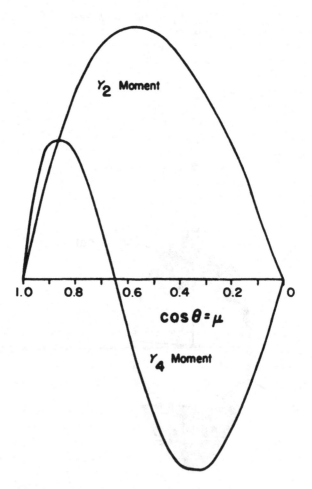

Fig. 9-36 Plots of $\int_{\mu_0}^{1} P_L(\mu) \, d\mu$ as a function of μ_0 for $L = 2$ and $L = 4$. (After Bertsch [48].)

proportional to the integral $\int_{\mu_0}^1 P_L(\mu)\,d\mu$. At the beginning of the shell, added nucleons are placed in orbits as close to the 3-axis as possible for prolate orbitals. Eventually, the equatorial orbitals are filled to make a spherically symmetric distribution again with shell closure. The quantity μ_0 thus varies from 1 to 0 as more and more nucleons are added. In Fig. 9–36, the quantities $\int_{\mu_0}^1 P_L(\mu)\,d\mu$ are plotted as functions of μ_0 for $L = 2$ and 4. We see from this plot that, whereas β_2 and Q_2 are of the same sign and have a maximum in the middle of the shell, β_4 and Q_4 undergo a change in sign as shown, in agreement with the experimental findings of Hendrie et al. [47].

In a microscopic calculation, to obtain a hexadecapole (P_4) moment in the density distribution of a nucleus, we must generalize the form V_h of the deformed one-body potential, given by Eq. 9–159. Nilsson et al. [49] have used the form

$$V_h = \tfrac{1}{2}\hbar\omega_0(\varepsilon,\varepsilon_4)\rho^2[1 - \tfrac{2}{3}\varepsilon P_2(\cos\theta_t) + 2\varepsilon_4 P_4(\cos\theta_t)] \qquad (9\text{–}170)$$

in the stretched coordinates. The oscillator constant ω_0 is now a function of ε and ε_4 to satisfy the volume-conservation condition $\omega_3\omega_\perp{}^2 = \mathring{\omega}_0{}^3$. Note

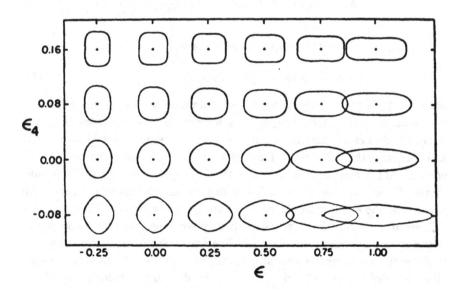

Fig. 9-37 The nuclear equipotential shapes for V_h given by Eq. 9–170 as a function of ε and ε_4. (After Nilsson et al. [49].)

that in Eq. 9–170 we have only P_2 and P_4 degrees of freedom, but the
potential could easily be generalized to include the mass asymmetry parameter P_3 and higher multipoles like P_6. These additional modes of distortion
are of importance in the calculation of fission barriers. With only the P_2-type
of deformation, we found in Section 9–8 that no major-shell mixing took
place in stretched coordinates; this is no longer true for a potential of the
form 9–170. Figure 9–37 shows the nuclear equipotential shapes as functions
of ε and ε_4. For the special case of $\varepsilon = \varepsilon_4 = 0$, the shape reduces to a sphere;
for spheroids, $\varepsilon_4 = 0$. We see that, even for large positive values of ε and ε_4,
the shape does not develop a prefission neck but instead remains cylinder-
like. This parametrization of the shape is therefore not very suitable for
calculation of fission barriers, as it is unable to describe the later stages of
fission. For calculation of fission barriers, other parametrizations of the
nuclear shape are also made [50].

In Section 9–8 we introduced a term proportional to λ^2 to correct the
oscillator shape. Nilsson et al. [49] choose to use instead a term ρ^4, whose
matrix elements within one oscillator shell are the same as those of $-\frac{1}{2}\lambda^2$,
apart from additive constants. Thus the generalized single-particle Hamiltonian of Nilsson in stretched coordinates is

$$
H_{SP} = \tfrac{1}{2}\hbar\omega_0(\varepsilon, \varepsilon_4)\left[(-\nabla_\zeta^2 + \rho^2) + \tfrac{1}{3}\varepsilon\left(2\frac{\partial^2}{\partial\xi_3^2} - \frac{\partial^2}{\partial\xi_2^2} - \frac{\partial^2}{\partial\xi_1^2}\right)\right.
$$
$$
\left. - \tfrac{2}{3}\varepsilon\rho^2 P_2(\cos\theta_t) + 2\varepsilon_4\rho^2 P_4(\cos\theta_t)\right]
$$
$$
- 2\kappa\hbar\mathring{\omega}_0[\lambda\cdot s - \mu(\rho^4 - \langle\rho^4\rangle_N)], \tag{9–171}
$$

where $\langle\rho^4\rangle_N$ is the average value of ρ^4 within a shell. The actual numerical
values of the parameters μ, κ, and $\mathring{\omega}_0$ for protons and neutrons are given
in ref. 49. As before, one may obtain the nuclear energy $E = \langle\sum_i(t_i + \tfrac{1}{2}V_i)\rangle$
and minimize this with respect to deformation. To refine the calculation,
one can add pairing correlations and the Coulomb energy of the deformed
shape. Möller [51] has done such equilibrium calculations in the rare-earth
nuclei, and the equilibrium values of ε and ε_4 that he finds are in reasonable
agreement with experimental values if only couplings between major shells
N and $N \pm 2$ are taken into account. The agreement for ε_4 (when converted
to β_4) with experiment is partially destroyed, however, if coupling between
all major shells is considered. This may be due to the inadequacy of the
volume-conservation condition, which is not applied to the angular-momen-
tum-dependent part of the one-body potential.

Finally, we mention that by calculating E in this one-body model as a function of deformation, although we obtain reasonable values of the equilibrium ground-state deformation, the value of E itself is found to be very unrealistic. Furthermore, this method of finding the deformation dependence of E is not accurate for large deformations and hence is not useful for the calculation of fission barriers, which involve delicate cancellation of the surface and Coulomb energies. The fission barriers of nuclei can be found accurately if the one-body Hamiltonian H_{SP} of Eq. 9–171 (or a more general form) is used only to calculate the shell correction in the total nuclear energy, which fluctuates with mass number and deformation, while the smooth part is taken directly from the semiempirical mass formula discussed in Chapter 6. This method is also capable of reproducing the equilibrium $(\varepsilon, \varepsilon_4)$ values and the binding energy of the ground state accurately, and has been used by Möller [51] with the H_{SP} of Eq. 9–171 and by Götz et al. [34] with a one-body Woods-Saxon potential. We shall describe this method in some detail in Section 12–4.

9-10 LEVELS OF DISTORTED ODD-A NUCLEI

Examination of the levels in Figs. 9–28 through 9–32 shows that at reasonable distortions there is a fairly uniform level distribution with spacing from 100 to 200 keV. We can therefore expect low-lying particle excitations, well below any vibrational levels. Such states are not found in even-even nuclei because of the energy required to break a correlated nucleon pair. Almost all the low-lying levels of distorted odd-A nuclei can be understood as single-particle states and the rotational bands built on them. A thorough study is given by Mottelson and Nilsson [5].

The most obvious check of the theory is the prediction of ground-state spins. Knowing the distortion from the quadrupole moment, one can count up the occupied levels in the level diagrams and determine the orbit of the last nucleon. This gives Ω; for the lowest state, $J = K = \Omega$ (unless $\Omega = \frac{1}{2}$), and one can compare the predicted and experimental J-values. There is often some doubt as to the predicted J because of the density of the levels and inversions of order which occur for small changes of ε. However, in all cases, the experimental value is one of those predicted.

We shall consider in detail three examples in different parts of the periodic table, rather than attempt to summarize all the data. The nuclides ^{25}Al and ^{25}Mg have been carefully studied by Litherland et al. [52]. Quadrupole moments in this region suggest $\varepsilon \simeq 0.3$. For this distortion, the $\frac{1}{2}^+$ [202]

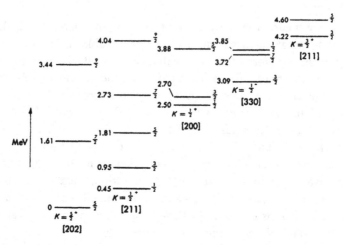

Fig. 9-38 The levels of ^{25}Al, displayed to show the various rotational bands.

level and the $\frac{1}{2}^+$ [211] level are very close (Fig. 9–28). Experimentally the ground state is $\frac{5}{2}^+$, and there is an excited state of $\frac{1}{2}^+$ at 450 keV. The other states are shown in Fig. 9–38; it is seen that they can be assigned to different states of the last particle, together with rotations of the intrinsic structure. The highest band shown corresponds to a hole in the $\frac{3}{2}^+$ [211] level and a pair in, presumably, the next state, $\frac{5}{2}^+$ [202]. We should not be content that these levels are correctly interpreted until the various consequences of the theory for electromagnetic properties, β-decay, and so forth, are examined.

One simple necessary condition is the level spacing in the various alleged rotational bands. The levels of the ground-state band satisfy quite well the $E \propto J(J + 1)$ rule, particularly if the small rotation-vibration correction is included. The $K = \frac{1}{2}$ bands, however, require more careful study, because of the Coriolis term. Equation 9–142 gives the relative energies in a given band as

$$E_J = \left(\frac{\hbar^2}{2\mathscr{I}}\right)[J(J + 1) + \delta_{K,(1/2)}a(-)^{J+(1/2)}(J + \tfrac{1}{2})]. \qquad (9–172)$$

In the special case of the single-particle χ's calculated by Nilsson,

$$\chi_\Omega = \psi_{\Omega\alpha} = \sum_{l\Lambda} a_{l\Lambda}|Nl\Lambda\Sigma\rangle, \qquad (9–162)$$

equations 9–141 and 9–138 give for the decoupling parameter

$$a = (-)^l \sum_l [a_{l0}^2 + 2(l(l+1))^{1/2} a_{l0} a_{l1}]. \tag{9-173}$$

The values of a depend on the distortion. To obtain exact fit for the levels of ^{25}Al, slightly different values of ε have to be used for the three $K = \frac{1}{2}$ bands, but this is neither unexpected nor physically unreasonable. Indeed, different equilibrium distortions arise from different configurations.

Further detailed consideration of other properties shows that the $A = 25$ nuclei are very well described by the rotations of intrinsic *distorted shell-model* states. There are indications that other nuclides from $A = 19–31$ also display collective properties, although there are nuclides in this region to which the collective model does not seem to apply [45]. In addition to ^{25}Mg and ^{25}Al, the model is apparently successful in ^{19}F [53], ^{24}Mg [54], ^{28}Al [55], ^{29}Si [56], and ^{31}P [57].

In the rare-earth region, which includes A from about 150 to 190, we shall study the $A = 183$ isobars as typical examples. The level schemes of $^{183}_{74}$W and $^{183}_{75}$Re are shown in Fig. 9–39. The spins and parities of the low-lying levels of these two nuclides are well established; our simplified figure does not show all known states. If we count up from 90 neutrons in Fig. 9–30, we find that the 109th neutron is expected to be in the $\frac{1}{2}^-$ [510] state for $\varepsilon \gtrsim 0.2$. This result is consistent with the ground-state assignments, and the levels labeled α can be considered to constitute a rotational band. Referring again to Fig. 9–30, we may expect a state caused by excitation to the $\frac{3}{2}^-$ [512] level at about 200 keV ($\hbar\omega_0 \approx 7$ MeV). It is natural, therefore, to associate the 209-keV level with this state. Two other states of its band appear; their energies above 209 keV are in the ratio 2.44, compared with the value 2.40 predicted by the $J(J+1)$ law. Finally, it is reasonable to assign the 453-keV level to the $\frac{7}{2}^-$ [503] band. While we are still examining the odd-neutron diagram, let us consider ^{183}Os. We expect its ground state to be $\frac{9}{2}^+$[624], and this appears to be confirmed. The $\frac{1}{2}^-$ level at 171 keV must then be the $\frac{1}{2}^-$ [510] state; the fact that it is only 171 keV above the ground state may suggest that ^{183}Os is less distorted than ^{183}W, with $\varepsilon \sim 0.15$.

The theoretical expectations for the odd-proton nuclides Ta and Re are found from Fig. 9–29. Counting up from the 50-shell, we find that for $\varepsilon \sim 0.3$ the 73rd proton would be either $\frac{9}{2}^-$ [514] or $\frac{5}{2}^+$ [402]. However, the ground states of all the Ta isotopes are $\frac{7}{2}^+$, as are those of the $_{71}$Lu isotopes. It would appear that we have here an example of extra pairing energy in the state $\frac{9}{2}^-$ ($l = 5$), compared with the state $\frac{7}{2}^+$ [404] ($l = 4$). If $\varepsilon \sim 0.3$, the two levels are very close, and the three protons 71–73 are presumably in the configuration $(\frac{7}{2}^+)(\frac{9}{2}^-)^2$. We would then expect the first particle excitation in Ta

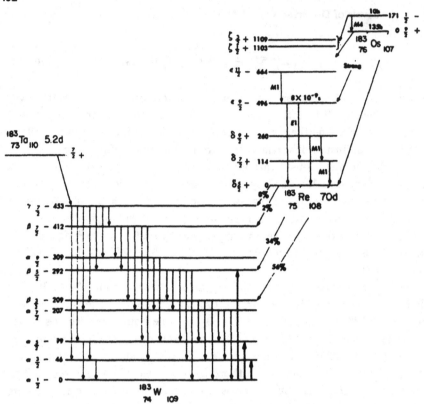

Fig. 9-39 The levels of the $A = 183$ nuclides (energies in keV). The Greek letters identify different rotational bands.

to be $(\frac{7}{2}+)^2(\frac{9}{2}-)$, with the configuration $(\frac{5}{2}+ [402])(\frac{9}{2}-)^2$ slightly higher. These expectations are borne out in ^{181}Ta. Also in ^{181}Ta, there is a $\frac{1}{2}+$ state at 612 keV; this is presumably due to a hole in the $\frac{1}{2}+$ [411] state, so that the last five protons are $(\frac{1}{2}+)(\frac{7}{2}+)^2(\frac{9}{2}-)^2$. Turning to $_{75}$Re and consistently assuming high pairing energy for the $\frac{9}{2}-$ state, we expect the last five protons to be $(\frac{7}{2}+)^2(\frac{5}{2}+)(\frac{9}{2}-)^2$. The ground states of the Re isotopes are indeed $\frac{5}{2}+$. In ^{183}Re we find also the expected particle excitations $(\frac{7}{2}+)^2(\frac{5}{2}+)^2(\frac{9}{2}-)$ and $(\frac{1}{2}+ [411])(\frac{7}{2}+)^2(\frac{5}{2}+)^2(\frac{9}{2}-)^2$. Presumably there should be a $\frac{7}{2}+$ [404] excitation; if there is, it is not clear why it has not been observed in the Os decay. In Fig. 9–39, the level-spacing ratio $[E(\frac{9}{2}) - E(\frac{5}{2})]/[E(\frac{7}{2}) - E(\frac{5}{2})]$ in the band labeled δ is 2.28, compared with the theoretical value $\frac{16}{7} = 2.29$. We may also remark that the moments of inertia of excited bands differ from the value in the ground-state band.

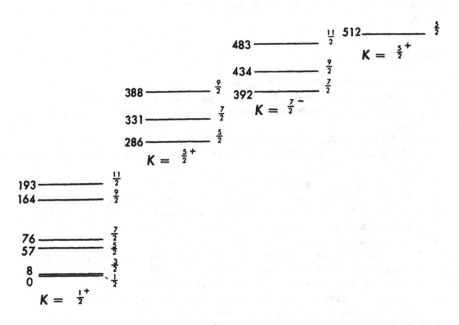

Fig. 9-40 The energy levels of ^{239}Pu, displayed to show the rotational bands (energies in keV).

The elements with $A > 224$ have rotational spectra. Figure 9–40 shows the levels of ^{239}Pu. It is left to the reader to make the detailed comparison with the level sequences deduced from Fig. 9–32. We mention only that the decoupling parameter $a = -0.58$ is consistent with the ground-state assignment $\frac{1}{2}^+$ [631] and that reasonable assignments for the 286-, 392-, and 512-keV levels are $\frac{5}{2}^+$ [622], $\frac{7}{2}^-$ [743], and $\frac{5}{2}^+$ [633], respectively. The moments of inertia are, of course, greater than in the rare-earth region, with correspondingly smaller level spacing.

A noteworthy feature of the heavy-element region is the appearance of a "magic" number at 152 neutrons. Such a number cannot arise for spherical nuclei, but we see that at distortions $0.2 < \varepsilon < 0.3$ there is a marked gap in the level spacing in Fig. 9–32. The $N = 152$ nuclides may then be magic, although they will not have the specially simple properties due to the spherical nuclei of the other magic nuclides. Experimental evidence for this energy gap is supplied by the unusually great α-decay energies of the $N = 154$ nuclides and the apparent instability against fission of the nuclides above the energy gap. A gap of about 300 keV is indicated between $N = 151$ and $N = 153$.

REFERENCES

1. J. Rainwater, *Phys. Rev.* 79, 432 (1950).
2. A. Bohr, *K. Danske Vidensk. Selsk. Mat.-Fys. Medd.* 26, No. 14 (1952).
3. A. Bohr and B. R. Mottelson, *K. Danske Vidensk. Selsk. Mat.-Fys. Medd.* 27, No. 16 (1953).
4. S. G. Nilsson, *K. Danske Vidensk. Selsk. Mat.-Fys. Medd.* 29, No. 16 (1955).
5. B. R. Mottelson and S. G. Nilsson, *K. Danske Vidensk. Selsk. Mat.-Fys. Skr.* 1, No. 8 (1959).
6. J. de Boer and J. Eichler, *Advan. Nucl. Phys.* I, 1 (1968).
7. Rayleigh, Third Baron, *The Theory of Sound*, Vol. II, Macmillan, London, 1877, p. 364.
8. O. Nathan and S. G. Nilsson, *Alpha-, Beta- and Gamma-Ray Spectroscopy*, Vol. I, edited by Kai Siegbahn, North-Holland, Amsterdam, 1965, p. 601.
9. T. Tamura and T. Ugadawa, *Phys. Rev.* 150, 783 (1966).
10. G. DoDang, R. Dreizler, A. Klein, and C. S. Wu, *Phys. Rev. Lett.* 17, 709 (1966).
11. M. Baranger and K. Kumar, *Nucl. Phys.* A122, 241 (1968).
12. K. Kumar and M. Baranger, *Nucl. Phys.* A122, 273 (1968).
13. A. Faessler, W. Greiner, and R. K. Sheline, *Nucl. Phys.* 70, 33 (1965).
14. S. Das Gupta and A. B. Volkov, *Phys. Rev.* C6, 1893 (1972).
15. F. M. Bernthal, J. O. Rasmussen, and J. M. Hollander, *Phys. Rev.* C3, 1294 (1971).
16. A. Johnson, H. Ryde, and S. A. Hjorth, *Nucl. Phys.* A179, 753 (1972).
17. A. S. Davydov and G. F. Filippov, *Nucl. Phys.* 8, 257 (1958).
18. A. S. Davydov and V. S. Rostovsky, *Nucl. Phys.* 12, 58 (1959).
19. A. S. Davydov and A. A. Chaban, *Nucl. Phys.* 20, 499 (1960).
20. M. A. J. Mariscotti, Gertrude Scharff-Goldhaber, and Brian Buck, *Phys. Rev.* 178, 1864 (1969).
21. B. R. Mottelson and J. G. Valatin, *Phys. Rev. Lett.* 5, 511 (1960).
22. B. R. Mottelson, *Cours de l'Ecole d'Eté de Physique Théorique des Houches*, 1958, Dunod, Paris, 1959; R. M. Rockmore, *Phys. Rev.* 116, 469 (1959); G. E. Brown, *Unified Theory of Nuclear Models and Forces*, North-Holland, Amsterdam, 1967, p. 72.
23. C. K. Ross, private communication.
24. A. B. Volkov, *Phys. Lett.* 35B, 299 (1971).
25. T. K. Das, R. M. Dreizler, and A. Klein, *Phys. Lett.* 34B, 235 (1971).
26. See, for example, J. D. Immele and G. L. Struble, *Nucl. Phys.* A187, 459 (1972).
27. P. H. Stelson and L. Grodzins, *Nucl. Data* A1, 21 (1965).
28. M. A. Preston and D. Kiang, *Can. J. Phys.* 41, 742 (1963).
29. E. R. Marshalek, *Phys. Rev.* 139B, 770 (1965).
30. A. K. Kerman, *Ann. Phys.* 12, 300 (1961).
31. B. L. Birbrair, L. K. Peker, and L. A. Sliv, *JETP* 9, 566 (1959).
32. S. Das Gupta and M. A. Preston, *Nucl. Phys.* 49, 401 (1963).
33. K. Kumar and M. Baranger, *Nucl. Phys.* A110, 529 (1968).
34. U. Götz, H. C. Pauli, K. Alder, and K. Junker, *Nucl. Phys.* A192, 1 (1972).
35. M. Sakai, *Nucl. Data* A8, 323 (1970).

36. H. S. Gertzman, D. Cline, H. E. Gove, and P. M. S. Lesser, *Nucl. Phys.* **A151**, 282 (1971).

37. R. J. Pryor and J. X. Saladin, *Phys. Rev.* **C1**, 1573 (1970).

38. W. T. Milner, F. K. McGowan, R. L. Robinson, P. H. Stelson, and R. O. Sayer, *Nucl. Phys.* **A177**, 1 (1971).

39. K. Kumar and M. Baranger, *Nucl. Phys.* **A92**, 608 (1967).

40. L. S. Kisslinger and R. A. Sorensen, *K. Danske Vidensk. Selsk. Mat.-Fys. Medd.* **32**, No. 9 (1960).

41. M. R. Gunye, S. Das Gupta, and M. A. Preston, *Phys. Lett.* **13**, 246 (1964).

42. L. Funke, K. H. Kaun, P. Kemnitz, H. Sodan, and G. Winter, *Nucl. Phys.* **A190**, 576 (1972); *Phys. Lett.* **39B**, 179 (1972).

43. A. K. Kerman, *K. Danske Vidensk. Selsk. Mat.-Fys. Medd.* **30**, No. 15 (1956).

44. T. D. Newton, *Can. J. Phys.* **38**, 700 (1960); *Atomic Energy of Canada Limited Report No.* CRT-886, 1960. See also *Can. J. Phys.* **37**, 944 (1959).

45. A. J. Rassey, *Phys. Rev.* **109**, 949 (1958).

46. D. Bes and Z. Szymanski, *Nucl. Phys.* **28**, 42 (1961).

47. D. L. Hendrie, N. K. Glendenning, B. G. Harvey, O. N. Jarvis, H. H. Duhm, J. Saudinos, and J. Mahoney, *Phys. Lett.* **26B**, 127 (1968).

48. G. F. Bertsch, *Phys. Lett.* **26B**, 130 (1968).

49. S. G. Nilsson, C. F. Tsang, A. Sobiczewski, Z. Szymanski, S. Wycech, C. Gustafson, I.-L. Lamm, P. Möller, and B. Nilsson, *Nucl. Phys.* **A131**, 1 (1969).

50. J. R. Nix, *Ann. Rev. Nucl. Sci.* **22**, 65 (1972).

51. P. Möller, *Nucl. Phys.* **A142**, 1 (1970).

52. A. E. Litherland, H. McManus, E. B. Paul, D. A. Bromley, and H. E. Gove, *Can. J. Phys.* **36**, 378 (1958).

53. E. B. Paul, *Phil. Mag.* **2**, 311 (1957).

54. R. Batchelor, A. J. Ferguson, H. E. Gove, and A. E. Litherland, *Nucl. Phys.* **16**, 38 (1960).

55. R. K. Sheline, *Nucl. Phys.* **2**, 382 (1956).

56. D. A. Bromley, H. E. Gove, and A. E. Litherland, *Can. J. Phys.* **35**, 1057 (1957).

57. A. E. Litherland, E. B. Paul, G. A. Bartholomew, and H. E. Gove, *Can. J. Phys.* **37**, 53 (1959).

PROBLEMS

9–1. Prove Eq. 9–26.

9–2. The electric multipole operator has been defined as

$$Q_{\lambda\mu} = e \sum_{k=1}^{Z} r_k{}^{\lambda} Y_{\lambda}{}^{\mu}(\theta_k, \phi_k) \equiv \sum_{k=1}^{Z} Q_{\lambda\mu}(\mathbf{r}_k), \tag{1}$$

where (r_k, θ_k, ϕ_k) are the spherical polar coordinates of the kth proton, the sum in Eq. 1 extending over all the Z protons in the nucleus.

(a) Consider an even-even nucleus with ground-state angular momentum zero. Show, using Eq. 9–22, that

$$B(E\lambda, 0 \to f) = (2\lambda + 1)|\langle \lambda 0 | Q_{\lambda 0} | 0 \rangle|^2, \tag{2}$$

where $|0\rangle$ is the ground state and $|f\rangle$ the final state with angular momentum λ.

(b) Define the energy-weighted sum

$$S_\lambda = \sum_f (E_f - E_0) B(E\lambda, 0 \to f),\tag{3}$$

the sum going over all final states $|f\rangle$ with angular momentum λ, and E_f and E_0 being the energies of $|f\rangle$ and $|0\rangle$. Show that

$$S_\lambda = \tfrac{1}{2}(2\lambda + 1)\langle 0|[Q_{\lambda 0}, [H, Q_{\lambda 0}]]|0\rangle,\tag{4}$$

where H is the Hamiltonian of the nucleus and a bracket denotes a commutator.
  (c) Let

$$H = -\frac{\hbar^2}{2M}\sum_{i=1}^A \nabla_i^2 + V,$$

where V is assumed to contain no velocity-dependent or exchange interactions. It then follows that

$$[H, Q_{\lambda 0}] = -\frac{\hbar^2}{2M}\sum_{k=1}^Z [\nabla_k^2, Q_{\lambda 0}(\mathbf{r}_k)].$$

Prove that

$$S_\lambda = \frac{\hbar^2}{2M}(2\lambda + 1)\,\langle 0|\sum_{k=1}^Z [\nabla_k Q_{\lambda 0}(\mathbf{r}_k)]^2|0\rangle.\tag{5}$$

(d) Consider the particular case of $\lambda = 2$. Note, from Eq. 3–8, that

$$Q_{20}(\mathbf{r}_k) = \sqrt{\frac{5}{16\pi}}\,e(3z_k^2 - r_k^2),$$

and show that

$$[\nabla_k Q_{20}(\mathbf{r}_k)]^2 = 4r_k^2 + 12r_k^2 \cos^2\theta.$$

Hence prove, assuming a spherically symmetric charge distribution $\rho(r_k)$, that

$$S_2 = \frac{Ze^2\hbar^2}{2M}\frac{25}{2\pi}\langle r^2\rangle,\tag{6}$$

where

$$\langle r^2\rangle = \frac{1}{Z}\int \rho(r_k)r_k^2 4\pi r_k^2\,dr_k.$$

For the more general λ-pole transition, Eq. 5 may be evaluated to give

$$S_\lambda = \frac{Ze^2\hbar^2}{2M}\frac{\lambda(2\lambda + 1)^2}{4\pi}\langle r^{2\lambda-2}\rangle,\tag{7}$$

which is an energy-weighted sum rule and includes both the $T = 0$ and $T = 1$ contributions. For further discussion of Eq. 7 relating to relevant experimental data, see ref. 8, p. 641.

9-3. Show that the vanishing of the products of inertia of a nucleus with isodensity surfaces

$$R = R_0(1 + \Sigma a_{2\mu} Y_2^\mu)$$

implies that

$$a_{22} = a_{2,-2} \quad \text{and} \quad a_{21} = a_{2,-1} = 0.$$

9-4. In the simple vibrational model of a spherical nucleus, simplify Eq. 9-24 to show that the reduced $E2$-transition probability from the ground state to the first excited 2^+ state is given by

$$B(E2, 0^+ \to 2^+) = \frac{45\hbar^2 e^2 r_0^4}{32\pi^2} \frac{Z^2 A^{4/3}}{E_2 B_2}, \tag{1}$$

where E_2 is the excitation energy of the first excited state, $R_0 = r_0 A^{1/3}$, and B_2 is the quadrupole inertial parameter.

Next, for the rotational model, use Eqs. 9-95 and 9-38 to show that the expression for the same quantity is given by

$$B(E2, 0^+ \to 2^+) = \frac{9\hbar^2 e^2 r_0^4}{16\pi^2} \frac{Z^2 A^{4/3}}{E_2 B_2}. \tag{2}$$

In both models we see that the inertial parameter B_2 is proportional to $[B(E2) \cdot E_2]^{-1}$.

Finally, for the ground-state band of a rotational nucleus, use Eqs. 9-38, 9-72, 9-73, 9-91, and 9-94 to show that

$$B(E2, J \to J - 2) = \frac{3}{32\pi^2} \frac{\hbar^2 e^2 r_0^4 (4J - 2)}{B_2(E_J - E_{J-2})} Z^2 A^{4/3} (J200|J - 2\,0)^2. \tag{3}$$

The $B(E2, J \to J - 2)$-value across the ground-state rotational band has been recently measured for ^{158}Er in the range $J = 2$ to $J = 16$ by Ward et al. [Phys. Rev. Lett. **30**, 493 (1973)]. We list their data in the following table:

Transition	$E_J - E_{J-2}$, keV	$B(E2)$, $e^2 b^2$
$2 \to 0$	192.7	0.55 ± 0.03
$4 \to 2$	335.7	0.87 ± 0.04
$6 \to 4$	443.8	1.14 ± 0.19
$8 \to 6$	523.8	1.08 ± 0.34
$10 \to 8$	578.9	1.12 ± 0.50
$12 \to 10$	608.1	$\gtrsim 0.90$
$14 \to 12$	510.0	0.77 ± 0.18
$16 \to 14$	473.2	1.35 ± 0.48

Assuming the validity of Eq. 3 above, calculate the inertial parameter B_2 for each of these transitions and comment on the peculiarity of its behavior.

Comment also on the assumptions that have gone into the derivation of Eq. 3. In your opinion, should this equation be applied in analyzing the experimental data given for ^{158}Er?

9-5. Prove Eq. 9–140, given

$$(JMK|J_\pm|JMK \pm 1) = \sqrt{(J \mp K)(J \pm K + 1)},$$

$$(j\Omega|j_\mp|j\,\Omega \pm 1) = \sqrt{(j \mp \Omega)(j \pm \Omega + 1)}.$$

The difference in sign in these two relations arises because $J_\pm$ refer to the components of $\mathbf{J}$, the total angular momentum, along moving axes, whereas j is defined as an angular momentum in the moving coordinate system.

9-6. The levels of ^{25}Mg and ^{25}Al have been carefully studied by Litherland et al. [Can. J. Phys. **36**, 378 (1958)]. Tables 9–4 and 9–5 list the rotational

TABLE 9-4

$K = \frac{5}{2}; \pi = +$			$K = \frac{1}{2}; \pi = +$		
J	E_{Mg}, MeV	E_{Al}, MeV	J	E_{Mg}	E_{Al}
$\frac{5}{2}$	0	0	$\frac{1}{2}$	0.58	0.45
$\frac{7}{2}$	1.61	1.61	$\frac{3}{2}$	0.98	0.95
$(\frac{9}{2})$		3.44	$\frac{5}{2}$	1.96	1.81
			$\frac{7}{2}$	2.74	2.73

TABLE 9-5

$K = \frac{1}{2}; \pi = +$			$K = \frac{1}{2}; \pi = -$		
J	E_{Mg}, MeV	E_{Al}, MeV	J	E_{Mg}	E_{Al}
$\frac{1}{2}$	2.56	2.50	$\frac{3}{2}$	3.40	3.09
$\frac{3}{2}$	2.80	2.70	$\frac{7}{2}$	3.97	3.72
$\frac{5}{2}$	3.90	3.88	$\frac{1}{2}$	4.26	3.85

bands, each built on a different Nilsson orbit, that have been identified. The orbits for the ground state of each band are the ones that Nilsson labels 5, 9, 11, 14, respectively. With the notation

$$\chi_\Omega = \sum_l (A_{l+}|N, l, \Omega - \tfrac{1}{2}, +\tfrac{1}{2}) + A_{l-}|N, l, \Omega + \tfrac{1}{2}, -\tfrac{1}{2})),$$

the wave functions for various distortions η are as follows:

Orbit 5: $N = 2$, $\Omega = \frac{3}{2}$, $A_{2+} = 1$ for all η;

Orbit 9: $N = 2$, $\Omega = \frac{1}{2}$;

η	2	4	6
A_{2+}	1.000	1.00	1.000
A_{0+}	-15.696	15.901	6.635
A_{2-}	11.489	-25.472	-16.609

Orbit 11: $N = 2$, $\Omega = \frac{1}{2}$;

η	2	4	6
A_{2+}	1.000	1.000	1.000
A_{0+}	-0.717	-1.143	-1.287
A_{2-}	-1.066	-0.675	-0.454

Orbit 14: $N = 3$, $\Omega = \frac{1}{2}$;

η	2	4	6
A_{3+}	1.000	1.000	1.000
A_{1+}	0.359	0.637	0.809
A_{3-}	0.777	0.632	0.500
A_{1-}	0.164	0.198	0.183

Also listed is orbit 7: $N = 2$, $\Omega = \frac{3}{2}$, $\eta = 2$, $A_{2+} = 1.000$, $A_{2-} = 0.351$.

(a) The four given bands all fit the formula

$$E = A[J(J + 1) + a(-)^{J+1/2}(J + \tfrac{1}{2})]$$
$$+ B[J(J + 1) + a(-)_{J}^{+1/2}(J + \tfrac{1}{2})]^2,$$

where the respective bands have

$A =$ 272 keV	176 keV	160 keV	117 keV
$a = 0$	-0.03	-0.58	-3.16
$B = -1.69$	-0.70	-0.55	-0.21

Explain this formula and calculate the effective moments of inertia.

(b) (i) Find the coefficients $c_{j\Omega}$ necessary to express orbit 9 (for $\eta = 4$) in the form $\chi_\Omega = \sum_j c_{j\Omega}\chi_{j\Omega}$.

(ii) Calculate the decoupling parameter for $\eta = 4$, for orbit 9. Compare with experiment.

(iii) Determine the level order of the band built on level 14, for $\eta = 2$ and for $\eta = 4$.

9-7. According to Nilsson, the $\frac{1}{2}^-$ [510] level has the following wave functions:

$$\eta' = 4, \quad \psi = \quad 1.000|550 +\rangle - 2.730|530 +\rangle + 2.440|510 +\rangle$$
$$- 0.606|551 -\rangle + 0.780|531 -\rangle + 1.431|511 -\rangle;$$

$$\eta = 6, \quad \psi = \quad 1.000|550 +\rangle - 2.359|530 +\rangle + 1.981|510 +\rangle$$
$$- 0.469|551 -\rangle + 0.420|531 -\rangle + 0.993|511 -\rangle.$$

Find the decoupling constant for $\eta = 4$ and $\eta = 6$ and compare the results with the experimental value for ^{183}W (Fig. 9-39).

9-8. The following level schemes are given. What would you expect the character of each state to be? At what energy would you look for a 10+ state in ^{234}U? an $\frac{11}{2}$+ level in ^{237}Np?

	MeV			MeV
		0+ ———————		0.810
4+ ———————	1.542	8+ ———————		0.499
2+ ———————	1.476			
? ———————	1.473	6+ ———————		0.296
2+ ———————	0.658	4+ ———————		0.143
0+ ———————	0	2+ ———————		0.0435
^{110}Cd		0+ ———————		0
		^{234}U		

	MeV
$\frac{9}{2}-$ ———————	0.159
$\frac{7}{2}-$ ———————	0.103
$\frac{9}{2}+$ ———————	0.076
$\frac{5}{2}-$ ———————	0.060
$\frac{7}{2}+$ ———————	0.033
$\frac{5}{2}+$ ———————	0
^{237}Np	

	MeV
$\frac{1}{2}-$ ———————	0.747
$\frac{9}{2}+$ ———————	0
^{97}Nb	

9-9. An odd-odd nucleus with $\Omega = 0$ and $K = 0$ has two possible intrinsic wave functions:

$$\chi(\pm) = \psi_\omega(n)\phi_{-\omega}(p) \pm \psi_{-\omega}(n)\phi_\omega(p),$$

where ψ_ω and ϕ_ω are Nilsson orbits with $j_3 = \omega$. Using Eqs. 9-131 and 9-132 and the relations following 9-138, show that, if the parities of ψ and ϕ are opposite, $\chi(-)$ is the ground state of a rotational band of odd J's while $\chi(+)$ is the ground state of a band of even J's; and that, if the parities are equal, the roles of $\chi(-)$ and $\chi(+)$ are reversed.

Similarly, show that there are only even spins in a $K = 0$ band in an even-even nucleus in which all particles are paired in Nilsson orbits.

Chapter 10

Hartree-Fock and Particle-Hole Calculations

10-1 REFINEMENTS ON THE SPHERICAL-SHELL MODEL

We noted in Chapter 7 the successes of the extreme single-particle model in explaining the ground-state properties of spherical or near-spherical nuclei. In Chapter 9, this scheme was broadened to generate the intrinsic states of deformed nuclei, where the average potential well in which the particles move was taken to be deformed. Although it was understood that the average field was being generated by the two-body interaction between the nucleons, this field was chosen in an empirical manner to reproduce the experimental single-particle levels, rather than calculating it from first principles. At first sight it seems very puzzling that the realistic two-nucleon potential, which is so strong and may be singular at short ranges, is able to produce a smooth average one-body potential. This puzzle was essentially resolved in Chapter 8, where it was shown that the effective interaction (the so-called G-matrix) between the nucleons in a nucleus is well behaved and smooth. It becomes meaningful, then, to formulate a scheme for generating a self-consistent average field for nucleons interacting with a smooth two-body force. This is called the Hartree-Fock method and we shall outline it in Section 10–3, where the meanings of the vague terms "smooth" and "self-consistent" will be clarified.

Apart from the ground-state properties, we are also interested in the excited states of a nucleus and in the decay modes and transition probabilities between the various nuclear states. Ideally, starting with a smooth, density-dependent effective interaction which has been obtained from the realistic two-nucleon potential, we should be able to reproduce the properties of the nucleus in the ground and excited states. In an earlier chapter we concentrated on how the effective interaction could be calculated starting from the fundamental two-nucleon force. In this chapter we shall concern ourselves primarily with the many-body problem of calculating the nuclear spectra, assuming the existence of a smooth effective interaction between the nucleons. In the next section, we shall first outline how the excited states of a nucleus can be calculated in a very simple-minded shell-model picture, and then compare the results with experimental observations. We shall find that some of the basic assumptions of this simple model have to be modified to explain the experimental results. In order to fix our ideas, we shall concentrate on the oxygen region, although the methods outlined are quite general.

10-2 THE SHELL-MODEL APPROACH TO CALCULATING NUCLEAR SPECTRA

To illustrate the method, we take the example of ^{18}O, which has two neutrons outside a closed shell of ^{16}O. The simplest empirical approach is to assume

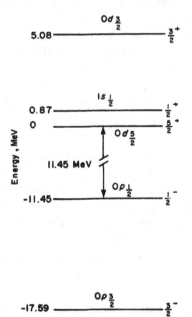

Fig. 10-1 Experimentally determined single-particle neutron levels in ^{16}O for the filled p-shell and the empty s-d shell. The zero of energy has been arbitrarily taken at the $1d_{5/2}$-level.

that the ^{16}O core is inert, in the sense that it provides an average field for the two neutrons, blocks states that would otherwise be available to the neutrons, but does not otherwise affect their motion. In the empirical approach, the single-particle levels generated by this potential are found from the experimental data. The energies of the levels in the s-d shell are obtained in experiments in which a neutron is dropped into unfilled levels through the stripping reaction ^{16}O(d, p)^{17}O. Correspondingly, the energies of the levels in the p-shell may be obtained by pulling out a neutron in a "pick-up" reaction like ^{16}O(d, t)^{15}O. These experimental single-particle states are shown in Fig. 10–1, where it will be seen that there is a gap of 11.45 MeV between the $0p_{1/2}$ and the $0d_{5/2}$ levels. This is the basis of the assumption that the ^{16}O core is inert, because it should cost 11.45 MeV to excite a particle from the core.

To calculate the energy spectrum of ^{18}O, we now consider that the two valence neutrons in the s-d shell interact through a two-body effective interaction, which may be completely phenomenological or may be derived from a realistic force. For simplicity, we ignore the continuum beyond the s-d shell and consider virtual scattering of the two neutrons within the s-d shell only. For example, a 0^+ shell state in ^{18}O could be composed of the following configurations:

$$|(0d_{5/2})^2 0\rangle, \qquad |(1s_{1/2})^2 0\rangle, \qquad \text{and} \qquad |(0d_{3/2})^2 0\rangle.$$

The two-neutron interaction forms a 3×3 matrix within the space spanned by these three states; the eigenvalues of the matrix are the energies of distinct 0^+ states. The choice of appropriate single-particle wave functions is a matter of some subtlety. In calculations with phenomenological interactions they are taken to be harmonic-oscillator wave functions with the oscillator constant adjusted to fit the rms radius of the nucleus.

Similarly, the 2^+ state may be formed by five different unperturbed configurations, and it is necessary to diagonalize a 5×5 matrix. Figure 10–2 shows the experimental spectrum of ^{18}O, together with a typical calculated spectrum in this scheme, due to Kahana and Tomusiak [1]. The particular spectrum is obtained by assuming the two-neutron force to be an S-state free-reaction matrix, which fits only the low-energy phase-shift data. It is a characteristic feature of any short-range attractive interaction to push the 0^+ ground state far below the unperturbed configuration—a manifestation of the pairing effect.

Although we see that the correct level ordering has been reproduced and the energy spacings are qualitatively right, we cannot obtain, in this model,

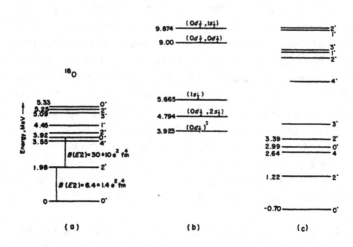

Fig. 10-2 (a) Experimental energy levels of ^{18}O. Also shown are the $B(E2)$-values for the transitions from 2^+ to 0^+ and $0^{+'}$ to 0^+ states. (b) The unperturbed energy levels due to two noninteracting neutrons in the s-d shell. (c) The calculated spectrum in the inert-core model, using the Kahana-Tomusiak force.

the observed electric quadrupole transition rates between certain states. The electromagnetic transition rate between two states of a nucleus is directly proportional to the square of the matrix element of the electromagnetic operator between the initial and final states. A measure of the quadrupole transition is thus the reduced matrix element $B(E2)$. Two values are: $B(E2, 2^+ \rightarrow 0^+) = 6.4 \pm 1.4\ e^2\ \mathrm{fm}^4$ and $B(E2, 0^{+'} \rightarrow 2^+) = 30 \pm 10\ e^2\ \mathrm{fm}^4$. At first sight, since the two particles outside the closed shell are neutral, it would seem that in the simple shell model there can be no electric quadrupole transition in ^{18}O. But an extracore neutron is in a non-spherical orbit and it induces a small eccentricity of order A^{-1} in the motion of each core nucleon. Hence the Z protons produce a quadrupole moment of order Z/A times a single particle moment. One can simulate this effect by assigning an "effective charge" $\sim Ze/A$ to each extracore neutron. Distortion of an orbit is equivalent to admixture of excited states; hence Fig. 10–3 is a diagrammatic representation of this effect. With $e_{\mathrm{eff}} = 0.5\ e$ one can explain the value of $B(E2, 2^+ \rightarrow 0^+)$, but simple shell-model wave functions yield $B(E2, 0^{+'} \rightarrow 2^+) = 0.8\ e^2\ \mathrm{fm}^4$ [2], nearly 2 orders of magnitude too small. Large $B(E2)$ values indicate collective effects. This clearly shows that the simple two-neutron model of ^{18}O must be unrealistic and that the core must be taking an active

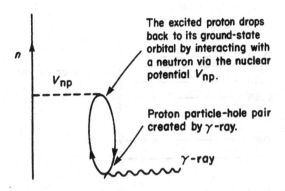

The excited proton drops back to its ground-state orbital by interacting with a neutron via the nuclear potential V_{np}.

Proton particle-hole pair created by γ-ray.

Fig. 10-3 The electric field couples to the extracore neutron through a proton particle-hole pair, giving rise to an effective charge of the neutron. If there is an extracore proton, its effective charge will similarly be increased. A calculation of the effective charge involves summing over all p-h proton states.

rather than a passive role in the description of these states. Even in the description of the energy levels in ^{18}O from the simple model, there are some important discrepancies. The experimental spectrum (see Fig. 10–2) shows a 2^+ and a 0^+ state at 5.25 and 5.33 MeV, respectively. The stripping experiments show that both these states contain a large amount of two-neutron configurations. However, in the shell-model calculations, nearly all the two-particle strengths are absorbed in the five lowest states below 5 MeV. The rest go into states at about 8 MeV and higher.

Further clues that the ^{16}O core is not inert can be found by examining the experimental spectrum of ^{17}O in more detail, as shown in Fig. 10–4. It will be seen that there are three odd-parity levels around 4-MeV excitation, which cannot be generated at these energies by the one-body potential of our model. These must arise from the excitation of an odd number of particles from the p-shell of the core into the s-d shell, leaving the appropriate number of holes in the core[†]. We should try to understand why these excitations from the core cost so little in energy.

Further evidence of core excitation can be obtained by examining the energy spectrum of ^{16}O, as shown in Fig. 10–5. In the simple shell-model picture, an odd-parity state can come only at an excitation energy of about

---

[†] These odd-parity states are primarily of two-particle, one-hole nature. See B. Margolis and N. de Takacsy, *Phys. Lett.* **15**, 329 (1965).

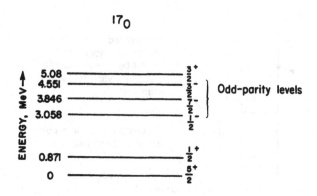

Fig. 10-4 Experimental spectrum of ^{17}O for the first few levels.

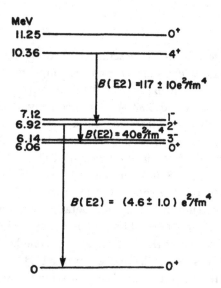

Fig. 10-5 Experimental spectrum of ^{16}O. There are strong $E2$-transitions from the 6.92-MeV 2^+ to the 6.06-MeV 0^+ state, and from the 10.36-MeV 4^+ to the 6.92-MeV 2^+ state.

12 MeV, whereas experimentally a 3^- level is seen at 6.14 MeV. One might suspect that this might be an octupole vibration state, which could show

that the core is far from inert. More surprising is the appearance of an
0+ excited state at only 6.06 MeV, whereas in the spherical single-particle
model it should come at about $\sim$ 24 MeV. Large $B(E2)$-transition rates
between the 6.92-MeV 2+ state and the 6.06-MeV 0+ state further suggest a
collective nature for these excited states. In Section 10–5 we shall try to
see how there can be low-lying even-parity excited states in ^{16}O, and will
then give a more realistic picture of ^{18}O. In Section 10–6, we shall try to
understand the odd-parity states of ^{16}O within the framework of an extended
shell model.

We have chosen the oxygen region to illustrate a simple shell model
calculation and also to show the failings of the inert-core picture. Similar
discrepancies are noted in the doubly closed-shell Ca and Ni regions as well.
The shell model is more successful, however, when we consider the Pb region,
where we have the doubly closed-shell nucleus ^{208}Pb. Figure 10–6 shows
plots of the experimentally determined low-lying levels of ^{209}Pb from the
reaction ^{208}Pb$(d, p)^{209}$Pb [3] and also the single-particles levels as generated
by a suitably chosen Wood-Saxon potential with spin-orbit force [4]. It will

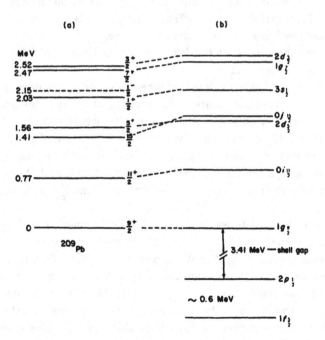

Fig. 10-6   (a) Experimental energy levels of the low-lying states in ^{209}Pb. (b) Some
levels generated by a neutron moving in an appropriate Woods-Saxon well with a spin-
orbit potential [4].

be seen that seven out of eight excited states up to 2.52 MeV can be explained in the shell-model picture, which is a much better situation than that for ^{17}O, where there were three low-lying odd-parity states (Fig. 10–4) which could not be generated in the extreme single-particle model. Moreover, Mukherjee and Cohen [3] have shown that the experimental angular distribution of the proton in $^{208}Pb(d, p)^{209}Pb$ and the average cross sections over scattering angles 45–90° as functions of excitation energy could be reproduced fairly well by assigning to these excited states just the pure shell-model configurations, as shown in Fig. 10–6.

This is a classic example of the success of the simple model, where the core does seem to play a passive role. We naturally seek to know why this is so, since the shell gap between the $2p_{1/2}$- and $1g_{9/2}$-levels is only about 3.4 MeV. It should be realized, however, that the size of the energy gap should be considered in relation to the spacings of the occupied single-particle levels, which are only about 0.5 MeV for the top few levels in ^{208}Pb. In the spectrum of ^{208}Pb, all the low-lying excited states up to 4 MeV have odd parity. It is conceivable that in some of the excited states of ^{208}Pb, for example, the core of ^{208}Pb will be excited too. Within the framework of our model, we could describe such an excited state of ^{209}Pb by coupling the $g_{9/2}$- or $i_{11/2}$-neutron to one of the low-lying excited states of the core.

Why is it, then, that all the low-lying excited states of ^{209}Pb seem to be almost pure single-particle states, with no admixture of core-coupled states? The answer lies in the fact that when a $g_{9/2}$- or $i_{11/2}$-neutron is coupled to any of the excited core states, the resulting state has also odd parity. Therefore it can mix only with the $j_{15/2}$ single-particle state of odd parity (see Fig. 10–6). All the other low-lying states of ^{209}Pb are of positive parity and therefore remain relatively "pure." Even the mixing in the odd-parity $j_{15/2}$-state is small, since the first excited state of ^{208}Pb is at 2.61 MeV, which is much greater than the experimental position of the $\frac{15}{2}-$ state (1.41 MeV) in ^{209}Pb.

Looking at Fig. 10–6, we notice that the experimental positions of all but the first excited state are somewhat lower in energy than the single-particle states generated by the Wood-Saxon potential. This signifies that small admixtures of other configurations are present in these states which result in the lowering of energy. In particular, there is an inversion of the order of the $\frac{5}{2}+$ and $\frac{15}{2}-$, with the $\frac{15}{2}-$ coming down, implying that the $\frac{5}{2}+$ state has less admixture of higher configurations than the $\frac{15}{2}-$. This is consistent with our earlier remarks. More specifically, we can describe the excited $\frac{15}{2}-$ state in terms of the following configurations:

$$\left|\tfrac{15}{2}-\right\rangle = C_1 \left|\text{core ground state } 0^+ \otimes j_{15/2}; \tfrac{15}{2}-\right\rangle$$
$$+ C_2 \left|\text{core } 3^- \otimes g_{9/2}; \tfrac{15}{2}-\right\rangle + C_3 \left|\text{core } 3^- \otimes i_{11/2}; \tfrac{15}{2}-\right\rangle + \cdots$$

with $\sum_i |C_i|^2 = 1$.

The quantity $|C_1|^2$ gives the strength of the pure single-particle configuration and is called the "spectroscopic factor" in analyses of d-p reactions. It is possible to extract the spectroscopic factors of the single-particle configurations of the various excited states by analyzing the cross-section data for d-p reactions. These analyses confirm that the spectroscopic factors for all the even-parity excited states of ^{209}Pb are between 0.95 and 1.0, while for the $\tfrac{15}{2}-$ state values range from 0.5 to 0.9 [5]. A great deal of experimental work has been done in the lead region [5], especially to examine how good the shell-model wave functions are, particularly when there are two or more valence particles or holes. It is found that in these nuclei the shell model is fairly successful.

Shell-model calculations escalate in complications with increasing number of particles outside the closed shell, because of the complexity of the angular-momentum algebra and the increasing dimension of the matrices. However, calculations with three or four particles outside the core are quite feasible and have been done [7]. Generally it is necessary to resort to more approximations in the calculation of spectra of nuclei with four or more extracore particles—a topic that we shall presently discuss.

10-3 THE HARTREE-FOCK (HF) EQUATIONS

In the shell model we have seen that many of the ground-state properties of nuclei can be described by making the simple assumption of independent particle motion in an average field subject to the exclusion principle. In this section we examine how the average one-body field of force is generated through a given two-nucleon effective interaction. This may be the reaction matrix calculated from the actual two-body force, or it may be a phenomenological interaction. In this section, we consider the theory when the effective interaction is local. In realistic models, the interaction may be nonlocal and density dependent, and these introduce some additional complications.

In the second quantized notation (see Appendix C), let us take the Hamiltonian to be

$$H = \sum_{\alpha,\beta} \langle \alpha|t|\beta \rangle a_\alpha{}^+ a_\beta + \tfrac{1}{2} \sum_{\alpha,\beta,\gamma,\delta} \langle \alpha\beta|V|\gamma\delta \rangle a_\alpha{}^+ a_\beta{}^+ a_\delta a_\gamma, \qquad (10\text{-}1)$$

where V denotes the effective two-body interaction operator, and the single-particle states $|\alpha\rangle$, $|\beta\rangle$, etc., form an arbitrary complete set. The operator a_α^+ creates a particle in the state $|\alpha\rangle$, while a_β destroys a particle in the state $|\beta\rangle$. The sums in Eq. 10-1 are unrestricted, going over the complete set of states. Essentially, we seek to obtain the best possible ground-state wave function and energy of an A-body system in the independent-particle approximation. In this picture, each particle is moving in some average one-body potential U, which generates a set of single-particle states $|a\rangle$, $|b\rangle$,..., $|l\rangle$, $|m\rangle$,.... Of these single-particle states, A are singly occupied by the particles and we denote them by $|l\rangle$, $|m\rangle$, $|n\rangle$,..., while the unoccupied states are $|a\rangle$, $|b\rangle$, $|c\rangle$,.... Our problem is to find U and the single-particle orbitals that are occupied in the ground state. Since the average field U is generated through the interaction V, we would expect its matrix element between two orbitals r and s to be

$$\langle r|U|s\rangle = \sum_{m=1}^{A} [\langle rm|V|sm\rangle - \langle rm|V|ms\rangle]. \qquad (10\text{-}2)$$

The average field U thus depends on which orbitals m are occupied, and the orbitals in turn are generated by the single-particle Hamiltonian $H_{SP} = t + U$. In second-quantized form, we can write

$$H_{SP} = \sum_{k,i} \left[\langle k|t|i\rangle + \sum_{m=1}^{A} (\langle km|V|im\rangle - \langle km|V|mi\rangle)\right] a_k^+ a_i, \qquad (10\text{-}3)$$

where the sums over k and i are unrestricted. The solution of the problem demands self-consistency: choosing a trial set of orbitals and selecting those which are occupied, we can calculate the one-body potential U by Eq. 10-2, and then solve the Schrödinger equation with U to obtain a new set of orbitals. The procedure is repeated until the process converges.

The procedure that we have described can be put on a firmer basis by using the variational principle. In the independent-particle picture, the many-body wave function $|\Phi\rangle$ is an antisymmetrized product of the single-particle orbitals. The optimum choice of these orbitals is such that the expectation value of the energy is stationary, that is,

$$\delta \frac{\langle \Phi_0|H|\Phi_0\rangle}{\langle \Phi_0|\Phi_0\rangle} = 0, \qquad (10\text{-}4)$$

where by $|\delta\Phi_0\rangle$ we mean a first-order change (i.e., change in a single orbital) in $|\Phi_0\rangle$, such that $\langle \Phi_0|\delta\Phi_0\rangle = 0$. The determinantal ground state that satisfies Eq. 10-4 is

$$|\Phi_0\rangle = \prod_{m=1}^{A} a_m{}^+|0\rangle, \tag{10-5}$$

where $|0\rangle$ denotes the true vacuum state with no particles. For $|\delta\Phi_0\rangle$, we choose

$$|\delta\Phi_0\rangle = \eta a_b{}^+ a_m|\Phi_0\rangle, \tag{10-6}$$

where η is a constant, $b > A$, and $m \leqslant A$. Then a straightforward but lengthy calculation to simplify Eq. 10-4 yields the condition (see Appendix C-3)

$$\langle m|t|b\rangle + \sum_{n=1}^{A} (\langle mn|V|bn\rangle - \langle nm|V|bn\rangle) = 0, \tag{10-7}$$

which must be obeyed for the optimum choice of orbitals, with $b > A$ and $m \leqslant A$. If we use the definition of H_{SP} as given in Eq. 10-3, it follows that

$$\langle m|H_{SP}|b\rangle = 0, \tag{10-8}$$

which implies that with this choice of H_{SP} there are no transitions between the occupied states $|m\rangle$ and the empty states $|b\rangle$. This, in fact, justifies our choice of U (Eq. 10-2), because it means that this U accounts for all the one-particle one-hole ($1p$-$1h$) transitions from the ground state that can be brought about by V. In other words, the matrix elements of the residual interaction $(V - U)$ link the HF ground state only to states in which two particles are excited, that is, $2p$-$2h$ states. Since U is a one-body potential, this is the best that can be achieved. The inclusion of the multiple particle-hole state constitutes higher-order corrections to the HF ground state. The minimal condition (Eq. 10-8), obtained from the variational principle, implies that the occupied and empty states are completely decoupled so far as H_{SP} is concerned, and this means that H_{SP} may be diagonalized in these spaces separately. This is achieved if the HF orbitals are generated by H_{SP}:

$$(t + U)|m\rangle = e_m|m\rangle, \tag{10-9a}$$

that is,

$$\langle l|H_{SP}|m\rangle = e_l\delta_{lm}, \quad l, m \leqslant A. \tag{10-9b}$$

Since the occupied orbitals are distinct from the unoccupied ones, it also follows that Eq. 10-8 is satisfied.

It is worth while to write Eq. 10-9a in coordinate space to emphasize the nonlocal nature of U. Using Eq. 10-2, we find

$$-\frac{\hbar^2}{2M}\nabla_1^2\phi_m(\mathbf{r}_1) + \left(\sum_{l=1}^{A}\int \phi_l^*(\mathbf{r}_2)V(|\mathbf{r}_1-\mathbf{r}_2|)\phi_l(\mathbf{r}_2)\,d^3r_2\right)\phi_m(\mathbf{r}_1)$$

$$-\sum_{l=1}^{A}\int \phi_l^*(\mathbf{r}_2)V(|\mathbf{r}_1-\mathbf{r}_2|)\phi_l(\mathbf{r}_1)\phi_m(\mathbf{r}_2)\,d^3r_2 = e_m\phi_m(\mathbf{r}_1),$$

where $\langle \mathbf{r}|m\rangle = \phi_m(\mathbf{r})$ is a single-particle state.

Writing

$$U_H(\mathbf{r}_1) = \sum_{l=1}^{A}\int \phi_l^*(\mathbf{r}_2)V(|\mathbf{r}_1-\mathbf{r}_2|)\phi_l(\mathbf{r}_2)\,d^3r_2 \qquad (10\text{--}10)$$

and

$$U_F(\mathbf{r}_1,\mathbf{r}_2) = \sum_{l=1}^{A}\phi_l^*(\mathbf{r}_2)V(|\mathbf{r}_1-\mathbf{r}_2|)\phi_l(\mathbf{r}_1),$$

we obtain

$$-\frac{\hbar^2}{2M}\nabla_1^2\phi_m(\mathbf{r}_1) + U_H(\mathbf{r}_1)\phi_m(\mathbf{r}_1) - \int U_F(\mathbf{r}_1,\mathbf{r}_2)\phi_m(\mathbf{r}_2)\,d^3r_2 = e_m\phi_m(\mathbf{r}_1).$$

$$(10\text{--}11)$$

Here we have denoted the local part of U by U_H and the nonlocal part by U_F, the latter arising because of the antisymmetrization of the HF wave function.

We should mention here the modifications that take place in the HF equation 10–11 when the effective V is density dependent. For simplicity, we assume that V is still local and has the form

$$\langle \mathbf{r}_1,\mathbf{r}_2|V|\mathbf{r}_1',\mathbf{r}_2'\rangle = V[r_{12},\rho(\mathbf{R}_{12})]\,\delta(\mathbf{r}_{12}-\mathbf{r}_{12}')\,\delta(\mathbf{R}_{12}-\mathbf{R}_{12}'), \qquad (10\text{--}12)$$

where $\mathbf{R}_{12} = (\mathbf{r}_1+\mathbf{r}_2)/2$, $\mathbf{r}_{12} = \mathbf{r}_1 - \mathbf{r}_2$, and similarly for the primed variables. We are assuming that the interaction depends on the density ρ at the center of mass of the two particles. Since

$$\rho(\mathbf{R}) = \sum_{m=1}^{A}\phi_m^*(\mathbf{R})\phi_m(\mathbf{R}),$$

an additional term arises when we minimize the ground-state energy with respect to the single-particle orbitals $\phi_m^*(\mathbf{r})$ at any point $\mathbf{r}$; in fact,

$$\frac{\delta V}{\delta \phi_m^*(\mathbf{r})} = \frac{\delta V}{\delta \rho}\frac{\delta \rho(\mathbf{R})}{\delta \phi_m^*(\mathbf{r})} = \frac{\delta V}{\delta \rho}\phi_m(\mathbf{r})\,\delta(\mathbf{R}-\mathbf{r}).$$

Consequently, for a density-dependent local V, the HF equation 10–11 is generalized to

$$-\frac{\hbar^2}{2M}\nabla_1^2\phi_m(\mathbf{r}_1) + U_H\phi_m(\mathbf{r}_1) - \int U_F(\mathbf{r}_1, \mathbf{r}_2)\phi_m(\mathbf{r}_2)\, d^3r_2$$
$$+ U_R(\mathbf{r}_1)\phi_m(\mathbf{r}_1) = e_m\phi_m(\mathbf{r}_1), \tag{10-13}$$

where $U_R(\mathbf{r}_1)$ is called the rearrangement potential. Its presence, of course, results in modifications of the values of the single-particle energies e_m and the total single-particle potential U. Explicit calculations show [8, 9] that U_R can be written as

$$U_R(\mathbf{r}_1) = \frac{1}{2}\sum_{l,m}\int \phi_l{}^*(\mathbf{r}_2)\phi_m{}^*(\mathbf{r}_3)\frac{\partial V(r_{23}, \rho(\mathbf{R}_{23}))}{\partial\rho(\mathbf{R}_{23})}\,\delta(\mathbf{R}_{23} - \mathbf{r}_1)$$
$$\cdot [\phi_l(\mathbf{r}_2)\phi_m(\mathbf{r}_3) - \phi_l(\mathbf{r}_3)\phi_m(\mathbf{r}_2)]\, d^3r_2\, d^3r_3. \tag{10-14}$$

In actual calculations, the HF orbitals are generally expressed in terms of some conveniently chosen basic set (denoted here by Greek symbols)

$$|l\rangle = \sum_\beta x_\beta{}^l|\beta\rangle, \tag{10-15}$$

with expansion coefficients $x_\beta{}^l$, where the basic set $|\beta\rangle$, for example, is generated by a spherically symmetric or axially symmetric harmonic-oscillator potential. In principle, the sum over β in Eq. 10–15 should be over all states; in practice, the space in truncated. In principle, of course, the choice of the space determines the effective interaction V (see Section 8–4). For example, for a calculation in the $0p$-shell region like that of carbon, the basic states $|\beta\rangle$ may include all states up to the $1p$-$0f$ levels of the harmonic oscillator. For convenience in notation, let us write

$$\langle ln|V|mn\rangle - \langle ln|V|nm\rangle = \langle ln|V|mn\rangle_A \tag{10-16}$$

and define the HF single-particle density operator

$$\rho = \sum_{m=1}^{A}|m\rangle\langle m|. \tag{10-17}$$

Then

$$\rho_{\gamma\delta} \equiv \langle\gamma|\rho|\delta\rangle = \sum_{m=1}^{A}(x_\gamma{}^m)^*(x_\delta{}^m). \tag{10-18}$$

Equation 10–9b may be rewritten as

$$\sum_l \langle l|t|m\rangle|l\rangle + \sum_l \sum_{n=1}^{A} \langle ln|V|mn\rangle_A|l\rangle = e_m|m\rangle. \qquad (10\text{-}19)$$

Using Eqs. 10–15 and 10–18, we reduce this to

$$\sum_{\beta} \langle\alpha|t|\beta\rangle x_\beta{}^m + \sum_{\beta,\gamma,\delta} \langle\alpha\gamma|V|\beta\delta\rangle_A \rho_{\gamma\delta} x_\beta{}^m = e_m x_\alpha{}^m. \qquad (10\text{-}20)$$

Initially, a set of coefficients $x_\beta{}^m$ for each orbital $|m\rangle$ is guessed, and $\rho_{\gamma\delta}$ is calculated by deciding which of these are filled. The left side of Eq. 10–20 can then be calculated in the given (truncated) space and diagonalized. This generates a new set of orbitals, and the procedure is repeated until successive diagonalizations produce the same set $x_\beta{}^m$.

The ground-state energy of the system with the HF state $|\Phi\rangle$ is

$$E_{\text{HF}} = \langle\Phi_0|H|\Phi_0\rangle$$

$$= \sum_{l=1}^{A} \langle l|t|l\rangle + \tfrac{1}{2}\sum_{l,m=1}^{A} \langle lm|V|lm\rangle_A \qquad (10\text{-}21\text{a})$$

or

$$E_{\text{HF}} = \sum_{l=1}^{A} e_l - \tfrac{1}{2}\sum_{l,m=1}^{A} \langle lm|V|lm\rangle_A \qquad (10\text{-}21\text{b})$$

$$= \sum_{l=1}^{A} \langle l|t + \tfrac{1}{2}U|l\rangle = \sum_{l=1}^{A} [e_l - \tfrac{1}{2}\langle l|U|l\rangle], \qquad (10\text{-}21\text{c})$$

where U is defined by Eq. 10–2.

It will be seen from Eq. 10–21 that the total ground-state energy is not simply the sum of the occupied eigenvalues e_l of H_{SP}. Nevertheless, we can interpret the e_l's as the approximate excitation energies of the levels and call them the self-consistent single-particle energies. This can be seen by finding the expectation value of H with respect to a state $a_b{}^+a_m|\Phi_0\rangle$, which is a configuration in which a particle in orbital $m \leqslant A$ has been excited to an orbital $b > A$. This is found to be $[E_{\text{HF}} + e_b - e_m - \langle bm|V|bm\rangle_A]$, implying that the excitation energy for this configuration is $e_b - e_m - \langle bm|V|bm\rangle_A$. For a system with large A, it is clear from Eq. 10–2 that the last term, $\langle bm|V|bm\rangle_A$, is of order $1/A$ compared to $e_b - e_m$ and hence may be ignored. If this is done, it follows that for the ground state the self-consistent energies of all the occupied orbitals should be less than those of the empty ones, so that only the lowest A HF orbitals are filled in the ground state.

Having established a variational principle for calculating the ground-state energy E_{HF}, we look for corrections to this expression in a perturbation scheme. Essentially, we started with a Hamiltonian for the A-nucleon system,

$$H = \sum_i t_i + \sum_{i<j} V_{ij}, \tag{10-22}$$

which we rewrote as

$$H = \sum_i (t_i + U_i) + (\sum_{i<j} V_{ij} - \sum_i U_i) \tag{10-23}$$

by introducing the one-body potential operator U defined in Eq. 10-2. The independent-particle A-body problem was then solved to obtain an expression for

$$E_{HF} = \sum_{l=1}^{N} \langle l|t + \tfrac{1}{2}U|l\rangle;$$

the factor $\tfrac{1}{2}$ appeared in U because it is being generated, by definition, by a two-body potential V_{ij} subject to the condition that the pairs i-j should not be double counted. For the HF approximation to be a good one, the contribution to the ground-state energy of the residual interaction $V - U$ appearing in Eq. 10-23 should be small compared to E_{HF}. In principle, the residual interaction, being a two-body operator, can cause transitions from the HF ground state $|\Phi_0\rangle$ to states $|\Phi_0'\rangle$, which differ only in one orbital, or to states $|\Phi_0''\rangle$, in which two orbitals are changed from $|\Phi_0\rangle$. Using perturbation theory, we find that the energy shift in the ground state due to the residual interaction is

$$\Delta E = \sum_{\Phi_0'} \frac{|\langle \Phi_0|V - U|\Phi_0'\rangle|^2}{E_{HF} - E'} + \sum_{\Phi_0''} \frac{|\langle \Phi_0|V|\Phi_0''\rangle|^2}{E_{HF} - E''}. \tag{10-24}$$

The first term on the right side of Eq. 10-24, which is due to the one-particle one-hole transitions, drops out automatically because of the self-consistent definition of U, as given in Eq. 10-2. This is the bonus obtained by doing a self-consistent calculation. If we had chosen U arbitrarily, rather than through Eq. 10-2, the one-particle one-hole contribution could be substantial. Let us now consider the last term in Eq. 10-24. Let

$$|\Phi_0''\rangle = a_b{}^+ a_d{}^+ a_m a_n |\Phi_0\rangle; \quad b \neq d > A, \quad m \neq n \leqslant A.$$

Then

$$\langle \Phi_0 | V | \Phi_0'' \rangle = \langle mn | V | bd \rangle_A \quad \text{and}$$

$$\sum_{\Phi_0''} \frac{|\langle \Phi_0 | V | \Phi_0'' \rangle|^2}{E_{\mathrm{HF}} - E''} = \sum_{\substack{m,n \leqslant A \\ b,d > A}} \frac{|\langle mn | V | bd \rangle_A|^2}{e_m + e_n - e_b - e_d}, \tag{10-25}$$

where e_m, etc., are the self-consistent HF single-particle energies. In order for ΔE to be small, the effective interaction V should be weak, so that its matrix elements connecting two occupied orbitals to two unoccupied ones are small. In other words, the effective interaction should be smooth, in the sense that it may not have high Fourier components in momentum. Moreover, if there is a substantial gap in energy between the occupied single-particle levels and the unoccupied ones, the energy denominator $e_m + e_n - e_b - e_d$ will be large and ΔE small. Estimates of such second-order corrections to the HF energy have been made, for example, by Kerman and Pal [10] in ^{16}O and ^{40}Ca, using the Tabakin separable potential as the effective interaction V. They find that, whereas the first-order potential energies per particle in ^{16}O and ^{40}Ca are 21.8 and 25.6 MeV, respectively, the second-order corrections in the potential energy per particle in these nuclei are 4.3 and 7.2 MeV. These result in substantial corrections to the binding-energy values but indicate that higher-order corrections are probably small. Essentially, in this perturbation scheme, corrections to the HF ground state $|\Phi_0\rangle$ are introduced by adding 2p-2h correlations. Denoting the correlated ground state by $|\tilde{\Phi}_0\rangle$, we can write

$$|\tilde{\Phi}_0\rangle = |\Phi_0\rangle + \sum_{\Phi_0''} \frac{\langle \Phi_0'' | V | \Phi_0 \rangle}{E_{\mathrm{HF}} - E''} |\Phi_0''\rangle. \tag{10-26}$$

We shall see in Section 10–6 that inclusion of 2p-2h correlations may have important effects on the equilibrium shape of the lowest intrinsic state of the nucleus. In the so-called random-phase approximation (RPA), a correlated ground state is introduced that has admixtures of 0p-0h, 2p-2h, 4p-4h, and higher-order configurations. We shall not discuss this method in this book, and the reader is referred to Rowe [11] for further development.

10-4 BINDING ENERGIES OF CLOSED-SHELL NUCLEI IN THE HARTREE-FOCK APPROXIMATION

The HF method has been extensively applied in nuclear calculations. A basic project is the calculation of the binding energy, the density distribution, and the equilibrium density of nuclei—in particular, of closed-shell spherical nuclei, for which the calculations are simpler. The binding energy and the saturation density of a system are very sensitive to the effective interaction and also to the nature of the approximations in the calculation. The first realistic attempt to do such a calculation on ^{16}O, ^{40}Ca, and ^{90}Zr was made by Brueckner, Lockett, and Rotenberg [12] and on ^{208}Pb by Masterson and Lockett [13]. They used the G-matrix from nuclear matter as the effective interaction, with the local-density approximation (see Chapter 8). Since the effective interaction itself is density dependent, there appears the extra rearrangement potential term with U_R in the HF equation 10–13. This extra term is of great importance in giving the right binding energy and saturation density. For an effective interaction that has no density dependence, we saw from Eq. 10–21 that the ground-state energy E_0 can be written as half the sum of the kinetic and the single-particle energies. If the radius of the nucleus is known, the kinetic energies are roughly known, and all the single-particle energies in a light nucleus like ^{16}O are also known. Adding these yields only about half the experimental binding of ^{16}O. In doing HF calculations for binding energy and saturation density, it is necessary, therefore, to take into account properly the density dependence of the effective interaction. The calculation of Brueckner et al. [12] included only the density dependence due to the repulsive core, omitting the strong density dependence brought about by the tensor force. This resulted in too little saturation, leading to unrealistically high central densities. In many subsequent calculations (see, e.g., Davies et al. [14]), the Brueckner G-matrix has been calculated in the harmonic-oscillator basis and the HF calculations have been performed with the matrix elements so obtained. In these calculations, the effect of the rearrangement potential and the contribution of higher order clusters have been omitted, resulting in too little binding.

The construction of a local, density-dependent interaction for finite nuclei, starting from the nuclear-matter G-matrix, was discussed in Chapter 8. Negele [15], Sprung and Banerjee [16], and Nemeth and Ripka [17] have constructed such interactions in the local-density approximation for HF calculations. As we have seen, realistic two-nucleon interactions that yield the correct equilibrium density of nuclear matter give a binding energy which is a few MeV less than the empirical value of 16 MeV. This additional binding may arise from higher-order clusters, from modifications in the tensor component of the force, and from a three-nucleon force [18]. The higher-order cluster contributions depend sensitively on the tensor component of the force. Negele, as well as Sprung and Banerjee, has accounted for these uncertainties of the nuclear-matter G-matrix by slightly ormalizing it (mainly in the $^3S_1$-state) to make it reproduce the correct ding and equilibrium density of nuclear matter. It has then been dem- trated [15, 19] that the effective density-dependent local interaction ived from such a G-matrix gives good results for finite spherical nuclei the HF calculations.

In the following treatment, rather than describing in detail the more mplicated calculations of Negele [15] or Campi and Sprung [19], we shall tline a HF calculation with a particularly simple, phenomenological teraction, due to Vautherin and Brink [20], which is density dependent. irst, consider for simplicity the S-state diagonal matrix element of a local otential $V(r)$ in momentum space:

$$\langle k|V|k\rangle = \int V(r) j_0{}^2(kr)\, d^3r,$$

where k is the magnitude of the relative momentum of two nucleons inside the Fermi sea. If $V(r)$ is of very short range, we can expand $j_0{}^2(kr)$ in a power series, obtaining

$$\langle k|V|k\rangle = \int V(r)\, d^3r - \tfrac{1}{3}k^2 \int V(r) r^2\, d^3r + \cdots, \qquad (10\text{--}27a)$$

where the higher-order terms can be neglected if the higher moments of $V(r)$ are very small. In coordinate space, the interaction may then be written as

$$V(p, r) = t_1\, \delta(r) + t_2[p^2\, \delta(r) + \delta(r) p^2], \qquad (10\text{--}27b)$$

where t_1, t_2 are constants and $\mathbf{p}$ is the momentum operator. Following this idea, Brink and Vautherin take the effective interaction to be

$$V = \sum_{i<j} V_{ij}^{(2)} + \sum_{i<j<k} V_{ijk}^{(3)}, \qquad (10\text{--}28)$$

where the last term in this equation is a phenomenological three-body interaction, whose role we shall presently discuss. The matrix elements of the two-body interaction between plane-wave states are of the most general form, bilinear in relative momenta:

$$\langle \mathbf{k}|V^{(2)}|\mathbf{k}'\rangle = t_0(1 + x_0 P_\sigma) + \tfrac{1}{2}t_1(k^2 + k'^2) + t_2\mathbf{k}\cdot\mathbf{k}' + iW_0(\boldsymbol{\sigma}_1 + \boldsymbol{\sigma}_2)\cdot\mathbf{k}\times\mathbf{k}'.$$

$$(10\text{--}29)$$

In Eq. 10–29, x_0 is the coefficient of the spin-exchange operator P_σ; t_0, t_1, t_2, and W_0 are constants; and the last term is a zero-range spin-orbit force.

A phenomenological effective interaction of this form was first proposed by Skyrme [21], and a very similar form has been used by Moszkowski [22] for the gross properties of nuclei. Note that the quadratic terms in k involving t_1 and t_2 not only make the interaction in coordinate space of finite range, but also give extra density dependence. This is so because the contribution to potential energy of a central, finite-range, static interaction goes dominantly like k_F^3, whereas for a quadratically momentum-dependent interaction the contribution goes as k_F^5. The three-body effective interaction is also taken to be of zero range:

$$V^{(3)}_{123} = t_3 \, \delta(\mathbf{r}_1 - \mathbf{r}_2) \, \delta(\mathbf{r}_1 - \mathbf{r}_3) \qquad (10\text{--}30)$$

and represents empirically the effect of higher-order clusters. It may be regarded as a density-dependent, two-body interaction, since

$$\int V^{(3)}_{123} \rho(\mathbf{r}_3) \, d^3r_3 = \rho(\mathbf{r}_1) t_3 \, \delta(\mathbf{r}_1 - \mathbf{r}_2). \qquad (10\text{--}31)$$

Delta-function interactions in HF calculations give rise to great simplicity because the direct and exchange terms have similar structures, and the self-consistent HF potential U turns out to be local.

The expectation value of the Hamiltonian with respect to a Slater determinant Φ can be expressed as

$$E = \langle \Phi | T + V | \Phi \rangle = \int \varepsilon_\rho(\mathbf{r}) \, d^3r, \qquad (10\text{--}32)$$

where $\varepsilon_\rho(\mathbf{r})$, the spatial-energy density, is a function of the single-particle density ρ. The simplicity of interaction 10–29 makes it possible to express $\varepsilon_\rho(\mathbf{r})$ in terms of $\rho(\mathbf{r})$, the kinetic-energy density $\tau(\mathbf{r})$, and the spin density $J(\mathbf{r})$:

$$\rho(\mathbf{r}) = \sum_i |\phi_i(\mathbf{r})|^2, \qquad \tau(\mathbf{r}) = \sum_i |\nabla \phi_i(\mathbf{r})|^2,$$

$$J(\mathbf{r}) = \frac{1}{r} \sum_i [\phi_i^*(\mathbf{l} \cdot \boldsymbol{\sigma}) \phi_i], \qquad (10\text{--}33)$$

where the sum i is over occupied orbitals only. For simplicity in algebra, we are assuming equal numbers of neutrons and protons and ignoring the Coulomb force. The results can be generalized by defining neutron and proton densities separately, as has been done by Brink and Vautherin [20]. Evaluating Eq. 10–32, we find

$$\varepsilon_\rho(\mathbf{r}) = \frac{\hbar^2}{2M} \tau + \tfrac{3}{8} t_0 \rho^2 + \tfrac{1}{16} t_3 \rho^3 + \tfrac{1}{16}(3t_1 + 5t_2)\rho\tau \; +$$

$$+ \tfrac{1}{16}(9t_1 - 5t_2)(\nabla\rho)^2 + \tfrac{3}{4}W_0\left(J\frac{d\rho}{dr}\right). \tag{10-34}$$

We should mention at this point that, although Eq. 10–34 has been obtained from the simplified interaction 10–29, Negele and Vautherin [23] have shown that form 10–34 for ε_ρ may also be obtained by starting from the nuclear-matter G-matrix derived from a realistic two-nucleon force, although in that case the coefficients t_i in Eq. 10–34 are somewhat density dependent.

The HF equation is obtained by varying Eq. 10–32 with respect to the single-particle states ϕ_i^* and has the form

$$\left\{-\nabla\left[\frac{\hbar^2}{2M^*(r)}\right]\cdot\nabla + U(\mathbf{r})\right\}\phi_i(\mathbf{r}) = e\phi_i(\mathbf{r}), \tag{10-35}$$

where, using Eq. 10–34, we find that

$$\frac{\hbar^2}{2M^*(r)} = \frac{\hbar^2}{2M} + \tfrac{1}{16}(3t_1 + 5t_2)\rho(r) \tag{10-36}$$

and

$$U(\mathbf{r}) = \tfrac{3}{4}t_0\rho + \tfrac{3}{16}t_3\rho^2 + \tfrac{1}{16}(3t_1 + 5t_2)\tau$$

$$+ \tfrac{1}{32}(5t_2 - 9t_1)\nabla^2\rho + \frac{3}{2r}W_0\frac{d\rho}{dr}\mathbf{l}\cdot\mathbf{s}. \tag{10-37}$$

Note that now the HF equation is for a local $U(\mathbf{r})$, which itself depends on the orbitals $\phi_i(\mathbf{r})$ through the density. Assuming a form for $\rho(\mathbf{r})$, one can obtain U from Eq. 10–37 and then solve Eq. 10–35 to find the single-particle states ϕ_i. This procedure can be iterated to obtain self-consistency.

It will be seen from Eq. 10–36 that the parameter $3t_1 + 5t_2$ determines the density dependence of the effective mass M^*, which is important for the single-particle level spacings. On the other hand, the coefficient $5t_2 - 9t_1$ of $\nabla^2\rho$ is important in determining the surface thickness. The parameter t_3, which is the coefficient of the three-body term in Eq. 10–30, is effective in giving saturation. The free parameters of the effective interaction V are fixed by fitting the binding energy and the equilibrium density of nuclear matter, and also the binding energies, radii, and separation energies of spherical nuclei. The parameters of one such set, S3, developed at Orsay [6] are as follows:

$$t_0 = -1128.75 \text{ MeV fm}^3, \qquad t_1 = 395.0 \text{ MeV fm}^5,$$

$$t_2 = -95.0 \text{ MeV fm}^5, \qquad t_3 = 14{,}000.0 \text{ MeV fm}^6,$$

$$x_0 = 0.45, \quad \text{and} \quad W_0 = 120 \text{ MeV fm}^5. \tag{10-38}$$

The numerical values of these parameters are by no means unique, and there are other sets of values that yield equally good results. The search for better parameters is a continuing process.

These parameters have also been evaluated by Negele and Vautherin [23], starting from a realistic nuclear-matter G-matrix. All the coefficients except t_2 are in qualitative agreement with the empirical set 10–38.

Vautherin and Brink [20] are able to reproduce, with good accuracy, the experimental binding energies and rms radii of ^{16}O, ^{40}Ca, ^{48}Ca, ^{90}Zr, and ^{208}Pb. We noted in Chapter 4 that the rms charge radius of ^{48}Ca, according to electron scattering and muonic X-ray data, is almost equal to (or is slightly smaller than) that of ^{40}Ca. This particular feature is not reproduced in HF calculations in general, and even with a density-dependent interaction some discrepancy remains. The single-particle levels calculated from the density-dependent interactions in the HF framework are much less bound and in much better agreement with experimental results than the corresponding density-independent calculations, because much of the binding comes from the rearrangement potential.

A detailed comparison of the calculated and experimental single-particle energies is given by Campi and Sprung [19]. In Fig. 10–7a we reproduce, from ref. 19, the perspective plots of the calculated density distributions of some spherical nuclei; the proton density is plotted on the left, and the neutron density on the right. The small oscillations in the density distribution, with a wavelength of the order of π/k_F, which arise because of shell effects,

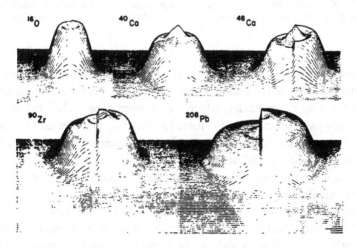

Fig. 10-7a Perspective plots of the calculated density distributions of some spherical nuclei. (After Campi and Sprung [19].)

are characteristic of HF calculations. In Fig. 10–7b we show, from ref. 15, the one-body self-consistent potential for ^{40}Ca obtained with the density-dependent effective interaction. This will be seen to be very state dependent, giving rise to charge form factors in reasonable agreement with electron-scattering data.

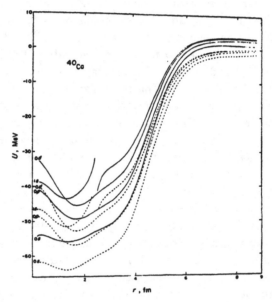

Fig. 10-7b Self-consistent single-particle proton (solid line) and neutron (dashed line) potentials obtained from the density-dependent effective interaction of Negele [15].

Summarizing these results, we can say that the local density approximation of the nuclear-matter G-matrix gives rise to a smooth density-dependent effective force in finite nuclei, which, for convenience in HF calculations, can be simply parameterized by the form 10–29 in k-space. Actual calculations show that the density dependence is of crucial importance in yielding the correct binding energy and saturation density in nuclei, and at the same time reproducing the experimental single-particle energies. Other phenomenological calculations [24] also bear out the last point.

Finally, it should be noted that the effective interaction (Eq. 10–29) has been used by Vautherin [25] to perform HF calculations in deformed nuclei, both in the s-d shell and in the rare-earth region, with satisfactory results.

10-5 DEFORMED HARTREE-FOCK ORBITALS

In Section 10–4 we saw that a complete HF calculation can be done, starting with a given effective interaction V between all the particles in a nucleus, to obtain a self-consistent potential U. In this full HF approach, the single-particle states are obtained in minimizing the energy by adjusting the single-particle wave functions ϕ_i with respect to their dependence on all possible radial and angular variables. For spectroscopic calculations, however, it is often sufficient to perform a more restricted HF calculation in which the ϕ_i are taken from one major shell; their radial dependence is that of the states of a harmonic-oscillator potential which fits the size of the nucleus, and only the angular-momentum part of the orbitals is varied in order to minimize the energy. In such calculations, one is not attempting to study the saturation property of the nucleus, which is taken for granted, but is only treating the configuration interactions in a major shell, just as in a shell-model calculation.

We know that the assumption of a spherical nucleus is a reasonable approximation for the ground states of doubly closed-shell nuclei. But we have seen that, when a few particles are added to a closed core, deformation sets in and the average potential is no longer spherically symmetric. Even for the doubly closed-shell case, we cannot discount the possibility that the nuclei are deformed in some excited states. We have seen in the Nilsson model that the single-particle states can be generated in a deformed one-body potential which is chosen empirically. In the HF calculations that we will describe now, the deformed orbitals in a major shell are generated in a self-consistent manner, and an intrinsic determinantal state can then be constructed with these orbitals. After determining which shape of the nucleus is energetically most favorable, the results are compared with those of a Nilsson-type calculation. To find the nuclear spectra, the various states of total angular momentum J are then projected from such an intrinsic state.

As an example of the method, we shall describe a simple calculation by Kelson [26] for some nuclei in the s-d shell which are known to have deformed shapes. At the end of this section, we shall refer to more involved and realistic calculations that have been done extensively by other groups.

In doing these calculations, it is convenient to take as a reference a closed-shell nucleus—in this particular case, ^{16}O. In the ground state of ^{16}O, it is assumed that the four 0s-shell orbits and twelve 0p-shell orbits are filled. Then ^{20}Ne, for example, can be considered to be a nucleus where four particles (two neutrons and two protons) are added in the 1s-0d shell orbits. In the hypothetical situation of four noninteracting particles, these orbitals would be generated by the spherical field of the ^{16}O core; in the actual case there is some effective interaction between the four nucleons, which will modify the average field, and the orbitals should be determined self-consistently. This method is quite versatile; for example, the negative-parity states of ^{19}F can be formed by putting four nucleons in the $1s$-$0d$ shell orbitals, with a hole in the $0p$-shell. To start with, we consider the simple case of ^{20}Ne.

In this model, there are only the four nucleons outside the closed core interacting with the Hamiltonian:

$$H = \sum_{\alpha,\beta} \langle \alpha|H_0|\beta \rangle a_\alpha{}^+ a_\beta{}^+ + \tfrac{1}{2} \sum_{\alpha,\beta,\gamma,\delta} \langle \alpha\beta|V|\gamma\delta \rangle a_\alpha{}^+ a_\beta{}^+ a_\delta a_\gamma. \tag{10-39}$$

Here H_0 is a one-body operator that represents the kinetic plus the potential energy of a particle moving in a field generated by the ^{16}O core. The single-particle j being a good quantum number in such a field, we have

$$\langle j'm'|H_0|jm \rangle = e_j \delta_{jj'} \delta_{mm'}, \tag{10-40a}$$

where by e_j we denote the single-particle energy generated by H_0. Since we are taking ^{16}O as the reference nucleus, the e_j's may be considered to be the separation energies of the last neutron in ^{17}O from the various orbits, which are known to be

$$e_{d_{5/2}} = -\,4.14 \text{ MeV}, \qquad e_{s_{1/2}} = -\,3.27 \text{ MeV}, \qquad \text{and} \qquad e_{d_{3/2}} = 0.94 \text{ MeV}. \tag{10-40b}$$

In this calculation, only the e_j's appear and the form of H_0 need not be specified. In our example, the effective interaction V is assumed to be simply a central force with Rosenfeld exchange mixture:

$$V = V_0 \tfrac{1}{3}(\boldsymbol{\tau}_1 \cdot \boldsymbol{\tau}_2)(0.3 + 0.7\,\boldsymbol{\sigma}_1 \cdot \boldsymbol{\sigma}_2) f(r), \tag{10-41}$$

with $f(r) = e^{-r/a}/(r/a)$. The parameter a is taken as 1.37 fm, and the quantity V_0 remains adjustable. Of course, V should, in principle, be a given interaction, but in view of the numerous approximations, particularly the treatment of the ^{16}O core as inert, some flexibility in V is necessary in order to fit the experimental energy spacings. To obtain the HF orbitals and the single-particle energies, we have to apply the condition $\delta\langle \Phi_0|H|\Phi_0 \rangle = 0$, where Φ_0 is a 4×4 determinant. As we have seen, this is equivalent to diagonalizing the single-particle operator H_{SP} in the occupied orbitals ϕ_m, ϕ_n:

$$\langle \phi_m|H_{SP}|\phi_n \rangle = \left\{ \langle \phi_m|H_0|\phi_n \rangle + \sum_{p=1}^{A} \langle \phi_m\phi_p|V|\phi_n\phi_p \rangle_A \right\} \delta_{mn}. \tag{10-42}$$

The HF orbitals ϕ_i are, in general, deformed, and consequently the single-particle j is no longer a good quantum number (see Sections 9–7 and 9–8). If axial symmetry is assumed, the projection Ω of j along the body-fixed symmetry axis is still a conserved quantity, and we may denote each HF orbital by $\phi_{\lambda,\Omega}$, where λ distinguishes various orbitals with the same Ω. We can then write

$$|\phi_{\lambda,\Omega}\rangle = \sum_j C^j_{\lambda\Omega}|j\Omega\rangle, \qquad (10\text{–}43)$$

where the coefficients $C^j_{\lambda\Omega}$ are determined self-consistently.

In Eq. 10–43, the sum over j is restricted to the 1s-0d shell, and the radial dependence of the states $|j\Omega\rangle$ is taken to be that of harmonic-oscillator states, with the oscillator constant chosen to reproduce the rms radius of the nucleus. The states $|\phi_{\lambda,\Omega}\rangle$ and $|\phi_{\lambda,-\Omega}\rangle$ are degenerate in energy. Thus, in each self-consistent single-particle state with energy e_Ω, we can place two neutrons and two protons. Kelson [26], on doing the calculations for ^{20}Ne, found that the four extracore particles occupied the lowest deformed orbital ($\lambda = 1$), with $\Omega = \pm \frac{1}{2}$. For $V_0 = 50$ MeV, the structure of the HF orbital is

$$|\phi_{1,1/2}\rangle = 0.71|0d_{5/2}; \tfrac{1}{2}\rangle - 0.39|0d_{3/2}; \tfrac{1}{2}\rangle + 0.58|1s_{1/2}; \tfrac{1}{2}\rangle, \qquad (10\text{–}44)$$

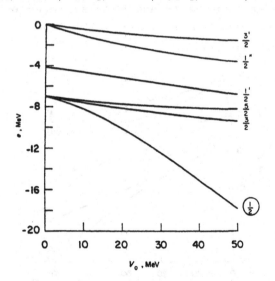

Fig. 10-8 The calculated single-particle self-consistent energies for ^{20}Ne, as a function of the strength V_0 of the attractive effective interaction. The single-particle energies e_j obtained from H_0 (Eq. 10–40) are taken to be somewhat different from those given in the text. Only the lowest $\Omega = \frac{1}{2}$ level is occupied by four nucleons. (After Kelson [26].)

which is quite similar to the Nilsson wave function. Figure 10–8 shows a plot of the self-consistent energies e_Ω for the HF orbitals as functions of the strength V_0 of the effective interaction. These energies are eigenvalues of a single-particle operator $H_{SP} = t + U$, where U is the self-consistent potential with the lowest $\Omega = \pm \frac{1}{2}$ levels occupied by four particles. It will be seen that, as V_0 increases, a marked gap appears between the occupied and the unoccupied levels. For a reasonable strength $V_0 = 50$ MeV, the gap is nearly 10 MeV. Qualitatively, this large gap indicates that it becomes increasingly difficult to promote particles from the occupied to the empty orbitals, making the deformed intrinsic state $|\Phi_{1,1/2}\rangle$ stable against vibrations as V_0 increases. We should remember, however, that if two particles are promoted to the $\Omega = \frac{3}{2}$ orbit the self-consistent potential U is no longer the same, with the result that the spectrum displayed in Fig. 10–8 will be modified. In general, the gap in this configuration between the $\Omega = \frac{1}{2}$ and the $\Omega = \frac{3}{2}$ levels will be less than that indicated by Fig. 10–8.

The energy levels shown in Fig. 10–8 have some similarity with the Nilsson levels of Fig. 9–28, which are plotted against the deformation parameter η. For $\eta = 2$, we get the same level ordering in the Nilsson levels as in Fig. 10–8, but the gap between the occupied and unoccupied levels in this case is less than 2 MeV. This has important effects in the calculation of the moment-of-inertia parameter, and improves the results of the cranking model if self-consistent energies are used instead of the energies of the Nilsson orbitals [27].

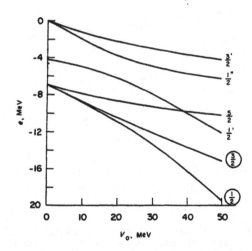

Fig. 10-9 The calculated self-consistent single-particle energies for ^{24}Mg as a function of the strength V_0 of the effective interaction, for the axially symmetric solution. Now the lowest $\Omega = \frac{1}{2}$ and $\frac{3}{2}$ levels are occupied. (After Kelson [26].)

Another important difference from the Nilsson model becomes apparent on examining the axially symmetric solution of ^{24}Mg. In Fig. 10–9 we have plotted the self-consistent single-particle energies of ^{24}Mg as a function of the strength V_0 of the effective interaction. In this case, the lowest two levels, with $\Omega = \pm \frac{1}{2}$ and $\Omega = \pm \frac{3}{2}$, are occupied by the eight extracore particles. For $V_0 = 50$ MeV, the $\Omega = \frac{1}{2}$ level is depressed only by about 2 MeV from its ^{20}Ne value, but the now-occupied $\Omega = \frac{3}{2}$ level has changed considerably, from about -8 MeV in ^{20}Ne to nearly -15 MeV in ^{24}Mg. This shows a very important characteristic of the self-consistent calculations, namely, the energy of a single-particle level is depressed markedly as it is occupied by more and more particles. This is due to the strong mutual attraction between particles in the same orbit. In the Nilsson model, this effect is nearly completely neglected, and the levels shift only slightly because of small changes in the equilibrium deformation of the nucleus.

Another important feature of the axially symmetric ^{24}Mg case is that the gap between the last occupied $\Omega = \frac{3}{2}$ level and the next unoccupied $\Omega = \frac{1}{2}$

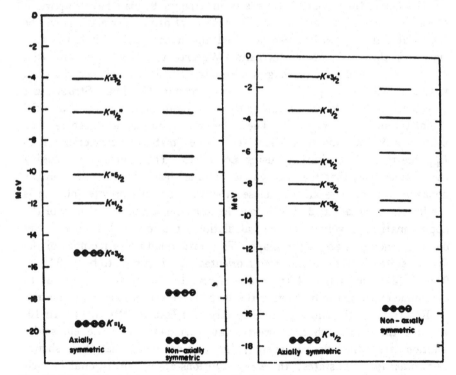

Fig. 10-10 A comparison of the calculated self-consistent single-particle energies for the axially symmetric and the nonaxial solutions: (a) ^{20}Ne, and (b) ^{24}Mg. (After Bar-Touv and Kelson [28].)

level is only about 3 MeV, in contrast to the gap of 10 MeV in ^{20}Ne. This suggests that the axially symmetric HF solution for ^{24}Mg is less stable than the one obtained for ^{20}Ne, since in mixing states of different Ω we create a state that does not have axial symmetry. Bar-Touv and Kelson [28], on doing a HF calculation without imposing the restriction of axial symmetry, found that a solution slightly lower in energy was obtained for an ellipsoidal shape with unequal axes.

In Fig. 10–10 we show the single-particle self-consistent spectra of ^{20}Ne and ^{24}Mg for both the axially symmetric and ellipsoidal shapes. Whereas for ^{20}Ne the axially symmetric shape is energetically favored, it is reversed for ^{24}Mg. There is also a larger gap between the occupied and empty single-particle levels in the triaxial solution than in the symmetric one. From this type of calculation, it would be concluded that the ground state of ^{24}Mg is asymmetric.

A note of caution must be added at this point about taking these results literally. We shall see shortly that HF calculations leave out the pairing type of correlations which, as we saw in Chapter 9, may have important effects in equilibrium deformation. The effect of pairing correlations will be appreciable if the gap between occupied and unoccupied orbitals is small, since it is then easier to scatter pairs. We expect, by this argument, that the symmetric solution of ^{24}Mg will be lowered more in energy than the triaxial solution when pairing is taken into account. Goodman, Struble, and Goswami [29] have found, on taking this effect into consideration, that axial symmetry in ^{24}Mg is restored. The same conclusion has also been reached by Gunye, Warke, and Khadkikar [30], who estimated the correction ΔE to E_{HF} due to $2p$-$2h$ excitations, using Eq. 10–25. They found, on including these corrections, that the two symmetric (one oblate, one prolate) and the nonaxial solutions are all very close in energy, but the prolate solution is the lowest, lying about 2 MeV below the nonaxial solution. In view of the approximations involved in the calculation, it is difficult to reach any definite conclusion about the shape. The static quadrupole moment of the first 2$^+$ state of ^{24}Mg is found from reorientation experiment to be $-$ 0.24 $\pm$ 0.06 eb [31], showing that the shape is prolate, but it is not possible to determine experimentally whether the shape is triaxial or axially symmetric.

Returning to the single-particle spectrum of ^{20}Ne or ^{24}Mg in Fig. 10–10, we note that, although the lowest axially symmetric and the ellipsoidal solutions are rather close in energy, this does not mean that there should be two *low-lying* 0$^+$ states; these two solutions are not orthogonal to each other, having been generated by two different single-particle Hamiltonians.

In fact, the close proximity of two such solutions may indicate that there is considerable overlap between these states. It is possible, however, to do "shape-mixing" calculations with nonorthogonal states, a point that we shall discuss in the next section.

Bar-Touv and Kelson [28] have done the HF calculations for all even-even nuclei in the s-d shell by removing the restriction of axial symmetry. Usually, more than one stationary intrinsic state exists, and there is competition between the lowest axially symmetric and nonaxially symmetric

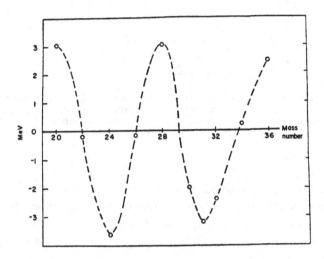

Fig. 10-11 The difference between the lowest axially symmetric and nonaxial intrinsic state energies, ΔE, as a function of mass number. The calculated points, shown by open circles, are joined by a dotted line to display the regularity of the curve. (After Bar-Touv and Kelson [28].)

intrinsic states. In Fig. 10–11 we plot the difference in the expectation values of H:

$$\Delta E = \langle \Phi_{nonax} | H | \Phi_{nonax} \rangle - \langle \Phi_{ax} | H | \Phi_{ax} \rangle$$

against the mass number A. Negative values of ΔE on the curve indicate that an axially asymmetric state is energetically favored. Two regions of axial asymmetry, around ^{24}Mg and ^{32}S, are clearly seen. An interesting result is shown in Fig. 10–12, where the expectation value of the mass quadrupole moment operator Q, with respect to the energetically lowest intrinsic state, is plotted against the mass number. This, of course, does not correspond to

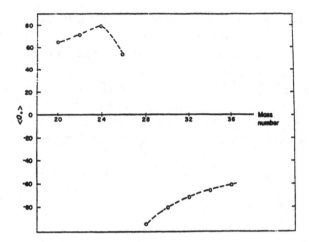

Fig. 10-12 The dependence of the quadrupole moment $\langle Q_0 \rangle$ of the lowest intrinsic state on the mass number A. The units of $\langle Q_0 \rangle$ are arbitrary. Note the abrupt change of sign around mass number 28, showing prolate to oblate transition. (After Bar-Touv and Kelson [28].)

any experimentally measurable quantity, but a positive $\langle Q_0 \rangle$ indicates a prolate ground-state band, while a negative $\langle Q_0 \rangle$ represents an oblate band. There is an abrupt change in shape near $A = 28$, for which both the prolate and oblate solutions are very close in energy, with the oblate shape only slightly lower.

It should be realized that in HF calculations of this type, in which only the particles in the s-d shell are assumed to interact with an effective *central* interaction V (Eq. 10–41), it is not very reasonable to take the spin-orbit splitting between the $d_{5/2}$- and $d_{3/2}$-states from ^{17}O, as given in Eq. 10–40b. This spin-orbit splitting is likely to be modified by the noncentral interactions between the pairs of particles in the s-d shell; therefore, if one takes a central interaction V as in Eq. 10–41, the $d_{5/2}$-$d_{3/2}$ input energies can be varied within reasonable limits to obtain the best result. In the case of ^{28}Si, the prolate-oblate energy difference is found to be very sensitive to the strength of the one-body spin-orbit force, the oblate solution being slightly lower in energy with reasonable values of the spin-orbit parameter. With more realistic effective interactions V containing noncentral components of force (like the Kuo-Brown G-matrix elements), it is reasonable to take the values of e_i given in Eq. 10–40b for the field of the core nucleons, and the oblate solution is then nearly 5 MeV lower in energy than the prolate one [30a]. Das Gupta and Harvey [30b] have pointed out that the lowest oblate and prolate solutions in ^{28}Si are orthogonal to each other, and that the ground state and the lowest few states of ^{28}Si are generated from rotations or particle-hole excitations of the oblate intrinsic state, while the 0^+ state at 6.69 MeV is the band head of the prolate intrinsic state. The static quadrupole moment Q_{2+} of the first excited 2^+ state of ^{28}Si has been measured to be $0.17 \pm 0.05 \, eb$ [31, 32], corresponding to an oblate shape,[†] in agreement with this calculation.

The position of the Fermi level in a nucleus can be obtained from experimental data by measuring the separation energy of a neutron from the top level. This is found to be substantially constant throughout the s-d shell, a result in agreement with the HF calculations of Bar-Touv and Kelson. For an even-even nucleus with mass number A, the gap between the last occupied orbital and the next empty level is just the difference between the Fermi levels of nuclei with mass numbers $A + 1$ and A. Experimentally, this gap is about 10 MeV for ^{20}Ne, and decreases steadily for even-even nuclei as more particles are added in the s-d shell [33]. This feature is also reproduced in these HF calculations.

It will now be interesting to go back and consider the low-lying even-parity states of ^{16}O. In Section 10-2, we pointed out that the first 0^+ excited state comes at 6.06-MeV excitation, which is very surprising from the spherical-shell-model viewpoint. In Fig. 10–13, we show the experimental evidence that certain of these low-lying excited states form rotational bands [34]. As will be observed from this figure, the 6.06-MeV 0^+ state is the band head of an even-parity rotational band, suggesting a deformed rotating nucleus in the excited states, although the ground state is spherical. If we turn back to the Nilsson diagram of Fig. 9–28, we find that, although there is a large gap of about 11.5 MeV between the occupied $0p_{1/2}$- and the empty $0d_{5/2}$-levels for the spherical shape, this gap decreases sharply as deformation sets in. To generate the even-parity excited states of ^{16}O, it is necessary to promote an even number of nucleons from the filled to the empty orbits. From the HF point of view, it is conceivable that, when this is done, the self-consistent field becomes highly deformed and much less energy is required to generate these excited states.

It is natural, therefore, to attempt a HF calculation in which the positive-parity intrinsic states are built from $2p$-$2h$ (two particle-two hole) and $4p$-$4h$ states. We have already noted a characteristic feature of HF calculations, in that a self-consistent orbital (in this case the lowest $\Omega = \frac{1}{2}^+$ orbit

---

[†] Note, from Eq. 9–110, that the intrinsic and static quadrupole moments have opposite signs for a given shape in the rotation-model approximation.

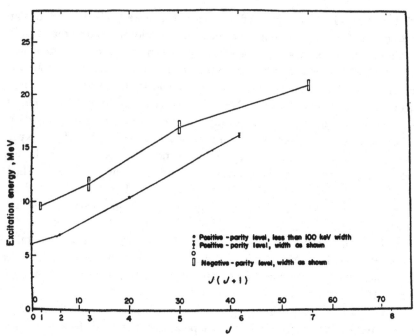

Fig. 10-13   States in ^{16}O excited strongly in the elastic scattering of α-particles by ^{12}C lie on a rotational band.   (After Carter et al. [34].)

in the s-d shell) is sharply depressed as it is occupied by more and more particles.   It may be possible, then, that the 4p-4h states are energetically more favored than the 2p-2h ones, since the gap between the occupied and empty orbitals is decreased drastically (see Fig. 9–28) for positive η.   This has been demonstrated in a HF calculation by Bassichis and Ripka [35], assuming axially symmetric shape.   In their calculation, they assumed an effective interaction V as given in Eq. 10–41, with a Gaussian form for $f(r)$. This calculation is very similar to the one on ^{20}Ne that we have described, except that instead of four particles in the s-d shell there are now two particles in the s-d shell and two holes in the p-shell with respect to the ground state ^{16}O configuration.

The lowest 4p-4h intrinsic state is the one in which the four particles are in the lowest $\Omega = \frac{1}{2}^+$ orbit and the four holes are in the highest $\Omega = \frac{1}{2}^-$ orbit, and this has isospin $T = 0$.   In Fig. 10–14, we plot the energies of this intrinsic 4p-4h state and one of the 2p-2h intrinsic states obtained from the HF calculation, relative to the ground state, as functions of the strength V_0

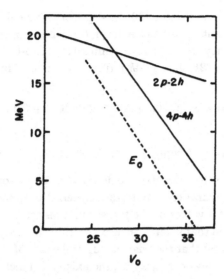

Fig. 10-14 The energies of the lowest $4p$-$4h$ and one of the $2p$-$2h$ intrinsic states of ^{16}O, relative to the ground state, plotted as a function of V_0, the effective interaction. The position of the projected band head of the even-parity rotational band is shown by the dotted line. (After Bassichis and Ripka [35].)

of the effective interaction. The dotted line shows the position of the head of the rotational band based on the $4p$-$4h$ intrinsic state. It has been obtained by projecting out the 0^+ state from the intrinsic state, and experimentally its position is at 6.06 MeV. We shall indicate presently how this projection can be done in a simple-minded manner. It will be seen that, for $V_0 = 32$ MeV, the experimental position of the band head is reproduced. It is interesting to note that with a Gaussian interaction this value of V_0 is also close to what would be required to reproduce the ^{20}Ne spectrum. Thus the $4p$-$4h$ intrinsic state, which is well isolated from the $2p$-$2h$ ones, gives rise to a rotational band.

In the simplified analysis just described, the impression was given that the 6.06-MeV 0^+ state is the band head of a pure $4p$-$4h$ intrinsic HF state. If one goes one step further than the HF approximation, the residual interaction $(V - U)$ will cause some mixing between the $0p$-$0h$, $2p$-$2h$, and $4p$-$4h$ configurations in all the low-lying even-parity states of ^{16}O. That there is some admixture of deformed orbitals even in the ground state is seen from the rather large $B(E2)$-transition value of 4.6 ± 1 e^2 fm^4 between the 6.92-MeV 2^+ state and the 0^+ ground state (see Fig. 10-5). To estimate the amount of this mixing, Brown and Green [36] have performed a calculation in which the single-particle orbitals are taken from the Nilsson model, rather than self-consistently. They diagonalize a semirealistic two-body interaction in the space of $0p\text{-}0h$, $2p\text{-}2h$, and $4p\text{-}4h$ intrinsic states and find, for example, that

$$|0^+, 0 \text{ MeV}\rangle = 0.87|(0p\text{-}0h)\rangle + 0.47|(2p\text{-}2h)\rangle + 0.13|(4p\text{-}4h)\rangle$$

and

$$|0^+, 6.06 \text{ MeV}\rangle = -0.26|(0p\text{-}0h)\rangle + 0.23|(2p\text{-}2h)\rangle + 0.94|(4p\text{-}4h)\rangle.$$

Similarly, states of other angular momenta also have admixtures of the various intrinsic configurations. With this model, Brown and Green are able to fit the experimental values of the transition rates reasonably well.

We should mention that in later calculations Hayward [37] and Stephenson and Banerjee [38] found that the triaxial elliptical shape of the $4p\text{-}4h$ intrinsic state has considerably lower energy than the axially symmetric case discussed above. The reason for the triaxial shape may be understood in a simple manner by the following argument. If only four particles in the $s\text{-}d$ shell (as in ^{20}Ne) are considered, an axially symmetric prolate shape is obtained. Likewise, by considering four-hole states (as in ^{12}C), an axially symmetric oblate shape is obtained. In $4p\text{-}4h$ states of ^{16}O, these two intrinsic states are locked together at right angles to each other by the attractive particle-hole interaction, giving rise to maximum overlap between the prolate and oblate shapes, and yielding a triaxial intrinsic state which generates the rotational band. These assertions are confirmed by the detailed calculations of Engeland and Ellis, as well as de Takacsy and Das Gupta [38].

10-6 PROJECTION OF ANGULAR-MOMENTUM STATES FROM THE INTRINSIC STATE

The HF wave function $|\Phi\rangle$, being a product of single-particle orbitals, is not in general an eigenstate of the operator J^2, where J is the total angular momentum of the state. For example, we have seen that, when there are some extra particles outside the closed core, the self-consistent potential U is deformed rather than spherically symmetric in space. This implies that the shape of the HF density distribution is also deformed, that is, it has a preferred direction in space with reference to a space-fixed system of axes. Thus rotational invariance has been lost in the HF approximation, and the deformed solution $|\Phi\rangle$ cannot be an eigenstate of J^2. This is not at all satisfactory, since the total Hamiltonian

$$H = \sum_i t_i + \sum_{i<j} V_{ij}$$

is certainly rotationally invariant, and a physical state should have good J. In the case of axial symmetry, although J is not conserved in the HF approximation, its projection along the symmetry axis is still conserved; we denote this quantum number by

$$K = \sum \Omega_i = \langle J_3 \rangle,$$

where Ω_i is the projection of the angular momentum of the ith particle along the symmetry axis. For example, in ^{20}Ne, four nucleons are placed in the $\Omega = \pm \frac{1}{2}$ orbits, and $K = 0$. The HF intrinsic state $|\Phi\rangle$ here represents a superposition of states with various values of J, but each of these has $K = 0$. Mathematically, it is tempting to project out of $|\Phi\rangle$ the states with good J and to identify these as the physical states of a given K-band. How can this procedure be justified on variational grounds (on which the HF approximation is based); moreover, under what circumstances can we expect to get a rotational band?

These questions have been examined by Hill and Wheeler [39] and by Peierls and Yoccoz [40]. The point to note is that for a deformed well U there is a directional degeneracy in the HF energy E_{HF}, since any orientation of the well in space generates a HF wave function with the same value of E_{HF}. In Chapter 9, we saw that the orientation of a body-fixed frame with respect to a space-fixed axis can be specified by three Euler angles θ_i, with $i = 1, 2$, and 3. Let $|\Phi_K\rangle$ be the HF solution for a given orientation of U; then, for a different orientation, $R(\theta_i)|\Phi_K\rangle$ will be the HF solution with the same energy E_{HF}, where $R(\theta_i)$ is a rotation operator defined by Eq. A–47. We can take advantage of this directional degeneracy by constructing a trial wave function that is a linear combination of these degenerate wave functions with a given K, the coefficients being determined by the variational procedure of minimizing the energy. More specifically, we define a function $|\Psi_K{}^\alpha\rangle$, with an as yet unspecified quantum number α:

$$|\Psi_K{}^\alpha\rangle = \int f_K{}^\alpha(\theta_i) R(\theta_i)|\Phi_K\rangle \sin \theta_2 \, d\theta_2 \, d\theta_1 \, d\theta_3,$$

where the limits of θ_2 are from 0 to π, and θ_1, θ_3 extend from 0 to 2π. The trial wave function $|\Psi_K{}^\alpha\rangle$ is a superposition of all the degenerate HF solutions over all possible orientations, with the coefficients $f_K{}^\alpha$ to be determined. Since the state $|\Psi_K{}^\alpha\rangle$ has no preferred direction in space, it is rotationally invariant. It is therefore possible for this to be an eigenstate of J^2 and J_z of the space-fixed axes, and we may write

$$|\Psi^J_{MK}\rangle = \int f^J_{MK}(\theta_i) R(\theta_i) |\Phi_K\rangle \, d\Omega, \tag{10-45}$$

where, in shorthand notation, we have put $d\Omega$ for $\sin \theta_2 \, d\theta_2 \, d\theta_1 \, d\theta_3$, with the limits of integration as specified above. The coefficients $f^J_{MK}(\theta_i)$ are to be determined by demanding that $\langle \Psi^J_{MK} | H | \Psi^J_{MK} \rangle / \langle \Psi^J_{MK} | \Psi^J_{MK} \rangle$ be stationary with respect to variations δf^J_{MK}, where H is the total Hamiltonian. It can be shown† that, to satisfy this requirement, f^J_{MK} is simply $\mathscr{D}^J_{MK}(\theta_i)$, as defined in Eq. A–53. We thus obtain

$$|\Psi^J_{MK}\rangle = \int \mathscr{D}^J_{MK}(\theta_i) R(\theta_i) |\Phi_K\rangle \, d\Omega. \tag{10-46}$$

To understand the physical significance of Eq. 10–46, consider the intrinsic wave function $|\Phi_K\rangle$ in the body-fixed axes, which may be expressed as a linear combination of good angular momentum states $|\Phi_K{}^J\rangle$:

$$|\Phi_K\rangle = \sum_{J'} C_{J'} |\Phi_K{}^{J'}\rangle. \tag{10-47}$$

If we want to express the same wave function in a rotated frame, we apply an appropriate rotation operator $R(\theta_i)$, obtaining, from Eq. A–50,

$$R(\theta_i) |\Phi_K\rangle = \sum_{J'} C_{J'} \sum_{M'} \mathscr{D}^{J'*}_{M'K}(\theta_i) |\Phi^{J'}_{M'}\rangle.$$

Substituting this in Eq. 10–46 yields

$$|\Psi^J_{MK}\rangle = \sum_{J',M'} C_{J'} |\Phi^{J'}_{M'}\rangle \int \mathscr{D}^{J*}_{MK}(\theta_i) \mathscr{D}^{J'}_{M'K}(\theta_i) \, d\Omega$$

$$= \frac{8\pi^2}{2J+1} C_J |\Phi_M{}^J\rangle, \tag{10-48}$$

showing that the Peierls-Yoccoz wave function is essentially the projection of $|\Phi_K\rangle$ with angular momentum J. We have thus justified the projection procedure on a variational basis. Since the total Hamiltonian H commutes with J^2, we can write the energy of the state with (J, K) as

$$E_{JK} = \frac{\langle \Psi^J_{MK} | H | \Psi^J_{MK} \rangle}{\langle \Psi^J_{MK} | \Psi^J_{MK} \rangle} = \frac{\langle \Phi_K | H | \Psi^J_{MK} \rangle}{\langle \Phi_K | \Psi^J_{MK} \rangle}. \tag{10-49}$$

---

† For the gory details in a simple two-dimensional model, see G. E. Brown, *Unified Theory of Models and Forces*, North-Holland, Amsterdam, 1967, p. 64.

Noting that the energy cannot depend on the choice of the space-fixed z-axis (i.e., the quantum number M), we can write

$$E_{JK} = \frac{\langle \Phi_K | H | \Psi_{KK}^J \rangle}{\langle \Phi_K | \Psi_{KK}^J \rangle}$$

$$= \frac{\int_0^\pi d\theta_2 \sin \theta_2 \, d_{KK}^J(\theta_2) \langle \Phi_K | H \exp(-i\theta_2 J_2) | \Phi_K \rangle}{\int_0^\pi d\theta_2 \sin \theta_2 \, d_{KK}^J(\theta_2) \langle \Phi_K | \exp(-i\theta_2 J_2) | \Phi_K \rangle}, \qquad (10\text{-}50)$$

where we have used Eqs. 10–46 and A–53.

The next question to answer is whether the E_J's in a given K-band form a rotational spectrum. If the wave function $|\Phi_K\rangle$ belongs to a well of large anisotropy, its overlap with $\exp(-i\theta_2 J_2)|\Phi_K\rangle$, an intrinsic state of a different orientation, is small unless the directions are nearly the same, that is, $\theta_2 = 0$ or π. Thus the overlap function $\langle \Phi_K | \exp(-i\theta_2 J_2) | \Phi_K \rangle$ as a function of θ_2 is expected to peak sharply about $\theta_2 = 0$ and π for a well-deformed well. A similar property may be assumed to be valid for the energy-overlap integral $\langle \Phi_K | H \exp(-i\theta_2 J_2) | \Phi_K \rangle$, since H operates on at most two particles simultaneously, so that for the remaining particles we have just the overlap of the intrinsic parts at different orientations. Under these conditions, Yoccoz [41] and Verhaar [42] have shown[†] that Eq. 10–50 gives rise to a rotational band for a given K, including the special case for $K = \frac{1}{2}$.

Bassichis, Giraud, and Ripka [43] applied Eq. 10–50 for the calculation of E_J for the light s-d shell nuclei, using the lowest $K = 0$ intrinsic state and evaluating the overlap integrals numerically. In Fig. 10–15 we show their calculated overlap function $\langle \Phi_{K=0} | \exp(-i\theta_2 J_2) | \Phi_{K=0} \rangle$ as a function of θ_2 for ^{20}Ne and note that it does peak at $\theta_2 = 0$ and $\theta_2 = \pi$. Figure 10–16 reproduces the spectrum of ^{20}Ne as obtained by Macfarlane and Shukla [44]; column (a) is the result of projection from the lowest $K = 0$ intrinsic state. In this calculation, the interaction V given by Eq. 10–41 was used with a Yukawa form for $f(r)$, with range parameter $a = 1.37$ fm, $V_0 = -50$ MeV, and the oscillator constant $b = \hbar/M\omega = 1.65$ fm for the basis states. The input single-particle energies were taken to be $e(d_{5/2}) = -7.0$ MeV, $e(s_{1/2}) = -4.2$ MeV, and $e(d_{3/2}) = 0$. In column (d) of Fig. 10–16, the experimental positive-parity states of predominant configuration $(0d, 1s)^4$ in this energy range are shown for comparison. It will be seen that the projection from

---

[†] Note that for θ_2 close to zero one can write

$$d_{KK}^J(\theta_2) = 1 - \frac{1}{4}[J(J+1) - K^2]\theta_2^2 + \cdots,$$

which, when inserted into Eq. 10–50, gives rise to a rotational spectrum to leading order.

the lowest $K = 0$ intrinsic state yields only five of the eight experimental levels, whereas a complete shell-model calculation with the same interaction, whose spectrum is shown in Fig. 10–16c, does not suffer from this deficiency. This indicates that other low-lying intrinsic states play a part in the description of these levels. We then inquire as to how to include the effects of the other intrinsic states in a more complete calculation.

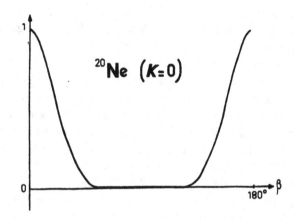

Fig. 10-15 The overlap function $\langle \Phi_K | \exp(- i\theta_2 J_2) | \Phi_K \rangle$ for ²⁰Ne, using the HF intrinsic state for $K = 0$, as a function of the Euler angle θ_2. (After Ripka [33].)

We have already seen that for a given nucleus many solutions (i.e., intrinsic states) of the HF equation can be obtained by filling up different sets of orbitals, or by imposing different symmetries (i.e., axial or nonaxial shapes). For ²⁰Ne, five solutions of the HF equation are known [45]. Of these, four have axial symmetry—two are prolate and two oblate—and one is axially asymmetric. The lowest intrinsic state has a prolate shape and gives rise to the spectrum shown in Fig. 10–16a. In ²⁸Si the situation is even more complicated. Ripka [33] lists eight different solutions lying between $E_{\mathrm{HF}} = -123$ and -124.44 MeV. In ²⁴Mg, we have seen that there are solutions corresponding to axially symmetric and triaxial ellipsoidal shapes, which are close in energy. In addition to these, other intrinsic states can be generated by filling up different orbitals, e.g. taking $1p$-$1h$ configurations based on the lowest intrinsic state. The point to note is that all these different intrinsic states of a nucleus need not be orthogonal to each other, because they are not eigenstates of the same Hamiltonian.† This is so since the HF Hamiltonian

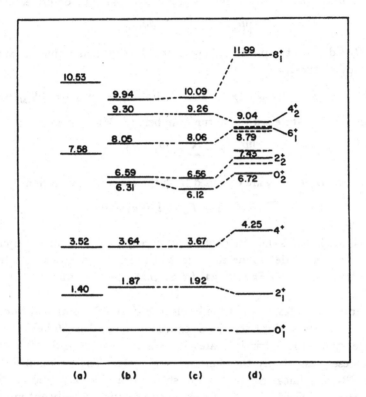

Fig. 10-16 Energy levels of ^{20}Ne: (a) the spectrum obtained from the lowest single intrinsic HF state; (b) the result of a shape-mixing calculation; (c) the spectrum of an exact shell-model calculation in the $(s, d)^4$ configuration; (d) the experimental spectrum. (After Macfarlane and Shukla [44].)

changes, through the one-body potential U, as the occupied orbitals are changed. Hence different intrinsic states, when they correspond to various local minima in a narrow energy region, cannot be interpreted as the different

† Under special conditions, of course, it is possible for two different intrinsic states to be strictly orthogonal. For example, in ^{19}F, the ground state is the band head of a positive-parity $K = \frac{1}{2}$ band, while the $\frac{1}{2}^-$ state at 0.110 MeV originates from the intrinsic state of negative parity with $K = \frac{1}{2}$.

excited states of the nucleus. In this case, there will be appreciable mixing of the intrinsic states via the effective interaction. We now illustrate this phenomenon by considering ^{20}Ne [44].

Let us denote the ith intrinsic state by $|\Phi(i)\rangle$. We can generate the states of good angular momentum J by the projection technique described earlier:

$$|\Psi_M{}^J(i)\rangle = P_M{}^J|\Phi(i)\rangle,$$

where $P_M{}^J$ denotes the projection operator. We can then define the overlap and energy matrices,

$$N_J(i, j) = \langle\Psi_M{}^J(i)|\Psi_M{}^J(j)\rangle \quad \text{and} \quad H_J(i, j) = \langle\Psi_M{}^J(i)|H|\Psi_M{}^J(j)\rangle.$$

Then the nuclear state with quantum numbers (J, M) is given by

$$|\Psi_M{}^J\rangle = \sum_i a_i|\Psi_M{}^J(i)\rangle,$$

where the coefficients a_i are obtained from the eigenvalue equation

$$\sum_j [H_J(i, j) - E_J N_J(i, j)]a_j = 0.$$

In this equation, E_J is the energy of the state $|\Psi_M{}^J\rangle$. Thus a matrix equation has to be solved to determine the extent of mixing of the various intrinsic states. Macfarlane and Shukla [44] found that when the above calculation was done for ^{20}Ne, taking only the four symmetric (two prolate and two oblate) intrinsic states, none of the levels in Fig. 10–16a (obtained from just the lowest prolate intrinsic state) was shifted by more than 20 keV. In fact, all these symmetric intrinsic states have high overlap with each other; they represent essentially the same physical state. The nonaxial intrinsic state in ^{20}Ne is a linear combination of states with $K = 0$, 2, and 4. When these states are mixed with the lowest prolate solution, significant improvement to the energy spectrum is obtained. Figure 10–16b shows the spectrum obtained by doing a mixing calculation with the lowest prolate, the nonaxial state ($K = 0$ and $K = 4$ components only), and five $1p$-$1h$ intrinsic states with $K = 0$ and 2, based on the lowest prolate solution. Now the agreement with the shell-model calculation is excellent, and all the experimental levels believed to be coming from the $(0d, 1s)^4$ configuration have been reproduced.

We can now see that, in addition to the lowest intrinsic HF state, some others may also play an important part in the description of the nuclear states. This type of shape-mixing calculation may be of particular importance in ^{28}Si, where the projected spectrum from the lowest oblate intrinsic state is far too compressed in comparison with the experimental values.

We shall now make some comments about the even-parity states of ^{18}O, which we discussed in the simple shell-model picture in Section 10–2. It was noted that the experimental data indicated the ^{16}O core to be playing an active role in the description of even the low-lying states. We now realize that in the HF picture there should be two distinct intrinsic states describing these low-lying levels: one having two particles in the s-d shell, and the other having four particles in the s-d shell, with two holes in the p-shell. The latter intrinsic state, because of the increased deformation of the orbitals with higher occupancy, may not cost too much in energy. There is, of course, some mixing between the 2p and 4p-2h intrinsic states. On doing the calculation, it is found that the ground state 0$^+$, the first excited 2$^+$ (1.98 MeV), and the second 0$^+$ (3.63 MeV) are mainly of 2p-nature, with some admixture of 4p-2h states. This admixture is sufficient to yield $B(E2)$-transition rates close to the experimental values if an effective charge of 0.5 e is taken for the extracore neutrons [46].

The HF scheme, which gives rise to a self-consistent one-body potential, cannot take account of pairing correlations adequately. We may appreciate this deficiency by considering the schematic pairing interaction, with constant matrix element $G = \langle ii|V|jj \rangle$. Since G can contribute to the potential U only when $i = j$, the HF method cannot adequately describe pairing. This is not the case for other components of the two-nucleon interaction; for example, the quadrupole-quadrupole part contributes appreciably to U, giving rise to a deformation. A formalism for generating a self-consistent potential that treats the pairing part of the two-nucleon potential on an equal footing with the other field-producing parts of the interaction has been developed; this is called the Hartree-Fock-Bogolubov (HFB) formalism.†

Without going into any details of this method, we would like to draw attention to two effects of pairing correlations that are particularly relevant in affecting the HF results. First, depending on the gap between the occupied and empty single-particle states, some solutions of the HF equation may gain more energy from the pairing correlations than others, altering the symmetry of the ground-state band. This is possible because the various HF intrinsic states may differ very little in energy. As an example, we have noted that pairing correlations seem to restore axial symmetry in the ground-state band of ^{24}Mg. Second, when a HF calculation is done without taking pairing effects into account, it is often found that the projected spectra, although

---

† See, for example, the article "Theory of Finite Nuclei," by M. Baranger, in *1962 Cargése Lectures in Theoretical Physics*, Benjamin, New York, 1963.

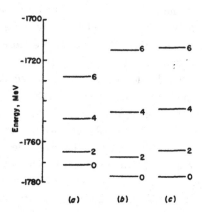

Fig. 10-17 The experimental energy spectrum of ^{22}Ne is shown in column (c). The energy spectra shown in columns (a) and (b) are computed in a space of the first four major oscillator shells by projecting good angular momentum states from the intrinsic HF and HFB states, respectively. The absolute energy scale (in MeV) is given at the left of the figure. (After Gunye and Khadkikar [48].)

reproducing the right level ordering, are too compressed in energy in comparison to the experimental levels. Some examples of this in the s-d shell are ^{22}Ne, ^{24}Mg, and ^{28}Si, but this deficiency is even more prominent in heavier nuclei of the p-f shell [47]. To the extent that we may use the concepts of the rotational model, the omission of the pairing correlations will result in the moment of inertia $\mathscr{I}$ being too large in simple HF calculations. This is borne out by a calculation [48] of the energy spectra of ^{22}Ne. The HF and HFB calculations were done with all 22 nucleons, using a slightly modified version of the effective interaction of Elliot et al. [49]. The basis space consisted of the first four major shells of the oscillator potential. The energy spectrum of ^{22}Ne, when projected from the lowest intrinsic HF state, is shown in column (a) of Fig. 10–17 and is seen to be quite compressed when compared to the experimental values listed in column (c) of the same figure. There is a marked improvement in the spectrum shown in column (b), however, where the effect of pairing is taken into account in a HFB calculation and the states are then projected out.

10-7 PARTICLE-HOLE (TAMM-DANCOFF) CALCULATIONS

For simplicity, we continue with the example of ^{16}O, although the physical ideas that we develop here are quite general and may be used for other closed-shell nuclei like ^{40}Ca or ^{208}Pb, and even, with some modifications, in other regions of the periodic table. In this section, we shall concentrate on the odd-parity excited states of ^{16}O. In particular, a sharp 3^- level (with $T = 0$) is seen at 6.13 MeV in reactions like the inelastic electron-scattering experiment ^{16}O$(e, e')^{16}$O*, brought about by the electromagnetic interaction between the incident electron and ^{16}O. Although a host of other excited states are identified in this reaction at excitation energies of up to 30 MeV or more, no other 3^- state is found. In some other reactions like ^{12}C$(\alpha, \alpha)^{12}$C other 3^- ($T = 0$) states at 11.63 and 13.14 MeV are detected; these are rather weak, in the sense that it is difficult to excite the nucleus to these states.

A similar situation prevails with respect to the 1^- excited states. These are most commonly studied by incident γ-rays on the nucleus in reactions like ^{16}O$(\gamma, \gamma)^{16}$O*, ^{16}O$(\gamma, n)^{15}$O, and ^{16}O$(\gamma, p)^{15}$N. When the incident γ-ray energy is small compared to the excited states of the nucleus, a photon can only shake the whole nuclear charge, producing Thomson (i.e., elastic) scattering. At higher excitations, but below the particle threshold, individual levels may be excited by the reaction ^{16}O$(\gamma, \gamma)^{16}$O*. A reaction like ^{16}O$(\gamma, n)^{15}$O, involving particle emission, can occur only when the incident γ-ray energy is sufficient to provide the separation energy of the neutron and the difference between the ground-state energies of ^{16}O and ^{15}O; these require a minimum γ-ray energy of 15.7 MeV.

For higher energies, many other reaction modes are possible, involving multiparticle emission. Because so many decay modes are possible at higher energies, the γ-ray absorption cross section is no longer sharp but shows broad structures. For example, measurement of the ^{16}O$(\gamma, \gamma)^{16}$O* cross section shows very broad resonance peaks located at about 22 and 25 MeV with spin 1^-, $T = 1$, indicating electric dipole absorption. These are called giant dipole resonances. From the shell-model point of view, the simplest odd-parity excited states in ^{16}O should result from the excitation of an odd number of nucleons from the filled p-orbits to the vacant s-d shell. The simplest intrinsic states, in the HF picture, describing such excited states differ from the ground-state configuration in $1p$-$1h$ orbits only. One should then expect such excited states to have energies ≈ 11 MeV, close to the shell gap between the $0p_{3/2}$- and the $0d_{5/2}$-orbits (see Fig. 10-1). Experimentally, as we have mentioned, the 3^- ($T = 0$) state occurs only at about half this excitation, while the 1^- ($T = 1$) state is at about twice the shell gap. Moreover, from the single-particle point of view, there should be many such states, whereas experimentally the strength is concentrated only in a single 3^- and in a broad 1^- with some structure in it. In this section we shall try to understand this behavior of these excited states within the framework of the single-particle model.

First we shall have to formulate more precisely, in terms of angular-momentum algebra, the meaning of expressions like "hole states" and "particle-hole configuration." We consider the special case of a filled j-shell in the spherical basis, and denote this closed-core state by $|c\rangle$:

$$|c\rangle = a_{jj}^+ a_{jj-1}^+ \cdots a_{jm}^+ \cdots a_{j-j}^+ |0\rangle, \tag{10-51}$$

where $|0\rangle$ is the vacuum with no particles present. If a particle in the orbit (jm) is missing from this core, we call this vacancy a hole, and denote the state as $|h\rangle_{jm}$:

$$|h\rangle_{jm} = a_{jj}^+ a_{jj-1}^+ \cdots a_{jm+1}^+ a_{jm-1}^+ \cdots a_{j-j}^+ |0\rangle. \tag{10-52}$$

Since the a's are Fermion operators, a pair like $(a_{jm}a_{jm}^+)$ commutes with every a or a^+ and hence can be inserted into Eq. 10–52 without altering its value:

$$|h\rangle_{jm} = a_{jj}^+ a_{jj-1}^+ \cdots a_{jm+1}^+ (a_{jm}a_{jm}^+) a_{jm-1}^+ \cdots a_{j-j}^+ |0\rangle$$
$$= (-)^{j-m} a_{jm} |c\rangle. \tag{10-53}$$

It can be readily shown that the hole state $|h\rangle_{jm}$ transforms under rotation as if it has an angular momentum j and azimuthal quantum number $-m$. We can therefore put

$$|h\rangle_{jm} \equiv (-)^{j-m} |j, -m\rangle.$$

We can now construct particle-hole configurations of a given angular momentum. Let there be a particle in an unfilled shell of angular momentum j_a, coupling to a hole of angular momentum j_n to give a state with resultant J, M:

$$|an^{-1}; JM\rangle = \sum_{m_a, m_n} (-)^{j_n - m_n} (j_a j_n m_a - m_n | JM) a_{j_a m_a}^+ a_{j_n m_n} |c\rangle. \tag{10-54}$$

Since there is a two-body effective interaction V between the nucleons,

$$V = \tfrac{1}{2} \sum_{\alpha, \beta, \gamma, \delta} \langle \alpha\beta | V | \gamma\delta \rangle a_\alpha^+ a_\beta^+ a_\delta a_\gamma, \tag{10-55}$$

it follows that there will be matrix elements of V between particle-hole states of type 10–54. We have seen that, if the ground state $|\Phi_0\rangle$ has been constructed from self-consistent orbitals, the residual interaction $(V - U)$ cannot cause $1p$-$1h$ transitions, so that the ground state would have no admixture of $1p$-$1h$ states. However, here we are trying to describe odd-parity excited states through the $1p$-$1h$ configuration, which may be formed, for example, because of the absorption of a photon of a certain multipolarity. The unperturbed $1p$-$1h$ states would then become mixed through the interaction V. We want to see whether it is possible for certain excited states to have a collective character, with excitation energies very different from their unperturbed values. For simplicity, we suppress the angular-momentum coupling in Eq. 10–54 and denote the $1p$-$1h$ state by $|an^{-1}\rangle$; then an excited state $|\Psi_1\rangle$ of the nucleus contains admixtures of such states

$$|\Psi_1\rangle = \sum_{n,a} C_{an}|an^{-1}\rangle, \qquad (10\text{–}56)$$

where, if we ignore multiparticle excitations, the coefficients are such that

$$\sum_{n,a} |C_{an}|^2 = 1. \qquad (10\text{–}57)$$

We can rewrite Eq. 10–23 for the Hamiltonian as

$$H = H_0 + V', \qquad (10\text{–}58)$$

where $H_0 = \sum_i (t_i + U_i)$, and

$$V' = \sum_{i<j} V_{ij} - \sum_i U_i$$

is the residual interaction. Then, for an excited state $|\Psi_1\rangle$ with energy E_1, we must have

$$(E_1 - H_0)|\Psi_1\rangle = V'|\Psi_1\rangle. \qquad (10\text{–}59)$$

Noting that a state like $|an^{-1}\rangle$ is an eigenstate of H_0 with energy $e_a - e_n \equiv e_{an}$ relative to the ground state, where the e's are the HF single-particle orbitals of the self-consistent ground state, we immediately obtain

$$(E_1 - e_{bm})C_{bm} = \sum_{n,a} \langle bm^{-1}|V'|an^{-1}\rangle C_{an}. \qquad (10\text{–}60)$$

In order to obtain the eigenvalues E_1 of the excited states, we therefore must evaluate p-h matrix elements of the type $\langle bm^{-1}|V'|an^{-1}\rangle$. To avoid cumbersome notation, let us denote

$$a_{j_n m_n} \equiv a_n, \qquad (10\text{–}61)$$

and similarly for the creation operators. Using Eqs. 10–54 and 10–55, we obtain

$$\langle bm^{-1}|V|an^{-1}\rangle = \tfrac{1}{2} \sum_{\alpha,\beta,\gamma,\delta} \langle c|a_m{}^+ a_b a_\alpha{}^+ a_\beta{}^+ a_\delta a_\gamma a_a{}^+ a_n|c\rangle \langle \alpha\beta|V|\gamma\delta\rangle. \qquad (10\text{–}62)$$

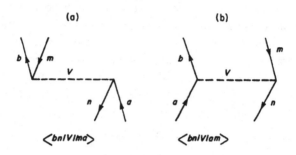

Fig. 10-18   The particle-hole matrix elements occurring in Eq. 10–63.

In Appendix C (Eq. C–38), we show how such matrix elements can be simplified in a systematic manner. It is shown there that the result is simply

$$\langle bm^{-1}|V|an^{-1}\rangle = \langle bn|V|ma\rangle - \langle bn|V|am\rangle, \qquad (10\text{–}63)$$

which is displayed diagramatically in Fig. 10–18. By convention, the first term on the right side of Eq. 10–63 is called the direct and the second the exchange term. Substitution of expression 10–63 for $\langle bm^{-1}|V|an^{-1}\rangle$ in Eq. 10–60 and subsequent diagonalization should yield the excited state energies E_1 and the corresponding wave functions.

The earliest calculations of this type, with phenomenological residual interaction, were due to Elliott and Flowers [50]. Their calculation agreed well with the experimental energies and lifetimes of the odd-parity levels of ^{16}O. They found, for example, that out of all the 1^- ($T = 1$) states one linear combination of the unperturbed p-h states separates markedly in energy and is pushed up. Furthermore, it has collective properties in the sense that this is the state most easily excitable by the electric dipole absorption of the γ-ray. Similarly, for the 3^- ($T = 0$) states, one particular state again has most of the collective strength, but it is pushed down in energy by the p-h residual interaction. These results are in agreement with the experimental observations.

To understand these results physically, we shall follow a model suggested by Brown and Bolsterli [51]. In this model, the exchange term in Eq. 10–63 is neglected, resulting in great simplicity. Some justification for this omission is based on the fact that the exchange term is almost identical to the direct one for a very short-range force, except that it has a different spin-isospin weight factor. This results in the right side of Eq. 10–63 having approximately just the direct term with a certain weight. In this model, then,

$$\langle bm^{-1}|V|an^{-1}\rangle \approx \langle b(1)n(2)|V|m(1)a(2)\rangle. \tag{10–64}$$

A general expansion of the two-body operator V in terms of the spherical harmonics is

$$V(|\mathbf{r}_1 - \mathbf{r}_2|) = \sum_{J'} (2J' + 1)V_{J'}(r_1, r_2)P_{J'}(\cos\theta_{12}),$$

where J' can take only integral values, and $V_{J'}(r_1, r_2)$ is a function of the scalars r_1, r_2. Using Eq. A–70, we obtain

$$V(|\mathbf{r}_1 - \mathbf{r}_2|) = 4\pi \sum_{J',M'} V_{J'}(r_1, r_2)Y_{J'}^{M'}(\hat{r}_1)Y_{J'}^{M'*}(\hat{r}_2). \tag{10–65}$$

Considerable simplification takes place if a separable assumption is made for $V_{J'}(r_1, r_2)$:

$$4\pi V_{J'}(r_1, r_2) = \chi f_{J'}(r_1)f_{J'}(r_2), \tag{10–66}$$

where $f_{J'}$ is some function still to be specified, and χ is a constant. Substituting Eqs. 10–65 and 10–66 into 10–64, we obtain the p-h matrix element as

$$\langle bm^{-1}|V|an^{-1}\rangle \approx \chi \sum_{J',M'} \langle b(1)|f_{J'}(r_1)Y_{J'}^{M'}(\mathbf{r}_1)|m(1)\rangle\langle n(2)|f_{J'}(r_2)Y_{J'}^{M'*}(\mathbf{r}_2)|a(2)\rangle.$$

$$\tag{10–67}$$

If we now consider an excited state of angular momentum J, M, the unperturbed p-h states $|bm^{-1}\rangle$ and $|an^{-1}\rangle$ are each coupled to angular momentum J. We show in Appendix C (Eq. C–39) that, for any one-body operator defined by

$$O_{J'} = \sum_{\alpha,\beta} \langle \alpha|f(r)Y_{J'}^{M'}|\beta\rangle a_\alpha^+ a_\beta, \tag{10–68}$$

$$\langle b|f(r)Y_{J'}^{M'}|m\rangle = \langle bm^{-1}|O_{J'}|c\rangle. \tag{10–69}$$

If $|c\rangle$ has angular momentum zero, the above matrix element is nonzero only for J', $M' = J$, M, since $|bm^{-1}\rangle$ has angular momentum J, M. This results in only one term contributing in the sum in Eq. 10–67. We saw in Chapter 3 that the Hamiltonian for the electromagnetic field can be expanded in multipole operators. Defining the transition matrix element of the electric multipole operator of order J by

$$D_{bm}^J = \langle b|r^J Y_J^0|m\rangle, \tag{10–70}$$

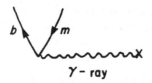

γ - ray

Fig. 10-19 The incident γ-ray creates a particle-hole pair as shown. For the Jth electric multipole of the electromagnetic field, the matrix element is denoted by D^J_{bm}.

we see that it can be represented graphically by Fig. 10–19, which has the same structure as the direct p-h matrix element of Fig. 10–18, cut in half. If it were possible to write $f_J(r) = r^J$, Eqs. 10–67 and 10–70 would yield

$$\langle bm^{-1}|V|an^{-1}\rangle \approx \chi D^J_{bm} D^{J*}_{an}, \tag{10–71}$$

thus relating the direct p-h matrix element to the transition amplitudes of the multipole operator. Even though $V_J(r_1, r_2)$ cannot be expected to have the simple form $r_1{}^J r_2{}^J$, Eq. 10–71 should be an order-of-magnitude approximation. The special property of this matrix element, namely, that only the Jth multipole of the interaction is involved in the mixing of p-h states coupled to J, is characteristic of the direct term only—other multipoles can certainly contribute in the exchange term. Substitution of Eq. 10–71 in Eq. 10–60 yields

$$(E_1 - e_{bm})C_{bm} = \chi D^J_{bm} \sum_{n,a} C_{an} D^{J*}_{an}.$$

Using the fact that

$$\chi \sum_{n,a} C_{an} D^{J*}_{an} = N_J, \quad \text{a constant,} \tag{10–72}$$

we obtain

$$C_{bm} = \frac{N_J D^J_{bm}}{E_1 - e_{bm}},$$

where these coefficients must obey the normalization condition 10–57.

Multiplying the left-hand side of the above equation by D_{bm}^{J*} and summing over b, m yields

$$\sum_{b,m} C_{bm}D_{bm}^{J*} = N_J \sum_{b,m} \frac{|D_{bm}^{J}|^2}{E_1 - e_{bm}}.$$

Using Eq. 10–72, we obtain

$$\frac{1}{\chi} = \sum_{b,m} \frac{|D_{bm}^{J}|^2}{E_1 - e_{bm}}. \tag{10–73}$$

For a given χ, which is determined from the nature of the p-h interaction, the excited states E_1 may be obtained from a graphical solution of Eq. 10–73, as shown in Fig. 10–20. All the solutions except one are trapped between the

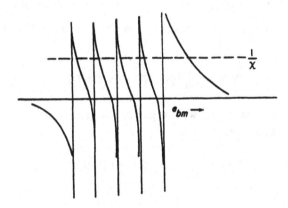

Fig. 10-20 Graphical solution of Eq. 10–73 for a positive χ, that is, a repulsive particle-hole interaction.

unperturbed energies. For a positive χ (i.e., a repulsive p-h interaction), one solution is pushed high up in energy and becomes collective, in the sense that it acquires a large share of the electromagnetic multipole strength. It is clear from Fig. 10–20 that for an attractive p-h interaction (negative χ) the collective solution will come down in energy. An extreme example is the degenerate limit, where all the e_{bm}'s equal e. Then, from Eq. 10–73, we obtain

$$\frac{1}{\chi} = \frac{D^2}{E_1 - e}, \quad \text{with} \quad D^2 = \sum_{b,m} |D_{bm}^j|^2. \tag{10-74}$$

In this case, all the solutions except one have the unperturbed energy e. The collective state has the eigenvalue $E_1 = e + \chi D^2$, and it contains all the multipole strength D^2.

Experimentally, we have seen that the $T = 1$ collective states are pushed up in energy, indicating a repulsive p-h interaction, whereas the $T = 0$ collective state comes down, showing the effect of an attractive p-h interaction. We would like to understand this behavior of the p-h interaction, starting with a short-range attractive particle-particle force that is independent of spin and i-spin. The results can then be generalized for a more realistic nucleon-nucleon force. In the i-spin space, we denote a proton state by $|\gamma\rangle = |\tfrac{1}{2}\tfrac{1}{2}\rangle$ and a neutron state by $|\delta\rangle = |\tfrac{1}{2} -\tfrac{1}{2}\rangle$. Then, using the phase convention defined in Eq. 10-53, a proton hole $|\bar{\gamma}\rangle$ and a neutron hole $|\bar{\delta}\rangle$ are given by

$$|\bar{\gamma}\rangle = |\tfrac{1}{2} -\tfrac{1}{2}\rangle \equiv |\delta\rangle, \qquad |\bar{\delta}\rangle = -|\tfrac{1}{2}\tfrac{1}{2}\rangle \equiv -|\gamma\rangle. \tag{10-75}$$

As we noted in Chapter 1, a neutron-proton state with $T = 1$, $T_3 = 0$ is given by

$$^3(\tau)_0 = \frac{1}{\sqrt{2}} [\gamma(1)\delta(2) + \gamma(2)\delta(1)].$$

Let us construct a state $^3(\tau)_0$ with a particle (1) and a hole (2). Making use of relation 10-75, we write

$$^3(\tau)_0 = \frac{1}{\sqrt{2}} [\gamma(1)\bar{\gamma}(2) - \delta(1)\bar{\delta}(2)]. \tag{10-76}$$

Similarly,

$$^3(\tau)_1 = -\gamma(1)\bar{\delta}(2), \qquad ^3(\tau)_{-1} = \delta(1)\bar{\gamma}(2)$$

and

$$^1(\tau)_0 = \frac{1}{\sqrt{2}} [\gamma(1)\bar{\gamma}(2) + \delta(1)\bar{\delta}(2)]. \tag{10-77}$$

In a given nucleus like ^{16}O, all the states should have $T_3 = 0$; hence the excited p-h configurations may be described by $^3(\tau)_0$ or $^1(\tau)_0$. Schematically, therefore, the p-h states can be denoted by

$$|ph\rangle = \frac{1}{\sqrt{2}} (\gamma\bar{\gamma} \pm \delta\bar{\delta}) \tag{10-78}$$

with the positive sign for $T = 0$ and negative sign for $T = 1$. With a spin-isospin-independent potential V, it can be easily seen that

$$\langle \gamma\bar{\gamma}|V|\gamma\bar{\gamma}\rangle = \text{(direct)} - \text{(exchange)} \quad \text{and} \quad \langle \gamma\bar{\gamma}|V|\delta\bar{\delta}\rangle = \text{(direct)}, \quad (10\text{--}79)$$

where (direct) and (exchange) stand for the matrix elements shown in Figs. 10-18a, 10-18b, respectively. It then follows that

$$\langle ph|V|ph\rangle = 2\text{(direct)} - \text{(exchange)} \quad \text{for} \quad T = 0,$$

$$= - \text{(exchange)} \quad \text{for} \quad T = 1. \quad (10\text{--}80)$$

For a short-range V, we have (direct) $\approx$ (exchange); and if the particle-particle interaction is attractive, it follows that the p-h matrix elements of V are attractive in $T = 0$ and repulsive in $T = 1$ states. This general result still holds for any reasonable exchange mixture.

To understand these results in physical terms, let us examine the form of the dipole operator

$$Q_1 = e \sum_p \mathbf{r}_p,$$

where the subscript p denotes a proton, and the sum is over all protons in the nucleus. The above equation can be rewritten as

$$Q_1 = \frac{e}{2} \sum_i (1 + \tau_{3i})\mathbf{r}_i,$$

where the sum i is over all nucleons. Defining a center-of-mass $\mathbf{R} = (1/A)\sum_i \mathbf{r}_i$, we obtain

$$Q_1 = \frac{e}{2} \sum_i \tau_{3i}\mathbf{r}_i + \frac{e}{2} A\mathbf{R}, \quad (10\text{--}81)$$

which implies that Q_1 consists of a $T = 1$ part and a $T = 0$ component which is spurious in the sense that it only affects the center-of-mass motion. For a nucleus with $T_3 = 0$ $(N = Z)$, this component can be eliminated by redefining the dipole operator with respect to the center of mass:

$$Q_1' = \frac{e}{2} \sum_i \tau_{3i}\mathbf{r}_i = e \sum_p (\mathbf{r}_p - \mathbf{R}). \quad (10\text{--}82)$$

If we now define the centers of mass of protons and neutrons separately,

$$\mathbf{R}_Z = \frac{1}{Z} \sum_p \mathbf{r}_p, \quad \mathbf{R}_N = \frac{1}{N} \sum_n \mathbf{r}_n,$$

a little algebra shows that

$$Q_1' = \frac{eNZ}{A}\,\mathbf{r},\qquad\qquad(10\text{–}83)$$

where $\mathbf{r} = (\mathbf{R}_Z - \mathbf{R}_N)$.

In Eq. 10–83, we have expressed the dipole operator in a collective coordinate, which is the relative distance between the proton and neutron mass centers. Thus absorption of dipole radiation ($T = 1$) can be interpreted as setting up relative motion between the protons and neutrons as a whole, which is opposed by the attractive n-p interaction. This requires additional energy for the $T = 1$ collective states, which are therefore pushed up from their unperturbed values. On the other hand, in a $T = 0$ excitation the neutrons and protons move in phase, bringing the collective state down in energy.

These ideas are fully exploited in the hydrodynamical model [64, 65], where giant dipole resonance results from the vibrations of a proton fluid against a neutron fluid. The resonance frequency in this model is inversely proportional to the linear dimension of the nucleus along which the vibrations take place, that is, to $A^{-1/3}$. An excellent and detailed treatment of this model, which is applicable for medium-mass and heavy nuclei, is given by

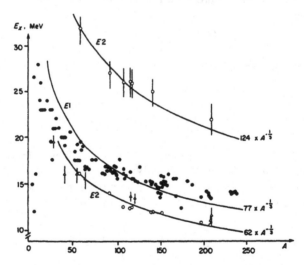

Fig. 10-21 Giant resonance energies as a function of A. The dark dots are for giant dipole $E1$-resonances, and the curve $77A^{-1/3}$ is the prediction of the hydrodynamical model. The upper curve $E2$ is through the $T = 1$ giant $E2$-resonances, while the lower curve $E2$ is through the $T = 0$ $E2$-resonances. The open circles are from (e, e') experiments; the triangles, from (p, p') experiments [67]. (After Walcher [68].)

Eisenberg and Greiner [65]. In Fig. 10–21, the experimentally found giant-dipole-resonance energies are shown by dark dots as a function of A. It will be seen that the smooth curve $E1$ given by the hydrodynamical model fits the gross experimental trend for $A > 100$. In the hydrodynamical model, one expects two peaks in the giant dipole resonance for axially symmetric deformed nuclei, corresponding to the two different frequencies of oscillation along the major and minor axes of the nucleus. This is found to be so experimentally; for example, in Nd isotopes, ^{150}Nd, which is deformed, exhibits two peaks, whereas the lighter isotopes show only one peak [66].

Finally, we should mention that (e, e') and (p, p') experiments have also revealed giant $E2$- (both $T = 0$ and $T = 1$) and giant $M1$-resonances [67]. A review of these findings is given by Walcher [68]. In Fig. 10–21, the upper $E2$ curve goes through the $T = 1$ giant quadrupole resonances, while the lower $E2$ curve is through the $T = 0$ resonances. Note that again the $T = 0$ resonances are lowered in energy, whereas the $T = 1$ resonances are pushed up, the dependence on mass number in both cases being $A^{-1/3}$. These characteristics for the giant $E2$-resonances were predicted by Bohr and Mottelson [69], whose theory fits the experimental data very well.

10-8 CALCULATION OF INERTIAL PARAMETERS

In the discussion of the rotational motion of well-deformed nuclei, we derived expression 9–37 for the moment of inertia $\mathscr{I}$ in the hydrodynamical model. We see in Fig. 10–22 that this expression fares very poorly when compared with the experimental data. The situation does not improve if we take the other extreme model, that of a rigid deformed shape, and modify Eq. 9–40 accordingly. In Section 10–6, we saw how states of good angular momenta J could be projected from an intrinsic wave function in a body-fixed system. In this section, we shall first attempt to find how the moment-of-inertia parameter $\mathscr{I}$ can be calculated for well-deformed nuclei, and then briefly discuss the microscopic basis of calculating the inertial parameter B_2 for quadrupole vibrations of spherical nuclei.

Consider a nucleus that is rotating with an angular velocity ω about a body-fixed 1-axis and has an axis of symmetry about the 3-axis. The same nucleus could be described in a space-fixed system, where, without any loss of generality, we may assume that the spaced-fixed x-axis coincides with the body-fixed 1-axis. If we denote the Hamiltonian in the space-fixed frame by H, the state $|\Psi\rangle$ will be a solution of the time-dependent wave equation,

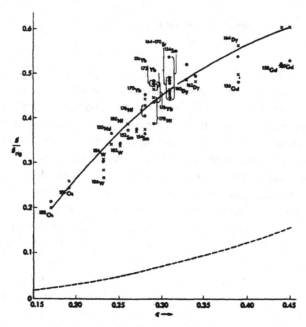

Fig. 10-22 The dependence of moments of inertia on deformation for even-even rare-earth nuclides. The circles indicate experimental values, the crosses are values calculated by superfluid theory [58], and the dashed line shows the prediction for irrotational fluid flow. The experimental values of $\mathscr{I}$ are deduced from the level spacing in rotational bands; the values of the deformation parameter are deduced from the electric quadrupole moments. Units of $\mathscr{I}$: $\mathscr{I}_{\text{rig}} = \frac{2}{5}AMR_0^2(1 + 0.31\beta + \cdots)$.

$$H|\Psi\rangle = i\hbar \frac{\partial |\Psi\rangle}{\partial t}. \tag{10-84}$$

The wave function in the rotating frame, $|\Phi_\omega\rangle$, is related to $|\Psi\rangle$ by the transformation

$$|\Psi\rangle = R_1(\omega t)|\Phi_\omega\rangle, \tag{10-85}$$

where R_1 is the rotation operator about the 1-axis,

$$R_1(\omega t) = \exp(-iJ_1\omega t). \tag{10-86}$$

Substituting Eq. 10-85 into 10-84, we obtain

$$H_\omega|\Phi_\omega\rangle = i\hbar \frac{\partial}{\partial t}|\Phi_\omega\rangle \tag{10-87}$$

with

$$H_\omega = R_1^{-1}\left(HR_1 - i\hbar \frac{\partial R_1}{\partial t}\right). \tag{10-88}$$

Recalling the simple form of R_1 from Eq. 10-86, we note that H_ω simplifies to

$$H_\omega = \exp(+ iJ_1\omega t)\, H \exp(- iJ_1\omega t) - \hbar\omega J_1. \tag{10-89}$$

If we demand stationary solutions of Eq. 10-87, we can put

$$H_\omega|\Phi_\omega\rangle = E_\omega|\Phi_\omega\rangle, \tag{10-90}$$

obtaining the relation between the energy eigenvalues in the two frames:

$$E = \langle\Psi|H|\Psi\rangle = E_\omega + \hbar\omega\langle\Phi_\omega|J_1|\Phi_\omega\rangle. \tag{10-91}$$

Note that, if we had started with the total Hamiltonian

$$H = \sum_i t_i + \sum_{i<j} V_{ij},$$

which is rotationally invariant, Eq. 10-89 would simply reduce to

$$H_\omega = H - \hbar\omega J_1, \tag{10-92}$$

showing that in a rotating frame an extra term $- \hbar\omega J_1$ arises in the Hamiltonian. If, on the other hand, we take the simple-minded approach of the independent-particle model with neglect of all residual interaction, we can assume that the quantity $\exp(iH_1\omega t)\, H \exp(- iJ_1\omega t)$ occurring in Eq. 10-89 is simply the sum of the time-independent one-body operators in the body-fixed frame, and denote this by H_0:

$$H_0 \equiv \exp(iJ_1\omega t)\, H \exp(- iJ_1\omega t) \approx \sum_i (t_i + U_i). \tag{10-93}$$

In the absence of any cranking (or rotation) the energy of the system is given by

$$H_0|\Phi_0\rangle = E_0|\Phi_0\rangle, \tag{10-94}$$

while in the presence of rotation the system energy is given by E of Eq. 10-91.

The quantity $E - E_0$ is the rotational energy and is given by

$$E - E_0 = \tfrac{1}{2}\mathscr{I}\omega^2. \tag{10-95}$$

If we assume ω to be small, E may be evaluated by taking $- \hbar\omega J_1$ as a perturbation. To do this, we note from Eqs. 10-89 and 10-93 that

$$H_\omega = H_0 - \hbar\omega J_1,$$

so that from perturbation theory

$$|\Phi_\omega\rangle = |\Phi_0\rangle - \hbar\omega \sum_{\Phi' \neq \Phi_0} \frac{\langle\Phi_0|J_1|\Phi'\rangle}{E_0 - E'}, \tag{10-96}$$

where $|\Phi'\rangle$ is a state connected through the operator J_1 with $|\Phi_0\rangle$, so that $H_0|\Phi'\rangle = E'|\Phi'\rangle$. Note further that, in the independent-particle model with axial symmetry, $|\Phi_0\rangle$ is a determinantal state with a definite value of $J_3 = \Omega$; and since J_1 is a linear combination of the raising and lowering operators J_+ and J_-, which alter Ω by ± 1, the diagonal matrix element

$$\langle \Phi_0|J_1|\Phi_0\rangle = 0. \tag{10-97}$$

When Eqs. 10–96 and 10–97 are used, it follows immediately that

$$\langle \Phi_\omega|J_1|\Phi_\omega\rangle = 2\hbar\omega \sum_{\Phi' \neq \Phi_0} \frac{|\langle \Phi'|J_1|\Phi_0\rangle|^2}{E' - E_0}. \tag{10-98}$$

Likewise, using perturbation theory up to second order and noting that the first-order term (10–97) is zero, we obtain

$$E_\omega = E_0 - (\hbar\omega)^2 \sum_{\Phi' \neq \Phi_0} \frac{|\langle \Phi'|J_1|\Phi_0\rangle|^2}{E' - E_0}. \tag{10-99}$$

We can now substitute the expressions for E_ω and $\langle \Phi_\omega|J_1|\Phi_\omega\rangle$ obtained in Eqs. 10–99 and 10–98 into expression 10–91 for E to get

$$E = E_0 + (\hbar\omega)^2 \sum_{\Phi' \neq \Phi_0} \frac{|\langle \Phi'|J_1|\Phi_0\rangle|^2}{E' - E_0}. \tag{10-100}$$

Comparing this expression with Eq. 10–95, we find

$$\mathscr{I} = 2\hbar^2 \sum_{\Phi \neq \Phi_0} \frac{|\langle \Phi'|J_1|\Phi_0\rangle|^2}{E' - E_0}. \tag{10-101}$$

The same expression for $\mathscr{I}$ could be obtained by recalling that

$$\langle \Phi_\omega|\hbar J_1|\Phi_\omega\rangle = \mathscr{I}\omega \tag{10-102}$$

and equating this to expression 10–98. Since the operator J_1 is a sum of one-body angular-momentum operators j_1 about the 1-axis, $|\Phi'\rangle$ can differ from $|\Phi_0\rangle$ only in one of the orbitals: namely, a hole is created in a single-particle orbital m and a particle in an originally unoccupied orbital b. Equation 10–101 thus simplifies to

$$\mathscr{I} = 2\hbar^2 \sum_{\substack{m \\ \text{(holes)}}} \sum_{\substack{b \\ \text{(particles)}}} \frac{|\langle b|j_1|m\rangle|^2}{e_b - e_m}. \tag{10-103}$$

This formula, originally derived by Inglis [52], is known as the cranking formula. In order to evaluate $\mathscr{I}$ by Eq. 10–103, it is necessary to choose an appropriate H_0 whose single-particle energies e_i are known—this could be, for example, the Nilsson Hamiltonian. In a more rigorous approach to the problem, however, one should calculate the self-consistent field in a rotating frame by performing a Hartree-Fock calculation on the Hamiltonian $(H - \hbar\omega J_1)$, with the constraint 10–102. This would take into account the changes brought about in the intrinsic state because of the cranking of the nucleus.[†] This self-consistent approach to cranking was formulated by Thouless [53], and a simplified formula developed by Thouless and Valatin [54]. A review of these methods is given by Rowe [55].

It has been shown [56] that, if the cranking formula 10–103 is applied to a rotating anisotropic harmonic-oscillator well, the rigid moment of inertia is obtained. As will be seen from Fig. 10–21, the experimental values are only 20–50% of the rigid values. Although the nucleons in the deformed well are moving independently of each other, they are frozen in their orbits because of the Pauli principle, and no scattering can take place in the absence of two-body interactions. In Eq. 10–103, we have assumed the occupancy of the orbitals to be either zero or one; furthermore, the equation also reveals that $\mathscr{I}$ should be very sensitive to the energy gap between the occupied and unoccupied orbitals. These considerations suggest that the two-body pairing force, which smears the occupancy of the orbitals near the Fermi energy and also introduces a gap of at least $2\varDelta$ (in even-even nuclei) between the ground state and the excited states, may modify $\mathscr{I}$ drastically from its rigid value. Consider an even-even nucleus, where we are taking into account the pairing force between identical nucleons. On solving the pairing problem in the BCS approximation (see Chapter 8), the ground state $|\Phi_0\rangle$ is described by a state with no quasiparticles, and the excited states $|\Phi'\rangle$ contain two quasiparticles. Thus we write

$$|\Phi'\rangle = \alpha_\mu{}^+\alpha_\nu{}^+|\Phi_0\rangle, \qquad (10\text{–}104)$$

where μ (or ν) stands for the quantum numbers required to specify a single-particle state in an axially symmetric well, that is, in a Nilsson potential, $\mu \equiv (N, \varOmega)$, N being the principal quantum number and $\varOmega$ the 3-component of the angular momentum. The operator $\alpha_\mu{}^+$ creates a quasiparticle in the state μ. We may now apply the cranking formula 10–101 for the moment of inertia, obtaining

$$\mathscr{I} = \hbar^2 \sum_{\substack{\mu,\nu \\ (\text{all})}} \frac{|\langle\Phi_0|J_1\alpha_\mu{}^+\alpha_\nu{}^+|\Phi_0\rangle|^2}{E_\nu + E_\mu}, \qquad (10\text{–}105)$$

[†] A self-consistent calculation of this type, with pairing taken into account, can reproduce the "back-bending" of the moment of inertia against ω^2 for high spins, described in Section 9–5. Details are given in ref. 63.

where $E_\mu = \sqrt{(e_\mu - \lambda)^2 + \Delta^2}$ is the quasiparticle energy. Note that the factor 2 is not present in front of the right-hand side of Eq. 10–105; it is canceled by the factor $\frac{1}{2}$, introduced to avoid double-counting μ and ν. On simplifying this expression, we obtain the formula [57]

$$\mathscr{I} = 2\hbar^2 \sum_{\substack{\mu > 0 \\ \nu \text{ all}}} \frac{|\langle \mu|j_1|\nu\rangle|^2}{E_\nu + E_\mu} (V_\mu U_\nu - V_\nu U_\mu)^2. \tag{10–106}$$

In this equation, $\mu > 0$ means that no time-reversed states of μ are included; while for ν both $|N, \Omega_\nu\rangle$ and the time-reversed state $|N, -\Omega_\nu\rangle$ are taken in the sum. When this formula is used for the evaluation of $\mathscr{I}$ for rare-earth nuclei with Nilsson wave functions, good agreement with experimental values is obtained [58, 59], as will be seen from Fig. 10–22.

Other collective parameters may also be calculated in the cranking approximation. Equation 10–101 for the moment of inertia is only a special case of a more general theorem when the adiabatic approximation for the collective mode is invoked. If any collective variable α which changes slowly compared to the intrinsic motion provides a term $\frac{1}{2}B\dot{\alpha}^2$ in the kinetic energy, the inertial parameter B is given by

$$B = 2\hbar^2 \sum_{\Phi' \neq \Phi_0} \frac{|\langle \Phi'|\partial/\partial\alpha|\Phi_0\rangle|^2}{E' - E_0}. \tag{10–107}$$

The analogy to Eq. 10–101 is clear if we recall that an angular-momentum component is the canonically conjugate variable to a rotation angle, and that the quantum-mechanical operator conjugate to any variable α is $-i\hbar\,\partial/\partial\alpha$. In applying formula 10–107 to obtain the quadrupole inertial vibrational parameter B_2 of spherical nuclei, we should remember that the first excited 2+ state is often in the range 1–2 MeV, and that the adiabatic approximation is not as good in this case as for the low-lying rotational states.

From the microscopic point of view, the first excited 2+ state of a spherical nucleus may be considered to be a coherent combination of one-particle, one-hole states, brought down in energy by the quadrupole-quadrupole part of the two-nucleon force. The mechanism by which a particular combination of $1p$-$1h$ states may show collective effects was discussed in a schematic model in Section 10–7, with emphasis on the giant dipole state. In the presence of pairing force, a $1p$-$1h$ state goes over to a two-quasi-particle state with a minimum energy of $\sim 2\Delta$. But a coherent combination of these two-quasi-particle states can be appreciably lowered in energy through the attractive quadrupole-quadrupole interaction. In this pairing plus quadrupole-quadrupole model, not only the energy of the 2^+ state but also its wave function can be calculated. It is then possible to calculate quantities like $B(E2)$ or the inertial parameter B_2 in a microscopic manner in this model. The quadrupole-quadrupole potential V_Q was defined earlier in Eq. 9–115, and an equation of form 10–73 can be obtained to determine the energy of the excited 2^+ state, which we denote by E_2. Modification due to pairing is straightforward, and the detailed formula is given by Bes and Sorensen [60]. One point in this modification is worth noting. We found in Eq. 10–106 for the moment of inertia an occupancy factor $(V_\mu U_\nu - V_\nu U_\mu)^2$, where the negative sign arose because the operator j_1 changes sign with time reversal. When Eq. 10–73 for the quadrupole force is modified because of pairing, however, the occupancy factor is $(V_\mu U_\nu + V_\nu U_\mu)^2$, due to the fact that $r^2 Y_2$ does not change sign with time reversal.

These particle-hole calculations can be refined by including the possibility that in the ground state, which was previously taken to be a HF determinant, there can be present $2p$-$2h$ excitations caused by residual interactions. This is called the random-phase approximation (RPA), and in the schematic model of Section 10–7 the resulting equations are only slightly more complicated. In this RPA calculation for E_2, it is found that, if $(E_2)^2 \ll (2\Delta)^2$ (i.e., the collective effect of lowering the energy of the 2^+ state is appreciable), the resulting equations for the inertial parameter B_2 becomes identical to the cranking-model expression 10–107. Kisslinger, Sorensen, and Uher [61, 62] have calculated the $B(E2)$-transition rates from the ground state 0^+ to the first excited 2^+ state of even-even spherical nuclei with the RPA equations in the pairing plus quadrupole-quadrupole model, obtaining reasonable agreement with experiment. For well-deformed nuclei, the same physical ideas can be applied for the calculation of the excitation energy and transition rate, *not* for the first excited 2^+ state (which is obtainable from the rotation of the same intrinsic state), but for the γ-vibrational state.

REFERENCES

1. S. Kahana and E. L. Tomusiak, *Nucl. Phys.* **71**, 402 (1965).
2. T. Engeland, *Nucl. Phys.* **72**, 68 (1965).
3. P. Mukherjee and B. L. Cohen, *Phys. Rev.* **127**, 1284 (1962).
4. J. Blomquist and S. Wahlborn, *Ark. Fysik* **16**, 545 (1960).
5. N. Stein, *Proceedings of the International Conference on Properties of Nuclear States, Montreal, 1969*, University of Montreal Press, 1969; see Fig. 4 on p. 342.
6. H. Flocard, M. Beiner, P. Quentin and D. Vautherin, Proc. Int. Conf. on Nucl. Phys. Vol. 1, Munich, 1973; North-Holland, Amsterdam, 1974; p. 40.

7. For a detailed list of references to these and other shell-model calculations, see the review by J. M. Irvine, *Rep. Progr. Phys.* **21**, Part I, 1 (1968).

8. K. A. Brueckner and D. Goldman, *Phys. Rev.* **116**, 424 (1959).

9. Kailash Kumar, *Perturbation Theory and the Nuclear Many-Body Problem*, North-Holland, Amsterdam, 1962, p. 172.

10. A. K. Kerman and M. K. Pal, *Phys. Rev.* **162**, 970 (1967).

11. D. J. Rowe, *Nuclear Collective Motion*, Methuen, London, 1970, Chap. 14.

12. K. A. Brueckner, A. M. Lockett, and M. Rotenberg, *Phys. Rev.* **121**, 255 (1961).

13. K. A. Masterson and A. M. Lockett, *Phys. Rev.* **129**, 776 (1963).

14. K. T. R. Davies, M. Baranger, R. M. Tarbutton, and T. T. S. Kuo, *Phys. Rev.* **177**, 1519 (1969).

15. J. W. Negele, *Phys. Rev.* **C1**, 1260 (1970).

16. D. W. L. Sprung and P. K. Banerjee, *Nucl. Phys.* **A168**, 273 (1971).

17. J. Nemeth and G. Ripka, *Nucl. Phys.* **A194**, 329 (1972).

18. B. A. Loiseau, Y. Nogami, and C. K. Ross, *Nucl. Phys.* **A165**, 601 (1971).

19. X. Campi and D. W. L. Sprung, *Nucl. Phys.* **A194**, 401 (1972).

20. D. Vautherin and D. M. Brink, *Phys. Lett.* **32B**, 149 (1970); *Phys. Rev.* **C5**, 626 (1972).

21. T. H. R. Skyrme, *Phil. Mag.* **1**, 1043 (1956).

22. S. A. Moszkowski, *Phys. Rev.* **C2**, 402 (1970).

23. J. W. Negele and D. Vautherin, *Phys. Rev.* **C5**, 1472 (1972).

24. H. Meldner, *Phys. Rev.* **178**, 1815 (1969); A. B. Volkov, unpublished results.

25. D. Vautherin, *Phys. Rev.* **C7**, 296 (1973).

26. I. Kelson, *Phys. Rev.* **132**, 2189 (1963).

27. I. Kelson and C. A. Levinson, *Phys. Rev.* **134**, B269 (1964).

28. J. Bar-Touv and I. Kelson, *Phys. Rev.* **138**, B1035 (1965).

29. A. L. Goodman, G. L. Struble, and A. Goswami, *Phys. Lett.* **26B**, 260 (1968).

30. M. R. Gunye, C. S. Warke, and S. B. Khadkikar, *Phys. Rev.* **C3**, 1936 (1971).

30a. S. K. M. Wong, J. Le Tourneux, N. Quang-Hoc, and G. Saunier, *Nucl. Phys.* **A137**, 318 (1969).

30b. S. Das Gupta and M. Harvey, *Nucl. Phys.* **A94**, 602 (1967).

31. D. Schwalm, A. Bamberger, P. G. Bizzeti, B. Povh, G. A. P. Engelbertink, J. W. Olness, and E. K. Warburton, *Nucl. Phys.* **A192**, 449 (1972).

32. O. Hausser, B. W. Hooton, D. Pelte, T. K. Alexander, and H. C. Evans, *Phys. Rev. Lett.* **22**, 359 (1969); *Can. J. Phys.* **48**, 35 (1970).

33. G. Ripka, *Advan. Nucl. Phys.* **1** (1968); see Table VI on p. 209.

34. E. B. Carter, G. Mitchell, and R. Davies, *Phys. Rev.* **133**, B1421 (1964).

35. W. H. Bassichis and G. Ripka, *Phys. Lett.* **15**, 320 (1966).

36. G. E. Brown and A. M. Green, *Phys. Lett.* **15**, 320 (1966). Also see the article by G. E. Brown in the *Proceedings of the International School of Physics*, "Enrico Fermi" Course, Vol. 36, p. 524.

37. J. Hayward, *Nucl. Phys.* **81**, 193 (1966).

38. G. J. Stephenson, Jr., and M. K. Banerjee, *Phys. Lett.* **24B**, 209 (1967); T. Engeland and P. J. Ellis, *Phys. Lett.* **25B**, 57 (1967); N. de Takacsy and S. Das Gupta, *Nucl. Phys.* **A152**, 657 (1970).

39. D. L. Hill and J. A. Wheeler, *Phys. Rev.* **89**, 1102 (1953).
40. R. E. Peierls and J. Yoccoz, *Proc. Phys. Soc. (London)* **70**, 381 (1957).
41. J. Yoccoz, *Proc. Phys. Soc. (London)* **70**, 388 (1957).
42. B. J. Verhaar, *Nucl. Phys.* **54**, 641 (1964).
43. W. H. Bassichis, B. Giraud, and R. Ripka, *Phys. Rev. Lett.* **15**, 980 (1965).
44. M. H. Macfarlane and A. P. Shukla, *Phys. Lett.* **35B**, 11 (1971).
45. J. C. Parikh, *Phys. Lett.* **25B**, 181 (1967).
46. H. G. Benson and J. M. Irvine, *Proc. Phys. Soc. (London)* **89**, 249 (1966); A. Bottino and B. Carazza, *Nuovo Cimento* **46B**, 137 (1967).
47. S. B. Khadkikar and M. R. Gunye, *Nucl. Phys.* **A110**, 472 (1968).
48. M. R. Gunye and S. B. Khadkikar, *Phys. Rev. Lett.* **24**, 910 (1970).
49. J. P. Elliott, A. D. Jackson, H. A. Mavromatis, E. A. Sanderson, and B. Singh, *Nucl. Phys.* **A121**, 241 (1968).
50. J. P. Elliott and B. H. Flowers, *Proc. Roy. Soc. (London)* **242A**, 57 (1957).
51. G. E. Brown and M. Bolsterli, *Phys. Rev. Lett.* **3**, 472 (1959).
52. D. Inglis, *Phys. Rev.* **96**, 1059 (1954). For higher-order corrections to the Inglis formula, see S. M. Harris, *Phys. Rev.* **138**, B509 (1965).
53. D. J. Thouless, *Nucl. Phys.* **21**, 225 (1960).
54. D. J. Thouless and J. G. Valatin, *Nucl. Phys.* **31**, 211 (1962).
55. D. J. Rowe, *Nuclear Collective Motion*, Methuen, London, 1970. Chapter 10.
56. B. R. Mottelson, *Rendiconti della Scuola Internationale di Fisica, Varenna 1960*, Zanichelli, Bologna, 1962, p. 44.
57. S. T. Belyaev, *K. Danske Vidensk. Selsk. Mat.-Fys. Medd.* **31**, No. 11 (1959).
58. J. J. Griffin and M. Rich, *Phys. Rev.* **118**, 850 (1960).
59. S. G. Nilsson and O. Prior, *K. Danske Vidensk. Selsk. Mat.-Fys. Medd.* **32**, No. 16 (1961).
60. D. R. Bes and R. A. Sorensen, *Advan. Nucl. Phys.* **2**, 129 (1969).
61. L. S. Kisslinger and R. A. Sorensen, *K. Danske Vidensk. Mat.-Fys. Medd.* **32**, No. 9 (1960).
62. R. A. Uher and R. A. Sorensen, *Nucl. Phys.* **86**, 1 (1966).
63. P. C. Bhargava, *Nucl. Phys.* **A207**, 258 (1973); B. Banerjee, H. J. Mang, and P. Ring, *Nucl. Phys.* **A215**, 366 (1973).
64. M. Goldhaber and E. Teller, *Phys. Rev.* **74**, 1946 (1948); H. Steinwedel and J. H. D. Jensen, *Z. Naturforsch.* **5a**, 413 (1950).
65. J. M. Eisenberg and W. Greiner, *Nuclear Models*, Vol. 1, North-Holland, Amsterdam, 1970, Chap. 10.
66. P. Carlos, H. Beil, R. Bergère, A. Leprêtre, and A. Veyssière, *Nucl. Phys.* **A172**, 437 (1972).
67. R. Pitthan and Th. Walcher, *Phys. Lett.* **36B**, 563 (1971); *Z. Naturforsch.* **27a**, 1683 (1972); R. Pitthan, *Z. Physik* **260**, 283 (1973); S. Fukuda and Y. Torizuka, *Phys. Rev. Lett.* **29**, 1109 (1972); M. B. Lewis and F. E. Bertrand, *Nucl. Phys.* **A196**, 337 (1972).
68. Th. Walcher, Proc. Int. Conf. on Nucl. Phys., Vol. 2, Munich, 1973; North-Holland, Amsterdam, 1974, p. 509.
69. A. Bohr and B. R. Mottelson, *Nuclear Structure*, Vol. 2, Addison-Wesley, Reading, Mass. (in press).

PROBLEMS

10-1. Very detailed and thorough experimental studies of ^{18}F were recently performed by two groups at Strasbourg and Toronto [see *Nucl. Phys.* **A199**, 232–398 (1973)]. The level scheme of ^{18}F up to the excitation energy of 5 MeV, shown in Fig. 10–23, is taken from the paper of C. Rolfs, A. M. Charlesworth, and R. E. Azuma [*Nucl. Phys.* **A199**, 257 (1973)].

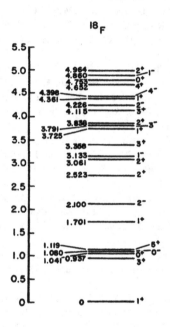

Fig. 10-23   Energy levels of ^{18}F.

(a) Comparing this spectrum of ^{18}F with that of ^{18}O, given in Fig. 10–2a, we find that far more levels (within about 5 MeV excitation) are present in ^{18}F. Explain why you expect this on theoretical grounds.

(b) If two nucleons are put in the $(1d_{5/2})^2$ configuration, then, for the $T = 0$ states, the resultant $J = 1^+, 3^+, 5^+$ only, while for $T = 1$ states $J = 0^+, 2^+, 4^+$ only are possible. Prove this.

(c) As discussed in the text, a simple model for ^{18}O consists of an inert ^{16}O-core and two neutrons in the s-d shell with $T = 1$. In ^{18}F, however, there are one neutron and one proton outside the ^{16}O-core, and both $T = 0$ and $T = 1$ states are possible. By comparing the given spectrum of ^{18}F with that of ^{18}O (Fig. 10–2a), identify the first five $T = 1$ excited states in ^{18}F.

Also, look up the low-lying levels of ^{18}Ne given by Ajzenberg-Selove, *Nucl. Phys.* **A190**, 1 (1972), and comment.

10-2. Two identical spin-$\frac{1}{2}$ particles in the spin singlet state are each moving in an average harmonic potential and further interacting via a two-body harmonic potential. The Hamiltonian of the system is given by (for convenience we take the mass M of each particle to be unity and the oscillator constant $\omega = 1$)

$$H = \tfrac{1}{2}(p_1^2 + r_1^2) + \tfrac{1}{2}(p_2^2 + r_2^2) + \chi \left[\frac{1}{\sqrt{2}} (\mathbf{r}_1 - \mathbf{r}_2) \right]^2,$$

where χ is a measure of the interaction strength.

(a) Define, for this problem, the relative and center-of-mass coordinates

$$\mathbf{r} = \frac{\mathbf{r}_1 - \mathbf{r}_2}{\sqrt{2}}; \qquad \mathbf{R} = \frac{\mathbf{r}_1 + \mathbf{r}_2}{\sqrt{2}}.$$

Show that the exact ground-state wave function for the system separates into relative and CM coordinates, and is given by (normalization = 1)

$$\Psi_{ex}(r, R) = \frac{1}{(\pi\hbar)^{3/4}} \exp(-R^2/2\hbar) \left(\frac{\sqrt{2\chi+1}}{\pi\hbar} \right)^{3/4} \exp - \left(\frac{\sqrt{2\chi+1}}{2\hbar} r^2 \right), \quad (1)$$

and the ground-state energy is

$$E_{ex} = \tfrac{3}{2}\hbar + \tfrac{3}{2}\hbar\sqrt{2\chi+1} \tag{2}$$

(b) Since the spin part of the wave function is asymmetric, the spatial wave function is symmetric, and the Hartree-Fock solution for the spatial part is the same as the Hartree solution in the s-state. Write down the Hartree equation and show that it can be solved analytically. Show that the normalized Hartree solution is

$$\Psi_H(r_1, r_2) = \left(\frac{\sqrt{\chi+1}}{\hbar\pi} \right)^{3/2} \exp\left(-\frac{\sqrt{\chi+1}}{2\hbar} r_1^2 \right) \exp\left(-\frac{\sqrt{\chi+1}}{2\hbar} r_2^2 \right), \tag{3}$$

and the ground-state energy in this approximation is

$$E_H = 3\hbar\sqrt{\chi+1}. \tag{4}$$

(c) Show that the overlap of the exact solution (1) and the Hartree solution (3) is simply given by

$$\langle \Psi_{ex} | \Psi_H \rangle = 8(\sqrt{2\chi+1})^{3/4} \left[\frac{\sqrt{\chi+1}}{(1+\sqrt{\chi+1})(\sqrt{2\chi+1}+\sqrt{\chi+1})} \right]^{3/2}. \tag{5}$$

Plot the overlap function (5) and the fractional error in the energy $(E_{ex} - E_H)/E_{ex}$ as a function of the interaction strength χ, varying χ from 0 to 3 in steps of 0.5. What is your conclusion from these plots?

If you have difficulty in solving this problem, you may consult the article by M. Moshinsky, *Am. J. Phys.* **36**, 52 (1968), and the erratum, *Am. J. Phys.* **36**, 763 (1968).

10–3. L. Zamick [*Phys. Lett.* **45B**, 313 (1973)] has proposed a simple model to determine the dependence of the nuclear incompressibility K on the density dependence of the effective interaction and the binding energy of the nucleus. Assume that the effective interaction is such that it yields the observed binding energy E_B at the right saturation density, and that it is parameterized by the expression

$$V = -\alpha\, \delta(\mathbf{r}) + \gamma \rho^\sigma(R)\, \delta(\mathbf{r}), \tag{1}$$

where the parameters α, γ, and σ are chosen to fit the experimental data, and **r** is the relative and **R** the center-of-mass coordinate of the two interacting nucleons. Assume further that the single-particle wave functions of the finite nucleus may be well approximated by harmonic-oscillator wave functions (see Problem 4–10), characterized by the oscillator length $b = (\hbar/M\omega)^{1/2}$.

(a) Show, on dimensional grounds, that the energy of the nucleus may be written in the form

$$E = -E_B = \frac{A'}{b^2} + \frac{B'}{b^3} + \frac{C'}{b^{3+3\sigma}}, \tag{2}$$

where A', B', and C' are independent of b. In this model, find an expression for the kinetic energy $\langle T \rangle$ of the nucleons and show that it may be identified with the first term in Eq. 2, that is,

$$\langle T \rangle = \frac{A'}{b^2}. \tag{3}$$

At saturation, we must have

$$\frac{\partial E}{\partial b} = 0. \tag{4}$$

Using relations 2, 3, and 4, verify that

$$B' = b^3 \left(E_B \frac{1+\sigma}{\sigma} + \langle T \rangle \frac{1+3\sigma}{3\sigma} \right)$$

and

$$C' = \frac{b^{3+3\sigma}}{\sigma} (E_B + \tfrac{1}{3}\langle T \rangle). \tag{5}$$

In this model, the nuclear incompressibility K may be defined as

$$K = \left(b^2 \frac{\partial^2}{\partial b^2} \frac{E}{A} \right)_{\text{equil.}}, \tag{6}$$

which is consistent with the definition given in Problem 8–7.

(b) Show that in this model the incompressibility K is

$$K = \frac{1}{A} [\langle T \rangle + 9E_B + \sigma(9E_B + 3\langle T \rangle)]. \tag{7}$$

For a medium-heavy or heavy nucleus it is not unreasonable to take $\langle T \rangle / A \approx 20$ MeV and $E_B/A \approx 8$ MeV; thus Eq. 7 yields

$$K = (92 + 132\sigma) \text{ MeV for a finite nucleus.}$$

For nuclear matter, $\langle T \rangle / A \approx 24$ MeV and $E_B/A \approx 16$ MeV, and the incompressibility expression 7 extrapolated to very large A is

$$K = (168 + 216\sigma) \text{ MeV.}$$

Unfortunately, since no experimental value of K is known, the parameter σ cannot be determined. Nuclear-matter calculations with reasonable forces yield a value for K in the range of 80–150 MeV, showing that a simple interaction of type 1 overestimates K.

10–4. The distribution of angular-momentum states in the deformed intrinsic state $|\Phi_K\rangle$, according to Eq. 10–47, is determined by the quantities $|C_J|^2$.

(a) Using the notation of the text, show that

$$|C_J|^2 = \frac{2J+1}{8\pi^2} \int [\mathscr{D}_{KK}^J(\theta_i)]^* \langle \Phi_K | R(\theta_i) | \Phi_K \rangle \, d\Omega. \tag{1}$$

For the ground-state $K = 0$ band of an even-even nucleus, show that Eq. 1 reduces to the form

$$|C_J|^2 = \tfrac{1}{2}(2J+1) \int_0^\pi \sin \theta_2 \, d\theta_2 \, P_J(\cos \theta_2) \langle \Phi_0 | \exp(-i\theta_2 J_2) | \Phi_0 \rangle. \tag{2}$$

(b) The overlap function $\langle \Phi_0 | \exp(-i\theta_2 J_2) | \Phi_0 \rangle$ is symmetric about $\theta_2 = \pi/2$ (see Fig. 10–15) and is well approximated, for a well-deformed nucleus, by the expression

$$\langle \Phi_0 | \exp(-i\theta_2 J_2) | \Phi_0 \rangle \approx \exp\left(-\frac{\theta_2^2 \langle J^2 \rangle}{4}\right), \quad 0 \leqslant \theta_2 \leqslant \frac{\pi}{2}, \tag{3}$$

where $\langle J^2 \rangle = \langle \Phi_0 | J^2 | \Phi_0 \rangle$ is typically 15–20 for s-d shell nuclei and of the order of 100 for rare-earth nuclei. Making approximation (3) in the exact expression (2) for $|C_J|^2$, one obtains

$$|C_J|^2 \approx (2J+1) \int_0^{\pi/2} \sin \theta_2 \, d\theta_2 \, P_J(\cos \theta_2) \exp\left(-\frac{\theta_2^2 \langle J^2 \rangle}{4}\right). \tag{4}$$

Since the dominant contribution to this integral comes from small values of θ_2, one can make the following small-angle approximation, valid for all J:

$$P_J(\cos \theta_2) \approx J_0\left[2\sqrt{J(J+1)}\ \sin\frac{\theta_2}{2}\right], \tag{5}$$

where J_0 is the cylindrical Bessel function of order zero.

(a) Show that, with this approximation,

$$|C_J|^2 \approx \frac{2(2J+1)\exp[-J(J+1)/\langle J^2\rangle]}{\langle J^2\rangle}. \tag{6}$$

In deducing Eq. 6, you will have to make use of the relation

$$\int_0^\infty \sqrt{x}\exp(-\alpha x^2)\,J_0(xy)\sqrt{xy}\,dx = \frac{\sqrt{y}}{2\alpha}\exp(-y^2/4\alpha).$$

(b) Using expression 6, find which angular-momentum states contribute importantly to $|\Phi_0\rangle$ for (i) s-d shell nuclei, and (ii) rare-earth nuclei.

Part III

Alpha Disintegration and Fission of Nuclei

Chapter 11

Alpha Radioactivity

11-1 INTRODUCTION

The fact that certain of the naturally occurring radioactive elements emit α-particles was one of the first discoveries in the early days of modern physics; by 1908 Rutherford [1] had conclusively demonstrated that α-particles are ^{4}He nuclei. There are 30 α-emitters in three naturally occurring radioactive chains, known as the uranium, actinium, and thorium series. These start at ^{238}U, ^{235}U (AcU), and ^{232}Th and, after a number of radioactive transformations, finish at the stable nuclides ^{206}Pb, ^{207}Pb, and ^{208}Pb, respectively. In addition, the great majority of the artificially created isotopes of the elements above Pb are α-active; so are some lighter nuclides, notably certain isotopes of Sm, Cm, Hf, Nd, Eu, Gd, Tb, Dy, Au, and Hg [2, 3]. Altogether there are over 150 α-active nuclides, which have been studied to some extent. In this section we shall set forth the main features of the pattern that can be seen in the experimental data.

In the earliest days of nuclear experimentation, it was thought that, in general, each nuclide emitted α-particles of a distinct and characteristic energy. In 1930 Rosenblum [4], analyzing the charged α-particles in a strong magnetic field, discovered that many nuclides emit several groups of α-particles of distinct energy. For example, $^{224}_{88}$Ra emits two main groups of α-particles: 95% of the emitted particles have energy 5.681 MeV; 5% have 5.445 MeV. Thus it is apparent that the decay need not leave the final nucleus in its ground state; in our example, when the 5.681-MeV α-particles are emitted, the daughter nucleus is $^{220}_{86}$Rn in its ground state, but when a

5.445-MeV α-particle is emitted, the ^{220}Rn is in a state excited by 226 keV.
The pattern for ^{224}Ra is typical of even-even nuclides; the feature to be
noted at the moment is that the ground-state transition is considerably more
frequent than the modes of decay which produce lower-energy particles.
On the other hand, the odd-A and odd-odd nuclei have a different pattern,
of which ^{223}Ra and ^{241}Am are typical representatives. Their spectra are shown
in Table 11-1, which also includes the spectra of a number of other nuclides.

TABLE 11-1

Nuclide	E_{α}, MeV	Abundance, %	Half-life	Partial half-life, years	Hindrance factor
$^{241}_{95}$Am	5.535	0.34	470y	1.38×10^5	560
	5.503	0.21		2.24×10^5	560
	5.476	84.4		5.60×10^2	1
	5.433	13.6		3.46×10^3	4
	5.379	1.4		3.36×10^4	16
	5.314	0.015		$3 \quad \times 10^6$	
$^{223}_{88}$Ra	5.860	Weak	11.2d	$\gtrsim 2$	$\sim 10^5$
	5.735	9		0.340	65
	5.704	53		0.0580	8
	5.592	24		0.128	5
	5.525	9		0.340	6
	5.487	2		1.5	17
	5.419	3		1.0	5
$^{224}_{88}$Ra	5.681	95.1	3.64d	1.053×10^{-2}	
	5.445	4.9		0.218	
	5.15	0.009		$1.0 \quad \times 10^2$	
$^{211}_{84}$Po	7.434	99	0.52s	1.6×10^{-8}	530
	6.88	0.50		3.3×10^{-6}	1700
	6.56	0.53		3.1×10^{-6}	100

It will be seen that in these non-even-even cases the α-particles of highest
energy are not usually the most abundant, but may be quite weak relative
to a group of somewhat smaller energy. Also, a comparison of the absolute
half-lives shows a characteristic difference. In fact, one of the most remarkable things about α-radioactivity is the great range of half-lives. If λ is the
probability per second that a radioactive nucleus will decay, the half-life
$\tau_{1/2} = (\ln 2)/\lambda$. We shall often call these λ's *decay constants*. Each mode

has its own decay constant λ_i, and the total decay constant of the nuclide is $\lambda_{tot} = \sum \lambda_i$. The *partial half-life* of a mode is

$$\tau_{1/2,i} = \frac{\ln 2}{\lambda_i}.$$

The great range of decay probabilities in α-activity is illustrated by the fact that the half-life of ^{213}Po, emitting 8.336-MeV α-particles, is 4.2×10^{-6} sec or 1.33×10^{-13} yr, whereas the half-life of ^{232}Th, which emits 3.98-MeV particles, is 1.39×10^{10} yr. A table of energies and half-lives illustrates clearly that the greater decay rates correspond to greater energies. As early as 1911, Geiger and Nuttall [5] showed that a good straight line was obtained by plotting the logarithm of the half-life against the logarithm of the range of the α-particles in air. This range is connected with the energy. Now that we have more data, of greater accuracy, we prefer to draw our curves only through the points belonging to the same element. Figure 11–1 shows the experimental points for even-even nuclides, and it is seen that joining the

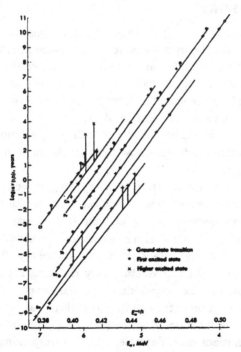

Fig. 11-1 Half-life versus energy relation for even-even α-emitters. The points appreciably off the empirical curves are either transitions to second excited states or pertain to isotopes of Em and Po whose decay involves a magic number.

points of the same Z produces a very regular family of lines. Moreover, with the exception of two isotopes of Rn and three of Po, the points for all the ground-state components of the α-spectra are seen to lie very close to the empirical lines.

If we were to plot the corresponding data for even-Z, odd-N nuclides on this diagram, we would see immediately that there are great discrepancies. In most cases, the half-life is 100–1000 times that for an even-even nuclide of the same Z and same energy; in no case is the half-life shorter than the corresponding even-even value. The ratio of the actual half-life to that predicted by the empirical even-even line is called the *hindrance factor*. There is no apparent correlation between hindrance factor and energy; it will be one of our tasks to try to "explain" the hindrance factor and to see what information concerning nuclear structure it can suggest to us. The pattern for odd-even and odd-odd nuclides is similar. The average hindrance factor for the odd-odd nuclides is perhaps greater than that for the odd-even cases.

11-2 BASIC THEORY

One of the first successful applications of quantum mechanics to a problem in nuclear physics was the theory of α-decay, presented in 1928 independently by Gamow [6] and by Condon and Gurney [7]. Their theory accounts for the very sensitive dependence of half-life on energy and predicts a relationship like that described by the smooth curves in Fig. 11–1. As we shall see, this theory essentially concerns the motion of the α-particle outside the nucleus, and consequently it is really not the overall λ-E_α relationship, but only the hindrance factors, that can throw light on the more interesting details of nuclear structure.

The half-lives of all α-emitters are very long indeed (10^{-6} to 10^{17} s) compared with the times of typical nuclear motions (10^{-21} s). Thus the average time before decay occurs is at least 10^{15} nuclear periods and may be as many as 10^{28}; the nucleus has plenty of opportunity to establish a pattern of motion, and the state is practically stationary. We might expect the wave functions of such quasi-stationary decaying states to differ only slightly from those of true bound states, at least over time intervals that are not too great a fraction of the half-life. It is clear that over longer intervals the part of the wave function which describes the probability of finding the α-particle in the nucleus must decrease, with a corresponding increase in the probability of finding that the particle has reached "infinity," that is, has been emitted from the nucleus.

The concept of the slowly decaying, almost stationary, state arises in many problems and is handled in a variety of ways. In β- and γ-transitions one uses first-order time-dependent perturbation theory, but in cases like α-decay, which involve a leakage through a potential barrier of particles going from one region of space to another, one normally deals with the problem as a time-independent one, ignoring the very slow change in the flux onto the barrier. Alternatively, one explicitly takes account of the time rate of change by allowing the energy to have a small imaginary part, so that the factor $\exp(- iEt/\hbar)$ introduces a decaying exponential $\exp(- \lambda t)$ into the otherwise steady-state wave function. In some ways the introduction of a complex energy provides a neat formulation of the theory, and it has been applied extensively to α-decay (e.g., refs. 6, 11). Here, however, we shall derive the results by the perhaps more readily understood WKB formulation of barrier penetration.

Our picture is that from time to time, as the nucleons move in the nucleus, an α-particle is formed near the surface and moving outward. It is subject to strong attractive nuclear forces while it is close to the rest of the nucleus; but once the α-particle has left the region in which nuclear forces act between it and the product nucleus, the interaction is electromagnetic. If the nucleus were a static spherically symmetric distribution of charge, the potential energy would be just the Coulomb potential $V_C = 2Ze^2/r$, where Z is the charge number of the *daughter* nucleus and r is the distance between the center of the α-particle and the charge center of the nucleus. Actually, since nuclear-charge distributions are nonstatic and, in general, nonspherical, higher electric moments must be considered. The most important is the quadrupole, which contributes a potential energy

$$V_Q = \frac{2Qe^2}{r^3} P_2(\cos \omega), \qquad (11\text{-}1)$$

where Q is the nuclear quadrupole moment, and ω is the angle between $\mathbf{r}$ and the nuclear axis. Since Q is of the order of the square of nuclear radii, and since Eq. 11-1 applies only for r greater than the nuclear radius, $V_Q < V_C$, and it is reasonable first to solve the problem with the Coulomb potential only, and to use these solutions as a basis for the solution of the more complete problem.

What can be said about the nuclear force felt by the α-particle near the nucleus or in its surface region? Just as nucleon scattering yields the potential between nucleons, and electron scattering reveals the nuclear-charge distribution, it seems reasonable to attempt to discover the actual force on an α-particle in a nucleus by studying the scattering of α-particles from nuclei. In addition to elastic scattering, it is possible to have inelastic scattering or nuclear reactions in which the α-particle is not re-emitted. The cross sections for all these processes involve a single potential, called the *optical* potential for the α-particle. Optical potentials are discussed at length in Chapter 18 of M. A. Preston, *Physics of the Nucleus* (Addison-Wesley, 1962) and in other standard accounts of nuclear reactions. For the present discussion, we need to remark only that this potential is complex: the real part gives directly the force causing elastic scattering, whereas the imaginary part is present to take account of the "disappearance" of α-particles as they initiate nuclear reactions. The imaginary part therefore yields information on how far an α-particle can penetrate nuclear matter without breaking up, or at any rate ceasing to be a recognizable α-particle cluster. The real part represents the potential barrier that the α-particle must traverse.

Data for these potentials have been given by Igo [8]. However, the optical potential is accurately known only in the extreme outer region of the nucleus. Igo's analysis shows that in the uranium nucleus no appreciable number of elastically scattered α-particles gets closer to the center than 9.9 fm. Consequently the potential is known accurately only in this surface region, which is in the very diffuse nuclear tail; in uranium, the density has fallen to one-tenth its central value at about 8.5 fm.[†] This very fast absorption of incident α-particles might suggest a very short lifetime for α-particle clusters formed in the nucleus. However, we must remember that we expect an important amount of granular structure in the region outside the half-density radius, with the possibility of α-particle clustering being favored; hence it is difficult to relate directly the fate of externally incident α-particles to the α-particle flux incident on the barrier of a decaying nucleus.

On the other hand, although the optical potential does not give information about the motion of α-particles in the nuclear interior, it does yield the barrier which must be penetrated. The nuclear potential, to which must be added the Coulomb energy, is [8]

$$V(r) = -1100 \exp\left[-\left(\frac{r - 1.17A^{1/3}}{0.574}\right)\right] \text{MeV}, \qquad (11\text{–}2)$$

with r measured in femtometers. This expression is valid in the region in which $|V(r)| \lesssim 10$ MeV and yields a potential barrier with a very sharp inner surface. For example, the ground-state α-particles of the U isotopes,

---

[†] Since the α-particle has a rms radius of 1.6 fm, some of the matter of an α-particle at 9.9 fm reaches a short way inside the one-tenth-density shell.

with energies ranging from 4.2 to 6.7 MeV, all strike the inner side of the barrier at radii between 9.38 and 9.42 fm.

Consequently, the potential felt by the α-particle is of the nature shown by the solid line in Fig. 11-2. This figure also shows, as a dashed line, the "traditional" potential, a spherical well of sharp radius R and constant potential inside R. Of course, the assumption of constant potential inside the nucleus is an oversimplification, but it is of some value in understanding the process since it permits analytic solutions for the decay rate. On the other hand, it would seem more reasonable to assume some potential that fits smoothly onto Eq. 11-2. Such an expression, for example, would be a Woods-Saxon form

$$V_{\text{int}} = - V_0 \left[1 + \exp\left(\frac{r - R_0}{a}\right)\right]^{-1} \tag{11-3}$$

with appropriate constants. We shall discuss later the sensitivity of the results to the form of V_{int}.

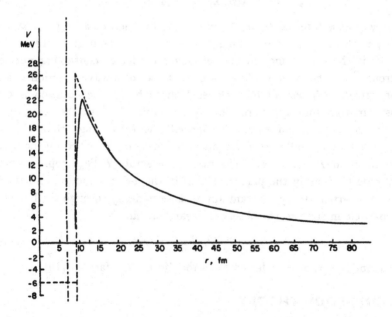

Fig. 11-2 Models of α-nucleus potential. The solid line is the optical potential of Eq. 11-2. The dashed line is the traditional model and the dot-dash line is the "surface well" model (see text).

So far the formulation ignores the imaginary part of the optical potential or, more physically, the fact that α-particles in a nucleus have a finite mean free path against losing their identity. One simple way of representing this phenomenon is to insist that α-particles can be found only within a short distance of the nuclear radius R, and to achieve this by postulating an infinite repulsion at a small distance from R. This has been called the surface-well model [9] and is represented in Fig. 11-2. It is not a very precise representation of the situation in which α-particles disappear, since the repulsive barrier is not absorptive but indeed reflects all particles striking it.

It is more satisfactory to consider the probability of finding an α-particle near the surface from the microscopic point of view of the dynamics of the nucleons. If an α-particle is to be considered as subject to a single-particle potential V similar to those in Fig. 11-2, its motion as it moves within the nucleus and through the barrier is described by a wave function $u(\mathbf{R})$, where $\mathbf{R}$ is the position of the center of mass of the α-particle referred to the center of the nucleus and u is a suitable eigenfunction of V. From the nucleonic point of view the wave function to describe this state of an escaping α-particle is

$$\Psi_\alpha = \mathscr{A}\{u(\mathbf{R})\psi_\alpha(\mathbf{x}_1, \mathbf{x}_2, \mathbf{x}_3)\Psi_f(\mathbf{r}_5, \ldots, \mathbf{r}_A)\}, \tag{11-4}$$

where ψ_α is the internal (ground-state) wave function of an α-particle, containing nucleons 1–4, x_i are internal coordinates (9 spatial degrees of freedom), and Ψ_f is the wave function of the daughter nucleus containing nucleons 5 through A. The symbol $\mathscr{A}$ indicates that the total wave function is antisymmetrized with respect to particles 1 through A; the notation is compressed in that spin and i-spin indices are omitted. But the state of the parent nucleus does not equal Ψ_α; indeed, the α-decaying "channel" may constitute only a small part of the nucleonic motions described by the actual wave function $\Psi_0(1, \ldots, A)$. The overlap integral $\langle \Psi_0 | \Psi_\alpha \rangle$ represents the amplitude for finding the parent nucleus in the α-decay mode; and if λ^0 is the single-particle decay constant appropriate to decay through the barrier V of a particle in state u, the observed decay constant is

$$\lambda = \lambda^0 |\langle \Psi_0 | \Psi_\alpha \rangle|^2 = \lambda^0 p. \tag{11-5}$$

The quantity p is often referred to as the "formation factor" [10].

11-3 ONE-BODY THEORY

We first consider the simplest one-body form of the theory, in which the barrier is entirely Coulomb and the potential inside the barrier is constant, that is, the "traditional" potential of Fig. 11-2. We make use of the WKB approximation described in Appendix D in order to write the barrier penetration. It is $e^{-2\omega}$, where

$$\omega = \int_R^{r_E} \left[\frac{2\eta k}{r} + \frac{(l+\frac{1}{2})^2}{r^2} - k^2 \right]^{1/2} dr. \qquad (11\text{-}6)$$

Here we are considering an emission in which E is the total kinetic energy of the α-particle and residual nucleus,[†] $k^2 = 2mE/\hbar^2$, m is the reduced mass:

$$m = \frac{m_\alpha m_Z}{m_\alpha + m_Z},$$

$$\eta = \left(\frac{2Ze^2}{\hbar}\right)\left(\frac{m}{2E}\right)^{1/2}, \qquad (11\text{-}7)$$

Z is the charge of the daughter nucleus, and in the centrifugal-energy term $l(l+1)$ has been replaced by $(l+\frac{1}{2})^2$ since this gives greater numerical accuracy in the first-order WKB approximation. The range of integration in Eq. 11-6 is between the two points where the total energy equals the energy of the potential barrier; in this case R is the "nuclear radius" and r_E is the outer classical turning point, that is,

$$\frac{2Ze^2}{r_E} + \frac{(l+\frac{1}{2})^2\hbar^2}{2mr_E^2} = E.$$

Note that ω in this chapter is the quantity K of Eq. D-10. When the wave function is normalized to one particle "in the nucleus," the decay constant λ^0 is equal to the flux of outgoing α-particles at a large distance, since this is then the number of decays per second per parent nucleus, in each of which there is precisely one α-particle. When the wave function ψ is an outgoing wave, the flux is $(\hbar/m)k|\psi|^2$, so that

$$\lambda^0 = \frac{\hbar k}{m} P|\psi(R)|^2, \qquad (11\text{-}8)$$

where P is the factor relating the value of the probability density at infinity with that at the radius R. In terms of the WKB approximation described in Eq. D-11, P is given as

$$P = \left| \frac{k(R)}{k(\infty)} \right| \exp(-2\omega), \qquad (11\text{-}9)$$

[†] More precisely, the total energy should be reduced by approximately $65.3(Z+2)^{7/5}$ eV to allow for electron screening [11].

where $k(r)$ is the local wave number, that is,

$$k^2(r) = \frac{2m|E - V|}{\hbar^2}. \qquad (11\text{--}10)$$

Hence, for the "traditional" model,

$$P = \left[\frac{2Ze^2}{RE} + \frac{(l + \frac{1}{2})^2\hbar^2}{2mR^2E} - 1\right]^{1/2} e^{-2\omega}$$

$$= \left[\frac{2\eta}{Rk} + \frac{(l + \frac{1}{2})^2}{k^2R^2} - 1\right]^{1/2} e^{-2\omega}. \qquad (11\text{--}11)$$

It is convenient to define a number α by the relationship

$$\tan^2\alpha = \frac{2\eta}{kR} + \frac{(l + \frac{1}{2})^2}{k^2R^2} - 1. \qquad (11\text{--}12)$$

Then it is readily seen that

$$\omega = \eta(2\alpha - \sin 2\alpha). \qquad (11\text{--}13)$$

To obtain this last result, we have ignored the square of the ratio of the centrifugal term to the Coulomb term, which is small compared with unity. Assembling these results, we find

$$\lambda^0 = \frac{\hbar k}{m} \tan\alpha\, e^{-2\omega}|\psi_0(R)|^2, \qquad (11\text{--}14)$$

where ψ_0 is the wave function for the α-particle inside the barrier.

The normalization of ψ_0 can be found to a good approximation from the condition that there be one α-particle in the nucleus, namely,

$$\int_0^R |\psi_0(r)|^2\, dr = 1. \qquad (11\text{--}15)$$

It follows that we can write

$$\lambda^0 = \frac{2v}{R} \tan\alpha\, e^{-2\omega}\frac{1}{F} \qquad (11\text{--}16)$$

where

$$F = \frac{2}{R}\frac{\int_0^R |\psi_0|^2\, dr}{|\psi_0(R)|^2}. \qquad (11\text{--}17)$$

Note that F is a dimensionless quantity dependent only on the wave function *inside* the nucleus.[†]

The factors in Eq. 11–16 for λ^0 can be interpreted physically. The number of times per second that an α-particle in the nucleus will hit the boundary of the nucleus is of the order of the velocity of the particle in the nucleus divided by the nuclear radius. Now, the velocity in the nucleus is not the same as the velocity v which the particle has at infinity, but the two are of the same order of magnitude. The last factor in Eq. 11–16 depends on the inside wave function. Hence it is reasonable to regard $2v/RF$ as the frequency with which the particle hits the wall. The other two terms depend only on the outside behavior and can be interpreted as the probability that the particle striking the wall will penetrate the barrier. For the energies of actual α-particles, η is about 20–25. Thus the variation in λ^0 comes essentially from the exponential. Also, because of the large value of η, the absolute value of λ^0 is very sensitive to the value of R used in determining α. In actual practice E and λ are known experimentally, and the equations have been used to determine the nuclear radius and a parameter in F. This was one of the earliest methods of finding the radii of heavy nuclei.

For example, with the traditional model of Fig. 11–2 and for the special case $l = 0$, the explicit form of the interior solution is

$$u_i = c_i \sin Kr,$$

where $K^2 = (2m/\hbar^2)(E - V_0)$ and V_0 is the constant interior potential. In order to fix two parameters we need a relationship in addition to Eq. 11–16. This arises from the realization that K is not arbitrary, since the wave function and its derivative must be continuous everywhere, in particular at the nuclear radius; we express this as the equality of the logarithmic derivative du/udr. Inside the well this is simply $K \cot KR$. To obtain its value just inside the Coulomb barrier we have resorted to the WKB approximation, which indicates that the solution under the barrier is

---

[†] One can generally demonstrate for the regular solution of a one-body Schrödinger equation that F can be alternatively expressed as

$$F = \frac{\hbar^2}{mR} \frac{d}{dE} \left(\frac{1}{\psi_0} \frac{d\psi_0}{dr} \right)_{r=R}. \tag{11-17a}$$

In other words, it is essentially the derivative with respect to the energy of the state of the logarithmic derivative of the wave function. This form of F arises naturally in the alternative derivation, which assigns the energy an imaginary part, and the fact that expressions 11–17 and 11–17a are equal provides the connection between the two derivations. See, for example, M. A. Preston, *Physics of the Nucleus*, Addison-Wesley, 1962, p. 362.

$$u(r) = c[k(r)]^{-1/2}(e^{\omega} + \tfrac{1}{2}ie^{-\omega}),$$

of which only the first term is significant at $r = R$. It is straightforward to see that the logarithmic derivative is $- k \tan \alpha$. Thus the two equations governing the simplest case of one-body α-decay are

$$K \cot KR = - k \tan \alpha \qquad (11\text{--}18a)$$

and

$$\lambda^0 = \frac{2v}{R}\frac{\tan \alpha}{F}e^{-2\omega} \qquad (11\text{--}18b)$$

with F given by Eq. 11-17 as

$$F = \operatorname{cosec}^2 KR - (\cot KR)/KR. \qquad (11\text{--}18c)$$

In these equations certain terms in $1/\eta$ have been omitted, causing an error of about 2%. The procedure used in a number of papers [12, 13] has been to solve Eq. 11-18a for K with an assumed value of R, and then to calculate λ^0 using this value of K. If the experimental value of λ is not obtained, a different choice of R is made, and the procedure is repeated until agreement is reached. Since λ^0 is the decay constant after the α-particle has formed (i.e., the λ^0 of Eq. 11-5), the procedure is equivalent to assuming that the formation factor p has a value of 1. This is certainly an overestimate of p; λ is an underestimate of λ^0. If more correct (i.e., higher) values of λ^0 were used, higher values would be obtained for R, since increasing the radius makes the barrier thinner and increases the penetration probability. However, some correction to R for the finite size of the α-particle is presumably necessary. Thus we say that a precise correlation of R with nuclear radius is not possible.

Nevertheless, for the ground-state transitions in even-even nuclei, the values of R calculated from Eq. 11-18 with the assumption that $p = 1$ are in excellent agreement with $R = r_0 A^{1/3}$, where $r_0 = 1.57 \pm 0.015$ fm. Also, the values of V_0 are found to satisfy the equation $E + V_0 = 0.52 \pm 0.01$ MeV. Although the physical significance of V_0 is not very clear, this result for $E + V_0$ has a simple meaning: it implies that KR is always just a little less than π, so that the interior wave function is very small at the boundary, as indicated in Fig. 11-3. (Actually $KR = 2.986 \pm 0.005$.)

The above value for r_0 is higher than that found from experiments involving electron and neutron scattering, for which the interpretation is

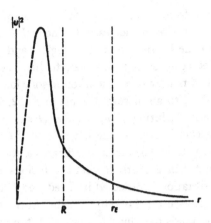

Fig. 11-3 The α-decay probability density.

much less ambiguous. These experimental results suggest $r_0 < 1.2$ fm. Let us assume as a simple model that R is made up of a true nuclear radius R_0 and an α-particle radius, $R = R_0 + R_\alpha$. If we demand that $R_0 = 1.2 \, A^{-1/3}$ fm, and that the λ^0's from Eq. 11–18 agree with the experimental decay constants, we find that $R_\alpha = 2.2$ fm. In other words, with $r_0 = 1.2$ fm and $R_\alpha = 2.2$ fm, we find that the formation factor p is unity. The rms radius of ^{4}He measured by electron scattering is 1.6 fm, which gives an equivalent square-well value of 2.1 fm. To decrease p it would be necessary to increase R_α to unreasonable values. The great regularity of the empirical values for R and $E + V_0$ suggests that, whatever the value of p, it is fairly constant for most even-even nuclides. This is the significance of the close agreement of the experimental data with the lines in Fig. 11–1. Indeed, we see that Eqs. 11–5, 11–13, and 11–18b yield

$$\ln \lambda = \ln p + \ln \left(\frac{2v}{RF} \tan \alpha \right) - (2m)^{1/2} \frac{Ze^2}{\hbar} (2\alpha - \sin 2\alpha) E^{-1/2}. \quad (11\text{–}19)$$

For the observed disintegration energies, the values of α turn out to be in the region of $\pi/3$, and the function $(2\alpha - \sin 2\alpha)$ does not vary too greatly over the observed energy·range. Since the second logarithmic term in Eq. 11–19 also turns out to be fairly insensitive to the energy, we see that $\log \lambda$ is approximately a linear function of $E^{-1/2}$ apart from fluctuations in p.

Apart from variations in p_i, experimental values of λ_i may be expected to depart from the smooth curves of Eq. 11–19 if the relative angular momentum is different from zero. The effect of angular momentum on the outside solution is such as to decrease λ, since $l(l + 1)/r^2$ may be considered an addition to the barrier, making it higher and thicker. At first, this was the only effect considered; however, to be consistent one should also include the effect of nonzero l on the interior solution, and such considerations can actually lead to an increased value of λ [12, 14]. In the traditional model the interior solution is $u_i = c_i Kr j_l(Kr)$. If the logarithmic derivative of this function is used as the left side of Eq. 11–18a, it will be found that K increases with l. The increase indicates a greater kinetic energy inside the nucleus and hence more frequent collisions with the surface. Mathematically, this situation is reflected in a reduced value of F, which in turn tends to increase λ^0 (see Eq. 11–18b). On the other hand, the increased barrier reduces $e^{-\omega}$ and tends to reduce λ. For $l \lesssim 4$, the result is an increase in λ^0; for $l > 4$, the exponential decrease becomes the deciding factor.

As we have already remarked, the traditional model is seen in Fig. 11–2 to be a not unreasonable approximation to the optical potential with, say, a Woods-Saxon potential inside the nucleus. But our results, typified by the exponential dependence in Eq. 11–18, show a very marked sensitivity to R or, more precisely, to the behavior of the wave function near the nuclear surface. The result for λ^0 is clearly affected by both the shape of the potential barrier and the approximations arising in the calculation. Both factors have been considered [15, 16]. By replacing a constant potential by a more physically reasonable Woods-Saxon shape joining smoothly onto the Igo optical potential, the quantity RF can be changed by 50–75%. Replacing the WKB approximation by numerical integration through the barrier is a time-consuming procedure, since the accurate calculation of an exponentially decreasing function requires great numerical accuracy; nevertheless, it has been done and, for reasonable values of R, causes about 10% effects.

Another consideration is the effect of abandoning the assumption that the one-body potential is local. One would, in fact, expect an effective α-nucleus potential to be momentum dependent or nonlocal, at least at short distances. Such potentials have been considered [17, 18]. Unfortunately, α-particle scattering has not been done at sufficient energies to sample much of the nuclear interior, so that considerable ambiguity is available to the theorist. One should, of course, use potentials that give the correct scattering cross sections; on the basis of the results of refs. 17 and 18, it is probably reasonable to consider that the use of nonlocal potentials restricted in this way will produce changes in the one-body factor ranging over about the same values as are covered by a sequence of acceptable Woods-Saxon potentials.

To summarize the above, we may note that the major uncertainty of the one-body model is connected with the potential and wave function at distances inside about 9 fm. We recall that the well-known part of the optical potential of Eq. 11-2 is outside this distance, and that the inner turning point R is near 9.4 fm for the observed range of α-particle energies. Consequently, it is reasonable to calculate the penetrability P of the barrier and to interpret λ/P as the single-body α-particle probability at R, subsuming in λ/P all the internal effects. Of course, P must be calculated with the most realistic possible barrier, that is, the nuclear potential of Eq. 11-2, together with the Coulomb and angular-momentum barriers.

These calculations have been made by Rasmussen [19]. He lists penetration factors for most well-known transitions and expresses the flux in terms of the *reduced width* δ^2 (MeV), defined as

$$\delta^2 = \frac{\lambda}{P} h, \qquad (11\text{--}20)$$

where h is Planck's (uncrossed) constant. The quantity δ^2/h is the hypothetical decay rate in the absence of a barrier. It is, of course, related to the formation factor p. From the point of view of nuclear dynamics, the interesting aspect of α-decay is the determination of δ^2. Rasmussen's values of δ^2 have been calculated with a realistic barrier and therefore may be considered good empirical values. Since they are based on the WKB approximation, they are subject to a correction of the order of magnitude noted above. The values might also change slightly if the Igo potential were replaced by a nonlocal one; but, as we have suggested, values of P are likely to be quite close for formally different potentials that are equivalent in elastic α-nucleus scattering.

Of course, the single-particle models, including the optical model, do not provide an understanding of the values of δ^2 beyond the rough order of magnitude v/R; one must find p from microscopic theory.

11-4 HIGHER ELECTRIC MOMENTS

The quadrupole potential (Eq. 11-1) involves the angle between the radius to the α-particle and the axis of the nucleus; it is a particular case of noncentral interactions for which the relative orbital angular momentum is not constant. A particle can leave the nucleus with angular-momentum quantum number l_1, and later, through the intermediary of such noncentral potentials, it can acquire a different value of l, say l_2, with the nucleus going to a state such that the total angular momentum and parity of the system are conserved. Of course, there is also an exchange of energy. For example, an even-even nucleus might undergo its ground-state transition, emitting an α-particle with $l = 0$. Then, before the particle had moved too far away, the quadrupole interaction could raise the nucleus to an excited state with $J = 2$, simultaneously reduce the kinetic energy of the particle, and change l to 2. Since the barrier would then have been penetrated partially at one energy and partially at another, the λ's of the two groups of particles would not have the values obtained by the theory discussed in Section 11-3. Moreover, we also see that the observed ratio of the intensity of the l_1-group to that of the l_2-group is not the same at infinity as at the nuclear radius, for the number of particles going from l_1 to l_2 is not the same as the number of particles involved in the reverse transition. In other words, the quantities $w_1 = \lambda_1/\lambda_1^0$ and $w_2 = \lambda_2/\lambda_2^0$, which, according to the previous theory, would be the formation factors p_1 and p_2, now differ from them because of the effect of the noncentral forces. In addition to the change in the amplitude of different angular momenta as the α-particles move out from the nucleus, a nonspherical shape may cause a different angular distribution of α-particles over a sphere *near* the nucleus.

Let us consider a spheroidal even-even nucleus, emitting α-particles in transitions to both the ground state and the various rotational states of its daughter. The angular distribution of a wave is closely related to the relative amplitudes of the constituent angular-momentum states. A com-

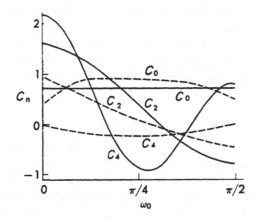

Fig. 11-4 Components of distribution functions peaked at angle ω_0.

pletely isotropic distribution is entirely an $l = 0$ wave. As the distribution becomes more peaked, the higher components become more important, and the relative amounts of various l's present depend on the angle about which the distribution is peaked. For example, the relative amounts of $l = 0$, 2, and 4 are shown for various angles and for two different distributions in Fig. 11-4, in which $|C_n|^2$ is the relative proportion of the state with $l = n$. If $|f(\omega)|^2$ is the angular distribution of α-particles, the solid lines represent the very peaked distribution

$$f(\omega) = \tfrac{1}{2} \delta(\omega - \omega_0) + \tfrac{1}{2} \delta(\omega - \pi - \omega_0),$$

and the dotted lines the broad distribution

$$f(\omega) = \cos^2(\omega - \omega_0), \qquad \omega \leqslant \frac{\pi}{2}.$$

Now, for a nucleus of positive quadrupole moment, the greatest excursions of nuclear matter from the center are in the directions 0 and π, with the angles measured from the symmetry axis. Consequently, to reach the same distance from the center of the nucleus, particles traveling out in different directions have to cover varying distances within the Coulomb barrier; and, assuming that the encounters of nascent α-particles with the nuclear surface are uniformly distributed over the surface, we might therefore expect that near the nucleus the distribution of α-particles would be peaked around the angles 0 and π. On the other hand, the quadrupole potential V_Q is greatest in the directions 0 and π, so that, although particles moving in these directions have a head start through the barrier, they also encounter a higher barrier. Thus during the progress through the barrier the intensity in these directions will suffer a relative decrease, or, in other words, the effect of the quadrupole moment is to shift the peak in the angular distribution toward angles 90° from the maximum of V_Q. It is clear that the same considerations apply also to negative Q's, where the initial maximum at $\pi/2$ would be swung toward 0 and π.

Figure 11-4 shows that, as the peak of the distribution moves toward $\pi/2$, the proportions of the $l = 4$ and $l = 2$ components decrease with respect to the $l = 0$ wave. Consequently, for positive Q, the waves with $l = 2$ and $l = 4$ will lose intensity to the $l = 0$ wave as the particles move out. Thus the ratio of the apparent formation factors w_2/w_0 will be less than p_2/p_0. If we assume that all even-even nuclei have about the same ratio of actual formation factors (p_2/p_0), we expect w_2/w_0 to decrease as Q increases. Since the natural α-emitters are located in the part of the periodic table where increasing Z corresponds to the filling of proton and neutron shells with increasing Q, w_2/w_0 should decrease if Z increases. If $l = 4$ states are involved, Fig. 11–4 would indicate that the ratio w_4/w_0 should exhibit this decrease even more markedly. Both effects are found.

These considerations have been formulated quantitatively [20]. The electrostatic interaction of the α-particle with the nucleus is expanded in spherical harmonics,

$$V(\mathbf{r}) = \sum_{p=1}^{A} \frac{2ee_p}{|\mathbf{r} - \mathbf{r}_p|} = 2e \sum_p \sum_\lambda e_{\lambda p} r_p{}^\lambda P_\lambda [\cos(\mathbf{r}, \mathbf{r}_p)] r^{-(\lambda+1)}, \quad (11\text{–}21)$$

where $e_{\lambda p}$ is the effective nucleon charge. The similarity between this expression and the multipole expansion of the electromagnetic field in appendix E suggests that the same nuclear matrix elements $B(E\lambda)$ will enter the calculation. We may write the wave function of the final state as a superposition of waves with the angular momentum of the initial nucleus shared in various ways between the α-particle and the residual nucleus:

$$\Psi = \sum_{l, I_f, \tau} r^{-1} u_\nu(r) \chi_\nu{}^M(\theta, \phi, \xi),$$

$$\chi_\nu{}^M(\theta, \phi, \xi) = \sum_m (l J_f m\, M - m | J_i M) Y_l{}^m(\theta, \phi) \psi_{J_f \tau}^{M-m}(\xi), \quad (11\text{–}22)$$

where ν stands for (l, J_f, τ), τ represents all unspecified quantum numbers of the residual nucleus, ξ represents all coordinates of the residual nucleus, and $\mathbf{r}$ is the coordinate of the α-particle. (Since the α-particle is supposed to be in the same state initially and finally, its internal wave function is not included.) Equation 11–22 is substituted in the Schrödinger equation for the whole system,

$$\left[H_\xi - \frac{\hbar^2}{2m} \nabla^2 + V(\mathbf{r}) - E \right] \Psi = 0,$$

the scalar product is taken with $\chi_\nu{}^M$, use is made of the fact that

$$H_\xi \psi_\nu(\xi) = E_\nu{}^{\text{nuc}} \psi_\nu(\xi),$$

and, after some simplification of the angular-momentum coefficients, we obtain [11]

$$-\frac{\hbar^2}{2M} \frac{d^2 u_\nu}{dr^2} + \left[\frac{l(l+1)\hbar^2}{2mr^2} + E_\nu{}^{\text{nuc}} - E + \frac{2Ze^2}{r} \right] u_\nu$$

$$= \sum_{\lambda=1}^{\infty} \frac{2e}{r^{\lambda+1}} \sum_{\nu'} u_{\nu'}(r) [B_{\nu' \to \nu}(E\lambda)]^{1/2} c(l, l', J_f, J_f'; \lambda). \quad (11\text{–}23)$$

Here

$$c(l, l', J_f, J_f'; \lambda) = \pm (-)^{J_f - 1} \frac{[4\pi(2l + 1)(2l' + 1)(2J_f' + 1)]^{1/2}}{2\lambda + 1}$$

$$\times (ll'00|\lambda 0) W(lJ_f l'J_f'; J_i\lambda),$$

where W is a Racah function. The uncertainty in sign comes from taking the square root of $B(E\lambda)$ at a certain point in the calculation; it relates to the two possible phases of the corresponding γ-transition. Sometimes angular γ-distributions will fix the sign experimentally; in other cases, it is fixed by a sufficiently detailed theoretical model.

The left side of Eq. 11-23 is the equation for the radial function u_ν in the absence of noncentral interaction, which we have already studied in detail. The solutions for the different angular-momentum states are coupled by the right side. The coupling gives mathematical expression to the physical idea that an α-particle could start at the nucleus with a single specific angular momentum and energy [$u_{\nu'}(R)$ finite, $u_\nu(R) = 0$ for $\nu \neq \nu'$] and, after traveling to some distance r, would have exchanged energy and angular momentum with the nucleus with a certain probability related to the now nonzero amplitude $u_\nu(r)$. The experimentally measured intensities of the different groups E_ν in the decay of a nuclide are proportional to $|u_\nu(\infty)|^2$. Internal nuclear dynamics determine $u_\nu(R)$, and the set of coupled differential equations 11-23 provides the link between the relative intensities at the nucleus and at infinity.

It is clear that the extent to which the population of one group (ν) is increased at the expense of another (ν') as the particles travel out is dependent on the size of the reduced matrix element $B_{\nu' \to \nu}$. The magnitude of the effect also depends on the energy difference $|E_{\nu'} - E_\nu|$, and is greater for smaller energy differences.

It has been shown that electric dipole oscillations are unimportant. Two groups are emitted by ^{241}Am decaying to levels of ^{237}Np separated by only 60 keV and connected by an $E1$-transition. With matrix elements of single-proton magnitude, the coupling effect is of the order of 10^{-2} of the intensity of the strong transition; but when the actual retarded value for $M(E1)$ is used, the result is reduced to 10^{-7} [11].

The noncentral effect is appreciable only in transitions to different states of the same rotational band in spheroidal nuclei. Here the quadrupole matrix element $B(E2)$ is large, and the energy differences are reasonably small. Since most natural α-emitters are spheroidal nuclei, the effects of quadrupole coupling have been extensively studied [11, 21, 22, 23, 24, 25].

The wave functions u_ν in Eq. 11–23 are known at large distances, for as r tends to infinity, the coupling terms become negligible, and u_ν is simply a Coulomb wave function for a particle of energy $E - E_\nu{}^{nuc}$ and charge $2e$, moving out from the charge Ze. When F_ν and G_ν are the regular and irregular Coulomb functions discussed in Appendix B, then

$$u_\nu \sim A_\nu e^{i\delta_\nu}(G_\nu + iF_\nu). \tag{11–24}$$

The magnitudes A_ν are fixed by the experimentally measured intensities of the different groups, since the intensity $I_\nu = v_\nu A_\nu{}^2$, where v_ν is the velocity of the particles in group ν. However, the phases δ_ν are not fixed by this consideration. To determine them experimentally, one would have to measure some quantity that depends on the interference between the various terms of Ψ. Such experiments (e.g., the measurement of the angular distribution of α-particles from oriented nuclei) are difficult to perform. Fortunately the phases must also satisfy a condition imposed by theory. Since the α-particles are emitted by quasi-stationary states, the α-particle flux at the nuclear surface must be due entirely to the imaginary part of the energy. In other words, in a calculation using only the real part of the energy, there must be no current of α-particles at the nuclear surface; this condition requires that the imaginary part of u_ν vanish on the surface.

One can then envisage obtaining the physical solution of the set of equations 11–23 by starting at some large value of r with form 11–24, in which the A_ν's are known. By numerical computation, the equations are integrated inward for various values of δ_ν, and those values of δ_ν are selected which make the imaginary parts vanish on the nuclear surface. It turns out that this condition permits two solutions for each δ_ν, one near zero and one near π. (Of course, one of the δ_ν's, say δ_0, can always be equated to zero, because of the arbitrary phase of the total wave function.) Consider

TABLE 11-2 Relative Amplitudes of the Waves of α-Particles of Different Angular Momenta on the Surface of the ^{242}Cm Nucleus. After Chasman and Rasmussen [25].

Case:	A	B	C	D	E	F	G	H
$\dfrac{u_2}{u_0}$	1.208	1.175	− 1.013	− 1.080	1.204	1.171	− 1.005	− 1.071
$\dfrac{u_4}{u_0}$	0.423	0.221	− 0.276	− 0.626	0.348	0.145	− 0.148	− 0.496
$\dfrac{u_6}{u_0}$	0.336	0.261	− 0.431	− 0.562	− 0.188	− 0.264	0.461	0.334

an even-even decay. The quadrupole moment couples the transitions to the ground state and to the first excited state $J = 2$ of the rotational band, that is, u_0 and u_2 are coupled. Also u_2 is coupled to u_4, the radial function for decay to the $J = 4$ rotational state. In turn u_4 is coupled to u_6 as well as to u_2, and so on. However, the number of α-particles feeding the higher levels rapidly decreases, so that the effect of u_8 on the value of u_4 may be ignored. It is therefore a reasonable procedure to consider only the first four equations of the set 11–23 and include only u_0, u_2, u_4, and u_6. The ambiguity in δ_2, δ_4, and δ_6 means that there are eight possible sets of phases and eight different solutions. In general, if we keep n wave functions, there are 2^{n-1} solutions, and the selection of the correct one can be made satisfactorily only by an interference experiment.

The large number of numerical results may be typified by a single case, ^{242}Cm, for which we assume the quadrupole moment $Q_0 = 11$ eb. We show in Table 11-2 the relative values of u_0, u_2, u_4, and u_6 on a sphere of radius 9 fm, which lies in the nuclear surface. The squares of these numbers are

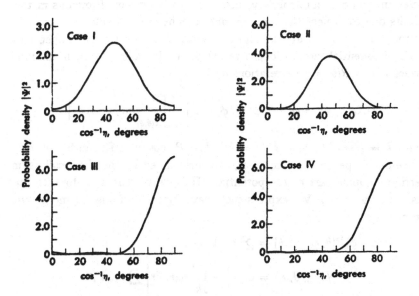

Fig. 11-5 Distribution of α-particles over the spheroidal surface of ^{242}Cm, based on the assumption that $Q_0 = 11.3$ eb. The abscissa is $\cos^{-1} \eta$, where η is the difference of the distances between the point on the spheroid and the two foci divided by the interfocal distance: $\eta = (r_1 - r_2)/a$. The polar angle θ is approximately $\cos^{-1} \eta$. This graph is approximate, being based on calculations [23] which ignored the $l = 6$ wave.

therefore the relative probabilities of forming the four different groups of α-particles. It is seen that among the eight cases there are some quite different solutions, particularly in the amplitude of the particles with $l = 4$. These results demonstrate that the noncentral effect may be quite important, but that we cannot calculate surface amplitudes uniquely without knowing the phases at infinity as well as the intensities.

Another way of presenting the results is to show the distribution of α-particles over the nuclear surface. Figure 11–5 illustrates the angular variation of $|\Psi|^2$ on the spheroidal surface of ^{242}Cm for four of the eight typical cases. In the cases labeled III and IV, α-particle formation is confined to a band of approximately 30° on either side of the equator. On the basis of our ideas of nuclear structure this seems improbable, and hence we incline toward accepting one of the other solutions.

It would be tedious and expensive to perform numerical integration of the barrier penetration for each case of decay of a nonspherical nucleus. One looks, therefore, for an extension of the WKB method that will deal with penetration of a barrier containing the noncentral quadrupole force. Such a technique has been developed by Fröman [21]. Given an arbitrary wave function over an arbitrary surface near the nucleus, Fröman's method permits one to project the function onto a sphere R_0 outside the barrier by multiplying the value at each surface point (θ, ϕ) by the one-dimensional WKB exponential factor along a radial path. In other words, if the inner surface is $r = R(\theta, \phi)$, the relationship is

$$\psi(R_0, \theta, \phi) = \psi[R(\theta, \phi), \theta, \phi] \exp\left[-\int_{R(\theta,\phi)}^{R_0} k(r, \theta, \phi)\, dr\right],$$

where $k = (2m/\hbar^2)^{1/2}[V - E]^{1/2} = \sum_L k_L(r) P_L(\cos\theta)$. The order of magnitude of the penetration factor is, of course, fixed by the spherically symmetric monopole part of the potential. It is convenient, therefore, to take this out as a factor. We express each wave function in angular-momentum components

$$\psi[R(\theta, \phi), \theta, \phi] = \sum c_{lm} Y_{lm},$$

$$\psi(R_0, \theta, \phi) = \exp\left[-\int_{R}^{R_0} k_0(r)\, dr\right] \sum d_{lm} Y_l{}^m,$$

where R is the radius of a spherical surface near the nuclear surface $R(\theta, \phi)$ — indeed, a convenient one is the sphere enclosing the same volume. We then define the matrix k by

$$d_{lm} = \sum_{LM} c_{LM} k_{lL}^{mM}.$$

If there were spherical symmetry, k would be the unit matrix, and in the general case it gives the angular-momentum distribution of the α-particles at a large distance R_0 in terms of their distribution on the nuclear surface. Of course, there is no change in the angular momentum once α-particles have traveled sufficiently far from the nucleus that the quadrupole field is negligible. Consequently, to evaluate the matrix k we need only calculate WKB integrals through the barrier. Fröman's initial calculations have been extended by Poggenburg, Mang, and Rasmussen [26].

11-5 FORMATION FACTORS

In addition to nonzero values of l and noncentral interactions, there is still a third factor which produces departures of the empirical lifetimes from the smooth values for even-even nuclides, namely, variation of the formation factor. It is the only one of the three that seems capable of producing factors of 10 or more in the half-life, and its value is potentially most informative about nuclear structure since it is the only one that depends sensitively on the properties of the interior of the nucleus. It seems certain that the formation factors play the leading role in causing the longer lifetimes which we saw in Section 11-1 to be characteristic of nuclides with odd protons or neutrons. The hindrance factors of Table 11-1 are not directly related to the formation factors, since they include also the effect of higher angular momenta and noncentral interactions. Moreover, since the formation factor for even-even nuclides is not unity, a hindrance factor is merely a number that is of the order of magnitude of the ratio of the formation factor of an even-even nucleus to that of the actual nucleus in question. Our problem is to understand the pattern of the formation factors in terms of nuclear structure.

First, we recognize that odd-A nuclides contain an unpaired nucleon which is less tightly bound than the others. It may reasonably be assumed that, when an α-particle forms, it is not likely to incorporate the odd nucleon, but will probably consist of protons and neutrons that are already paired. For transitions in which the odd nucleon does not change its quantum state, the formation factor would be expected to be about the same as for even-even nuclei, since the nucleons forming the α-particle would have very similar wave functions in the two cases. Consequently, if the ground state of the daughter nucleus differs from that of the parent, while an excited state has a similar wave function, we expect the ground state to show a considerable departure factor, whereas the excited state may have essentially the lifetime appropriate for an even-even case. This would explain the typical pattern of odd-A nuclides, which we discussed in Section 11–1.

As an example of a considerable body of similar data, consider the decay of ^{243}Cm, shown in Fig. 11–6. The various quantum numbers of the single-particle levels in a spheroidal potential are shown. It is seen that only 13% of the transitions go to the ground-state band A of ^{239}Pu, whereas 86% go to the two levels of the band labeled B. It is not hard to find the

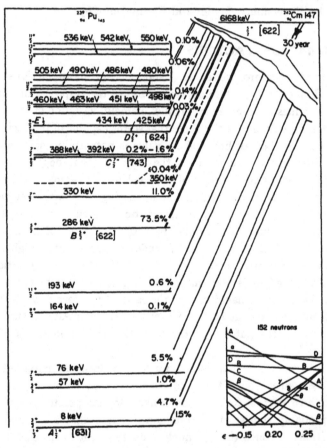

Fig. 11-6 Decay of ^{243}Cm. About 30 α-groups are known; the 24 of highest energy are shown. The notation for a band is $K\Pi[N, n_z, \Lambda]$. Based on *Nuclear Data Sheets* B3, No. 2, 1969 and S. A. Baranov et al., *Yadern Fiz.* 4, 1108 (1966). The insert of a portion of a level diagram has the notation: $A\ \frac{1}{2}^+[631]$, $B\ \frac{5}{2}^+[622]$, $C\ \frac{7}{2}^-$ [743], $D\ \frac{7}{2}^+[624]$, $\alpha\ \frac{9}{2}^-[734]$, $\beta\ \frac{5}{2}^+[633]$, $\gamma\ \frac{1}{2}^-[501]$, $\delta\ \frac{5}{2}^-[503]$, $\varepsilon\ \frac{3}{2}^-[501]$, $\theta\ \frac{13}{2}^+[606]$.

reason: the transitions to band A have $\Delta K = 2$ and $\Delta \Lambda = 1$, whereas there are no changes in quantum numbers in the transitions to band B. Another very typical example is provided by ^{241}Am, 84% of whose decays go to a $\frac{5}{2}^-$ level at 60 keV rather than a $\frac{5}{2}^+$ ground state; the quantum numbers of the decaying state of Am ($J = \frac{5}{2}$, $\pi = -1$, $N = 5$, $n_z = 2$, $\Lambda = 3$, $K = \frac{5}{2}$) are identical with those of the Np state preferentially fed, whereas the ground state of ^{237}Np is $\frac{5}{2}^+$ (6, 4, 2, $\frac{5}{2}$). Of course, the reason that it is an excited state of the daughter which resembles the ground state of the parent is simply that, with the addition of two more neutrons (or protons), the odd particle in the ground state has moved to the next Nilsson orbit. The atomic number of Np is 93; and, as Fig. 9–31 shows, for $\eta \sim 5$ the $\frac{5}{2}^+$ (6, 4) state and the $\frac{5}{2}^-$ (5, 2) state are, respectively, the ground and first excited states for the 93rd proton, and the latter is the ground state for the 95th proton.

Within a given rotational band, the states have the same internal structure, but in some transitions the α-particle may carry off zero angular momentum, whereas in others it must remove an appreciable amount. The decays requiring higher l-values for the α-particle are expected to be less probable, if only because of the increasingly greater radius at which the particle must be formed. This distinction is present in the cases cited in the last paragraph (note that all the levels which are appreciably fed have $\Delta J = 0$ or 1), and the additional hindrance with higher l's is shown also in even-even decays. The decays to the first excited 2^+ states are not delayed; in fact, in many cases the reduced width δ^2 is actually greater for the $l = 2$ transition. However, all the $l = 4$ transitions to 4^+ rotational states are slow. A quantitative expression is readily given in terms of the *reduced hindrance factor*, which is the ratio δ_l^2 to δ_0^2, where δ_l^2 is the reduced width $h\lambda/P_l$, obtained with a barrier penetrability including the centrifugal barrier, and δ_0^2 is the reduced width of the ground state. The reduced hindrance factors for $l = 2$ range from 0.8 to 1.2, while those for $l = 4$ increase with nuclear distortion, ranging from 2 at $Z = 90$ to a maximum of 150 in ^{244}Cm, and then decrease [19]. However, these numbers do not measure the relative difficulty of forming an α-particle with $l = 4$, for they are based on the intensity of such particles at infinity, not at the nuclear surface, and we know that the quadrupole interaction can seriously affect the relative intensities.

We see that the α-decay of spheroidal nuclei is consistent with the collective model, both in its general features and in assignments of quantum numbers to specific states.

The spherical nuclei near ^{208}Pb also provide some interesting confirmations of the single-particle model. One of the simpler facts to understand is that the decay of all Bi isotopes $(Z = 83)$ is strongly forbidden. The unpaired proton of Bi is in the $h_{9/2}$ state, while the 81st proton in the daughter is $s_{1/2}$; the α-particle must carry off at least four units of angular momentum and presumably must incorporate the $h_{9/2}$ proton. Both these factors make its formation difficult.

The shell model was tested by Mang [27] in a quantitative calculation of the formation factors of the Po and At isotopes. The most striking feature of the experimental data is the sharp drop by a factor of about 5 in the reduced width of α-emitters with 126 or fewer neutrons. Mang is able to explain this phenomenon in terms of a simple spherical shell-model nuclear structure.[†] The essential idea of the calculation is that the probability of formation of an α-particle is proportional to the overlap of the wave function of the final state with that of the initial state.

Let the initial state be $\Psi_0(\mathbf{r}_1, \mathbf{r}_2, \ldots, \mathbf{r}_A)$. Suppose that nucleons 1, 2, 3, and 4 form an α-particle of angular momentum l. If we write $\mathbf{R}$ for the center of mass of the α-particle and $\mathbf{x}_1$, $\mathbf{x}_2$, and $\mathbf{x}_3$ for its internal coordinates (they are functions of $\mathbf{r}_1, \ldots, \mathbf{r}_4$), the final-state wave function is of the form

$$\Psi_\alpha = \mathscr{A}\{R^{-1}u_l(R)\chi_{lJ_f}^M(\mathbf{x}_1, \mathbf{x}_2, \mathbf{x}_3, \mathbf{r}_5, \ldots, \mathbf{r}_A)\},$$

where

$$\chi = \sum_m (lJ_f m\, M - m|J_i M)\, Y_l^m(\Theta, \Phi)\Psi_f^{M-m}(\mathbf{r}_5, \ldots, \mathbf{r}_A)\psi_\alpha(\mathbf{x}_1, \mathbf{x}_2, \mathbf{x}_3);$$

in this sum ψ_α is the internal wave function of an α-particle, and Ψ_f is the state of the daughter nucleus. With a nucleus in state Ψ_0, the amplitude for finding this α-particle with its center at radius R is

$$A = \int d\mathbf{x}_1\, d\mathbf{x}_2\, d\mathbf{x}_3\, d\mathbf{r}_5 \cdots d\mathbf{r}_A \sin\Theta\, d\Theta\, d\Phi\, \Psi_0{}^*\chi.$$

Factors must be introduced to take account of the other possible choices of four nucleons to constitute the particle, but A is the fundamental quantity in the formation factor. Mang obtained the general expression for the reduced width δ in terms of A, and then evaluated δ for the Po and At isotopes. It is necessary to assume wave functions for the various nuclei involved. For the α-particle Mang used a wave function that had been

---

[†] An earlier explanation, which invoked a discontinuous change in nuclear radius at $N = 126$, is apparently incorrect.

successfully employed to predict the binding energy and the charge distribution, namely,

$$\psi_\alpha = {}^1\sigma_0(1, 2){}^1\sigma_0(3, 4) \exp[-\tfrac{1}{2}\beta(x_1{}^2 + x_2{}^2 + x_3{}^2)],$$

where (1, 2) are the protons, (3, 4) are the neutrons, and

$$x_1 = 2^{-1/2}(r_1 - r_2), \qquad x_2 = 2^{-1/2}(r_3 - r_4),$$

$$x_3 = \tfrac{1}{2}(r_1 + r_2 - r_3 - r_4).$$

For heavy nuclei, simple shell-model wave functions were used, for example,

$${}^{204}\text{Po}: \quad \text{protons } h_{9/2}^2, \text{ neutrons } p_{1/2}^{-2}f_{5/2}^{-4};$$

$${}^{200}\text{Pb}: \quad \text{protons closed shell, neutrons } p_{1/2}^{-2}f_{5/2}^{-4}p_{3/2}^{-2}.$$

The radial wave functions of the single-particle states were taken as harmonic-oscillator functions, with the one adjustable parameter fixed to give a suitable rms radius for all nuclei. The values of the formation factor were calculated for $R = 9$ fm, a radius inside which most α-particle formation occurs; the penetration factors required to obtain empirical values of δ^2 have been calculated for a Coulomb barrier starting at the same radius.

A comparison of the calculated and empirical values of reduced widths is shown in Fig. 11–7. The absolute values of the calculated δ^2 can be altered by changing R and the parameters in the wave function, but the relative values are insensitive to such changes; the calculated data have been normalized by fitting the value for ^{210}Po to experiment.

It is seen that even the simple shell model provides very satisfactory agreement with the observed trend for these spherical nuclei. Mang's early calculations have been greatly extended in the direction of using Nilsson wave functions appropriate to the heavier nuclei and simultaneously introducing the configuration mixing caused by the pairing force [28]. One might expect that the superfluid smearing of the Fermi surface would be important. Consider the case of ^{243}Cm, discussed earlier (Fig. 11–6). Transitions to band A are strongly forbidden, while those to band B are unhindered; consequently, if the BCS quasiparticles should mix states in these bands to even a small extent, one might expect marked effects.

Mang and Rasmussen did find that the theoretical α-decay rates are quite sensitive to details of the pairing-force wave functions. In their later work [26] they therefore used a wave function which strictly conserves particle number; furthermore, they assumed an attractive δ-force interaction between like nucleons in Nilsson orbitals, rather than using the standard

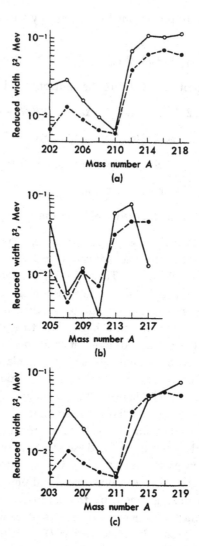

Fig. 11-7 Reduced widths of Po and At isotopes: (a) even-even Po, (b) even-odd Po, and (c) odd-even At. (After Mang [27].) The closed circles are Mang's calculated values, the open circles are empirical values. The calculated values are normalized at ^{210}Po.

assumption of a constant matrix element G between all pairing states. These calculations are rather lengthy, and it does not seem desirable to give all details here. These workers have calculated α-intensities for most of the actinide region, using the wave functions mentioned to obtain formation factors $\langle \Psi_0 | \Psi_\alpha \rangle$ and employing the Igo potential and Fröman's method to calculate barrier penetrabilities.

The quality of the agreement of these calculations with experiment is well illustrated by the case of ^{243}Cm, shown in Fig. 11–6 and Table 11–3. The rotational bands labeled A, B, C, and D are consistent with the systematics of neighboring nuclei, correspond to neutron orbitals in this part of the Nilsson diagram, and have reasonable values of the moments of inertia and of the decoupling parameter for the A-band.

TABLE 11-3 Comparison of Calculated and Experimental Values of Relative Intensities in the α-Decay of ^{243}Cm. Calculated values from Ref. 26. Experimental values from *Nucl. Data Sheets* B3-2-8 (1969), based mostly on S. A. Baranov, V. M. Kulakov, V. M. Shatinsky, *Nucl. Phys.* **56**, 252 (1964).

Daughter level		Calculated,	Experimental,
Band	Energy	%	%
A	0	0.43	1.5
A	8	2.2	4.7
A	57	0.89	1.0
A	76	1.9	5.5
A	164	0.20	0.1
A	193	0.21	0.6
A	320[a]	0.013	
B	286	76	73.5
B	330	11.5	11.0
B	388	1.8	1.6
B	463	0.022	0.01
B	436	0.004	0.002
C	392	0.068	0.2
C	434	0.028	0.14
C	486	0.008	0.02
C	542	0.00006	0.006
E	460	0.00028	0.06 or 0.03
E	490	0.00056	0.009
E	505	0.0005	0.007

[a] Calculated energy of $\frac{13}{2}+$ state. Not seen in α-decay experiments.

It is seen that the agreement is impressive, particularly since the intensities vary over 4 orders of magnitude. This constitutes a sensitive test of the wave functions, since it is to their structure that one must attribute the very marked departure from a regular energy dependence of the intensities.

The band labeled E may be a $K = \frac{1}{2}$ band, containing the 451-, 460-, 490-, and 505-keV levels. However, none of the nearby Nilsson orbitals seems to have a decoupling parameter near the value necessary to fit the level scheme. The calculations for band E in Table 11-3 are based on the assumption that the three levels listed form the $\frac{1}{2}^-[501]$ band; since the results are 1 or 2 orders of magnitude too small, this assumption is apparently wrong. Similarly the assignment of the 542-keV level to band C may be questioned because the calculated and experimental intensities differ by 2 orders of magnitude.

Although their relative intensities are in good agreement with experiment, the absolute values of decay rates calculated by Poggenburg, Mang, and Rasmussen [26] are too small by almost an order of magnitude. As we saw in Section 11-3, this is probably due to deficiencies in the α-nucleus potential and in the WKB approximation.

We may summarize this chapter by two remarks. First, we understand reasonably well the phenomena of α-decay, which occur "after the formation" of the α-particle, although great accuracy of calculation of barrier penetration may require tedious computation. Second, "formation of the α-particle" occurs continuously in the sense that the overlap $\langle \Psi_\alpha | \Psi_0 \rangle$ is nonzero, that is to say, the correlations of nucleons in the nuclear wave function Ψ_0 are such that "some of the time" four of the nucleons are an α-cluster. Moreover, the intrinsic correlations need not be known in great detail, for the wave functions of the Nilsson model with pairing-force configuration mixing are found to be quite adequate. This fact should not be very surprising; the only important aspects that such a wave function overlooks are the short-range two-body correlations which "wound" the single-particle wave functions, and these depend primarily on density, which in turn is essentially constant for nuclei throughout the α-active region. Hence, at the most, the neglect of short-range correlations in both the nuclear wave functions and the intrinsic α-particle wave functions will introduce a common normalization factor.

Alpha decay, at one time a source of puzzling phenomena, is now seen to provide a convincing confirmation of many aspects of the collective model.

REFERENCES

1. E. Rutherford and H. Geiger, *Proc. Roy. Soc. (London)* **A81**, 141, 162 (1908); also E. Rutherford and T. Royds, *Phil. Mag.* **17**, 281 (1909).
2. J. O. Rasmussen, S. G. Thompson, and A. Ghiorso, *Phys. Rev.* **89**, 33 (1953).
3. R. D. Macfarlane and T. P. Kohman, *Phys. Rev.* **121**, 1758 (1961).
4. S. Rosenblum, *J. Phys. Radium* **1**, 438 (1930), and many later papers.
5. H. Geiger and J. M. Nuttall, *Phil. Mag.* **22**, 613 (1911).
6. G. Gamow, *Z. Phys.* **51**, 204 (1928).
7. E. U. Condon and R. W. Gurney, *Nature* **122**, 439 (1928); *Phys. Rev.* **33**, 127 (1929). In addition to refs. 6 and 7, a number of accounts on α-decay theory was published in the years 1930–1937, of which the most commonly quoted are H. Bethe, *Rev. Mod. Phys.* **9**, 161 (1937), and F. Rasetti, *Elements of Nuclear Physics*, Prentice-Hall, Englewood Cliffs, N.J., 1936. Others are listed in ref. 12.
8. G. Igo, *Phys. Rev.* **115**, 1665 (1959).
9. G. H. Winslow, *Phys. Rev.* **96**, 1032 (1954).
10. I. Perlman, A. Ghiorso, and G. T. Seaborg introduced the concept of a formation factor in *Phys. Rev.* **77**, 26 (1950).
11. I. Perlman and J. O. Rasmussen, *Encyclopedia of Physics*, Vol. XLII, Springer-Verlag, Berlin, 1957, p. 109.
12. M. A. Preston, *Phys. Rev.* **71**, 865 (1947).
13. I. Kaplan, *Phys. Rev.* **81**, 962 (1951).
14. M. A. Preston, *Phys. Rev.* **83**, 475 (1951).
15. Gy. Bencze and A. Sandulescu, *Phys. Lett.* **22**, 473 (1966).
16. L. Scherk and E. Vogt, *Can. J. Phys.* **46**, 1119 (1968).
17. Gy. Bencze, *Phys. Lett.* **23**, 713 (1967).
18. M. L. Chaudhury, *J. Phys. A: Gen. Phys.* **4**, 328 (1971); *Nucl. Phys.* **76**, 181 (1966).
19. J. O. Rasmussen, *Phys. Rev.* **113**, 1593, and **115**, 1675 (1959).
20. M. A. Preston, *Phys. Rev.* **75**, 90 (1949), and **82**, 515 (1951).
21. P. O. Fröman, *K. Danske Vidensk. Selsk. Mat.-Fys. Skr.* **1**, No. 3 (1957).
22. R. R. Chasman and J. O. Rasmussen, *Phys. Rev.* **112**, 512 (1958).
23. E. M. Pennington and M. A. Preston, *Can. J. Phys.* **36**, 944 (1958).
24. J. O. Rasmussen and E. R. Hansen, *Phys. Rev.* **109**, 1656 (1958).
25. R. R. Chasman and J. O. Rasmussen, *Phys. Rev.* **115**, 1257 (1959).
26. J. K. Poggenburg, H. J. Mang, and J. O. Rasmussen, *Phys. Rev.* **181**, 1697 (1969).
27. H. J. Mang, *Phys. Rev.* **119**, 1069 (1960); *Z. Phys.* **148**, 572 (1957).
28. H. J. Mang and J. O. Rasmussen, *K. Danske Vidensk. Selsk. Mat.-Fys. Skr.* **2**, No. 3 (1962); H. J. Mang, *Ann. Rev. Nucl. Sci.* **14**, 1 (1964).

PROBLEMS

11-1. In 1964 (*Phys. Lett.* **13**, 73) Flerov et al. reported the synthesis of the nuclide $^{260}104$ in the reaction ^{242}Pu ($^{22}Ne, 4n$). The species was identified by spontaneous fission products and assigned a half-life of 0.3 s. The name kurchatovium was proposed. The nuclide ^{260}Ku should decay by α-emission to ^{256}No. From binding-energy systematics the Q-value for α-decay is estimated at 8.99 MeV.

Justify this estimate, calculate the half-life for α-decay, and comment on whether or not it should be observed.

11-2. The nuclide ^{253}Fm has a half-life of 3.0 days, decaying 88% by electron capture and 12% by α-decay. The energies and intensities of the α-groups are as follows:

E_α, MeV	I_α, %
7.092	1.3
7.032	6.7
6.952	42.7
6.910	9.8
6.876	0.9
6.855	8.4
6.682	23.2
6.659	2.4
6.639	2.6
6.550	1.5
6.496	0.3

Use these results to suggest Nilsson levels and rotational band structure for ^{249}Cf. You may find it useful to know that the ground state of ^{251}Cf has $J\pi = \frac{1}{2}^+$ and that the ground state of ^{252}Es is thought to have $J\pi = 7^+$.

Chapter 12

Fission

12-1 THE PHENOMENON OF FISSION

Hahn and Strassmann [1] were the first to establish that uranium, when bombarded with neutrons, splits into lighter elements with about half the atomic number of uranium. The phenomenon of a nucleus undergoing a division into two more or less equal fragments, either spontaneously or induced by some projectile, is termed binary fission. Less frequently, division into three fragments has also been observed [2]; this is called ternary fission. It is easy to see, from the binding-energy versus mass-number curve shown in Fig. 6–1, why fission is an energetically favorable process for heavy nuclei. The binding energy per particle, B/A, is maximum for ^{56}Fe, being about 8.8 MeV, and falls off rather slowly with increasing A. The quantity B/A in the mass region $A = 240$ is about 7.6 MeV, while it is nearly 8.5 MeV for half this mass number. The mass of the atomic species (A, Z) is simply

$$M(A, Z) = ZM_H + (A - Z)M_n - B(A, Z). \qquad (12\text{–}1)$$

It follows that, if a heavy nucleus of $A = 240$ splits into two nearly equal parts, the sum of the masses of the two fragments will be less than the mass of the fissioning nucleus by $240(8.5 - 7.6) = 216$ MeV, which should be released in the form of energy. At the instant that the division takes place (the scission point), the two fragments are in highly deformed excited states, and most of this energy is stored in the form of the deformation and excitation energy of the fragments and the mutual Coulomb energy between them, the latter being the dominant term. As a crude estimate, if we assume that the charge centers of the two fragments ($Z_1 = 50$, $Z_2 = 44$) are separated by $D = 15$ fm at the time of scission, the Coulomb energy $(Z_1 Z_2 e^2)/D \approx 210$ MeV. In the postscission stages, the two fragments recede from each other because of Coulombic repulsion, picking up velocity and at the same time collapsing to their equilibrium shapes. By the time the separation D between the charge centers has become 150 fm, the mutual Coulomb energy has reduced to a tenth of its original value, the rest being converted to kinetic energy.[†] A very simple calculation shows that this implies a fragment velocity of the order of 10^9 cm/sec, and that each fragment has covered a distance of about 70 fm in roughly 10^{-20} s, by this time already attaining 90% of the maximum kinetic energy. Moreover, each of the intermediate-mass primary fragments is neutron rich, since the neutron excess increases with atomic number.

About two to four neutrons are emitted from the fragments after these have attained their full kinetic energy, within a time of 10^{-15} to 10^{-18} s [3] after scission. In the laboratory frame the angular distribution of these neutrons is strongly peaked in the directions of fragment motion; in the frame of the moving fragment the neutron distribution is very nearly isotropic. This supports the view that these neutrons are emitted from the fully accelerated fragments themselves and not at the time of scission. These prompt neutrons have a Maxwellian distribution in energy, and on the average each neutron carries away about 2 MeV of energy. In addition, of course, it costs energy for these neutrons to "evaporate" from the fragments—about 8 MeV for each neutron. When the fragment-excitation energy has been reduced below the neutron-emission threshold, the fragments de-excite by γ-emission ($< 10^{-11}$ s) to reach their ground state. These secondary fragments (also called primary products) are far removed from the β-stability line because their charges have not been readjusted. The primary products thus undergo the slow process of β-decay ($\sim 10^{-2}$ s or more), forming stable end products. Occasionally it may happen that certain β-decay paths lead to a level that is neutron unstable, causing the emission of delayed neutrons. The delayed neutron yield is only about 1% of the total neutron emission.

Now, using mass formula 6–21, we shall investigate at what point of the periodic table it becomes energetically more favorable for a nucleus to undergo binary fission. As we have seen, the mass formula with the parameters given in Eq. 6–24 reproduces the binding energies of nuclei very well, although small corrections like those for shell effects and deformation have

---

[†] Noncentral components of the electric field can also exchange energy with the internal degrees of freedom of the fragments.

been ignored in it. For simplicity, we consider only the most stable even-even isobar with mass $M(A, Z)$ undergoing symmetric fission into two identical fragments, each of mass $M(A/2, Z/2)$. This is an oversimplification of the actual situation, since very often the fragments have unequal masses. Using Eqs. 6–21 and 12–1, we find that

$$\Delta M = M_0(A, Z) - 2M\left(\frac{A}{2}, \frac{Z}{2}\right)$$

$$= a_s(1 - 2^{1/3})A^{2/3} + \kappa_c(1 - 2^{-2/3})\frac{Z^2}{A^{1/3}}$$

$$- a_{\text{surf sym}}(1 - 2^{1/3})I^2 A^{-4/3} + a_{\text{pair}}(1 - \delta 2^{3/2})A^{-1/2}, \qquad (12-2)$$

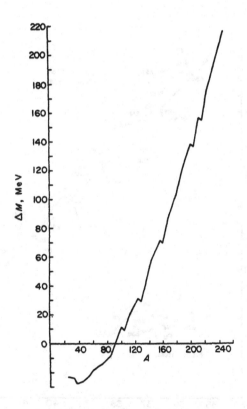

Fig. 12-1 The energy release ΔM in the symmetric fission of even-even nuclides, using Eq. 12-2. Only those even-even nuclides (A, Z) are considered which are the most stable isobars, and which yield fragments with $(A/2, Z/2)$ even-even, so that δ in Eq. 12-2 is $+ 1$. The kinks in the curve arise from discontinuous changes in $I = N - Z$ for the most stable isobars.

where $\delta = +1$, -1, or 0, according as each fragment is even-even, odd-odd, or odd, while I as before is $N - Z$. In Fig. 12-1 we plot ΔM against A, and it will be seen that ΔM becomes positive at about $A \approx 90$, indicating that binary fission becomes energetically favorable.

Using similar arguments, we find that for heavier nuclei even more energy should be released for ternary fission into three equal fragments. The experimental observations that most nuclei are stable against spontaneous fission and that even the half-life of a heavy nucleus like ^{238}U is of the order of 10^{16} years, as well as the fact that ternary fission is rather infrequent, all point to the conclusion that there is an energy barrier which the fragments must either penetrate or surmount before fission can occur. We shall discuss the origin and implications of the fission barrier in the next section.

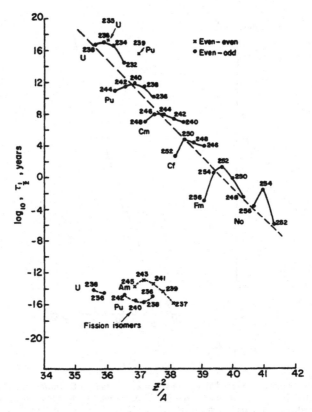

Fig. 12-2 The spontaneous fission half-lives of elements in the actinide region plotted against Z^2/A.

One can estimate the height of this barrier either from the spontaneous half-lives of heavy nuclei by making some assumptions about its shape, or more directly by inducing fission by the use of some external projectile (photon, neutron, deuteron, etc.) and obtaining the fission threshold energy. In Fig. 12-2 we plot on a logarithmic scale the spontaneous half-lives of some elements in the actinide region against Z^2/A. It will be seen that the general trend of the even-even nuclei can be roughly reproduced by a straight line, although the various even-even isotopes of a particular nuclide seem to fall on a smooth curve. Note that odd nuclides possess half-lives which are longer by a few orders of magnitude. Also shown are the lifetimes of the short-lived fission isomers, whose existence has led to new information about the barrier shapes, and which we shall discuss at some length in Section 12-5.

For the moment, we want a rough estimate of the barrier height, using the even-even nuclides as a guide. Seaborg [4] noted that the trend in the half-lives of even-even nuclides could be reproduced by the empirical formula[†]

$$\tau_{1/2} = (10^{178-3.75Z^2/A}) \cdot 10^{-21} \text{ s}. \qquad (12-3)$$

The lifetime in a spontaneous fission process will be determined by the probability of a fragment penetrating the energy barrier, which in turn depends on the shape and height of the barrier. On general grounds, we expect the probability factor for penetration to be of the approximate form $\exp(-\kappa E_b)$, where E_b is the height of the barrier with respect to the ground state of the fissioning nucleus, and κ depends on the shape of the barrier. Since the half-life $\tau_{1/2}$ is inversely proportional to the probability of penetration, we expect, using Eq. 12-3, that E_b should have a linear dependence on Z^2/A. Seaborg [4] has given an empirical formula

$$E_b = 19.0 - \frac{0.36Z^2}{A} \text{ MeV} \qquad (12-4)$$

based on such an analysis, which is approximately valid for even-even nuclei over a limited range of Z^2/A. We emphasize that Eq. 12-4 is just a rough estimate based on a one-humped barrier. We also find, from Eqs. 12-3 and 12-4, that each factor-of-10 increase in $\tau_{1/2}$ corresponds to an increase of about 0.13 MeV in the barrier height, showing the extreme sensitivity of half-life to barrier height.

[†] This form of the formula, in a simple barrier model, is derived in Section 12-3. See, in particular, Eqs. 12-26 to 12-29.

A more direct way of determining barrier height is to impart enough excitation energy to the nucleus by a projectile so that the barrier can be surmounted. A basic difficulty in this method is that there is no unique value of threshold energy, since, even if the excitation energy E_{ex} is less than E_b, fission can take place by barrier penetration. From our previous formula, below-the-barrier fission yield would be expected to rise exponentially with the increase in excitation energy, while for $E_{ex} > E_b$ the yield should level off until new reaction channels become available with increasing energy (see, e.g., Fig. 12–6).

One of the most direct methods of estimating the barrier height is by exciting the nucleus with γ-rays, although it is rather difficult to obtain a monochromatic source. As the γ-ray energy is increased from a low value, the fission yield increases exponentially and then levels off rather sharply but in a smooth fashion [5]. The bending of the fission-yield curve is also partly due to the opening up of (γ, n) channels, and it is not possible to determine E_b unambiguously. Typical uncertainties in E_b thus determined are in the range of 0.2–0.3 MeV.

The other commonly used method of determining E_b is to excite the target nucleus by a monoenergetic neutron beam. When a low-energy neutron is captured by the target nucleus, it forms a compound nucleus, with the energy of the incident neutron redistributed by many collisions over a time period very long ($\sim 10^{-14}$ s) compared to the time in the nuclear scale ($\sim 10^{-22}$ s). Above a certain threshold of neutron energy fission becomes dominant, while below this energy the compound nucleus de-excites by electromagnetic processes. A zero-energy incident neutron imparts an excitation energy B_n to the compound nucleus, where B_n is the binding energy of the neutron in the compound nucleus. Consequently, if the threshold kinetic energy of the neutron is E_n, the barrier height $E_b = E_n + B_n$. It is necessary, therefore, to know B_n from other sources in order to estimate E_b by this method. Furthermore, if the compound nucleus has an even number of neutrons, B_n may be larger than E_b and even thermal neutrons can induce fission—in such cases this method is not applicable. On the other hand, the method works for a target nucleus like ^{234}U, where the compound nucleus ^{235}U has an odd number of neutrons, and $B_n = 5.3$ MeV. In this case, the threshold energy E_n is ~ 0.5 MeV, giving the barrier height in ^{235}U as 5.8 MeV. For most nuclei in the actinide region, E_b is between 5 and 6 MeV; for Cf and Fm isotopes the fission barrier is smaller, in the range 2.5–4 MeV.

For the case in which $E_b < B_n$, an ingenious method of measuring the threshold has been devised by Northrup, Stokes, and Boyer [6], using the (d, pf) reaction. Fission thresholds are obtained by measuring the energy spectrum of protons in coincidence with fission events induced by deuterons of known kinetic energy E_d. If the outgoing proton energy is E_p, and the deuteron binding energy is ε_d, the captured neutron has an equivalent kinetic energy E_n, given by

$$E_d - E_p = \varepsilon_d + E_n. \qquad (12\text{-}5)$$

The excitation energy of the compound nucleus, as before, is

$$E_{ex} = E_n + B_n = (E_d - E_p - \varepsilon_d) + B_n. \qquad (12\text{-}6)$$

The point to note is that E_n now may be negative in cases where nearly all the kinetic energy is taken away by the proton, so that $E_d - E_p < \varepsilon_d$. In this case the excitation energy $E_{ex} < B_n$, and by measuring the E_d and E_p at fission threshold it is possible to estimate the barrier height from Eq. 12-6. We shall discuss the results of such experiments later in this section while considering the variation of fission cross section with excitation energy.

We have already noted that in actinide nuclei with an odd number of neutrons (e.g., ^{233}U, ^{235}U, ^{239}Pu) a large fission rate can be induced by thermal

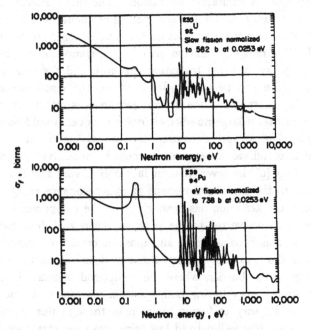

Fig. 12-3 Fission cross section σ_f as a function of neutron energy in the energy range 10^{-3} to 10^4 eV, incident on (a) ^{235}U and (b) ^{239}Pu. (After Hyde [7].)

neutrons, a fact which makes them technologically important. Some insight into the fission process is gained by examining the variation of the fission cross section with increasing neutron energy in this range. In Figs. 12–3a and 12–3b we plot the fission cross section σ_f of ^{235}U and ^{239}Pu against the incident neutron energy in the energy range 10^{-3} to 10^4 eV [7]. It will be seen that there is a drastic drop in σ_f from about 1000 b at the lower energies to less than 10 b in the keV region.

From 0 to 0.2 eV, the cross section drops inversely with the velocity v of the incoming neutron. This reflects the fact that the slower neutrons spend a larger fraction of time in the vicinity of the target nucleus and therefore have a greater probability of being captured and imparting the excitation energy for fission.

Beyond 0.2 eV, there are many sharp resonances in the cross section. These occur when the incident kinetic energy plus the binding energy of the neutron equals the energy of one of the excited states of the compound nucleus, the density of such states being very large at an excitation energy of about 6 MeV. If the width of a resonance in the fission cross section is denoted by Γ_f, then $\Gamma_f/\hbar$ is the probability per unit time that the compound nucleus will lose its excitation by fission. The many resonances that are found in this region have an average spacing of about 1 eV, and the widths, which vary from resonance to resonance, are a few tenths of an electron volt or less. In this range of neutron energy, the compound nucleus can also de-excite by γ-ray emission (radiative capture), and the ratio of radiative capture to fission can be substantial (0.69 for the resonance at 0.296 eV in ^{239}Pu). The sharpness of the fission resonances is rather surprising, since it would seem that a large number of fission channels would be open to the compound nucleus at such excitation energies, giving rise to a large fission width, and that individual resonances would be unresolved. It was pointed out by A. Bohr [8], however, that in fact only a very few fission channels are available to the compound nucleus at each resonance energy. Although about 6 MeV of excitation has been given to the compound nucleus, most of this energy is converted to the potential energy of deformation as the compound nucleus becomes more and more deformed on its way to fission. Bohr visualized a "transition state" of the nucleus: an unstable equilibrium near the top of the barrier, where the compound nucleus is "cold," most of its excitation energy having been spent in reaching this deformed state. A nucleus, on its way to fission, must pass through this transition state, which has only a few well-defined low-lying quantum states available to it, because of its low excitation (see Fig. 12–4). Further restrictions on the

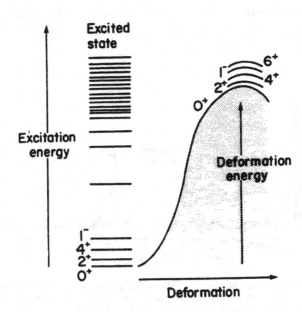

Fig. 12-4 Schematic view of the transition states of a nucleus in the liquid-drop picture. (After Hyde [7].)

levels of the transition state through which fission can take place are imposed by the conservation of angular momentum, a point that we shall discuss at some length later.

An interesting phenomenon is observed when low-energy neutrons (0.2–8 keV) bombard an even-neutron nucleus like ^{240}Pu. In this case, the fission is in the subthreshold region, since the excitation energy imparted is less than the barrier height. The plot of fission cross section in Fig. 12–5 demonstrates that the fission widths are abnormally large (~ 50 eV) for some approximately equally spaced groups of resonances (at intervals of ~ 650 eV), and are extremely small between them, instead of being randomly distributed. This phenomenon, known as an intermediate structure effect, cannot be explained by the conventional one-humped barrier. We shall discuss the implications of this finding in Section 12–5 when we consider the barrier shape in detail.

In Fig. 12–6, we have plotted the fission cross section of ^{238}U in the MeV region, as an example of a nuclide that has a threshold energy in this range. In the subthreshold range of energies, the cross section rises exponentially, until neutron emission becomes possible. The competition from neutron re-emission gives rise to kinks in the curve below the fission threshold.

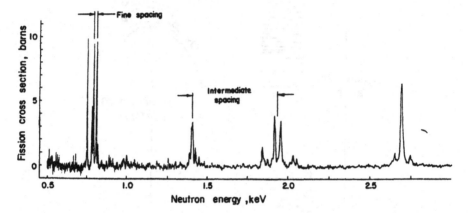

Fig. 12-5 The fission cross section σ_f of ^{240}Pu as a function of incident neutron energy, showing the grouping of resonances at regular intervals with large widths. [After E. Migneco and J. P. Theobald, *Nucl. Phys.* **A112**, 603 (1968).]

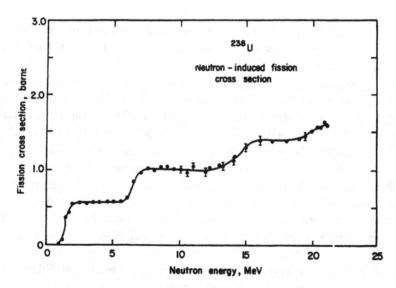

Fig. 12-6 Fission cross section σ_f for n-induced fission of ^{238}U as a function of energy. (After Hyde [7].)

The neutron-fission cross section in the MeV region is governed by the ratio of the widths Γ_f/Γ_n for fission as compared with neutron emission. The curve beyond the threshold is flat up to about 6 MeV, where there is again a sharp rise. At this point, the excitation energy is large enough to permit a neutron emission as well as fission over the barrier, so that a second reaction channel (n, nf) has opened up, giving increased cross section. The rises in the cross section at 14 and 20 MeV correspond to the opening up of "third-try" $(n, 2nf)$ and "fourth-try" fission.

An interesting phenomenon is observed when ^{230}Th is bombarded by neutrons in the energy range 690–1230 keV [9]. The fission threshold in this case is about 1 MeV. As shown in Fig. 12–7, a well-developed maximum in σ_f is observed in the subthreshold region around 750 keV, and careful analysis shows [10] that this cannot be attributed to neutron-emission competition. One is forced to assume that the fission width Γ_f has a nonmonotonic

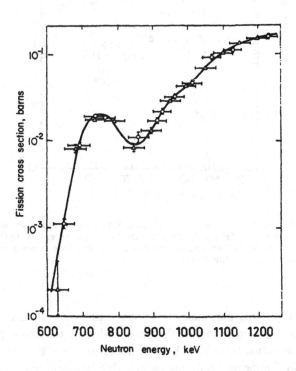

Fig. 12-7 Neutron-induced fission cross section of ^{230}Th, showing a marked resonance in the subthreshold neutron energy, around 750 keV. (After Vorotnikov et al. [9].)

dependence on the excitation energy, a behavior not allowed by a simple one-humped barrier. This anomaly, which is believed to be due to vibrational resonances in a double-humped barrier, will be discussed later. Further examples of such vibrational resonances have been found in the (d, pf) reactions in the compound nuclei ^{234}U, ^{236}U, ^{240}U, and ^{242}Pu [11]. In Fig. 12–8,

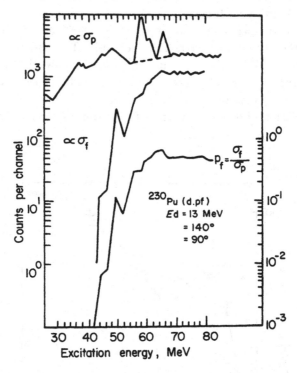

Fig. 12-8 The proton yield, fission yield, and fission probability P_f when 13-MeV deuterons are incident on ^{239}Pu, plotted against the excitation energy. Note the resonance-like structure around 5-MeV excitation, where neutron emission is not possible. (After J. Pedersen and B. P. Kuzminov [11].)

as an example of this phenomenon, we have plotted the curves of proton yield, coincident fission yield, and their ratio, which is the fission probability, against the excitation energy for 13-MeV deuterons on ^{239}Pu. The resonance at $E_{ex} \approx 5$ MeV appears below the neutron binding energy of 6.4 MeV and therefore cannot be explained by neutron competition. We shall see later that these resonances have a natural explanation when the barrier shape has two minima, giving rise to vibrational states. The peculiar barrier shapes arise because of shell effects, which persist even for large deformations, and which we shall discuss after describing the simple liquid-drop picture of fission.

12-2 FISSION-FRAGMENT CHARACTERISTICS

In this section we discuss the mass and energy distributions of the fragments in binary fission and examine in more detail some properties of the emitted prompt neutrons. It is important, while discussing the experimental data, to distinguish between mass distributions before and after neutron emission—the difference between the two gives us information about the number of prompt neutrons emitted. Radiochemical and mass-spectroscopic analyses of the fission fragments are done with the final fission products or the cumulative yields. The post-neutron-fragment kinetic energies can also be obtained by detecting the fission fragments with solid-state detectors after the fragments have traversed a path length of a meter or two (i.e., after about 10^{-7} s from the instant of fission). The solid-state detectors have good counting efficiency, and the pulse height for a given ion mass is linear with energy; thus an analysis of the pulse heights with a calibrated detector enables one to obtain the kinetic energies E_{k_1} and E_{k_2} of the fragments *after* the prompt neutrons have been emitted.

Let the pre-neutron-emission fragment masses be m_1^* and m_2^*, with kinetic energies $E_{k_1}^*$ and $E_{k_2}^*$. The corresponding unstarred symbols refer to post-neutron-emission quantities. Then momentum and mass conservation imply

$$m_1^* E_{k_1}^* = m_2^* E_{k_2}^*, \tag{12-7}$$

$$m_1^* + m_2^* = A, \tag{12-8}$$

and

$$m_i = m_i^* - \nu_i, \qquad i = 1, 2, \tag{12-9}$$

where A is the mass of the fissioning nuclide, and ν_i is the number of neutrons emitted from the ith fragment. It follows immediately from Eqs. 12–7 to 12–9 that

$$m_1^* = \frac{A E_{k_2}^*}{E_{k_1}^* + E_{k_2}^*}, \tag{12-10}$$

and a similar relation exists for m_2^*.

To obtain the initial fragment masses m_1^* and m_2^*, it is therefore necessary to know the pre-neutron-emission kinetic energies. Experimentally, however, it is only possible to measure the post-neutron energies E_{k_1} and E_{k_2}. Schmitt, Neiler, and Walter [12] have developed a widely used approximate procedure to deduce m_1^* and m_2^* from the experimentally measured quantities. First, it is assumed that, although the kinetic energy of each fragment changes because of neutron emission, its velocity remains almost unaltered, that is, $v_i^* = v_i$. The plausibility of this assumption relies on the conservation of momentum, coupled with the fact that the neutrons evaporate isotropically from the fragment in its own reference frame and hence have an average velocity in the laboratory frame that is the same as the fragment velocity. It then follows that

$$E_{k_i}^* = E_{k_i}\left(\frac{m_i^*}{m_i}\right) \tag{12-11}$$

and

$$m_i^* = \frac{AE_{k_2}}{E_{k_2} + E_{k_1}(1 + \xi_1)}, \tag{12-12}$$

where

$$\xi_1 = \frac{1 + v_1/m_1}{1 + v_2/m_2} - 1 \approx \frac{v_1}{m_1} - \frac{v_2}{m_2}. \tag{12-13}$$

One could, therefore, obtain m_1^* and m_2^* from a measurement of E_{k_1} and E_{k_2} provided that the values of v_1 and v_2 were known as a function of fragment mass and energy. However, usually experiments have been able to determine only the average number $\bar{v}$ of neutrons emitted from a fragment of known mass, but distributed over the available energies. Consequently, some approximations must be made. The simplest comes from putting the small number $\xi_1 = 0$; the resulting m_i^* are referred to as "provisional" masses in some papers in the literature. Another approximation, which is sufficiently accurate for most purposes, is to replace v_i by $\bar{v}_i$. Fission-fragment double energy measurements have been made by several groups, using silicon-surface-barrier detectors and the fission yield $N(m^*)$ deduced from the above analysis. Work along these lines on several different nuclides has been done by Schmitt et al. [12], Bennet and Stein [13], Pleasonton [14], and John et al. [15]. Before discussing these results, we outline the other commonly used method of determining the pre-neutron-emission fragment distribution.

The most direct method of obtaining the primary fragment characteristics is to determine the velocities of the two fragments by time-of-flight measurements, the flight path of each fragment being about 2 meters. The fragments attain their full velocity when their separation is a few thousand fm, that is, within a time period of about 10^{-19} s, after which they should travel with uniform velocities through vacuum, provided their motion is not altered by subsequent neutron emission or through scattering from the walls of the flight tube. If, as before, it is assumed that $v_i = v_i{}^*$, then, since momentum conservation implies that

$$m_1{}^* v_1{}^* = m_2{}^* v_2{}^*, \qquad (12\text{--}14)$$

the measured velocities v_1 and v_2, together with relation 12–8, immediately

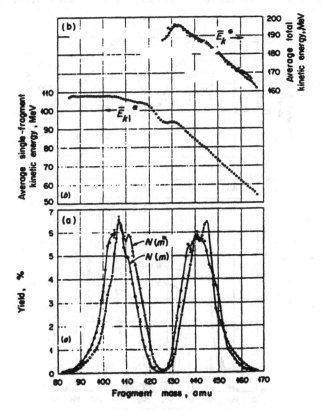

Fig. 12-9 Spontaneous fission of ^{252}Cf. (a) Pre-neutron-emission mass distribution $N(m^*)$ and post-neutron-emission mass distribution $N(m)$ as functions of fragment mass. (b) Average single-fragment and total pre-neutron-emission kinetic energy as functions of mass. (After Schmitt et al. [12].)

give the pre-neutron-emission fragment masses $m_1{}^*$ and $m_2{}^*$. The double velocity measurements were pioneered by Stein [16] and Milton and Fraser [17, 18]. The latter workers studied the spontaneous fission of ^{252}Cf [17] and the thermal-neutron-induced fission of ^{233}U, ^{235}U, and ^{239}U [18]. Their results were somewhat inaccurate [12] because of the effects of fragment scattering from the walls of the flight tubes. The subsequent experiment of Whetstone [19] on the spontaneous fission of ^{252}Cf was free of such errors, and more accurate measurements on ^{235}U $+ n$ have been made by Signarbieux, Ribrag, and Nifenecker [20].

We now present a few experimental results. A very large body of experimental data is now available [21], but we shall confine our discussion largely to spontaneous fission and fission induced by low-energy neutrons. The most striking characteristic in these cases is the mass asymmetry of the

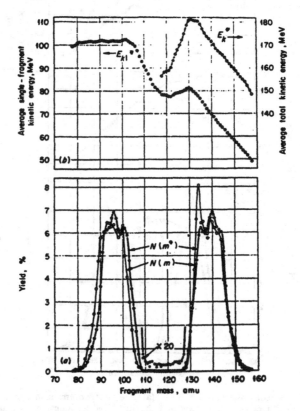

Fig. 12-10 Thermal-neutron-induced fission of ^{235}U. Captions for (a) and (b) are the same as in Fig. 12-9. (After Schmitt et al. [12].)

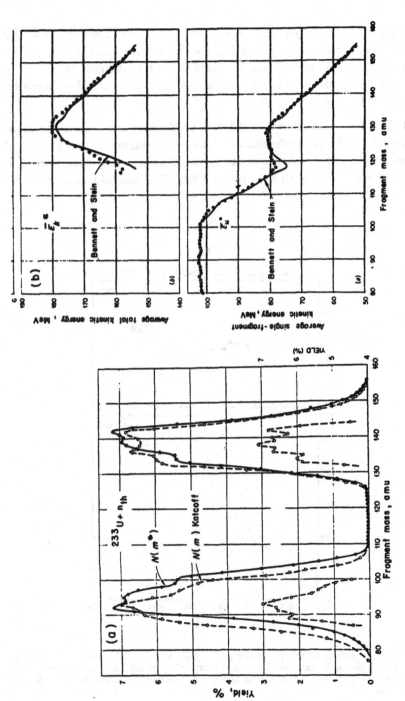

Fig. 12-11 Thermal-neutron-induced fission of ^{233}U. Captions for (a) and (b) are the same as in Fig. 12-9. (After F. Pleasonton [14].)

fragments. This is illustrated in Figs. 12–9 to 12–11, where the pre- and post-neutron-emission fission yields, $N(m^*)$ and $N(m)$, are shown for the spontaneous fission of ^{252}Cf and the thermal-neutron-induced fission of ^{235}U and ^{233}U, respectively. The following characteristics should be apparent from these figures:

1. The most probable mode of fission in all three cases is with an asymmetric mass split of the fragments. The heavy-fragment mass peak in all appears around $A \sim 140$, while the light mass peak shifts from $A \sim 95$ in ^{233}U $+ n$ and ^{235}U $+ n$ to $A \sim 108$ in ^{252}Cf. This general feature is further

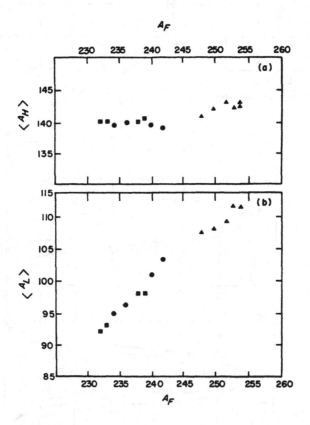

Fig. 12-12 Positions of the (a) heavy peak $\langle A_H \rangle$ and (b) light peak $\langle A_L \rangle$ in initial mass distribution as a function of the mass number A_F of the fissioning nucleus. (After Pappas et al. [21].) ▲: spontaneous fission; ●: thermal neutron fission; ■: fast neutron fission.

illustrated in Fig. 12–12, where we plot the positions of the heavy peak, $\bar{A}_H$, and the light peak, $\bar{A}_L$, in the initial mass distribution as a function of the mass number A_F of the fissioning nuclide. Whereas the heavy peak remains substantially at the same location, $\bar{A}_H \sim 140$, the position of the light peak keeps shifting.

2. There is considerable fine structure in the peaks of both $N(m^*)$ and $N(m)$, with $N(m)$ shifted slightly to the left of $N(m^*)$, showing the effect of neutron emission.

3. The fine-structure peaks of $N(m^*)$ in $^{233}U + n$ appear at heavy-fragment masses $A \sim 134$, 139, and 142, at roughly the same mass numbers as in $^{235}U + n$. For ^{252}Cf, the peaks are again at $A \sim 140$ and 145. These indicate a periodicity of about 5 mass units in the fine-structure peaks. We shall discuss the implications of this observation shortly.

In Figs. 12–9 to 12–11, we have also plotted the average single-fragment and average total pre-neutron-emission kinetic energies as functions of the fragment mass. The lighter fragment is seen to have more kinetic energy, and its value for a particular fissioning nuclide is nearly a constant. The heavy-fragment kinetic energy falls steadily, with increasing fragment mass. The total pre-neutron-emission kinetic energy, $\bar{E}_k^*$, has a dip near symmetric fission in all three cases, although for ^{252}Cf this dip is only about 8 MeV, while for the two lighter fissioning nuclides it is approximately 22 MeV. In Fig. 12–13 we plot the average number of neutrons emitted per fragment and the average number of neutrons emitted per fission (i.e., by both the fragments) as functions of the fragment mass for thermal-neutron-induced fission of ^{233}U, ^{235}U, and ^{239}Pu [22]. In Fig. 12–14 the average neutron yield per fragment as a function of fragment mass is shown for the spontaneous fission of ^{252}Cf [23]. In all these curves of sawtooth shape, it will be noted that (a) the average number of neutrons emitted is minimum when the fragment mass is about 132, and (b) the maximum number of neutrons is emitted near symmetric fission.

It is worth while to discuss now the above experimental results in more detail. First, we make some comments on the asymmetry in the mass distribution of the fragments. From the empirical liquid-drop mass formula 6–21, there is no reason to expect this asymmetry, as the calculated energy release from this formula is found to be maximum for symmetric fission. We may then suspect that the mass asymmetry may be arising because of shell effects, implying thereby that shell effects persist even at the scission point, when the nucleus is in a highly deformed configuration.

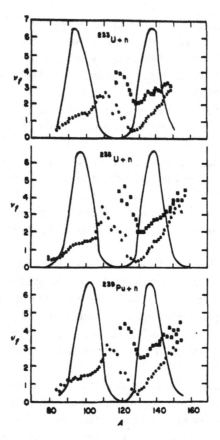

Fig. 12-13 Neutron yields of thermal-neutron-induced fission of ^{233}U, ^{235}U, and ^{239}Pu as a function of fragment mass. ● : The average number of neutrons emitted per fragment; ◆ : the average number of neutrons emitted per fission as a function of the heavy-fragment mass. The mass distributions are shown for comparison. (After Apalin et al. [22].)

As will be seen from Figs. 12–9 to 12–11, appreciable asymmetric fission yield starts building up only when one of the fragments has at least 82 neutrons and 50 protons (two magic numbers), that is, beyond the heavy-fragment mass number 132.

As already discussed in Chapter 6, shell effects in the binding energy of a nucleus arise because of the bunching of single-particle quantum levels, due to the motion of a nucleon in an average potential well. This average potential well should change only slowly in a time compared to the time required for a nucleon to cross the nuclear diameter, which is of the order

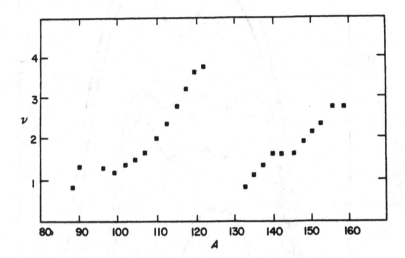

Fig. 12-14 Average neutron yield per fragment as a function of the fragment mass in the spontaneous fission of ^{252}Cf. (After Bowman et al. [23].)

of 10^{-22} s. During the time of scission, the nuclear shape is changing rapidly, and a neck connecting the two fragments is formed at this stage and snapped at scission. Shell effects in the fragments should show up in fission if the scission time (the time from when the nucleus is in the quasi-stationary state at the top of the barrier to the snapping of the neck) is long compared to $\sim 10^{-22}$ s, so that the nucleonic motion in each fragment can be regarded as being in a slowly changing average potential. The evidence that there are shell effects in the fragments points to the conclusion that the scission time may be long compared to 10^{-22} s. Further evidence of this is the near isotropy of the neutrons emitted from a fragment in its frame of reference, indicating that the neutrons are emitted after scission. If the snapping of the neck is very sudden, some neutrons would be expected to acquire enough excitation energy to escape at the point of scission. Fuller [24] has estimated that, if complete severance of the neck takes place within 2.5×10^{-22} s, it would result in the ejection of about 1.5 neutrons per fission in ^{252}Cf at the time of scission. Assuming that the angular distribution and energy of such neutrons would not by accident be such as to make them indistinguishable from evaporation neutrons, we conclude that the severance of the neck must be more gentle. We should point out that the

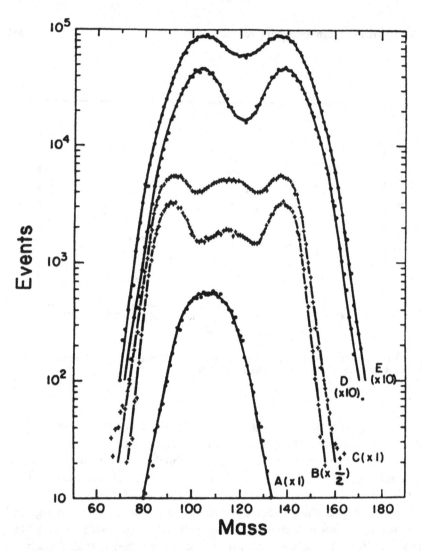

Fig. 12-15 Total initial mass yield distribution. For convenience of display, the number of events for each system has been multiplied by a scale factor given in parentheses. (A) ^{209}Bi (42.0-MeV α, f); (B) ^{226}Ra (30.8-MeV α, f); (C) ^{226}Ra (38.7-MeV α, f); (D) ^{238}U (29.4-MeV α, f); (E) ^{238}U (42.0-MeV α, f). (After J. P. Unik and J. R. Huizenga [29].)

experimental data are not completely consistent with isotropy of the emitted neutrons, and it is believed that about 10% of the emitted neutrons in ^{252}Cf originate from the fissioning nucleus rather than from the fragments [25].

Although the scission time may be longer than 10^{-22} s, as indicated here, it is doubtful whether a statistical equilibrium is established at scission, as claimed by Fong [26]. Up to the present time, there is no satisfactory *dynamical* theory of fission that describes the asymmetric mass split of the fissioning nucleus, but one may obtain some insight by plotting the static energy surfaces of the distorted nucleus, including the shell correction, as functions of the deformation parameters [27, 28], which we discuss in Section 12–5.

It is not always the case that the mode with asymmetric mass splitting is more probable than the symmetric one. When fission is induced by imparting enough excitation energy to target nuclei with $Z \leqslant 83$, the mass distribution in general is symmetric. Since the height of the fission barrier in these ,nuclei is rather large, considerable excitation energy has to be supplied. In Fig. 12–15, we show the fragment mass distribution obtained with 42-MeV ^{4}He ions incident on a ^{209}Bi target, leading to the fission of ^{213}At, and it is seen to be symmetric. When fission is induced in nuclides in the actinide region by imparting large excitation energy, the fine structures in the mass distribution disappear, and symmetric fission becomes nearly as frequent as asymmetric fission. This is also illustrated in Fig. 12–15 for ^{226}Ra and ^{238}U for ^{4}He-ion-induced fission [29].

Another interesting finding in regard to the mass distribution of the fragments has been reported by John et al. [15]. They find that thermal-neutron-induced fission of ^{257}Fm is strongly symmetric, as shown in Fig. 12–16. On the other hand, the spontaneous fission of ^{257}Fm still exhibits asymmetry, as shown in Fig. 12–17, although symmetric fission is also quite frequent. These mass distributions were derived from double energy measurements on coincident fission fragments by solid-state detectors, and the masses correspond to provisional masses (i.e., those with $\xi_1 = 0$ in Eq. 12–13). The most interesting point is that symmetric mass distribution in the low-energy fission of heavy actinides seems to appear abruptly at ^{257}Fm, since ^{256}Fm

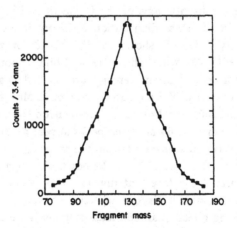

Fig. 12-16 Mass distribution for the thermal-neutron-induced fission of ^{257}Fm. (After John et al. [15].)

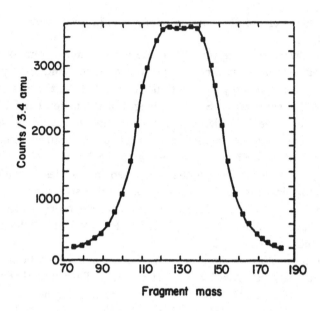

Fig. 12-17 Mass distribution for the spontaneous fission of ^{257}Fm. (After John et al. [15].)

fissions asymmetrically. This observation lends support to the occurrence of double-humped fission-barrier shapes due to shell effects and will be further discussed in Section 12–5.

We now make a few comments on the fine structure observed in the initial and final mass distributions of the fragments. The fine structure is more pronounced in the mass distribution of heavy fragments having the same kinetic energy. This is shown in Fig. 12–18 for thermal-neutron-induced fission of ^{235}U [30], where the initial and final fission yields of heavy fragments having a certain kinetic energy are plotted against the heavy-fragment mass number A_H. For a kinetic energy of 72 MeV there are distinct peaks at $A_H \approx 130$, 135, and 140 in the pre-neutron-emission distribution, while for a kinetic energy of 64 MeV the peaks appear at 140, 146, and 151. The distributions of the final masses are shifted somewhat to the left because of neutron emission. Thomas and Vandenbosch [31] correlated the peaks in the fine structure to the fact that division into even-even fragments is energetically more favorable in fission, since these have more binding energy, because of the pairing effect, than odd-mass fragments. Furthermore, for a given charge of the even-even heavy fragment, the binding energy reaches

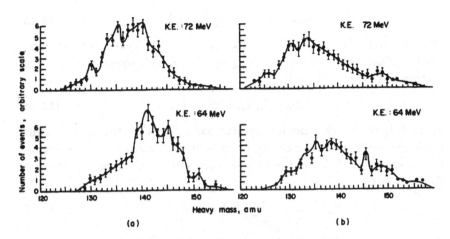

Fig. 12-18 The initial (a) and final (b) mass distributions for fixed kinetic energies in the thermal-neutron-induced fission of ^{235}U. Indicated errors are statistical ones. (After G. Andritsopoulos [30].)

a maximum for some value of A_H. For example, this maximum for $Z_H = 58$ comes near $A_H = 150$, for $Z_H = 56$ at $A_H \approx 145$, and for $Z_H = 54$ at $A_H \approx 140$ [31]. The peaks in the fine structure thus appear at the same values of A_H at which the even-even mass surface has peaks. Hence it appears that pairing and symmetry energy effects (which are the most rapidly varying terms in the mass formula when Z is fixed) are still operative at the scission point. Since both these effects are related to the concept of a Fermi surface, this lends credence to a description of the nucleus in terms of single-particle orbitals, even at this stage in the fission process. It is amusing that historically the fine structure in the fission-fragment mass distribution was better understood than the gross structure itself with its accompanying mass asymmetry.

Figures 12–9 to 12–11 also show the variation of the total pre-neutron-emission kinetic energy $E_k{}^*$ as a function of the fragment mass for ^{252}Cf, ^{233}U + n, and ^{235}U + n. If we simply equated $E_k{}^*$ to $e^2 Z_1 Z_2/D$, where D is the separation of the charge centers at scission, $E_k{}^*$ would be a maximum for symmetric fission with $Z_1 = Z_2$ (assuming that D is approximately constant and $Z_1/A_1 = Z_2/A_2$). We find from the experimental data, however, that $E_k{}^*$ has a marked dip for symmetric fission and is maximum for $A_H \approx 132$. Milton and Fraser [18] correlated this behavior of $E_k{}^*$ with the energy that goes into fragment excitation. If E_R is the energy release in fission, then

$$E_R = E_k{}^* + E_{ex,H}^* + E_{ex,L}^*, \qquad (12\text{--}15)$$

where the subscripts H and L refer, as before, to the heavy and light fragments. The excitation energy $E_{ex,H}^* + E_{ex,L}^*$ is subsequently released in neutron and γ-emission, so that

$$E_{ex,i}^* = \bar{\nu}_i(\bar{B}_{ni} + \bar{E}_{ni}) + \bar{E}_{\gamma i}, \qquad (12\text{--}16)$$

where $\bar{\nu}_i$ is the average number of neutrons emitted by the ith fragments, $\bar{B}_{ni}$ is the average binding energy of an emitted neutron, $\bar{E}_{ni}$ is its average kinetic energy, and $\bar{E}_{\gamma i}$ is the average energy released through γ-emission. We thus obtain

$$E_k{}^* = E_R - \sum_{i=H,L} [\bar{\nu}_i(\bar{B}_{ni} + \bar{E}_{ni}) + \bar{E}_{\gamma i}]. \qquad (12\text{--}17)$$

For given A_H, A_L, Milton and Fraser [18] took for E_R the maximum energy released, averaged over the various charge distributions Z_H, Z_L. The quantities $\bar{E}_n$ and $\bar{E}_\gamma$ were assumed to be independent of the fragment mass and were given values of 1.2 and 7.5 MeV, respectively, for ^{233}U as well as ^{235}U. By taking reasonable values of $\bar{B}_n$, Milton and Fraser could reproduce the variation of $E_k{}^*$ with A_H. For symmetric fission, as will be seen from Fig. 12–14, ν is maximum, giving a dip in $E_k{}^*$. For $A_H \sim 132$, ν is minimum, and this gives rise to the peak in $E_k{}^*$. For this value of A_H, the heavy fragment is doubly magic ($N_H = 82$, $Z_H = 50$) and is not easily deformable; consequently it does not take up much excitation energy and the neutron emission is minimum.

In the preceding discussion, we arbitrarily assumed at one point that the charge division in the fragments is such that $Z_1/A_1 = Z_2/A_2 = Z/A$. Experimentally, this is found to be a good first approximation. More precisely, for a given fragment mass A_1, the charge distribution of the fragment is found to be Gaussian, with its peak at $Z_1 = (Z/A) \cdot A_1$.

We end this section with a few comments about ternary fission, for which a fair number of experimental data exist [32]. According to Feather [2], ternary fission is a process in which three fragments appear within a time period of 10^{-18} s after scission. By far the most frequent third fragment is observed to be an α-particle, the other two being heavy fragments. The fractional yield of ternary fission in induced fission of the actinide nuclei is found to be about 2.1×10^{-3} for all of the species and is independent of the excitation energy between 6 and 40 MeV. This means that about 2 ternary fissions are observed, with one fragment an α-particle, in 1000 fission processes. The energy distribution of these emitted α-particles has been studied experimentally, although the necessity of shielding the detectors from natural α-particles and other background effects has resulted, most often, in the omission of the low-energy part of the spectrum. The energy spectrum of the α-particle has a Gaussian shape, centered at ∼ 16 MeV. Measurements of the angular distribution of the emitted α-particles reveal a peak at about 82° with respect to the direction of the lighter of the two heavy fragments. The variation of the α-energy with angle of emission has also been measured [33]. Using these input data, computer calculations have been made [33] to back-track the trajectory of the α-particle, obtaining the initial dynamical parameters that yield the observed distribution. The data are consistent with the following initial conditions: the α-particle is emitted isotropically at the instant of scission or within 10^{-21} s of it, in the region between the two heavy fragments, and has an initial kinetic energy of 2–3 MeV.

Clark and Kugler [34] have made a careful study of the energy spectrum of the α-particles emitted from $^{235}U + n$, including the low-energy part of the spectrum. In addition to the peak at ∼ 16 MeV for the long-range α-particles, they find almost an equal number of short-range α-particles with energy < 8 MeV. This is a most puzzling phenomenon, since if an α-particle was initially in the Coulomb field of the two fragments, it should pick up appreciable kinetic energy at the detection point. It may be, as suggested by Kugler [35], that these α-particles are emitted at a very late stage of the scission, so that the distances between them and the fragment charge centers are much larger than the corresponding values for the long-range α-particles. More work on the low-energy distribution of the emitted α-particles may help to shed further light on this problem.

12-3 THE LIQUID-DROP MODEL

From a microscopic point of view, the phenomenon of fission is extremely complicated. As the fissioning nucleus is torn apart in a very short time, the motion of each individual nucleon is radically altered. The entire interval from the instant that some energy is imparted to the nucleus (e.g., by a neutron capture) to the time at which it is torn apart (scission) may be divided into two parts.

During the first interval of time, the imparted energy is redistributed among the nucleons over a comparatively large period ($\sim 10^{-15}$ s) involving many nucleonic collisions. During this time, the energy that goes into the collective degrees of freedom causes increased deformation of the nucleus, and this change in deformation takes place very slowly compared to the time of individual nucleonic motion. If none of the individual nucleons attains enough energy to escape the nucleus (e.g., by neutron emission), the imparted energy may be nearly all spent in causing a large critical deformation, beyond which the system is unstable. This metastable state is reached at the saddle point, at the end of the first interval of time. Once the nucleus is over the saddle point, shape instability takes place, and the nucleus can distort very easily to reach the point of scission.

The second time interval, from the saddle point to scission, is very short ($\sim 10^{-21}$ s), and the separation of collective and intrinsic coordinates may not be meaningful at this stage, although there is evidence for the persistence of shell effects even here, as discussed in Section 12-2. (See the introduction of Chapter 9 for a discussion of the adiabatic approximation.)

In this section, we shall concentrate our attention on the collective motion of the nucleus up to the saddle point, where we can still meaningfully describe the motion in terms of collective variables. There are two aspects of the problem: the statics (i.e., the potential-energy surfaces of the nucleus with increasing deformation) and the dynamics—its equation of motion. In the absence of any complete theory, it is instructive to construct intuitive models of fission to see whether the salient experimentally observed characteristics can be extracted, and the liquid-drop model is an attempt in this direction. Immediately after the discovery of fission, Meitner and Frisch [36] emphasized the analogy of the process concerned with the division of a charged liquid drop into two smaller droplets due to deformation caused by an external impact. Bohr and Wheeler [37] gave an account of the mechanism of fission based on the statics of the model by calculating the potential-energy surfaces of a charged deformed liquid drop. A more ambitious project was undertaken by Nix and Swiatecki [38] and Nix [39], who traced out in detail both the static and the dynamic characteristics of the division of an idealized (nonviscous, irrotational) liquid drop whose size, surface tension, and charge are those of a nucleus, and compared the results with what is observed experimentally in the fission of real nuclei. This simple model is reasonably successful in describing the fission of nuclei lighter than radium, where the mass distribution of the fragments is symmetric, but fails for heavier nuclei. Despite its failure in the most interesting region of nuclei, the model has two important factors in its favor. The first is its conceptual simplicity, although the actual calculations (particularly the dynamics) are quite complicated. The second factor is that its failure in the heavy-mass region indicates the importance of the missing ingredients (e.g., single-particle effects like shell correction) in the simple model. Recent progress in the understanding of fission phenomena has consisted mainly in the modification of the statics to account for shell effects, and has led to better agreement between the predictions of theory and the findings of experiment.

Before doing any calculation with the model, we should first inquire as to why a charged liquid drop should behave in any way like a nucleus [38]. We shall be particularly concerned with the variation of the potential energy with deformation. There are two important characteristics of the nucleus that we must keep in mind. First, it is reasonable to assume that the nucleus is incompressible, so that the total nuclear volume remains unaltered in fission, and the average nuclear density ρ is identical in the primary fragments and in the fissioning nucleus. Second, although the two-nucleon force is rather complicated, its range r_n is small compared to the nuclear radius R_0 of the heavy nuclei being considered. Consequently, the nuclear part of the potential energy can be expressed as the sum of a volume term proportional to A (independent of the shape of the nucleus), a surface term proportional to $A \cdot (r_n/R_0)$ (shape dependent), and terms containing higher powers of r_n/R_0, which are smaller in magnitude (see Eq. 6–9). There are also the symmetry terms (see Eq. 6–21) proportional to $(N - Z)^2$. The volume part of the symmetry energy is shape independent, while the surface-symmetry term has the same shape dependence as the main surface term. The symmetry effects can be taken into account by replacing the coefficients a_v and a_s in Eq. 6–21 with C_v and C_s, as is done in Eq. 6–22. For a nucleus in the actinide region, the volume term is typically a few thousand MeV, the surface term (including the surface-symmetry energy) is several hundred MeV, and the volume-symmetry term is a few hundred MeV. Of these three, only the variation of the surface term with deformation need be considered, the other two remaining almost unchanged with deformation and through fission. The other large shape-dependent contribution is the Coulomb energy, which is of the order of 1000 MeV.

If we decide to neglect the finer corrections to the binding energy arising from pairing and shell effects ($\sim$ a few MeV, but deformation dependent), we see that the *change* in the nuclear potential energy with deformation comes about through the variation of the surface and Coulomb terms only. Therefore, for considering the statics (i.e., the dependence of potential energy on deformation of the nucleus), a good first approximation should be just to consider the surface and Coulomb energies. Furthermore, although the nuclear force is rather complicated in nature, the surface term arises because a nucleon at the surface has fewer neighbors to interact with, and the proportionality constant C_s (see Eqs. 6–22 and 6–23) is accurately known.

The situation is completely analogous to that of a liquid drop. As the nucleus or the liquid drop is deformed, its surface area increases, resulting in an increase in the surface energy and a decrease in the Coulomb energy. Thus there is some cancellation of the two effects. For lighter elements this cancellation is only partial, but for elements of higher charge in the actinide region it is nearly complete, with the result that the neglected shell corrections of a few MeV can no longer be ignored. It is not surprising, then, that the simple liquid-drop model works reasonably well for the lighter nuclei but fails in the actinide region. A logical improvement to the model would be to incorporate the single-particle effects on the potential-energy surfaces of the liquid drop. In this section, we shall consider only the liquid-drop part of the energy, and examine the predictions of this simple model. In the next two sections shell correction will be treated in detail.

Let us now consider the potential energy of a charged drop as a function of the deformation parameters, with the size, surface tension, and charge of the drop the same as those of a nucleus. Let E_s and E_C be the surface and Coulomb energies of the deformed drop, and E_{s0} and E_{C0} the corresponding quantities for the undistorted spherical shape. Then

$$E_{s0} = 4\pi R_0^2 S = C_s A^{2/3}, \qquad (12\text{--}18)$$

where S is the surface tension of the liquid, related to the coefficient C_s of the surface term in the mass formula 6–22, with $C_s = 18.56\,[1 - 1.79(N - Z)^2/A^2]$ MeV. The Coulomb energy of a spherical drop is given by

$$E_{C0} = \frac{3}{5}\frac{Z^2 e^2}{R_0} = \frac{\kappa_C Z^2}{A^{1/3}}, \qquad (12\text{--}19)$$

where $\kappa_C = 0.717$ MeV is the coefficient of the shape-dependent part of the Coulomb term in mass formula 6–22. For a deformed shape,

$$E_s = S\sigma \quad \text{and} \quad E_C = \frac{1}{2}\int\frac{\rho(\mathbf{r}_1)\rho(\mathbf{r}_2)}{|\mathbf{r}_1 - \mathbf{r}_2|}\,d\tau_1\,d\tau_2, \qquad (12\text{--}20)$$

where σ is the surface area of the drop, and $\rho(\mathbf{r})$ is the charge density at point $\mathbf{r}$. As the drop deforms, the surface energy increases ($E_s - E_{s0} > 0$) while the Coulomb energy decreases ($E_C - E_{C0} < 0$). If we take the spherical drop as the reference, the potential energy V of the drop up to second order in α is given by

$$V = (E_s - E_{s0}) + (E_C - E_{C0}) = \frac{1}{2}\sum_{\lambda,\mu} C_\lambda |\alpha_{\lambda\mu}|^2, \qquad (12\text{--}21)$$

where C_λ is defined by Eqs. 9–9 to 9–11. A detailed derivation of this expression may be found in a book by Wilets [40].

We will consider, for simplicity, only small deformations of the type $\lambda = 2$. Using Eqs. 12–18 and 12–19, we find that

$$V = \frac{1}{2\pi} (E_{s0} - \tfrac{1}{2}E_{c0}) \sum_\mu |\alpha_{2\mu}|^2. \qquad (12\text{–}22)$$

If V is positive, it means that there is an effective energy barrier for deformation and fission, while a negative or zero V indicates instability of the spherical drop toward deformation. From Eq. 12–22 we conclude that a charged spherical drop is unstable toward small deformations of the $\lambda = 2$ type if

$$E_{s0} \leqslant \tfrac{1}{2}E_{c0} \quad \text{or} \quad x = \frac{E_{c0}}{2E_{s0}} \geqslant 1. \qquad (12\text{–}23)$$

The dimensionless quantity x is called the fissionability parameter, and, using Eqs. 12–18 and 12–19, we find

$$x = \frac{\kappa_C Z^2/A^{1/3}}{2C_s A^{2/3}} = \frac{Z^2/A}{51.77[1 - 1.79(N - Z)^2/A^2]}. \qquad (12\text{–}24)$$

According to this formula, $x = 0.6991$ for ^{210}Po, increasing to the values 0.7597 for ^{235}U and 0.8274 for ^{254}Fm. As x approaches unity, we expect the height of the barrier to decrease. For $x < 1$, spontaneous fission is still possible through barrier penetration.

Although nuclei with $x < 1$ are stable with respect to small arbitrary deformations, a larger deformation will give the long-range repulsion more advantage over the short-range attraction, and it should be possible for a suitably deformed nucleus to divide spontaneously in this case. For this to happen, however, the nucleus will have to reach a critical stage of deformation, at which it will be on the verge of division. This will be a state of unstable equilibrium, and it can be found by plotting the potential-energy contours of the liquid drop as a function of the deformation parameters. For axially symmetric deformations, we can use the expansion

$$R(\theta) = \frac{R_0}{\xi} \left[1 + \sum_{\lambda=2}^\infty a_\lambda P_\lambda(\cos\theta) \right], \qquad (12\text{–}25)$$

where $a_\lambda = [(2\lambda + 1)/4\pi]^{1/2}\alpha_{\lambda 0}$, and can then write down an expression for the deformation energy $V(x, a_2, a_3, \ldots, a_\lambda, \ldots)$, in a power series of the a_λ's. Cohen and Swiatecki [41] have done this calculation, taking terms up to $\lambda = 18$.

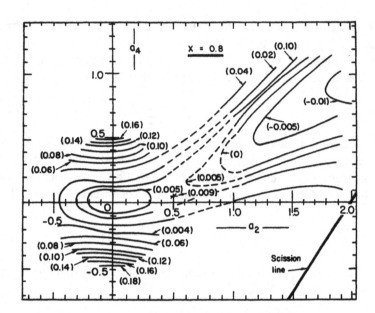

Fig. 12-19 Contour map of the potential-energy surfaces in the liquid-drop model as functions of a_2 and a_4, all other a_λ's being set equal to zero. The fissionability parameter x is fixed at 0.8. The labeled energies are in units of E_{s0}. The saddle point is shown by a cross. (After Wilets [40].)

For the purpose of orientation, let us consider only deformations of types a_2 and a_4, putting all other $a_\lambda = 0$. The potential-energy contours for $x = 0.8$ are plotted in the a_2-a_4 plane in Fig. 12–19, each contour being a line of constant potential energy labeled by values of V/E_{s0}. For $x < 1$, as we have seen, the spherical shape with $a_2 = a_4 = 0$ is always a local minimum. We note from the figure that for large distortions there are other contours with potential energy equal to or less than that of the spherical configuration. In fact, these successive contours of lower and lower potential energy form a deep valley which favors more and more deformation of the nucleus, ultimately leading to scission. However, in order to go down this valley, starting from the spherical shape, the nucleus has to pass through some contours of higher potential energy.

The potential barrier that the nucleus must surmount will depend on the path that it follows in the a_2-a_4 plane. There is, of course, a particular path of least resistance leading to the potential hollow on the other side, given by the values of a_4 (in terms of a_2 and x) for which the potential is minimal, that is, $\partial V/\partial a_4 = 0$ and $\partial^2 V/\partial a_4{}^2 > 0$. On this fission path, there will be a particular point in the a_2–a_4 plane where the potential energy is maximum, and this is the barrier height that the nucleus has to surmount to go down the valley of fission. This point is called the saddle point because the potential energy increases in the direction in the a_2–a_4 plane perpendicular to the fission path, and it is a configuration of unstable equilibrium for the nucleus. It is the low point or pass in the potential-energy ridge that separates the spherical drop from the valley leading to two-fragment division. In the general case, there may appear other passes leading to even deeper hollows in the potential that correspond to three- or many-fragment divisions of the drop, but these passes lie at considerably higher energy, hindering such divisions. In Fig. 12–19, where we have plotted the potential-energy contours, the position of the saddle point is shown by a cross.

It is thus possible in this simple model not only to obtain the height of the fission barrier but also to find the shape of the nucleus at the saddle point, since the deformation parameters at this point are known. In a more general treatment of the model, both even and odd a_λ's may be considered, including the possibility of asymmetric mass division. It is found, however, that this gives rise to higher fission barriers, implying thereby that only symmetric division is favored in the liquid-drop model.

In Cohen and Swiatecki's work [41], all the even a_λ's up to $\lambda = 18$ in expansion 12–25 were considered. As before, there is a fission path in the multidimensional a_λ-space leading to the saddle point. The shape of the liquid drop at the saddle point has been numerically computed for various values of the fissionability parameter x, and this is shown in Fig. 12–20. For $x = 1$, as we have seen, the spherical shape itself is an unstable equilibrium, while for the other extreme, $x = 0$, there is effectively no Coulomb repulsion—hence the saddle-point configuration consists of two touching spheres.

Calculation of the height of the fission barrier in the liquid-drop model gives a pronounced and smooth dependence of E_b on x, in contradiction with the experimental results. Since the barrier in the liquid-drop model for the actinide nuclides is only a few MeV, it is clear that the cancellation of the surface and Coulomb terms is nearly complete, and that the neglected shell effects (which show a rapid variation with the parameter x) can play a dominant role. To stress this point, we have plotted the fission barrier E_b against the mass number A in the liquid-drop model for nuclei along the stability line in Fig. 12–21, and the result is seen to be a smooth curve. The effect of including the shell correction (following the method of Myers

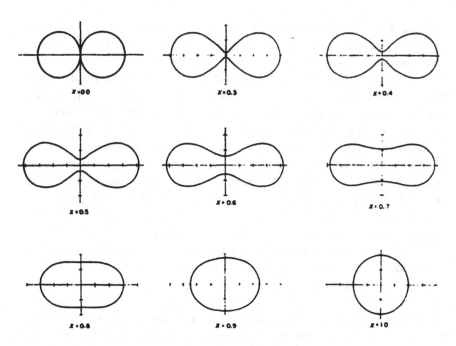

Fig. 12-20 Numerically computed saddle-point shapes of the liquid drop for various values of the fissionability parameter x. (After Cohen and Swiatecki [41].)

and Swiatecki [42]) is shown in the same figure by the dotted curve. It is clear that, although the liquid-drop model can be taken as a first approximation to describe the fission process, it is absolutely necessary to include shell effects in the theory. Moreover, the liquid-drop model predicts only a one-humped barrier, while shell effects, which are deformation dependent, may change the shape radically, introducing more than one hump, with important consequences. We shall devote the next section to an explanation of how the shell effect is evaluated in a single-particle picture.

We outlined some results of the dynamics of an oscillating liquid drop in Chapter 9. It was noted that for small deformations there are β-vibrations about the equilibrium shape which preserve the symmetry axis. The inertial parameter B associated with such oscillations is not given realistically in the hydrodynamical model. Without going into finer details, it is still instructive to obtain a crude estimate of the fission half-life of a nucleus from the simple ideas developed so far. In the multidimensional α-space, there is, as we have seen, a curve of least resistance to fission, which we call

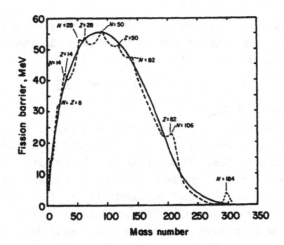

Fig. 12-21 The fission-barrier height E_b (in MeV) against the mass number A in the liquid-drop model is shown by the continuous curve. The effect of including the shell correction is shown by the dotted line. (After Myers and Swiatecki [42].)

the fission path. Along this fission path, the liquid-drop model predicts a single hump around the saddle point. We may imagine parameterizing this path by a single parameter α, which is a measure of the distortion of the nucleus along this path. If we denote the value of this parameter by α_s at the saddle point, the potential of the charged drop near this point along the fission path may be written as

$$V(\alpha) = V(\alpha_s) - C(\alpha - \alpha_s)^2 + \cdots , \qquad (12\text{-}26)$$

where C is a measure of the curvature of the barrier, and $V(\alpha_s) = E_b$, the height of the barrier. It is convenient to set $\alpha - \alpha_s = \alpha'$. The kinetic energy of the nucleus around the saddle point is $\frac{1}{2}B\dot{\alpha}'^2$, where B is the associated inertial parameter evaluated at the saddle point. When this model is applied to the nucleus, the Hamiltonian near the saddle point is

$$H(\alpha') = \frac{1}{2}B(\dot{\alpha}')^2 - C(\alpha')^2 + E_b. \qquad (12\text{-}27)$$

To emphasize the resemblance of this Hamiltonian to that of an inverted harmonic oscillator, we can put $C = \frac{1}{2}B\omega_b^2$, where ω_b has the dimensions of frequency.

In order for fission to take place, the nucleus as a whole has to tunnel through this barrier. The problem of penetration through one-dimensional barriers in the WKB approximation is dealt with in detail in Appendix D.

For an inverted harmonic-oscillator barrier, the WKB result is exact. If an energy E ($< E_b$) has been imparted to the nucleus, the penetration probability is given by (see Eqs. D–11, D–12)

$$P = (1 + \exp 2K)^{-1} \approx \exp(-2K) \tag{12–28}$$

with $2K = 2\pi(E_b - E)/\hbar\omega_b$.

The curvature of the barrier along the fission path at the saddle point can be obtained from liquid-drop statics. In order to estimate $\hbar\omega_b$, it is also necessary to know $B(\alpha_s)$. Detailed microscopic calculations indicate that the inertial parameter B of a nucleus with large deformation (corresponding to the saddle point) is about ten times larger than the corresponding irrotational hydrodynamical value [43]. We can, however, estimate $\hbar\omega_b$ from a knowledge of the spontaneous half-lives of fissioning nuclei. The vibrating nucleus is making a certain number of assaults n on the barrier per second. This number is usually equated to the frequency of β-vibrations, setting $\hbar\omega_{vib} \approx 1$ MeV. This corresponds to $n = 10^{20.38}$ assaults per second on the barrier. The fission rate is then $n \cdot P$, and the half-life in seconds is

$$\tau_{1/2} = \frac{\ln 2}{n \cdot P} = 10^{-20.54} P^{-1} \text{ s}, \tag{12–29}$$

where P is given by Eq. 12–28. Empirical data give a value of E_b fluctuating in the range 4.5–6.5 MeV for A going from 230 to 250. Knowing the half-lives $\tau_{1/2}$ for various nuclei, we can deduce the value of $\hbar\omega_b$ from Eqs. 12–28 and 12–29; $\hbar\omega_b \approx 300$–400 keV for these nuclei. It is to be noted that in the presence of a double-humped barrier the problem has to be reanalyzed, although the basic ideas are unaltered.

12-4 SHELL CORRECTION TO THE LIQUID-DROP MASS FORMULA

In Chapter 6, we found that the variation of the binding energy per nucleon with mass number A could be fairly well reproduced by the semiempirical mass formula. This formula yielded a smooth dependence of B/A on A for the most stable isobars. As was pointed out in Section 6–3, the experimental binding energies revealed small deviations from this smooth curve, and we could write (cf. Eq. 6–22)

$$B(A, Z) = B_{dist}^{LD}(A, Z) + \delta B(A, Z), \tag{12–30}$$

where $B(A, Z)$ is the experimental binding energy of a nucleus (A, Z), B_{dist}^{LD} is its value predicted by the distorted liquid-drop mass formula 6–22, and $\delta B(A, Z)$ arises because of shell correction. The important point to note is that, when δB is plotted against A, it has a structure which is *not* reproducible by adjusting the coefficients of the mass formula. In Chapter 6, the form of the semiempirical mass formula was obtained by considering a Fermi gas model of the nucleus, where the single-particle density of states was a smooth function of energy. Although terms like symmetry energy do have their origin in quantum effects (the Pauli principle), the specific effect of fluctuation of single-particle level density due to shell structure was ignored in the mass formula. Thus the mass formula may be regarded as a semiclassical approximation for the energy of the nucleus, where the single-particle levels, instead of being discrete and bunched together in shells, have somehow been smoothed to yield an average level density. This approximation, together with the assumption of the incompressibility of the nucleus, is sufficient to yield a smooth dependence of the energy on A.

It is important to realize that the semiempirical mass formula already contains the important two-body correlation terms (e.g., the volume, surface, Coulomb, and pairing terms all arise from some two-body effective interaction), but a correction of specific single-particle origin needs to be incorporated in it to obtain better agreement with experimental results. Although this correction is a small fraction of one per cent of the total binding, it is of great importance in determining the stability of superheavy nuclei and the shapes of fission barriers [44–46]. Because of the single-particle origin of the shell effect, we may attempt to extract it from a simple independent-particle model, and add this correction to B_{dist}^{LD} to obtain the shape of the fission barriers as well as to investigate the stability of superheavy nuclei.

Shell effects do not arise just from the discreteness of the levels but come about because of the degeneracy or of the bunching of states in the energy space. To appreciate this simple fact, consider one species of noninteracting fermions occupying singly the energy levels of a one-dimensional harmonic oscillator. The single-particle energies are

$$e_N = (N + \tfrac{1}{2})\hbar\omega, \qquad N = 0, 1, 2, \ldots, \tag{12-31}$$

and, considering A fermions, the Fermi energy is

$$\lambda = (A - \tfrac{1}{2})\hbar\omega. \tag{12-32}$$

The sum of the occupied single-particle energies E, is given by

$$E = [\tfrac{1}{2} + \tfrac{3}{2} + \cdots + (A - \tfrac{1}{2})]\hbar\omega$$
$$= \frac{A^2}{2}\hbar\omega. \tag{12-33}$$

We shall presently show that it is the sum of the single-particle energies of the occupied orbitals that contains shell effects; if this quantity E is a smooth function of A, as in Eq. 12–33, no shell correction is present. For a more realistic three-dimensional isotropic harmonic oscillator, although the energies are equally spaced, there is a degeneracy $\frac{1}{2}(N+1)(N+2)$ in a given major shell N, and it is no longer possible to express E as a smooth function of A. Degeneracy, of course, is an extreme limit of bunching of states with respect to energy, and it may be thought that with increasing deformation of the potential well, as the levels become more uniformly distributed, the shell effects should disappear. That this is not so is illustrated in Fig. 12–22 by plotting the energy levels of an axially symmetric harmonic-oscillator potential (Eq. 9–147) against ε, for $\omega_1 = \omega_2 = \omega_\perp = \omega_0(1 + \varepsilon/3)$ and $\omega_3 = \omega_0(1 - 2\varepsilon/3)$. It will be seen that the shell structure persists even at large deformations, and regular energy gaps with large degeneracy appear for $\omega_3/\omega_\perp = \frac{2}{1}, \frac{1}{1}$ or $\frac{1}{2}$, that is, at $\varepsilon = -0.75$, 0, and 0.6, respectively.

In regard to the empirical mass formula 6–22, the liquid-drop part $B_{\text{dist}}^{LD}(A, Z)$ always prefers a spherical shape of the nucleus. It is the effect described by the shell-correction term δB which causes certain nuclei to have deformed shapes with extra stability. To see this in more detail, examine the Nilsson diagram (Fig. 9–32) of neutron levels for $N > 126$. The closed shells for spherical shapes in this region occur at $N = 126$ and 184.

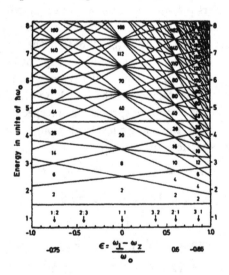

Fig. 12-22 Single-particle levels of an anisotropic harmonic oscillator (with $\omega_1 = \omega_2 = \omega_\perp$) as a function of $\varepsilon = (\omega_\perp - \omega_3)/\omega_0$. [After R. K. Sheline, I. Ragnarsson, and S. G. Nilsson, *Phys. Lett.* 41B, 115 (1972).]

If we consider a nucleus with $N = 152$, then for a spherical shape the $g_{9/2}$ and $i_{11/2}$ levels are fully occupied and the $j_{15/2}$ levels partly so. Since the density of single-particle states near the Fermi energy is very large in this case, most of the valence nucleons are near this energy and the nucleus has less binding than in the situation (which occurs for $\varepsilon \sim 0.05$) where the single-particle states are more uniform. The most favorable configuration is at $\varepsilon \approx 0.25$, when the density of states near the Fermi level is very small, and most of the valence neutrons occupy deeper and more bound states. For such a deformation, we expect the shell correction δB to be positive, giving a deformed ground state. As is clear from Fig. 12–22, a closed shell may appear near $N \sim 140$ for $\varepsilon \approx 0.6$, which is around the liquid-drop saddle point, thus completely altering the shape of the liquid-drop barrier.

Recognizing the point that shell correction is basically a single-particle effect and arises from the fluctuations in the single-particle level density, we shall now discuss a method of theoretically estimating δB. In Chapter 10, we saw that in the Hartree-Fock theory we can write the energy of a nucleus as

$$E_0 = \sum_m e_m^{(\mathrm{HF})} - \frac{1}{2} \int \rho(\mathbf{r})\rho(\mathbf{r}') V(\mathbf{r}, \mathbf{r}') \, d^3r \, d^3r',$$

where $V(\mathbf{r}, \mathbf{r}')$ is the effective two-body interaction, the sum runs over the occupied HF orbitals m with self-consistent energies $e_m^{(\mathrm{HF})}$, and the exchange term has been neglected. Strutinsky [47] has shown that E_0 can be approximated by the following expression:

$$E_0 \approx \sum_m e_m^{(sm)} - \frac{1}{2} \int \tilde{\rho}(\mathbf{r})\tilde{\rho}(\mathbf{r}') V(\mathbf{r}, \mathbf{r}') \, d^3r \, d^3r',$$

where $\tilde{\rho}(\mathbf{r})$ is the "smooth" component of $\rho(\mathbf{r})$, and $e_m^{(sm)}$ are the single-particle energies derived from the one-body potential $U = \int \tilde{\rho}(\mathbf{r}') V(\mathbf{r}, \mathbf{r}') \, d^3r'$, which we may identify as a phenomenological shell-model type of potential. Let us denote $\sum_m e_m^{(sm)}$ by E; this quantity contains the shell effects. Also, let us assume that there is a prescription for extracting the part of E that smoothly varies with nucleon number and deformation, and designate this as $\tilde{E}$. Then we may write

$$E_0 = (E - \tilde{E}) + \tilde{E} - \frac{1}{2} \int \tilde{\rho}(\mathbf{r})\tilde{\rho}(\mathbf{r}') V(\mathbf{r}, \mathbf{r}') \, d^3r \, d^3r'$$

The last two terms on the right side of the equation have no shell effects present, and we can identify these as the terms representing the liquid-drop expression, E^{LD}. Hence we finally obtain

$$E_0 = E^{LD} + (E - \bar{E}). \qquad (12\text{-}34)$$

According to Eq. 12–34, the liquid-drop energy should be "renormalized" by the shell-correction term $(E - \bar{E})$ to obtain the energy of the many-nucleon system. We shall now describe the method of Strutinsky for obtaining $\bar{E}$ from a given shell-model potential (e.g., the Nilsson potential). The method works particularly well for a one-body potential in which all the levels are discrete, as in the Nilsson potential. There are some complications, however, for a finite U like the Woods-Saxon potential, where some bound-state levels are followed by a continuum; this we shall discuss later. In the following we shall also drop the superscript "sm" from the single-particle energies for simplicity.

The single-particle density of states is given by

$$g(e) = \sum_i \delta(e - e_i), \qquad (12\text{-}35)$$

where the e_i's are the eigenvalues of a properly chosen one-body potential, and the sum over i is taken over all states, taking account of the degeneracy of the levels. If we consider a nucleus with N neutrons, then, for the neutrons,

$$E = \sum_{\substack{m \\ \text{(occupied)}}} e_m. \qquad (12\text{-}36)$$

In this model, there is a similar expression for E for the protons. Equation 12–36 takes into account the quantum nature of the levels and their uneven fluctuations due to shell effects. In order to obtain the smooth part $\bar{E}$ from E, we should either average it over a large number of nuclides, or smooth the quantum fluctuations of the level density $g(e)$ in a single nucleus. Strutinsky [47] chooses the latter procedure and gives a well-defined method for smoothing $g(e)$, from which $\bar{E}$ can be obtained. To do this, it is necessary to smooth the individual delta functions occurring in Eq. 12–35 in such a manner that these overlap sufficiently with each other to yield a smooth $\bar{g}(e)$. A simple way to do this is to replace $\delta(e - e_i)$ by a Gaussian centered about e_i, with appropriate normalization to preserve the strength of the level and a width γ which is at least of the same order as the shell spacings ($\sim \hbar\omega$ in the case of the harmonic oscillator). In other words, we could replace

$$\delta(e - e_i) \rightarrow \frac{1}{\gamma\sqrt{\pi}} \exp[- (e - e_i)^2/\gamma^2] \qquad (12\text{-}37)$$

to obtain a smooth $\bar{g}(e)$.

Even though this replacement preserves the normalization of each level (when the e-space extends from $-\infty$ to ∞), it does not reproduce the higher moments of the delta function. We can preserve all the moments of the delta function from 0 to $2S + 1$ by making the following replacement:

$$\delta(e - e_i) \rightarrow \frac{1}{\gamma\sqrt{\pi}} \exp(-u_i^2)\, L_S^{1/2}(u_i^2) \qquad (12\text{–}38)$$

with $u_i = (e - e_i)/\gamma$. The function $L_n^{1/2}(x)$ is a polynomial of order n in x,

$$L_n^{1/2}(x) = \sum_{m=0}^{n} \frac{1}{(n-m)!\,m!} \frac{\Gamma(n + \frac{3}{2})}{\Gamma(m + \frac{3}{2})} (-x)^m, \qquad (12\text{–}39)$$

where $\Gamma(n)$ is the gamma function.

According to the Strutinsky prescription, then,

$$\tilde{g}(e) = \frac{1}{\gamma\sqrt{\pi}} \sum_{i} \exp(-u_i^2)\, L_S^{1/2}(u_i^2), \qquad (12\text{–}40)$$

where u_i has been defined in Eq. 12–38. The function $L_S^{1/2}$ which modulates the Gaussian is often called the curvature function of the $(2S)$th order. Note that, since each level is in principle being spread also into the negative ranges of e (even when the bottom of the well is at $e = 0$), $\tilde{\lambda}$ and $\tilde{E}$ are defined by

$$N = \int_{-\infty}^{\tilde{\lambda}} \tilde{g}(e)\, de, \qquad \tilde{E} = \int_{-\infty}^{\tilde{\lambda}} \tilde{g}(e)e\, de. \qquad (12\text{–}41)$$

If the Strutinsky smoothing procedure is applied to an already smooth density distribution $\eta(e)$ [as can be done by replacing the sum $\sum_i$ in Eq. 12–40 by an integral $\int \eta(e)\, de$], self-consistency demands that the same $\eta(e)$ be reproduced. This condition is satisfied with a $(2S)$th-order curvature function if $\eta(e)$ is expressible in a polynomial of order $2S + 1$ or less. In order that the shell correction δE_{th} be unique for a given nuclear spectrum, one could require that the result be independent of γ over a wide range and not alter with increasing order, but less restrictive conditions have been given [62].

To illustrate the last two points about the uniqueness of δE for a discrete spectrum, we have calculated it for the 126 neutrons of ^{208}Pb. The single-particle spectrum of a modified oscillator type was taken following Krappe and Wille [48], and is given for completeness:

$$e_{j=l+1/2} = (2n + l + \tfrac{3}{2}) - \kappa(N)l - \kappa(N)\mu_0[l(l + 1) - \tfrac{1}{2}N(N + 3)],$$

$$e_{j=l-1/2} = (2n + l + \tfrac{3}{2}) + \kappa(N)(l + 1) - \kappa(N)\mu_0[l(l + 1) - \tfrac{1}{2}N(N + 3)].$$

$$(12\text{--}42)$$

Here the energy is expressed in units of $\hbar\omega$; n, l are the usual oscillator quantum numbers, with $N = 2n + l$. For a reasonable empirical fit of the neutron levels, the following values for the parameters occurring in Eq. 12–42 are chosen:

$$\mu_0 = 0.308, \qquad \kappa(N) = 0.21[\tfrac{1}{2}(N + 1)(N + 2)]^{-1/3}. \qquad (12\text{--}43)$$

Taking $\hbar\omega = 41/A^{1/3} \approx 6.92$ MeV, we show the results for $\delta E = (E - \tilde{E})$, using second- to tenth-order Strutinsky calculations, in Fig. 12–23 as a function of the smearing parameter γ. As $\gamma \to 0$, each Gaussian goes over to the delta-function limit, and $\tilde{E} \to E$. The second-order calculation does not yield a unique δE, because $\tilde{g}(e)$ is more complicated than a third-order polynomial. Note that calculations in all higher orders give the same result. As the order $(2S)$ of the curvature function is increased, the value of γ for which δE becomes independent of γ also increases slightly. As $S \to \infty$, the Gaussian modulated by the curvature function again reproduces the delta function exactly for any finite γ, and $\delta E \to 0$. For any finite order, however, a unique δE can be found for a discrete spectrum provided that γ is chosen to be sufficiently large. We should mention that care must be taken to include the contribution to $\tilde{E}$ coming from the negative regions of e in the

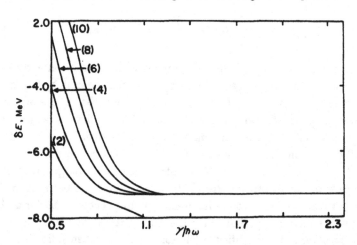

Fig. 12-23 Shell correction (in MeV) for the neutrons in ^{208}Pb as a function of γ for the discrete spectrum defined by Eq. 12–42. The smearing parameter γ is expressed in units of $\hbar\omega$ (1 $\hbar\omega$ = 6.92 MeV), and the shell correction is shown for various curvature functions.

integrals of Eq. 12–41, where $\tilde{g}(e)$ is oscillatory in form. Furthermore, for a given γ, all states i in Eq. 12–40 must be included that may contribute finitely to $\tilde{g}(e)$ for $e < \tilde{\lambda}$. For $\gamma \approx 2.5\ \hbar\omega$, this implies the inclusion of states as high as $N = 20$. In practice, $\gamma \approx 1.2\ \hbar\omega$ yields the right shell correction.

The Strutinsky method of renormalizing the semiempirical mass formula has been used by Nilsson et al. [44] in an investigation of the stability of superheavy nuclei and also for the determination of the shapes of fission barriers. We saw in Section 9–9 that the same method has been used successfully to calculate the deformation of rare-earth nuclei. The form of the single-particle Hamiltonian chosen by Nilsson et al. for the calculation of shell corrections in actinide and superheavy nuclei is given by Eq. 9–171. The sum of the occupied single-particle energies, E, is replaced in the presence of pairing by the expression

$$E = \sum_\nu e_\nu V_\nu^2 - \frac{\Delta^2}{G} - G(\sum_\nu V_\nu^4 - \sum_\nu{}' 1),\qquad(12\text{–}44)$$

where e_ν are the single-particle energies, and the other symbols were defined in Eq. 8–113. The sums are taken separately over neutrons and protons with pairing matrix elements G_n and G_p, respectively. The last term, with the summation $\sum'$ over occupied energy levels, represents the subtraction of the diagonal pairing energy, so that only the correlation energy due to pairing remains. The theoretical estimate of the shell correction to binding energy, which we denote by $\delta B_{\text{th}}(A, Z)$, and by which the liquid-drop mass formula should be renormalized, is given by

$$\delta B_{\text{th}} = -\ \delta E_{\text{th}} = -\ (E - \tilde{E} - \langle E_{\text{pair}}\rangle),\qquad(12\text{–}45)$$

where E is defined by Eq. 12–44, and $\tilde{E}$ by Eq. 12–41, and $\langle E_{\text{pair}}\rangle$ is the average pairing energy, taken simply as $-\ 2.3$ MeV for nuclei in this region.

An estimate of the binding energy of a nucleus (A, Z) as a function of the deformation parameters ε and ε_4 can now be made by adding this theoretical estimate of shell correction δB_{th} to the distorted liquid-drop part of the mass formula, $B_{\text{dist}}^{LD}(A, Z)$:

$$B_{\text{th}}(A, Z) = B_{\text{dist}}^{LD}(A, Z) + \delta B_{\text{th}}(A, Z).\qquad(12\text{–}46)$$

Nilsson et al. [44] have compared the results of their calculation with experimental values for the rare-earth and actinide regions and extrapolated these estimates into the superheavy region. When Eq. 6–1 is used, the comparison can as well be made in terms of mass M rather than of the binding B of a nucleus. It is convenient, for this purpose, to make the

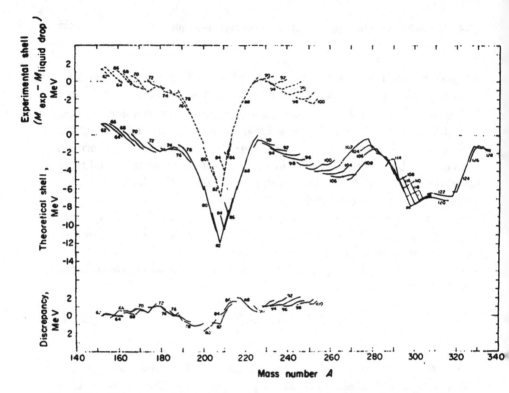

Fig. 12-24 Experimental and theoretical mass values (including shell correction) for $150 < A < 130$ plotted relative to the spherical liquid-drop mass. (After Nilsson et al. [44].)

comparisons with respect to the spherical liquid-drop mass, M_{LD}, defined in terms of $B_{\mathrm{sph}}(A, Z)$ of Eq. 6-21. In Fig. 12-24 we plot the experimental and theoretical mass values, $(M_{\exp} - M_{LD})$ and $(M_{\mathrm{th}} - M_{LD})$, versus A, following Nilsson et al., and show the discrepancy in the known region in the lower scale. This discrepancy is within ± 2 MeV throughout the range shown, displaying excellent agreement with experiment. It should be noted that in deriving M_{th} the Myers-Swiatecki mass formula has been used, whose parameters (given in Eq. 6-24) were determined assuming a special form for δB_{th}. In principle, these parameters should be redetermined using δB_{th} estimated by the Strutinsky procedure.

It will be seen from Fig. 12-24 that the estimated masses for superheavy nuclei beyond the present experimental range show a broad shell structure at $Z = 114$ and $N = 184$–196. Although this is not as pronounced as the ^{208}Pb shell, it is the main reason for the hope that there may exist in this region an island of relative stability, which may be experimentally detectable.

We have seen that the Strutinsky method of calculating δB_{th} is unambiguous when applied to a single-particle spectrum with all discrete levels, as in a Nilsson-type potential. As pointed out by Lin [49], the method runs into difficulties, however, when applied to a finite potential well (e.g., the Woods-Saxon well) which has a single-particle spectrum of a finite number of discrete levels, followed by a continuum. If the continuum is completely ignored, and the Strutinsky prescription is applied to the bound levels only, δB_{th} turns out to depend both on the order and on γ. The results of such a calculation for the neutrons, using a Woods-Saxon potential well which fits the experimentally known top few single-particle levels of ^{208}Pb reasonably well, are shown in Fig. 12–25.

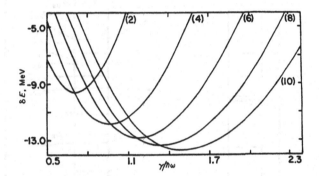

Fig. 12-25 Shell correction (in MeV) for the neutrons in ^{208}Pb as a function of γ for a Woods-Saxon spectrum, ignoring continuum effects. The smearing parameter γ is expressed in units of $\hbar\omega$ ($1\ \hbar\omega = 6.92$ MeV), and the shell correction is shown for various curvature functions. (After Ross and Bhaduri [50].) The parameters of the Woods-Saxon potential are given by J. Blomquist and S. Wahlborn, *Ark. Fysik* **16**, 545 (1960).

The reason for this ambiguity is not hard to understand. The Strutinsky method is tailor-made for a spectrum with an infinite number of discrete levels. In the process of smoothing such a spectrum, individual level strengths are distributed in the entire e-space from $-\infty$ to $+\infty$. As the Gaussian level width γ is increased, the low-lying levels lose more and more of their strength. This loss, however, is compensated for by the higher-lying discrete levels, more and more of which begin to contribute to the low-lying energies as γ increases. Thus the single-particle smoothed level density $\tilde{g}(e)$ remains γ-independent, provided that γ is of the order of the shell spacing or greater. When, however, only a finite number of discrete levels are being smoothed, this situation no longer exists, and the result becomes γ-dependent. To overcome this difficulty in applying the Strutinsky method to the spectrum of a finite well, one should consider not only the discrete bound states but also the modifications in the continuum-level density brought about by the potential.

This point is worth appreciating because it also clarifies the meaning of "resonances" in the continuum. For this purpose, we show in Fig. 12–26 the single-particle levels that are occupied by the neutrons in the topmost

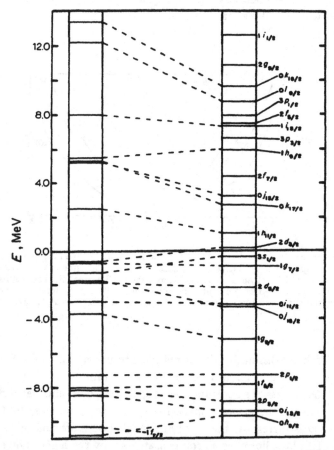

Fig. 12-26 Comparison of the single-particle spectra for the neutrons in ^{208}Pb, obtained using a Woods-Saxon potential (left) and an oscillator-type potential (right). The "levels" lying above zero energy in the Woods-Saxon spectrum correspond to peaks in the function $\Delta g(e)$ of Eq. 12–47 and are associated with continuum resonances. (The widths of these are not indicated.) (After Ross and Bhaduri [50].)

shell and the empty levels beyond $N = 126$, as generated by the Woods-Saxon potential and an appropriate spherical Nilsson potential. Although there are some differences in the level orderings in the two potentials, the few unoccupied bound levels in the Woods-Saxon potential are the same as those generated by the Nilsson potential. Comparing the two spectra in Fig. 12–26, we infer that the Woods-Saxon potential is only slightly too weak to bind the $1h_{11/2}$, $0k_{17/2}$, $0j_{13/2}$ levels. In the Strutinsky calculation with the Nilsson potential, these bound states do play a vital role, as is made evident by the fact that, if the spectrum is cut off at $2d_{3/2}$ (the last bound state of the Woods-Saxon), the results become quite γ-dependent. The bound states that are thus missed in the Woods-Saxon potential nevertheless appear as "bumps" in the continuum-level density (instead of delta functions) and are termed resonances. In the absence of any potential, the single-particle level density $g(e)$ is simply proportional to $\sqrt{e}$ for $e > 0$. In the presence of a central potential $U(r)$, the modification to this density, $\Delta g(e)$, is given by

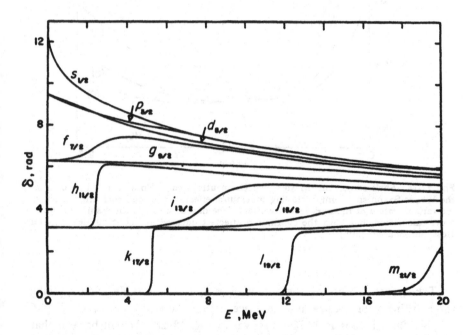

Fig. 12-27 Plot of some of the phase shifts generated by the Woods-Saxon potential for various l $(= j - \frac{1}{2})$ values against energy e. Note the steep rise in some of these through $\pi/2$ at certain energies corresponding to resonances. (After Ross and Bhaduri [50].)

$$\Delta g(e) = \frac{1}{\pi} \sum_l (2l + 1) \frac{d}{de} \delta_l(e), \tag{12-47}$$

where $\delta_l(e)$ is the phase shift of the lth partial wave at energy e due to the potential $U(r)$ (see Problem 5–7). In the presence of a spin-orbit potential, the effective potential

$$U_{\text{eff}}(r) = U(r) + \frac{\hbar^2}{2M} \frac{l(l+1)}{r^2} \tag{12-48}$$

is different for the two possible values of $l = j \pm \frac{1}{2}$ for a given j, and the phases have to be calculated separately in these two channels. In Fig. 12–27, we plot some of the phases for $l = j - \frac{1}{2}$ as a function of e. At a resonance, $\delta(e)$ rises sharply, contributing a bump in the level density given by Eq. 12–47. We find that a level like $1h_{11/2}$, which was missed as a bound state, appears as a resonance at $e \approx 2.5$ MeV.

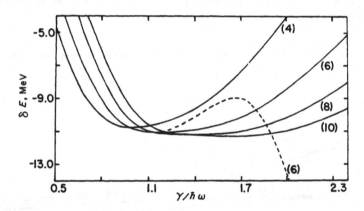

Fig. 12-28 Shell correction (in MeV) for the neutrons in ^{208}Pb as a function of γ for the Woods-Saxon spectrum, including continuum effects. The shell correction is shown for various values of the curvature function. The dashed curve is the result of a sixth-order calculation, but with the continuum effects truncated at about 20 MeV. (After Ross and Bhaduri [50].)

Taking into account all these resonances up to $e \approx 100$ MeV and applying the Strutinsky procedure as before, we obtain the shell correction δE [50] for the 126 neutrons in ^{208}Pb as shown in Fig. 12–28. It will be seen that fourth- to tenth-order calculations all yield a result of about -11 MeV, in contrast to the ambiguous results of Fig. 12–25, where only the bound levels of the finite well were smoothed. This calculation, although rather cumbersome for practical application (especially with deformed potentials), demonstrates the necessity to take into account the modifications in the continuum due to the finite potential.

Bolsterli et al. [51] have applied the Strutinsky method to finite potential wells. They diagonalize the deformed single-particle Hamiltonian in a large basis of deformed harmonic-oscillator states, and smooth the delta functions at not only the bound but also the artificial unbound eigenstates. With the optimum choice of the matrix size, this procedure ensures the overall smooth behavior of the density of states above the Fermi level. The method is practical and is widely used.

12-5 THE TRIUMPH OF THE DOUBLE HUMP

In this section we shall study how the shape of the fission barrier is affected by shell correction and the effect of this on various aspects of fission phenomena [46]. We noted before that the simple liquid-drop model energetically prefers a spherical ground state and gives rise to a one-humped barrier, as shown by the dotted curve in Fig. 12–29. Addition of the shell correction for energy to the liquid-drop expression can have drastic effects on the fission-barrier shape, especially for the actinide nuclides, where this correction is comparable in magnitude to the liquid-drop barrier itself. As we noted before, the shell effect may give rise to a deformed ground state, as shown

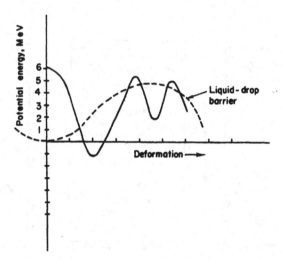

Fig. 12-29 A schematic diagram of the fission barrier after the inclusion of shell effects, for nuclides with neutron number $N \sim 142$–148. For a discussion of the relative heights of the barrier, see text. The liquid-drop barrier is shown by the dotted line.

in Fig. 12–29. One important characteristic of the shell correction is its modulation of the deformation energy as a function of deformation, due to the alternate compression and thinning out of single-particle levels near the Fermi surface with increasing deformation (Fig. 12–22). This may result in a second potential minimum at a much larger value of deformation, as shown in Fig. 12–29. For the modulation to be effective in giving rise to a second minimum, it should appear at a deformation close to that of the liquid-drop saddle point, where the liquid-drop profile of the potential energy

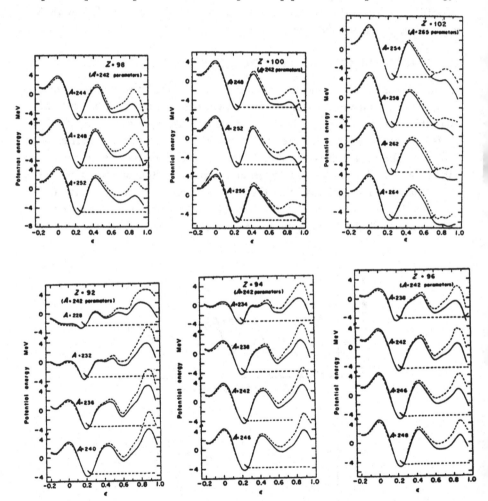

Fig. 12-30 Calculated fission barriers of some even-even nuclides, including shell correction. The calculations were done using a modified Nilsson-type potential. (After Nilsson et al. [44].)

is flat. This happens to be the case for nuclides with neutron numbers in the vicinity of 142–148, where theoretical calculations indeed predict a second minimum [44].

In Fig. 12–30, we show the fission barriers of some even-even nuclides as calculated by Nilsson et al. [44], using the deformed Nilsson-type potential to evaluate the shell correction. These and other calculations reveal a general trend toward a weak and shallow second minimum in Th nuclides and a greater height of the second barrier than of the first. In the U nuclides the second barrier height is still greater, with the second minimum increasing in depth, a trend that continues in Pu and Np, where the second barrier is only slightly higher. For Am and heavier nuclei ($Z \geqslant 95$) the second minimum is shallower again, and the height of the second barrier is less than the first. The second minimum of the fission barrier almost disappears for the isotopes of $_{100}$Fm and is absent in the neighboring heavier nuclides. All the ground states are found to be prolate. We also show, in Fig. 12–31, the stabilizing effect of the shell correction on some nuclides in the superheavy region, this being particularly evident for the spherical ground state of the nuclide with $Z = 114$ and $N = 184$. A number of subsequent calculations performed by other authors [51] with more realistic finite wells have produced

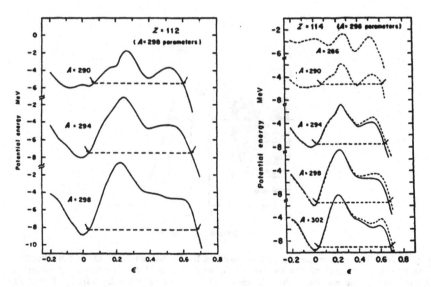

Fig. 12-31 Calculated fission barriers of some superheavy nuclides, including shell correction. (After Nilsson et al. [44].)

similar results. It is difficult to assess the significance of the quantitative differences obtained from these finite-well calculations, because of the uncertainty in shell-correction results mentioned in Section 12–4. It is found that in most cases the second minimum (when it exists) is about 2–3 MeV higher than the first one.

   Before trying to correlate the results of experiments with these double humps along the fission path, we shall make a few relevant comments about the possible states in the two wells. As we mentioned before, when some excitation energy is given to a nuclide by the capture of a low-energy neutron, this energy is dissipated into many degrees of freedom by many nucleonic collisions and a compound nucleus is formed. Now, since there are two minima in potential energy, there are two possibilities for the compound-nucleus formation with different deformations. The level spacings between the compound nuclear states are very small, because the energy can be dissipated in very many degrees of freedom, not just in the fission degree of freedom along the fission path. At about 5-MeV excitation, for example, the compound nuclear levels of the first well (class I levels) may have a typical spacing of about 10 eV. The second compound nuclear configuration occurs at a greater deformation, where the energy available for distribution in the different degrees of freedom is about 3 MeV less (the difference between

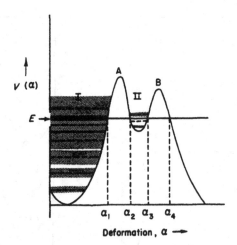

Fig. 12-32 Penetration through a two-humped fission barrier at an excitation energy E. The vibrational states of the two potential wells I and II are indicated by horizontal dashed lines. The widths of these levels are shown schematically by hatched lines. The class I states at this excitation energy form a continuum, while the class II states may be regarded as distinct.

the second and first minima). Consequently, the compound nuclear levels of the second well (class II levels) have a much greater spacing at the same excitation, of the order of a few hundred eV. At higher excitations, the class I and class II compound nuclear states can mix thoroughly. In addition to these compound nuclear levels, which arise basically because of the abundance of the nonfission degrees of freedom, there are certainly some vibrational levels in the two potential wells that correspond to vibrations in the fission degree of freedom. The vibrational states are really the respective bound states of the two potential wells shown in Fig. 12–32. The lowest vibrational states, especially in the second well, may be of narrow width, but the higher ones are very broad because of damping into the compound nuclear states. In analyzing the experimental results, which we now examine, we shall draw upon the above physical ideas.

As will be seen from Fig. 12–2, the spontaneous fission lifetimes display some systematic trends when plotted against Z^2/A. It was therefore a considerable surprise to discover an odd-odd isotope of americium $(A = 242)$

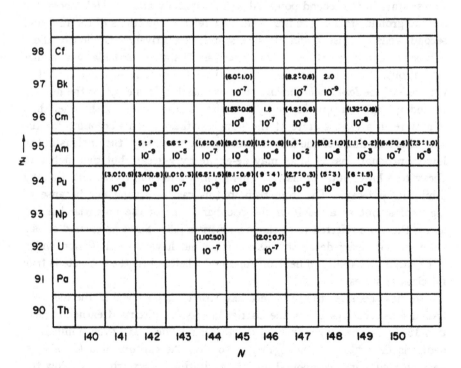

Fig. 12-33 A chart of the shape isomers. Half-life entries in seconds indicate nuclides for which these isomers have been observed.

decaying by a spontaneous fission lifetime of the order of milliseconds [52], shorter by a factor of 10^{20} than the lifetimes of more stable species. This fission isomer and some others were discovered before the existence of two-humped barriers had been realized. Subsequently, over 25 such short-lived isomers have been discovered, some with lifetimes of 10^{-8} s or less [53]. Figure 12-33 shows a table of these isomers.

In the conventional liquid-drop picture, such tremendous shortening of fission lifetimes requires the isomeric states in question to be excited by several MeV above the ground state. It is then difficult to explain the long lifetime of such a state against γ-decay, which is certainly not much stronger than spontaneous fission. The stability against γ-decay of these states would require them to have very high spins, which were experimentally not found to be present [54]. The mystery is easily solved by taking into account the modified two-humped fission-barrier shape, which is expected to prevail in the region of neutron numbers 142–148, just where the fission isomers are found. Such a two-humped fission barrier is shown in Fig. 12-32, where the lowest state in the second potential well is typically about 3 MeV above the normal ground state of the first well. We refer to the states of the first and second wells as class I and class II states, respectively. Obviously there will be some mixing between class I and class II states, particularly at higher excitations. Consider now the deformation of a nuclide to be such that it is in one of the *lowest vibrational* states of class II. It is easy for the nuclide to undergo spontaneous fission by tunneling to the right—much easier than if it were in a class I state at the same excitation. This accounts for the very short lifetimes of the fission isomers. Alternatively, the nucleus in the low vibrational class II state may de-excite by γ-decay, but then it has to decay to a lower state of class I. This implies tunneling to the left, thereby inhibiting γ-decay. The isomerism in this case is arising not because of high spins, but as a result of the peculiar shape of the potential barrier, and is therefore referred to as shape isomerism [55]. Since an excited class I state cannot easily decay by fission, it should have a small fission width but a large probability of neutron emission, whereas just the opposite is true for class II states.

We next consider the resonance-like structures observed in the induced-fission cross sections below the barrier energy, as discussed earlier in Section 12-1. In these experiments, the target nuclides have even numbers of neutrons (like $^{230}_{90}\text{Th}_{140}$ and $^{240}_{94}\text{Pu}_{146}$), so that the capture of a low-energy neutron results in a compound nuclear excitation energy which is below the fission-barrier height. Resonance structures are observed in two distinct energy ranges. The grouping of the fission resonances is observed in a high-resolution experiment with neutrons having energies of a few hundred eV or a few keV energy incident on a ^{240}Pu target (see Fig. 12–5). This is an example of intermediate structure in fission cross section, with a fine spacing of about 15 eV within a group and an intermediate spacing of about 0.7 keV between them. On the other hand, the so-called vibrational resonances are observed with incident neutrons of a few hundred keV range, being captured by ^{230}Th and leading to subbarrier fission (see Fig. 12–7). Here the experimental resolution is a few tens of keV. We outline below, in a qualitative manner, how both these types of resonances can arise naturally in a double-humped barrier. The detailed quantitative theory is given in Lynn's article [56].

We first consider the phenomenon of vibrational resonances in subbarrier fission cross sections (Figs. 12–7 and 12–8). As we noted before, it is not possible to explain this in the liquid-drop picture, where the fission cross section should increase steeply with increasing excitation in the subbarrier region. We cannot postulate that this decrease in the fission cross section is due to the opening of channels leading to neutron emission, since in the (d, pf) reactions the excitation energy imparted is less than the neutron binding energy. Consider Fig. 12–32, where the (compound) nucleus has been given an excitation energy E which is close in magnitude to one of the lowest vibrational class II states. Consider also the special situation in which the lowest one or two vibrational states have widths less than the spacing between the compound nuclear class II states at this excitation—a situation likely to occur if the second well is rather shallow with small zero-point energy (as in Th nuclei). At such excitations, the width of class I vibrational states is very large because of damping into the nearly continuous class I compound nuclear states, so that we can assume a continuous spectrum. The low-lying class II vibrational states at these excitations, on the other hand, can be considered to be discrete, and their decay to class I states is hindered by the potential barrier. The situation is thus analogous to the quantum-mechanical problem of penetration of a free particle (continuous spectrum) through a double-humped barrier. We briefly outline the treatment of Ignatyuk et al. [57] to demonstrate that under such conditions there will be resonances in the penetration function when the excitation energy E coincides with one of the well-defined bound states of the second well. These, in turn, will show up in the fission cross section.

Referring to Fig. 12–32, let P_A and P_B be the quasi-classical penetration probabilities of the separate barriers A and B at energy E. Then, by the WKB approximation (see Appendix D),

$$P_A = \exp\left[-2\int_{\alpha_1}^{\alpha_2}|\kappa(\alpha)|\,d\alpha\right],$$

$$P_B = \exp\left[-2\int_{\alpha_3}^{\alpha_4}|\kappa(\alpha)|\,d\alpha\right], \tag{12-49}$$

with $\kappa = \hbar^{-1}\{2B[E - V(\alpha)]\}^{1/2}$, where the α_i's are shown in Fig. 12-32, and B is the inertial parameter of the nucleus. The phase associated with the second well is defined by

$$\phi(E) = \int_{\alpha_2}^{\alpha_3}\kappa(\alpha)\,d\alpha, \tag{12-50}$$

and the bound states of the second well are located at energies E_n, for which

$$\phi(E_n) = (n + \tfrac{1}{2})\pi. \tag{12-51}$$

Ignatyuk et al. [57] show that the penetration probability through the two barriers is given approximately by

$$P(E) = \tfrac{1}{4}P_A P_B\left(\frac{(P_A + P_B)^2}{16}\sin^2\phi + \cos^2\phi\right)^{-1}. \tag{12-52}$$

When E coincides with one of the bound states E_n of the second well, Eq. 12-51 is satisfied, and $\cos\phi = 0$. In this case, the penetrability is maximum:

$$P_{\max} = \frac{4P_A P_B}{(P_A + P_B)^2}. \tag{12-53}$$

This value is unity for a symmetrical two-humped barrier, where $P_A = P_B = \tfrac{1}{2}$. The minimum penetrability is given by the condition that $\sin\phi = 0$; then

$$P_{\min} = \tfrac{1}{4}P_A P_B, \tag{12-54}$$

which is just $\tfrac{1}{16}$ for the symmetric case. In any case,

$$\frac{P_{\max}}{P_{\min}} = \frac{16}{(P_A + P_B)^2} \gg 1, \tag{12-55}$$

thus showing a structure in the subbarrier penetration function for each bound state of the second well. Such a resonance, for example, is seen at a neutron excitation of 715 keV in the fission cross section of ^{230}Th and in the (d, pf) reactions, as discussed earlier (see Figs. 12-7 and 12-8).

We next discuss briefly the intermediate structure in the subbarrier fission cross-section resonances as shown, for example, in Fig. 12-5. Let us envisage a situation in which the excitation is low enough for both the compound nuclear class I and class II states to be considered as discrete, but the vibrational states (in both the wells) to have widths greater than the spacing between the compound nuclear levels. For a given excitation, as mentioned before, the class I compound nuclear levels will be much more closely packed than the levels of the same nature in the second well. It is natural to expect a fine-grained resonance structure in the fission cross section due to the class I compound nuclear levels. Two distinct physical processes are taking place. First, the neutron is captured to form a compound nucleus, and then the compound nucleus undergoes fission. If the incident neutron excitation energy coincides with one of the class I compound nuclear levels, there is an enhancement in the cross section. Since these level spacings are of the order of 10–15 eV only, a group of them can have energy in the neighborhood of a class II compound nuclear level, whose width is typically a few tens of eV, with a level spacing of a few hundred eV. When this is the case, there is a strong coupling between class I and class II compound nuclear levels through the intermediate barrier. This will result in groups of finely grained resonances being strongly enhanced, as shown in Fig. 12–5, with a spacing between these groups corresponding to the level spacing between the class II compound nuclear states.

We now proceed to discuss, again in a qualitative manner, the effect of the double hump of the fission barrier on the angular distribution of the fission fragments. To do this, we shall first formulate the role of the "transition state" at the saddle point in the fission process, concentrating for simplicity on a single-humped barrier. We shall then point out what modifications may be expected in this picture because of the presence of a double barrier, and correlate these with the experimental observations.

Consider the situation in which the excitation energy imparted to the fissioning nucleus (by incident photons or neutron capture) is low enough to be near the threshold. In order to fission, the nucleus has to pass through the saddle point, where most of the excitation energy has been expended in deformation and the nucleus spends a relatively long time in a state of unstable equilibrium. These (decaying) states at the saddle point, called transition states, form a spectrum of excited states analogous to that near the ground state (Fig. 12–4). It is assumed that the nucleus has an axis of symmetry in its ground state and one in the transition state, although the orientation of these axes need not be the same, in general. Although the total angular momentum J and its projection M along a space-fixed axis must be conserved throughout the fission process, the projection K along the symmetry axis may be different in the ground and transition states, depending on the orientation of the symmetry axis. Since the passage from the saddle to scission point is very fast, the assumption is that it is the direction of the symmetry axis at the transition state along which the fission fragments separate. Since J is fixed, this direction is determined by the quantum number K of the transition state through which the nucleus passes. The different transition states at the saddle point with well-defined values of energy, J and K are called the fission channels; at low excitation energies only a few of these are available.

These ideas are quite successful in the explanation of the observed angular distribution in a number of cases. For example, in the photofission of even-even nuclei like ^{232}Th and ^{238}U near threshold, there is a pronounced anisotropy of the fission fragments perpendicular to the direction of the photon beam [58]. In this case, all the nucleonic spins are paired off and $K = 0$ for both the ground and the lowest transition states. At low energies, it is the dipole absorption, which imparts an angular momentum of $1\hbar$ to the nucleus with negative parity, that is important. The fission channels at the saddle point must therefore have $J = 1^-$. The low-lying spectrum of the transition states corresponds (as for an even-even nucleus) to the $K = 0$ rotational band, with possibly only one 1^- low-lying octupole vibrational state (see Chapter 9). Thus the low-energy photofission has to go through this $K = 0$, $J = 1^-$ transition state, where the symmetry axis is perpendicular to the direction of angular momentum **J**. Since a photon beam carries angular momentum only along and against the beam direction, the symmetry axis of the transition state in this fission channel is perpendicular to the beam direction. From our previous considerations, it then follows that the fragments will fly perpendicularly to the beam direction. As the excitation energy is increased, many more 1^- fission channels will become available with other values of K, and the anisotropy should be less drastic, in accordance with observations.

Similar reasoning would lead us to expect pronounced anisotropy in the angular distribution of fragments in neutron-induced fission near the threshold. For the excitation energy imparted through neutron capture to be near threshold, it is necessary, as discussed before, for the target nucleus to have an even number of neutrons, so that the binding energy of the odd neutron added to form the fissioning compound nucleus is small. Anisotropies have been experimentally observed in many such cases, some of which are shown in Fig. 12–34. If $W(\theta)$ denotes the angular distribution of a fission fragment at an angle θ from the incident beam direction, a measure of the anisotropy

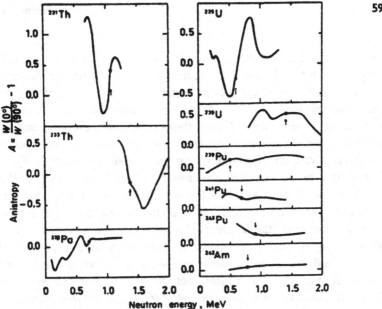

Fig. 12-34 Anisotropies in the angular distribution of a fission fragment in neutron-induced fission in the MeV region. The neutron energy corresponding to the fission barrier is shown by an arrow and a dot on each curve. Note that the structure in the anisotropy disappears in the heavier specimens. (After Bjornholm and Strutinsky [46].)

is $A = [W(0)/W(\pi/2)] - 1$. The approximate location of the fission barrier is shown by an arrow in these plots.

It will be seen from Fig. 12-34 that in the lighter actinides (up to ^{235}U) pronounced anisotropy appears in the angular distribution. Large (negative) dips in A near the threshold in these nuclides imply more fragments flying off in a direction perpendicular to the beam, as expected. However, the anisotropy is very weak or nearly absent in the heavier actinides, particularly in ^{243}Pu and ^{242}Am. This peculiarity cannot be explained by a one-humped barrier, but all these observations find a natural explanation on the basis of double fission barriers when we remember the relative variation of heights of the two barriers with increasing mass number. When two barriers are present, the threshold should be determined by the higher one. In the Th, Pa, and U nuclides, where the second barrier was estimated to be higher, the quantum numbers K of the few available transition states associated with this barrier should determine the angular distribution, resulting in anisotropy. However, in the nuclides ^{241}Pu, ^{243}Pu, and ^{242}Am, the first barrier is 1 or 2 MeV higher than the second, and therefore determines the threshold. The second well in these nuclides is still deep enough for the nucleus to spend a considerable time there, in the process "forgetting" the quantum numbers K associated with the transition states of the first barrier.

However, the fact that the excitation energy available to the nucleus is nearly equal to the threshold or the height of the first barrier makes it scan the transition states of the second barrier at an excitation of 1 or 2 MeV, where many fission channels with different values of K are available, and not just the lowest ones. Under these conditions, the resulting angular distribution is a superposition of many channels, averaging out peculiarities of individual distributions and giving only weak anisotropy [46]. These experimental results confirm the result obtained from theory that for americium and heavier actinides the second barrier is lower than the first.

Finally, we should mention that static calculations of the potential-energy surfaces with the inclusion of shell correction have led to a partial understanding of mass asymmetry in low-energy fission in the actinide nuclides. As discussed in an earlier section, even low-energy fission is found to be mass symmetric for lighter nuclei like ^{210}Po and heavier nuclei like ^{257}Fm, but has marked mass asymmetry for nuclei between these. In the liquid-drop model, one would expect this to imply the inclusion of P_3, P_5, and other odd Legendre polynomials in the angular dependence of the distortion, leading to a saddle-point shape that had no mirror symmetry about an axis perpendicular to the symmetry axis. As was pointed out earlier, however, the liquid-drop barrier was found to be stable against the inclusion of deformations of odd multipoles; this meant that such deformations only increased the barrier height. In the liquid-drop picture, therefore, the nucleus, on its path to fission, would deform in such a manner as to exclude these modes, and its shape at the saddle point would give rise to mass-symmetric fission only. As in the case of photofission of even-even nuclei near the threshold, exceptions would exist where the fission channel goes through the pear-shaped 1^- octupole state, resulting in mass asymmetry (see Fig. 9–14). Nevertheless, for the vast majority of observed cases, mass asymmetry remained an outstanding puzzle which defied even qualitative explanation. Following a conjecture of Swiatecki, however, Möller and Nilsson [27] calculated the potential-energy surfaces of ^{210}Po, ^{236}U, ^{242}Pu, and ^{252}Fm, including the shell correction to energy, by using the Strutinsky procedure, and taking into account P_3- and P_5-modes of distortion in the modified Nilsson-type well. They found that, whereas in ^{210}Po the barrier remained stable against deformations of P_3- and P_5-types, in the other isotopes the height of the second barrier decreased when these deformations were included. This decrease was particularly large for ^{236}U, where there was a reduction in height of more than 2 MeV for the second barrier due to these modes, resulting in the saddle-point shape shown in Fig. 12–35.

Nuclear shape at $\epsilon = 0.85$,
$\epsilon_3 = -0.16$, $\epsilon_4 = 0.12$, $\epsilon_5 = 0.08$

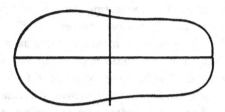

Fig. 12-35 The saddle-point shape of ^{236}U, including, in addition to the even modes, P_3- and P_5-modes of distortion in the nuclear shape. These odd modes actually decrease the height of the second barrier by about 2 MeV. (After Möller and Nilsson [27].)

One can expect unequal division of masses from such a shape. This instability of the second barrier against the inclusion of odd multipole distortions was much reduced in ^{252}Fm, where the decrease in height of the barrier peak was less than 0.5 MeV. This trend toward smaller asymmetry in going from uranium to fermium agrees with experimental observation. One would expect, on this basis, no asymmetry for the heavier isotopes of fermium (since the second barrier disappears), and this is found to be the case in the spontaneous fission of ^{257}Fm (see Fig. 12-16).

If a nuclear reaction leads to the formation of a nucleus in an excited state of the secondary minimum, this should result in electromagnetic transitions within the levels of the secondary well preceding isomeric fission. Specifically, in an even-even nucleus, the final decay in such a situation should proceed via $E2$-transitions within the rotational band built on the isomeric level. The de-excitation of the nucleus via $E2$-transitions may result either in the emission of γ-rays or in the emission of conversion electrons. The latter process, which is more important for nuclei with high Z and large deformation, is due to the sudden change during a transition in the electromagnetic field experienced by the atomic electrons, resulting in their acceleration and, in particular, the emission of some electrons on acquiring sufficient kinetic energy. The internal conversion process, as it is called, is thus the conversion of nuclear excitation energy to kinetic energy of the electrons through the direct electromagnetic interaction between the nucleons and electrons. The conversion electrons have been measured preceding the spontaneous fission of the ^{240}Pu isomer, which has a half-life of about 4×10^{-9} s, by Specht et al. [59]. This resulted in the identification of the rotational band in the secondary minimum of ^{240}Pu, up to the 8$^+$ level. Specht et al. used an α-beam of 25 MeV in the reaction ^{238}U(α, 2n)^{240}Pu, and identified the rotational band by measuring the delayed coincidence between the conversion electrons and the fission fragments of ^{240}Pu. This rotational band may be fitted by choosing the moment of inertia $\mathscr{I}$ so that $\hbar^2/2\mathscr{I} =$ 3.331 keV, which is very different from the corresponding value of 7.156 keV for the ground-state band. The increased value of $\mathscr{I}$ for the rotational band in the secondary minimum is due to the increased deformation. This experiment may be regarded as direct evidence for the existence of the secondary well in a fission isomer.

We may summarize this section by noting that diverse facets of fission phenomena find natural explanation in the existence of a fission barrier with a double hump, which comes about because of shell effects. More detailed treatment of the subject is given by Brack et al. [60] and by Swiatecki and Bjørnholm [61].

REFERENCES

1. O. Hahn and F. Strassmann, *Naturwissenschaften* 27, 11 (1939); 27, 89 (1939).
2. For a review on ternary fission, see I. Halpern, *Ann. Rev. Nucl. Sci.* 21, 245 (1971) and N. Feather, in *Proceedings of the Second International Symposium on the Physics and Chemistry of Fission, Vienna, 1969*, International Atomic Energy Agency, Vienna, 1969, p. 83.
3. The time for prompt neutron emission was estimated to be less than 4×10^{-14} s from the instant of fission by J. S. Fraser and J. C. D. Milton, *Phys. Rev.* 93, 818 (1954); J. E. Gindler and J. R. Huizenga quote a time range of 10^{-15} to 10^{-18} s for prompt neutron emission. See Fig. 1 in their article "Nuclear Fission," in *Nuclear Chemistry*, Vol. 2, edited by L. Yaffe, Academic Press, New York-London, 1968.
4. G. T. Seaborg, *Phys. Rev.* 88, 1429 (1952).
5. R. A. Schmitt and R. B. Duffield, *Phys. Rev.* 105, 1277 (1957).
6. J. A. Northrup, R. G. Stokes, and K. Boyer, *Phys. Rev.* 115, 1277 (1959).
7. These figures are taken from E. K. Hyde, *The Nuclear Properties of the Heavy Elements*, Vol. III, Prentice-Hall, Englewood Cliffs, N.J., 1964. This book gives a detailed account of fission phenomena up to 1963.
8. A. Bohr, *Proceedings of the United Nations Conference on the Peaceful Uses of Atomic Energy, Geneva, 1955*, Vol. 2, United Nations, New York, 1956, p. 151.
9. P. E. Vorotnikov, S. M. Dubrovina, G. A. Otroschenko, and V. A. Shigin, *Sov. J. Nucl. Phys.* 5, 210 (1967).

10. J. E. Lynn, *The Theory of Neutron Resonance Reactions*, Clarendon Press, Oxford, 1968, Chap. 8, Secs. 11 and 12.
11. J. Pedersen and B. P. Kuzminov, *Phys. Lett.* **29B**, 176 (1969).
12. H. W. Schmitt, J. H. Neiler, and F. J. Walter, *Phys. Rev.* **141**, 1146 (1966).
13. M. J. Bennet and W. E. Stein, *Phys. Rev.* **156**, 1277 (1967).
14. F. Pleasonton, *Phys. Rev.* **174**, 1500 (1968).
15. W. John, E. K. Hulet, R. W. Lougheed, and J. J. Wesolowski, *Phys. Rev. Lett.* **27**, 45 (1971).
16. W. E. Stein, *Phys. Rev.* **108**, 94 (1957).
17. J. C. D. Milton and J. S. Fraser, *Phys. Rev. Lett.* **7**, 67 (1961).
18. J. C. D. Milton and J. S. Fraser, *Can. J. Phys.* **40**, 1626 (1962).
19. S. L. Whetstone, Jr., *Phys. Rev.* **131**, 1232 (1963).
20. C. Signarbieux, M. Ribrag, and H. Nifenecker, *Nucl. Phys.* **A99**, 41 (1967).
21. See, for example, the review lecture by A. C. Pappas, J. Alstad, and E. Hageb, "Mass, Energy, and Charge Distribution in Fission," in *Proceedings of the Second International Symposium on the Physics and Chemistry of Fission, Vienna, 1969* (ref. 2).
22. V. F. Apalin, N. Yu. Gritsyuk, I. E. Kutikov, V. Lebedev, and L. A. Mikaelyan, in *Proceedings of the Symposium on the Physics and Chemistry of Fission, Salzburg, Austria, 1965*, Paper SM-60/92, International Atomic Energy Agency, Vienna, 1965.
23. H. R. Bowman, J. C. D. Milton, S. G. Thompson, and W. J. Swiatecki, *Phys. Rev.* **129**, 2133 (1963).
24. R. W. Fuller, *Phys. Rev.* **126**, 684 (1962).
25. See the article "Nuclear Fission" by J. E. Gindler and J. R. Huizenga (ref. 3), p. 155.
26. P. Fong, *Phys. Rev.* **135**, B1338 (1965).
27. P. Möller and S. G. Nilsson, *Phys. Lett.* **31B**, 283 (1970).
28. V. V. Pashkevich, *Nucl. Phys.* **A169**, 275 (1971).
29. J. P. Unik and J. R. Huizenga, *Phys. Rev.* **134**, B90 (1964).
30. G. Andritsopoulos, *Nucl. Phys.* **A94**, 537 (1967).
31. T. D. Thomas and R. Vandenbosch, *Phys. Rev.* **133**, B976 (1964).
32. See the section on ternary fission in *Proceedings of the Second International Symposium on the Physics and Chemistry of Fission, Vienna, 1969* (ref. 2).
33. G. M. Raisbeck and J. D. Thomas, *Phys. Rev.* **172**, 1272 (1968).
34. G. Kugler and W. B. Clarke, *Phys. Rev.* **C5**, 551 (1972).
35. G. Kugler, unpublished thesis on "Ternary Fission Studies of ^{235}U," McMaster University, Hamilton, 1970.
36. L. Meitner and O. R. Frisch, *Nature* **143**, 239 (1939).
37. N. Bohr and J. A. Wheeler, *Phys. Rev.* **56**, 426 (1939).
38. J. R. Nix and W. J. Swiatecki, *Nucl. Phys.* **71**, 1 (1965).
39. J. R. Nix, *Nucl. Phys.* **130**, 241 (1969).
40. L. Wilets, *Theories of Nuclear Fission*, Clarendon Press, Oxford, 1964, Chap. 2.
41. S. Cohen and W. J. Swiatecki, *Ann. Phys.* **22**, 406 (1963).
42. W. D. Myers and W. J. Swiatecki, *Nucl. Phys.* **81**, 1 (1966).
43. A. Sobiczewski, Z. Szymanski, S. Wycech, S. G. Nilsson, J. R. Nix, Chin Fu Tsang, C. Gustafson, P. Möller, and B. Nilsson, *Nucl. Phys.* **A131**, 67 (1969).
44. S. G. Nilsson, C. F. Tsang, A. Sobiczewski, Z. Szymanski, S. Wycech, C. Gustafson, I. Lamm, P. Möller, and B. Nilsson, *Nucl. Phys.* **A131**, 1 (1969).
45. T. Johansson, S. G. Nilsson, and Z. Szymanski, *Ann. Phys. (France)* **5**, 377 (1970).

46. S. Bjørnholm and V. M. Strutinsky, *Nucl. Phys.* **A136**, 1 (1969).
47. V. M. Strutinsky, *Nucl. Phys.* **A95**, 420 (1967); **A122**, 1 (1968).
48. H. J. Krappe and U. Wille, in *Proceedings of the Second International Symposium on the Physics and Chemistry of Fission, Vienna, 1969* (ref. 2), p. 197.
49. W.-F. Lin, *Phys. Rev.* **C2**, 871 (1970).
50. C. K. Ross and R. K. Bhaduri, *Nucl. Phys.* **A188**, 566 (1972).
51. Y. A. Muzychka, V. V. Pashkevich, and V. M. Strutinsky, *Sov. J. Nucl. Phys.* **8**, 417 (1969); V. V. Pashkevich, *Nucl. Phys.* **A169**, 275 (1971); M. Bolsterli, E. O. Fiset, J. R. Nix, and J. L. Norton, *Phys. Rev. Lett.* **27**, 681 (1971); *Phys. Rev.* **C5**, 1050 (1972).
52. S. M. Polikanov, V. A. Druin, V. A. Karnaukhov, V. L. Mikheev, A. A. Pleve, N. K. Skobelev, V. G. Subbotin, G. M. Ter-Akop'yan, and V. A. Fomichev, *Sov. Phys. JETP* **15**, 1016 (1962).
53. J. Borggreen, Yu. P. Gangrsky, G. Sletten, and S. Bjørnholm, *Phys. Lett.* **25B**, 402 (1967); N. N. Lark, G. Sletten, J. Pedersen, and S. Bjørnholm, *Nucl. Phys.* **A139**, 481 (1969).
54. G. N. Flerov, Yu. P. Gangrsky, B. N. Markov, A. A. Pleve, S. M. Polikanov, H. Jungclaussen, *Sov. J. Nucl. Phys.* **6**, 12 (1968).
55. Two excellent nontechnical articles on this subject are the following: D. D. Clark, *Phys. Today*, December 1971, p. 23, and D. H. Wilkinson, *Comments Nucl. Particle Phys.* **2**, 131 (1968).
56. J. E. Lynn, in *Proceedings of the Second International Symposium on the Physics and Chemistry of Fission, Vienna, 1969* (ref. 2), p. 249.
57. E. B. Gai, A. B. Ignatyuk, N. C. Rabotnov, and G. N. Smyrenkin, in *Proceedings of the Second International Symposium on the Physics and Chemistry of Fission, Vienna, 1969* (ref. 2), p. 337.
58. E. J. Winhold and I. Halpern, *Phys. Rev.* **103**, 990 (1956); A. P. Baerg, R. M. Bartholomew, F. Brown, L. Katz, and S. B. Kowalski, *Can. J. Phys.* **37**, 1418 (1959).
59. H. J. Specht, J. Weber, E. Konecny, and D. Heunemann, *Phys. Lett.* **41B**, 43 (1972).
60. M. Brack, J. Damgaard, A. S. Jensen, H. C. Pauli, V. M. Strutinsky, and C. Y. Wong, *Rev. Mod. Phys.* **44**, 320 (1972).
61. W. J. Swiatecki and S. Bjørnholm, *Phys. Rept.* **4C**, No. 6 (1972).
62. M. Brack and H. C. Pauli, *Nucl. Phys.* **A207**, 401 (1973).

PROBLEMS

12-1. *Prompt neutron spectrum in fission.* It is generally assumed that these neutrons are emitted from moving fission fragments after the fragments have attained their maximum velocities (see Section 10–1).

(a) Let a neutron be emitted from the fragment at a center-of-mass angle θ_{CM}, with energy in the *CM*-frame given by E_{CM}. Show that the energy E of the neutron in the laboratory frame is given by

$$E = E_f + E_{CM} + 2(E_f E_{CM})^{1/2} \cos \theta_{CM}, \qquad (1)$$

where E_f is the fragment kinetic energy per nucleon, that is, the energy $\frac{1}{2}Mv_f^2$ of a neutron moving with the velocity of the fragment.

(b) Assume that the emission of neutrons in the *CM*-frame is (i) isotropic; (ii) anisotropic but symmetric about $\theta_{CM} = \pi/2$. In both situations, prove that the average energies obey the relation

$$\bar{E} = \bar{E}_f + \bar{E}_{CM}. \tag{2}$$

(c) Let the energy distribution of the emitted neutrons in the laboratory frame be $N(E)$, normalized, for convenience, to unity. For given E_f and E_{CM}, and assuming isotropic emission, show that

$$N(E) = \tfrac{1}{4}(E_f E_{CM})^{-1/2}, \quad (\sqrt{E_{CM}} - \sqrt{E_f})^2 < E < (\sqrt{E_{CM}} + \sqrt{E_f})^2,$$

$$= 0, \qquad \text{otherwise.} \tag{3}$$

If the energy distribution of the neutrons in the CM-frame is given by $N'(E_{CM})$, it follows that the resulting $N(E)$ for a given E_f is

$$N(E) = \int_{(\sqrt{E} - \sqrt{E_f})^2}^{(\sqrt{E} + \sqrt{E_f})^2} \frac{N'(E_{CM})\, dE_{CM}}{4(E_f E_{CM})^{1/2}}. \tag{4}$$

From Eq. 4 show that, for very low energies $(E \to 0)$, $N(E) \propto \sqrt{E}$. If the distribution $N'(E_{CM})$ is taken to be Maxwellian, $N(E)$ can be obtained from Eq. 4. The resulting expression for $N(E)$, which can be written in a closed form, is known as the Watt formula. Empirically, however, the experimental data for thermal neutron capture of ^{233}U, ^{235}U, and ^{239}Pu are as well fitted by a Maxwellian form for $N(E)$:

$$N(E) \propto \sqrt{E} \exp\left(-\frac{E}{\tau}\right),$$

with $\tau \approx 1.3$ MeV. See, for more details, the paper by J. Terrell, *Phys. Rev.* **113**, 527 (1959).

12–2. *Smoothed single-particle density of states and shell correction.* (a) Consider the idealized single-particle spectrum given by Eq. 12–31, and assume that the lowest A states are each singly occupied by A fermions of one kind (we ignore the spin and i-spin degeneracy for simplicity here.) In the notation of the text, the single-particle density of states is given by

$$g(e) = \sum_{N=0}^{\infty} \delta(e - (N + \tfrac{1}{2})\hbar\omega).$$

Since, on the average, there is one state in the interval $\hbar\omega$, the smoothed density of states should simply be

$$\bar{g}(e) = 1/\hbar\omega, \quad e \geqslant 0,$$

$$= 0, \qquad e < 0.$$

Find the quantities $\tilde{\lambda}$ and $\bar{E}$, using Eq. 12–41 for this case, and show that $\bar{E} = A^2/2\hbar\omega$; it thus follows from Eq. 12–33 that the shell correction to energy is zero.

(b) The smoothed density $\bar{g}(e)$ of the spectrum in (a) may also be found by the method outlined in Problem 6–6, which you should do before attempting this part. Using the notation of Problem 6–6, show that the single-particle partition function $Z(\beta)$ is given by

$$Z(\beta) = \tfrac{1}{2} \operatorname{cosech} \frac{\hbar\omega\beta}{2} .$$

Making the expansion in powers of $\hbar$ (i.e., a semiclassical expansion), we obtain

$$Z(\beta) = \frac{1}{\hbar\omega\beta} - \frac{1}{24}\hbar\omega\beta + 0(\beta^3) + \cdots .$$

Taking the Laplace inverse term by term, we get

$$g(e) = \frac{1}{\hbar\omega} - \frac{1}{24}\hbar\omega\,\delta'(e) + 0[\delta'''(e)] + \cdots .$$

The first two terms in the series contribute smoothly to the energy.

Show that, if we include the δ'-term in the density of states, $\bar{E} = (A^2/2)\hbar\omega + \tfrac{1}{24}\hbar\omega$, so that the shell correction $(E - \bar{E}) = -\tfrac{1}{24}\hbar\omega$ for this case. An identical shell correction is found numerically when the Strutinsky method is applied to the single-particle spectrum given in (a).

(c) Consider the energy spectrum of an isotropic three-dimensional harmonic oscillator, for which

$$e_N = (N + \tfrac{3}{2})\hbar\omega, \quad N = 0, 1, 2, \ldots ,$$

with a degeneracy of $\tfrac{1}{2}(N + 1)(N + 2)$. We have again ignored the spin and i-spin degeneracy. Using the partition function method of (b), show that in this case the density of states contributing to the smoothed energy is

$$\bar{g}(e) = \frac{e^2}{2(\hbar\omega)^3} - \frac{1}{8}\frac{1}{\hbar\omega} + \frac{17}{1920}\hbar\omega\,\delta'(e).$$

For a discussion of this method and more solvable examples, see the paper by R. K. Bhaduri and C. K. Ross, *Phys. Rev. Lett.* **27**, 606 (1971).

12–3. As was mentioned in Section 12–2, the most frequent third particle in the ternary fission of a spontaneously fissioning nucleus is an α-particle. Kugler and Clarke [34] used the stacked foil assembly shown in Fig. 12–36a of dimensions such that essentially all the emitted α-particles from fissioning ^{235}U were stopped in the foils. The number N_i of α-particles stopped in the ith foil from the source (see Fig. 12–36b) was measured by mass-spectrometric techniques; knowledge of the range-energy relation of the α-particles in lead then made it possible to determine the energy distribution of the emitted α-particles.

(a) Let p_i be the probability that an α-particle of energy E and range $r(E)$, emitted in an arbitrary direction, is stopped in the ith foil (Fig. 12–36b).

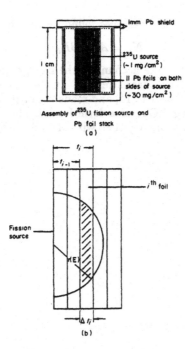

Fig. 12-36 (a) Assembly of ^{235}U fission source and lead foil stack. (b) The schematic geometry of fission source and foil stack. (After G. Kugler [35].)

Let Δt_i be the thickness of the ith foil, and t_i the total foil thickness up to and including the ith foil. Show that

$$p_i = \frac{\Delta t_i}{r(E)}, \qquad r(E) \geqslant t_i,$$

$$= \frac{r(E) - t_{i-1}}{r(E)}, \quad t_{i-1} < r(E) < t_i,$$

$$= 0, \qquad r(E) \leqslant t_{i-1}.$$

(b) Let the number of emitted α-particles between energies E and $E + dE$ be $n(E)\,dE$; then show that the total probability of finding a particle in the ith foil is

$$P_i = \Delta t_i \int_{E(t_i)}^{\infty} \frac{n(E)\,dE}{r(E)} + \int_{E(t_{i-1})}^{E(t_i)} \frac{r(E) - t_{i-1}}{r(E)} n(E)\,dE,$$

where $E(t_i)$ is the energy of an α-particle with range t_i.

The total number N_i of α-particles found in the ith foil is therefore

$$N_i = A + BP_i,$$

where A is a constant to take account of the background, and B is another constant depending on the source strength. In their experiment, Kugler and Clarke measured N_i for $i = 1$ to 11, and made use of the known range-energy function $r(E)$. Assuming a Gaussian form for $n(E)$, one could determine the parameters of the Gaussian, together with A and B, by a least-squares fit of the data.

Appendixes

Angular Momentum, Spherical Tensors, and Dirac Matrices

The quantum theories of angular momentum and of the relativistic electron are notoriously plagued by a variety of notations and of different authors' personal choices of arbitrary phases. It therefore seems desirable to collect a number of important formulas in this appendix, to show the notation used in this book, and to provide a convenient single summary. The appendix is not an exposition of these parts of quantum mechanics, although it does contain most of the basic equations.

Throughout this appendix, angular momenta are measured in units of $\hbar$; in other words, $\hbar$ is set equal to 1.

A-1 ANGULAR MOMENTUM

(a) *Fundamental matrix elements.* Let any angular momentum **J** have components J_x, J_y, J_z along spaced-fixed axes. Let

$$J_\pm = J_x \pm iJ_y. \tag{A-1}$$

The fundamental relationships that identify the vector **J** as an angular momentum are the commutators:

$$[J_x, J_y] = iJ_z \quad \text{and cyclic permutations,} \tag{A-2}$$

$$[J_z, J_\pm] = \pm J_\pm, \tag{A-3}$$

$$[J^2, J_z] = 0. \tag{A-4}$$

Using the basic relationship $[x, p_x] = i$, we may prove these results for the orbital angular momentum $\mathbf{l} = \mathbf{r} \times \mathbf{p}$ of a single particle and for any sum of orbital angular momenta. Equations A–2 are postulated to hold for spin angular momentum $\mathbf{s}$, for which there is no classical analog. It can then be proved that they hold for any angular momentum

$$\mathbf{J} = \sum_i (\mathbf{l}_i + \mathbf{s}_i).$$

From the commutation relations, it follows that J^2 and J_z are simultaneously observable, that is, we can specify states in which they both have given values. These values are $J(J+1)$ and M, respectively, where J is integral or half-integral and nonnegative and $-J \leqslant M \leqslant J$. In symbols, we denote the wave function of such a state by $|JM\rangle$, and

$$J^2|JM\rangle = J(J+1)|JM\rangle, \tag{A-5}$$

$$J_z|JM\rangle = M|JM\rangle. \tag{A-6}$$

Of course the states are orthogonal and normalized, that is,

$$\langle JM|J'M'\rangle = \delta_{JJ'}\delta_{MM'}. \tag{A-7}$$

The components J_x and J_y cannot be given definite values simultaneously with J_z, but the combinations $J_\pm$ have the important property of operating on the state $|JM\rangle$ and converting it into a state with an M-value one greater or one less:

$$J_\pm|JM\rangle = [J(J+1) - M(M \pm 1)]^{1/2}|JM \pm 1\rangle \tag{A-8}$$

$$= [(J \mp M)(J \pm M + 1)]^{1/2}|JM \pm 1\rangle. \tag{A-9}$$

Consequently,

$$\langle JM|J_\pm|J'M'\rangle = \delta_{JJ'}[(J \mp M)(J \pm M + 1)]^{1/2}\delta_{M,M'\pm 1}. \tag{A-10}$$

(b) *Orbital angular momentum.* In the Schrödinger picture of quantum mechanics, which uses differential operators, single-particle wave functions depend on the coordinates $\mathbf{r}$ of a particle, and the momentum operator is $\mathbf{p} = -i\nabla$. For the angular-momentum operators it is natural to use spherical polar coordinates, and from $\mathbf{L} = \mathbf{r} \times \mathbf{p}$ we find

$$L_\pm = e^{\pm i\phi}\left(\pm \frac{\partial}{\partial\theta} + i\cot\theta\frac{\partial}{\partial\phi}\right), \tag{A-11a}$$

$$L_z = -i\frac{\partial}{\partial\phi}, \tag{A-11b}$$

$$L^2 = -\frac{1}{\sin^2\theta}\frac{\partial^2}{\partial\phi^2} - \frac{1}{\sin\theta}\frac{\partial}{\partial\theta}\left(\sin\theta\frac{\partial}{\partial\theta}\right). \qquad \text{(A–11c)}$$

The expression for L^2 is essentially the angular terms of the Laplacian. The eigenvalues of L^2 and L_z are $l(l+1)$ and m, respectively, where l takes on only nonnegative integral values and $m = -l, -l+1, \ldots, l-1, l$. The eigenfunctions $|lm\rangle$ are given in this representation by the spherical harmonics

$$|lm\rangle = Y_l{}^m(\theta, \phi). \qquad \text{(A–12)}$$

The definition of spherical harmonics most widely used in nuclear physics is that given by Condon and Shortley [1]; it is the one adopted in this book:

$$Y_l{}^m = \Theta_l{}^m(\theta)(2\pi)^{-1/2}e^{im\phi}, \qquad \text{(A–13a)}$$

$$\Theta_l{}^m(\theta) = (-)^m\left[\frac{2l+1}{2}\frac{(l-m)!}{(l+m)!}\right]^{1/2}\sin^m\theta\,\frac{d^m}{(d\cos\theta)^m}P_l(\cos\theta), \quad m>0,$$
$$\text{(A–13b)}$$

$$\Theta_l{}^{-m}(\theta) = (-)^m\Theta_l{}^m(\theta), \quad m>0, \qquad \text{(A–13c)}$$

$$P_l(\cos\theta) = \frac{(-)^l}{2^l l!}\frac{d^l}{(d\cos\theta)^l}\sin^{2l}\theta. \qquad \text{(A–13d)}$$

Another way of writing $\Theta_l{}^m$, which is valid for either sign of m, is

$$\Theta_l{}^m = (-)^l\left[\frac{2l+1}{2}\frac{(l+m)!}{(l-m)!}\right]^{1/2}\frac{1}{2^l l!}\frac{1}{\sin^m\theta}\frac{d^{l-m}}{(d\cos\theta)^{l-m}}\sin^{2l}\theta. \quad \text{(A–14)}$$

Useful explicit forms are as follows:

$$\Theta_0{}^0 = \sqrt{\tfrac{1}{2}};$$

$$\Theta_1{}^0 = \sqrt{\tfrac{3}{2}}\cos\theta, \qquad \Theta_1{}^{\pm1} = \mp\sqrt{\tfrac{3}{4}}\sin\theta;$$

$$\Theta_2{}^0 = \sqrt{\tfrac{5}{8}}(3\cos^2\theta - 1),$$

$$\Theta_2{}^{\pm1} = \mp\sqrt{\tfrac{15}{4}}\cos\theta\sin\theta, \qquad \Theta_2{}^{\pm2} = \sqrt{\tfrac{15}{16}}\sin^2\theta;$$

$$\Theta_3{}^0 = \sqrt{\tfrac{7}{8}}(5\cos^3\theta - 3\cos\theta), \qquad \Theta_3{}^{\pm1} = \mp\sqrt{\tfrac{21}{32}}(5\cos^2\theta - 1)\sin\theta,$$

$$\Theta_3{}^{\pm2} = \sqrt{\tfrac{105}{16}}\cos\theta\sin^2\theta, \qquad \Theta_3{}^{\pm3} = \mp\sqrt{\tfrac{35}{32}}\sin^3\theta.$$

(c) *Spin.* Since the spin of a fermion is a purely quantum-mechanical concept, there is no differential operator suitable to describe it, but an explicit representation of the operators s is provided by any mathematical entities that satisfy Eqs. A–2 to A–9. The standard procedure is to use the Pauli spin matrices σ, defined by

$$s = \tfrac{1}{2}\sigma, \tag{A–15}$$

$$\sigma_x = \begin{pmatrix} 0 & 1 \\ 1 & 0 \end{pmatrix}, \qquad \sigma_y = \begin{pmatrix} 0 & -i \\ i & 0 \end{pmatrix}, \qquad \sigma_z = \begin{pmatrix} 1 & 0 \\ 0 & -1 \end{pmatrix}. \tag{A–16}$$

The commutation relations A–2 through A–4 are satisfied in the form

$$[\sigma_x, \sigma_y] = 2i\sigma_z, \qquad [\sigma_z, \sigma_\pm] = \pm\, 2\sigma_\pm. \tag{A–17}$$

In addition, the operators anticommute, that is,

$$\sigma_i\sigma_j + \sigma_j\sigma_i = 2\delta_{ij}. \tag{A–18}$$

In particular,

$$\sigma_x^2 = \sigma_y^2 = \sigma_z^2 = 1.$$

These results imply the cyclic relations typified by

$$\sigma_x\sigma_y = i\sigma_z. \tag{A–19}$$

We see that $s^2 = \tfrac{1}{4}\sigma^2 = \tfrac{3}{4}$, which shows that these are appropriate operators to describe an angular momentum $J = \tfrac{1}{2}$.

The "wave functions" representing the eigenstates of s^2 and s_z are the two-component vectors

$$\alpha = \begin{pmatrix} 1 \\ 0 \end{pmatrix}, \qquad \beta = \begin{pmatrix} 0 \\ 1 \end{pmatrix}. \tag{A–20}$$

Since

$$\sigma_z\alpha = \alpha, \qquad \sigma_z\beta = -\,\beta, \tag{A–21}$$

we see that α describes a state with spin "up" and β a state with spin "down," that is, $m = \tfrac{1}{2}$ and $-\tfrac{1}{2}$, respectively. Also

$$\sigma_-\alpha = 2\beta, \qquad \sigma_+\beta = 2\alpha. \tag{A–22}$$

The wave function $\begin{pmatrix} a \\ b \end{pmatrix}$ represents a state with probability $|a|^2$ of finding the spin up and probability $|b|^2$ of finding it down. The state is normalized if $|a|^2 + |b|^2 = 1$.

The Hermitian adjoint of the vector

$$|\psi\rangle = \begin{pmatrix} a \\ b \end{pmatrix} \quad \text{is} \quad \langle\psi| = (a^*b^*),$$

and the product $\langle \psi_1 | \psi_2 \rangle$ is explicitly

$$(a_1{}^*b_1{}^*) \begin{pmatrix} a_2 \\ b_2 \end{pmatrix} = a_1{}^*a_2 + b_1{}^*b_2. \tag{A-23}$$

The eigenstates α and β are of course orthogonal,

$$(\alpha | \beta) = 0.$$

(d) *Isospin.* The essential feature of the three Pauli matrices and their associated eigenvectors is their ability to describe states distinguished from each other by the value of a quantity (like s_z) which has two possible values. It is therefore possible to use the same representation for the i-spin of nucleons. With

$$t_i = \tfrac{1}{2}\tau_i, \tag{A-24}$$

$$\tau_1 = \begin{pmatrix} 0 & 1 \\ 1 & 0 \end{pmatrix}, \qquad \tau_2 = \begin{pmatrix} 0 & -i \\ i & 0 \end{pmatrix}, \qquad \tau_3 = \begin{pmatrix} 1 & 0 \\ 0 & -1 \end{pmatrix},$$

$$\gamma = \begin{pmatrix} 1 \\ 0 \end{pmatrix}, \qquad \delta = \begin{pmatrix} 0 \\ 1 \end{pmatrix},$$

we have

$$(\tfrac{1}{2} + t_3)\gamma = \gamma, \qquad (\tfrac{1}{2} + t_3)\delta = 0, \tag{A-25}$$

$$t_+\delta = (t_1 + it_2)\delta = \gamma, \qquad t_-\gamma = \delta. \tag{A-26}$$

Equation A–25 shows that γ and δ provide representations for two charge states, γ being the proton and δ the neutron, with the operator for charge being $(\tfrac{1}{2} + t_3)e$. Equations A–26 show that t_+ and t_- are operators which, respectively, convert neutron states into proton states and vice versa. They are therefore useful in describing β-decay.

(e) *Addition of two angular momenta.* Two angular momenta, $\mathbf{J}_1$ and $\mathbf{J}_2$, may be added to form a resultant,

$$\mathbf{J} = \mathbf{J}_1 + \mathbf{J}_2.$$

The possible values of J are those that can be formed by a vector triangle, that is,

$$J = J_1 + J_2, J_1 + J_2 - 1, \ldots, |J_1 - J_2| + 1, |J_1 - J_2|. \tag{A-27}$$

The state in which J^2 and J_z have fixed values $J(J + 1)$ and M is not, in general, a state in which J_{1z} and J_{2z} have fixed values, but is a combination of all the states for which

$$M_1 + M_2 = M. \tag{A-28}$$

Specifically, if we denote the state with quantum numbers J_1, J_2, J, and M by $|J_1J_2JM\rangle$, we have

$$|J_1J_2JM\rangle = \sum_{M_1,M_2} (J_1J_2M_1M_2|JM)|J_1M_1\rangle|J_2M_2\rangle. \tag{A-29}$$

In Eq. A–29, $|J_1M_1\rangle|J_2M_2\rangle$ is the state in which both angular momenta $\mathbf{J}_1$ and $\mathbf{J}_2$ have fixed magnitudes and z-components, and $(J_1J_2M_1M_2|JM)$ is the *Clebsch-Gordan vector-addition coefficient*.

These important coefficients have been given a wide variety of notations. There has also been some variation in the choice of phases, but most authors now use the conventions adopted by Condon and Shortley, namely, that all coefficients are real, those with $M = J$ are positive, and all angular-momentum states satisfy Eq. A–10. The Clebsch-Gordan coefficients are closely related to Wigner's *3-j symbols*, which refer to the addition of three angular momenta to a zero resultant. Precisely,

$$\begin{pmatrix} J_1J_2J \\ M_1M_2M \end{pmatrix} = \frac{(-)^{J_1-J_2-M}}{(2J+1)^{1/2}} (J_1J_2M_1M_2|J - M). \tag{A-30}$$

The Clebsch-Gordan coefficients satisfy the symmetry relations:

$$(J_1J_2M_1M_2|JM) = (-)^{J_1+J_2-J}(J_2J_1M_2M_1|JM), \tag{A-31a}$$

$$= (-)^{J_1-J+M_2}\left(\frac{2J+1}{2J_1+1}\right)^{1/2} (JJ_2M - M_2|J_1M_1), \tag{A-31b}$$

$$= (-)^{J_2-J-M_1}\left(\frac{2J+1}{2J_2+1}\right)^{1/2} (J_1J - M_1M|J_2M_2), \tag{A-31c}$$

$$= (-)^{J_1+J_2-J}(J_1J_2 - M_1 - M_2|J - M). \tag{A-31d}$$

Also,

$$\sum_{M_1,M_2} (J_1J_2M_1M_2|JM)(J_1J_2M_1M_2|J'M') = \delta_{JJ'}\delta_{MM'}, \tag{A-32a}$$

$$\sum_{JM} (J_1J_2M_1M_2|JM)(J_1J_2M_1'M_2'|JM) = \delta_{M_1M_1'}\delta_{M_2M_2'}. \tag{A-32b}$$

A state of fixed M_1 and M_2 is not, in general, a state of fixed J, but a linear combination of states of all J from $|J_1 - J_2|$ to $J_1 + J_2$, namely,

$$|J_1M_1\rangle|J_2M_2\rangle = \sum_{J} (J_1J_2M_1M_2|JM_1 + M_2)|J_1J_2JM_1 + M_2\rangle. \tag{A-33}$$

There is a closed form (due to Racah) for Clebsch-Gordan coefficients:

$$(J_1 J_2 M_1 M_2 | JM)$$
$$= \delta_{M, M_1 + M_2}[(2J + 1)(J_1 + J_2 - J)!(J + J_1 - J_2)!(J + J_2 - J_1)!$$
$$\times (J_1 + M_1)!(J_1 - M_1)!(J_2 + M_2)!(J_2 - M_2)!(J + M)!(J - M)!]^{1/2}$$
$$\times [(J + J_1 + J_2 + 1)!]^{-1/2}$$
$$\times \sum_s (-)^s[s!(J_1 + J_2 - J - s)!(J_1 - M_1 - s)!(J_2 + M_2 - s)!$$
$$\times (J - J_2 + M_1 + s)!(J - J_1 - M_2 + s)!]^{-1}. \qquad (A-34)$$

The summation over s includes all values for which none of the factorials have negative arguments ($0! = 1$). This formula is cumbersome to use, and tables of Clebsch-Gordan coefficients have been prepared [1, 2, 3, 4, 5]. The coefficients are all square roots of rational fractions; some of the tables list the separate factors of these fractions, whereas others give simply numerical values. Some of the coefficients for low values of J_2 are given in Table A-1. It is also worth noting that

$$(J_1 J_2 J_1 J_2 | J_1 + J_2 J_1 + J_2) = 1, \qquad (A-35)$$

$$(J0M0 | JM) = 1. \qquad (A-36)$$

TABLE A-1   Values of $\Phi(J_2, M_2, J)^a$.

	$J_2 = 0$:	$\Phi(0, 0, 0) = 1$	
	$J_2 = \frac{1}{2}$:	$\Phi(\frac{1}{2}, \frac{1}{2}, J_1 + \frac{1}{2}) = 1$	

$J_2 = 1$	$J = J_1$	$J = J_1 + 1$	
$M_2 = 0$	$- 2M_1$	$- \sqrt{2(J_1 - M_1 + 1)}$	
$M_2 = 1$	$- \sqrt{2(J_1 - M_1)}$	1	

$J_2 = \frac{3}{2}$	$J = J_1 + \frac{1}{2}$	$J = J_1 + \frac{3}{2}$	
$M_2 = \frac{1}{2}$	$J_1 - 3M_1$	$- \sqrt{3(J_1 - M_1 + 1)}$	
$M_2 = \frac{3}{2}$	$- \sqrt{3(J_1 - M_1)}$	1	

$J_2 = 2$	$J = J_1$	$J = J_1 + 1$	$J = J_1 + 2$
$M_2 = 0$	$2[3M_1^2 - J_1(J_1 + 1)]$	$2M_1\sqrt{6(J_1 - M_1 + 1)}$	$\sqrt{6(J_1 - M_1 + 2)(J_1 - M_1 + 1)}$
$M_2 = 1$	$(2M_1 + 1)\sqrt{6(J_1 - M_1)}$	$2(J_1 - 2M_1)$	$- 2\sqrt{J_1 - M_1 + 1}$
$M_2 = 2$	$\sqrt{6(J_1 - M_1)(J_1 - M_1 - 1)}$	$- 2\sqrt{J_1 - M_1}$	1

^{a} Defined in Eq. A-37.

The quantities in Table A-1 are $\Phi(J_2, M_2, J)$, defined by

$$(J_1 J_2 M_1 M_2 | JM) = (-)^{M_2-J_2}(2J+1)^{1/2}$$

$$\times \left[\frac{(J_1+J-J_2)!(J+M)!}{(J_1+J_2+J+1)!(J_1+M_1)!} \right]^{1/2} \Phi(J_2, M_2, J).$$

$$(A-37)$$

This equation holds only for $J \geqslant J_1$ and $M_2 \geqslant 0$. For $J < J_1$ and/or $M_2 < 0$, Eqs. A-31 must be used first to obtain a coefficient to which Eq. A-37 can be applied.

In particular, we may note the addition of orbital and spin angular momenta of a single nucleon:

$$\mathbf{j} = \mathbf{l} + \mathbf{s}.$$

The state $j = l + \frac{1}{2}, m$ is

$$(2l+1)^{-1/2}[(l+m+\tfrac{1}{2})^{1/2}Y_l^{m-(1/2)}\alpha + (l-m+\tfrac{1}{2})^{1/2}Y_l^{m+(1/2)}\beta], \quad (A-38a)$$

and the state $j = l - \frac{1}{2}, m$ is

$$(2l+1)^{-1/2}[-(l-m+\tfrac{1}{2})^{1/2}Y_l^{m-(1/2)}\alpha + (l+m+\tfrac{1}{2})^{1/2}Y_l^{m+(1/2)}\beta]. \quad (A-38b)$$

Another important case is the addition of the spins of two particles. The singlet state ($j = 0$) and the triplet states ($j = 1$) are given by

$$^1(\sigma)_0 = (\tfrac{1}{2}\tfrac{1}{2}\tfrac{1}{2}-\tfrac{1}{2}|0\ 0)\alpha(1)\beta(2) + (\tfrac{1}{2}\tfrac{1}{2}-\tfrac{1}{2}\tfrac{1}{2}|0\ 0)\beta(1)\alpha(2)$$

$$= (1/\sqrt{2})(\alpha(1)\beta(2) - \beta(1)\alpha(2)),$$

$$^3(\sigma)_1 = (\tfrac{1}{2}\tfrac{1}{2}\tfrac{1}{2}\tfrac{1}{2}|1\ 1)\alpha(1)\alpha(2) = \alpha(1)\alpha(2),$$

$$^3(\sigma)_{-1} = (\tfrac{1}{2}\tfrac{1}{2}-\tfrac{1}{2}-\tfrac{1}{2}|1\ -1)\beta(1)\beta(2) = \beta(1)\beta(2),$$

$$^3(\sigma)_0 = (\tfrac{1}{2}\tfrac{1}{2}\tfrac{1}{2}-\tfrac{1}{2}|1\ 0)\alpha(1)\beta(2) + (\tfrac{1}{2}\tfrac{1}{2}-\tfrac{1}{2}\tfrac{1}{2}|1\ 0)\alpha(2)\beta(1)$$

$$= (1/\sqrt{2})(\alpha(1)\beta(2) + \alpha(2)\beta(1)).$$

Also, the two-particle states with quantum numbers $SLJM$ are given by

$$\mathcal{Y}_{JSL}^M = \sum_{M_S, M_L} (SLM_S M_L | JM)^{2S+1}(\sigma)_{M_S} Y_L^{M_L}.$$

For example,

$$\mathcal{Y}_{112}^1 = \frac{1}{\sqrt{10}}\,^3(\sigma)_1 Y_2^0 - \sqrt{\frac{3}{10}}\,^3(\sigma)_0 Y_2^1 + \sqrt{\frac{3}{5}}\,^3(\sigma)_{-1} Y_2^2.$$

(f) *Addition of three angular momenta.* Suppose that $\mathbf{J} = \mathbf{J}_1 + \mathbf{J}_2 + \mathbf{J}_3$. We can form two kinds of states, for both of which the magnitude and orientation of $\mathbf{J}$ are fixed. First, we can have states in which the magnitude of $\mathbf{J}_{12} = \mathbf{J}_1 + \mathbf{J}_2$ is fixed; these states are

$$|(J_1 J_2) J_{12}, J_3; JM\rangle = \sum_{M_3} (J_{12} J_3 M_{12} M_3 | JM) |J_{12} M_{12}\rangle |J_3 M_3\rangle. \quad \text{(A-39a)}$$

Alternatively, we can take states in which $\mathbf{J}_{13} = \mathbf{J}_1 + \mathbf{J}_3$ has a definite magnitude:

$$|(J_1 J_3) J_{13}, J_2; JM\rangle = \sum_{M_2} (J_{13} J_2 M_{13} M_2 | JM) |J_{13} M_{13}\rangle |J_2 M_2\rangle. \quad \text{(A-39b)}$$

Of course, the states $|J_{12} M_{12}\rangle$ and $|J_{13} M_{13}\rangle$ are themselves complex. For example,

$$|J_{12} M_{12}\rangle = \sum_{M_1} (J_1 J_2 M_1 M_2 | J_{12} M_{12}) |J_1 M_1\rangle |J_2 M_2\rangle.$$

One class of states, say with J_{12} fixed, can be expressed in terms of the other set with J_{13} fixed. The coefficients of the relationship are called Racah's W-functions:

$$|(J_1 J_2) J_{12}, J_3; JM\rangle$$
$$= \sum_{J_{13}} W(J_1 J_2 J_3 J; J_{12} J_{13})[(2J_{12} + 1)(2J_{13} + 1)]^{1/2} |(J_1 J_3) J_{13}, J_2; JM\rangle.$$
$$\text{(A-40)}$$

The W-functions are essentially the same as Wigner's *6-j symbols*, defined by

$$\begin{Bmatrix} J_1 J_2 J_{12} \\ J J_3 J_{13} \end{Bmatrix} = (-)^{J_1 + J_2 + J_3 + J} W(J_1 J_2 J_3 J; J_{12} J_{13}). \quad \text{(A-41)}$$

It is not surprising that W-functions arise in discussions of the individual-particle model, in angular correlations, and in differential-reaction cross sections involving three angular momenta. However, because we have not exhibited the mathematical details of such calculations in the text, we shall not summarize the various orthogonality relations of the W's or all their connections with the Clebsch-Gordan coefficients. We shall simply quote one expression for W and give the definitions of the functions Z and F_λ, which are referred to in some sections on angular correlations:

$$W(J_1 J_2 J_3 J; J_{12} J_{13})$$
$$= \sum_{\text{all } M} (-)^S \times$$

$$\times \begin{pmatrix} J_1 J_2 J_{12} \\ M_1 M_2 - M_{12} \end{pmatrix} \begin{pmatrix} J_1 J_3 J_{13} \\ M_1 M_3 - M_{13} \end{pmatrix} \begin{pmatrix} J_2 J_{13} J \\ M_2 M_{13} - M \end{pmatrix} \begin{pmatrix} J_3 J_{12} J \\ M_3 M_{12} - M \end{pmatrix}$$

$$= \sum_{\text{all } M} (-)^{S+T}$$

$$\times [(2J_{12} + 1)(2J_{13} + 1)]^{-1/2}(2J + 1)^{-1}(J_1 J_2 M_1 M_2 | J_{12} M_{12})$$

$$\times (J_1 J_3 M_1 M_3 | J_{13} M_{13})(J_2 J_{13} M_2 M_{13} | JM)(J_3 J_{12} M_3 M_{12} | JM), \quad \text{(A-42)}$$

where $S = J_1 + J_2 - J_{12}$ and $T = J_{12} + M_{12} + J_{13} + M_{13}$. When L and L' are integral, we define

$$Z(LJL'J'; j\lambda) = i^{L'-L+\lambda}[(2L + 1)(2L' + 1)(2J + 1)(2J' + 1)]^{1/2}$$

$$\times (LL'00|\lambda 0)W(LL'JJ'; \lambda j), \quad \text{(A-43)}$$

$$F_\lambda(LL'J_1 J_2) = (-)^{J_1 - J_2 + 1}[(2L + 1)(2L' + 1)(2J_2 + 1)]^{1/2}(LL'1 - 1|\lambda 0)$$

$$\times W(LL'J_2 J_2; \lambda J_1). \quad \text{(A-44)}$$

A-2 SPHERICAL TENSORS

(a) *Rotation matrices $\mathscr{D}$.* It is important to be able to connect the wave function referred to one system of axes with the wave functions in a different frame obtained by a set of rotations of the original frame.

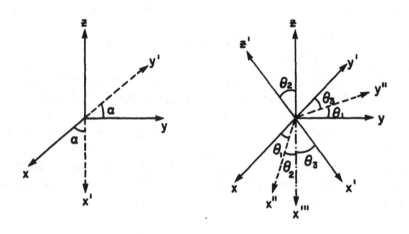

Fig. A-1 Rotations of Axes.

Consider the rotation of a right-handed frame of reference (x, y, z) about the z-axis by an angle α, as shown in Fig. A–1. Let P be a fixed point (x, y, z) in the old frame; then in the rotated frame the same point P is (x', y', z'). It is simpler to use polar coordinates (r, θ, ϕ) in the old frame, and (r', θ', ϕ') in the rotated frame. In the simple example shown in Fig. A–1, the rotation operation by α about the z-axis implies that $r' = r$, $\theta' = \theta$, and $\phi' = \phi - \alpha$. Thus the rotation operation $R_z(\alpha)$ on a function $f(r, \theta, \phi)$ gives

$$R_z(\alpha)f(r, \theta, \phi) = f(r, \theta, \phi - \alpha)$$

$$= f(r, \theta, \phi) - \alpha \frac{\partial}{\partial \phi} f(r, \theta, \phi) + \frac{\alpha^2}{2!} \frac{\partial^2}{\partial \phi^2} f(r, \theta, \phi) - \cdots$$

$$= \exp\left(-\alpha \frac{\partial}{\partial \phi}\right) f(r, \theta, \phi).$$

Using Eq. A–11b, we can rewrite the above equation as ($\hbar = 1$),

$$R_z(\alpha)f(r, \theta, \phi) = \exp(-i\alpha L_z)\, f(r, \theta, \phi). \tag{A–45}$$

Similarly, the rotation operator in the configuration space for rotation α about an arbitrary unit vector $\mathbf{n}$ is given by $\exp[-i\alpha(\mathbf{n} \cdot \mathbf{L})]$. Generalizing the space to include both the configuration space and the spin space, we can write the rotation operator for a rotation α about the unit vector $\mathbf{n}$ as

$$R_\mathbf{n}(\alpha) = \exp[-i\alpha(\mathbf{n} \cdot \mathbf{J})], \tag{A–46}$$

where $\mathbf{J} = \mathbf{L} + \mathbf{S}$. Note that the operator $R_\mathbf{n}(\alpha)$ is unitary.

The rotations taking one coordinate system into another are conveniently specified by the Euler angles $(\theta_1, \theta_2, \theta_3)$, which denote three successive rotations: a rotation θ_1 about the z-axis, followed by a rotation θ_2 about the new y-axis y'', followed by a rotation θ_3 about the final z-axis, z'. The ranges of values taken by the θ_i's are $0 \leqslant \theta_1 \leqslant 2\pi$, $0 \leqslant \theta_2 \leqslant \pi$, and $0 \leqslant \theta_3 \leqslant 2\pi$. We can thus write the general rotation operator in terms of the Euler angles θ_i as

$$R(\theta_1, \theta_2, \theta_3) = \exp(-i\theta_3 J_{z'}) \exp(-i\theta_2 J_{y''}) \exp(-i\theta_1 J_z).$$

Using the relations

$$\exp(-i\theta_2 J_{y''}) = \exp(-i\theta_1 J_z) \exp(-i\theta_2 J_y) \exp(i\theta_1 J_z)$$

and

$$\exp(-i\theta_3 J_{z'}) = \exp(-i\theta_2 J_{y''}) \exp(-i\theta_3 J_z) \exp(i\theta_2 J_{y''}),$$

we finally obtain

$$R(\theta_i) = R(\theta_1, \theta_2, \theta_3) = \exp(-i\theta_1 J_z) \exp(-i\theta_2 J_y) \exp(-i\theta_3 J_z). \quad (A\text{-}47)$$

Since, from the above definition, $[R(\theta_i), J^2] = 0$, it follows that

$$J^2 R(\theta_i)|JM\rangle = R(\theta_i)J^2|JM\rangle$$
$$= J(J+1)R(\theta_i)|JM\rangle,$$

so that rotations cannot change the total angular momentum of a state. However, unless the axis of rotation is along the z-axis, the z-component of angular momentum will change, so that

$$R(\theta_i)|JK\rangle = \sum_M \langle JM|R(\theta_i)|JK\rangle|JM\rangle. \quad (A\text{-}48)$$

We define† the $\mathscr{D}$-functions as

$$[\mathscr{D}^J_{MK}(\theta_i)]^* = \langle JM|R(\theta_i)|JK\rangle \quad (A\text{-}49)$$

to rewrite Eq. A-48:

$$R(\theta_i)|JK\rangle = \sum_M [\mathscr{D}^J_{MK}(\theta_i)]^*|JM\rangle. \quad (A\text{-}50)$$

Thus, for the rotation operator $R(\theta_i)$, we can associate a $(2J+1) \times (2J+1)$ matrix $\mathscr{D}^J(\theta_i)$, whose elements are given by Eq. A-49. By combining Eqs. A-47 and A-49, we obtain

$$[\mathscr{D}^J_{M'M}(\theta_i)]^* = \exp(-i\theta_1 M') \langle JM'|\exp(-i\theta_2 J_y)|JM\rangle \exp(-i\theta_3 M)$$
$$= \exp(-i\theta_1 M') \, d^J_{M'M}(\theta_2) \exp(-i\theta_3 M), \quad (A\text{-}51)$$

with

$$d^J_{M'M}(\theta_2) = \langle JM'|\exp(-i\theta_2 J_y)|JM\rangle. \quad (A\text{-}52)$$

We have chosen the representation in which J_z is diagonal and J_y pure imaginary; it can then be shown that $d^J_{M'M}(\theta_2)$ is real. It follows from Eq. A-51 that

$$\mathscr{D}^J_{M'M}(\theta_i) = \exp(i\theta_1 M') \, d^J_{M'M}(\theta_2) \exp(i\theta_3 M). \quad (A\text{-}53)$$

We now list some useful properties of the rotation matrices. The unitarity relations are

---

† The reader must be warned that there are several different sets of conventions in the literature. There are different conventions for the definition of Euler angles, handedness of axes, and phases of the $\mathscr{D}$-functions. The phase used here is different from that in the earlier book by M. A. Preston. References 6, 7, and 8 differ from each other.

$$\sum_{M} [\mathscr{D}^J_{MK'}(\theta_i)]^* \mathscr{D}^J_{MK}(\theta_i) = \delta_{K'K} \tag{A-54}$$

and

$$\sum_{K} [\mathscr{D}^J_{M'K}(\theta_i)]^* \mathscr{D}^J_{MK}(\theta_i) = \delta_{M'M}. \tag{A-55}$$

The d-functions defined by Eq. A–52 obey the following relations:

$$d^J_{MK}(\theta_2) = (-)^{M-K} d^J_{-M,-K}(\theta_2), \tag{A-56}$$

$$= (-)^{M-K} d^J_{KM}(\theta_2), \tag{A-57}$$

$$= d^J_{KM}(-\theta_2); \tag{A-58}$$

$$d^J_{MK}(\pi - \theta) = (-)^{J-M} d^J_{M,-K}(\theta), \tag{A-59}$$

$$= (-)^{J+K} d^J_{-M,K}(\theta). \tag{A-60}$$

Also,

$$d^J_{MK}(0) = \delta_{MK}, \qquad d^J_{MK}(\pi) = (-)^{J-M} \delta_{M,-K}. \tag{A-61}$$

Using Eqs. A–56 to A–58, we can easily prove that

$$[\mathscr{D}^J_{MK}(\theta_i)]^* = (-)^{M-K} \mathscr{D}^J_{-M,-K}(\theta_i), \tag{A-62}$$

$$= (-)^{M-K} \mathscr{D}^J_{M,K}(-\theta_i), \tag{A-63}$$

since

$$\mathscr{D}^J_{M,K}(-\theta_i) = \mathscr{D}^J_{-M,-K}(\theta_i). \tag{A-64}$$

Using Eq. A–58 and definitions A–51 and A–53, it also follows that

$$[\mathscr{D}^J_{MK}(\theta_1, \theta_2, \theta_3)]^* = \mathscr{D}^J_{KM}(-\theta_3, -\theta_2, -\theta_1). \tag{A-65}$$

Important special cases are the $\mathscr{D}$'s with a zero subscript; these are spherical harmonics:

$$\mathscr{D}^J_{M0}(\theta_1, \theta_2, \theta_3) = \left(\frac{4\pi}{2J+1}\right)^{1/2} Y_J{}^M(\theta_2, \theta_1), \tag{A-66}$$

$$\mathscr{D}^J_{0K}(\theta_1, \theta_2, \theta_3) = (-)^K \left(\frac{4\pi}{2J+1}\right)^{1/2} Y_J{}^K(\theta_2, \theta_3), \tag{A-67}$$

$$\mathscr{D}^J_{00}(\theta_1, \theta_2, \theta_3) = P_J(\cos\theta_2). \tag{A-68}$$

The spherical-harmonic addition theorem follows from these results. If a certain vector $\mathbf{r}$ has polar angles (θ, ϕ) with respect to the (x, y, z)-axes, and

(θ', ϕ') with respect to the (x', y', z')-axes obtained by the rotation operator $R(\theta_i)$, it follows from Eq. A–50 that

$$Y_l^m(\theta', \phi') = \sum_{m'} [\mathscr{D}_{m'm}^l(\theta_1, \theta_2, \theta_3)]^* Y_l^{m'}(\theta, \phi). \tag{A–69}$$

For $m = 0$, since

$$Y_l^0(\theta', \phi') = \sqrt{\frac{2l+1}{4\pi}}\, P_l(\cos \theta'),$$

we obtain, using Eq. A–66,

$$P_l(\cos \theta') = \frac{4\pi}{2l+1} \sum_{m'} [Y_l^{m'}(\theta_2, \theta_1)]^* Y_l^{m'}(\theta, \phi). \tag{A–70}$$

Since (θ_2, θ_1) are the polar angles of the z'-axis with respect to the (x, y, z)-axes and θ' is the angle between the z'-axis and $\mathbf{r}$, Eq. A–70 is the addition theorem for the Legendre polynomial of the angle between two vectors, namely, the ones we have called $\mathbf{r}$ and the z'-axis.

Consider again a frame (x, y, z) and a rotated frame (x', y', z'), obtained as before by applying the rotation operator $R(\theta_i)$ on the (x, y, z)-axes. Let $\mathbf{J}$ be an angular-momentum vector with z-component M and z'-component K. It can then be shown [6, 7, 8] that

$$J^2 \mathscr{D}_{MK}^J = J(J + 1)\mathscr{D}_{MK}^J,$$

$$J_z \mathscr{D}_{MK}^J = M \mathscr{D}_{MK}^J,$$

$$J_{z'} \mathscr{D}_{MK}^J = K \mathscr{D}_{MK}^J, \tag{A–71}$$

and

$$J_\pm \mathscr{D}_{MK}^J = [(J \pm K)(J \mp K + 1)]^{1/2} \mathscr{D}_{M,K \mp 1}^J, \quad \text{with} \quad J_\pm = J_{x'} \pm iJ_{y'}.$$

In the above situation, we can use Eq. A–50 to write

$$\langle x, y, z | R(\theta_i) | JK \rangle = \sum_M [\mathscr{D}_{MK}^J(\theta_i)]^* \langle x, y, z | JM \rangle$$

or

$$\langle x', y', z' | JK \rangle = \sum_M [\mathscr{D}_{MK}^J(\theta_i)]^* \langle x, y, z | JM \rangle$$

or

$$\Psi_{JK}(x', y', z') = \sum_M [\mathscr{D}_{MK}^J(\theta_i)]^* \Psi_{JM}(x, y, z). \tag{A–72}$$

Using Eq. A–55, we can invert Eq. A–72 to obtain

$$\Psi_{JM}(x, y, z) = \sum_K \mathscr{D}^J_{MK}(\theta_i)\Psi_{JK}(x', y', z'). \tag{A-73}$$

The following coupling relationships can then be derived, using Eqs. A-72 and A-73, where the $\mathscr{D}$'s have the same arguments $(\theta_1, \theta_2, \theta_3)$:

$$\mathscr{D}^{J_1}_{M_1K_1}\mathscr{D}^{J_2}_{M_2K_2} = \sum_J \mathscr{D}^J_{MK}(J_1J_2M_1M_2|JM)(J_1J_2K_1K_2|JK), \tag{A-74}$$

$$\mathscr{D}^J_{MK} = \sum_{M_1,K_1} (J_1J_2M_1M_2|JM)(J_1J_2K_1K_2|JK)\mathscr{D}^{J_1}_{M_1K_1}\mathscr{D}^{J_2}_{M_2K_2}, \tag{A-75}$$

$$\sum_{M_1,M_2} \mathscr{D}^{J_1}_{M_1K_1}\mathscr{D}^{J_2}_{M_2K_2}\mathscr{D}^{J_3*}_{M_3K_3}(J_1J_2M_1M_2|J_3M_3) = (J_1J_2K_1K_2|J_3K_3). \tag{A-76}$$

The normalization factor for $\mathscr{D}^J_{MK}$ is $[(2J + 1)/8\pi^2]^{1/2}$:

$$\frac{[(2J_1 + 1)(2J_2 + 1)]^{1/2}}{8\pi^2} \int_0^\pi \sin\theta_2\, d\theta_2 \int_0^{2\pi} d\theta_1 \int_0^{2\pi} d\theta_3\, \mathscr{D}^{J_1*}_{M_1K_1}\mathscr{D}^{J_2}_{M_2K_2}$$

$$= \delta_{M_1M_2}\delta_{K_1K_2}\delta_{J_1J_2}. \tag{A-77}$$

Equations A-74 and A-77 yield

$$\int \mathscr{D}^{J_1}_{M_1K_1}\mathscr{D}^{J_2}_{M_2K_2}\mathscr{D}^{J_3*}_{M_3K_3} \sin\theta_2\, d\theta_2\, d\theta_1\, d\theta_3$$

$$= \frac{8\pi^2}{2J_3 + 1} (J_1J_2M_1M_2|J_3M_3)(J_1J_2K_1K_2|J_3K_3). \tag{A-78}$$

By using Eq. A-66, this result can be specialized to give

$$\int Y^{m_1}_{l_1}(\theta_2, \theta_1) Y^{m_2}_{l_2}(\theta_2, \theta_1) Y^{m_3*}_{l_3}(\theta_2, \theta_1) \sin\theta_2\, d\theta_2\, d\theta_1$$

$$= \left[\frac{(2l_1 + 1)(2l_2 + 1)}{4\pi(2l_3 + 1)}\right]^{1/2} (l_1l_2m_1m_2|l_3m_3)(l_1l_200|l_30). \tag{A-79}$$

(b) *Spherical tensors.* A spherical tensor of rank n is defined as $2n + 1$ quantities $T_n{}^\mu(\mu = -n,\dots, n)$ which transform in the same way as angular-momentum matrices under a rotation of axes. In symbols,

$$T_n{}^{\mu'}(c') = \sum_\mu T_n{}^\mu(c)\mathscr{D}^n_{\mu\mu'}(\Omega), \tag{A-80}$$

where c and c' indicate two different coordinate systems to which quantities are referred, and the argument Ω of $\mathscr{D}^n$ represents the Euler angles carrying c into c'.

Any vector operator A_x, A_y, A_z can be written as a spherical tensor of rank 1 by means of the definitions

$$T_1^{\pm 1} = \mp 2^{-1/2}(A_x \pm iA_y), \qquad T_1^0 = A_z. \tag{A-81}$$

A scalar invariant is a tensor of rank 0.

All tensor operators relevant to a given system have the same commutation relations with the total angular momentum of the system:

$$[J_\pm, T_n^\mu] = [n(n+1) - \mu(\mu + 1)]^{1/2} T_n^{\mu \pm 1}, \tag{A-82a}$$

$$[J_z, T_n^\mu] = \mu T_n^\mu. \tag{A-82b}$$

A fundamental property of spherical tensors is expressed in the *Wigner-Eckart theorem*, which gives a simple relationship for matrix elements between states with different orientations:

$$\langle \tau' J'M' | T_n^\mu | \tau J M \rangle = (\tau' J' \| T_n \| \tau J)(JnM\mu | J'M'). \tag{A-83}$$

The left side of this equation is the matrix element of T_n^μ between any two states; τ represents all quantum numbers other than J and M. As the notation indicates, the quantity $(\tau' J' \| T_n \| \tau J)$ does not depend on M, M', or μ; it is called the *reduced matrix element* and is defined by Eq. A-83.[†]

An immediate consequence of Eq. A-83 is the following important selection rule, which provides the basis for the angular-momentum selection rules in electromagnetic transitions and β-decay: a tensor of rank n has nonzero matrix elements between two states only if $\Delta J \leqslant n$ and $J + J' \geqslant n$.

A result which is often required is the reduced matrix element of the vector operator **J** itself:

$$(\tau' J' \| \mathbf{J} \| \tau J) = \delta_{\tau \tau'} \delta_{J J'} [J(J+1)]^{1/2}.$$

Spherical tensors of various ranks can be constructed from two tensors. Suppose that T_n^μ, U_k^κ are two tensors; their tensor product of rank N is

$$V_N^Q = \sum_{\mu, \kappa} (nk\mu\kappa | NQ) T_n^\mu U_k^\kappa. \tag{A-84}$$

Examples of such product tensors are found in the theory of forbidden β-decay. If T_n and U_k both act on the same dynamical variable (e.g., if both are particle operators like **r** and **p** acting on the same particle), then

$$(\tau' J' \| V_N^Q \| \tau J) = (-)^{J+J'+N} \sum_{\tau'' J''} \begin{Bmatrix} nkN \\ JJ'J'' \end{Bmatrix}$$
$$\times (\tau' J' \| T_n \| \tau'' J'')(\tau'' J'' \| U_k \| \tau J)[(2N+1)(2J''+1)]^{1/2}. \tag{A-85}$$

[†] Defined by some authors to be $(-)^{2n}(2J'+1)^{1/2}$ times our value.

However, if T_n and U_k act on different spaces, the formula is more complicated and involves 9-j symbols; such cases arise when T_n acts on spatial and U_k on spin coordinates, or when T_n and U_k are particle operators acting on different particles.

We shall illustrate the application of Eq. A–85 by using it to derive the relationship employed in Eq. 5–2 to obtain a nuclear magnetic moment. If $\mathbf{A}$ is any vector operator that acts in the same space as the angular momentum $\mathbf{J}$, then

$$\langle JM|A_z|JM\rangle = \frac{M}{J(J+1)} \langle JM|\mathbf{J}\cdot\mathbf{A}|JM\rangle. \qquad \text{(A–86)}$$

We begin the proof of this result by noting that the scalar product is

$$\mathbf{J}\cdot\mathbf{A} = -\sqrt{3}\, V_0^0,$$

where V_0^0 is the tensor product of rank 0, given by

$$V_0^0 = \sum_\mu (1\,1\,\mu\,-\mu|0\,0)J_1{}^\mu A_1{}^{-\mu}.$$

Now

$$\begin{aligned}
\langle JM|V_0^0|JM\rangle &= (J0M0|JM)(J||V_0^0||J) \\
&= (J||V_0^0||J) \\
&= (-)^{2J}\sum_{J'}\begin{Bmatrix}110\\JJJ'\end{Bmatrix}(J||J_1||J')(J'||A_1||J)(2J'+1)^{1/2} \\
&= (-)^{2J}\begin{Bmatrix}110\\JJJ\end{Bmatrix}(J(J+1))^{1/2}(J||A_1||J)(2J+1)^{1/2}.
\end{aligned}$$

The 6-j symbol with a zero is a particularly simple one:

$$\begin{Bmatrix}J_1J_10\\J_2J_2J\end{Bmatrix} = \frac{(-)^{J+J_1+J_2}}{\sqrt{(2J_1+1)(2J_2+1)}}.$$

Therefore

$$\langle JM|\mathbf{J}\cdot\mathbf{A}|JM\rangle = \sqrt{J(J+1)}\,(J||A_1||J).$$

But

$$\begin{aligned}
\langle JM|A_z|JM\rangle &= \langle JM|A_1{}^0|JM\rangle = (J1M0|JM)(J||A_1||J) \\
&= \frac{M}{\sqrt{J(J+1)}}(J||A_1||J),
\end{aligned}$$

and Eq. A–86 follows immediately.

The tensor nature of various electromagnetic operators is important in the process of deducing transition rates in the collective model.

A-3 DIRAC EQUATION

We shall present a summary of the Dirac equation for relativistic fermions, without any detailed discussion of the physical background. As Dirac first suggested, a free relativistic particle of spin $\frac{1}{2}$ may be described by a wave function satisfying a wave equation linear in momentum. The Dirac equation can be written as

$$i\frac{\partial}{\partial t}\psi_p(\mathbf{r}, t) = H\psi_p(\mathbf{r}, t), \tag{A-87}$$

where the Hamiltonian for the free particle is

$$H = \boldsymbol{\alpha} \cdot \mathbf{p} + \beta m. \tag{A-88}$$

The quantities $\boldsymbol{\alpha}$ and β are Hermitian matrices; $\mathbf{p}$ is the momentum operator. We have set $\hbar = c = 1$, but we keep the mass m explicit. The energy and momentum are related relativistically by

$$E^2 = m^2 + p^2. \tag{A-89}$$

We assume a plane-wave solution of Eq. A-87,

$$\psi_p(\mathbf{r}, t) = \psi(\mathbf{p})\exp[i(\mathbf{p}\cdot\mathbf{r} - Et)]; \tag{A-90}$$

then, from Eqs. A-87 and A-88, we obtain

$$(\boldsymbol{\alpha}\cdot\mathbf{p} + \beta m)\psi(\mathbf{p}) = E\psi(\mathbf{p}). \tag{A-91}$$

For this equation to be consistent with

$$(p^2 + m^2)\psi(\mathbf{p}) = E^2\psi(\mathbf{p}),$$

β and each of the three components of $\boldsymbol{\alpha}$ are required to have square unity and all four must anticommute:

$$\{\alpha_i, \alpha_j\} = 2\delta_{ij}, \tag{A-92a}$$

$$\{\alpha_i, \beta\} = 0, \tag{A-92b}$$

$$\alpha_1{}^2 = \alpha_2{}^2 = \alpha_3{}^2 = \beta^2 = 1, \tag{A-92c}$$

where $\{A, B\} = AB + BA$.

In order that there be four matrices satisfying Eq. A-92, it is necessary that they be of order 4 at least. Many explicit representations are possible, but we select

$$\alpha = \begin{pmatrix} 0 & \sigma \\ \sigma & 0 \end{pmatrix}, \qquad \beta = \begin{pmatrix} 1 & 0 \\ 0 & -1 \end{pmatrix}, \qquad \text{(A-93)}$$

where σ represents the 2×2 Pauli spin matrices defined in Eq. A-16, and 1 is the 2×2 unit matrix. We also introduce the 4×4 spin matrices, denoted by Σ:

$$\Sigma = \begin{pmatrix} \sigma & 0 \\ 0 & \sigma \end{pmatrix}, \qquad \text{(A-94)}$$

which are to be interpreted as the spin operators in relativistic calculations.

The fact that the matrices are 4×4 implies that the wave function ψ must have four components. Just as the two components of the spin wave function

$$\alpha = \begin{pmatrix} 1 \\ 0 \end{pmatrix} \quad \text{and} \quad \beta = \begin{pmatrix} 0 \\ 1 \end{pmatrix}$$

are connected with the two spin states in nonrelativistic mechanics, the four components of ψ also have physical significance. In order to see this in more detail, let us write the four-component "spinor" $\psi(\mathbf{p})$ as

$$\psi = \begin{pmatrix} \phi \\ \chi \end{pmatrix}, \qquad \text{(A-95)}$$

where ϕ and χ are each two-component spinors. Then, writing out Eq. A-91 in detail by using A-93 and A-95, we obtain

$$(\sigma \cdot \mathbf{p})\chi + m\phi = E\phi, \qquad \text{(A-96a)}$$

$$(\sigma \cdot \mathbf{p})\phi - m\chi = E\chi, \qquad \text{(A-96b)}$$

where

$$E = \pm |E|, \qquad |E| = \sqrt{m^2 + p^2}. \qquad \text{(A-97)}$$

From Eq. A-96, we see that

$$\chi = \frac{\sigma \cdot \mathbf{p}}{E + m} \phi \qquad \text{(A-98)}$$

and

$$\phi = \frac{\sigma \cdot \mathbf{p}}{E - m} \chi. \qquad \text{(A-99)}$$

From these equations, it is clear that, if E is positive, that is, $E = |E|$, then χ is very small compared to ϕ in the nonrelativistic limit, whereas for negative energies, $E = -|E|$, the opposite is true. Thus, for positive and negative energies, we can write the solutions ψ as

$$\psi_+ = \begin{pmatrix} \phi \\ \dfrac{\boldsymbol{\sigma} \cdot \mathbf{p}\phi}{E+m} \end{pmatrix}, \qquad E > 0,$$

$$\psi_- = \begin{pmatrix} \chi \\ -\dfrac{(\boldsymbol{\sigma} \cdot \mathbf{p})\chi}{|E|+m} \end{pmatrix}, \qquad E < 0. \tag{A--100}$$

By choosing $\begin{pmatrix} 1 \\ 0 \end{pmatrix}$ and $\begin{pmatrix} 0 \\ 1 \end{pmatrix}$ for ϕ as well as χ, we thus find the four independent solutions of the free Dirac equation:

$$\psi_+{}^{\mathrm{I}} = \begin{pmatrix} 1 \\ 0 \\ \dfrac{p_z}{E+m} \\ \dfrac{p_x + ip_y}{E+m} \end{pmatrix}, \qquad \psi_+{}^{\mathrm{II}} = \begin{pmatrix} 0 \\ 1 \\ \dfrac{p_x - ip_y}{E+m} \\ \dfrac{-p_z}{E+m} \end{pmatrix}, \qquad E > 0,$$

$$\psi_-{}^{\mathrm{I}} = \begin{pmatrix} \dfrac{-p_z}{|E|+m} \\ \dfrac{-p_x - ip_y}{|E|+m} \\ 1 \\ 0 \end{pmatrix}, \qquad \psi_-{}^{\mathrm{II}} = \begin{pmatrix} \dfrac{-p_x + ip_y}{|E|+m} \\ \dfrac{p_z}{|E|+m} \\ 0 \\ 1 \end{pmatrix}, \qquad E < 0. \tag{A--101}$$

Note that the four solutions above are not eigenfunctions of the spin operator Σ_z unless p_x and p_y are set to zero. For the special case in which $p = p_z$,

$$\Sigma_z \psi_\pm{}^{\mathrm{I}} = \psi_\pm{}^{\mathrm{I}}, \qquad \Sigma_z \psi_\pm{}^{\mathrm{II}} = -\psi_\pm{}^{\mathrm{II}}.$$

In this case (and also in the zero momentum limit) $\psi_+{}^{\mathrm{I}}$ is the positive-energy spin-up solution, $\psi_-{}^{\mathrm{II}}$ is the negative-energy spin-down solution, and so on. More precisely, although an arbitrary component of $\boldsymbol{\Sigma}$ is not a constant of motion in relativistic mechanics, the component of $\boldsymbol{\Sigma}$ in the direction of the motion is conserved. Formally, Σ_z does not commute with the Hamiltonian, but $S(p) = \mathbf{\Sigma} \cdot \mathbf{p}/p$ does. This important component of the spin is called the "helicity." In the above example, $\psi_\pm^{\,I}$ have helicity $+1$ and $\psi_\pm^{\,II}$ have helicity -1. The spinors in Eq. A–101 are not normalized to unity in a unit volume. To do this, we should multiply each spinor by a normalizing factor N:

$$N = \left[1 + \frac{p^2}{(|E| + m)^2}\right]^{-1/2} = \left(\frac{|E| + m}{2|E|}\right)^{1/2}. \tag{A-102}$$

Before proceeding, we point out that there is a simple interpretation for the matrices α_k ($k = 1, 2, 3$), for a free Dirac particle. Consider

$$\dot{x}_k = i[H, x_k] = i[\alpha_j p_j, x_k] = \alpha_k, \tag{A-103}$$

which says that $\boldsymbol{\alpha}$ is the velocity. Complications arise in the relativistic case, however, because $\boldsymbol{\alpha}$ (and therefore the velocity) is not a constant of motion, even for a free particle. In the nonrelativistic limit, however, it is quite proper to treat $\boldsymbol{\alpha}$ as $\mathbf{p}/m$. Similarly, the nonrelativistic limit of β is $+1$.

For many purposes, it is convenient to write the Dirac equation in a more symmetric form. If we write the coordinates as x_1, x_2, x_3, and $x_4 = it$, the time-dependent equation A–87 takes the form

$$\left[\frac{\partial}{\partial x_4} + \frac{1}{i}(\boldsymbol{\alpha} \cdot \boldsymbol{\nabla}) + \beta m\right]\Psi = 0.$$

Premultiplying the above equation by β, and putting

$$\gamma_k = -i\beta\alpha_k, \quad k = 1, 2, 3,$$

$$\gamma_4 = \beta, \tag{A-104}$$

we obtain

$$\left(\sum_{\mu=1}^{4} \gamma_\mu \frac{\partial}{\partial x_\mu} + m\right)\Psi = 0. \tag{A-105}$$

Since β and α_k are Hermitian, γ_k is also Hermitian. It follows directly from the definition that the γ-matrices anticommute:

$$\{\gamma_\mu, \gamma_\nu\} = 2\delta_{\mu\nu}. \tag{A-106}$$

We also define the important quantity

$$\gamma_5 = \gamma_1\gamma_2\gamma_3\gamma_4 = i\alpha_1\alpha_2\alpha_3, \tag{A-107}$$

which satisfies

$$\gamma_5^2 = 1, \quad \{\gamma_5, \gamma_\mu\} = 0, \quad [\gamma_5, \alpha_k] = 0. \tag{A-108}$$

In this representation,

$$\gamma_k = -i \begin{pmatrix} 0 & \sigma_k \\ -\sigma_k & 0 \end{pmatrix} \qquad \text{(A–109)}$$

and

$$\gamma_5 = \begin{pmatrix} 0 & -1 \\ -1 & 0 \end{pmatrix}. \qquad \text{(A–110)}$$

Also,

$$\alpha = -\Sigma\gamma_5, \qquad \Sigma = -\alpha\gamma_5, \qquad \text{(A–111)}$$

and

$$[\Sigma, \gamma_5] = 0. \qquad \text{(A–112)}$$

It must be pointed out that several notations are used for the γ's. Our notations are the same as those of Rose [10] and Wu and Moszkowski [11]. Some arise naturally from keeping the coordinate x_4 real and distinguishing contravariant and covariant vectors; then Eq. A–105 may look slightly different. In all notations, however, the commutation relations are the same. In our notation, the γ's are Hermitian, as are α, Σ, and β.

Let us briefly study the Dirac equation in the presence of a central one-body potential $V(r)$, which, for example, may be the Coulomb field produced by the nucleus at the site of the electron. In the relativistic case, even in a central potential, the orbital angular-momentum $\mathbf{l} = \mathbf{r} \times \mathbf{p}$ and Σ are not constants of motion, but $\mathbf{j}^2$ and j_z are, where $\mathbf{j} = (\mathbf{l} + \frac{1}{2}\Sigma)$ is the total angular momentum. There is another operator

$$K = \beta(\Sigma \cdot \mathbf{l} + 1), \qquad \text{(A–113)}$$

which also commutes with the Hamiltonian. From the definitions of K and $\mathbf{j}$, it follows that

$$K^2 = \mathbf{j}^2 + \tfrac{1}{4}. \qquad \text{(A–114)}$$

Let us denote the eigenvalues of $\mathbf{j}^2$, j_z, and K by $j(j+1)$, μ, and $-\kappa$, respectively. It follows from Eq. A–114 that

$$\kappa = \pm (j + \tfrac{1}{2}). \qquad \text{(A–115)}$$

We write the equation $K\psi = -\kappa\psi$, using the two-component spinors ϕ and χ of Eq. A–95,

$$\begin{pmatrix} \boldsymbol{\sigma} \cdot \mathbf{l} + 1 & 0 \\ 0 & -\boldsymbol{\sigma} \cdot \mathbf{l} - 1 \end{pmatrix} \begin{pmatrix} \phi \\ \chi \end{pmatrix} = -\kappa \begin{pmatrix} \phi \\ \chi \end{pmatrix},$$

which can be written as

$$(\boldsymbol{\sigma} \cdot \mathbf{l} + 1)\phi = -\kappa\phi, \qquad (\boldsymbol{\sigma} \cdot \mathbf{l} + 1)\chi = +\kappa\chi. \tag{A-116}$$

Thus the two-component spinors ϕ and χ have opposite signs of κ. Note that ϕ and χ are separately eigenfunctions of $\boldsymbol{\sigma} \cdot \mathbf{l}$, with different eigenvalues. Since

$$l^2 = j^2 - \boldsymbol{\sigma} \cdot \mathbf{l} - \tfrac{3}{4}, \tag{A-117}$$

it follows that ϕ and χ are separately eigenfunctions of l^2, although the four-component spinor ψ is not. For a fixed κ, the eigenvalues l_A of ϕ and l_B of χ are completely determined. For example, if $\kappa = j + \tfrac{1}{2}$, then $l_A = j + \tfrac{1}{2}$ and $l_B = j - \tfrac{1}{2}$. For $\kappa = -(j + \tfrac{1}{2})$, the above values of l_A and l_B become interchanged.

Noting this, let us define two-component spinors $\mathscr{Y}_{jl}^\mu$ which are eigenfunctions of j^2, j_z, and l^2:

$$\mathscr{Y}_{jl}^\mu = (l\,\tfrac{1}{2}\,\mu - \tfrac{1}{2}\,\tfrac{1}{2}|j\,\mu)\,Y_l^{\mu-(1/2)} \begin{pmatrix} 1 \\ 0 \end{pmatrix}$$

$$+ (l\,\tfrac{1}{2}\,\mu + \tfrac{1}{2}\,-\tfrac{1}{2}|j\,\mu)\,Y_l^{\mu+(1/2)} \begin{pmatrix} 0 \\ 1 \end{pmatrix}. \tag{A-118}$$

Then we can write

$$\psi = \begin{pmatrix} \phi \\ \chi \end{pmatrix} = \begin{pmatrix} g_\kappa(r)\mathscr{Y}_{jl_A}^\mu \\ if_\kappa(r)\mathscr{Y}_{jl_B}^\mu \end{pmatrix}, \tag{A-119}$$

where $g_\kappa(r)$ and $f_\kappa(r)$ are radial wave functions to be determined. The factor i is introduced to make $f_\kappa(r)$ real.

If we substitute this form of ψ in the Dirac equation, we find

$$(\boldsymbol{\sigma} \cdot \mathbf{p})\chi = [E - V(r) - m]\phi, \qquad (\boldsymbol{\sigma} \cdot \mathbf{p})\phi = [E - V(r) + m]\chi. \tag{A-120}$$

To proceed, we have to know what $\boldsymbol{\sigma} \cdot \mathbf{p}$ does operating on ϕ and χ. Using the identity

$$\boldsymbol{\sigma} \cdot \mathbf{p} = \frac{\boldsymbol{\sigma} \cdot \mathbf{r}}{r^2}\left(-ir\frac{\partial}{\partial r} + i\boldsymbol{\sigma} \cdot \mathbf{l}\right) \tag{A-121}$$

and the relation

$$\frac{\boldsymbol{\sigma} \cdot \mathbf{r}}{r} \, \mathcal{Y}^{\mu}_{j \, l = j \pm (1/2)} = - \, \mathcal{Y}^{\mu}_{j \, l = j \mp (1/2)},\qquad (A\text{-}122)$$

we get

$$(\boldsymbol{\sigma} \cdot \mathbf{p})\chi = -\frac{df}{dr}\mathcal{Y}^{\mu}_{jl_A} - \frac{1-\kappa}{r}f\mathcal{Y}^{\mu}_{jl_A}$$

and a similar expression for $(\boldsymbol{\sigma} \cdot \mathbf{p})\phi$. Thus the radial form of the Dirac equation is

$$-\frac{df}{dr} - \frac{1-\kappa}{r}f = (E - V - m)g,$$

$$\frac{dg}{dr} + \frac{1+\kappa}{r}g = (E - V + m)f. \qquad (A\text{-}123)$$

There are two linearly independent solutions to this system, one regular at the origin and one irregular. When V is the Coulomb potential $- Ze^2/r$ for an electron and a point nucleus, the calculation is straightforward. Rose [9] provides the regular solutions to these relativistic Coulomb functions, normalized thus:

$$\int r^2 \, dr\{f_\kappa(E')f_\kappa(E) + g_\kappa(E')g_\kappa(E)\} = \delta(E' - E), \qquad (A\text{-}124)$$

so that

$$\int d\mathbf{r} \; \psi^H_{\tau'}(E')\psi_\tau(E) = \delta_{\tau'\tau}\delta(E' - E), \qquad (A\text{-}125)$$

where the superscript H represents the Hermitian adjoint, and τ stands for all quantum numbers, excluding energy. This normalization corresponds to one particle per unit energy interval in all space. Asymptotically,

$$f_\kappa \sim -\frac{1}{r}\left(\frac{E-m}{\pi p}\right)^{1/2} \sin(pr + \delta_c),$$

$$g_\kappa \sim \frac{1}{r}\left(\frac{E+m}{\pi p}\right)^{1/2} \cos(pr + \delta_c), \qquad (A\text{-}126)$$

where p is the asymptotic momentum $(E^2 - m^2)^{1/2}$ and δ_C is the Coulomb phase shift:

$$\delta_C = \frac{\alpha Z E}{p}\ln 2pr - \arg \Gamma\!\left(\gamma + i\frac{\alpha Z E}{p}\right) + \eta - \frac{\pi}{2}\gamma, \qquad (A\text{-}127)$$

$$\gamma = \kappa^2 - \alpha^2 Z^2, \tag{A-128}$$

and

$$\exp(2i\eta) = -\frac{\kappa - i\alpha Z/p}{\gamma + i\alpha ZE/p}. \tag{A-129}$$

An irregular solution is formed by introducing any constant phase into the arguments of the trigonometric functions in Eq. A-126.

Inside a nucleus of nonzero radius, the solutions will not be Coulomb functions, but outside they will be linear combinations of regular and irregular functions, although the normalization will have changed.

The solution for $Z = 0$ yields the angular-momentum eigenstates of a free particle. This solution is

$$f_\kappa = \frac{\kappa}{|\kappa|} \left(\frac{E - m}{\pi p}\right)^{1/2} p j_{l_B}(pr),$$

$$g_\kappa = \left(\frac{E + m}{\pi p}\right)^{1/2} p j_{l_A}(pr), \tag{A-130}$$

where the j-functions are spherical Bessel functions. In β-decay, Eqs. A-130 are the low-Z approximation for the electron functions, and, with $E = p$, $m = 0$, they are the neutrino functions.

A-4 STRONG, ELECTROMAGNETIC, AND WEAK INTERACTION HAMILTONIANS

In relativistic mechanics, in going from one inertial frame to another, the velocity of light must be preserved. This is achieved by what is called a Lorentz transformation:

$$x_\mu' = \sum_{\nu=1}^{4} a_{\mu\nu} x_\nu, \tag{A-131}$$

such that

$$\sum_{\mu=1}^{4} x_\mu^2 = \sum_{\mu=1}^{4} x_\mu'^2,$$

which implies the orthogonality condition

$$\sum_\nu a_{\mu\nu} a_{\lambda\nu} = \delta_{\mu\lambda}.$$

Therefore

$$\det a = \pm 1. \tag{A-132}$$

When $\det a = +1$, it is called a proper Lorentz transformation. This includes ordinary spatial rotations, but not space inversion or time reversal. Space inversion implies changing x_k to $-x_k$ ($k = 1, 2, 3$), while keeping x_4 the same. This changes the "handedness" of the spatial axes.

A quantity that remains invariant under a proper Lorentz transformation as well as space inversion is called a scalar. A pseudoscalar is one that changes sign when the axes change handedness, but is otherwise invariant. Quantities that transform as in Eq. A-131 and whose spatial components change sign under space inversion (like ordinary vectors) are vectors. A pseudovector (also called an axial vector) does not change sign under space inversion.

It turns out that $\psi^{+}\psi$ is not invariant under Lorentz transformation, and hence we define an "adjoint" wave function

$$\bar{\psi} = \psi^{+}\gamma_4, \tag{A-133}$$

which has the property that $\bar{\psi}\psi$ is a scalar. By inserting appropriate γ's between $\bar{\psi}$ and ψ, we can also construct other quantities that behave like pseudoscalars, vectors, axial vectors, or tensors under proper Lorentz transformation and space inversion. Both ψ and $\bar{\psi}$ are four-component spinors; hence sixteen bilinear combinations can be made out of them. These can be grouped into five different classes:

$\bar{\psi}\psi$: scalar (S);

$i\bar{\psi}\gamma_\mu\psi$: vector with four components (V);

$\bar{\psi}\sigma_{\mu\nu}\psi$: antisymmetric tensor (T) containing six different components, where $\sigma_{\mu\nu} = (1/2i)(\gamma_\mu\gamma_\nu - \gamma_\nu\gamma_\mu)$;

$i\bar{\psi}\gamma_\mu\gamma_5\psi$: axial vector (A) with four components;

$i\bar{\psi}\gamma_5\psi$: pseudoscalar (P). $\tag{A-134}$

The factor i has been introduced in the appropriate quantities to ensure hermiticity. Note that there is no pseudotensor form because the sixteen quantities $\gamma_\mu\gamma_\nu\gamma_5$ are the same sixteen as $\gamma_\mu\gamma_\nu$ but arranged in different order.

In many applications, it is necessary to construct matrix elements of the interaction Hamiltonian involving the above quantities. In the case of strong and electromagnetic interactions, since parity is conserved, the interaction Hamiltonian has to be a scalar. For weak interactions like β-decay, on the other hand, the Hamiltonian must have a pseudoscalar part which violates parity conservation. We now indicate briefly how such matrix elements may be constructed in each of the above cases.

For the electromagnetic case, the current operator of a Dirac particle is $ie\gamma_4\gamma_\mu$, where e is the electric charge. In the nonrelativistic limit, the three spatial components give rise to the ordinary electric current $\mathbf{j}$ and the fourth component is the charge. The interaction of this current with the electromagnetic field is $ie\gamma_4\gamma_\mu A_\mu$, where A_μ is the four-potential of the electromagnetic field. The matrix element of this interaction therefore takes the form

$$ie(\psi_f{}^+\gamma_4\gamma_\mu A_\mu\psi_i) = ie(\bar{\psi}_f\gamma_\mu A_\mu\psi_i), \tag{A-135}$$

where ψ_f and ψ_i are the final and initial states, respectively, of the Dirac particle. For a free particle, expression A–135 contains the interaction of the normal magnetic moment and the charge of the Dirac particle with the electromagnetic potentials. The anomalous magnetic moment arises from an operator $\sigma_{\alpha\beta}q_\beta$, where q_β is the four-momentum transferred to the electromagnetic field.

For strong interactions, there are many ways of constructing the interaction matrix element. Consider the case of pions interacting with nucleons. Now, instead of the vector field A_μ of the electromagnetic case, we have the pionic field ϕ_π, which is a pseudoscalar. This is so since the pion has spin zero and negative intrinsic parity. In order to construct a scalar for the interaction energy of the pion field with nucleons, we recall that $i\bar{\psi}\gamma_5\psi$ is a pseudoscalar, and therefore $i\gamma_4\gamma_5\phi_\pi$ is an operator with scalar matrix elements between different states of the nucleon. The matrix element of the pseudoscalar interaction thus takes the form

$$i\sqrt{4\pi}\,g(\psi_f{}^+\gamma_4\gamma_5\phi_\pi\psi_i) = i\sqrt{4\pi}\,g(\bar{\psi}_f\gamma_5\phi_\pi\psi_i), \tag{A-136}$$

where g is the pseudoscalar pion-nucleon coupling constant. There are other ways of coupling the pion field with the nucleonic "current." For example, we may use the "gradient coupling" with matrix elements

$$i\sqrt{4\pi}\,f\left(\bar{\psi}_f\gamma_\mu\gamma_5\,\frac{\partial}{\partial x_\mu}\,\phi_\pi\psi_i\right), \tag{A-137}$$

where f is the appropriate coupling constant.

For a vector meson with four components ϕ_μ, the interaction matrix elements take the same form as the electromagnetic case (cf. Eq. A–135):

$$i\sqrt{4\pi}\,g_v(\bar{\psi}_f\gamma_\mu\phi_\mu\psi_i), \tag{A-138}$$

where g_v is the corresponding coupling constant. Finally, for a vector meson, there can be the gradient coupling also:

$$\sqrt{4\pi}\, f_v \left[\bar{\psi}_f \sigma_{\mu\nu} \left(\frac{\partial \phi_v}{\partial x_\mu} - \frac{\partial \phi_\mu}{\partial x_\nu} \right) \psi_i \right].$$ (A–139)

In the case of weak interactions, the important point is that the Hamiltonian must contain both scalar and pseudoscalar parts. In addition, we must distinguish between the wave functions of the baryons and the leptons. A general way of writing the interaction matrix element is

$$(\bar{\Psi}_f O \Psi_i)(\bar{\psi}_f O_L \psi_i),$$ (A–140)

where Ψ_f and Ψ_i are the final and initial baryonic states, while ψ_f and ψ_i are the corresponding leptonic ones. The operator O acts on the baryonic states, and O_L on the leptonic ones only. Different possible forms of O and O_L are used in β-decay theory. For example, we may take

$$O = C_V \gamma_\mu + C_A \gamma_\mu \gamma_5 \quad \text{and} \quad O_L = \gamma_\mu (1 + \gamma_5),$$

with C_V and C_A as constants to be determined by experiments.

REFERENCES

1. E. U. Condon and G. H. Shortley, *The Theory of Atomic Spectra*, Cambridge University Press, London, 1935: Formulas of Clebsch-Gordan coefficients for $J_3 = \frac{1}{2}, 1, \frac{3}{2}, 2$.
2. D. L. Falkoff, G. S. Holladay, and R. E. Sells, *Can. J. Phys.* **30**, 253 (1952). Formulas of Clebsch-Gordan coefficients for $J_3 = 3$.
3. M. A. Melvin and N. V. J. Swamy, *Phys. Rev.* **107**, 186 (1957). Formulas of Clebsch-Gordan coefficients for $J_3 = \frac{5}{2}$.
4. A. Simon, *Numerical Table of the Clebsch-Gordan Coefficient*, U.S.A.E.C. Rept. ORNL-1718, 1954. Given as decimal fractions for any $J \leqslant \frac{9}{2}$.
5. M. Rotenberg, R. Bivins, N. Metropolis, and J. K. Wooten, *The 3-j and 6-j Symbols*, The Technology Press, Massachusetts Institute of Technology, Cambridge, Mass. 1959. All Clebsch-Gordan coefficients with all three $J \leqslant 8$ are given as rational fractions.
6. A. R. Edmonds, *Angular Momentum in Quantum Mechanics*, Princeton University Press, Princeton, N.J., 1957.
7. M. E. Rose, *Elementary Theory of Angular Momentum*, John Wiley, New York, 1957.
8. E. P. Wigner, *Group Theory and Its Applications to Quantum Mechanics of Atomic Spectra*, Academic Press, New York, 1958.
9. M. E. Rose, *Phys. Rev.* **51**, 484 (1937).
10. M. E. Rose, *Relativistic Electron Theory*, John Wiley, New York, 1961.
11. C. S. Wu and S. A. Moszkowski, *Beta Decay*, John Wiley, New York, 1966.

Some Results Used in Scattering Theory

B-1 RADIAL WAVE FUNCTIONS

For the scattering of two particles, the radial wave equation outside the region of nuclear interaction is

$$\frac{d^2u}{d\rho^2} + \left[1 - \frac{2\eta}{\rho} - \frac{l(l+1)}{\rho^2}\right]u = 0, \tag{B-1}$$

where

$$\rho = kr. \qquad k = \frac{mv}{\hbar}, \qquad \eta = \frac{Z_1 Z_2 e^2}{\hbar v}, \tag{B-2}$$

and the various quantities are the reduced mass, relative velocities, and so forth, appropriate to the center-of-mass system.

It is easy to check by differentiation that, for ρ large enough to ignore ρ^{-2} in comparison with unity, Eq. B-1 is satisfied by taking u as either the sine or the cosine of the argument $(\rho - \eta \ln \rho + \text{constant})$. The constant is selected to make the solution $\sin(\rho - \eta \ln \rho + \text{constant})$ become the asymptotic form of the solution of Eq. B-1, which is regular at $\rho = 0$. The equation is the confluent hypergeometric equation, which has been thoroughly studied; we shall merely quote results. If one particle is neutral, $\eta = 0$, and the equation is a form of Bessel's equation.

We shall denote the regular solution of Eq. B-1 by F_l; it is defined by insisting that $F_l(0) = 0$ and that its asymptotic oscillations have unit amplitude. Then it can be shown that

$$F_l \sim \sin\left(\rho - \eta \ln 2\rho - \tfrac{1}{2}l\pi + \sigma_l\right), \tag{B-3}$$

where

$$\sigma_l = \arg \Gamma(l+1+i\eta) = \sigma_0 + \sum_{s=1}^{l} \tan^{-1}\frac{\eta}{s}. \tag{B-4}$$

For the other independent solution, we choose the one that is asymptotically out of phase by $\pi/2$:

$$G_l = \cos\left(\rho - \eta \ln 2\rho - \tfrac{1}{2}l\pi + \sigma_l\right). \tag{B-5}$$

Then the Wronskian is

$$G_l \frac{dF_l}{d\rho} - F_l \frac{dG_l}{d\rho} = 1. \tag{B-6}$$

Complete series expansions about $\rho = 0$ may be given, but we shall simply note that F_l is a series of ascending powers beginning with a term in ρ^{l+1}, and that G_l consists of two series, one beginning with a term in ρ^{-l} and the other starting with ρ^{l+1} and being multiplied by $\ln 2\rho$. For $l = 0$,

$$F_0(\rho) = C_0\rho(1 + \eta\rho + \cdots), \tag{B-7a}$$

$$G_0(\rho) = \frac{1}{C_0}\{1 + 2\eta\rho[\ln 2\eta\rho + 2\gamma - 1 + h(\eta)] + \cdots\}, \tag{B-7b}$$

where

$$C_0^2 = 2\pi\eta[\exp(2\pi\eta) - 1]^{-1}, \tag{B-8a}$$

$$\gamma = \text{Euler's constant} = 0.57722\ldots,$$

$$h_0(\eta) = \eta^2 \sum_{n=1}^{\infty} \frac{1}{n(n^2 + \eta^2)} - \ln\eta - \gamma. \tag{B-8b}$$

For $l > 0$, the behavior for small ρ is

$$F_l = C_l\rho^{l+1}\left(1 + \frac{\eta\rho}{l+1} + \cdots\right), \tag{B-9a}$$

$$G_l = \frac{1}{(2l+1)C_l}\rho^{-l}\left[1 + 0\left(\frac{\eta\rho}{l}\right)\right], \tag{B-9b}$$

where

$$C_l = C_0 \frac{1}{(2l+1)!!} \prod_{s=1}^{l}\left(1 + \frac{\eta^2}{s^2}\right)^{1/2}$$

$$= 2^l \exp(-\tfrac{1}{2}\pi\eta) \frac{|\Gamma(l + 1 + i\eta)|}{(2l + 1)!}$$

and $(2l + 1)!! = 1 \cdot 3 \cdot 5 \cdots (2l + 1)$.

In a scattering experiment, the incoming and outgoing Coulomb waves are taken as

$$I_l = \exp(i\omega_l)\,(G_l - iF_l) \sim \exp[-i(\rho - \eta \ln 2\rho - \tfrac{1}{2}l\pi + \sigma_0)], \quad \text{(B-10a)}$$

$$O_l = \exp(-i\omega_l)\,(G_l + iF_l) \sim \exp[i(\rho - \eta \ln 2\rho - \tfrac{1}{2}l\pi + \sigma_0)]. \quad \text{(B-10b)}$$

For uncharged particles, the functions F_l and G_l are replaced by spherical Bessel functions, $j_l(\rho)$ and $n_l(\rho)$, according to

$$F_l \to \rho j_l(\rho), \qquad G_l \to -\rho n_l(\rho). \quad \text{(B-11)}$$

These Bessel functions in turn behave as follows:

$$j_l(\rho) \sim \rho^{-1} \sin(\rho - \tfrac{1}{2}l\pi), \quad \text{(B-12a)}$$

$$n_l(\rho) \sim -\rho^{-1}\cos(\rho - \tfrac{1}{2}l\pi), \quad \text{(B-12b)}$$

$$n_l \frac{dj_l}{d\rho} - j_l \frac{dn_l}{d\rho} = -\frac{1}{\rho^2}. \quad \text{(B-13)}$$

At small r,

$$j_l = \frac{\rho^l}{(2l + 1)!!}\left[1 - \frac{\rho^2}{2(2l + 3)} + \cdots\right], \quad \text{(B-14a)}$$

$$n_l = -\frac{(2l - 1)!!}{\rho^{l+1}}\left[1 + \frac{\rho^2}{2(2l - 1)} + \cdots\right]. \quad \text{(B-14b)}$$

The spherical Bessel functions are related to the Bessel functions by

$$j_l(\rho) = \left(\frac{\pi}{2\rho}\right)^{1/2} J_{l+(1/2)}(\rho), \quad \text{(B-15a)}$$

$$n_l(\rho) = (-)^{l+1}\left(\frac{\pi}{2\rho}\right)^{1/2} J_{-l-(1/2)}(\rho). \quad \text{(B-15b)}$$

The incoming wave is

$$I_l = -i\rho h_l^{(2)}(\rho) = -i\rho(j_l - in_l) \sim \exp[-i(\rho - \tfrac{1}{2}l\pi)], \quad \text{(B-16a)}$$

and the outgoing wave is

$$O_l = i\rho h_l^{(1)}(\rho) = i\rho(j_l + in_l) \sim \exp[i(\rho - \tfrac{1}{2}l\pi)], \quad \text{(B-16b)}$$

where we have introduced the spherical Hankel functions $h_l^{(1)}$ and $h_l^{(2)}$.

We also note the explicit forms

$$j_0(\rho) = \frac{\sin \rho}{\rho}, \qquad n_0(\rho) = -\frac{\cos \rho}{\rho};$$

$$j_1(\rho) = \frac{\sin \rho}{\rho^2} - \frac{\cos \rho}{\rho}, \qquad n_1(\rho) = -\frac{\cos \rho}{\rho^2} - \frac{\sin \rho}{\rho}. \qquad \text{(B–17)}$$

The functions F_l and ρj_l begin to increase at small ρ like ρ^{l+1}, then they increase more quickly, and after the point

$$\rho = \eta + [\eta^2 + l(l+1)]^{1/2}$$

they oscillate, the period and amplitude of the oscillations gradually tending to the asymptotic forms.

B-2 EFFECTIVE-RANGE FORMULA

Consider s-wave proton-proton scattering with the radial function in the center-of-mass system given by

$$\frac{d^2 u_1}{dr^2} - \left[\frac{1}{Rr} + V(r)\right] u_1 = -k_1^2 u_1, \qquad \text{(B–18)}$$

where $R = \eta_1/k_1 = \hbar^2/Me^2$ as in Eq. 2–50, and $V(r)$ is the nuclear potential multiplied by $M/\hbar^2$. We write the same equation for a different energy:

$$\frac{d^2 u_2}{dr^2} - \left[\frac{1}{Rr} + V(r)\right] u_2 = -k_2^2 u_2. \qquad \text{(B–19)}$$

If we multiply Eq. B–18 by u_2, B–19 by u_1, subtract, and integrate from r to r', we obtain

$$\left(u_2 \frac{du_1}{dr} - u_1 \frac{du_2}{dr}\right)\Bigg|_r^{r'} = (k_2^2 - k_1^2) \int_r^{r'} u_1 u_2 \, dr. \qquad \text{(B–20)}$$

We denote the asymptotic form of u by ϕ; then ϕ satisfies Eq. B–18 with $V(r)$ omitted, which is the Coulomb equation.

A similar procedure yields

$$\left(\phi_2 \frac{d\phi_1}{dr} - \phi_1 \frac{d\phi_2}{dr}\right)\Bigg|_r^{r'} = (k_2^2 - k_1^2) \int_r^{r'} \phi_1 \phi_2 \, dr. \qquad \text{(B–21)}$$

By subtracting Eq. B–20 from B–21, we obtain

$$\phi_2 \phi_1' - \phi_1 \phi_2' - u_2 u_1' + u_1 u_2'\big|_r^{r'} = (k_2^2 - k_1^2) \int_r^{r'} (\phi_1 \phi_2 - u_1 u_2) \, dr. \qquad \text{(B–22)}$$

We now choose the limit r' so large that $u(r')$ has reached its asymptotic value. Consequently the upper limit on the left-hand side of Eq. B–22 is zero, and, since the integrand on the right vanishes for $r > r'$, we may write

$$(\phi_2\phi_1' - \phi_1\phi_2' - u_2u_1' + u_1u_2')_r = (k_2^2 - k_1^2)\int_r^\infty (\phi_1\phi_2 - u_1u_2)\, dr. \qquad \text{(B–23)}$$

Since u_1 and u_2 are solutions for the physical problem, they vanish at $r = 0$. Since the ϕ's are solutions of the Coulomb equation and are the asymptotic forms of the u's, we have

$$\phi \propto \sin(\rho - \eta \ln 2\rho + \sigma_0 + \delta)$$

when δ is the nuclear phase shift. The proportionality factor determines the wave-function normalization, but clearly divides through Eq. B–23. We therefore choose

$$\phi_i = C_{0i}[G_0(k_i r) + \cot \delta_i\, F_0(k_i r)].$$

Equations B–7 show that, as r tends to zero, the limit of ϕ is unity. There is therefore no problem of convergence in the integral in Eq. B–23 if we let r approach zero.

We still require the derivative of ϕ for small r. Again using Eq. B–7, we see that

$$\phi_i' = \frac{d\phi_i}{dr} = \frac{1}{R}\left(\ln \frac{r}{R} + 2\gamma\right) + \frac{1}{R} h(\eta_i) + C_{0i}^2 k_i \cot \delta_i.$$

Therefore, the limit of Eq. B–23 as r tends to zero is

$$\frac{1}{R}[h(\eta_2) - h(\eta_1)] + C_{02}^2 k_2 \cot \delta_2 - C_{01}^2 k_1 \cot \delta_1$$

$$= (k_2^2 - k_1^2)\int_0^\infty (\phi_1\phi_2 - u_1u_2)\, dr. \qquad \text{(B–24)}$$

We now let k_1 tend to zero and define

$$-a_p^{-1} \equiv \lim_{k \to 0} C_0^2 k \cot \delta + \frac{1}{R} h(\eta), \qquad \text{(B–25)}$$

which is a generalization of Eq. 2–33.

We may now rewrite Eq. B–24 as

$$C_0^2 k \cot \delta + \frac{h(\eta)}{R} = -\frac{1}{a_p} + k^2 \int_0^\infty (\phi\phi_0 - uu_0)\, dr. \qquad \text{(B–26)}$$

This equation is exact; we obtain the effective-range expansion, Eq. 2–57, by defining

$$r_0 = 2 \int_0^\infty (\phi_0{}^2 - u_0{}^2)\, dr, \tag{B–27}$$

and by noting that at low energies

$$\phi(k;r) = \phi_0(r) + 0(k^2).$$

The simpler derivation of the effective-range formula for neutron-proton scattering is left to the reader.

B-3 DERIVATION OF THE PROTON-PROTON SCATTERING CROSS SECTION

For particles incident along the z-axis, the total wave function is

$$\psi = \sum c_l r^{-1} u_l(r) P_l(\cos \theta), \tag{B–28}$$

where u_l satisfies

$$\frac{d^2 u_l}{dr^2} + \left[k^2 - \frac{1}{Rr} + V(r) - \frac{l(l+1)}{r^2} \right] u_l = 0.$$

Outside the nuclear potential, u_l is a linear combination of the Coulomb functions F_l and G_l, so that

$$u_l(r) \sim \sin \left(kr - \eta \ln 2kr - \tfrac{1}{2} l\pi + \sigma_l + \delta_l \right), \tag{B–29}$$

where δ_l is the change in the phase shift due to the potential $V(r)$. Note that δ_l is *not* the phase shift that $V(r)$ would produce in the absence of a Coulomb force.

Now the coefficients c_l in Eq. B–28 must be chosen so that the only incoming waves are those in the incident beam. The presence of the Coulomb force somewhat complicates the problem, since a plane wave $\exp(ikz)$ is not a solution of

$$\left(\nabla^2 + k^2 - \frac{1}{Rr} \right) \psi = 0,$$

even for large r. However, it can be verified by differentiating that $\exp\{i[kz + \eta \ln k(r - z)]\}$ is a solution for large r to order $1/r^2$; this expression is *the Coulomb distorted plane wave*, and it can be shown that

$$\exp\{i[kz + \eta \ln k(r - z)]\} \sim \sum_{l=0}^{\infty} (2l + 1)i^l(kr)^{-1}$$

$$\times \sin (kr - \eta \ln 2kr - \tfrac{1}{2}l\pi)P_l(\cos \theta).$$

Hence, if we set $c_l = (2l + 1)k^{-1}i^l \exp[i(\delta_l + \sigma_l)]$ in Eq. B–28, we have

$$\psi = \exp\{i[kz + \eta \ln k(r - z)]\} - f(\theta)\frac{1}{r} \exp[i(kr - \eta \ln 2kr)], \quad (B\text{–}30)$$

where

$$f(\theta) = -\frac{1}{2ik} \sum_{l=0}^{\infty} (2l + 1)\{\exp[2i(\delta_l + \sigma_l)] - 1\}P_l(\cos \theta). \quad (B\text{–}31)$$

For pure Coulomb scattering ($\delta_l = 0$), it can be shown from the properties of the Coulomb functions[†] that the amplitude $f(\theta)$ can be summed to give

$$f_C(\theta) = -\frac{1}{4k^2R} \operatorname{cosec}^2 \frac{\theta}{2} \exp\left(- 2i\eta \ln \sin \frac{\theta}{2} + i\pi + 2i\sigma_0\right). \quad (B\text{–}32)$$

Hence $f = f_C + f_N$, where, as in Eq. 5–27,

$$f_N = -\frac{1}{2ik} \sum (2l + 1)e^{2i\sigma_l}(e^{2i\delta_l} - 1)P_l(\cos \theta).$$

To find the proton-proton cross section we must take account of the identity of the two particles. Interchange of their coordinates is equivalent to replacing θ by $\pi - \theta$. Hence the spatial wave function must be taken in singlet spin states as

$$\psi(r, \theta) + \psi(r, \pi - \theta),$$

and in triplet states as

$$\psi(r, \theta) - \psi(r, \pi - \theta).$$

Since $P_l(\cos \theta)$ has the parity of l, Eq. B–31 shows that these combinations automatically eliminate both singlet odd and triplet even states.

A wave function of form B–30 describes a unit incident proton flux and an outgoing flux through any large sphere of $|f(\theta)|^2$ particles per unit solid angle per second. In other words, $\sigma(\theta) = |f(\theta)|^2$. Therefore the scattering cross section is given in a singlet state as

$$|f(\theta) + f(\pi - \theta)|^2,$$

[†] For example, L. Schiff, *Quantum Mechanics*, 1st ed., McGraw-Hill, New York, 1949, p. 116.

and in a triplet state as

$$|f(\theta) - f(\pi - \theta)|^2.$$

In experiments with unpolarized beams, the relative weights of triplet and singlet states are 3 to 1; moreover, their contributions add incoherently. We thus obtain the final result,

$$\sigma(\theta) = \tfrac{3}{4}|f(\theta) - f(\pi - \theta)|^2 + \tfrac{1}{4}|f(\theta) + f(\pi - \theta)|^2$$

$$= |f(\theta)|^2 + |f(\pi - \theta)|^2 - \mathrm{Re}\,[f^*(\theta)f(\pi - \theta)], \qquad \text{(B-33)}$$

which is Eq. 5–36. The first term can be considered to describe direct scattering, the second exchange scattering, and the third the interference between the direct and exchange waves. By substituting the expression in Eq. B–32 for $f(\theta)$, we obtain by direct calculation from Eq. B–33 the Mott scattering due only to the Coulomb force:

$$\sigma_{\text{Mott}} = \left(\frac{e^2}{4E}\right)^2 \left[\operatorname{cosec}^4 \frac{\theta}{2} + \sec^4 \frac{\theta}{2} - \operatorname{cosec}^2 \frac{\theta}{2} \sec^2 \frac{\theta}{2} \cos\left(\eta \ln \tan \frac{\theta}{2}\right)\right], \qquad \text{(B-34)}$$

where all quantities are measured in the center-of-mass system.

When nuclear forces are present, it is easy to see that Eq. B–33 may be written as

$$\sigma(\theta) = \sigma_{\text{Mott}}(\theta) + \sigma_N(\theta) + \sigma_{NC}(\theta), \qquad \text{(B-35)}$$

where $\sigma_N(\theta)$, which involves only the nuclear phase shifts δ_l, and $\sigma_{NC}(\theta)$, which involves Coulomb-nuclear interference terms, are given by

$$\sigma_N(\theta) = |f_N(\theta)|^2 + |f_N(\pi - \theta)|^2 - \mathrm{Re}\,[f_N^*(\theta)f_N(\pi - \theta)], \qquad \text{(B-36)}$$

$$\sigma_{NC}(\theta) = \mathrm{Re}\,\{f_C^*(\theta)[2f_N(\theta) - f_N(\pi - \theta)]$$

$$+ f_C^*(\pi - \theta)[2f_N(\pi - \theta) - f_N(\theta)]\}. \qquad \text{(B-37)}$$

It is left to the reader as a straightforward exercise to obtain the following result for the differential cross section when two protons scatter at an energy low enough so that the nuclear force is entirely in the S-state:

$$\sigma(\theta) = \left(\frac{e^2}{4E}\right)^2 \left[\operatorname{cosec}^4 \frac{\theta}{2} + \sec^4 \frac{\theta}{2} - \operatorname{cosec}^2 \frac{\theta}{2} \sec^2 \frac{\theta}{2} \cos\left(\eta \ln \tan \frac{\theta}{2}\right)\right]$$

$$- \left(\frac{e^2}{4E}\right) \frac{1}{k} \sin \delta \left[\operatorname{cosec}^2 \frac{\theta}{2} \cos\left(\delta + \eta \ln \sin \frac{\theta}{2}\right)\right.$$

$$\left. + \sec^2 \frac{\theta}{2} \cos\left(\delta + \eta \ln \cos \frac{\theta}{2}\right)\right] + \left(\frac{1}{k^2}\right) \sin^2 \delta.$$

B-4 COHERENT-SCATTERING LENGTH

The possibility of coherent scattering arises when a slow neutron is scattered by a target containing several scattering centers, such as a molecule or a crystal. For very slow neutrons, the neutron wavelength may be comparable to or greater than the relative distances between the individual scatterers. In this case, the incident neutron wave interacts with many scattering centers at the same time. The scattered waves from individual centers may have a phase relationship if the individual scatterers have similar scattering characteristics, thus giving rise to interference and coherent scattering. On the other hand, random phase differences between individual scatterers will give rise to incoherent scattering. In coherent scattering all the scattering centers participate in a collective way, whereas they act independently in incoherent scattering.

The simplest (and historically the most important) example is the scattering of slow neutrons ($\lambda \sim 7 \,\text{Å}$) from molecular hydrogen. The two protons in H_2 can have a total spin $S_H = 1$ (orthohydrogen) or 0 (parahydrogen). The interproton distance is less than 1 Å, and the two scattering centers are very similar; hence the conditions for coherent scattering are satisfied. In coherent scattering, the amplitudes of the scattered waves should be added. The scattering amplitude f_k is given by Eq. 2–31; and in the limit $k \to 0$, using Eq. 2–33, $f_k = -a$, where a is the scattering length.

We denote the spin operator of the incident neutron by $s_n = \frac{1}{2}\sigma_n$ and, similarly, the spins of the two protons in H_2 by s_{p_1} and s_{p_2}. Then the general amplitude for the scattering of a neutron by a proton is

$$\hat{a} = \tfrac{1}{4}(3a_t + a_s) + (a_t - a_s)s_n \cdot s_p. \qquad (\text{B-38})$$

In this equation, $\hat{a}$ is to be regarded as an operator, which yields the correct scattering length when its expectation value is taken with respect to the appropriate wave function. This is so because, by using the equations immediately preceding Eq. 2–6, $\langle s_n \cdot s_p \rangle = -\tfrac{3}{4}$ for the singlet and $+\tfrac{1}{4}$ for the triplet state. The total amplitude for the scattering of the neutron from the two protons in H_2 is therefore

$$\hat{a}_H = \tfrac{1}{2}(3a_t + a_s) + (a_t - a_s)s_n \cdot S_H, \qquad (\text{B-39})$$

where $S_H = s_{p_1} + s_{p_2}$.

The cross section of scattering from the hydrogen molecule with a given S_H and S_{zH} by an unpolarized neutron beam is

$$\sigma_{\text{mol}} = \tfrac{1}{2} \sum_{S_{zn}} \langle s_{zn} S_{zH} S_H{}^2 | 4\pi \hat{a}_H{}^2 | s_{zn} S_{zH} S_H{}^2 \rangle. \qquad (\text{B-40})$$

Since

$$\sum_{s_{zn}} \langle s_n \cdot S_H \rangle = 0, \qquad \tfrac{1}{2} \sum_{s_{zn}} \langle (s_n \cdot S_H)^2 \rangle = \tfrac{1}{4} S_H(S_H + 1), \qquad \text{(B-41)}$$

Eq. B-40 yields,

$$\sigma_{\text{ortho}} = \pi(3a_t + a_s)^2 + 2\pi(a_t - a_s)^2,$$

$$\sigma_{\text{para}} = \pi(3a_t + a_s)^2. \qquad \text{(B-42)}$$

If the nuclear force were spin independent, then $a_t = a_s$, giving $\sigma_{\text{ortho}} = \sigma_{\text{para}}$. Early experiments showed that $\sigma_{\text{ortho}} \gg \sigma_{\text{para}}$. Since a_t has a positive sign, this indicates that a_s is large and negative. For parahydrogen, since $S_H = 0$, the coherent-scattering amplitude is

$$a_H = \tfrac{1}{2}(3a_t + a_s). \qquad \text{(2-46)}$$

Interference between neutron waves scattered coherently by nuclei gives rise to optical phenomena like total reflection. Denote the wave number of the neutron wave by k in vacuum; then this is altered to some value k' inside the scattering medium because of interference between the waves. The refractive index of the medium is defined as

$$n = \frac{k'}{k}, \qquad \text{(B-43)}$$

and if this is less than 1, there can be total reflection. The refractive index is simply related to the coherent-scattering amplitude of the scattering centers in the medium. If it is assumed for simplicity that there are N molecules per unit volume, all of the same kind with coherent-scattering amplitude A_{mol},

$$n = 1 - \frac{\lambda^2}{2\pi} N A_{\text{mol}}. \qquad \text{(B-44)}$$

A simple derivation of this formula can be given, following Lord Rayleigh [1]. Let a plane neutron wave $\exp(ikz)$ be incident on a layer of thickness l of the medium, with $l \gg \lambda$, the wavelength of the neutron. When the wave emerges from the layer, it is given on the z-axis by

$$\exp[ik'l + ik(z - l)] = \exp[inkl + ik(z - l)]$$

$$= \exp(ikz) \exp[ikl(n - 1)]. \qquad \text{(B-45)}$$

The number of scattering centers in the elementary volume shown in Fig. B-1 is $(yl\, dy\, d\phi)N$; hence, when the transmitted beam is written as a sum of the incident plus the coherently scattered waves, it takes the form

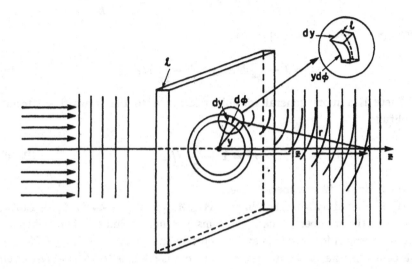

Fig. B-1 Neutron Refractive Index. The slab of scattering matter is of thickness $l \gg \lambda$. A neutron beam coming from the left parallel to the z-axis has plane wave fronts suggested by the perpendicular lines. Neutrons scattered in the indicated volume element constitute a spherical wave whose wave fronts are suggested by the arcs along the ray of length r.

$$e^{ikz} + 2\pi l N f_k \int_0^\infty \frac{e^{ikr}}{r} \, y \, dy, \tag{B-46}$$

where $r = \sqrt{y^2 + z^2}$. Equation B–46 may be rewritten in the form

$$\exp(ikz) - 2\pi l N A_{\text{mol}} \int_z^\infty \exp(ikr) \, dr. \tag{B-47}$$

Using the relation

$$\int_z^\infty e^{ikr} \, dr = \lim_{p \to 0} \int_z^\infty e^{ikr} e^{-p^2 r} \, dr = -\frac{\exp(ikz)}{ik},$$

the transmitted wave on the z-axis is given by

$$e^{ikz} \left(1 - i \, \frac{2\pi l N A_{\text{mol}}}{k} \right). \tag{B-48}$$

Equating this with the alternative form B–45, we can write

$$\exp[ikl(n-1)] \approx 1 + ikl(n-1) = 1 - \frac{i2\pi lNA_{\text{mol}}}{k},$$

since $k(n-1) \ll 1$. Consequently,

$$n = 1 - \frac{2\pi}{k^2}NA_{\text{mol}} = 1 - \frac{\lambda^2}{2\pi}NA_{\text{mol}}. \qquad (B\text{--}49)$$

There is a more general formula relating n to the forward scattering amplitude,

$$n^2 = 1 + \frac{\lambda^2}{\pi}Nf_k, \qquad (B\text{--}50)$$

of which Eq. B–49 is a special case.

With a liquid hydrocarbon like $C_{12}H_{18}$, A_{mol} in Eq. B–44 is $A_C + 1.5A_H$, and N is the number of carbon atoms per unit volume. The coherent-scattering length for carbon is experimentally known ($A_C = 6.63 \pm 0.03$ fm). This compound has n slightly less than 1, so it is possible to do total reflection experiments. For the grazing angle θ_c for total reflection, using Snell's law, we can write

$$\cos\theta_c = 1 - \frac{\theta_c^2}{2} = n = 1 - \frac{\lambda^2}{2\pi}N(A_C + 1.5A_H)$$

and

$$\theta_c = \lambda\left[\frac{N(A_C + 1.5A_H)}{\pi}\right]^{1/2}. \qquad (B\text{--}51)$$

Accurate determination of θ_c therefore determines A_H, since A_C is known from other sources.

Fermi and Marshall [2] first used this method to determine the coherent-scattering lengths of various elements. More recent extensive work was done by Koester [3], who measured the critical angle θ_c for the total reflection of neutrons from the "mirror" surface of various hydrocarbons.

REFERENCES

1. Lord Rayleigh, *Phil. Mag.* **47**, 375 (1899).
2. E. Fermi and L. Marshall, *Phys. Rev.* **71**, 666 (1947); **72**, 408 (1947).
3. L. Koester, *Z. Phys.* **198**, 187 (1967); **208**, 515 (E) (1967).

The Notation of Second Quantization

C-1 THE WAVE FUNCTION OF A MANY-FERMION SYSTEM

For a system of identical fermions in the independent-particle model, the Slater determinantal wave function ensures the indistinguishability of the particles and the antisymmetry under the exchange of any two particle coordinates. Since it is not meaningful, however, to determine which particle is in which single-particle state, the particle indices are really superfluous. This suggests the adoption of an occupation number representation, in which only the state labels are used, and one requires to know only which states are occupied. Since what is being exploited is the indistinguishability of the particles, this method is quite general and can be used for bosons as well, provided that one indicates in some way the number of particles in each state.

In the so-called occupation number representation, also called second quantization, algebraic manipulations are easily done in a systematic and concise manner. In what follows, we shall explain this notation, concentrating specifically on fermions.

Let us denote a complete set of single-particle states by $\{\alpha\}$: $\alpha_1, \alpha_2, \alpha_A$, etc. Here α stands for a whole set of quantum numbers which uniquely defines a single-particle state. Then an A-fermion determinantal wave function is denoted by

$$\frac{1}{\sqrt{A!}} \begin{vmatrix} \phi_{\alpha_1}(1) & \cdots & \phi_{\alpha_1}(A) \\ \vdots & & \vdots \\ \phi_{\alpha_A}(1) & \cdots & \phi_{\alpha_A}(A) \end{vmatrix} \equiv \frac{1}{\sqrt{A!}} \det\{\alpha_1 \cdots \alpha_A\}. \qquad \text{(C--1)}$$

In this equation, $(1) \cdots (A)$ are particle indices. In the second quantized notation, we define a set of operators a_α^+ and a vacuum state $|0\rangle$, such that

$$a_{\alpha_1}^+ a_{\alpha_2}^+ \cdots a_{\alpha_A}^+ |0\rangle \tag{C-2}$$

is identical to C-1 and represents a many-body state in which $\alpha_1, \alpha_2, \ldots, \alpha_A$ are singly occupied by particles. For the equivalence between C-1 and C-2 to hold, the operators a_α^+ have to obey certain algebraic relations. Since

$$a_{\alpha_1}^+ a_{\alpha_2}^+ \cdots a_{\alpha_A}^+ |0\rangle \equiv \frac{1}{\sqrt{A!}} \det \{\alpha_1 \alpha_2 \cdots \alpha_A\}$$

$$= -\frac{1}{\sqrt{A!}} \det \{\alpha_2 \alpha_1 \cdots \alpha_A\}$$

$$= -a_{\alpha_2}^+ a_{\alpha_1}^+ \cdots a_{\alpha_A}^+ |0\rangle,$$

it follows that

$$a_{\alpha_1}^+ a_{\alpha_2}^+ = -a_{\alpha_2}^+ a_{\alpha_1}^+$$

or

$$\{a_{\alpha_1}^+, a_{\alpha_2}^+\} = a_{\alpha_1}^+ a_{\alpha_2}^+ + a_{\alpha_2}^+ a_{\alpha_1}^+ = 0. \tag{C-3}$$

Similarly, defining the Hermitian conjugate of these operators, we obtain

$$\{a_{\alpha_1}, a_{\alpha_2}\} = 0. \tag{C-4}$$

Furthermore, we know that the overlap of two determinantal wave functions

$$\frac{1}{A} \int \det{}^* \{\alpha_1 \cdots \alpha_A\} \det \{\beta_1 \cdots \beta_A\} \, d^3r_1 \cdots d^3r_A \tag{C-5}$$

= 0 unless the sets $\{\alpha\}$ and $\{\beta\}$ are the same,
= + 1 if the sets are the same but differ by an even number of permutations,
= − 1 if the sets differ by an odd number of permutations.

Writing C-5 in the new notation, we obtain

$$\langle 0 | a_{\alpha_A} a_{\alpha_{A-1}} \cdots a_{\alpha_1} a_{\beta_1}^+ a_{\beta_2}^+ \cdots a_{\beta_A}^+ | 0 \rangle. \tag{C-6}$$

To satisfy the above conditions with C-6, we must have

$$\langle 0 | 0 \rangle = 1, \tag{C-7}$$

$$a_\alpha | 0 \rangle = 0 \quad \text{for all } \alpha, \tag{C-8}$$

and

$$\{a_\alpha, a_\beta{}^+\} = \delta_{\alpha\beta}. \tag{C-9}$$

Using Eqs. C-7 to C-9, one can prove that the two-particle wave function $a_{\alpha_1}^+ a_{\alpha_2}^+ |0\rangle$ is normalized to 1, that is,

$$\langle 0 | a_{\alpha_2} a_{\alpha_1} a_{\alpha_1}^+ a_{\alpha_2}^+ |0\rangle = 1.$$

The operators $a_\alpha{}^+$ and a_α can be interpreted as the creation and annihilation operators for a particle in state α. To see this, consider a state

$$|\psi\rangle = a_{\alpha_1}^+ a_{\alpha_2}^+ a_{\alpha_3}^+ |0\rangle$$
$$= |1, 1, 1, 0, 0, \ldots\rangle. \tag{C-10}$$

This means that in $|\psi\rangle$ there is one particle each in states $\alpha_1, \alpha_2, \alpha_3$, while all other states are empty. If a_{α_1} is an annihilation operator, its application to $|\psi\rangle$ should result in a state $|0, 1, 1, 0, 0, \ldots\rangle$. Using Eq. C-9, one finds that

$$a_{\alpha_1}|\psi\rangle = a_{\alpha_1} a_{\alpha_1}^+ a_{\alpha_2}^+ a_{\alpha_3}^+ |0\rangle$$
$$= (1 - a_{\alpha_1}^+ a_{\alpha_1}) a_{\alpha_2}^+ a_{\alpha_3}^+ |0\rangle$$
$$= a_{\alpha_2}^+ a_{\alpha_3}^+ |0\rangle - a_{\alpha_1}^+ a_{\alpha_2}^+ a_{\alpha_3}^+ a_{\alpha_1} |0\rangle$$
$$= a_{\alpha_2}^+ a_{\alpha_3}^+ |0\rangle, \text{ using Eq. C-8}$$
$$= |0, 1, 1, 0, 0, \ldots\rangle. \tag{C-11}$$

A similar operation by a_{α_2} on $|\psi\rangle$ reveals a difference:

$$a_{\alpha_2}|\psi\rangle = a_{\alpha_2} a_{\alpha_1}^+ a_{\alpha_2}^+ a_{\alpha_3}^+ |0\rangle$$
$$= - a_{\alpha_1}^+ a_{\alpha_3}^+ |0\rangle$$
$$= - |1, 0, 1, 0, 0, \ldots\rangle. \tag{C-12}$$

Thus there is an overall minus sign in Eq. C-12, although a_{α_2} still behaves like an annihilation operator. The minus sign appears because there is an odd number of operators (here just $a_{\alpha_1}^+$) between a_{α_2} and $a_{\alpha_2}^+$. More generally,

$$a_\alpha |\ldots, n_\alpha, \ldots\rangle = (-)^p \sqrt{n_\alpha} |\ldots, 1 - n_\alpha, \ldots\rangle,$$
$$a_\alpha{}^+ |\ldots, n_\alpha, \ldots\rangle = (-)^p \sqrt{1 - n_\alpha} |\ldots, 1 + n_\alpha, \ldots\rangle, \tag{C-13}$$

where any occupation number n_γ in a state γ is either 0 or 1, and

$$p = \sum_{\gamma=1}^{\alpha-1} n_\gamma.$$

It can also be shown that the operator

$$\hat{n} = \sum_{\text{all } \alpha} a_\alpha{}^+ a_\alpha \tag{C-14}$$

behaves like a number operator on a many-body state $|\psi\rangle$, so that

$$\hat{n}|\psi\rangle = \sum_\gamma n_\gamma |\psi\rangle = A|\psi\rangle. \tag{C-15}$$

C-2 THE OPERATORS

We now formulate the procedure for writing down operators in this notation. For simplicity, we consider first a one-body operator, such as kinetic energy,

$$T = \sum_{i=1}^{A} t(i), \tag{C-16}$$

where the index i refers to a particle. A standard result with the determinantal wave function C–1 for the following diagonal matrix element is

$$\frac{1}{A!} \int \det{}^* \{\alpha_1 \cdots \alpha_A\} T \det \{\alpha_1 \cdots \alpha_A\} \, d^3r_1 \cdots d^3r_A$$

$$= \sum_{\alpha_i=1}^{A} \int \phi_{\alpha_i}^*(\mathbf{r}_j) t(\mathbf{r}_j) \phi_{\alpha_i}(\mathbf{r}_j) \, d^3r_j = \sum_{\alpha_i=1}^{A} \langle \alpha_i | t | \alpha_i \rangle, \tag{C-17}$$

where the sum on α_i is over occupied states only, and the subscript on t is irrelevant. Similarly, the matrix element between two determinantal states in which all the single-particle states except one are identical is given by

$$\frac{1}{A!} \int \det \{\alpha_1 \alpha_2 \cdots \alpha_A\}^* T \det \{\alpha_M \alpha_2 \cdots \alpha_A\} \, d^3r_1 \cdots d^3r_A$$

$$= \langle \alpha_1 | t | \alpha_M \rangle, \quad M > A. \tag{C-18}$$

If the single-particle states in the two determinants differ by more than one, then the matrix element is zero.

The above results in the occupation number notation are as follows:

$$\langle \underbrace{1\,1\,1 \cdots 1\,0\,0 \cdots}_{A} | T | \underbrace{1\,1\,1 \cdots 1\,0\,0 \cdots}_{A} \rangle = \sum_{j=1}^{A} \langle \alpha_j | t | \alpha_j \rangle \tag{C-19}$$

and

$$\langle \underbrace{1\,1\,1 \cdots 1\,0\,0 \cdots}_{A} | T | \underbrace{0\,1\,1 \cdots 1\,0 \cdots}_{A} n_M = 1, 0 \cdots \rangle = \langle \alpha_1 | t | \alpha_M \rangle. \tag{C-20}$$

These equations follow if we define the one-body operator as

$$T = \sum_{\beta\alpha} \langle \beta|t|\alpha\rangle a_\beta^+ a_\alpha, \tag{C-21}$$

where the indices β, α run over the complete set.

Next we consider the two-body operator,

$$v = \sum_{i<j} v(i,j), \tag{C-22}$$

where i and j refer to particles. Again all the usual matrix elements of v can be obtained by working with the second-quantized wave function C-2 and writing the operator v as

$$v = \tfrac{1}{2} \sum_{\alpha\beta\gamma\delta} \langle \alpha\beta|v|\gamma\delta\rangle a_\alpha^+ a_\beta^+ a_\delta a_\gamma, \tag{C-23}$$

where the sums run over the complete set, with $\alpha \neq \beta$ and $\gamma \neq \delta$. Note the order of the two annihilation operators.

As a simple example, assume that two fermions occupy the two states η and ζ with wave function

$$\Psi = \frac{1}{\sqrt{2}} \det\{\eta, \zeta\} = \frac{1}{\sqrt{2}} [\phi_\eta(\mathbf{r}_1)\phi_\zeta(\mathbf{r}_2) - \phi_\eta(\mathbf{r}_2)\phi_\zeta(\mathbf{r}_1)]. \tag{C-24}$$

The diagonal matrix element of an operator $v(1, 2)$ with this wave function is simply

$$\langle \eta\zeta|v|\eta\zeta\rangle - \langle \eta\zeta|v|\zeta\eta\rangle, \tag{C-25}$$

where, for a local $v(1, 2)$,

$$\langle \alpha\beta|v|\gamma\delta\rangle = \int \phi_\alpha^*(\mathbf{r}_1)\phi_\beta^*(\mathbf{r}_2)v(\mathbf{r}_1, \mathbf{r}_2)\phi_\gamma(\mathbf{r}_1)\phi_\delta(\mathbf{r}_2)\, d^3r_1\, d^3r_2. \tag{C-26}$$

In the second-quantized notation, using the forms C-2 and C-23, the diagonal matrix element $\langle \Psi|v|\Psi\rangle$ is

$$\tfrac{1}{2} \sum_{\alpha\beta\gamma\delta} \langle \alpha\beta|v|\gamma\delta\rangle \langle 0|a_\zeta a_\eta a_\alpha^+ a_\beta^+ a_\delta a_\gamma a_\eta^+ a_\zeta^+|0\rangle. \tag{C-27}$$

The reader should verify that this expression reduces to C-25 on using the properties of the creation and annihilation operators given in Section C-1.

Now we can write down a Hamiltonian

$$H = \sum_{i=1}^{A} t(i) + \sum_{i<j} v(i,j), \tag{C-28}$$

where the second term on the right has $\tfrac{1}{2}A(A-1)$ terms, A being the number of particles. In the second-quantized version, we write this as

$$H = \sum_{\alpha\beta} \langle\alpha|t|\beta\rangle a_\alpha{}^+a_\beta + \tfrac{1}{2} \sum_{\alpha\beta\gamma\delta} \langle\alpha\beta|v|\gamma\delta\rangle a_\alpha{}^+a_\beta{}^+a_\delta a_\gamma, \qquad (C\text{-}29)$$

where there is no reference to the particle number A. The two expressions C–28 and C–29 are equivalent in the sense that their matrix elements with respect to many-body wave functions are identical.

In the independent-particle model, the ground state of a nucleus is described by a state $|c\rangle$ in which all states $\alpha_1, \alpha_2, \ldots, \alpha_A$ are occupied by nucleons and all others are empty. In our notation,

$$|c\rangle = a_{\alpha_1}^+ a_{\alpha_2}^+ \cdots a_{\alpha_A}^+ |0\rangle, \qquad (C\text{-}30)$$

where α_A defines the Fermi level of the system. Note that

$$a_\beta|c\rangle = 0 \quad \text{if} \quad \beta > \alpha_A$$

and

$$a_\beta{}^+|c\rangle = 0 \quad \text{if} \quad \beta \leqslant \alpha_A. \qquad (C\text{-}31)$$

It is often convenient to make a transformation such that $|c\rangle$ behaves like a vacuum state:

$$a_\beta{}^+ = b_\beta, \qquad a_\beta = b_\beta{}^+, \quad \text{for} \quad \beta \leqslant \alpha_A. \qquad (C\text{-}32)$$

The operator b^+ is called a "hole"-creation operator, while for $\beta > \alpha_A$ the a's are the usual particle operators. Now we obtain

$$a_\beta|c\rangle = 0, \quad \beta > \alpha_A,$$

$$b_\beta|c\rangle = 0, \quad \beta \leqslant \alpha_A, \qquad (C\text{-}33)$$

so that $|c\rangle$ behaves like the vacuum state with respect to particle and hole operators.

In order to evaluate the matrix elements of operators in a systematic manner, we now introduce a few definitions.

1. *The normal order of a product of operators* is such that all creation operators (of particles as well as holes) stand to the left of the annihilation operators. The sign is determined by whether an odd or an even number of permutations is required to obtain the normal ordering. For example,

$$N(a_{\alpha_1}a_{\alpha_2}^+a_{\alpha_3}) = -a_{\alpha_2}^+a_{\alpha_1}a_{\alpha_3} \quad \text{when} \quad \alpha_1, \alpha_2, \alpha_3 > \alpha_A$$

$$N(a_{\alpha_1}a_{\alpha_2}^+a_{\alpha_3}) = N(a_{\alpha_1}a_{\alpha_2}^+b_{\alpha_3}^+) \quad \text{when} \quad \alpha_1, \alpha_2 > \alpha_A, \alpha_3 < \alpha_A$$

$$= a_{\alpha_2}^+b_{\alpha_3}^+a_{\alpha_1}.$$

Note that the normal ordered product, operating on $|c\rangle$, gives zero identically.

2. *A contracted product of any two* (creation or annihilation) *operators A* and *B* is defined by

$$\underline{AB} = AB - N(AB). \tag{C-34}$$

Because of the commutation relations C-3 and C-4, the contracted product is simply a *c*-number, namely,

$$\underline{a_\alpha a_{\alpha'}^+} = \delta_{\alpha\alpha'}, \qquad \alpha, \alpha' > \alpha_A,$$

$$\underline{a_\alpha^+ a_{\alpha'}} = \underline{b_\alpha b_{\alpha'}^+} = \delta_{\alpha\alpha'}, \qquad \alpha, \alpha' \leqslant \alpha_A. \tag{C-35}$$

All other contractions are identically zero. Note that, because of definition 1,

$$\langle c|AB|c\rangle = \langle c|\underline{AB}|c\rangle = \underline{AB}.$$

3. *Wick's theorem*: Often we encounter a product of many creation and annihilation operators whose expectation value is required with respect to $|c\rangle$. Wick's theorem states that, if the total number of creation plus annihilation operators is odd, the expectation value of their product with respect to $|c\rangle$ vanishes. If, on the other hand, the total number is even, so that all of these operators can be contracted, the expectation value is just the sum of all possible fully contracted combinations. For example,

$$\langle c|a_{\alpha_1} a_{\alpha_2}^+ b_{\alpha_3}^+|c\rangle = 0,$$

$$\langle c|a_{\alpha_1} a_{\alpha_2}^+ b_{\alpha_3} b_{\alpha_4}^+|c\rangle = \underline{a_{\alpha_1} a_{\alpha_2}^+}\, \underline{b_{\alpha_3} b_{\alpha_4}^+} = \delta_{\alpha_1\alpha_2}\delta_{\alpha_3\alpha_4},$$

since all other contractions are zero, with $\alpha_1, \alpha_2 > \alpha_A$ and $\alpha_3, \alpha_4 \leqslant \alpha_A$.

C-3 EVALUATION OF MATRIX ELEMENTS

We can now apply the methods developed in Sections C-1 and C-2 to the calculation of matrix elements. A first-order change in Eq. 10-4 implies that

$$\langle c|H|\delta c\rangle + \langle \delta c|H|c\rangle = 0,$$

where $|c\rangle$ and $|\delta c\rangle$ are defined by Eqs. 10-5 and 10-6. In the above equation, the first term is simply the complex conjugate of the second. We thus get the minimal condition

$$\langle c|H a_b^+ a_m|c\rangle = 0,$$

with $b > A$ and $m \leqslant A$. Substituting for H from Eq. C-29, we can write

$$\sum_{\alpha,\beta} \langle \alpha|t|\beta\rangle\langle c|a_\alpha^+ a_\beta a_b^+ a_m|c\rangle + \tfrac{1}{2}\sum_{\alpha,\beta,\gamma,\delta} \langle \alpha\beta|v|\gamma\delta\rangle\langle c|a_\alpha^+ a_\beta^+ a_\delta a_\gamma a_b^+ a_m|c\rangle = 0. \tag{C-36}$$

We consider first the matrix element $\langle c|a_\alpha{}^+a_\beta a_b{}^+a_m|c\rangle$. Since $m \leqslant A$, $a_m = b_m{}^+$, from Eq. C–32. Also, for the matrix element to be nonzero, $\alpha \leqslant A$; consequently $a_\alpha{}^+ = b_\alpha$. Thus

$$\langle c|a_\alpha{}^+a_\beta a_b{}^+a_m|c\rangle = \langle c|b_\alpha a_\beta a_b{}^+b_m{}^+|c\rangle$$
$$= b_\alpha b_m{}^+a_\beta a_b{}^+ = \delta_{\alpha m}\delta_{\beta b}.$$

$$\therefore \sum_{\alpha,\beta} \langle \alpha|t|\beta\rangle\langle c|a_\alpha{}^+a_\beta a_b{}^+a_m|c\rangle = \langle m|t|b\rangle.$$

Next, we evaluate the matrix element

$$\langle c|a_\alpha{}^+a_\beta{}^+a_\delta a_\gamma a_b{}^+a_m|c\rangle, \quad \text{with} \quad \alpha \neq \beta, \quad \gamma \neq \delta.$$

Obviously, for a nonzero value, α, β, and $m \leqslant A$, so that

$$\langle c|a_\alpha{}^+a_\beta{}^+a_\delta a_\gamma a_b{}^+a_m|c\rangle = \langle c|b_\alpha b_\beta a_\delta a_\gamma a_b{}^+b_m{}^+|c\rangle.$$

(i) $\delta \leqslant A, \gamma > A$: We obtain

$$\langle c|b_\alpha b_\beta b_\delta{}^+a_\gamma a_b{}^+b_m{}^+|c\rangle = b_\alpha b_m{}^+b_\beta b_\delta{}^+a_\gamma a_b{}^+ - b_\alpha b_\delta{}^+b_\beta b_m{}^+a_\gamma a_b{}^+.$$

When substituted into the second term in Eq. C–36, this yields

$$\tfrac{1}{2}\left[\sum_{\delta=1}^{A} \langle m\delta|v|b\delta\rangle - \langle \delta m|v|b\delta\rangle\right].$$

(ii) $\delta > A, \gamma \leqslant A$: This case yields two more terms, identical to the above expression, except that now the dummy index is γ, going from 1 to A.

Summing (i) and (ii) and denoting the dummy index by n, we obtain the minimal condition given by Eq. 10–7.

Using these techniques, it is also straightforward to evaluate the quantity

$$\langle c|a_m{}^+a_b a_\alpha{}^+a_\beta{}^+a_\delta a_\gamma a_a{}^+a_n|c\rangle$$

occurring in the particle-hole matrix element $\langle bm^{-1}|v|an^{-1}\rangle$ of Eq. 10–62. We consider the nondiagonal matrix element when $b \neq a, m \neq n$, and we remember that, in our notation, m, $n \leqslant A$ while $a, b > A$. We then have

$$\langle c|b_m a_b a_\alpha{}^+a_\beta{}^+a_\delta a_\gamma a_a{}^+b_n{}^+|c\rangle. \tag{C–37}$$

Using Wick's theorem, we can reduce expression C–37 to the following sum of contracted pairs (noting that b_m can only be contracted with a_δ or a_γ):

$$b_m a_b a_\alpha{}^+ a_\beta{}^+ a_\delta a_\gamma a_a{}^+ b_n{}^+ + b_m a_b a_\alpha{}^+ a_\beta{}^+ a_\delta a_\gamma a_a{}^+ b_n{}^+$$

$$+ \, b_m a_b a_\alpha{}^+ a_\beta{}^+ a_\delta a_\gamma a_a{}^+ b_n{}^+ + b_m a_b a_\alpha{}^+ a_\beta{}^+ a_\delta a_\gamma a_a{}^+ b_n{}^+$$

$$= - \, \delta_{ab}\delta_{\beta n}\delta_{\gamma a}\delta_{\delta m} + \delta_{an}\delta_{\beta b}\delta_{\gamma a}\delta_{\delta m}$$

$$+ \, \delta_{ab}\delta_{\beta n}\delta_{\gamma m}\delta_{\delta a} - \delta_{an}\delta_{\beta b}\delta_{\gamma m}\delta_{\delta a}.$$

Substituting this value for expression C-37 in Eq. 10-62, we get

$$\langle bm^{-1}|v|an^{-1}\rangle = \tfrac{1}{2}(- \langle bn|v|am\rangle + \langle nb|v|am\rangle$$

$$+ \, \langle bn|v|ma\rangle - \langle nb|v|ma\rangle)$$

$$= \langle bn|v|ma\rangle - \langle bn|v|am\rangle, \tag{C-38}$$

which is Eq. 10-63 of the text. Furthermore, since we denote the matrix element $\langle \alpha\beta|v|\gamma\delta\rangle$ by the diagram

it follows that $\langle bn|v|ma\rangle$ becomes

while $\langle bn|v|am\rangle$ is equivalent to

We consider now a multipole operator of order J,

$$O_J = \sum_{\alpha,\beta} \langle \alpha|r^J Y_J{}^M|\beta\rangle a_\alpha{}^+ a_\beta.$$

Then, we define

$$D_{bm}^J = \langle bm^{-1}|O_J|c\rangle$$

$$= \sum_{\alpha,\beta} \langle bm^{-1}|a_\alpha{}^+ a_\beta|c\rangle \langle \alpha|r^J Y_J{}^M|\beta\rangle$$

$$= \sum_{\alpha,\beta} \langle c|a_m{}^+ a_b a_\alpha{}^+ a_\beta|c\rangle \langle \alpha|r^J Y_J{}^M|\beta\rangle$$

or

$$D_{bm}^J = \langle b|r^J Y_J{}^M|m\rangle. \tag{C-39}$$

Note that, if $|c\rangle$ has zero angular momentum, only the component of $|bm^{-1}\rangle$ that has angular momentum J contributes to D_{bm}^J.

The WKB Approximation

The WKB (Wentzel-Kramers-Brillouin) approximation is a solution to the Schrödinger equation which has validity if the local wave number varies sufficiently slowly. For the motion of a particle of energy E in a potential $V(x)$, the local wave number $k(x)$ is given by

$$k^2(x) = \frac{2m}{\hbar^2}[E - V(x)],$$

and the wave equation is simply

$$\frac{d^2u}{dx^2} + k^2(x)u = 0.$$

However, the considerations which follow are of a mathematical nature and apply equally well if the independent variable is the radial coordinate r, the coordinate β of the collective model, or any other coordinate for which the wave function can be separated. To emphasize this, we denote the variable by α and remark that, whatever the explicit form of the kinetic-energy operator, so long as the problem is separable, we can by transformation of variables put the wave equation into the form

$$\frac{d^2u}{d\alpha^2} + k^2(\alpha)u = 0 \tag{D-1a}$$

or

$$\frac{d^2u}{d\alpha^2} - \kappa^2(\alpha)u = 0, \tag{D-1b}$$

where k and κ are real functions given by

$$k^2 = C[E - V_e(\alpha)], \quad E > V_e, \tag{D-2a}$$

$$\kappa^2 = C[V_e(\alpha) - E], \quad E < V_e, \tag{D-2b}$$

where E is the energy of the system, $V_e(\alpha)$ is the effective potential energy, and C is a proportionality constant. When k is constant, the solutions of Eq. D-1 are exponentials $\exp(\pm ik\alpha)$, $\exp(\pm \kappa\alpha)$. Hence, if k is slowly varying, we might expect that the solutions

$$u_0 = \exp\left[\pm i \int^\alpha k(\alpha)\, d\alpha\right]$$

might have some validity. Indeed, we find that

$$\frac{d^2 u_0}{d\alpha^2} + k^2 u_0 = \pm \tfrac{1}{2}i \cdot \frac{dk}{d\alpha} u_0,$$

showing that, if $dk/d\alpha$ is small compared with k^2, the approximation is valid. This condition may be written as

$$\left(\frac{dk}{d\alpha}\right)\frac{1}{k} \ll k. \tag{D-3}$$

Since $1/k$ is the "local wavelength," this expression states that the result is valid if the change in k over a wavelength or, indeed, over several wavelengths is negligible compared with k itself. One situation in which this would arise would be in a region where $E \gg V_e(\alpha)$. It could also occur when $V_e \gg E$, provided that V_e is a slowly varying function. The one region where the condition certainly fails is that in which $V_e \approx E$, for there k is very small but $dk/d\alpha$ is large, being proportional to $k^{-1}(dV/d\alpha)$.

We can attempt to improve on the expression u_0 by finding the next approximation in the form $f(\alpha)u_0(\alpha)$, where $f(\alpha)$ remains near unity. It is easy to see, ignoring the second derivative of $f(\alpha)$, that $f(\alpha) = k^{-1/2}$. This second approximation is the one usually called the WKB solution,

$$u = Ak^{-1/2} \exp(\pm i\omega), \tag{D-4a}$$

or

$$u = B\kappa^{-1/2} \exp(\pm \omega), \tag{D-4b}$$

where

$$\omega = \int^\alpha k\, d\alpha \quad \text{or} \quad -\int^\alpha \kappa\, d\alpha, \tag{D-4c}$$

as the case may be.

The most common application of the WKB approximation in nuclear physics is to problems of barrier penetration. In considering a potential that goes to zero at large values of α, but rises well above the energy E for certain other values, as shown in Fig. D-1, it is seen that there are two regions where the WKB solution would be useful. In region 1, the solution of form D-4a

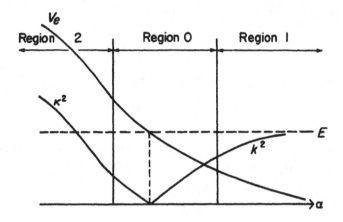

Fig. D-1 A potential $V_e(\alpha)$ that goes to zero for large α, but is greater than E for small α. In regions 1 and 2, where k^2 and κ^2 are not small, simple WKB solutions for the wave function can be found (see text).

with an imaginary exponential would be valid; in region 2, that of form D-4b. However, in the intermediate region, labeled region 0 in the figure, the approximation is not valid since k^2 is small, vanishing at the point where $E = V_e$. To obtain expressions for the penetration of particles in region 1 into region 2, it is necessary to make use of the wave function in region 0 only in order to connect the expressions in the other two regions. Explicitly, using an obvious notation, we can write the wave functions in regions 1 and 2, respectively, as

$$\psi(1) = A_- u_-(1) + A_+ u_+(1) \tag{D-5a}$$

and

$$\psi(2) = B_- u_-(2) + B_+ u_+(2). \tag{D-5b}$$

It is necessary to find the relationship between the B's and the A's in order to determine the relationship between the particle densities in the two asymptotic regions.

In the immediate vicinity of the point where $E = V_e$, we can treat k^2 as a linear function of α; that is, if we shift the origin of α to this turning point,

$$\frac{d^2u}{d\alpha^2} + a\alpha u = 0. \tag{D-6}$$

In this case,[†] if we take the hitherto unspecified lower limit on the integral D–4c defining ω to be $\alpha = 0$,

$$\omega = \tfrac{2}{3}a^{1/2}|\alpha|^{3/2},$$

and exact solutions[‡] of Eq. D–6 are

$$u_{\pm} = \omega^{1/2}|k|^{-1/2}Z_{\pm 1/3}(\omega), \tag{D-7}$$

where $Z_p(\omega)$ is a Bessel function and

$$Z_p(\omega) = J_p(\omega), \qquad\qquad \alpha > 0$$
$$= I_p(\omega) = i^{-p}J_p(i\omega), \quad \alpha < 0.$$

For small ω, $J_p(\omega) = I_p(\omega) = (\tfrac{1}{2}\omega)^p(p!)^{-1}$, and we can readily see that

$$u_+(\alpha \to 0^+) = u_+(\alpha \to 0^-) = 0,$$

$$u_-(\alpha \to 0^+) = u_-(\alpha \to 0^-) = \text{a nonzero number},$$

$$\frac{du_+}{d\alpha}(\alpha \to 0^+) = -\frac{du_+}{d\alpha}(\alpha \to 0^-) = \text{a nonzero number},$$

$$\frac{du_-}{d\alpha}(\alpha \to 0^+) = \frac{du_-}{d\alpha}(\alpha \to 0^-) = 0.$$

---

† We see the advantage of the minus sign in the definition D–4c of ω; ω increases with $|\alpha|$. If regions 1 and 2 are reversed so that region 2 lies to the right of region 1, the signs in Eq. D–4c are to be reversed.

‡ More generally, Eq. D–1 can be written, with $v = k^{1/2}u$, as

$$\frac{d^2v}{d\omega^2} + \left[1 + \frac{1}{4}\frac{1}{k^2}\left(\frac{dk}{d\omega}\right)^2 - \frac{1}{2}\frac{1}{k}\frac{d^2k}{d\omega^2}\right]v = 0,$$

and, if k is proportional to a power of ω, say ω^β, the solutions are $\omega^{1/2}Z_p(\omega)$, with $p = \tfrac{1}{2}(\beta - 1)$.

Hence, since the wave function and its derivative must be continuous at $\alpha = 0$, it follows that, if a solution for $\alpha > 0$ is

$$\psi = \omega^{1/2}k^{-1/2}[AJ_{1/3}(\omega) + BJ_{-1/3}(\omega)],$$

its value for $\alpha < 0$ must be

$$\psi = \omega^{1/2}\kappa^{-1/2}[-AI_{1/3}(\omega) + BI_{-1/3}(\omega)].$$

Now, for large ω,

$$J_p(\omega) \sim (\tfrac{1}{2}\pi\omega)^{-1/2} \cos\left[\omega - \left(p + \frac{1}{2}\right)\frac{\pi}{2}\right] \qquad \text{(D-8a)}$$

and

$$I_p(\omega) \sim (2\pi\omega)^{-1/2}\{e^\omega + e^{-\omega}\exp[-\pi i(\tfrac{1}{2} + p)]\}. \qquad \text{(D-8b)}$$

Of course, in the asymptotic expression D–8b, the term with the decreasing exponential can be written only when the solution taken is $u_+ - u_-$, for which the positive exponential vanishes.

The solutions valid for k small are thus seen to have asymptotic forms D–8 similar to the WKB solutions D–4. Hence, if in region 1 our solution is an outgoing wave, $u_+(1) = A_+k^{-1/2}e^{i\omega}$, we need to take in region 0 the solution with the same asymptotic form, namely,

$$A_+k^{-1/2}i\exp\left(i\,\frac{\pi}{4}\right)\mathrm{cosec}\,p\pi\,(\tfrac{1}{2}\pi\omega)^{1/2}\left[\exp\left(-ip\,\frac{\pi}{2}\right)J_p - \exp\left(\frac{ip\pi}{2}\right)J_{-p}\right].$$

To assure continuity at $\alpha = 0$, the solution for $\alpha < 0$ must, therefore, be

$$A_+\kappa^{-1/2}i\exp\left(i\,\frac{\pi}{4}\right)\mathrm{cosec}\,p\pi\,(\tfrac{1}{2}\pi\omega)^{1/2}\left[-\exp\left(-ip\,\frac{\pi}{2}\right)I_p - \exp\left(ip\,\frac{\pi}{2}\right)I_{-p}\right].$$

For large ω, this becomes $\exp(-i\pi/4)\,A_+\kappa^{-1/2}e^\omega$.

Thus we have seen that a solution in region 1 given by $Ak^{-1/2}e^{i\omega}$ is connected to a solution in region 2 given by $\exp(-i\pi/4)\,A\kappa^{-1/2}e^\omega$. However, this formula cannot be used in reverse. In other words, if we know that the wave function in region 2 is closely approximated by $\kappa^{-1/2}e^\omega$, we cannot deduce that the wave in region 1 is simply an outgoing wave, since the phase of the oscillating wave in region 1 is determined by the mixture of the increasing and decreasing exponentials in region 2, and, when the region-2 wave function contains the increasing exponential, our approximate solution, of course, tells us nothing about the amount of decreasing exponential present. However, it is possible in a similar way to establish the connection giving the amplitude and phase of the wave in region 1 if region 2 is known to contain only the decreasing exponential. This connection is

$$\kappa^{-1/2}e^{-\omega} \to 2k^{-1/2}\cos\left(\omega - \frac{\pi}{4}\right).$$ (D-9a)

Also, for a wave of general phase γ in region 1, the connection giving the magnitude in region 2 is

$$k^{-1/2}\cos\left(\omega + \gamma - \frac{\pi}{4}\right) \to \sin\gamma\,\kappa^{-1/2}e^{\omega}.$$ (D-9b)

We can see how to use these formulas to calculate transmission through a reasonably wide barrier as shown in Fig. D-2. Let us consider a wave incident from the left, so that region III contains both a wave moving in the positive α-direction and a reflected wave moving in the negative sense. The wave is transmitted through the classically forbidden zone of region II, and in region III there is only an "outgoing" wave, that is, one moving toward positive α. In short, the asymptotic form in region I is simply

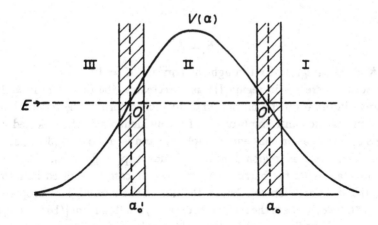

Fig. D-2 Transmission of a wave of incident energy E through a barrier $V(\alpha)$. The shaded regions are in the vicinity of the classical turning points O and O'.

$$u(\text{I}) \sim A k^{-1/2} e^{i\omega}.$$

We know, therefore, from the above discussion that in region II, when we are sufficiently far from the turning point O,

$$u(\text{II}) \sim A \exp(-i\pi/4)\, \kappa^{-1/2} e^{\omega}.$$

It is necessary that the barrier be sufficiently broad that the term in $e^{-\omega}$ will be negligible at points within region II near O'. This expression refers, of course, to an ω defined to vanish at the right-hand turning point. To make the connection between regions II and III, we must use relations in which we define our origin at the left-hand turning point. If we use primes to denote such quantities, we have for region II

$$\omega' = \int_0^{\alpha'} \kappa \, d\alpha' = \int_{\alpha_{\mathsf{o}}'}^{\alpha} \kappa \, d\alpha = \int_{\alpha_{\mathsf{o}}'}^{\alpha_{\mathsf{o}}} \kappa \, d\alpha - \int_{\alpha}^{\alpha_{\mathsf{o}}} \kappa \, d\alpha = \int_{\alpha_{\mathsf{o}}'}^{\alpha_{\mathsf{o}}} \kappa \, d\alpha - \omega.$$

Hence near the left side of region II we can write

$$u(\text{II}) \sim A \exp(-i\pi/4)\, \kappa^{-1/2} \exp\left(\int_{\alpha_{\mathsf{o}}'}^{\alpha_{\mathsf{o}}} \kappa \, d\alpha\right) e^{-\omega'},$$

and Eq. D–9a shows that the wave function in region III is

$$u(\text{III}) \sim 2A \exp(-i\pi/4) k^{-1/2} \exp\left(\int_{\alpha_{\mathsf{o}}'}^{\alpha_{\mathsf{o}}} \kappa \, d\alpha\right) \cos\left(\omega' - \frac{\pi}{4}\right)$$

$$= B k^{-1/2}(e^{-i\omega'} + i e^{i\omega'}),$$

where

$$B = A e^{K} \tag{D–10}$$

and K is the integral of κ through the barrier: $K = \int_{\alpha_{\mathsf{o}}'}^{\alpha_{\mathsf{o}}} \kappa \, d\alpha$.

Since ω' increases in region III as α decreases, the first term in $u(\text{III})$ is the wave incident on the barrier, and the second term is the reflected wave. It is seen that, for unit amplitude B of the incident wave, the reflected wave, to this approximation, has equal amplitude and is $\pi/2$ out of phase, and the transmitted wave in region I has amplitude $\exp(-\int_{\alpha_{\mathsf{o}}'}^{\alpha_{\mathsf{o}}} \kappa \, d\alpha)$. This is the fundamental result for barrier penetration. It can be phrased in terms of flux. If α is a Cartesian coordinate, the flux in the α-direction is proportional to $\text{Im}[u^*(du/d\alpha)]$; if α is the radial coordinate r, the flux is $\text{Im}\,[(1/r^2)u^*(du/dr)]$; and in general the flux is $g(\alpha)\,\text{Im}\,[u^*(du/d\alpha)]$, where $g(\alpha)$ is some "geometrical" factor. Hence the flux at a point α_1 in region I, compared to that at a point α_3 in region III, is

$$\frac{\text{Flux in I}}{\text{Flux in III}} = \left[\frac{g(\alpha_1)}{g(\alpha_3)}\right] \exp(-2K). \qquad \text{(D-11a)}$$

Equation D-10 gives the barrier transmission, that is, the intensity ratio $(A/B)^2$,

$$T = \exp(-2K). \qquad \text{(D-11b)}$$

If the barrier is so thin that the term $e^{-\omega}$ is not negligible near O', it can be shown† that a better approximation is

$$T \approx \frac{1}{1 + \exp(2K)}, \qquad \text{(D-12a)}$$

$$R \approx \frac{\exp(2K)}{1 + \exp(2K)}, \qquad \text{(D-12b)}$$

where R is the intensity of the reflected wave in region III. Notice that $R + T = 1$, as it should. The approximation that we have developed is, of course, the limit of Eq. D-12 for $K \gg 1$, and is valid when the turning points are well separated.

We may note that for one particular case, namely, a parabolic barrier, Eq. D-12 is exact.

Equation D-11 is valid if the points α_1 and α_3 are sufficiently remote from the barrier that $dk/d\alpha$ is small and satisfies Eq. D-3. If this is not the case, an extra factor must be introduced into Eq. D-11a, namely, the ratio of the value of $1 - (1/2k^2)(dk/d\alpha)$ at α_1 to its value at α_3.

---

† See, for example, N. Fröman and P. O. Fröman, *JWKB Approximation*, North-Holland, Amsterdam; 1965.

Electromagnetic Transitions

E-1 MULTIPOLE EXPANSION OF THE ELECTROMAGNETIC FIELD

We stated in Chapter 3 that the interaction energy of a system of charged particles with the electromagnetic field can be expanded in a series of multipole moments. The usefulness of Eq. 3–1 is subject to the provisos that the field changes only slightly over the nuclear volume, and that the nucleons move nonrelativistically. The first condition is equivalent to requiring that the wavelength of the radiation be large compared to nuclear dimensions, or that

$$R \ll \lambda, \quad \text{i.e.,} \quad \omega R \ll c,$$

where ω is the angular frequency of the radiation and R is the radius of the nucleus. This condition may be rewritten in terms of the energy of the corresponding photon as

$$\hbar \omega \ll \frac{\hbar c}{R} \approx \frac{197}{1.2 \, A^{1/3}} \approx 165 \, A^{-1/3} \text{ MeV.}$$

In other words, for energies of 25 MeV or more, the convergence of the series is much slower, and different approaches may be preferable.

The various multipole moments are tensors. In studying nuclei, it is more useful to employ spherical tensors than Cartesian tensors, because spherical tensors have special properties in connection with angular momentum. A summary of the properties of spherical tensors is given in Appendix A.

Of special usefulness in this connection is the Wigner-Eckart theorem, given by Eq. A–83.

Let us first write the electromagnetic field in free space in terms of spherical tensors. For a pure radiation field, we may put the electrostatic potential $\phi = 0$ and express the electric and magnetic fields $\mathcal{E}$ and $\mathcal{H}$ in terms of the vector potential $\mathcal{A}$ by

$$\mathcal{E} = -\frac{1}{c}\frac{\partial \mathcal{A}}{\partial t}, \tag{E-1a}$$

$$\mathcal{H} = \text{curl } \mathcal{A}. \tag{E-1b}$$

Maxwell's equations are satisfied if

$$\left(\nabla^2 - \frac{1}{c^2}\frac{\partial^2}{\partial t^2}\right)\mathcal{A} = 0, \tag{E-2a}$$

$$\text{div } \mathcal{A} = 0. \tag{E-2b}$$

A general field $\mathcal{A}$ can be expressed in terms of any complete orthogonal set of solutions of (E–2). If we put

$$\mathcal{A} = q_k e^{-i\omega t} \mathbf{A}_k(\mathbf{r}), \tag{E-3}$$

with $k = \omega/c$, Eqs. (E–2) are equivalent to

$$\text{curl curl } \mathbf{A}_k - k^2 \mathbf{A}_k = 0. \tag{E-4}$$

It is found that the following two expressions both satisfy this last equation for integral values of λ and μ:

$$\mathbf{A}_{\lambda\mu}^E = \left(\frac{-i}{k}\right)\text{curl } [\mathbf{r} \times \text{grad } (j_\lambda(kr) Y_\lambda{}^\mu(\theta, \phi))], \tag{E-5a}$$

$$\mathbf{A}_{\lambda\mu}^M = \mathbf{r} \times \text{grad } (j_\lambda(kr) Y_\lambda{}^\mu), \tag{E-5b}$$

where j_λ denotes the spherical Bessel functions. By allowing λ, μ, and k to take on all possible values, these solutions are made to span a complete orthogonal set, and consequently any $\mathcal{A}$ can be written as a series of them:

$$\mathcal{A} = \sum_{\lambda,\mu,\sigma}\int dk\, q_k{}^{\lambda\mu\sigma} e^{-i\omega t} \mathbf{A}_{\lambda\mu}^\sigma(k, \mathbf{r}). \tag{E-6}$$

The individual solutions are of considerable interest in themselves.

We define

$$\mathbf{E}_{\lambda\mu}^\sigma = ik\mathbf{A}_{\lambda\mu}^\sigma, \tag{E-7a}$$

$$\mathbf{H}_{\lambda\mu}^\sigma = \text{curl } \mathbf{A}_{\lambda\mu}^\sigma, \tag{E-7b}$$

and, using Eq. (E–4), we note that

$$\mathbf{H}^M = \mathbf{E}^E, \qquad \mathbf{H}^E = -\mathbf{E}^M.$$

Consequently,

$$\mathbf{r} \cdot \mathbf{E}^M = \mathbf{r} \cdot \mathbf{H}^E = 0, \tag{E–8}$$

which means that the electric field is transverse in the M-mode and the magnetic field is transverse in the E-mode. These are the properties of the fields of oscillating magnetic and electric multipoles, respectively. The fields with superscript M and E are therefore called magnetic and electric radiations. The energy in the field for one of these solutions is

$$\text{energy} = \frac{1}{4\pi} \int [|\mathbf{E}_{\lambda\mu}^{\sigma}|^2 + |\mathbf{H}_{\lambda\mu}^{\sigma}|^2] \, dV. \tag{E–9}$$

The energy flow density is given by Poynting's vector, and the momentum density in the field is just $1/c^2$ of this quantity: $(1/4\pi c)\, \mathbf{E} \times \mathbf{H}$. Consequently the angular momentum carried by the field is:

$$\text{angular momentum of field} = \frac{1}{4\pi c} \int \mathbf{r} \times \mathrm{Re}\,(\mathbf{E}_{\lambda\mu}^{*} \times \mathbf{H}_{\lambda\mu}) \, dV. \tag{E–10}$$

For the classical field, it is just a matter of juggling with vector relationships and integral theorems to show that the z-component of the angular momentum is (μ/ω) times the energy. If we introduce quantization simply by saying that the energy of the field is $\hbar\omega$ times the number of photons, we have demonstrated that the photons associated with a field of the type $\mathbf{A}_{\lambda\mu}^{\sigma}$ each have an angular-momentum z-component equal to $\mu\hbar$. More satisfactorily, we can introduce a quantized field by treating the quantities $q_k^{\lambda\mu\sigma}$ in Eq. E–6 as quantum-mechanical operators, which create and destroy photons of type λ, μ, k. Then Eqs. E–9 and E–10 are operator equations, and the eigenvalues of energy and angular momentum are readily determined, since q^*q is the operator whose eigenvalues are the numbers of photons present. In this way, the same result is obtained, together with the fact that the square of the angular momentum carried by one photon is $\lambda(\lambda + 1)\hbar^2$.

There are, of course, other sets of solutions of Eq. E–4, of which the simplest are sets of plane waves of various polarizations. These lead to a quantized description with photons of definite linear momentum, and fixed directions of $\mathbf{E}$ and $\mathbf{H}$. Such photons are suitable for discussing some problems, but for most nuclear problems photons of definite angular momentum are more convenient. Which of these two alternatives we use is merely a matter of convenience, since an arbitrary state can be expressed as a linear combination of the states of any complete set.

The interaction energy between a charged system and the electromagnetic field is

$$-\int\left[\frac{1}{c}\,\mathbf{j}(\mathbf{r})\cdot\mathcal{A}(\mathbf{r})+\mathbf{M}(\mathbf{r})\cdot\mathcal{H}(\mathbf{r})\right]d^3\mathbf{r}, \qquad (\text{E--11})$$

where $\mathbf{j}(\mathbf{r})$ is the current density and $\mathbf{M}(\mathbf{r})$ is the dipole moment per unit volume due to a distribution of spins. If we take the more fundamental viewpoint from which $\mathbf{j}$ and $\mathbf{M}$ are attributed to A particles, this energy becomes

$$H_{\text{int}}=-\sum_{i=1}^{A}\left[\frac{e_i}{m_ic}\,\mathbf{p}_i\cdot\mathcal{A}(\mathbf{r}_i)+\mu_i\mathbf{s}_i\cdot\mathcal{H}(\mathbf{r}_i)\right], \qquad (\text{E--12})$$

where μ_i is the intrinsic magnetic moment of the ith particle, μ_0 is the nuclear magneton, and the remaining notation is evident.

The transition probability per second for a process due to this interaction which takes the charged system and the field from a state I to a state F is

$$T=\frac{2\pi}{\hbar}\,|\langle F|H_{\text{int}}|I\rangle|^2\rho_E, \qquad (\text{E--13})$$

where ρ_E is the number of final states per unit energy interval. The coefficients $q_k{}^{\lambda\mu\sigma}$ and $q_k{}^{\lambda\mu\sigma*}$, treated as annihilation and creation operators for the quantized field, have nonzero matrix elements between states which differ by precisely one photon of type k, λ, μ, σ. For example, radiation may be discussed by taking an initial state with no photons and a final state with one photon, or absorption may be handled by reversing these states. In either case, it is clear that, in addition to the matrix element of q, an essential role is played by the matrix element $\langle f|H'|i\rangle$, where f and i refer only to the states of the charged system and

$$H'=-\sum_i\left[\frac{e_i}{m_ic}\,\mathbf{p}_i\cdot\mathbf{A}_{\lambda\mu}^{\sigma}(k,\mathbf{r})+\mu_i\mathbf{s}_i\cdot\mathbf{H}_{\lambda\mu}^{\sigma}(k,\mathbf{r})\right]. \qquad (\text{E--14})$$

Of course, $|E_i-E_f|=\hbar\omega=\hbar ck$, since $E_I=E_F$. In the special case of $kR\ll 1$, the Bessel functions satisfy

$$j_\lambda(kr)\simeq\frac{(kr)^\lambda}{(2\lambda+1)!!},$$

where $(2\lambda + 1)!! = 1 \cdot 3 \cdot 5 \cdots (2\lambda + 1)$. With this approximation, it can be seen by straightforward calculation that

$$\langle f|H'|i\rangle \quad \text{is proportional to} \quad \langle f|M_{\lambda\mu}|i\rangle \quad \text{or} \quad \langle f|Q_{\lambda\mu}|i\rangle$$

for $\sigma = M$ or E, respectively, where $M_{\lambda\mu}$ and $Q_{\lambda\mu}$ are magnetic and electric multipole operators,

$$Q_{\lambda\mu} = \sum_i [e_i r_i^\lambda Y_\lambda^{\mu*}(\Omega_i) - ig_{si}\mu_0 k(\lambda + 1)^{-1}\boldsymbol{\sigma}_i \times \mathbf{r}_i \cdot \text{grad} \, (r^\lambda Y_\lambda^{\mu*})_i], \qquad \text{(E–15)}$$

$$M_{\lambda\mu} = \mu_0 \sum_i \left(g_{si}\mathbf{s}_i + \frac{2}{\lambda+1}g_{li}\mathbf{l}_i\right) \cdot \text{grad} \, (r^\lambda Y_\lambda^{\mu*})_i, \qquad \text{(E–16)}$$

μ_0 is the nuclear magneton, and the g's are the gyromagnetic ratios.

These operators are closely related to the multipoles introduced in Chapter 3; apart from a normalizing factor, $M_{\lambda 0}$ is the magnetic 2^λ-pole moment of Eq. 3–20, and the first term of $Q_{\lambda 0}$ is the Q_λ of Eq. 3–6. (The matrix element of the second term of $Q_{\lambda\mu}$ is of the same order of magnitude as that of $M_{\lambda+1}$ and is usually ignored.) Thus the static electromagnetic moments are the diagonal matrix elements of the same operators that are involved in transitions.

When all the expressions are properly normalized (our expressions for **A** are not), the transition probability for emission of a photon of energy $\hbar\omega$, angular momentum λ, μ, and of electric or magnetic type, with the nucleus going from a state i to a state f, is

$$T_{if}(\sigma\lambda\mu) = \frac{8\pi(\lambda + 1)}{\lambda[(2\lambda + 1)!!]^2} \frac{k^{2\lambda+1}}{\hbar} |\langle f|\mathcal{O}_{\lambda\mu}|i\rangle|^2, \qquad \text{(E–17)}$$

where $\mathcal{O}_{\lambda\mu}$ stands for $Q_{\lambda\mu}$ or $M_{\lambda\mu}$. Usually we are not interested in the orientation of either the initial or the final nucleus, so we sum over M_f and average over M_i. We define the so-called reduced matrix element:

$$B(\sigma\lambda, J_i \to J_f) = (2J_i + 1)^{-1} \sum_{M_i,M_f} |\langle f|\mathcal{O}_{\lambda\mu}|i\rangle|^2, \qquad \text{(E–18a)}$$

$$T(\sigma\lambda) = \frac{8\pi(\lambda + 1)}{\lambda[(2\lambda + 1)!!]^2} \frac{k^{2\lambda+1}}{\hbar} B(\sigma\lambda). \qquad \text{(E–18b)}$$

Of course, only one value of μ, that is, $\mu = M_f - M_i$, contributes to each term.

We can make a very rough estimate of the matrix elements by replacing spherical harmonics and angular momentum vectors by unity, and noting that the initial and final wave functions vanish outside the nuclear radius R:

$$\langle f|Q_{\lambda\mu}|i\rangle \sim ZeR^{\lambda}, \tag{E-19a}$$

$$\langle f|M_{\lambda\mu}|i\rangle \sim A\left(\frac{e\hbar}{2Mc}\right)R^{\lambda-1}. \tag{E-19b}$$

We note that the ratio of the magnetic and electric moments of the same order is approximately $(\hbar/Mc)/R$, which varies from about 0.2 to 0.03. Since the square appears in Eq. E-17, the relative transition probabilities vary from 0.04 to 0.001, but in view of the crudity of the approximations, these last numbers might be in error by a factor of nearly ten. On the same basis of approximation, the second term in $Q_{\lambda\mu}$ contributes about $kA(e\hbar/2Mc)R^{\lambda}$ to $\langle f|Q|i\rangle$. Its ratio to the contribution E-19a is thus of the order of $\hbar\omega/Mc^2$, and this term can be neglected except in unusual circumstances.

The estimates E-19 substituted in Eq. E-17 show that the ratio of transition rates of successive multipoles λ and $\lambda + 1$ is of order $(kR)^2/(2\lambda + 3)^2$. So long as we are dealing with $kR \ll 1$ for which our theory has been developed, we may therefore consider only the lowest multipole which contributes to a given transition.

The selection rules are important. The parity of $Q_{\lambda\mu}$ is that of $Y_{\lambda}{}^{\mu}$, namely $(-)^{\lambda}$, but $M_{\lambda\mu}$ has parity $-(-)^{\lambda}$, since in addition to $Y_{\lambda}{}^{\mu}$ it contains a vector in the gradient operator and pseudo vectors s and l. Also, since $Q_{\lambda\mu}$ and $M_{\lambda\mu}$ are spherical tensors, the selection rule $J_i + J_f \geqslant \lambda \geqslant |J_i - J_f|$ applies. The lowest multipoles arising in specified γ-transitions are therefore as shown in Table E-1. (The square brackets indicate that these multipoles do not contribute to cases in which either J_i or J_f is zero.) If our estimates of the matrix elements are reliable, some of these transi-

TABLE E-1

ΔJ	$\Pi_i \Pi_f$	Multipoles
$0 \to 0$		(E0) special case
$\frac{1}{2} \to \frac{1}{2}$	$+1$	(E0) $M1$
$\frac{1}{2} \to \frac{1}{2}$	-1	$E1$
0	$+1$	(E0) $M1$ $E2$
0	-1	$E1$ $M2$
1	$+1$	$M1$ [$E2$]
1	-1	$E1$ [$M2$]
2	$+1$	$E2$ [$M3$]
2	-1	$M2$ [$E3$]
3	$+1$	$M3$ [$E4$]
3	-1	$E3$ [$M4$]

tions will emit pure multipole radiation. For example, a $\Delta J = 0$ transition with parity change may proceed by either $E1$- or $M2$-radiation. The relative intensities are expected to be in the ratio $(Mc^2/\hbar\omega)^2$, which is at least 10^4 or 10^5, and pure $E1$-radiation will result. However, when the lower-order radiation is magnetic, the situation is not so clear-cut. For $\Delta J = 0$ or $\Delta J = 1$ with no parity change, $M1$ and $E2$ may mix with relative intensity $[(\hbar/Mc)/R]^4(Mc^2/\hbar\omega)^2$. This ratio is still considerably greater than unity in most cases, but the first factor may be as small as 10^{-4}. Since quadrupole moments are considerably enhanced in distorted nuclei, we may expect to find appreciable $M1 - E2$ mixtures in their spectra.

The multipole $E0$ occurs in brackets in Table E–1. As a glance at Eqs. E–5 shows, there is no radiation field for $\lambda = 0$. This vanishing of the field is the mathematical form of the physical statement that each photon carries off at least one unit of angular momentum. Consequently, although the matrix element of Q_{00} may not vanish, there is no term $\lambda = 0$ in the expansion of the radiation field, and the $E0$-multipoles do not contribute to γ-transitions. In particular, $0 \rightarrow 0$ γ-transitions are completely forbidden.

How, then, does an excited state of spin zero decay, if there are no other states between it and a zero-spin ground state? It might go by β-decay, forming a different nuclide, but this is a very slow process and, although the excited state cannot emit a single photon, it can decay more quickly through the electromagnetic interaction between the nucleus and its atomic electrons. The excess nuclear energy is given to one of the electrons which is thereby expelled from 'the atom. This form of energy transfer is called *internal conversion*, and the emitted electron is a *conversion electron*. A $0 \rightarrow 0$ transition can also go by higher-order electromagnetic processes such as the emission of two photons, or one photon and a conversion electron, but for most nuclear states these processes would have a probability at least two orders of magnitude smaller than internal conversion. If a state has excitation energy greater than two electron masses, it can decay also by the electromagnetic creation of a positron-negatron pair. At high energies, this *internal pair creation* is more important than internal conversion. Besides providing a mechanism for $0 \rightarrow 0$ decays, internal conversion and internal pair creation often compete significantly with γ-emission for transitions of higher multipolarity.

E-2 SINGLE-PARTICLE TRANSITION RATES

In view of the sensitivity of the static moments to configuration mixing and collective effects, it is not surprising that the extreme single particle model does not yield realistic values of γ-decay probabilities. The formulas for lifetimes in this model are particularly simple, however, and are often used as reference values for comparing experimental data.

We shall first estimate the transition probability E-17 for electric λ-pole transition of a single proton in this model. For simplicity, we assume that the orbital angular momentum of the proton is 1 in the initial state, and zero in the final state. Let the initial intrinsic spin of the proton be spin up, i.e., parallel to 1, and remain so after the transition. Clearly, the electric multipole radiation emitted is of order $\lambda = l$. The initial and final wavefunctions of the proton may be written as

$$\phi_i = u_i(r) Y_l{}^m(\theta, \phi)\alpha, \quad \text{and} \quad \phi_f = u_f(r)(4\pi)^{-1/2}\alpha,$$

where α is the spin-up wave function, and u_i, u_f are functions of $r = |\mathbf{r}|$ only. In evaluating the matrix-element $\langle f|Q_{\lambda\mu}|i\rangle$, we shall ignore the second term in Eq. E-15. We then obtain $(\lambda = l)$,

$$\langle f|Q_{\lambda\mu}|i\rangle = (4\pi)^{-1/2} e \int_0^\infty r^\lambda u_f(r) u_i(r) r^2 \, dr.$$

For an order of magnitude estimate, we shall replace the radial integral by a rough approximation. Taking the u's constant over the nuclear radius R, we obtain

$$\int_0^\infty r^\lambda u_f(r) u_i(r) r^2 \, dr \approx \frac{3R^\lambda}{(\lambda + 3)},$$

so that

$$|\langle f|Q_{\lambda\mu}|i\rangle|^2 \approx \frac{e^2}{4\pi} R^{2\lambda} \left(\frac{3}{\lambda + 3}\right)^2. \tag{E-20}$$

Substituting this in Eq. E-17, we have the estimate for the single-proton transition probability, which is also called the Weisskopf unit,

$$T_W(E, \lambda \to 0) = \frac{2(\lambda + 1)}{\lambda[(2\lambda + 1)!!]^2} \left(\frac{3}{\lambda + 3}\right)^2 \frac{e^2}{\hbar c} \left(\frac{\omega R}{c}\right)^{2\lambda} \omega \, \text{sec}^{-1}. \tag{E-21a}$$

In a similar manner, by taking the expression E-16 for $M_{\lambda\mu}$, we may obtain the corresponding single-particle estimate for the λ-pole magnetic transition, called the Moszkowski unit,

$$T_M(M, \lambda \to 0) = \frac{2(\lambda + 1)}{\lambda[(2\lambda + 1)!!]^2} \left(\frac{3}{\lambda + 2}\right)^2 \frac{e^2}{\hbar c} \left(\frac{\hbar/Mc}{R}\right)^2$$

$$\times \left(\mu_p \lambda - \frac{\lambda}{\lambda + 1}\right)^2 \left(\frac{\omega R}{c}\right)^{2\lambda} \omega \, \text{sec}^{-1}. \tag{E-21b}$$

There is also a Weisskopf unit for magnetic transition; it is

$$T_W(M, \lambda \to 0) = 10\left(\frac{\lambda+2}{\lambda+3}\right)^2\left(\mu_p\lambda - \frac{\lambda}{\lambda+1}\right)^{-2} T_M(M, \lambda \to 0). \quad \text{(E-21c)}$$

These units are often used for all experimental transitions, without distinguishing neutron or proton, and regardless of any model of the states involved. In obtaining numerical values of these units, the nuclear radius R is taken to be $1.2A^{1/3}$ fm. It is of interest to quote these numerical results in the form of the width for γ-decay, defined as

$$\Gamma_\gamma = \hbar T. \quad \text{(E-22)}$$

If E_γ is the γ-ray energy, $\Gamma_{\gamma W}(E\lambda)$ is proportional to $A^{2\lambda/3}E_\gamma^{2\lambda+1}$, and $\Gamma_{\gamma W}(M\lambda)$ to $A^{2(\lambda-1)/3}E_\gamma^{2\lambda+1}$. For E_γ in MeV, the numerical coefficients for $\Gamma_{\gamma W}$ are given in Table E-2. Of course, in comparison with experiment, it is the

TABLE E-2 Coefficients for $\Gamma_{\gamma W}$ in sec^{-1}.

λ	$E\lambda$	$M\lambda$
1	6.8×10^{-2}	2.1×10^{-2}
2	4.9×10^{-8}	1.5×10^{-8}
3	2.3×10^{-14}	6.8×10^{-15}
4	6.8×10^{-21}	2.1×10^{-21}
5	1.6×10^{-27}	4.9×10^{-28}

partial γ-ray width which must be used after corrections are made for competing processes such as internal conversion. The single-particle estimate for $B(\sigma\lambda)$ may be easily obtained from Eqs. E-18 and E-20. For example,

$$B(E2, 0 \to 2) = \sum_{M_f} |\langle f|Q_{2\mu}|i\rangle|^2 = 5|\langle f|Q_{2\mu}|i\rangle|^2$$

$$= \frac{5}{4\pi}\left(\frac{3}{5}\right)^2 R^4 e^2$$

$$\approx 2.95 \times 10^{-5} A^{4/3}\, e^2 b^2. \quad \text{(E-23)}$$

This is the unit used in Eq. 9-93 of the text; the value for the downward transition $2 \to 0$ is, of course, reduced by a factor of 5.

Author Index

Author Index

Subject Index

Subject Index

Printed in the United States
by Baker & Taylor Publisher Services